英汉·汉英
化工工艺与设备
图解词典

陈国桓 蔡 晖 主编

ENGLISH-CHINESE
CHINESE-ENGLISH
PICTORIAL DICTIONARY
OF CHEMICAL TECHNICAL
AND EQUIPMENT

化学工业出版社
·北京·

本词典共包括三部分内容：图解词典部分、英汉对照词汇部分和汉英对照词汇部分。

图解词典部分以专业排列系统介绍了常用化工工艺流程及机械设备，工艺流程包括石油化工、煤化工、化工工艺、废物处理；机械设备包括传热设备、蒸馏与吸收设备、萃取、离子交换与吸附设备、蒸发器与结晶设备、干燥设备、过滤设备、离心机与压榨设备、气固分离设备、粉碎设备、固体的混合与分离设备、造粒设备、流化床系统和反应器、泵、压缩机与风机、真空设备、旋转轴密封、制冷设备、压力容器、阀门与管件、输送与称量设备、工业炉、动力装置等。给出了流程及设备涉及的物料、过程、部件的中英文名称。

英汉对照词汇部分以英文顺序排列，汉英对照词汇部分以词汇中文名称的汉语拼音顺序排列，两部分词汇每词后给出了此词在图解词典中的位置，可供索引之用。

本词典可供化工、炼油、轻工、环保、能源、食品等领域的技术人员参考，也可作为高校学生专业英语课程的辅助读物。

图书在版编目（CIP）数据

英汉·汉英化工工艺与设备图解词典/陈国桓，蔡晖主编．—2版．—北京：化学工业出版社，2017.4
ISBN 978-7-122-29141-7

Ⅰ.①英… Ⅱ.①陈… ②蔡… Ⅲ.①化学工业-生产工艺-图解词典-英、汉②化工设备-图解词典-英、汉 Ⅳ.①TQ-61

中国版本图书馆CIP数据核字（2017）第033988号

责任编辑：傅聪智　　　　　　　　　　装帧设计：刘丽华
责任校对：边　涛

出版发行：化学工业出版社（北京市东城区青年湖南街13号　邮政编码100011）
印　　刷：北京永鑫印刷有限责任公司
装　　订：三河市胜利装订厂
850mm×1168mm　1/32　印张24¼　字数684千字
2017年6月北京第2版第1次印刷

购书咨询：010-64518888（传真：010-64519686）　售后服务：010-64518899
网　　址：http://www.cip.com.cn
凡购买本书，如有缺损质量问题，本社销售中心负责调换。

定　　价：168.00元　　　　　　　　　　　　　版权所有　违者必究

前　言

本书自 2003 年 11 月出版以来,深受高校师生及工程技术人员欢迎。在此,感谢读者关注。

本次再版,删去了一部分化工设备的内容;新增了化工工艺流程,主要是煤化工的新技术,其余有废水处理、固体废物处理、废气处理等。这些资料来源于新版图书、最新期刊杂志、会议文集等。

本书图解的化工工艺与设备,其中有的展示基础知识,更多的是展示国内外新技术。读者在查阅英汉词汇的同时,还可以领略国内外新技术。

天津市华邦科技发展有限公司张丽娟老师为本书的图文处理做了大量工作,还有退休的朱美娥高级工程师,在校大学生陈相宜、陈宗岳为本书的完成做了许多具体工作,在此一并致谢。

本书的再版,与化学工业出版社的鼓励及支持分不开。我们的水平有限,书中难免有不妥之处,敬请广大读者批评指正。

编者
2016.12

第一版前言

随着改革开放的深入，国家经济建设的加速，国际合作及技术交流日益广泛。尤其是进入WTO，对外贸易进一步扩大，引进技术与装备，输出技术与设备的工作更加频繁，科技工作者手头有一些专业性很强、图文并茂、英汉对照的工具书是有益的。

本书的内容涉及化工与相关领域，如炼油、轻工、环保、能源和食品等，其工艺与设备可供科技工作者查阅，也可作为高校学生的专业英语课程的辅助读物。由于书中对某一技术有多样性的展示，读者在查阅英汉词汇的同时，还有可能获得某些收益。

参加本书编写的工作人员分工如下。陈国桓，第1、11、15～18、26～28章及附录；安钢，第2、12～14、23章；蔡晖，第3、20章；朱美娥第4、21、22、24章；李永辉，第5～7、14、15、19章；张健飞，第8～10、25章。主编陈国桓，朱美娥。

天津天大天久科技股份有限公司总经理张敏卿教授、公司原总工程师赵汝文教授对本书的编写给予大力支持，在此深表谢意。

有很多同事为本书提供资料，如宋正孝教授、高励成教授、柴诚敬教授、康勇教授、熊光楠教授、马永明博士研究生、张利伟博士研究生等，特此致谢。还有陈相宜、陈宗岳、洪征辉、余辉星、黎昌明等同志为本书的完成做了许多具体工作，在此一并致谢。

本书的编写完成，与化学工业出版社的鼓励及支持分不开。我们的水平有限，书中难免有不妥之处，敬请广大读者批评指正，提出宝贵意见，以便再版时改正及提高。

陈国桓

2003年8月于天津大学化工学院

凡 例

本词典分三部分：图解词典部分，英汉对照词汇部分和汉英对照词汇部分。

图解词典部分以专业排列。共分 25 章。

英汉对照词汇部分以英文顺序排列。

汉英对照词汇部分以词汇中文名称的汉语拼音顺序排列。

英汉对照词汇和汉英对照词汇每词后给出了此词在图解词典中的部分位置，可作为索引使用。

本词典中"[ ]"内为替换词，[ ] 内的词替换其前面的词，字数一一对应；有","的情况下，","前后的字或词均可作替换。"( )"表示其中的内容可有可无成为说明性文字。

本词典中使用了一些非法定计量单位，其换算关系如下：

$1\text{ft}=0.3048\text{m}$

$1\text{in}=2.54\text{cm}$

$x\text{°F}=\dfrac{5}{9}(x-32)\text{°C}$

$1\text{lb}=0.4536\text{kg}$

$1\text{Å}=0.1\text{nm}$

$1\text{mile}=1.609\text{km}$

$1\text{gal}=4.546\text{L}$

$1\text{long ton}=1016.05\text{kg}$

$1\text{sh ton}=907.18\text{kg}$

$1\text{hp}=745.7\text{W}$

$1\,泊=0.1\text{Pa}\cdot\text{s}$

$1\,沲=10^{-4}\text{m}^2/\text{s}$

$1\text{bar}=10^5\text{Pa}$

$1\,乇=133.322\text{Pa}$

$1\,石油桶=159\text{L}$

目　　录

图解词典部分 ·· 1
1　Chemical Engineering Major Terms　化工常用名词 ································· 1
2　Flow Diagram for Chemical Process　化工工艺流程图 ······························ 11
　2.1　Petrochemical processing　石油加工 ·· 11
　2.2　Chemical processing of coal　煤化工 ··· 19
　2.3　Chemical process　化工工艺 ·· 78
　2.4　水净化处理　Purification treatment of water ······························ 103
3　Waste Treatment　废物处理 ·· 106
　3.1　Wastewater treatment　废水处理 ··· 106
　3.2　Biochemical wastewater treatment systems　生物化学废水处理系统 ··· 128
　3.3　Solid waste treatment　固体废物处理 ······································ 134
　3.4　Waste gas treatment　废气处理 ··· 149
　3.5　Emissions measurement　排放物测定 ······································ 152
4　Heat-Transfer Equipment　传热设备 ·· 161
　4.1　Shell-and-tube heat exchangers　管壳换热器（列管换热器） ············ 161
　4.2　Other heat exchangers for liquids and gases　液体和气体用的
　　　　其他型式换热器 ·· 164
　4.3　Heat exchangers for solids　固体用换热器 ································· 166
　4.4　Waste-heat boiler　废热锅炉 ·· 170
　4.5　Air-cooled heat exchangers　空冷换热器 ··································· 171
　4.6　Cooling tower　凉水塔 ·· 172
5　Distillation and Absorption Equipments　蒸馏与吸收设备 ······················· 173
　5.1　Type of towers　塔器类型 ··· 173
　5.2　Plate column　板式塔 ·· 174
　5.3　Packed column　填料塔 ··· 179
　5.4　Absorption column　吸收塔 ·· 183
　5.5　Molecular distillation still　分子蒸馏釜 ··································· 185
6　Extraction, Ion-exchange and Adsorption Equipments
　　　萃取、离子交换与吸附设备 ·· 189

6.1	Extraction equipment 萃取设备	189
6.2	Ion-exchange equipment 离子交换设备	197
6.3	Adsorption equipment 吸附设备	201

7 Evaporators and Crystallization Equipment 蒸发器与结晶设备 ... 208

7.1	Evaporators 蒸发器	208
7.2	Crystallization equipment 结晶设备	212

8 Drying Equipment 干燥设备 218

8.1	Classification of dryers 干燥器分类	218
8.2	Tray and compartment dryer 厢式干燥器	221
8.3	Tunnel dryer and belt dryer 隧道式和带式干燥器	222
8.4	Pneumatic dryer 气流式干燥器	224
8.5	Fluidized bed dryer 流化床干燥器	229
8.6	Spray dryer 喷雾干燥器	232
8.7	Rotating drum dryer 滚筒干燥器	234
8.8	Rotary dryer 回转干燥器	235
8.9	Vertical type dryer 立式干燥器	242

9 Filtration Equipment, Centrifuges and Expression Equipment 过滤设备、离心机与压榨设备 ... 245

9.1	Plate-and-frame type filter press 板框压滤机	245
9.2	Pressure leaf filter 加压叶滤机	247
9.3	Rotary vacuum drum filter 转筒真空过滤机	250
9.4	Vacuum filter and others 真空过滤机及其他过滤机	255
9.5	Centrifuge 离心机	258
9.6	Expression equipment 压榨设备	268

10 Gas-solids Separation Equipments 气固分离设备 269

10.1	Cyclone separators 旋风分离器	269
10.2	Bag filters 袋滤器	272
10.3	Electrical precipitators 电除尘器	274
10.4	Other separators 其他分离器	277
10.5	Scrubber 涤气器	280

11 Size Reduction Equipments 粉碎设备 285

11.1	Crushing equipment 破碎设备	285
11.2	Grinding equipment 研磨设备	291
11.3	Non-rotary ball or bead mills 不旋转的球磨机或珠磨机	296
11.4	Dispersion and colloid mills 分散磨和胶体磨	297

11.5　Fluid-energy or jet mill　流能磨或气流粉碎机 ⋯⋯⋯⋯⋯⋯⋯⋯⋯⋯⋯⋯⋯⋯⋯⋯ 298

11.6　Crushing and grinding practice　破碎与研磨的实际应用 ⋯⋯⋯⋯⋯⋯⋯⋯⋯⋯⋯ 300

12　Mixing and Separation Equipments for Solids　固体的混合与分离设备 ⋯ 304

12.1　Solids mixing machines　固体物料混合机械 ⋯⋯⋯⋯⋯⋯⋯⋯⋯⋯⋯⋯⋯⋯⋯⋯⋯ 304

12.2　Screening machines　筛分机械 ⋯⋯⋯⋯⋯⋯⋯⋯⋯⋯⋯⋯⋯⋯⋯⋯⋯⋯⋯⋯⋯⋯ 307

12.3　Dry classification　干式分级 ⋯⋯⋯⋯⋯⋯⋯⋯⋯⋯⋯⋯⋯⋯⋯⋯⋯⋯⋯⋯⋯⋯⋯ 309

12.4　Wet classifiers　湿式分级器 ⋯⋯⋯⋯⋯⋯⋯⋯⋯⋯⋯⋯⋯⋯⋯⋯⋯⋯⋯⋯⋯⋯⋯⋯ 312

12.5　Dense-media separation　稠密介质分离 ⋯⋯⋯⋯⋯⋯⋯⋯⋯⋯⋯⋯⋯⋯⋯⋯⋯⋯⋯ 313

12.6　Magnetic separators　磁力分离器 ⋯⋯⋯⋯⋯⋯⋯⋯⋯⋯⋯⋯⋯⋯⋯⋯⋯⋯⋯⋯⋯⋯ 317

12.7　Electrostatic separator and optical separator　静电分离器和光学分离器 ⋯⋯⋯⋯ 320

12.8　Flotation　浮选 ⋯⋯⋯⋯⋯⋯⋯⋯⋯⋯⋯⋯⋯⋯⋯⋯⋯⋯⋯⋯⋯⋯⋯⋯⋯⋯⋯⋯⋯ 322

13　Granulation Equipments　造粒设备 ⋯⋯⋯⋯⋯⋯⋯⋯⋯⋯⋯⋯⋯⋯⋯⋯⋯⋯⋯⋯ 324

13.1　Prilling tower　造粒塔 ⋯⋯⋯⋯⋯⋯⋯⋯⋯⋯⋯⋯⋯⋯⋯⋯⋯⋯⋯⋯⋯⋯⋯⋯⋯⋯ 324

13.2　Extrusion pelleting equipment　挤压造粒设备 ⋯⋯⋯⋯⋯⋯⋯⋯⋯⋯⋯⋯⋯⋯⋯⋯ 326

13.3　Rotating dish granulators　转盘造粒机 ⋯⋯⋯⋯⋯⋯⋯⋯⋯⋯⋯⋯⋯⋯⋯⋯⋯⋯⋯ 330

13.4　Fluidized bed and spouted bed granulators　流化床与喷动床造粒机 ⋯⋯⋯⋯⋯⋯ 332

13.5　Pelletizing by solidification method　固化法造粒 ⋯⋯⋯⋯⋯⋯⋯⋯⋯⋯⋯⋯⋯⋯⋯ 334

14　Fluidized-bed Systems and Reactors　流化床系统和反应器 ⋯⋯⋯⋯⋯⋯⋯⋯ 336

14.1　Design of fluidized-bed systems　流化床系统的设计 ⋯⋯⋯⋯⋯⋯⋯⋯⋯⋯⋯⋯⋯ 336

14.2　Uses of fluidized-bed　流化床的应用 ⋯⋯⋯⋯⋯⋯⋯⋯⋯⋯⋯⋯⋯⋯⋯⋯⋯⋯⋯⋯ 341

14.3　Reactors　反应器 ⋯⋯⋯⋯⋯⋯⋯⋯⋯⋯⋯⋯⋯⋯⋯⋯⋯⋯⋯⋯⋯⋯⋯⋯⋯⋯⋯⋯ 345

14.4　Agitated reactors　搅拌式反应器 ⋯⋯⋯⋯⋯⋯⋯⋯⋯⋯⋯⋯⋯⋯⋯⋯⋯⋯⋯⋯⋯⋯ 350

15　Miscellaneous Processes　其他化工过程 ⋯⋯⋯⋯⋯⋯⋯⋯⋯⋯⋯⋯⋯⋯⋯⋯⋯⋯ 352

15.1　Sublimation　升华 ⋯⋯⋯⋯⋯⋯⋯⋯⋯⋯⋯⋯⋯⋯⋯⋯⋯⋯⋯⋯⋯⋯⋯⋯⋯⋯⋯⋯ 352

15.2　Membrane processes　膜分离过程 ⋯⋯⋯⋯⋯⋯⋯⋯⋯⋯⋯⋯⋯⋯⋯⋯⋯⋯⋯⋯⋯ 352

15.3　Dielectrophoresis　介电电泳 ⋯⋯⋯⋯⋯⋯⋯⋯⋯⋯⋯⋯⋯⋯⋯⋯⋯⋯⋯⋯⋯⋯⋯⋯ 354

15.4　Diffusional separation processes　扩散分离过程 ⋯⋯⋯⋯⋯⋯⋯⋯⋯⋯⋯⋯⋯⋯⋯ 354

15.5　Coalescence processes　聚并过程 ⋯⋯⋯⋯⋯⋯⋯⋯⋯⋯⋯⋯⋯⋯⋯⋯⋯⋯⋯⋯⋯⋯ 355

15.6　Sedimentation operations　沉降操作 ⋯⋯⋯⋯⋯⋯⋯⋯⋯⋯⋯⋯⋯⋯⋯⋯⋯⋯⋯⋯ 357

15.7　Forming machine for rubber and plastics, mixing machine　橡胶塑料
加工成型及混合机械 ⋯⋯⋯⋯⋯⋯⋯⋯⋯⋯⋯⋯⋯⋯⋯⋯⋯⋯⋯⋯⋯⋯⋯⋯⋯⋯⋯⋯ 360

16　Pumps　泵 ⋯⋯⋯⋯⋯⋯⋯⋯⋯⋯⋯⋯⋯⋯⋯⋯⋯⋯⋯⋯⋯⋯⋯⋯⋯⋯⋯⋯⋯⋯⋯⋯⋯ 363

16.1　Classification of pumps　泵分类 ⋯⋯⋯⋯⋯⋯⋯⋯⋯⋯⋯⋯⋯⋯⋯⋯⋯⋯⋯⋯⋯⋯ 363

16.2　Reciprocating pump　往复泵 ⋯⋯⋯⋯⋯⋯⋯⋯⋯⋯⋯⋯⋯⋯⋯⋯⋯⋯⋯⋯⋯⋯⋯⋯ 366

16.3　Rotating pump　回转泵 ⋯⋯⋯⋯⋯⋯⋯⋯⋯⋯⋯⋯⋯⋯⋯⋯⋯⋯⋯⋯⋯⋯⋯⋯⋯⋯ 371

16.4	Centrifugal pump 离心泵	373
16.5	Special pump 特种泵	391

17 Compressors and Fans 压缩机与风机 ... 394
- 17.1 Type of compressor 压缩机类型 ... 394
- 17.2 Reciprocating compressor 往复式压缩机 ... 395
- 17.3 Rotary blower 回转式鼓风机 ... 403
- 17.4 Centrifugal compressor 离心压缩机 ... 404
- 17.5 Fan 风机 ... 406
- 17.6 Compressed air installation 压缩空气站 ... 407

18 Vacuum Equipments 真空设备 ... 408
- 18.1 Rotary oil sealed mechanical pump 油封式旋转机械真空泵 ... 408
- 18.2 Diffusion pump 扩散泵 ... 409
- 18.3 Ejector pump 喷射泵 ... 412
- 18.4 Water-ring vacuum pump 水环真空泵 ... 415
- 18.5 Vacuum system 真空系统 ... 416

19 Sealing of Rotating Shafts 旋转轴密封 ... 417
- 19.1 Mechanical seal 机械密封 ... 417
- 19.2 Stuffing box 填料箱 ... 419
- 19.3 Gas seal 气体密封 ... 421
- 19.4 Vacuum seal 真空密封 ... 423

20 Refrigeration 制冷 ... 424
- 20.1 Refrigeration system 制冷系统 ... 424
- 20.2 Refrigerator 制冷机 ... 429

21 Pressure Vessel and Attachment 压力容器及附件 ... 432
- 21.1 Pressure vessel 压力容器 ... 432
- 21.2 Storage and process vessels 贮罐及工艺容器 ... 434
- 21.3 Support 支座 ... 436
- 21.4 Head 封头 ... 437

22 Valves and Fittings 阀门与管件 ... 438
- 22.1 Valves 阀 ... 438
- 22.2 Traps 疏水阀 ... 442
- 22.3 Fittings 管件 ... 444
- 22.4 Flanges 法兰 ... 445
- 22.5 Expansion joints 膨胀节 ... 446
- 22.6 Piping system 管系 ... 447

- 23 Conveying and Weighing Equipments 输送与称量设备 ········· 449
  - 23.1 Pneumatic conveying systems 气流输送系统 ········· 449
  - 23.2 Belt-conveyor systems 带式输送系统 ········· 454
  - 23.3 Vibrating conveyor and miscelleaceous 振动输送机及其他 ········· 457
  - 23.4 Packaging and handling of solid and liquid products 固体和液体产品的包装与运送 ········· 461
  - 23.5 Weighing of bulk solids 粉粒体的称量 ········· 462
- 24 Industrial Furnaces 工业炉 ········· 463
  - 24.1 Industrial furnace 工业炉 ········· 463
  - 24.2 Burner 燃烧器 ········· 468
  - 24.3 Coal gasifier 煤气化炉 ········· 473
- 25 Power Plant 动力装置 ········· 474
  - 25.1 Cogeneration systems 联合生产系统 ········· 474
  - 25.2 Nuclear reactor 核反应堆 ········· 475
  - 25.3 Gas-turbine 燃气轮机 ········· 478
  - 25.4 Electric parts and power plant 电工器件与发电厂 ········· 480
  - 25.5 Steam-generation system 蒸汽发生系统 ········· 485
  - 25.6 Heat transport 热的输送 ········· 487

主要参考文献 ········· 490

英汉对照词汇部分 ········· 493

汉英对照词汇部分 ········· 628

图解词典部分

1 Chemical Engineering Major Terms 化工常用名词

absorption 吸收
absorption refrigeration 吸收制冷
activated sludge 活性污泥
activity coefficient 活度系数
adiabtic process 绝热过程
adsorption 吸附
aerobic bacteria 好氧细菌
affinity 亲和势
anaerobic bacteria 厌氧细菌
anergy 烁；无效能
apparent density 视密度；表观密度
arithmetic mean temperature difference 算术平均温差
artificial intelligence (AI) 人工智能
availability 烟；有效能
azeotrope 共沸物；恒沸物
azeotropic distillation 共沸蒸馏，恒沸蒸馏
backmixing 返混
batch distillation 间歇蒸馏，分批蒸馏
Bernoulli equation 伯努利方程
biocatalytic reaction 生物催化反应
biocell 生物电池
biochemical engineering 生化工程
biochemical oxygen demand (BOD) 生化需氧量
biochemical separation 生化分离
bioengineering 生物工程
biological agent 生物制剂
biotechnology 生物技术
biotransformation 生物转化
boundary condition 边界条件
bubble point 泡点
bulk density 堆密度
by-product 副产物
Carnot cycle 卡诺循环
carrier 载体
catalyst 催化剂
cavitation 汽蚀
cell culture 细胞培养
chemical absorption 化学吸收
chemical engineering 化学工程
chemical engineering science 化学工程学
chemical engineering thermodynamics 化工热力学
chemicals from petroleum 石油化工产品
chemical equilibrium 化学平衡
chemical industry 化学工业
Chemical Industry and Engineering Society of China (CIESC) 中国化工学会
chemical machinery 化工机械
chemical process 化工工艺
chemical oxygen demand (COD) 化学需氧量，化学耗氧量

chemical towers 化工塔类
chemical reaction engineering 化学反应工程
chemical systems engineering 化工系统工程
chemical vapor deposition (CVD) 化学气相沉积
chemisorption 化学吸附
circulation 环流
classical fluidization 经典流态化
clone 克隆
coefficient of performance (COP) 性能系数
coalescence 聚并，凝并
cold-flow model experiment 冷模试验
collection efficiency 捕集效率
column washer 洗涤塔
composite flow model 组合流动模型
composite membrane 复合膜
compressibility factor 压缩因子
compression refrigeration 压缩制冷
computer aided process design (CAPD) 计算机辅助过程设计
concentration profile 浓度［分布］剖面［图］
conversion 转化率；转化
corrugated wire gauze packing 网波纹填料
critical exponent 临界指数
culture medium 培养基
deactivation 失活
decantation 倾析
decay of activity 活性衰减
decontamination factor (DF) 去污指数；净化指数
deoxyribonucleic acid (DNA) 脱氧核糖核酸
dew point 露点
dense phase 密相，浓相
dialysis 渗析；透析
diffusion coefficient 扩散系数
dilatant fluid 胀塑性流体
dilute phase 稀相
dimensional analysis 量纲分析；因次分析
dimensionless group 无量纲数群
disintegration 破碎
discharge coefficient 流量系数；孔流系数
disk centrifuge 碟式离心机
distillate 馏出液
distillation with chemical reaction 反应蒸馏
distribution plate 分布板
downstream processing 下游处理；后处理
dumped packing 散装填料；乱堆填料
dynamic simulation 动态模拟
dynamic viscosity 动力黏度
eddy diffusion 涡流扩散
eddy flow 涡流
effectiveness factor 有效因子
electrodialysis 电渗析
electrophoresis 电泳
elutriation 扬析
empirical model 经验模型
encapsulation 胶囊化
endothermic reaction 吸热反应
energy balance 能量衡算；能量平衡
enrichment 富集
entrainment 雾沫夹带
enthalpy 焓

enthalpy-entropy diagram 焓熵图
entropy generation 熵产生
enzymatic hydrolysis 酶法水解
enzymatic reaction kinetics 酶反应动力学
enzyme membrane 酶膜
equality constraint 等式约束
equation of state (EOS) 状态方程
equilibrium still 平衡釜
equilibrium constant 平衡常数
evolutionary operation (EVOP) 调优操作
evaporation 蒸发
excess enthalpy 超额焓;过量焓
exergy 㶲;有效能;可用能
exothermic reaction 放热反应
expert system (ES) 专家系统
expression rate 压榨速率
external diffusion 外扩散
extend aeration 延时曝气
extent of reaction 反应进度,反应程度
extraction 萃取
extractive distillation 萃取蒸馏
failure diagnosis 故障诊断
feedback control 反馈控制
feedstock 原料
fermentation 发酵
film heat transfer coefficient 传热膜系数
filtration 过滤
filtration membrane 滤膜
finite element method 有限元法
first law of thermodynamics 热力学第一定律
flash evaporation 闪蒸
floating head heat exchanger 浮头换热器
flexibility (操作)弹性;柔性;适应性
flocculation 絮凝
flooding point 泛点
flotation 浮选
flow diagram 流程图
flow sheet 流程图
fluidization 流态化
fluid dynamics 流体动力学
flux 通量
fouling 污垢;结垢
fugacity coefficient 逸度系数
freeze drying 冷冻干燥
free sedimentation 自由沉降
gas-liquid equilibrium (GLE) 气液平衡
generalized fluidization 广义流态化
genetic engineering 基因工程;遗传工程
golden section method 黄金分割法
gradient 梯度
grain size analyzer 粒度分析仪
graphical method 图解法
gross error 过失误差
growth factor 生长因子
group activity coefficient 基团活度系数
head 扬程
heat carrier 载热体
heat flux 热通量
heat of absorption 吸收热
heating medium 载热体
heat·of formation 生成热
heat-pipe 热管
heat transfer 传热;热量传递
heat pump 热泵

height 扬程
height equivalent of a theoretical plate (HETP) 等（理论）板高度；理论板当量高度
height of a (heat) transfer unit 传热单元高度
height of overall transfer unit 总传质单元高度
heterogeneous reaction 非均相反应
hollow-fiber module 中空纤维组件
homogenization 匀化
hydraulic radius 水力半径
hydrostatic head 液柱静压头
hypersorption 超吸附
ideal gas 理想气体
imbibition 浸润；吸液
immobilization technology 固定化技术；固相化技术
immobilized enzyme reactor 固定化酶反应器
inactivation 失活
information flow diagram 信息流程图
infusion 浸渍；浸泡
initial condition 初始条件
input-output 投入产出
intermediate-product 中间产物
interstitial velocity 空隙速度
internal diffusion 内扩散
ion exchange 离子交换
irreversible process 不可逆过程
isenthalpic process 等焓过程
isochoric process 等容过程
isentropic process 等熵过程
isobaric process 等压过程
isothermal process 等温过程
jet reactor 射流反应器
jet tray 舌形板

key component 关键组分
knowledge base 知识库
Knudsen number 克努森数
kinematic viscosity 运动黏度
kinetic head 动压头
laminar flow 层流；滞流
latent heat 潜热
Laval nozzle 拉瓦尔喷嘴
law of conservation of energy 能量守恒定律
leaching 浸取
lean phase 贫相
lift 扬程
Linde cycle 林德循环
lining 衬里
liquid holdup 持液量；持液率
liquid hourly space velocity (LHSV) 液态空速
liquid-solid extraction 液固萃取
liquid-liquid equilibrium (LLE) 液液平衡
loading point 载点
logarithmic mean temperature difference 对数平均温差
loop reactor 环流反应器
lyophilization 冷冻干燥
Mach number 马赫数
macrofluid 宏观流体
mass flow rate 质量流率；质量流量
manometer 液柱压力计
mass transfer zone (MTZ) 传质区
material balance 物料衡算；物料平衡
mean residence time 平均停留时间
maximum mixedness 最大混合度
membrane module 膜组件
metastable region 亚稳区；介稳区

microcapsule 微胶囊
membrane bioreactor 膜生物反应器
microfiltration 微（孔过）滤
microorganism 微生物
minimum fluidizing velocity 最小流化速度
microwave drying 微波干燥
minimum reflux ratio 最小回流比
mixing lenght 混合长
mockup experiment 冷模试验
molecular sieve 分子筛
molecular thermodynamics 分子热力学
mother liquor 母液
momentum transfer 动量传递
mould 霉菌
multicomponent mixture 多元混合物；多组分混合物
multiple-effect evaporation 多效蒸发
multiphase flow 多相流
multistage compressor 多级压缩机
nanofiltration 纳米过滤
net positive suction head (NPSH) 汽蚀余量；净正吸压头
Newtonian fluid 牛顿流体
non-equilibrium stage model 非平衡级模型
nitrogen fixation 固氮（作用）
noise level 噪声水平
non-equilibrium system 非平衡系统
non-Newtonian fluid 非牛顿流体
nucleic acid 核酸
number of (mass) transfer units (NTU) 传质单元数
Nusselt number 努塞特数

nucleate boiling 泡核沸腾
objective function 目标函数
one-dimensional model 一维模型
operational variable 操作变量
osmotic coefficient 渗透系数
overall heat transfer coefficient 总传热系数
on-line 在线
on-off control 通断控制
parallel feed 平行进料
partial molar quantity 偏摩尔量
particle size 粒度
particle size distribution 粒度分布
particulate fluidization 散式流态化
pelletizing 造粒
percolation 渗滤
perforated plate 多孔板
perfect mixing 全混
peripheral speed 圆周速度
PERT——project evaluation and review technique 项目评审技术
phase equilibrium 相平衡
phase diagram 相图
pilot plant 中间试验装置
Pitot tube 皮托管
point efficiency 点效率
plate efficiency （塔）板效率
plug flow 平推流；活塞流
powder technology 粉体技术；粉体工程
poly-fluid theory 多流体理论
Prandtl number 普朗特数
pressure-enthalpy diagram 压焓图
principle of entropy increase 熵增（加）原理

process optimization 过程优化
process simulation 过程模拟
process system engineering 过程系统工程；化工系统工程
prototype experiment 原型试验
process equipment engineering 化工机械工程
protein engineering 蛋白质工程
quantiment 图像分析仪
quantity meter ［累计］总量表
quantum effect 量子效应
quasilinearization 拟线性化
quasi-static process 准静态过程
quench 冷激，骤冷
R&D——research and development 研究与开发
radial flow reactor 径向反应器
radiation intensity 辐射强度
raffinate 萃余液；抽余液
random process 随机过程
Raoult's law 拉乌尔定律
Raschig ring 拉西环
raw material 原料
RDC——rotating disc contactor 转盘塔
reaction kettle 反应釜
reaction kinetics 反应动力学
reaction mechanism 反应机理
reaction order 反应级数
real gas 真实气体
reboiler 再沸器；重沸器
recombinant DNA 重组 DNA
recovery 回收（率）
reduced temperature 对比温度
rectification 精馏

re-entrainment 二次夹带
reflux ratio 回流比
refrigeration cycle 制冷循环
regeneration 再生
relative volatility 相对挥发度
regression analysis 回归分析
reliability 可靠性
renewable resources 可再生资源
residual enthalpy 残余焓，剩余焓
RTD——residence time distribution 停留时间分布
residual error 残差
residue 残液；釜液
resolution 分辨率
retentate 渗余物
retention 截留；保留
reverse extraction 反萃取
reverse osmosis 反渗透
reversible reaction 可逆反应
Reynolds number 雷诺数
rheological property 流变性质
RNA——ribonucleic acid 核糖核酸
ribose 核糖
Roots blower 罗茨鼓风机
rich phase 富相
ROI——rate of return on investment 投资收益率
root-mean-square error 均方根误差
roughness 粗糙度
rotating extractor 转盘萃取器
saccharification 糖化作用
saddle-point azeotropic mixture 鞍点共沸物
safety factor 安全系数
sand-bed filter 砂滤器

sampled data control system 采样控制系统
Sauter mean diameter 索特平均直径；当量比表面直径
scale 污垢；结垢
scale factor 标度因子
scale up 放大
scheduling of production 生产排序
Schmidt number 施密特数
screw pump 螺杆泵
screen analysis 筛析；筛分
scrubbing 洗涤；水洗
secondary reaction 二次反应
second-order phase transition 二级相变
second law of thermodynamics 热力学第二定律
sedimentation 沉降
seed crystal 晶种
sediment 沉积物
segregation 离析
selective control 选择性控制
semi-continuous process 半连续过程
selectivity coefficient 选择性系数
semi-empirical model 半经验模型
semi-ideal solution 半理想溶液
semipermeable membrane 半透膜
separation factor 分离因子
sequential modular approach 序贯模块法
serial correlation 序列关联
settling 沉降
shake-flask culture 摇瓶培养
shallow bed 浅床
shape factor 形状系数

shelf dryer 厢式干燥器
shell-and-tube heat exchanger 管壳换热器；列管换热器
shell [side] pass 壳程
shrinking core model 缩核模型
shortcut method 简捷法
side cooler 中间冷却器
side reaction 副反应
sieve tray 筛板
sieve diameter 筛孔直径
simple distillation 简单蒸馏
simulated annealing 模拟重结晶法
simulation 模拟；仿真
single crystal 单晶
single-phase flow 单相流
sintering 烧结
size classification 粒度分级
size reduction 粉碎，磨细
SLE——solid-liquid equilibrium 固液平衡
sludge 污泥；淤泥
slugging 节涌；腾涌
sluice separation 淘析
slurry reactor 浆料反应器
solid-liquid separation 固液分离
solubility 溶解度
solute 溶质
solution 溶液
solution polymerization 溶液聚合
solvent extraction 溶剂萃取
STY——space time yield 空时收率
SV——space velocity 空间速率；空速
specific liquid rate 喷淋密度
specific death rate 比死亡速率

specific speed 比转速
specific surface area 比表面积
spiral dryer 螺旋干燥器
sphericity 球形度
spin flash dryer 旋转闪蒸干燥器
splitter 分流器
spontaneous process 自发过程
spouted bed 喷动床
spray density 喷淋密度
stability analysis 稳定性分析
stable state 稳态,稳定状态
stage-by-stage method 逐级计算法
standard error 标准误差
standard state 标准态
state variable 状态变量
static head 静压头
static mixer 静态混合器
statistical model 统计模型
steady state 定态;稳态;定常态
steam jet ejector 蒸汽喷射泵
steam distillation 水蒸气蒸馏
steam stripping 汽提
sterile operation 无菌操作
stiff equation 刚性方程
sterilization filter 除菌滤器
stirred type crystallizer 搅拌结晶器
stirring 搅拌
stochastic control 随机控制
Stokes diameter 斯托克斯直径
stream 物流
streamline flow 层流;滞流
stripping 提馏;解吸;反萃取
stripping section 提馏段
stripping factor 解吸因子
structured packing 整装填料;规整填料
substrate 底物;基质
subsystem 子系统
sudden enlargement 骤扩;突然扩大
sudden contraction 骤缩;突然缩小
supercritical fluid extraction 超临界(流体)萃取
supernatant 上清液
superheating 过热
supporter 载体
surface aerator 表面曝气器
surface diffusion 表面扩散
surface reaction control 表面反应控制
surface renewal theory 表面更新理论
surface work 表面功
surge 喘振
surge tank 缓冲罐
suspension 悬浮液,悬浮
surroundings (热力学)环境
suspension polymerization 悬浮聚合
symmetric membrane 对称膜
tanks-in series model 多釜串联模型
TDH——transport disengaging height (输送)分离高度
temperature gradient 温度梯度
temperature profile 温度(分布)剖面(图)
terminal velocity 终端速度
thawing 融化
theoretical plate 理论(塔)板
thermal diffusivity 热扩散系数;导温系数
thermal efficiency 热效率
thermal insulation 隔热;保温

thermal stability 热稳定性
thermodynamic consistency test 热力学一致性检验
thermodynamic analysis of process 过程热力学分析
thermodynamic characteristic function 热力学特性函数
thermodynamic equilibrium 热力学平衡
thickener 增稠器；浓密机
thickness 稠度
thin-film evaporator 薄膜蒸发器
thixotropy 触变性
three-phase fluidization 三相流态化
throttling process 节流过程
throughput 通过量；产量
tip speed 桨尖速度
tortuosity 曲折因子
total reflux 全回流
transfer function 传递函数
trap 疏水器；汽水分离器
tray 塔板
tray column 板式塔
trickle bed 滴流床；涓流床
triple point 三相点
tube bundle 管束
TSA——temperature swing adsorption 变温吸附
tube (side) pass 管程
turbine agitator 涡轮搅拌器
tunnel dryer 隧道干燥器
turbogrid tray 穿流栅板
turbulent flow 湍流；紊流
two-film theory 双膜理论
turbulent fluidized bed 湍动流化床

two-phase flow 两相流
Tyler standard sieve 泰勒标准筛
ultracentrifuge 超速离心机
ultrafiltration 超滤
underflow 底流；下漏
unit operation 单元操作
uniform conversion model 均匀转化模型
unreacted core model 非反应核模型
unsteady state 非定态；非稳态
U-tube heat exchanger U形管换热器
vacuum distillation 真空蒸馏；减压蒸馏
vacuum pump 真空泵
vapor-liquid equilibrium ratio 汽液平衡比
velocity profile 速度（分布）剖面（图）
Venturi tube 文丘里管
VST——vertical sieve tray 垂直筛板
vibrated fluidized bed 振动流化床
vibrating screen 振动筛
viscoelastic fluid 黏弹性流体
viscoplastic fluid 黏塑性流体
viscosity 黏度
VLE——vapor-liquid equilibrium 汽液平衡
voidage 空隙率
volumetric oxygen transfer coefficient 容积传氧系数
volumetric flow rate 体积流率；体积流量；体积流速
volute 蜗壳
votator apparatus 套管冷却结晶器

wall effect 壁效应
washings 洗涤液
waste heat boiler 废热锅炉
water hammer 水锤
waste water treatment 废水处理
Weber number 韦伯数
weir height 堰高
weighted mean 加权平均

wet bulb temperature 湿球温度
wetted perimeter 润湿周边
wetted wall column 湿壁塔
yeast 酵母
yield 收率
zeolite catalyst 沸石催化剂
zero leakage 零泄漏

2 Flow Diagram for Chemical Process
化工工艺流程图

2.1 Petrochemical processing 石油加工

2.1.1 Diagrammatic flow sheet of petroleum refinery
石油加工流程图

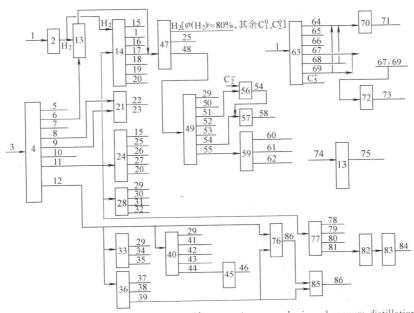

1. LPG (liquefied petroleum gas) 液化石油气
2. hydrogen making 制氢
3. crude oil 原油
4. atmospheric and vacuum distillation 常减压蒸馏
5. $C_3 \sim C_5$ LPG $C_3 \sim C_5$ 液化石油气
6. naphtha 石脑油

2.1 Petrochemical processing 石油加工

7. distilled gasoline 直馏汽油
8. virgin kerosene 直馏煤油
9. straight-run diesel oil 直馏柴油
10. straight-run heavy diesel fuel 直馏重柴油
11. vacuum diesel oil 减压柴油
12. vacuum residuum 减压渣油
13. hydrofining 加氢精制
14. hydrocracking 加氢裂化
15. dry gas 干气
16. light naphtha 轻石脑油
17. heavy naphtha 重石脑油
18. hydrocracking aviation fuel 加氢裂化航空燃料
19. hydrocracking diesel oil 加氢裂化柴油
20. tail oil 尾油
21. urea dewaxing 尿素脱蜡
22. low-freezing oil 低凝点油
23. liquid paraffin 液蜡
24. catalytic cracking 催化裂化
25. liquefied gas 液化气
26. catalytic cracked gasoline 催化裂化汽油
27. catalytic diesel oil 催化裂化柴油
28. thermal cracking 热裂化
29. petrogas 石油气
30. thermal gasoline 热裂化汽油
31. thermal diesel oil 热裂化柴油
32. residual oil 渣油
33. visbreaking 减黏裂化
34. raw gasoline 粗汽油
35. mazut 重油
36. solvent deasphalting 溶剂脱沥青
37. raw lubricating oil 润滑油料
38. raw materials for catalytic cracking 催化裂化原料
39. detarring asphalt 脱油沥青
40. delayed coking 延迟焦化
41. coking gasoline 焦化汽油
42. coking diesel oil 焦化柴油
43. coking waxy oil 焦化蜡油
44. coke 焦炭
45. calcination 煅烧
46. petrol coke 石油焦
47. catalytic reforming 催化重整
48. reformate 重整油
49. separation and reforming of aromatics 芳烃分离与转化
50. raffinate oil 抽余油
51. benzene 苯
52. methylbenzene 甲苯
53. dimethylbenzene 二甲苯
54. ethylbenzene 乙苯
55. heavy aromatic 重芳烃
56. hydrocarbylation 烃基化
57. dehydrogenation 脱氢
58. styrene 苯乙烯
59. heavy aromatics separation 重芳烃分离
60. blended gasoline 调和汽油
61. trimethylbenzene 三甲苯
62. durene 均四甲苯
63. gas separating 气体分离
64. propylene 丙烯
65. propagas 丙烷
66. normal butane 正丁烷
67. isobutane 异丁烷
68. isobutene 异丁烯
69. n-butene 正丁烯
70. polyforming 叠合重整
71. polybenzine 叠合汽油
72. alkylation 烷基化
73. gasoline alkylate 烷基化汽油
74. gasoline, kerosene or diesel fuel 汽油, 煤油或柴油
75. hydrogenated gasoline, hydrogenated kerosene or hydrogenated diesel fuel 加氢汽油、加氢煤油或加氢柴油
76. oxidized asphalt 氧化沥青
77. lubricating oil 润滑油
78. component solvent 混合溶剂
79. paraffin wax 石蜡
80. extractive oil 抽出油
81. slack wax 蜡膏
82. SDO (solvent deoiling) 溶剂脱油
83. fine purification 精制
84. paraffin wax or floor wax 石蜡或地蜡
85. blend 调和
86. asphalts 沥青

2.1.2　Crude oil drilling and oil production　石油钻探和采油

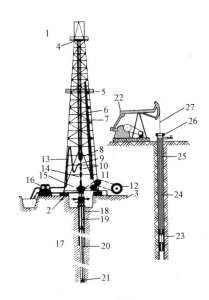

1　drilling rig　钻井机械

2　substructure　井架底座

3　crown safety platform　架顶安全平台

4　crown blocks　顶滑车，天车

5　working platform, an intermediate platform　工作平台，一个中间平台

6　drill pipes　钻杆

7　drilling cable　钻油井用的钢丝绳

8　travelling block　游动滑车

9　hook　大钩

10　swivel　旋转龙头

11　draw works　绞车

12　engine　发动机

13　standpipe and rotary hose　竖管和（旋转）泥浆管

14　kelly　方钻杆

15　rotary table　转盘

16　slush pump; mud pump　泥浆泵

17　well　钻井

18　casing　套管

19　drilling pipe　钻杆

20　tubing　井管

21　drilling bit　钻头

22　pumping unit　抽油装置

23　plunger　柱塞

24　tubing　井管，油管

25　sucker rods (pumping rods)　抽油杆（泵杆）

26　stuffing box　填料盒

27　polish (polished) rod　抛光杆

2.1.3 Crude unit with atmospheric and vacuum towers 具有常压塔和减压塔的原油蒸馏装置

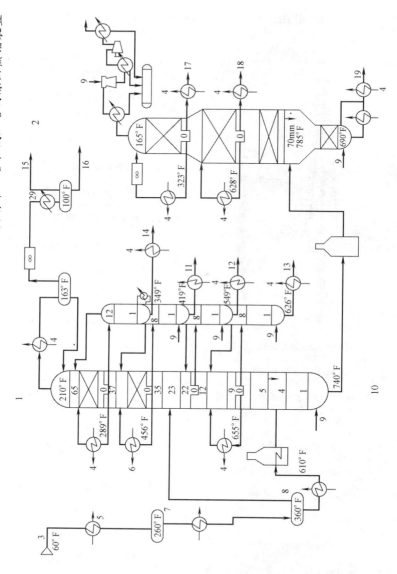

2.1.4 Crude atmospheric tower 原油常压塔

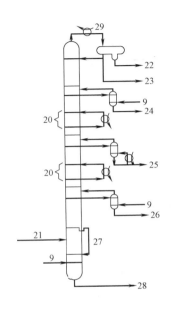

1 atmospheric tower 常压塔

2 vacuum tower 减压塔

3 crude charge 原油加入

4 crude 原油

5 heat exchange process 换热过程

6 crude and reboiler 原油与再沸器

7 desalter 脱盐设备

8 flash drum 闪蒸罐

9 STM（steam） 水蒸气

10 overflash set at 3 percent on feed 指定在进料3%的超闪蒸

11 kerosene 煤油

12 diesel 柴油

13 heavy gas oil 重柴油

14 HSR naphtha HSR 石脑油

15 gas 气体

16 LSR naphtha LSR 石脑油

17 LVGO（light vacuum gas oil） 轻减压粗柴油

18 HVGO（heavy vacuum gas oil） 重减压粗柴油

19 RESID（residuum） 残渣油

20 pump around 泵循环回流

21 crude oil 原油

22 water 水

23 overhead 塔顶馏出物

24 heavy naphtha 重石脑油

25 light distillate 轻馏出物

26 heavy distillate 重馏出物

27 overflash 超闪蒸

28 bottoms 塔底残留物

29 CW（cooling water） 冷却水

2.1.5 Catalytic cracking unit 催化裂化装置

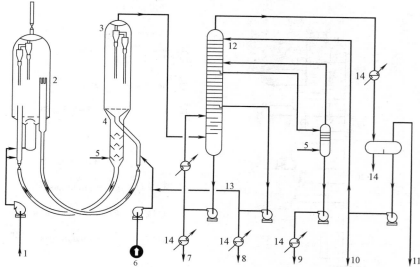

1. combustion air 氧化用空气
2. regenerator 再生器
3. reactor 反应器
4. catalyst stripper 催化汽提塔
5. steam 水蒸气
6. gas oil charge 加入粗柴油
7. heavy catalytic gas oil 重催化裂化粗柴油
8. intermediate catalytic gas oil 中间催化裂化粗柴油
9. light catalytic gas oil 轻催化裂化粗柴油
10. unstabilized gasoline 不稳定汽油
11. wet gas 湿气体
12. fractionator 分馏塔
13. recycle 再循环
14. water 水

2.1.6 Delayed-coking unit 延迟焦化装置

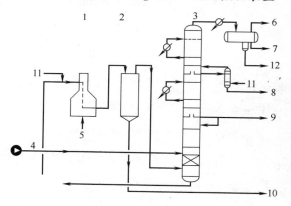

1. heater 加热器
2. coke drum 焦炭罐
3. main fractionator 主分馏塔
4. fresh feed 进料
5. fuel 燃料
6. gas 气体
7. coker naphtha 焦化石脑油
8. LCGO (light coking gas oil) 轻焦化粗柴油
9. MCGO (intermediate coking gas oil) 中间焦化粗柴油
10. coke 焦炭
11. steam 水蒸气
12. water 水

2.1.7 Oil-storage tank and main circulating loop
储油罐及主要循环回流管

1 supply and return lines for any number of branch circuits 供油与回油线,可供多个支线用
2 main-inlet shutoff valve for branch circuit 支线用主入口截止阀
3 low-pressure return line 低压回油线
4 check valve 止逆阀
5 shutoff valve 截止阀
6 for overhead tanks only 压力罐用
7 gauge well 表孔
8 manhole 人孔
9 vent 放空
10 ground level 地平
11 traps 阱
12 oil-storage tank 储油罐
13 fill pipe 加入管
14 tank suction heater (not required for light oil) 油罐吸入加热器（对轻油不需要）
15 suction line 吸入管
16 duplex oil filter 复式滤油器
17 duplicate pumps with integral pressure-relief valve 带内泄压阀的复式泵
18 pressure gauge 压力表
19 high-pressure supply line 高压供油线
20 diaphragm relief valve 膜式泄压阀
21 bypass valve 旁路阀

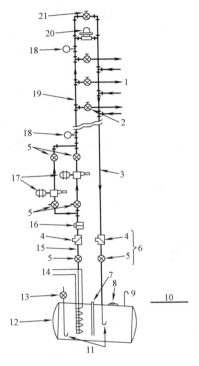

2.1.8 Koppers-Totzek oil gasfication process
Koppers-Totzek 油气化法

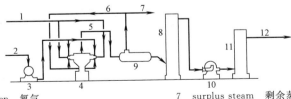

1 oxygen 氧气
2 oil 油
3 service bin 供油器
4 gasifire 气化炉
5 gas 气体
6 steam 水蒸气
7 surplus steam 剩余蒸汽
8 washer/cooler 洗涤塔/冷却塔
9 waste-heat boiler 废热锅炉
10 disintegrator 气体洗涤机
11 drop separator 液滴分离器
12 synthesis gas 合成气

2.1.9 Processing of oil sands 油砂加工

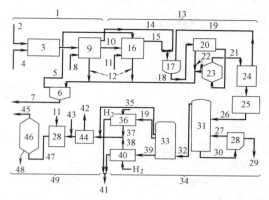

1. primary separation 一次分离
2. oil sand 油砂
3. tumbler mixer 转筒混合机
4. hot water/steam/caustic solution 热水、蒸汽、碱液
5. residual bitumen 剩余沥青
6. API separator 美国石油学会标准分离器
7. to tailings pond 去尾渣池
8. water 水
9. primary separation tank 一次分离罐
10. middlings 中间产物
11. air 空气
12. sand and water 砂和水
13. secondary separation 二次分离
14. primary froth 一次泡沫
15. secondary froth 二次泡沫
16. secondary separation tank 二次分离罐
17. dilution tank 稀释罐
18. diluted froth 稀释过的泡沫
19. naphtha 粗汽油
20. scroll-type centrifuge 涡旋式离心机
21. bitumen 沥青
22. solids 固体粒子
23. disk-type centrifuge 盘式离心机
24. naphtha recovery tank 粗汽油回收罐
25. hot-bitumen storage 热沥青贮罐
26. hot bitumen (sprayed in) 热沥青（喷入）
27. hot coke 热焦炭
28. burner 燃烧炉
29. product coke 焦炭产品
30. cold coke 冷焦炭
31. fluidized-bed coker 流化床焦化塔
32. vapor product 蒸气产品
33. fractionator 分馏塔
34. upgrading 改质
35. process gas 工艺气体
36. naphtha hydrotreater 粗汽油加氢处理装置
37. H_2 + light ends 氢气+轻馏分
38. naphtha + butane 粗汽油+丁烷
39. gas oil 瓦斯油
40. gas oil hydrotreater 瓦斯油加氢处理装置
41. synthetic crude (to pipeline) 合成的粗产物（去管线）
42. process gas (burned in utility plant) 工业废气（供公用电站燃烧）
43. H_2S (gas) H_2S（气体）
44. amine scrubber 脱胺塔
45. tailgas 尾气
46. Claus converter 克劳斯转化器
47. H_2S (gas) + SO_2 (gas) H_2S（气）+ SO_2（气）
48. S_8 (liquid) 单斜晶硫（液相）
49. desulfurization 脱硫

2.1.10 Resid hydrocracker reactor 渣油加氢裂化反应器

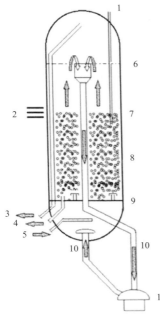

1　catalyst addition　催化剂加入
2　level detectors　液面测定仪
3　gas/liquid product to separators　气体/液体产品去分离器
4　catalyst withdrawal　催化剂出料
5　makeup H_2 and feed oil　补充氢气和油进料
6　max. LQ. level　最高液面
7　expanded level　膨胀液面
8　settled catalyst level　沉积催化剂液面
9　distributor grid plate　分布器栅板
10　recycling oil　循环油
11　ebullition pump　沸腾泵

2.2　Chemical processing of coal　煤化工

2.2.1　Technigues for chemical processing of coal　煤化工技术路线

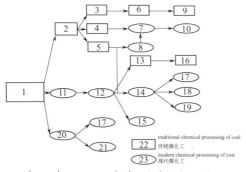

1　coal energy and chemical industrial chain　煤炭能源化工产业链
2　coking　煤焦化
3　coke　焦炭
4　coal tar　煤焦油
5　coke oven gas　焦炉气
6　calcium carbide　电石
7　crude benzol refining　粗苯精制
8　hydrogen production　制氢
9　acetylene　乙炔
10　BDO　丁二醇
11　coal gasification　煤气化
12　syngas　合成气
13　synthetic ammonia　合成氨
14　methanol　甲醇
15　indirect liquefaction　间接液化
16　nitrogen fertilizer　氮肥
17　olefin hydrocarbon　烯烃
18　dimethylether　二甲醚
19　acetic acid　乙酸
20　coal liquefaction　煤液化
21　gasoline and diesel oil　汽油和柴油

2.2.2 The production of clean fuels from coal 从煤制造洁净燃料

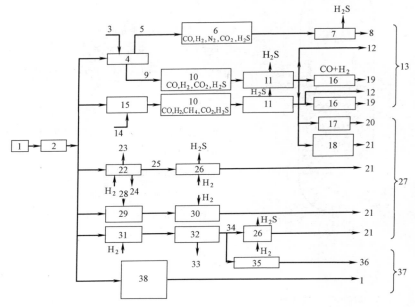

1. coal 煤
2. preparation 制备
3. H_2O + air or O_2 H_2O+空气或O_2
4. gasification 气化
5. when air 空气时
6. low-Btu 低热值
7. purification 净化
8. low-Btu 低热值
9. when O_2 氧气时
10. medium-Btu 中热值
11. shift conversion and purification 变换与净化
12. medium-Btu 中热值
13. clean gaseous fuels 洁净气体燃料
14. H_2, steam or synthesis H_2、蒸汽或合成气
15. hydro-gasification 加氢气化
16. methanation 甲烷化
17. methanol synthesis 甲醇合成
18. Fischer-Tropsch synthesis 费-托合成
19. high-Btu 高热值
20. methanol 甲醇
21. hydro-carbon 烃类
22. pyrolysis 热解
23. gas 气体
24. char 炭
25. oils 油
26. hydrotreating 加氢处理
27. clean liquid fuels 洁净液体燃料
28. coal-derived liquid 煤衍生液体
29. slurry preparation 煤浆制备
30. catalytic hydrogenation 催化加氢
31. dissolution in solvent 溶剂溶解
32. filtration and solvent removal 过滤与脱溶剂
33. air, pyritic sulfur 空气, 黄铁矿的硫
34. syn-crude 合成原油
35. solidification 固化成型
36. solvent refined coal 溶剂精制煤
37. clean solid fuels 洁净固体燃料
38. direct desulfurization by physical, chemical or thermal treatment 用物理、化学或热处理方法直接脱硫

2.2.3 Gasification and related technologies 气化和相关技术

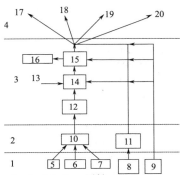

1 feedstocks 原料
2 gasification 气化
3 syngas processing 合成气加工
4 products 产品
5 coal 煤炭
6 petcoke 石油焦
7 biomass 生物质
8 natural gas 天然气
9 other gas 其他气体
10 gasification 气化
11 steam reforming 蒸汽重整
12 impurity removal 杂质脱除
13 steam 蒸汽
14 water gas shift 水气变换
15 CO_2 removal 二氧化碳脱除
16 sequestration 储存
17 hydrogen, electric power 氢气、电力
18 ammonia, nitrogen fertilizers 氨、氮肥
19 methanol, dimethyl ether, hydrocarbons 甲醇、二甲醚、碳氢化合物
20 substitute natural gas, Fischer-Tropsch hydrocarbons 替代天然气、费-托法合成烃类化合物

2.2.4 Lurgi-Ruhrgas process 鲁奇-鲁尔公司工艺

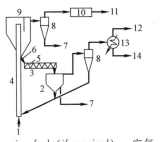

1 air, fuel (if required) 空气、燃料（如需要）
2 solids surge bin 固体颗粒缓冲仓
3 retort 干馏器
4 lift pipe 提升管
5 shale feed 油页岩进料
6 hot solids 热固体颗粒
7 waste solids 废固体颗粒
8 cyclone 旋风分离器
9 separation bin 分离仓
10 heat-recovery unit 热回收装置
11 gaseous waste 废气
12 gas product 气体产品
13 condenser 冷凝器
14 shale oil 页岩油

2.2.5　Tosco Ⅲ process　美国油页岩公司Ⅲ工艺

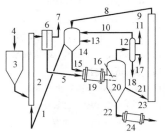

(TOSCO——the oil shale corporation of America)
1　hot flue gas　热烟道气
2　lift pipe　提升管
3　surge hopper　缓冲装料斗
4　shale feed　油页岩进料
5　preheated shale　已预热油页岩
6　separator　分离器
7　flue gas (to scrubber, atmosphere)　烟道气（去涤气器，放空）
8　balls　球状物，小球
9　ball elevator　小球提升机
10　gas　煤气
11　naphtha　石脑油
12　fractionator　分馏塔
13　air　空气
14　ball heater　小球加热器
15　hot balls　热球
16　trommel　滚筒回转筛
17　gas oil　瓦斯油
18　resid　残液
19　pyrolysis drum　裂解转筒
20　accumulator　收集器
21　warm balls　温热小球
22　hot spent shale　热的废页岩
23　spent shale (to disposal)　废页岩（去处置）
24　cooler　冷却器

2.2.6　IGT process　（美）煤气工艺研究所工艺

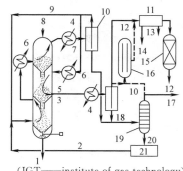

(IGT——institute of gas technology)
1　spent shale　用过的油页岩
2　makeup hydrogen　补充氢气
3　reactor　反应器
4　cooler　冷却器
5　product hydrocarbons　产物碳氢化合物
6　heater　加热器
7　light oil　轻油
8　shale feed　油页岩进料
9　hydrogen recycle　氢循环
10　separator　分离器
11　purification unit　提纯装置
12　gas product　气体产品
13　sulfur　硫黄
14　carbon dioxide　二氧化碳
15　methanator　甲烷转化器
16　hydrogasifier　加氢气化器
17　oil product　油产品
18　liquid product　液态产品
19　fractionator　分馏塔
20　heavy oil　重油
21　hydrogen plant　氢气生产装置

2.2.7 Petrosix process Petrosix 工艺

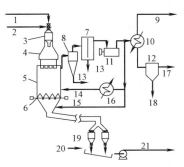

1　shale feed　油页岩进料
2　seal gas　密封气
3　feed hopper　加料斗
4　shale distributor　油页岩分布器
5　pyrolysis vessel　热解容器
6　discharge mechanism　卸料机械装置
7　electrostatic precipitator　静电除尘器
8　cyclone　旋风分离器
9　gas product　气体产品
10　condenser　冷凝器
11　compressor　压缩机
12　separator　分离器
13　heavy shale oil　重质页岩油
14　hot gas　热气体
15　cool gas　冷气体
16　heater　加热器
17　light shale oil　轻质页岩油
18　waste water　废水
19　seal system　密闭系统
20　water　水
21　retorted shale slurry (to disposal)　干馏过的页岩淤浆（去处置）

2.2.8 Lurgi process 鲁奇（煤气化）法

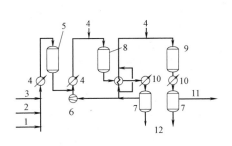

1　hydrogen　氢气
2　nitrogen　氮气
3　feed gas　原料气
4　steam　水蒸气
5　ZnO reactor　氧化锌反应器
6　compressor　压缩机
7　separator　分离器
8　main methanation reactor　主甲烷化反应器
9　final methanation reactor　最终甲烷化反应器
10　C.W (cooling water)　冷却水
11　product gas　成品气体
12　condensate　冷凝液

2.2.9　Chem systems process　化工系统过程

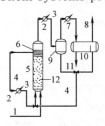

1　feed gas　原料气
2　boiler-feed water　锅炉给水
3　steam　水蒸气
4　inert liquid　惰性液体
5　reactor　反应器
6　liquid-vapor interface　液-气相界面
7　C.W（cooling water）　冷却水
8　product gas　产品气

2.2.10　Parsons process Parson工艺

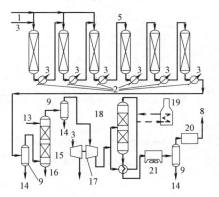

9　separator　分离器
10　decanter　倾析器
11　water　水
12　fluid catalyst bed　催化剂流化床
13　fresh liquor　新鲜溶液
14　condensate　冷凝液
15　CO_2 absorber　CO_2吸收塔
16　spent liquor　废液
17　compressor　压缩机
18　methanator　甲烷转化器
19　furnace　燃烧炉
20　glycol dryer　乙二醇干燥器
21　air cooler　空气冷却器

2.2.11　Lurgi gasifier　鲁奇煤气化炉

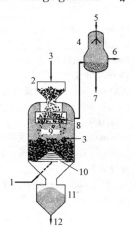

(uses a rotating grate underneath the coal bed for feeding oxygen and steam)
在煤床下设旋转凸栅并供入氧气及水蒸气

1　steam, air or O_2　水蒸气、空气或氧气
2　coal lock hopper　闭锁式煤斗
3　coal　煤
4　scrubber　涤气器
5　scrubbing liquid　洗涤液
6　raw synthesis gas　粗制合成气
7　dust and tar　粉尘与焦油
8　coal distributor　煤分布器
9　reactor　反应器
10　rotating grate　旋转凸形栅
11　ash lock hopper　闭锁式煤渣斗
12　ash　煤渣

2.2.12 Lurgi gasifier 鲁奇煤气化炉

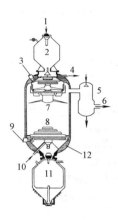

(coal enters the top of the gasifier and flows countercurrently to an ascending stream of air and steam 煤从炉顶加入，与上升的空气及蒸汽流作逆流流动)

1　coal　煤
2　coal lock　闭锁式煤斗
3　drive　传动装置
4　steam　水蒸气
5　scrubbing cooler　涤气冷却器
6　gas　煤气
7　distributor　布料器
8　grate　栅板
9　grate drive　转栅驱动器
10　steam+oxygen　水蒸气+氧气
11　ash lock　闭锁式灰斗
12　water jacket　水夹套

2.2.13 Lurgi dry ash gasifier 鲁奇干煤灰气化炉

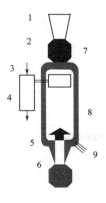

1　coal　煤
2　coal bunker　煤仓
3　quench liquid　急冷液
4　quench cooler　急冷室
5　grate　炉栅
6　ash lock　灰锁斗
7　lock hopper　煤锁斗
8　gasifier　气化炉
9　steam/oxygen　蒸汽/氧气

2.2.14　The slagging gasifier　熔渣气化炉

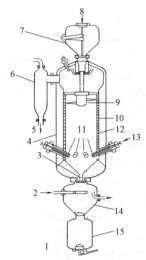

(uses a minimum of live steam　使用少量新鲜蒸汽)

1　source：British Gas Corp　来源：英国天然气协会
2　circulating quench-water　循环急冷水
3　slag tap　渣排口
4　pressure shell　受压外壳
5　gas outlet　煤气出口
6　gas quench　煤气急冷
7　coal lock-hopper　密封煤斗
8　feed coal　原料煤
9　coal distributor/stirrer　煤分布器/搅拌器
10　refractory lining　耐火材料衬里
11　tuyeres　吹风管嘴
12　water jacket　水夹套
13　steam/oxygen　水蒸气/氧气
14　slag quench-chamber　熔渣急冷室
15　slag lock-hopper　密封渣斗

2.2.15　Schematic of a typical GSP gasifier　典型的GSP气化炉示意

1　natural gas　天然气
2　tar/slurry　焦油浆
3　oil　油
4　cooling water　冷却水
5　quenching water　急冷水
6　gas　煤气
7　soot/water　煤灰水
8　slag　炉渣

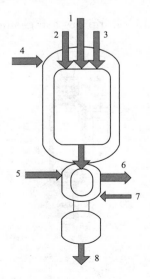

2.2.16 Schematic of a typical GSP plus gasifier 典型的 GSP+气化炉示意

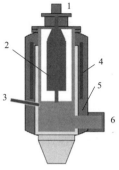

1. coal, oxygen, steam in 煤,氧气,蒸汽进入
2. suspended combustor 悬浮式燃烧器
3. quench water in 急冷水 进入
4. cooling wall 冷却壁
5. refractory 耐火材料
6. syngas out 合成气出口

2.2.17 Schematic of an ECUST gasifier ECUST 气化炉示意

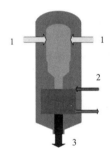

1. coal/water slurry, oxygen 水煤浆,氧气
2. quench water 急冷水
3. ash 灰分

2.2.18 The Winkler gasifier 温克勒气化炉

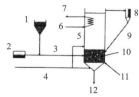

1. coal bunker 煤仓
2. screw feeder 螺旋给料机
3. coal 煤
4. steam + air or O_2 蒸汽+空气或氧气
5. free board 自由空间
6. water 水
7. steam 蒸汽
8. hydrocyclone 水力旋流分离器
9. fines return 细粉回流
10. bubbling fluidized bed 鼓泡式流化床
11. rotating grate 旋转炉栅
12. bottom ash 炉底灰

2.2.19 Schematic of a typical Winkler gasifier 典型的温克勒气化炉示意

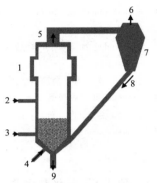

1　gasifier　气化炉
2　oxygen/air　氧气/空气
3　ground coal　碾碎的煤
4　steam, oxygen　蒸汽，氧气
5　raw gas　粗制煤气
6　clean gas　净化煤气
7　cyclone　旋风分离器
8　recovered fines　回收的粉末
9　ash removal　灰分去除

2.2.20 Pressurized feed system for the high temperature Winkler gasifier 用于高温温克勒气化炉的加压进料系统

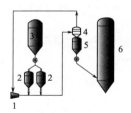

1　compressor　压缩机
2　lock hopper　闭锁式料斗
3　burner　燃烧炉
4　filter　过滤器
5　charge bin　加料料仓
6　gasifier　气化炉

2.2.21 ConocoPhillips gasifier 康菲气化炉

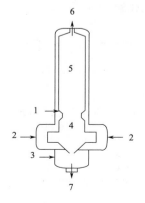

1　coal slurry　煤浆
2　coal slurry/O_2　煤浆/氧气
3　slag quench water　炉渣急冷水
4　first stage　一段
5　second stage　二段
6　product gas　成品气
7　slag/water slurry　水炉渣浆

2.2.22 ConocoPhillips E-Gas gasifier 科诺克飞利浦 E-Gas 气化炉

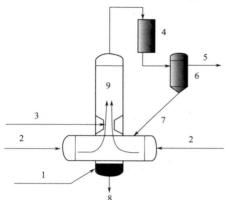

1　quench water　急冷水
2　1st stage coal/water slurry, O_2　一段水煤浆，氧气
3　2nd stage coal/water slurry　二段水煤浆
4　steam generator　蒸汽发生器
5　syngas　合成气
6　candle filter　烛型过滤器
7　char　灰渣
8　slag/water slurry　水炉渣混合浆
9　gasifier　气化炉

2.2.23 The Foster-Wheeler partial gasifier Foster-Wheeler 部分气化炉

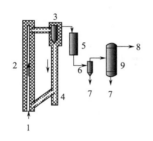

1　coal, air, steam　煤，空气，蒸汽
2　gasifier body　气化炉主体
3　recycle cyclone　循环旋风分离器
4　solid standpipe　固体煤粒立管
5　snygas cooler　合成气冷却器
6　pre-cleaner cyclone　预净化旋风分离器
7　char　灰渣
8　syngas　合成气
9　candle filter　烛型过滤器

2.2.24 The transport gasifier 输送床气化炉

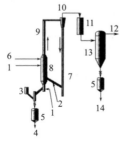

1　air, O_2, steam　空气，氧气，蒸汽
2　recycle syngas　循环合成气
3　startup burner　开工烧嘴
4　bottom ash　炉底灰
5　ash depressurization　灰分卸压
6　coal from lock hoppers　从煤锁斗来的煤
7　standpipe　立管
8　mixing zone　（气化炉）混合区
9　riser　提升器
10　solid separation unit　固体分离器
11　primary gas cooler　主煤气冷却器
12　syngas　合成气
13　particulate control device　微粒控制设备
14　fly ash　粉煤灰

2.2.25 The SHELL coal gasifier 壳牌煤气化炉

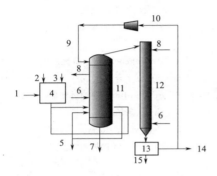

1　N_2　氮气
2　coal　煤炭
3　fly ash　粉煤灰
4　grinding, drying, & feeding　研磨，干燥，进料
5　O_2　氧气
6　boiler feed water　锅炉给水
7　slag　炉渣
8　steam　蒸汽
9　quench gas　急冷气
10　recycle compressor　循环压缩机
11　gasifier　气化炉
12　syngas cooler　合成气冷却器
13　dry solid removal　干固体去除
14　syngas　合成气
15　fly ash(recycled)　粉煤灰（循环的）

2.2.26 SIEMENS gasifier 西门子气化炉

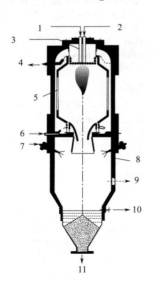

1　fuel　燃料
2　O_2/steam　氧气/蒸汽
3　burner　烧嘴
4　pressure, water outlet　压力水出口
5　cooling screen　冷却帘栅
6　pressure, water inlet　压力水进口
7　quench water　急冷水
8　cooling jacket　冷却夹套
9　gas outlet　煤气出口
10　water overflow　溢流水
11　granulated slag　粒状炉渣

2.2.27 Mono-ethanol-amina (MEA) based process for CO_2 removal from flue gas 一乙醇胺法脱除烟气中的 CO_2

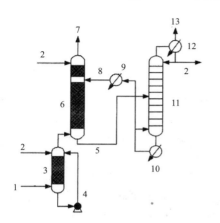

1. flue gas 烟气
2. water 水
3. direct contact cooler 直接接触式冷却器
4. water recycle 水循环
5. CO_2-loaded solution 含二氧化碳溶液
6. absorber 吸收塔
7. CO_2-free flue gas to atmosphere 无二氧化碳烟气排放到大气
8. MEA solution 一乙醇胺溶液
9. cooler 冷却器
10. reboiler 再沸器
11. stripper 解吸塔
12. condenser 冷凝器
13. CO_2 to compression 二氧化碳去压缩

2.2.28 Portion of a RECTISOL plant designed to process syngas prior to ammonia synthesis and subsequent conversion of ammonia and CO_2 to urea. 低温甲醇洗净化技术（RECTISOL）中，还有氨合成以及氨与二氧化碳制成尿素的工艺（未展示）

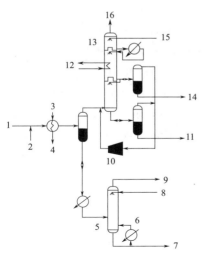

1. syngas feed 合成气进料
2. methanol 甲醇
3. cold product gas 冷成品气
4. warm product gas 热成品气
5. water splitter 水分离塔
6. steam 蒸汽
7. water 水
8. liquid methanol 液体甲醇
9. methanol vapor 甲醇蒸气
10. flash gas compressor 闪蒸气体压缩机
11. methanol + H_2S 甲醇 + 硫化氢
12. refrigerant 制冷剂
13. primary column 主塔
14. methanol + CO_2 甲醇 + 二氧化碳
15. chilled methanol 急冷甲醇
16. purified syngas 净化合成气

2.2.29 Schematic of a typical GE coal gasification system with quench chamber 典型的带急冷室的GE煤气化系统示意

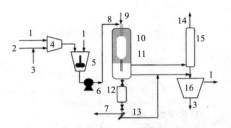

1 water 水
2 coal 煤
3 recycled solids 循环固体
4 mill 碾磨机
5 slurry tank 煤浆槽
6 slurry pump 煤浆泵
7 slag 炉渣
8 coal/water slurry 水煤浆
9 O_2 氧气
10 gasifier 气化炉
11 quench chamber 急冷室
12 lock hopper 灰锁斗
13 slag screen 炉渣筛
14 syngas 合成气
15 gas scrubber 气体洗涤塔
16 clarifier 澄清器

2.2.30 Schematic of typical GE gasification system with radiant syngas cooler 典型的带辐射合成气冷却器的GE煤气化系统示意

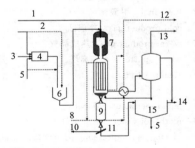

1 oxygen 氧气
2 makeup water 补充水
3 coal 煤
4 mill 碾磨机
5 recycled solids 循环固体
6 slurry tank 煤浆槽
7 gasifier 气化炉
8 boiler feed water 锅炉给水
9 lock hopper 灰锁斗
10 slag 炉渣
11 screen 筛子
12 steam 蒸汽
13 syngas 合成气
14 water 水
15 clarifier 澄清器

2.2.31 Schematic of typical SHELL gasification process MP. medium pressure 典型的壳牌气化工艺示意（MP，中压）

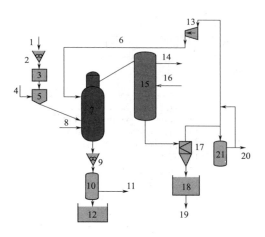

1　biomass　生物质
2　pulverizer　粉磨机
3　silo　筒仓
4　nitrogen　氮气
5　hopper　料斗
6　quench gas　急冷气
7　gasifier　气化炉
8　oxygen/steam　氧气/蒸汽
9　slag crusher　碎渣机
10　slag removal　除渣
11　slag　炉渣
12　slag water treatment　渣水处理
13　compressor　压缩机
14　superheated steam　过热蒸汽
15　syngas cooler　合成气冷却器
16　MP steam　中压蒸汽
17　dry solids removal　干固体去除
18　fly ash silo　粉灰筒仓
19　fly ash　粉灰
20　syngas　合成气
21　wet scrubbing　湿法涤气器

2.2.32 SELEXOL process for the selective removal of H_2S and CO_2 选择性脱除硫化氢和二氧化碳的 SELEXOL 工艺流程

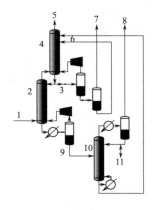

1　syngas　合成气
2　sulfur absorber　硫吸收塔
3　CO_2 loaded solvent　含有二氧化碳的溶剂
4　CO_2 absorber　二氧化碳吸收塔
5　purified syngas　净化合成气
6　lean solvent　贫溶剂
7　CO_2　二氧化碳
8　H_2S　硫化氢
9　H_2S loaded solvent　含有硫化氢的溶剂
10　H_2S stripper　硫化氢汽提塔
11　water　水

2.2.33 2-Stage Rectisol wash for ammonia/urea plant 用于合成氨/尿素厂的两级低温甲醇洗工艺流程

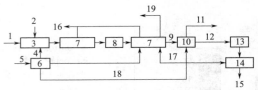

1 feed stock coal 原料煤
2 H_2O 水
3 partial oxidization 部分氧化
4 oxygen 氧气
5 air 空气
6 air separation unit 空分装置
7 rectisol wash 低温甲醇洗
8 conversion 变换
9 H_2 氢气
10 nitrogen wash 氮洗
11 nitrogen wash tail gas 氮洗尾气
12 N_2+H_2 氮气+氢气
13 ammonia synthesis 合成氨
14 urea synthesis 尿素合成
15 product urea 产品尿素
16 H_2S 硫化氢
17 CO_2 二氧化碳
18 N_2 氮气
19 N_2+CO_2 氮气+二氧化碳

2.2.34 Process of syngas conversion unit 合成气变换单元工艺流程

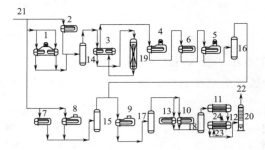

1 E01- water gas waste heat boiler Ⅰ E01-水煤气废热锅炉Ⅰ
2 E02- feed water heater for intermediate pressure boiler E02-中压锅炉给水加热器
3 E03- intermediate temperature heat exchanger/steam filter E03-中温换热器/蒸汽过滤器
4 E04- waste heat boiler of shift converter Ⅰ E04-变换炉废热锅炉Ⅰ
5 E05- waste heat boiler of shift converter Ⅱ E05-变换炉废热锅炉Ⅱ
6 E06- low pressure steam superheater E06-低压蒸汽过热器
7 E07- low pressure steam superheater E07-低压蒸汽过热器
8 E08- water gas waste heat boiler Ⅱ E08-水煤气废热锅炉Ⅱ
9 E09- low pressure waste heat boiler E09-低压废热锅炉

10　E11- feed water heater for low pressure boiler　E11-低压锅炉给水加热器

11　E12- desalinated water heater　E12-脱盐水加热器

12　E13- shift gas water cooling　E13-变换气水冷器

13　E17- feed water heater for intermediate pressure boiler　E17-中压锅炉给水加热器

14　V01- first water separation column　V01-第一水分离器

15　V02- second water separation column　V02-第二水分离器

16　V03- third water separation column　V03-第三水分离器

17　V04- fourth water separation column　V04-第四水分离器

18　V07- fifth water separation column　V07-第五水分离器

19　R01- shift converter　R01-变换炉

20　T01- ammonia washing tower　T01-洗氨塔

21　crude gas from coal gasification　来自煤气化装置粗煤气

22　to Rectisol process　去低温甲醇洗

23　cooling water　冷却水

24　CWR　冷却水回路

2.2.35　Schematic of coal slurry preparation unit　煤浆制备单元示意

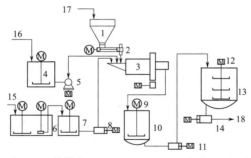

1　coal storage hopper　储煤斗
2　coal gravimetric feeder　煤称量给料机
3　pulverizer　磨煤机
4　water tank with grinder　研磨水槽
5　grinder pump with water　研磨水泵
6　underground tank for additives　添加剂地下槽
7　additive tank　添加剂槽
8　additive pump　添加剂泵
9　mixer for coal slurry discharge tank　煤浆出料槽搅拌器
10　coal slurry discharge tank　煤浆出料槽
11　pulverizer discharge tank pump　磨煤机出料槽泵
12　mixer for coal slurry tank　煤浆槽搅拌器
13　coal slurry tank　煤浆槽
14　coal slurry pump　煤浆泵
15　additive　添加剂
16　fresh water　新鲜水
17　feed stock coal　原料煤
18　to gasifier　去气化炉

2.2.36 Process of methane production from coal via syngas 煤经合成气制甲烷工艺流程

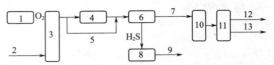

1. air separation unit 空分装置
2. coal 煤
3. coal gasification 煤气化
4. conversion 变换
5. adjust H_2/CO ratio 调节 H_2/CO 比值
6. desulfurization and decarbonization 脱硫脱碳
7. clean syngas 洁净合成气
8. sulfur recovery 硫回收
9. sulfur or sulfuric acid 硫或硫酸
10. methanation 甲烷化
11. separation 分离
12. natural gas to pipeline network 天然气去管网
13. water and naphtha, etc 水、石脑油等

2.2.37 Ammonia synthesis loop, showing a quench reactor 氨合成回路,显示了急冷反应器

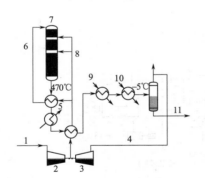

1. fresh feed gas 新鲜的原料气
2. feed compressor 进料压缩机
3. recycle compressor 循环压缩机
4. recycle gas 循环气
5. steam 蒸汽
6. feed gas (400℃) 原料气
7. synthesis reactor 合成反应器
8. quench gas 急冷气(150℃)
9. cooling water 冷却水
10. refrigerant 冷冻剂
11. liquid ammonia 液氨

2.2.38 Ammonia synthesis process 合成氨工艺流程

1. coal 煤
2. gasification agent 气化剂
3. gasification 造气
4. H_2, CO, CO_2, H_2S, N_2, etc 氢气、一氧化碳、二氧化碳、硫化氢、氮气等
5. conversion 变换
6. desulfurization 脱硫
7. CO_2, N_2, H_2, CO 二氧化碳,氮气,氢气,一氧化碳
8. decarbonization 脱碳
9. little amount of CO, CO_2, N_2, H_2 少量 CO, CO_2, N_2, H_2
10. cryogenic liquid ammonia washing 低温液氨洗
11. N_2, H_2 氮气和氢气
12. ammonia synthesis 氨合成
13. NH_3 氨
14. ammonia refrigeration 氨冷冻
15. liquid ammonia 液氨

2.2.39 Design of catalytic reactor 催化反应器的设计

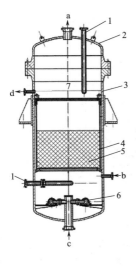

1 sleeves for thermocouples 热电偶套
2 cover 顶盖
3 reactor housing 反应器壳体
4 basket 催化剂筐
5 catalyst 催化剂
6 liquid dispenser 液体分配器
7 liquid level 液位
a unreacted gases, and methanol and water vapors 未反应的气体，以及甲醇和水蒸气
b cooled inert hydrocarbon liquid 冷却的惰性烃液体
c hydrogen and carbon dioxide 氢气和二氧化碳
d inert hydrocarbon liquid 惰性烃类液体

2.2.40 Techniques for ethanol synthesis from coal via syngas 煤制乙醇的主要技术路线

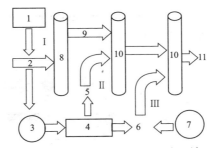

1 coal gasification 煤气化
2 syngas 合成气
3 methanol 甲醇
4 methanol carbonylation 甲醇羰基化
5 acetic acid 乙酸
6 acetate 乙酸酯
7 acetic acid esterification 乙酸酯化
8 synthesis reactor 合成反应器
9 C_2 oxygen containing compounds C_2 含氧化合物
10 hydrogenation reactor 加氢反应器
11 ethanol 乙醇

2.2.41 Process flow for ethanol synthesis from Syngas 合成气制乙醇工艺流程

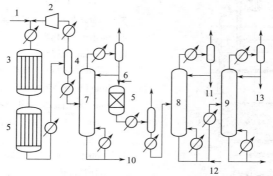

1　syngas　合成气
2　recycle compressor　循环压缩机
3　synthesis reactor　合成反应器
4　splitter　分离器
5　hydrogenation reactor　加氢反应器
6　hydrogen　氢气
7　first tower　1塔
8　second tower　2塔
9　third tower　3塔
10　H_2O，AcOH　水、乙酸
11　AcOEt，MeOH　乙酸乙酯、甲醇
12　NaOH　氢氧化钠　NaOH
13　95%EtOH　95%乙醇

2.2.42 Lurgi Company methanol synthesis reaction process Lurgi公司甲醇合成反应工艺流程

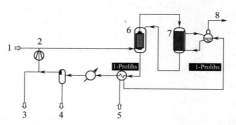

1　make-up gas　补充气体
2　recycle compressor　循环压缩机
3　sweeping gas　吹扫气
4　crude methanol　粗甲醇
5　boiler feed water　锅炉给水
6　gas-cooled reactor　气冷反应器
7　water-cooled reactor　水冷反应器
8　high pressure steam　高压蒸汽

2.2.43 Lurgi Company methanol synthesis and rectification process Lurgi 公司甲醇合成和精馏工艺流程

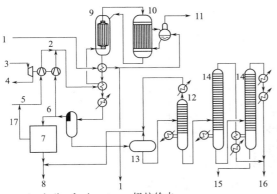

1 boiler feed water 锅炉给水
2 synthesis gas/recycle gas compressor 合成气/循环气压缩机
3 superheated low pressure steam 过热低压蒸汽
4 low pressure steam 低压蒸汽
5 syngas 合成气
6 released gas 放出气
7 pressure swing adsorption 变压吸附
8 fuel gas 燃料气
9 gas-cooled synthesis reactor 气冷式合成反应器
10 water-cooled synthesis reactor 水冷式合成反应器
11 high pressure steam 高压蒸汽
12 topping distillation column 拔顶蒸馏塔
13 expansion tank 膨胀罐
14 atmospheric methanol distillation column 常压甲醇蒸馏塔
15 to desaturation tower 至去饱和塔
16 methanol 甲醇
17 H_2 氢气

2.2.44 Davy methanol synthesis process 英国 Davy 公司甲醇合成工艺流程

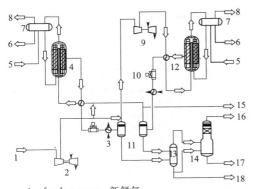

1 fresh syngas 新鲜气
2 syngas compressor 合成气压缩机
3 water cooler 水冷器
4 first synthesis column 第一合成塔
5 boiler feed water 锅炉给水
6 steam drum blow-off 汽包排污
7 steam drum 汽包
8 steam 蒸汽
9 recycle syngas compressor 循环气压缩机
10 air cooler 空冷器
11 splitter 分离器
12 second synthesis column 第二合成塔
13 flash tank 闪蒸槽
14 letdown tank 排放槽
15 purge gas 驰放气
16 flash steam 闪蒸汽
17 crude methanol to rectification 粗甲醇至精馏
18 crude methanol to stabilization tower 粗甲醇至稳定塔

2.2.45 Process flow for Davy Company improved low pressure methanol synthesis Davy 公司改进低压甲醇合成工艺流程

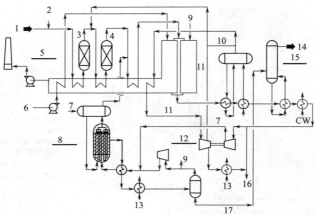

1　natural gas　天然气
2　recycle gas　循环气
3　desulfurization tower　脱硫塔
4　pre-reforming　预重整
5　reforming process　重整工序
6　air　空气
7　boiler feed water　锅炉给水
8　synthesis process　合成工序
9　fuel gas　燃料气
10　steam generation　发生蒸汽
11　high pressure steam　高压蒸汽
12　compression process　压缩工序
13　cooling water　冷却水
14　methanol　甲醇
15　distillation process　蒸馏工序
16　condensate water　凝结水
17　crude methanol　粗甲醇

2.2.46 Methanol synthesis using a double-tower series 串塔合成甲醇工艺流程示意

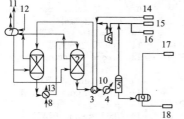

1, 2　methanol synthesis converter　甲醇合成塔
3　heat exchanger　换热器
4　water cooling tower　水冷塔
5　splitter　分离器
6　circulator　循环机
7　steam drum　汽包
8　waste heat recovery unit　废热回收装置
9　flash tank　闪蒸槽
10　CW　冷却水
11　intermediate pressure steam　中压蒸汽
12　boiler feed water　锅炉给水
13　steam　蒸汽
14　supplementary gas　补充气
15　purge gas　驰放气
16　return hydrogen　返氢气
17　flash steam　闪蒸汽
18　crude methanol　粗甲醇

2.2.47 Flow chart for two-stage synthesis of methanol 双级合成甲醇工艺流程示意

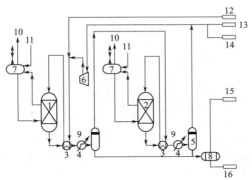

1, 2	methanol synthesis converter 甲醇合成塔	10	intermediate pressure steam 中压蒸汽
3	heat exchanger 换热器	11	boiler feed water 锅炉给水
4	water cooling tower 水冷塔	12	supplementary gas 补充气
5	splitter 分离器	13	purge gas 驰放气
6	circulator 循环机	14	return hydrogen 返氢气
7	steam drum 汽包	15	flash steam 闪蒸汽
8	flash tank 闪蒸槽	16	crude methanol 粗甲醇
9	CW 冷却水		

2.2.48 Methanol synthesis flowsheet based on a quench reactor 基于急冷反应器的甲醇合成工艺流程

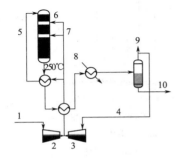

1　fresh feed gas　新鲜原料气
2　feed compressor　供料压缩机
3　recycle compressor　循环压缩机
4　recycle gas　循环气
5　feed gas, 200℃　原料气 (200℃)
6　synthesis reactor　合成反应器
7　quench gas　急冷气
8　cooling water　冷却水
9　purge gas　驰放气
10　crude methanol　粗甲醇

41

2.2.49 Tsinghua university one-step liquid phase dimethylether (DME) synthesis 清华大学液相一步法二甲醚工艺流程

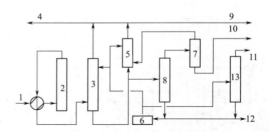

1 syngas 合成气
2 synthesis converter 合成塔
3 absorption tower 吸收塔
4 to decarbonization tower 去脱碳塔
5 scrubber 洗气塔
6 reserve liquid tank 储液罐
7 separation tank 分离罐
8 DME tower 二甲醚塔
9 purge gas 驰放气
10 dimethylether 二甲醚
11 refined methanol 精甲醇
12 discharge water 排放水
13 methanol tower 甲醇塔

2.2.50 NKK one-step liquid phase Dimethylether synthesis NKK液相一步法二甲醚工艺流程

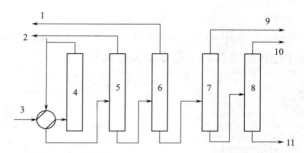

1 back to gasification 回造气
2 back to the absorption tower 回吸收塔
3 syngas 合成气
4 slurry bed reactor 浆态床反应器
5 separation tank 分离罐
6 decarbonization tower 脱碳塔
7 DME tower 二甲醚塔
8 methanol tower 甲醇塔
9 dimethylether 二甲醚
10 methanol 甲醇
11 discharge water 排放水

2.2 Chemical processing of coal 煤化工

2.2.51 Lurgi's MegaDME process 鲁奇公司大型二甲醚制备过程

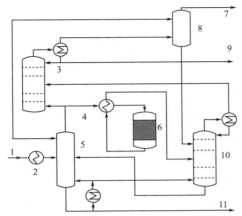

1　methanol feedstock　原料甲醇
2　methanol preheater　甲醇预加热器
3　DME (dimethyl ether) column　二甲醚塔
4　DME reactor feed/product HX　二甲醚反应器进料/产物氢交换
5　methanol vapouriser　甲醇汽化器
6　DME reactor　二甲醚反应器
7　light ends　轻馏分
8　scrubber　涤气塔
9　DME　二甲醚
10　reactor product splitter　反应产物分离塔
11　process water　工艺用水

2.2.52 Topsøe DME synthesis process (two pot synthesis) 托普索公司二甲醚合成工艺（两步合成法）

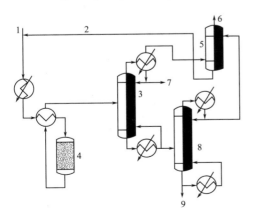

1　methanol　甲醇
2　recycle　循环
3　DME column　二甲醚塔
4　reactor　反应器
5　off gas absorber　尾气吸收塔
6　off gas　废气
7　DME product　二甲醚产品
8　water column　水塔
9　waste water　废水

43

2.2.53 Schematic diagram of pressure swing adsorption system 变压吸附系统示意图

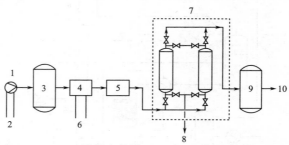

1. feed air compressor 进料空气压缩机
2. cooling 冷却
3. air receiver 储气罐
4. refrigerated dryer 冷冻式干燥机
5. air filters 空气过滤器
6. cooling 冷却
7. carbon sieve container 炭分子筛容器
8. oxygen enriched air 富氧空气
9. nitrogen receiver 氮储罐
10. dry nitrogen 干燥氮气

2.2.54 BASF process of acetic acid synthesis from methanol carbonylation BASF公司甲醇羰基化制乙酸工艺流程

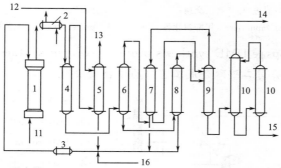

1. reactor 反应器
2. coolers 冷却器
3. preheater 预热器
4. low pressure separator 低压分离器
5. tail gas scrubber 尾气洗涤塔
6. degassing tower 脱气塔
7. separator column 分离塔
8. catalyst separator column 催化剂分离塔
9. azeotropic distillation column 共沸蒸馏塔
10. rectification column 精馏塔
11. carbon monoxide 一氧化碳
12. methanol 甲醇
13. tail gas 尾气
14. acetic acid 乙酸
15. by-product 副产品
16. dimethylether 二甲醚

2.2.55 Schematic of acetic acid synthesis from low pressure methanol carbonylation 低压甲醇羰基化制乙酸流程简图

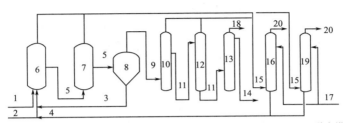

1. input material CO 原料 CO
2. input material methanol 原料甲醇
3. circulating mother liquid 循环母液
4. absorbed methanol 吸收甲醇
5. reaction fluid 反应液
6. reactor 反应釜
7. conversion reactor 转化釜
8. evaporator 蒸发器
9. flashed vapour 闪蒸气
10. off with light constituents tower 脱轻塔
11. crude acetic acid 粗乙酸
12. dehydration tower 脱水塔
13. finished tower 成品塔
14. product acetic acid 成品乙酸
15. tail gas 尾气
16. high pressure absorbing tower 高压吸收塔
17. fresh methanol 新鲜甲醇
18. vent 放空
19. low pressure absorbing tower 低压吸收塔
20. to torch 至火炬排放

2.2.56 Monsanto/BP process for acetic acid production Monsanto/BP 制乙酸工艺流程

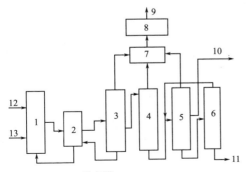

1. reactor 反应器
2. flash tank 闪蒸罐
3. light component removal column 脱轻组分塔
4. dehydration tower 脱水塔
5. paraffins removal column 脱烷塔
6. waste acid stripper 废酸汽提塔
7. buffer 缓冲
8. absorbing section 吸收工段
9. to torch 去往火炬
10. CH_3COOH final product 乙酸成品
11. liquid to be treated 待处理的料液
12. CO 一氧化碳
13. CH_3OH 甲醇

45

2.2.57 Process of continuous production of ethyl acetate 连续法生产乙酸乙酯工艺流程

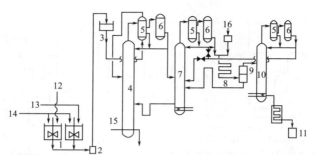

1 mixer   混合器
2 pump   泵
3 elevated tank   高位槽
4 esterification reaction tower   酯化反应塔
5 reflux dephlegmator   回流分凝器
6 complete condenser   全凝器
7 ester distilled tower   酯蒸出塔
8 mixing coil   混合盘管
9 splitter   分离器
10 ester drying tower   酯干燥塔
11 product storage tank   产品储槽
12 acetic acid   乙酸
13 sulfuric acid   硫酸
14 ethanol   乙醇
15 steam   蒸汽
16 water   水

2.2.58 Ethyl acetate production process 乙酸乙酯生产的工艺流程

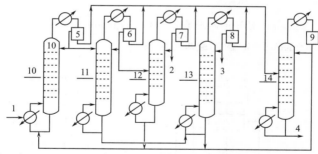

1 ethanol, acetic acid   乙醇、乙酸
2 low ester byproducts   低酯副产物
3 ethyl acetate products   乙酸乙酯产品
4 wastewater   废水
5 reflux tank of esterification tower   酯化塔回流罐
6 reflux tank of concentration tower   提浓塔回流罐
7 low ester recovery tower reflux tank   低酯回收塔回流罐
8 reflux tank of refining tower   精制塔回流罐
9 reflux tank of wastewater recovery tower   废水回收塔回流罐
10 esterification tower   酯化塔
11 concentration tower   提浓塔
12 low ester recovery column   低酯回收塔
13 refining tower   精制塔
14 wastewater recovery column   废水回收塔

2.2.59 Process of ethanol production from hydrogenation via ethyl acetate 经乙酸乙酯加氢制备乙醇的工艺流程

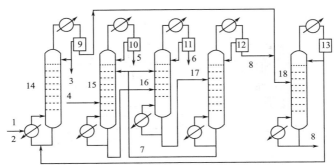

1 ethanol 乙醇
2 acetic acid 乙酸
3 crude ethyl acetate to hydrogenation process 粗乙酸乙酯到加氢工序
4 crude ethanol from hydrogenation process 粗乙醇来自加氢工序
5 ethyl acetate and ethanol 乙酸乙酯与乙醇
6 absolute alcohol product 无水乙醇产品
7 regenerated glycol 再生乙二醇
8 wastewater 废水
9 reflux tank of esterification tower 酯化塔回流罐
10 reflux tank of ethyl acetate tower 乙酸乙酯塔回流罐
11 reflux tank of ethanol tank 乙醇塔回流罐
12 reflux tank of regeneration tower 再生塔回流罐
13 reflux tank of wastewater recovery tower 废水回收塔回流罐
14 esterification tower 酯化塔
15 ethyl acetate tower 乙酸乙酯塔
16 ethanol tower 乙醇塔
17 regeneration tower 再生塔
18 wastewater recovery column 废水回收塔

2.2.60 Process of intermittent production of butyl acetate 间歇法生产乙酸丁酯工艺流程

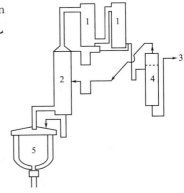

1 condenser 冷凝器
2 fractionator 分馏器
3 aqueous layer 水层
4 splitter 分离器
5 esterification reactor 酯化釜

2.2.61 Methyl formate production by dehydrogenation of methanol 甲醇脱氢制甲酸甲酯

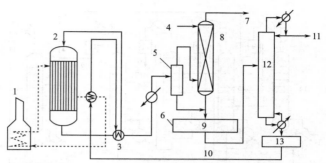

1　furnace　加热炉
2　reactor　反应器
3　evaporator　蒸发器
4　CH_3OH　甲醇
5　cooling unit　冷却器
6　separating unit　分离装置
7　H_2　氢气
8　absorber　吸收塔
9　crude methyl formate　粗甲酸甲酯
10　methanol recycle　甲醇循环
11　$HCOOCH_3$　甲酸甲酯
12　distillation column　精馏塔
13　methanol　甲醇

2.2.62 Leonard process for the indirect synthesis of formic acid via methyl formate 伦纳德工艺：甲醇经甲酸甲酯制甲酸

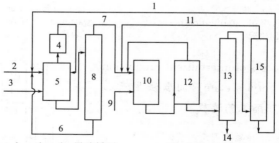

1　recirculated methanol　甲醇循环
2　methanol　甲醇
3　CO　一氧化碳
4　cooling unit　冷却装置
5　methyl formate reactor 42 bar　42巴甲酸甲酯反应器
6　methanol + catalyst　甲醇+催化剂
7　methyl formate　甲酸甲酯
8　distillation column　精馏塔
9　steam　蒸汽
10　hydrolysis reactor　水解反应器
11　recirculated methyl formate　甲酸甲酯循环
12　declamping tank　分层罐
13　product distillation　产品精馏
14　separation of water and 85% formic acid　分离的水和85%的甲酸
15　methanol-methyl formate separation　甲醇-甲酸甲酯分离

2.2.63 Methyl mercaptane production process 甲硫醇生产工艺

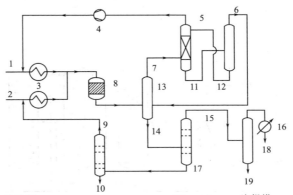

1 H_2S 硫化氢
2 CH_3OH 甲醇
3 heaters 加热器
4 compressor 压缩机
5 H_2S recycle 硫化氢循环
6 CH_3SH 甲硫醇
7 H_2S, CH_3SH 硫化氢，甲硫醇
8 reactor 反应器
9 CH_3OH recycle 甲醇循环
10 H_2O 水
11 absorber 吸收塔
12 stripper 汽提塔
13 stabiliser 稳定器
14 $CH_3SH, CH_3OH, (CH_3)_2S, H_2O$ 甲硫醇，甲醇，二甲硫醚，水
15 $CH_3SH, (CH_3)_2S$ 甲硫醇，二甲硫醚
16 cooling unit 冷却器
17 CH_3OH, H_2O 甲醇，水
18 CH_3SH product 甲硫醇产品
19 $(CH_3)_2S$ product or recycle 二甲硫醚产品或循环

2.2.64 Dimethyl terephthalate (DMT) production through p-xylene oxidation 对二甲苯氧化制对苯二甲酸二甲酯

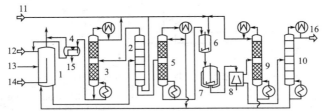

1 oxidation reactor 氧化反应器
2 esterification reactor 酯化反应器
3 methanol column 甲醇塔
4 separation vessel 分离器
5 ester column 酯塔
6 solvent container 溶剂容器
7 crystallsation 结晶
8 centrifuge 离心机
9 centrifugate distillation 离心液精馏
10 DMT column 对苯二甲酸二甲酯塔
11 methanol 甲醇
12 p-xylene 对二甲苯
13 catalyst 催化剂
14 air 空气
15 water 水
16 DMT 对苯二甲酸二甲酯

49

2.2.65 Evonik catalytic distillation methyl *t*-butyl ether (MTBE) process Evonik 催化蒸馏制甲基叔丁基醚（MTBE）工艺

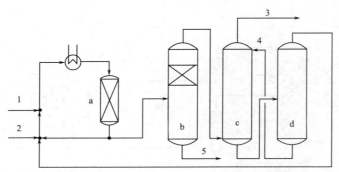

1　C_4- hydrocarbons　碳 4 烃
2　methanol　甲醇
3　raffinate　萃余液
4　H_2O　水
5　MTBE　甲基叔丁基醚

a　reactor　反应器
b　catalytic distillation column　催化蒸馏塔
c　methanol extraction　甲醇萃取
d　methanol column　甲醇塔

2.2.66 DuPont's ethylene glycol process 杜邦公司乙二醇生产工艺

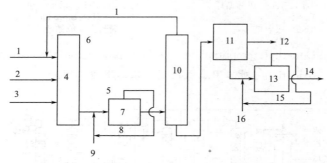

1　glycolic acid ＋ H_2O　乙醇酸＋水
2　diluted H_2SO_4　稀释的硫酸
3　formaldehyde ＋ water vapor from methanol-oxidation　甲醇-氧化生成甲醛和水蒸气
4　mixer　混合器
5　reactor　反应器
6　2 parts Glycolic acid　2 份乙醇酸
　　1 part formaldehyde　1 份甲醛
　　2 parts H_2O　2 份水
　　0.02 parts H_2SO_4　0.02 份 H_2SO_4
7　700 bar　700bar
　　200℃　200℃

　　5 min residence time　5min 停留时间
8　CO cycle　一氧化碳循环
9　fresh carbon oxide　新鲜的碳氧化物
10　vacuum distillation　真空精馏
11　esterification　酯化
12　waste　废弃物
13　hydrogenating　加氢反应
　　200℃　200℃
　　30 bar　30bar
　　Cu-chromite　铜铬铁矿
14　ethylene glycol　乙二醇
15　circulated water　循环水
16　fresh water　淡水

2.2.67 Process for a demonstration project of polyolefin produced from coal 煤制聚烯烃示范项目工艺流程

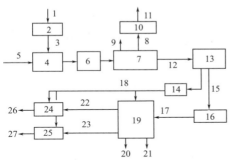

8	acid gas 酸性气
9	CO_2 二氧化碳
10	sulfur recovery 硫黄回收
11	sulfur 硫黄
12	$CO + H_2$ 一氧化碳+氢气
13	methanol synthesis 甲醇合成
14	PSA (pressure swing adsorption) 变压吸附
15	methanol 甲醇
16	methanol to olefins 甲醇制烯烃
17	reaction gas 反应气体
18	H_2 氢气
19	olefin separation 烯烃分离
20	mixed C_4 混合 C_4
21	mixed C_5 混合 C_5
22	ethylene 乙烯
23	propylene 丙烯
24	polyethylene 聚乙烯
25	polypropylene 聚丙烯
26	PE 聚乙烯产品
27	PP 聚丙烯产品

1　air　空气
2　air separation　空气分离
3　oxygen　氧气
4　gasification　煤气化
5　raw stuff coal　原料煤
6　CO conversion　CO变换
7　syngas purification (Rectisol process)
　　合成气净化（低温甲醇洗）

2.2.68 UOP/HYDRO MTO (methanol to olefins) process UOP/HYDRO MTO（甲醇制烯烃）工艺

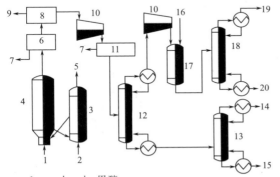

1　methanol　甲醇
2　air　空气
3　regenerator　再生器
4　MTO reactor　MTO反应器
5　CO_2, N_2　二氧化碳，氮气

6　water removal　脱水
7　water　水
8　CO_2 removal　脱二氧化碳
9　CO_2　二氧化碳
10　compressor　压缩机
11　dehydration　脱水
12　deethanizer　脱乙烷塔
13　depropanizer　脱丙烷塔
14　propylene　丙烯
15　C_4^+ product　4个C（含C_4）以上的烃类产品
16　H_2　氢气
17　acetylene saturator　乙炔饱和塔
18　demethanizer　脱甲烷塔
19　methane　甲烷
20　ethylene　乙烯

2.2.69 Methanol to olefins (MTO) process 甲醇制烯烃（MTO）工艺流程示意图

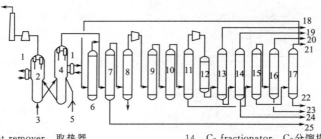

1 heat remover 取热器
2 regenerator 再生器
3 air 空气
4 reactor 反应器
5 methanol 甲醇
6 C_4 conversion reactor C_4转化反应器
7 quenching tower 急冷塔
8 water scrubber 水洗塔
9 caustic scrubber 碱洗塔
10 drying tower 干燥塔
11 de-C_2 tower 脱C_2塔
12 hydrogenation reactor 加氢反应器
13 de-C_1 tower 脱C_1塔
14 C_2 fractionator C_2分馏塔
15 de-C_3 column 脱C_3塔
16 C_3 fractionator C_3分馏塔
17 de-C_4 column 脱C_4塔
18 dry gas 干气
19 $C_2^=$ 乙烯
20 $C_3^=$ 丙烯
21 mixed C_4 混合C_4
22 C_{5+} 5个C（含C_5）以上的烃类
23 C_3^0 丙烷
24 C_2^0 乙烷
25 heavies 重馏分

2.2.70 SDTO process principle for methanol to low-carbon olefins 甲醇制低碳烯烃SDTO工艺原则流程

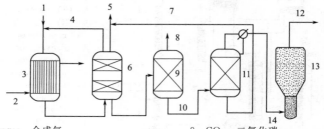

1 syngas 合成气
2 heat carrier 热载体
3 shell-and-tube reactor 列管式反应器
4 unreacted syngas 未反应之合成气
5 H_2O 水
6 absorbing tower 吸收塔
7 H_2O return to absorbing tower H_2O返回吸收塔
8 CO_2 二氧化碳
9 CO_2 removal tower 脱CO_2塔
10 DME aqueous solution 二甲醚水溶液
11 distillation column 蒸馏塔
12 low-carbon olefins 低碳烯烃
13 fluidized bed 流化床
14 DME 二甲醚

2.2.71 UOP/Hydro MTO process (production of polymer-grade olefins) UOP/Hydro MTO 工艺流程（生产聚合级烯烃）

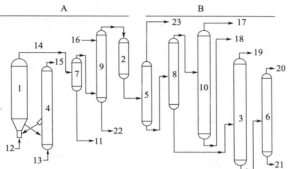

1　fluidized bed-reactor　流化床-反应器
2　dryer　干燥器
3　propylene fractionator　丙烯精馏塔
4　regenerator　再生器
5　demethanizer　脱甲烷塔
6　depropaniser　脱丙烷塔
7　quenching tower　急冷塔
8　deethanizer　脱乙烷塔
9　caustic washing tower　碱洗塔
10　ethylene fractionator　乙烯精馏塔
11　H_2O　水
12　methanol　甲醇
13　air　空气
14　product　产品
15　flue gas　烟气
16　alkali　碱
17　ethylene　乙烯
18　ethane　乙烷
19　propylene　丙烯
20　propane　丙烷
21　C_{4+}　4个C（含C_4）以上的烃类
22　CO_2　二氧化碳
23　CH_4　甲烷
A　reaction zone　反应区
B　product separation zone　产品分离区

2.2.72 Schematic of methanol to olefins (MTO) process 甲醇制烯烃（MTO）工艺装置简要流程

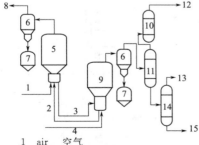

1　air　空气
2　spent catalyst　待生催化剂
3　regenerating catalyst　再生催化剂
4　gaseous methanol　气相甲醇
5　regenerator　再生器
6　separator　分离器
7　waste catalyst　废催化剂
8　flue gas to heat recovery system　烟气至热量回收系统
9　reactor　反应器
10　water scrubber　水洗塔
11　quenching tower　急冷塔
12　products to olefin separation　产品至烯烃分离
13　cooled stripping gas as concentrated water back to refining　汽提气冷却后作为浓缩水回炼
14　stripper　汽提塔
15　purified water to heat exchanger　净化水至换热器

2.2.73 Schematic drawing of oil product synthesis 合成油产品路线

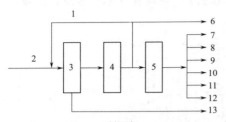

1　recycle gas　循环气
2　coal-based syngas　煤基合成气
3　synthesis　合成反应
4　product separation　产品分离
5　product refinery　产品精制
6　tail gas to reformer or fuel gas　尾气去重整或燃料气
7　$C_2 \sim C_4$ (olefin hydrocarbon and liquefied petroleum gas)　$C_2 \sim C_4$ (烯烃及液化气)
8　$C_5 \sim C_{12}$ (naphtha and gasoline)　$C_5 \sim C_{12}$ (石脑油及汽油)
9　$C_{13} \sim C_{19}$ (diesel fuel)　$C_{13} \sim C_{19}$ (柴油)
10　$C_{20} \sim C_{30}$ (paraffin level heavy oil)　$C_{20} \sim C_{30}$ (石蜡级重油)
11　$>C_{30}$ (solid wax)　$>C_{30}$ (固体蜡)
12　other fuels and chemicals　其他燃油及化学品
13　byproduct steam　副产蒸汽

2.2.74 Fischer-Tropsch Low temperature slurry bed oil synthesis process 低温浆态床费-托合成油工艺流程

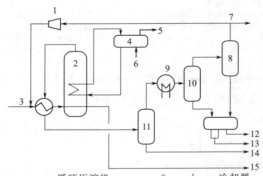

1　recycle compressor　循环压缩机
2　reactor　反应器
3　syngas　合成气
4　steam drum　汽包
5　steam　蒸汽
6　water　水
7　tail gas　尾气
8　demister　除沫器
9　coolers　冷却器
10　cryogenic separation tank　低温分离罐
11　hot separator tank　高温分离罐
12　synthetic light oil　合成轻油
13　polluted water　污水
14　soft wax　软蜡
15　hard wax　硬蜡

2.2.75 Schematic flow diagram of the fluid-bed methanol-to-gasoline process 甲醇制汽油的流化床工艺流程示意

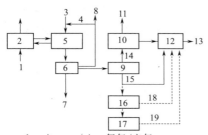

1　nitrogen/air　氮气/空气
2　regenerator　再生器
3　crude methanol　粗甲醇
4　recycle　循环
5　conversion reactor　转化反应器
6　product separator　产品分离器
7　water　水
8　light gas　轻质气
9　debutanizer　脱丁烷塔
10　alkylation　烷基化
11　propane　丙烷
12　blending　共混
13　gasoline　汽油
14　$C_{3,4}$　碳3碳4
15　C_{5+}　5个C以上（含C_5）的烃类
16　splitter　分离设备
17　HGT unit　重质汽油处理器
18　light gasoline　轻质汽油
19　treated heavy gasoline　处理后的重质汽油

2.2.76 Gasoline mode of the methanol-to-olefins gasoline plant at Wesseling, Germany (© Uhbe, ExxonMobil) 德国韦瑟灵公司甲醇制烯烃汽油厂的汽油模式（© Uhbe，埃克森美孚）

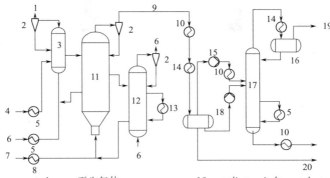

1　regenerated gas　再生气体
2　cyclone　旋风分离器
3　regenerator　再生器
4　nitrogen + air　氮气+空气
5　heating unit　加热器
6　nitrogen　氮气
7　methanol　甲醇
8　evaporator　蒸发器
9　crude gas　粗煤气
10　cooling unit　冷却器
11　reactor　反应器
12　cooling unit for catalyst　催化剂的冷却器
13　heat exchanger　热交换器
14　condenser　冷凝器
15　compressor　压缩机
16　reflux tank　回流罐
17　separating column　分离塔
18　pump　泵
19　gas　煤气
20　water　水

2.2.77 ExxonMobil MTG (methanol to gasoline) process. The feedstock is a mixture of dimethyl ether (DME) and water produced from methanol 埃克森美孚公司MTG（甲醇制汽油）工艺流程，原料是由甲醇制得的二甲醚（DME）和水的混合物

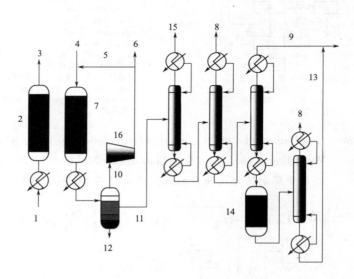

1 N_2, O_2 氮气，氧气
2 offline MTG reactor burning coke off catalyst 独立的MTG反应器，烧去催化剂表面附着的焦质
3 N_2, CO_2 氮气，二氧化碳
4 DME + H_2O 二甲醚＋水
5 recycle gas 循环气
6 purge 驰放
7 online MTG reactor 流程中的MTG反应器
8 LPG (liquefied petroleum gas) 液化石油气
9 light gasoline 轻质汽油
10 gas 气
11 hydrocarbon liquids 液烃
12 water 水
13 heavy gasoline 重质汽油
14 HGT (heavy gasoline treating) reactor 重质汽油处理反应器
15 C_2- 2个C（含C_2）以下的烃类
16 compressor 压缩机

2.2 Chemical processing of coal 煤化工

2.2.78 Gasoline synthesis process 合成油工艺流程

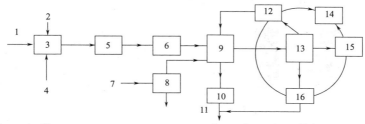

1. coal 煤
2. steam 蒸汽
3. gasification 气化
4. air separation 空分
5. purification 净化
6. conversion 变换
7. natural gas 天然气
8. partial oxidation 部分氧化
9. F-T synthesis 费-托合成
10. polyreaction 叠合
11. product gasoline and wax 产品油和蜡
12. reforming 重整
13. tail gas 尾气
14. ammonia or electricity or heat 氨或电或热
15. cryogenic separation 低温分离
16. oligomerization 低聚反应

2.2.79 Exxon Donor Solvent process for direct hydrogenation of coal to produce liquid fuels Exxon Donor 煤直接加氢制液体燃料的溶剂生产工艺

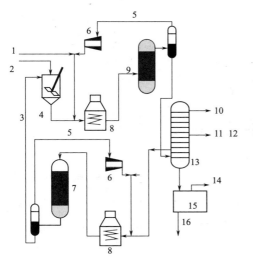

1. makeup H_2 补充氢气
2. coal 煤
3. recycle solvent 循环溶剂
4. slurry tank 煤浆槽
5. recycle H_2 循环氢气
6. recycle compressor 循环压缩机
7. trickle bed hydrotreater 滴流床加氢装置
8. preheat 预热
9. ebullating bed hydrotreater 沸腾床加氢装置
10. naphtha 石脑油
11. distillate 馏出液
12. fuel oil 燃料油
13. vacuum distillation 真空蒸馏
14. liquids 液体
15. flexicoker 灵活焦化器
16. coke to gasification 焦炭去气化

2.2.80 Simplified schematic diagram of chemical looping combustion system 化学循环燃烧系统简图

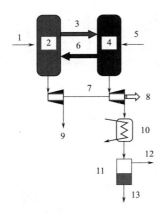

1. air 空气
2. MOR- metal oxidation reactor 金属氧化反应炉
3. MeO_x-metal oxide 金属氧化物
4. FOR-fuel oxidation reactor 燃料氧化反应炉
5. fuel 燃料
6. metal 金属
7. turbine 汽轮机
8. electricity 电力
9. O_2 depleted air 氧气耗尽的空气
10. cooling system 冷却系统
11. water condenser 水冷凝器
12. CO_2 二氧化碳
13. water 水

2.2.81 Scheme for high-pressure underground gasification 地下高压气化示意图

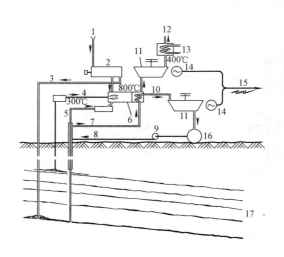

1. air 空气
2. compressor 压缩机
3. compressed air 压缩空气
4. rich gas 富煤气
5. lean gas 贫气
6. combustion chamber 燃烧室
7. steam(220℃) 水蒸气(220℃)
8. water 水
9. pump 泵
10. superheater 过热器
11. turbine 涡轮机
12. stack 烟囱
13. economizer 省煤器，废气预热器
14. alternator 交流发电机
15. electricity grid 电网
16. condenser 冷凝器
17. coal seams 煤层

58

2.2.82 Surface-direct retorting 地面直接干馏

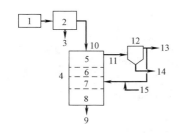

1 oil-shale mine 油页岩矿
2 crushing, feed preparation 粗碎，进料预处理
3 fines discard 废矿粉
4 retort 干馏
5 shale preheating zone 页岩预热段
6 shale retorting zone 页岩干馏段
7 combustion zone 燃烧段
8 spent-shale cooling zone 废页岩冷却段
9 spent-shale solids 废页岩固体颗粒
10 raw shale feed 未加工的页岩加料
11 vapors 气化物
12 separator 分离器
13 product gas 产品煤气
14 product oil (to upgrading) 产品油（去改质）
15 air 空气
16 fines discard 细颗粒排弃
17 heating medium 加热介质
18 spent shale 废页岩

2.2.83 Surface-indirect retorting 地面间接干馏

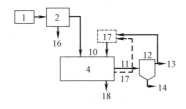

2.2.84 Oil-shale process uses a single multifunction retort 采用单台多功能干馏釜加工油页岩

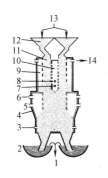

1 spent shale 用过的页岩
2 spent-shale discharge device 废页岩卸料装置
3 recycle gas (for cooling) 循环气（冷却用）
4 refractory lining 耐火层
5 gas burner 煤气燃烧器
6 recycle gas 循环气
7 gas burner 煤气燃烧器
8 recycle gas inlet 循环气入口
9 oil-vapor collection, evacuation chamber 油气收集、排气室
10 heat-carrier preparation chamber 热载体制备室
11 oil-shale semicoking chamber 油-页岩半焦化室
12 charging zone 装料区
13 oil shale 油页岩
14 oil vap 油气

2.2.85 Modified in-situ retorting 变型的就地干馏

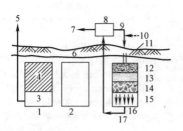

1. step1: mining 第一步：采矿
2. step2: rubblizing 第二步：粗碎
3. mined portion 坑道地段
4. shale pillar 油页岩矿柱
5. to surface retort 到地面干馏
6. overburden 表土层
7. oil (to upgrading) 油（去改质）
8. separation 分离
9. product gas 产品煤气
10. air 空气
11. ground surface 地表面
12. burned-out zone 燃尽段
13. combustion zone 燃烧段
14. retorting zone 干馏段
15. vapor condensation zone 蒸气冷凝段
16. oil and gas 油和煤气
17. step3: retorting 第三步：干馏
18. expected temperature profile 预期的温度分布剖面图
19. fractured shale zone 已破碎的页岩区
20. compressed air injection well 压缩空气注入井
21. oil, water and gas production well 油、水和煤气产出井
22. front movement 朝前移动
23. combustion gases, oil and water 已燃气、油和水
24. oil, water and gas drive zone 油、水和气推进区

2.2.86 True in-situ retorting 真实的就地干馏

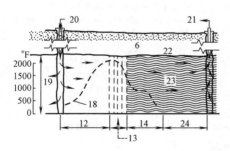

2.2.87 Retort 干馏器

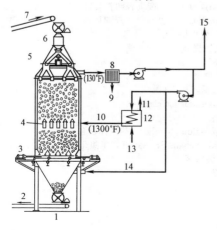

1. kiln (or retort) 窑（或者干馏器）
2. spent shale (200°F) 用过的油页岩（200°F）
3. discharge grate 出料格栅
4. gas injectors 进气喷射器
5. gas-disengaging plenum 气体释放空间
6. rotating segment feeder 旋转星形加料器
7. oil shale 油页岩
8. electrostatic precipitator 静电除尘器
9. shale oil 页岩油
10. heated recycle gas (1300°F) 已加热的循环气（1300°F）
11. flue gas 烟道气
12. furnace 加热炉
13. fuel+air 燃料+空气
14. cool recycle gas (130°F) 冷循环气（130°F）
15. high-Btu gas product 高热值气体产品

2.2.88 Galoter oil-shale retorting system 加劳特油页岩干馏装置

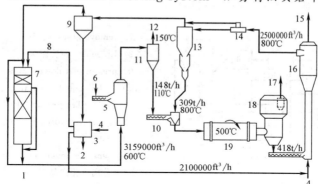

（uses shale as heat carrier 采用页岩作为载热体）
1　note: total capacity is 3760 tons/d
　　注：总产量为 3760 吨/天
2　spent shale 废油页岩
3　ash heat exchanger 炉灰热交换器
4　air 空气
5　dryer 干燥器
6　raw shale 153 tons/h 未加工的页岩 153 吨/时
7　waste-heat boiler 废热锅炉
8　air 615000 ft^3/h 空气 615000 英尺^3/时
9　ash separator 炉灰分离器
10　mixer 混合器
11　dry oil-shale separator 干的油页岩分离器
12　to stack 3954000ft^3/h 去烟囱 3954000 英尺^3/时
13　heat-carrier separator 热载体分离器
14　bypass 旁路
15　flue gas to electrostatic precipitator 烟道气去静电除尘器
16　air-blown firebox 鼓风式燃烧器
17　oil vapors to condensation section 油气去冷凝工段
18　dust-removal chamber 除尘室
19　reactor 反应器

2.2.89 General systems description: surface-coal gasification 地上煤气化的总流程

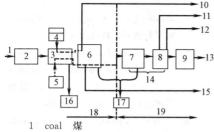

1　coal 煤
2　coal storage, preparation and handing 储煤仓，煤的准备与运送
3　gasifier 气化炉
4　air & steam 空气与水蒸气
5　oxygen & steam 氧与水蒸气
6　sulfur, particulate and tar removal 脱硫，除尘与除焦油
7　remaining acid gas removal 剩余酸性气体的脱除
8　shift 变换
9　methanation 甲烷化
10　low-Btu fuel gas 低热值燃料气
11　synthesis gas (for chemicals and indirect liquefaction) 合成气（制化学品与间接液化用）
12　hydrogen (for fertilizer or direct liquefaction) 氢（制肥料或直接液化用）
13　high-Btu pipeline gas 高热值管道气
14　note: the order of these processes may be reversed 注：这两个过程的次序可以互换
15　med-Btu fuel gas 中热值燃料气
16　ash disposal 排灰
17　waste disposal 排污
18　gasification 气化
19　product refinement 产品精制

2.2.90 Direct-coal-liquefaction generalized flow diagram 煤直接液化的一般流程

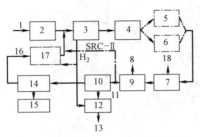

1　coal　煤
2　grind and dry　磨碎与干燥
3　slurry coal and solvent　煤浆与溶剂
4　preheat and dissolve　预热与溶解
5　catalytic (H-coal) liquefaction 催化 (H-coal) 液化
6　thermal (SRC & EDS) 热过程 (SRC & EDS)
7　remove gases　脱气
8　Lt. ends　轻质物
9　pressure-let-down　减压
10　solids liquid separation　固液分离
11　solids+heavy liquids　固体+重质液体
12　generate hydrogen　发生氢气
13　ash　灰
14　separate liquefaction products 分离液化产品
15　liquefaction products　液化产品
16　solvent　溶剂
17　solvent hydrogenation (EDS) 溶剂加氢 (EDS)
18　H_2S, NH_3, H_2O, CH_4　硫化氢、氨、水、甲烷

2.2.91 Solid SRC 固相溶剂精制煤

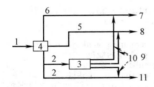

1　coal　煤
2　solid SRC (solvent refined coal) (as a petrochemical building block) 固体溶剂精制煤（作为石油化工基本原料）
3　LC-fining　液位控制-澄清
4　SRC-1　溶剂精制煤-1
5　naphtha, fuel oil　石脑油、燃料油
6　gas　煤气
7　fuel, ethylene, propylene　燃料、乙烯、丙烯
8　turbine fuel, crude gasoline stock 透平机燃料，粗汽油料
9　chemicals: benzene, naphthalene, phenol, cresols and xylenols, carbon-black feed 化学品：苯、萘、苯酚、混合甲酚和混合二甲酚、炭黑原料
10　low S, N products　低含硫、氮量产品
11　solid boiler fuel: anode coke, metallurgical coke 固体锅炉燃料：氧化焦炭、冶金焦炭

2.2.92 Flowsheet showing equipment and processing steps for solvent refined coal 溶剂精制煤加工步骤和装置的流程图

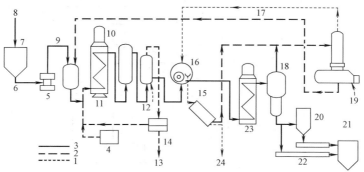

1. gas 气体
2. solvent 溶剂
3. coal 煤
4. hydrogen generation 发生氢气
5. coal pulverizer and dryer 煤粉碎和干燥机
6. 50 tons/d 50 吨/天
7. storage hopper 贮斗
8. raw coal 原煤
9. slurry blend 煤浆调和
10. slurry preheater dissolver 煤浆预热溶解器
11. fuel 燃料
12. flash 闪蒸
13. sulfur 硫
14. stretford desulfurization unit 蒽醌二磺酸钠法脱硫装置
15. residue dryer 滤渣干燥器
16. pressure flter 压力式过滤机
17. wash solvent 洗涤剂
18. vacuum flash 减压闪蒸
19. solvent-recovery distillation 溶剂回收蒸馏
20. prill tower 造粒塔
21. solvent-refined-coal product storage 溶剂精制煤产品仓
22. product flaking belt 片状产品输送带
23. preheater 预热器
24. mineral residue 矿物废渣

2.2.93 Integration of SRC and Clean-Coke demonstration plants 溶剂精制煤与洁净焦炭实验厂的整合

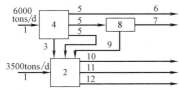

1. washed coal 洗过的煤
2. clean-coke demonstration plant 洁净焦炭实验厂
3. naphtha fraction 石脑油馏分
4. SRC demonstration plant 溶剂精制煤实验厂
5. SRC product 溶剂精制煤产品
6. power-plant fuel 动力装置燃料
7. coke for aluminum anodes 铝氧化所用焦炭
8. delayed coker 延迟焦化装置
9. coker distillate 焦化塔馏出物
10. chemical feedstocks 化学原料
11. liquid fuels 液体燃料
12. 2300 tons/d metullurgical coke 2300 吨/日冶金焦炭

2.2.94 SRC process yields clean coal 生产洁净煤的溶剂精制煤工艺

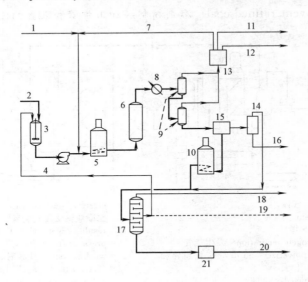

(the process as well as several byproducts 本工艺另产多种副产品)

1 makeup hydrogen （补充）氢气
2 coal 煤
3 slurry makeup tank 煤浆制备罐
4 solvent 溶剂
5 slurry preheater 煤浆预热器
6 dissolver 溶解器
7 recycle hydrogen 循环氢气
8 cooler 冷却器
9 gas/liquid separators 气液分离器
10 heater 加热器
11 hydrocarbon gases 碳氢化合物气体
12 hydrogen sulfide 硫化氢
13 gas separation and purification 气体分离和精制
14 dryer/solvent recovery 干燥器-溶液回收
15 filter 过滤器
16 ash and residue 灰渣
17 vacuum distillation column 真空精馏塔
18 light distillate 轻馏分油
19 middle distillate 中间馏分油
20 clean SRC 净化的溶剂精制煤
21 solidification 固化成型

2.2.95 SRC-I process scheme
SRC-I 过程的流程图

2.2.96 SRC-II process scheme
SRC-II 过程的流程图

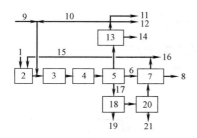
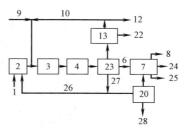

1 coal 煤
2 slurry-mix tank 煤浆混合槽
3 preheater 预热器
4 dissolver 溶解器
5 separator 分离器
6 oil 油
7 product fractionaton 产物分馏
8 naphtha 石脑油
9 makeup hydrogen 补充氢
10 recycle hydrogen 循环氢
11 HC gas 气体烃
12 purge gas 排出气
13 gas purification 气体净化
14 to sulfur recovery unit 去硫回收装置
15 recycle solvent 循环溶剂
16 process solvent 过程用溶剂
17 slurry 浆
18 solids separation 固体分离
19 residue 残渣
20 vacuum flash 减压闪蒸
21 SRC——solvent refined coal 溶剂精制煤
22 sulfur recovery 硫回收
23 separator and stripper 分离与汽提器
24 middle distillate 中馏分
25 heavy distillate 重馏分
26 recycle slurry 循环煤浆
27 stripped slurry 汽提后煤浆
28 vacuum residue 减压残渣

2.2.97 Flow scheme of the Exxon donor solvent process
Exxon 供氢溶剂过程的流程图

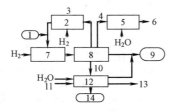

1 coal 煤
2 solvent hydrogenation 溶剂加氢
3 catalytic 催化
4 gas 气
5 gas purification and steam reforming 气体净化与蒸汽重整
6 hydrogen 氢
7 liquefaction 液化
8 distillation 蒸馏
9 liquid products 液体产品
10 heavy bottoms slurry 塔底重油浆
11 air 空气
12 flexicoking 灵活焦化
13 fuel gas 燃料气
14 ash residue 灰渣

2.2.98 H-coal process scheme 氢-煤过程流程图

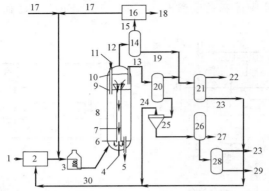

1. coal 煤
2. coal preparation 煤预处理
3. fired preheater, 371℃ 火焰预热器, 371℃
4. impeller shaft 叶轮轴
5. spent catalyst 废催化剂
6. distributor plate 分布板
7. recycle tube 回流管
8. ebullient-bed catalytic reactor, 20MPa, 455℃ 沸腾床催化反应器, 20MPa, 455℃
9. catalyst level 催化剂位
10. slurry level 浆面
11. regenerated catalyst 循环催化剂
12. vapor 蒸气
13. solid liquid 固体液体
14. condenser, 20MPa, 38℃ 冷凝器, 20MPa, 38℃
15. gas 气体
16. hydrogen recovery 氢回收
17. hydrogen 氢
18. hydrocarbon gas, H_2S, NH_3 气体烃, H_2S, NH_3
19. liquid 液体
20. flash separator, 0.1MPa 闪蒸分离器, 0.1MPa
21. still, 0.1MPa 蒸馏柱, 0.1MPa
22. light distillate 轻馏分
23. heavy distillate 重馏分
24. oil 油
25. hydroclone 水力旋流器
26. liquid-solid separator 固液分离器
27. residue 残渣
28. vacuum still 减压蒸馏
29. residual fuel 残渣燃料
30. recycle heavy distillate 循环重馏分

2.2.99 COED coal pyrolysis COED煤热解过程

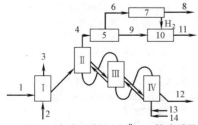

1. crushed coal (~⅛″) 粉碎后的煤 (~⅛″)
2. fluidizing gas 流化载气
3. to gas cleanup 去气体净化
4. pyrolysis gas 热解气
5. oil recovery and filtration section 油回收与过滤
6. COED gas COED气体
7. gas scrubbing and processing 气体洗涤与加工
8. product gas (ES) 产品气体 (ES)
9. COED oil COED油
10. fixed-bed hydrotreatment 固定床加氢处理
11. synthetic crude oil 合成原油
12. char product 产品炭
13. oxygen 氧气
14. steam 水蒸气

2.2.100 Simplified process diagram for the SASOL-I Fischer-Tropsch process SASOL-I 费-托合成过程的简化流程

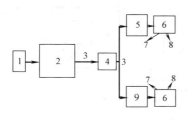

1 coal prep. 煤的准备
2 gasification reactor, high temperature and moderate pressure 气化反应器，高温与中压
3 synthesis gas 合成气
4 gas cleanup 气体净化
5 fixed bed synthesis 固定床合成
6 separator 分离器
7 liquid products 液体产品
8 gas products 气体产品
9 fluid bed synthesis 流化床合成

2.2.101 Process and alternatives for producing SNG from coal 煤生产合成天然气的工艺和替换方案

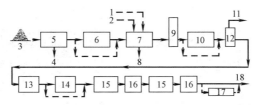

process (indicated by dark lines) 生产工艺用黑线表示

alternatives (indicated by broken lines) 替换工艺用虚线表示

1 hydrogen (or hydrogen-rich gas) 氢（或含氢富气）
2 oxygen 氧
3 coal from minemouth 来自矿井的煤
4 refuse 废料
5 coal preparation 煤预加工
6 pretreatment to prevent caking 防止结块处理
7 gasification 气化作用
8 ash 煤渣
9 scrubber 涤气器
10 shift converter 变换炉
11 H_2S to sulfur recovery H_2S 去硫回收
12 acid gas removal 脱除酸性气体
13 trace sulfur removed 脱除微量硫
14 trace organics removal 脱除微量有机物
15 methanation 甲烷化作用
16 drying 干燥
17 compression 压缩
18 SNG to pipeline 合成天然气去管线

(SNG——synthetic natural gas)

2.2.102 Flow scheme shows natural gas used for fuel and hydrogen 天然气生产燃料和氢气流程示意

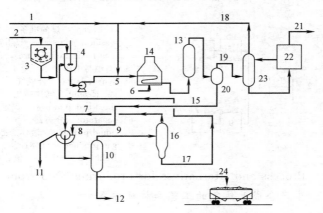

1   reformed natural gas   重整天然气
2   coal   煤
3   pulverizer   粉碎机
4   slurry mixing tank   煤浆混合罐
5   coal slurry   煤浆
6   natural gas   天然气
7   wash solvent   洗涤溶剂
8   filter   过滤器
9   solvent   溶剂
10 vacuum flash tank   真空闪蒸罐
11 residus   滤渣
12 pipeline fuel to power plant   燃料管道通发电站
13 dissolver（depolymerizer）溶解器
（解聚装置）
14 slurry preheater   煤浆预热器
15 product slurry   煤浆产品
16 solvent recovery still   溶剂回收蒸馏釜
17 recycle solvent   循环溶剂
18 recycle hydrogen   循环氢气
19 raw gas   未加工的气体
20 separator   分离器
21 sulfur   硫
22 sulfur plant   硫加工设备
23 absorber   吸收塔
24 solid product   固态产品

2.2.103 Basic SNG process 基本合成天然气工艺

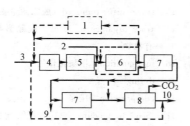

（using naphtha or LPG feedstocks is indicated dark lines. Process alternatives shown with broken lines 利用石脑油或者液化石油气为原料，工艺以黑线表示，替换方案以虚线表示）
1   gas reforming and conditioning
    气体转化与调理
2   steam   水蒸气
3   naphtha or LPG   石脑油或液化石油气
4   sulfur removal   脱硫
5   sulfur absorption   吸收硫
6   gasification   气化
7   methanation   甲烷化反应
8   CO_2 removal and gas drying   脱除 CO_2 和气体干燥
9   condensate   冷凝物
10 SNG（synthetic natural gas）
    合成天然气

2.2.104 Options for upgrading lean gas 贫气的提质方案

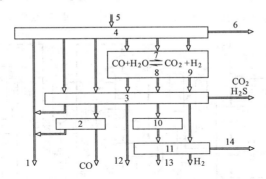

1. low-Btu fuel gas 低热值燃料气
2. gas separator 气体分离器
3. Rectisol acid-gas removal 甲醇法脱除酸性气体
4. gas cleaning unit 气体净化装置
5. raw gas 未加工的煤气
6. tars, solids 焦油、固体
7. shift reaction 变换反应
8. partial 部分的
9. complete 全部的
10. methanation 甲烷化反应
11. cryogenic separation 低温分离
12. synthesis gas 合成气
13. methane SNG 甲烷、合成天然气
14. reject gas N_2 废气氮气

2.2.105 Slurry feed system for coal processes 煤原料生产工艺的煤浆进料系统

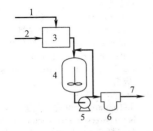

1. dry pulverized coal 干煤粉
2. solvent 溶剂
3. eductor mixer 喷射混合器
4. slurry blend tank 煤浆混合罐
5. recirculating pump 再循环泵
6. injection pump 注射泵
7. to process 去加工

2.2.106 Synthane process flowsheet for production of substitute natural gas 以煤代替天然气的合成天然气工艺流程图

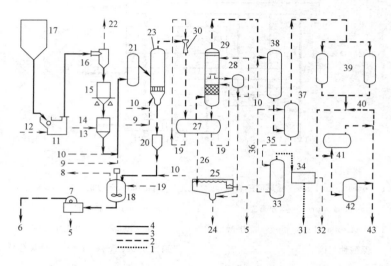

1   H_2S & sulfur   硫化氢和硫
2   fuel gas   燃料气体
3   char   炭
4   coal   煤
5   water to recycle   水去循环
6   char cake to storage   炭滤饼去贮存
7   filter   过滤机
8   vent   放空
9   oxygen   氧气
10   steam   水蒸气
11   flash drying mill   急速干燥粉碎机
12   hot air   热空气
13   injector   注射器
14   recycle CO_2 循环 CO_2
15   weigh hopper   称量斗
16   cyclone   旋风分离器
17   raw coal storage hopper   原煤贮仓
18   char slurry tank   炭料浆罐
19   water   水
20   lock hopper   闭锁式料斗
21   pretreater   预热器
22   air to recycle   空气去循环
23   gasifier   气化器
24   tar to gasifier   焦油去气化器
25   decanter   滗析器
26   waste liquors   废液
27   surge tank   缓冲罐
28   wash oil tank   洗油罐
29   scrubber   涤气器
30   Venturi scrubber   文丘里涤气器
31   sulfur to storage   硫去贮罐
32   CO_2 to recycle CO_2 去循环
33   absorbent regenerator   吸收剂再生器
34   Stretford desulfurizer   蒽醌二磺酸钠法脱硫器
35   rich absorbent   富吸收剂
36   lean absorbent   贫吸收剂
37   acid-gas absorber   酸性气体吸收器
38   shift converter   变换器
39   sulfur guard reactors   硫保护反应器
40   recycle   循环
41   methanator   甲烷转化器
42   hot gas recycle methanator (alternate)   热气循环甲烷转化器（交替使用）
43   high-Btu fuel gas   高热值燃料气

2.2 Chemical processing of coal 煤化工

2.2.107 Hydrogasification couples nuclear heat with methane reforming 加氢气化使核热与甲烷重整组合

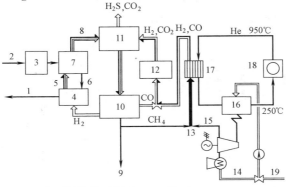

1 residual char 270tons/h 剩余炭 270 吨/小时
2 raw lignite 2200 tons/h 未加工的褐煤 2200 吨/小时
3 rotary tube dryer 旋转管束干燥器
4 gasfire 850℃ 气化炉 850℃
5 raw gas 未处理的煤气
6 lignite 褐煤
7 fluidized bed dryer 流化床干燥器
8 raw gas CH_4, H_2, CO, H_2S, CO_2 未加工的煤气甲烷、氢、一氧化碳、硫化氢、二氧化碳
9 CH_4-production $380 \times 10^3 m^3/h$ 甲烷产品 380×10^3 米³/小时
10 low temperature separation 低温分离
11 gas cleaning system 气体净化系统
12 shift conversion 变换炉
13 110MW (net power generation) 110 兆瓦（净发电量）
14 H_2O (water) 水
15 H_2O (steam) 水（蒸汽）
16 steam generator 蒸汽发生器
17 steam reformer 蒸汽重整炉
18 high-temperature nuclear reactor 3000 MW (thermal) 高温核反应堆 3000 兆瓦（热量）
19 H_2O (process water) 水（工业用水）

2.2.108 Plant combining HTGR and steam gasification of coal 气冷式高温反应堆与煤的气化联合工厂

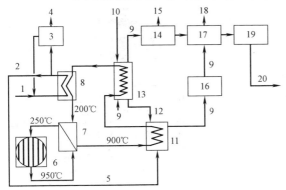

1 water 水
2 steam 水蒸气
3 power plant 发电站
4 electricity 电
5 process steam 工艺蒸汽
6 HTGR (high-temperature gas-cooled reactor) 气冷式高温反应堆
7 heat exchanger 换热器
8 steam generator 蒸汽发生器
9 gas 气体
10 coal 煤
11 gas generator 煤气发生器
12 char 炭
13 carbonizer 碳化塔
14 tar remover 脱焦油
15 tar, oil, etc 焦油、油等
16 dust remover 除尘
17 CO_2/H_2S remover 脱二氧化碳、硫化氢
18 CO_2, H_2S, etc 二氧化碳、硫化氢等
19 gas processing 煤气加工
20 gas to user 煤气去用户

2.2.109 Simplified block diagram of a modern Integrated Gasification Combined Cycle (IGCC) plant 现代整体煤气化联合循环 (IGCC) 工厂的简化框图

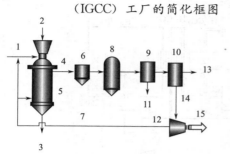

3 slag 炉渣
4 syngas 合成气
5 gasifier 气化炉
6 particulate removal 颗粒去除
7 steam 蒸汽
8 CO-shift reactor 一氧化碳变换反应炉
9 gas clean-up 煤气净化
10 CO_2 capture 二氧化碳捕集
11 sulfur 硫
12 power block 动力单元
13 CO_2 二氧化碳
14 H_2 氢气
15 electricity 电力

1 O_2 from ASU 从空分装置来的氧气
2 coal 煤

2.2.110 A Typical Integrated Gasification Combined Cycle Plant 典型的整体煤气化联合循环工厂

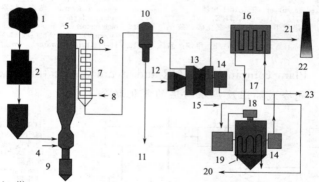

1 coal 煤
2 coal pulverizer 磨煤机
3 hopper 料斗
4 oxygen/air/steam 氧气/空气/蒸汽
5 gasifier 气化炉
6 to steam turbine 去蒸汽轮机
7 heat exchanger 换热器
8 feed water 给水
9 slag hopper 渣斗
10 gas cleaning unit 气体净化设备
11 sulphur, mercury and trace elements removal 硫, 汞和微量元素脱除
12 air 空气
13 gas turbine 燃气轮机
14 generator 发电机
15 from heat exchanger 来自换热器
16 HRSG 余热锅炉
17 steam 蒸汽
18 steam turbine 蒸汽轮机
19 cooling water 冷却水
20 to heat exchanger 去热交换器
21 to vent 去放空
22 stack 烟囱
23 electricity 电力

2.2.111 Combined cycle plant (Steam from syngas processing is not shown) 联合循环电厂(来自合成气处理过程的蒸汽未显示)

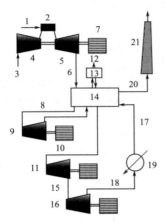

7 electric generator 发电机
8 high pressure steam 高压蒸汽
9 high pressure steam turbine & generator 高压蒸汽涡轮机和发电机
10 intermediate pressure steam 中压蒸汽
11 intermediate pressure steam turbine & generator 中压蒸汽涡轮机和发电机
12 NH_3 氨
13 SCR (selective catalytic reduction) 选择性催化还原
14 HRSG (heat recovery steam generator) 热回收蒸汽发生器
15 low pressure steam 低压蒸汽
16 low pressure steam turbine & generator 低压蒸汽涡轮机和发电机
17 boiler feed water 锅炉给水
18 exhaust steam 废蒸汽
19 condenser 冷凝器
20 cool flue gas 冷烟道气
21 stack 烟囱

1 fuel 燃料
2 burner box 箱式燃烧器
3 air 空气
4 compressor 压缩机
5 turbine 涡轮机
6 hot flue gas 热烟道气

2.2.112 A combination of oxy-combustion and IGCC with CCS (carbon capture and sequestration) 有氧燃烧与带有碳捕集和封存的整体煤气化联合循环的组合工艺

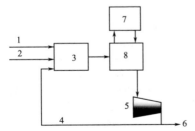

3 compressor/gas turbine/generator 压缩机/燃气轮机/发电机
4 recycle CO_2 循环二氧化碳
5 fan 风机
6 CO_2 to compression & sequestration 二氧化碳去压缩和封存
7 steam turbines/generator/condenser 蒸汽涡轮机/发电机/冷凝器
8 HRSG (heat recovery steam generator) 热回收蒸汽发生器

1 O_2 氧气
2 hot syngas 热合成气

73

2.2 Chemical processing of coal 煤化工

2.2.113 Schematic diagram of a triple combined cycle including SOFC (solid oxide fuel cell), gas, and steam turbines 固体氧化物燃料电池、燃气轮机和蒸汽轮机三结合联合循环示意图

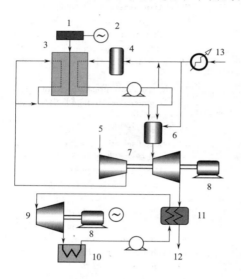

1 inverter （直流变交流）逆变器
2 AC electricity 交流电
3 SOFC 固体氧化物燃料电池
4 external reformer 外重整炉
5 air 空气
6 combustor 燃烧炉
7 gas turbine 燃气轮机
8 generator 发电机
9 steam turbine 汽轮机
10 condenser 冷凝器
11 regenerative heat exchanger 蓄热式换热器
12 exhaust 废气
13 NG (natural gas) 天然气

2.2.114 Simplified diagram of the advanced zero-emission power plant (AZEP) concept 创新的零排放电厂概念示意

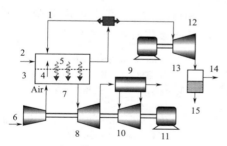

1 recycle 循环
2 NG (natural gas) 天然气
3 ITM (ion transport membranes) 离子迁移膜
4 O_2 氧气
5 heat 热
6 air 空气
7 O_2 depleted air 氧耗尽的空气
8 gas turbine 燃气轮机
9 HRSG (heat recovery steam generator) 热回收蒸汽发生器
10 steam turbine 汽轮机
11 generator 发电机
12 CO_2/steam turbine 二氧化碳/汽轮机
13 condenser 冷凝器
14 CO_2 二氧化碳
15 H_2O 水

2.2.115 Oxy-combustion power plant based on design by Haslbeck et al 哈斯贝克设计的氧燃烧动力装置

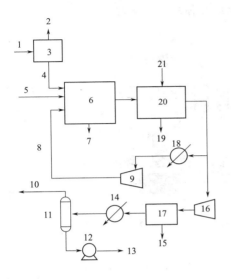

1 air 空气
2 N_2 氮气
3 air separation unit 空分装置
4 O_2 氧气
5 coal 煤
6 pulverized coal combustion 碎煤燃烧
7 ash 煤灰
8 flue gas recycle 烟气循环
9 fan 风机
10 vent gas 排出气
11 flash 闪蒸
12 pump 泵
13 flue gas to sequestration 烟气去储存待用
14 cool 冷却
15 water 水
16 compressor 压缩机
17 drier 干燥器
18 reheat 再加热
19 gypsum 石膏
20 flue gas desulfurization 烟气脱硫
21 limestone, water 石灰石，水

2.2.116 The Sargar process combines a pressurized fluidized bed combustion process with a post-combustion CO_2 removal process, here shown as the Benfield process 本菲尔德工艺——增压流化床燃烧与燃烧后脱碳联合的 Sargar 工艺流程

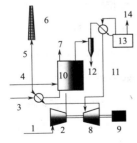

4 coal/limestone slurry 石灰石煤浆
5 flue gas 烟气
6 stack 烟囱
7 steam 蒸汽
8 turbine 涡轮机
9 generator 发电机
10 pressurized fluid bed combustion 增压流化床燃烧
11 CO_2-free flue gas 无二氧化碳烟气
12 fly ash 粉煤灰
13 benfield process 本菲尔德脱碳工艺
14 CO_2 二氧化碳

1 air 空气
2 compressor 压缩机
3 boiler feed water 锅炉给水

2.2.117 Simplified IGCC (integrated gasification combined cycle) with pre-combustion CO₂ capture process 带燃烧前二氧化碳捕集的整体煤气化联合循环简化流程

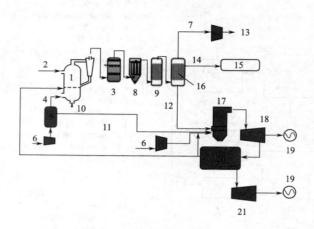

1 gasifier 气化炉
2 coal 煤炭
3 quench tower 骤冷塔
4 O₂ 氧气
5 air separation unit 空分装置
6 air 空气
7 CO₂ 二氧化碳
8 filter 过滤器
9 WGS 水煤气变换炉
10 ash 炉灰
11 N₂ 氮气
12 H₂ 氢气
13 CO₂ storage 二氧化碳存储
14 H₂S 硫化氢
15 clause process 克劳斯（二段脱硫）工艺
16 H₂S, CO₂ separation 硫化氢，二氧化碳分离
17 furnace 燃烧炉
18 gas turbine 燃气轮机
19 generator 发电机
20 heat exchange 热交换
21 steam turbine 蒸汽轮机

2.2.118 Schematic diagram of the oxyfuel combustion process for CO_2 capture 富氧燃烧工艺中二氧化碳捕集示意

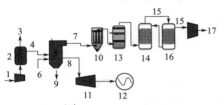

1 air 空气
2 air separation unit 空分装置
3 N_2 氮气
4 O_2 氧气
5 furnace 燃烧炉
6 coal 煤炭
7 flue gas 烟道气
8 steam 蒸汽
9 ash 炉灰
10 filter 过滤器
11 steam turbine 汽轮机
12 generator 发电机
13 wet scrubber 湿式涤气塔
14 CO_2 absorber 二氧化碳吸收塔
15 CO_2 二氧化碳
16 CO_2 stripper 二氧化碳汽提塔
17 CO_2 storage 二氧化碳存储

2.2.119 Summary of CO_2 capture strategies. (a) Pre-combustion capture, (b) post-combustion capture, and (c) oxyfuel combustion 二氧化碳捕集方法简介：(a) 燃烧前捕集；(b) 燃烧后捕集；(c) 富氧燃烧

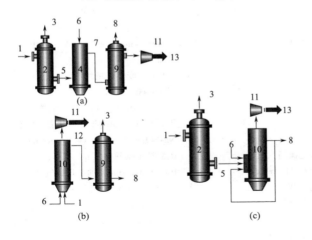

1 air 空气
2 ASU 空分装置
3 N_2 氮气
4 reformer 重整器
5 O_2 氧气
6 fuel 燃料
7 syngas 合成气
8 CO_2 二氧化碳
9 gas separation unit 气体分离装置
10 combustor 燃烧炉
11 turbine generator 涡轮发电机
12 fuel gas 燃气
13 electricity 电力

2.3 Chemical process 化工工艺

2.3.1 Flow diagram of the coal-to-ammonia plant
由煤生产合成氨工厂流程图

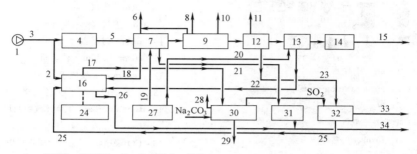

1. HHV = 11342Btu/lb, 4.2% S。高热值＝11342 英热单位/磅, 4.2%硫（HHV——high heat value）
2. 350 tons/d 350 吨/天
3. coal 2600 tons/d 煤, 2600 吨/天
4. coal preparation 煤的制备
5. coal 2250 tons/d 煤, 2250 吨/天
6. 950 psig steam（压力为）950 磅/英寸^2 蒸汽
7. gasification （煤）气化
8. 100 psig steam（压力为）100 磅/英寸^2 蒸汽
9. shift conversion and gas cooling 变换反应和气体冷却
10. 40 psig steam（压力为）40 磅/吋^2 的蒸汽
11. CO_2 to atm 二氧化碳通大气
12. Rectisol process for CO_2-removal 甲醇法脱（二氧化）碳
13. nitrogen wash 氮气洗涤
14. ammonia synthesis 氨合成
15. anhydrous ammonia 1500 tons/d 无水氨（1500 吨/天）
16. steam boiler and superheater 蒸汽锅炉和过热器
17. fluegas 烟道气
18. ash/soot 煤渣、烟灰
19. oxygen 1865 tons/d 氧气 1865 吨/天
20. nitrogen 氮气
21. ash 灰分
22. CO-rich gas 一氧化碳富气
23. H_2S-rich gas 硫化氢富气
24. power generation 14MW 发电 14 兆瓦
25. tailgas 废气, 尾气
26. boiler ash 炉灰
27. air separation plant 空分装置
28. fluegas to atmosphere 烟道气放空
29. $Na_2SO_4 + Na_2CO_3$ (4.0 tons/d) 硫酸钠＋碳酸钠（4.0 吨/天）
30. fluegas treating 加工烟道气
31. ash dewatering 灰分脱水
32. Claus sulfur plant 克劳斯制硫装置
33. 106.7 tons/d sulfur 106.7 吨/天硫黄
34. 550 tons/d wet ash (30% moisture) 550 吨/天湿灰（含湿量 30%）

2.3.2 Texaco coal-gasification process to an existing ammonia plant 德士古煤造气工艺连到现有的合成氨厂

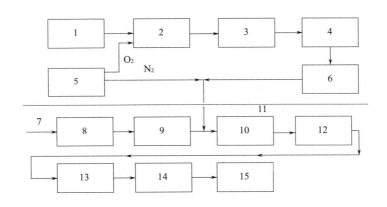

1 coal-handling and preparation 煤的加工制备

2 gasification unit 造气装置

3 intermediate-temperature CO-shift 中温一氧化碳变换

4 gas desulfurization 气体脱硫

5 air separation plant 空分装置

6 CO_2 removal 脱二氧化碳

7 natural gas 天然气

8 primary reformer 一段转化炉

9 secondary reformer 二段转化炉

10 high temperature shift 高温变换

11 existing plant 现有装置

12 low temperature shift 低温变换

13 CO_2 removal 脱二氧化碳

14 methanator 甲烷化反应器

15 ammonia synthesis 合成氨

2.3.3 Block diagram for ammonia plant 合成氨装置方框图

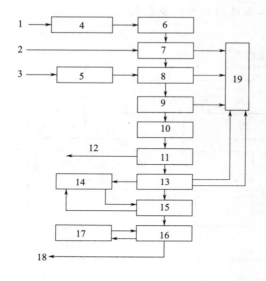

broken lines: heat
虚线：热量

1   natural gas   天然气

2   steam   水蒸气

3   air   空气

4   feed gas compressor   原料气压缩机

5   air compressor   空气压缩机

6   desulfuration   脱硫

7   primary reforming   一段转化

8   secondary reforming   二段转化

9   high temperature shift   高温变换

10   low temperature shift   低温变换

11   decarbonizing   脱碳

12   CO_2 by product CO_2 副产品

13   methanation   甲烷化

14   synthesis gas compressor   合成气压缩机

15   synthesis   合成

16   refrigerating   冷冻

17   ammonia compressor   氨压缩机

18   NH_3 product   氨成品

19   thermal recovery unit   热回收装置

2.3.4 Fixed-bed system divides the methanol-to-gasoline reaction 固定床反应器甲醇转化成汽油流程

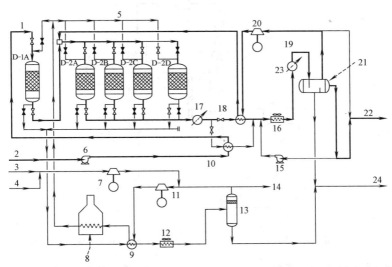

1. DME (dimethyl ether) reactor 二甲醚反应器
2. crude methanol 粗甲醇
3. air 空气
4. nitrogen 氮气
5. conversion reactors 转化反应器
6. feed pump 进料泵
7. air booster compressor 空气增压压缩机
8. startup/regen. furnace 开工/再生炉
9. regen./recycle exchanger 再生/循环换热器
10. reactor effluent vs methanol 反应器流出物与甲醇热交换
11. regen. recycle compressor 再生循环压缩机
12. regen. gas cooler 再生气冷却器
13. gas-liquid separator 气-液分离器
14. offgas to stack 废气去烟囱
15. recycle pump 循环泵
16. effluent air cooler 流出液空冷器
17. 600-psig steam 600磅/英寸^2 蒸汽
18. reactor effluent vs recycle gas 反应器流出液与循环气换热
19. effluent cooler 反应器流出液冷却器
20. gas recycle compressor 气体循环压缩机
21. product separator 产品分离器
22. raw gasoline 粗汽油
23. cooling water 冷却水
24. water 水

2.3.5 Pilot-plant fluid-bed reactor system 流化床反应器中试装置

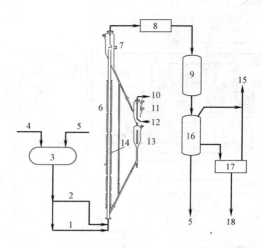

1 vapor feed 蒸气进料

2 liquid feed 液相进料

3 feed preparation 进料预处理

4 pure methanol 纯甲醇

5 water 水

6 reactor (10.2cm×7.6m) 反应器 (10.2cm×7.6m)

7 disengager (41cm×91cm) 释放器 (41cm×91cm)

8 filter 过滤器

9 three-stage condenser 三级冷凝器

10 flue gas 烟道气

11 regenerator (25.4cm×76cm) 再生器 (25.4cm×76cm)

12 air+N_2 空气+氮气

13 hopper (25.4cm×76cm) 贮液器 (25.4cm×76cm)

14 catalyst recirculation 催化剂循环

15 off gas 废气

16 separator 分离器

17 tank 贮罐

18 liquid hydrocarbons 液态碳氢化合物

2.3.6 Flowscheme relies on a solvent to remove phenolics
依靠溶剂除去酚醛树脂的流程

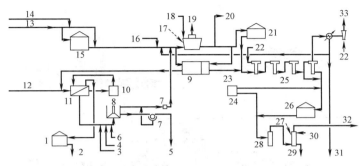

1   slop-oil tank   不合格石油产品贮槽
2   slop oil to boilers   废油去锅炉
3   acid   酸
4   caustic   氢氧化物
5   sludge cake to disposal   淤泥滤饼去处理
6   polymer and coagulant   多聚体和絮凝剂
7   filter   过滤机
8   clarifier   澄清器
9   cooling-tower feed surge pond   冷却塔进料缓冲池
10 oily-water holding pond   含油水储存池
11 separator   分离器
12 oily water   含油的水
13 syngas liquids from ammonia stripper   来自氨气提塔的合成气溶液
14 wastewater   废水
15 storage tank   贮槽
16 softened makeup water   补给软化水
17 process cooling towers   工艺冷却塔
18 cooling-water return   冷却水返回
19 evaporation   蒸发
20 cooling water to heat incinerators   冷却水去热焚烧炉
21 evaporator or feed surge tank   蒸发器或进料缓冲槽
22 steam   水蒸气
23 evaporator-feed surge pond   蒸发器-进料缓冲池
24 incinerator-feed surge pond   焚烧炉-进料缓冲池
25 multiple-effect evaporator   多效蒸发器
26 incinerator-feed surge tank   焚烧炉-进料缓冲槽
27 scrubber   涤气塔
28 incinerator   焚烧炉
29 effluent to ash conveyor   排放物去炉灰输送机
30 wastewater from boiler-feed water blowdown   来自锅炉给水排污的废水
31 excess condensate to deep well   剩余冷凝液去深井
32 flue gas to stack   烟道气去烟囱
33 vent   放空

2.3.7 Manufacture of synthetic fibres 合成纤维的生产

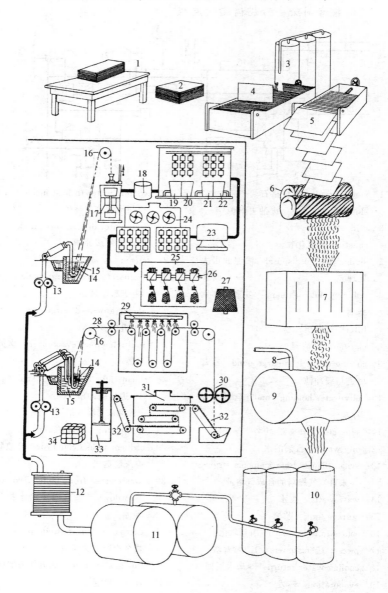

1	basic material 基本原料	19	washing 水洗
2	mixing cellulose sheets 混合纤维素板	20	desulphurizing 脱硫
		21	bleaching 漂白
3	caustic soda 苛性钠,烧碱	22	treating of cake to give filaments softness 使丝饼柔软的处理
4	steeping cellulose sheets in caustic soda 纤维素板浸入苛性钠中	23	hydro-extraction to remove surplus moisture 脱水以去除多余的水分
5	pressing out excess caustic soda 压出多余的苛性钠	24	drying in heated room 在烘房内烘干
6	shredding the cellulose sheets 粉碎纤维素板	25	winding yarn from cake into cone form 将丝饼绕成锥形丝筒
7	maturing of the alkali-cellulose crumbs 碱性纤维素碎片的熟化	26	cone-winding machine 锥形绕线机
8	carbon disulphide 二硫化碳	27	viscose rayon yarn on cone ready for use 卷于锥形筒上备用的黏胶丝
9	conversion of alkali-cellulose into cellulose xanthate 碱纤维素转变为纤维素黄酸酯	28~34	from viscose spinning solution to viscose rayon staple fibre 从黏胶纺丝液至黏胶短纤维(人造丝)
10	dissolving the xanthate in caustic soda for the preparation of the viscose spinning solution 把黄酸酯溶于苛性钠中以制备黏胶纺丝液	28	filament tow 拉出来的丝束
		29	overhead spray washing plant 喷淋水洗设备
11	vacuum ripening tanks 真空成熟槽	30	cutting machine for cutting filament tow to desired length 将丝切成所需长度的剪切机
12	filter press 压滤器		
13	metering pump 计量泵		
14	multi-holed spinneret 多孔纺纱头	31	multiple drying machine 多层烘干机
15	coagulating basin 凝固槽		
16	godet wheel 导丝轮	32	conveyor belt 输送带
17	Topham centrifugal pot (box) 托范式离心罐	33	baling press (压力)打包机
		34	bale of viscose rayon ready for dispatch 准备运送的黏胶纤维包
18	viscose rayon cake 黏胶丝饼		

2.3.8 Manufacture of polyamide fibres 聚酰胺纤维的制造

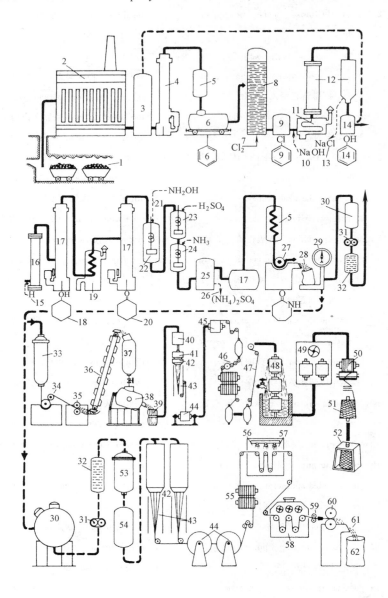

1	coal	煤
2	coking plant for dry coal distillation	煤干馏用的焦化设备
3	extraction of coal tar and phenol	焦油和酚的提取
4	gradual distillation of tar	焦油的分馏
5	condenser	冷凝器
6	benzene extraction and dispatch	苯的提取和运送
7	chlorine	氯
8	benzene chlorination	苯的氯化
9	monochlorobenzene	单氯苯
10	caustic soda solution	苛性钠溶液
11	evaporation of chlorobenzene and caustic soda	单氯苯和苛性钠的蒸发
12	autoclave	高压釜
13	sodium chloride, a by-product	氯化钠,一种副产品
14	phenol	(苯)酚
15	hydrogen inlet	氢气入口
16	hydrogenation of phenol to produce raw cyclohexanol	酚的氢化产生粗环己醇
17	distillation	蒸馏
18	pure cyclohexanol	纯环己醇
19	oxidation (dehydrogenation)	氧化(脱氢作用)
20	formation of cyclohexanone	环己酮的形成
21	hydroxylamine inlet	羟胺入口
22	formation of cyclohexanoxime	环己酮肟的形成
23	addition of sulphuric acid to effect molecular rearrangement	加硫酸使分子重新排列
24	ammonia to neutralize sulphuric acid	加氨中和硫酸
25	formation of caprolactam oil	己内酰胺油的形成
26	ammonium sulphate solution	硫酸铵溶液
27	cooling cylinder	冷却滚筒
28	caprolactam	己内酰胺
29	weighing apparatus	称重装置
30	melting pot	熔化罐
31	pump	泵
32	filter	过滤器
33	polymerization in the autoclave	在高压釜内聚合
34	cooling of the polyamide	聚酰胺的冷却
35	solidification of the polyamide	聚酰胺的固化
36	vertical lift	升降机,电梯
37	extractor	萃取器
38	drier	干燥机
39	dry polyamide chips	干燥的聚酰胺(碎)片
40	chip container	碎片容器
41	top of spinneret for melting the polyamide and forcing it through spinneret holes	熔化聚酰胺和从喷丝孔压挤出来的喷丝头
42	spinneret holes	喷丝孔
43	solidification of the polyamide filaments in the cooling tower	聚酰胺丝于冷却塔中凝固
44	collection of extruded filaments into thread form	将挤压出来的丝条集合成丝束
45	preliminary stretching	预拉伸
46	stretching of the polyamide thread to achieve high tensile strength	聚酰胺丝线(冷)拉伸达到高抗拉强度
47	final stretching	最后拉伸
48	washing of yarn packages	丝包的洗涤
49	drying chamber	干燥室
50	rewinding	复绕
51	polyamide cone	聚酰胺锥形筒
52	polyamide cone ready for dispatch	准备运送的聚酰胺筒
53	mixer	混合器
54	polymerization under vacuum	真空下聚合
55	stretching	拉伸
56	washing	水洗
57	finishing of tow for spinning	纺纱用丝束的整理
58	drying of tow	丝束的烘干
59	crimping of tow	丝束的起皱
60	cutting of tow into normal staple lengths	将丝束切成标准纤维长度
61	polyamide staple	聚酰胺(短)纤维
62	bale of polyamide staple	聚酰胺纤维包

2.3 Chemical process 化工工艺

2.3.9 Glass production 玻璃制造

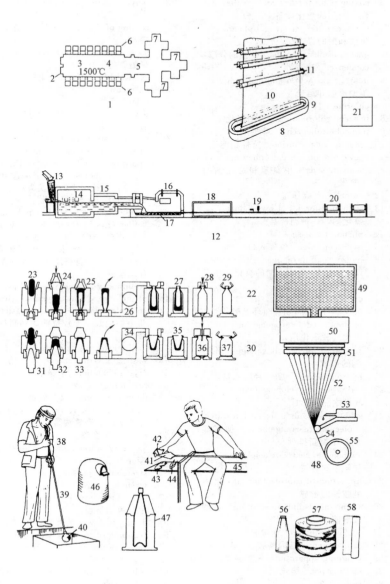

1~20	sheet glass production 平板玻璃制造	31	introduction of the gob of molten glass 加入熔融的玻璃料滴
1	glass furnace for the Fourcault process 弗克法玻璃窑	32	plunger 柱塞
2	filling end, for feeding in the batch 加料端，分批进料	33	pressing 压制
3	melting bath 熔化池	34	transfer from the press mould to the blow mould 从压制模转到吹制模
4	refining bath 精炼池	35	reheating 再加热
5	working baths 工作池	36	blowing (suction, final shaping) 吹制（吸气，最后成形）
6	burners 燃烧器	37	delivery of the completed vessel 成品容器的输送
7	drawing machines 拉制机	38~47	glassmaking 玻璃制造
8	Fourcault glass-drawing machine 弗克法玻璃拉制机	38	glassmaker 吹玻璃工人
9	slot 槽沟	39	blowing iron 吹玻璃用的铁管
10	glass ribbon being drawn upwards 正在引上的玻璃板	40	god 熔解的玻璃原料
11	rollers 滚子	41	hand-blown goblet 人工吹制的高脚杯
12	float glass process 浮法玻璃制造法	42	clappers for shaping the base of the goblet 高脚杯底成形用拍板
13	batch feeder 原料进给装置	43	trimming tool 修整工具
14	melting bath 熔化池	44	tongs 夹钳
15	cooling tank 冷却池	45	glassmaker's chair 玻璃工人坐椅
16	float bath in a protective inert-gas atmosphere 在惰性气体保护下的浮槽	46	covered glasshouse pot 有盖的熔化玻璃坩埚
17	molten tin 熔化的锡	47	mould, into which the parison is blown 吹制成形（长颈瓶）模型，玻璃半成品在内
18	annealing lehr 退火窑	48~55	production of glass fibre 玻璃纤维的生产
19	automatic cutter 自动截切机	48	continuous filament process 连续拉丝
20	stacking machines 堆置机		
21	a bottle-making machine 一种制瓶机	49	glass furnace 玻璃熔炉
22~37	blowing processes 吹制法	50	bushing containing molten glass 装熔化玻璃的套筒
22	blow-and-blow process 吹制过程	51	bushing tips 套筒端
23	introduction of the gob of molten glass 加入熔融的玻璃料滴	52	glass filaments 玻璃纤维丝
24	first blowing 第一次吹制	53	sizing 上涂料
25	suction 吸气	54	strand 玻璃纤维线
26	transfer from the parison mould to the blow mould 从玻璃瓶模转到吹制模	55	spool 卷线轴
		56~58	glass fibre products 玻璃纤维制品
27	reheating 再加热	56	glass yarn 玻璃丝
28	blowing 吹制	57	sleeved glass yarn 玻璃丝卷
29	delivery of the completed vessel 成品容器的输送	58	glass wool 玻璃棉
30	press-and-blow process 压-吹法		

2.3.10 Pulping 制浆

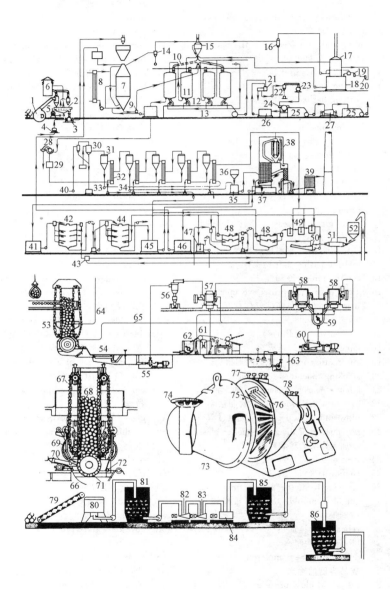

2.3 Chemical process 化工工艺

1~52 sulfate pulp mill 硫酸盐纸浆厂
1 chippers with dust extractor 带吸尘装置的削片机
2 rotary screen 旋转筛
3 chip packer 木片装料机
4 blower 鼓风机
5 disintegrator 粉碎机
6 dust-settling chamber 灰尘沉淀室
7 digester 蒸解锅
8 liquor preheater 液体顶热器
9 control tap 控制龙头
10 swing pipe 摆动管
11 blow tank 喷放罐
12 blow valve 喷放阀
13 blow pit 喷放池
14 turpentine separator 松节油分离器
15 centralized separator 集中分离器
16 jet condenser 喷射冷凝器
17 storage tank for condensate 凝结液贮存罐
18 hot water tank 热水槽
19 heat exchanger 热交换器
20 filter 过滤器
21 presorter 预选料器
22 centrifugal screen 离心筛
23 rotary sorter 旋转筛浆机
24 concentrator 浓缩器
25 vat 浆槽
26 collecting tank for backwater 回水收集槽
27 conical refiner 锥形磨浆机
28 black liquor filter 黑液过滤器
29 black liquor storage tank 黑液贮槽
30 condenser 冷凝器
31 separator 分离器
32 heater 加热器
33 liquor pump 液泵
34 heavy liquor pump 重液泵
35 mixing tank 混合槽
36 salt cake storage tank 盐饼贮槽
37 dissolving tank 溶解槽
38 steam heater 蒸汽加热器
39 electrostatic precipitator 静电集尘器
40 air pump 空气泵
41 storage tank for the uncleared green liquor 不洁绿液贮槽
42 concentrator 浓缩器
43 green liquor preheater 绿液预热器
44 concentrator for the weak wash liquor 稀洗液浓缩器
45 storage tank for the weak liquor 稀液贮槽
46 storage tank for the cooking liquor 蒸煮液贮槽
47 agitator 搅拌器
48 concentrator 浓缩器
49 causticizing agitators 苛化搅拌器
50 classifier 分级器
51 lime slaker 化灰器
52 reconverted lime 再生石灰
53~65 groundwood mill 磨木浆厂
53 continuous chain grinder 连续链式磨木机
54 strainer 滤筛
55 pulp water pump 纸浆水泵
56 centrifugal screen 离心筛
57 screen 筛浆机
58 secondary screen 二道筛
59 reject chest 废料池
60 conical refiner 锥形磨浆机
61 pulp-drying machine 纸浆干燥机
62 concentrator 浓缩器
63 waste water pump 废水泵
64 steam pipe 蒸汽管
65 water pipe 水管
66 continuous chain grinder 连续链式磨木机
67 feed chain 给料链
68 groundwood 磨木浆
69 reduction gear for the feed chain drive 进料链传动的减速齿轮
70 stone-dressing device 刻石装置
71 grinding stone 磨石
72 spray pipe 喷淋管
73 conical refiner 锥形磨浆机
74 handwheel for adjusting the clearance between the knives 调整刀片间隙的手轮
75 rotating bladed cone 旋转刀片的锥部
76 stationary bladed shell 固定刀片壳
77 inlet for unrefined cellulose or groundwood pulp 未精制纤维素或磨木纸浆的入口
78 outlet for refined cellulose or groundwood pulp 精制纤维素或磨木纸浆的出口
79~86 stuff preparation plant 备料设备
79 conveyor belt for loading cellulose or groundwood pulp 纤维素或磨木浆的输送带
80 pulper 碎浆机
81 dump chest 卸料池
82 cone breaker 锥形碎浆机
83 conical refiner 锥形磨浆机
84 refiner 精磨机
85 stuff chest 纸料池
86 machine chest 机械浆料池

2.3.11 Papermaking 造纸

2.3 Chemical process 化工工艺

1 stuff chest, a mixing chest for stuff 贮浆池，一种浆料混合池
2 Erlenmeyer flask 锥形（烧）瓶
3 volumetric flask 容量瓶
4 measuring cylinder 量筒
5 Bunsen burner 本生灯
6 tripod 三脚架
7 petri dish 有盖玻璃皿
8 test tube rack 试管架
9 balance for measuring basis weight 测重秤
10 micrometer 测微计
11 centrifugal cleaners ahead of the breast-box of a paper machine 造纸机料箱前的离心净浆器
12 standpipe 竖管
13~28 paper machine production line 造纸机生产线
13 feed-in from the machine chest with sand table and knotter 从带砂滤器和除节器的原料贮槽进料
14 wire 金属丝网
15 vacuum box 真空箱
16 suction roll 吸水滚轮
17 first wet felt 第一道湿毛布
18 second wet felt 第二道湿毛布
19 first press 第一道压榨
20 second press 第二道压榨
21 offset press 胶版压榨
22 drying cylinder 烘缸
23 dry felt 干毛布
24 size press 压榨机
25 cooling roll 冷却滚轮
26 calender rolls 压光滚轮
27 machine hood 机（器）罩
28 delivery reel 卷取纸轴
29~35 blade coating machine 刮刀涂布机
29 raw paper 粗纸
30 web 纸幅
31 coater for the top side 顶面涂布器
32 infrared drier 红外线干燥器
33 heated drying cylinder 加热烘缸
34 coater for the underside 底面涂布器
35 reel of coated paper 涂布的卷筒纸
36 calender 压光机
37 hydraulic system for the press rolls 压榨滚轮的液压系统
38 calender roll 压光滚轮
39 unwind station 退纸装置
40 lift platform 提升台
41 rewind station 回卷台
42 roll cutter 滚轮切纸机
43 control panel 控制盘
44 cutter 切纸机
45 web 卷筒纸
46~51 papermaking by hand 手工造纸
46 vatman 捞工
47 vat （纸）浆槽
48 mould 模盘
49 coucher 伏（辊）工
50 post ready for pressing 准备压制的位
51 felt 毛布，（毛）毡

2.3.12 Malting and Brewing I 麦芽制造和啤酒酿造 I

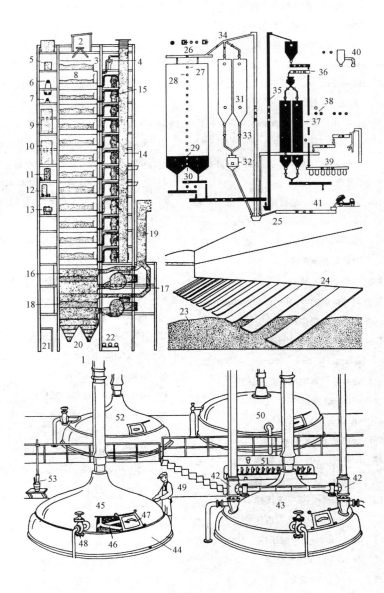

1~41	preparation of malt 制麦芽		29	drying kiln 烘干窑
1	malting tower 麦芽制造塔		30	curing kiln 熟化窑
2	barley hopper 大麦料斗		31	barley silo 大麦筒仓
3	washing floor with compressed air washing unit 有压缩空气洗涤装置的洗涤层		32	weighing apparatus 称量设备
			33	barley elevator 大麦提升机
			34	three-way chute 三通滑槽
4	outflow condenser 外流凝结器		35	malt elevator 麦芽提升机
5	water-collecting tank 集水箱		36	cleaning machine 清洗机
6	condenser for the steep liquor 浸液用的凝结器		37	malt silo 麦芽筒仓
			38	corn removal suction 吸除麦芽装置
7	coolant-collecting plant 冷却液收集装置		39	sacker 装袋器
			40	dust extractor 除尘器
8	steeping floor 浸渍层		41	barley reception 大麦的接收
9	cold water tank 冷水箱		42~53	mashing process in the mash-house 酿酒捣碎车间的捣浆过程
10	hot water tank 热水箱			
11	pump room （水）泵房			
12	pneumatic plant 气动设备		42	premasher for mixing grist and water 混合麦芽和水的初捣
13	hydraulic plant 液压设备			
14	ventilation shaft 通风管道		43	mash tub for mashing the malt 捣碎麦芽的捣浆槽
15	exhaust fan 排风机			
16~18	kilning floors 烘干层		44	mashing kettle 麦芽浆蒸煮锅
16	drying floor 干燥层		45	dome of the tun 桶顶
17	burner ventilator 燃烧送风机		46	propeller 搅拌桨
18	curing floor 熟化层		47	sliding door 拉门
19	outlet from the kiln 烘窑出气口		48	water supply pipe 供水管
20	finished malt collecting hopper 麦芽成品收集漏斗		49	brewer 啤酒酿造者
			50	lauter tun for settling the draff and filtering off the wort 沉淀残渣和过滤麦芽汁用的过滤槽
21	transformer station 变电所			
22	cooling compressor 冷冻压缩机			
23	green malt 绿麦芽		51	lauter battery for testing the wort for quality 检查麦芽汁品质用的过滤器组
24	turner 翻拌器			
25	central control room with flow diagram 有流程图的中央控制室			
			52	hop boiler for boiling the wort 蒸煮麦芽汁用的锅炉
26	screw conveyor 螺旋输送机			
27	washing floor 冲洗层			
28	steeping floor 浸渍层		53	ladle-type thermometer 勺式温度计

2.3 Chemical process 化工工艺

2.3.13 Brewhouse 啤酒厂

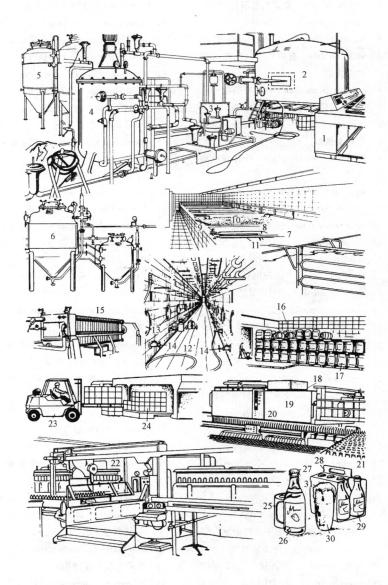

1～5　wort cooling and break removal　麦芽汁的冷却和残渣的去除

1　control desk　操纵台

2　whirlpool separator for removing the hot break　除去高温凝固物所用的涡流分离器

3　measuring vessel for the kieselguhr　板状硅藻土计量器

4　kieselguhr filter　硅藻土过滤器

5　wort cooler　麦芽汁冷却器

6　pure culture plant for yeast　酵母纯培养设备

7　fermenting cellar　发酵室

8　fermentation vessel　发酵槽

9　fermentation thermometer　发酵温度计

10　mash　麦芽浆

11　refrigeration system　制冷系统

12　lager cellar　（贮酒的）大地窖

13　manhole to the storage tank　贮酒罐的人孔

14　broaching tap　取汁龙头

15　beer filter　啤酒滤器

16　barrel store　酒桶仓库

17　beer barrel　啤酒桶

18　bottle-washing plant　洗瓶车间

19　bottle-washing machine　洗瓶机

20　control panel　操纵盘

21　cleaned bottles　已洗净的瓶子

22　bottling　装瓶

23　forklift truck　叉车

24　stack of beer crates　啤酒箱堆

25　beer can　罐装啤酒

26　beer bottle　瓶装啤酒

27　crown cork closure　冠状软木瓶盖

28　disposable pack　一次性包装

29　non-returnable bottle　不回收的瓶子

30　beer glass　啤酒杯

31　head　泡沫

2.3.14 Coking plant 焦化厂

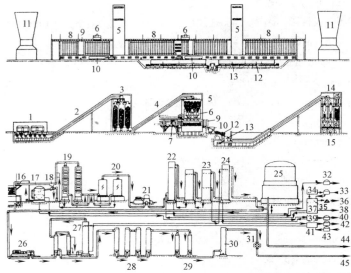

1 dumping of coking coal 焦煤卸料场
2 belt conveyor 带式输送机
3 service bunker 煤仓
4 coal tower conveyor 煤塔输送机
5 coal tower 煤塔
6 larry car 装煤车
7 pusher ram 推进溜板
8 battery of coke ovens 炼焦炉组
9 coke guide 焦炭导槽
10 quenching car, with engine 淬火车, 带发动机
11 quenching tower 淬火塔
12 coke loaking bay 凉焦台
13 coke side bench 放焦台
14 screening of lump coal and culm 块煤和屑煤筛分台
15 coke loading 焦炭装载
16～45 coke-oven gas processing 炼焦炉煤气的加工
16 discharge of gas from the coke ovens 炼焦炉煤气排出 (口)
17 gas-collecting main 煤气主管
18 coal tar extraction 焦油的提取
19 gas cooler 煤气冷却器
20 electrostatic precipitator 静电集尘器
21 gas extractor 煤气抽出器
22 hydrogen sulphide scrubber 硫化氢洗涤塔
23 ammonia scrubber 氨气洗涤塔
24 benzene scrubber 苯洗涤塔
25 gas holder 贮气罐
26 gas compressor 煤气压缩机
27 debenzoling by cooler and heat exchanger 冷却器和换热器的脱苯
28 desulphurization of pressure gas 加压煤气的脱硫
29 gas cooling 煤气冷却
30 gas drying 煤气干燥
31 gas meter 煤气表
32 crude tar tank 粗焦油罐
33 sulphuric acid supply 硫酸供应
34 production of sulphuric acid 硫酸产出
35 production of ammonium sulphate 硫铵产出
36 ammonium sulphate 硫酸铵
37 recovery plant for recovering the scrubbing agents 回收洗涤液的回收装置
38 waste water discharge 废水排出
39 phenol extraction from the gas water 从煤气中提取酚
40 crude phenol tank 粗酚罐
41 production of crude benzol 粗苯产出
42 crude benzol tank 粗苯罐
43 scrubbing oil tank 洗涤油罐
44 low-pressure gas main 低压煤气总管
45 high-pressure gas main 高压煤气总管

2.3.15 Flow diagram of the Sohio process for the manufacturing of acrylonitrile 美孚公司丙烯腈生产工艺流程图

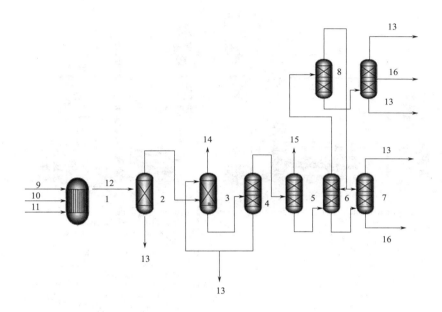

1   CSTR reactor   连续搅拌釜式反应器
2   neutralizer   中和塔
3   absorber   吸收塔
4   stripper   汽提塔
5   HCN column   氰化氢塔
6   extractive distillation column   萃取精馏塔
7   acetonitrile purification column   乙腈净化塔
8   acrylonitrile purification columns   丙烯腈净化塔
9   propylene   丙烯
10 ammonia   氨
11 air   空气
12 H_2SO_4 硫酸
13 waste   废物
14 tail gases   尾气
15 HCN   氰化氢
16 acetonitrile   乙腈

2.3.16 Carbon black process diagram 炭黑工艺流程图

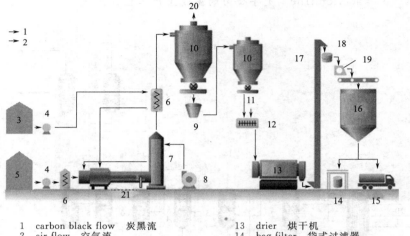

1 carbon black flow   炭黑流
2 air flow   空气流
3 feedstock oil   原料油
4 oil pump   油泵
5 fuel oil   燃油
6 oil preheater   油预热器
7 air preheater   空气预热器
8 air blower   鼓风机
9 pulverizer   粉碎机
10 filter bags   滤袋
11 granulation water   造粒用水
12 pelletizer   造粒机
13 drier   烘干机
14 bag filter   袋式过滤器
15 bulk transporting vehicle   散装运输车辆
16 product tank   产品罐
17 bucket elevator   斗式提升机
18 classifier   分选机
19 magnetic separator   磁力分离机
20 by-product gases (fuel for drier and power plant)   副产品气体（作为烘干机和电厂的燃料）
21 reactor   反应器

2.3.17 Furnace process (for carbon black production) 炉式工艺制炭黑流程

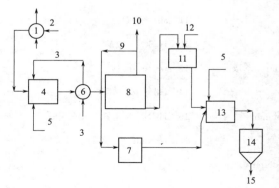

1 oil preheater   油预热器
2 oil   油
3 air   空气
4 reactor   反应器
5 natural gas   天然气
6 preheater   预热器
7 waste gas combustor   废气燃烧炉
8 bag filter   袋式过滤器
9 waste gas   废气
10 atm   大气
11 pelletizer   造粒机
12 water   水
13 dryer   烘干机
14 bulk STG   散装储存货架
15 product   产品

2.3.18 A process flow diagram to manufacture vinyl chloride monomer 氯乙烯单体生产工艺流程图

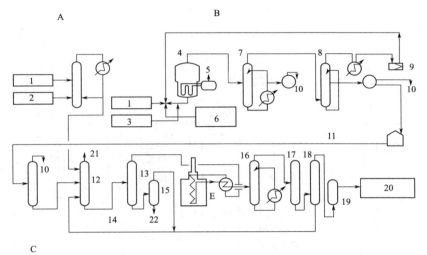

A	direct chlorination 直接氯化	10	water 水
B	oxychlorination 氧氯化	11	EDC (ethylene dichloride) 二氯乙烷
C	dehydrator 脱水器	12	EDC No.1 column EDC 一塔
D	EDC purification EDC 净化	13	EDC No.2 column EDC 二塔
E	EDC cracker EDC 裂解装置	14	recycle EDC 循环 EDC
F	VCM purification VCM 净化	15	EDC recovery column EDC 回收塔
1	ethylene 乙烯	16	EDC quencher EDC 急冷塔
2	chlorine 氯	17	HCl column 盐酸塔
3	oxygen 氧气	18	VCM column 氯乙烯单体塔
4	reactor 反应器	19	caustic tower 烧碱塔
5	steam 蒸汽	20	vinyl chloride monomer 氯乙烯单体
6	hydrogen chloride 氯化氢	21	lights 轻组分
7	quencher 急冷塔	22	heavies 重组分
8	caustic scrubber 碱洗涤器		
9	recycle gas compressor 循环气压缩机		

2.3.19 A process flow diagram to produce ethylene oxide 环氧乙烷生产工艺

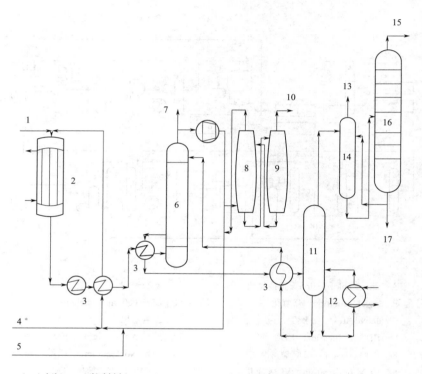

1 inhibitors 抑制剂
2 reactor 反应器
3 heat exchanger 热交换器
4 O_2 氧气
5 ethylene 乙烯
6 EO scrubber 环氧乙烷洗涤塔
7 vent A 放空 A
8 CO_2 absorber 二氧化碳吸收塔
9 CO_2 desorber 二氧化碳解吸塔
10 CO_2 二氧化碳
11 EO desorber 环氧乙烷解吸塔
12 reboiler 再沸器
13 vent B 放空 B
14 stripping column 汽提塔
15 ethylene oxide 环氧乙烷
16 distillation column 蒸馏塔
17 ethylene oxide with water 含水的环氧乙烷

2.4 水净化处理 Purification treatment of water

2.4.1 Essential features of a flow-through pasteurizer 流通式巴氏消毒器的基本特征

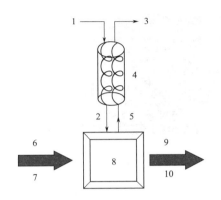

1. impure water T_0 不纯净的水（温度 T_0）
2. T_1 (temperature 1) 温度 T_1
3. pasteurized water T_2 巴氏消毒水（温度 T_2）
4. heat exchanger 热交换器
5. T_p (temperature P) 温度 T_P
6. solar power 太阳能电力
7. Q_s (sensible heat flow) 显热流
8. device (pasteurizer) 设备（巴氏消毒器）
9. heat loss 热损失
10. Q_L (latent heat flow) 潜热流

2.4.2 Schematic of a multistage flash distillation process （海水）多级闪速蒸馏工艺示意

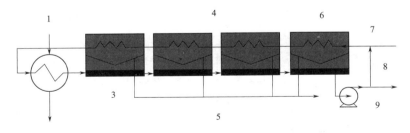

1. steam in 蒸汽进入
2. condensate out 冷凝水出
3. first stage 第一级
4. heat recovery stages 热回收级
5. product water 成品水
6. heat rejection stage 热耗损级
7. seawater in 海水进入
8. recycle brine 循环盐水
9. brine discharge 盐水排放

2.4.3 Schematic of a multiple-effect distillation (MED) plant 多效蒸馏法（海水淡化）厂示意

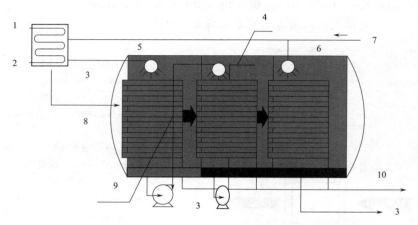

1 steam 蒸汽
2 condensate 冷凝水
3 brine 盐水
4 brine to other stages 盐水去其他级
5 heat recovery 热回收
6 heat rejection 热耗损
7 seawater 海水
8 vapor 蒸气
9 vapor flow 蒸气流
10 product water 成品水

2.4.4 Potable water production from a river 河水制备饮用水流程

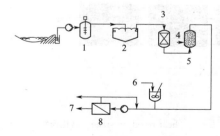

1 flocculation 絮凝
2 sedimentation 沉淀
3 granulated active carbon 粒状活性炭
4 O_3 臭氧
5 ozone treatment 臭氧处理
6 PAC (powdered activated carbon) 粉末活性炭
7 potable water to disinfection and distribution 饮用水去消毒和配送
8 UF capillary modules 超滤毛细管膜组件

2.4.5 An overview on water treatment process 水处理过程概述

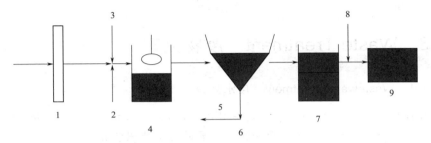

1   screens   筛网
2   coagulant   凝固剂
3   prechlorination   预加氯消毒
4   mixing/flocculation   搅拌/絮凝
5   sludge   污泥
6   settling   沉降
7   filtration   过滤
8   chlorination   加氯消毒
9   clear well   清水池

3　Waste Treatment　废物处理

3.1　Wastewater treatment　废水处理

3.1.1　Wastewater-treatment facility for crude-oil-topping petroleum refinery　炼油厂原油拔顶装置的污水处理装置

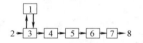

1　storm-water retention　雨水积池
2　oily water sewer　含油污水
3　separator　分离器
4　equalization basin　平衡水池
5　dissolved-air flotation　溶气浮选
6　aerated lagoon　曝气池
7　sand filters　砂粒过滤器
8　effluent　流出液

3.1.2　Typical wastewater treatment plant showing instrumentation　典型的污水处理厂所用装置

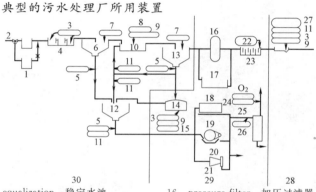

1　holding equalization　稳定水池
2　early warning　事先警告
3　pH　酸碱值
4　neutralization pond　中和池
5　% solids　固体百分率
6　primary clarifier　一次沉降池
7　bed depth　床深
8　DO——dissolved oxygen　溶解氧量
9　temperature　温度
10　aeration　曝气
11　flow　流量
12　thickener　增稠池
13　secondary clarifier　二次沉降池
14　digester　老化器
15　gas flow　气体流量
16　pressure filter　加压过滤器
17　gravity filter　重力过滤器
18　press filter　压滤机
19　vacuum filter　真空过滤机
20　centrifuge　离心机
21　dewatering　脱水
22　residual chlorine　残余氯
23　chlorine　氯
24　flue gas　烟道气
25　incinerator　焚烧炉
26　weight　质量
27　composite sample　复合样品
28　outfall　流出口
29　options　附加的系统
30　basic system　基本系统

3.1.3 Oil/water separator 油水分离器

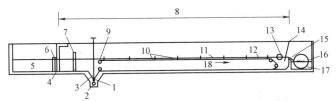

(built to standards of the American Petroleum Institute removes free oil from refinery wastewaters 美国石油学会标准,从炼油厂废水中清除浮油)

1. sludge-collecting hopper discharge with lead plug 带铅塞的集泥斗卸出口
2. sludge pump suction pipe 污泥泵吸入管
3. sludge-collecting hopper 集泥斗
4. slot for channel gate 流道口缝隙
5. forebay 前池
6. gateway pier 门框墙墩
7. diffusion device (vertical-slot baffle 直立挡板) 扩散装置
8. separator channel 分离器流道
9. flight scraper chain sprocket 刮泥板链轮
10. wood flights 木制刮板
11. flightscraper chain 刮泥板链条
12. water level 水平面
13. rotatable oil-skimming pipe 可旋转的撇油管
14. oil-retention baffle 截油挡板
15. effluent weir and wall 废水溢流堰板
16. effluent sewer 废水管
17. effluent flume 废水槽
18. flow 流向

3.1.4 Corrugated-plate interceptor 波纹板油水分离器

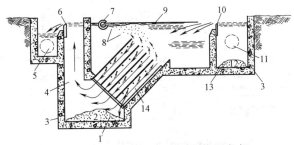

(for refinery wastewater 用于炼油厂废水)

1. sludge pit 污泥池
2. sludge 污泥
3. concrete 混凝土
4. clean-water outlet channel 净化水流出通道
5. outlet 排出口
6. adjustable outlet weir 可调出口堰
7. oil skimmer 撇油器
8. oil globules 油珠
9. oil layer 油层
10. adjustable inlet weir 可调进口堰
11. inlet 进口
12. sediment 沉淀物
13. sediment trap 沉淀物捕集器
14. plat assembly consisting of 24 or 48 corrugated parallel plates 由24~48块波纹平行板组成的部件

3.1.5 Solvent extraction of phenols from coke waste
从炼焦废液中溶剂萃取苯酚

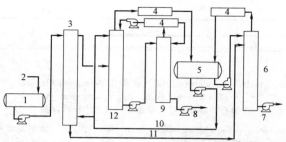

1 surge tank　缓冲罐
2 feed　进料
3 extraction column　萃取塔
4 condenser　冷凝器
5 solvent pumping tank　溶剂泵前贮罐
6 solvent stripping column　溶剂汽提塔
7 dephenolized effluent　脱酚后液体排出
8 crude phenol　粗酚
9 phenol still　苯酚蒸馏塔
10 recycle solvent　溶剂循环
11 raffinate　萃余液
12 solvent recovery column　溶剂回收塔

3.1.6 Batch-still system for recovery of waste solvents
回收废溶剂的间歇蒸馏系统

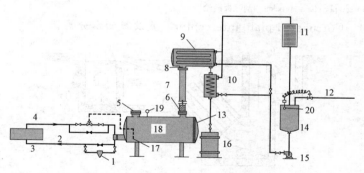

（plant capacity 50 gal/d　装置生产能力 50gal/d）

1 trap to drain　疏水阀排水
2 condensate　冷凝水
3 electric steam generator　电热蒸汽发生器
4 28 lb/h steam in　蒸汽进入，28磅/小时
5 rupture disk　防爆膜
6 packing　填料
7 4in dia. column　塔直径4英寸
8 TR——temperature recorder　温度记录仪
9 primary condenser　主冷凝器
10 secondary condenser　二次冷凝器
11 cooling tower　冷却塔
12 cold-water makeup　冷却水补充
13 hinged door for cleanout　清扫用的带铰链端盖
14 cold-water reservoir　冷水贮罐
15 cold-water recycle pump　冷水循环泵
16 product receiver　产品接收器
17 TRC——temperature recording controller　温度记录控制器
18 65-gal batch still　65加仑间歇蒸馏釜
19 PI——pressure indicator　压力表
20 LC——level controller　液位控制器

3.1.7 Stripping process 汽提过程

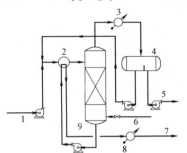

1. from waste water feed tank 来自废水料槽
2. feed-bottom exchanger 料液-塔底液换热器
3. condenser 冷凝器
4. decanter 倾析器
5. recovered organic chloride 回收的有机氯化物
6. L.P. steam 低压蒸汽
7. stripped water 汽提工艺水
8. stripped-water cooler 汽提工艺水冷却器
9. stripper 汽提塔

(useful for removing the lighter organic chlorides 可用于去除轻质有机氯化物)

3.1.8 An activated-sludge system 活性污泥系统

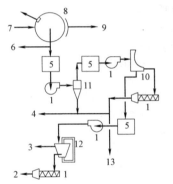

1. pump 泵
2. to blending tank 去混合槽
3. solids waste 固体废弃物
4. to grit removal 移走砂粒
5. wet well 湿井
6. aeration recycle 曝气循环
7. from aeration 来自曝气池
8. secondary clarifier 二次澄清器
9. secondary effluent 二次废水
10. separator 分离器
11. hydrocyclone 旋液分离器
12. centrifuge 离心机
13. to decantation tank 去滗析槽

(using a centrifuge 使用离心机)

3.1.9 Precoat filtration system 预涂过滤系统

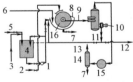

(for capturing fine solids that cannot be flocculated 可聚集不能絮凝的细颗粒)

1. precoat pump 预涂助滤剂泵
2. wash liquid 洗涤液
3. precoat material 预涂助滤剂
4. precoat mix tank 预涂助滤剂混合槽
5. precoat liquid 预涂助滤液
6. feed 进料
7. to drain 去排放
8. sludge discharge 滤渣卸料
9. vacuum line 真空管线
10. filtrate pump 滤液泵
11. filtrate 滤液
12. to process 去下道工序
13. to atmosphere 通大气
14. mist eliminator/silencer 捕沫器/消声器
15. vacuum pump 真空泵
16. vacuum filter 真空过滤机

3.1.10 Granular-media filter operates on a controlled cycle
控制循环下粒状介质过滤操作

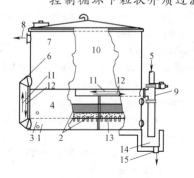

1 collection chamber 聚集室
2 filter media (coal, sand) 过滤介质 (煤，砂)
3 air 空气
4 filter compartment 过滤层段
5 inlet 供料口
6 transfer pipe 输液管
7 storage compartment 贮料层段
8 outlet 出口管
9 three-way valve 三通阀
10 water level 水平面
11 filter 过滤
12 backwash 反洗
13 nozzles 喷嘴
14 sump 污水槽
15 drain 排放口

3.1.11 Solids handling system 固体处理系统

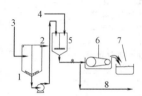

1 sludge thickening tank 污泥增稠罐
2 overflow to clarifier 溢流至澄清器
3 sludge from clarifier 来自澄清器污泥
4 conditioning agent 调节剂
5 sludge conditioning tank 污泥老化罐
6 rotary vacuum filter 旋转真空过滤机
7 sludge to ultimate disposal site 污泥至最终处置场地
8 water returned to clarifier 水返回至澄清器

3.1.12 Typical continuous-treatment system for heavy metals
典型重金属连续处理系统

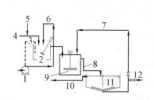

1 pump 泵
2 mixing, 2~3min 混合，2~3分钟
3 pH sensor pH值传感元件
4 raw waste 原废液
5 coagulant 絮凝剂
6 caustic soda 氢氧化钠
7 recycled sludge 循环污泥
8 polymer 多聚体
9 flocculation, 15~20min 絮凝作用，15~20分钟
10 treated effluent 处理后废水
11 clarifier, 2~4h 澄清器，2~4小时
12 sludge to waste 污泥至污水池

3.1.13 Induced-air flotation 导气浮选

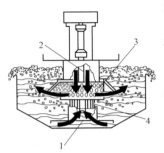

1 larger flotation units include false bottom to aid pulp flow 包括引导浆料流动的假底的较大浮选单元
2 upper portion of rotor draws air down the standpipe for thorough mixing with pulp 转子的上部使空气至立管下部流出与浆料充分混合
3 disperser breaks air into minute bubbles 分散器将空气碎裂为小气泡
4 lower portion of rotor draws pulp upward through rotor 转子较低的部位带动浆料通过转子朝上移动

(for oily wastes and suspended solids
用于含油废水和悬浮固体)

3.1.14 Dissolved-air flotation system 溶气浮选系统

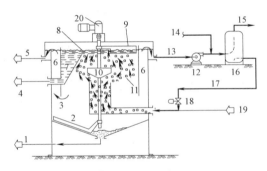

(has recycle-flow pressurization to separate
solids-laden or oily-water influents
在循环流动中增压以分离含油废水与所带的固体)

1 sludge 污泥
2 rotating scraper arm 旋转刮板臂
3 water 水
4 skimmed-oil discharge 撇出的油排出
5 clean water discharge 净化水排出
6 clean water 净化水
7 oil 油
8 skimmings hopper 撇油槽
9 rotating skimmer blade 旋转撇油器叶片
10 diffuser cone 锥形扩散器
11 rising air bubbles with attached oil 带着油上升的空气泡
12 recycle pump 循环泵
13 recycle water 循环水
14 compressed air 压缩空气
15 excess air 过剩空气
16 aeration tank 曝气罐
17 aerated recycle water 经曝气循环水
18 backpressure valve 止回阀
19 oily-water influent 含油废水进入
20 motor and gear 电机与减速机

3.1 Wastewater treatment 废水处理

3.1.15 Pressurization systems dissolve air (in feed liquids for the flotation of suspended solids or oily materials)
在进料液中加压溶气（以浮选悬浮固体或含油物质）

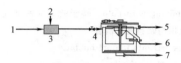

1　feed　进料
2　air　空气
3　pressurization system　加压系统
4　flotation unit　浮选装置
5　effluent　废水
6　float　浮选物
7　sludge　污泥

3.1.16 Hot air dries sludge directly in the fluidized bed
流化床中热空气直接干燥污泥

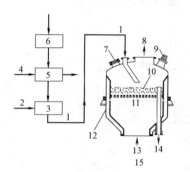

1　feed　加料
2　grinding recycled dry product　磨细的循环干物料
3　mixer/granulator　混合/造粒机
4　flocculant preparation　絮凝剂制剂
5　centrifugal decenter　离心式滗析器
6　sludge preparation　污泥制备
7　sight glass　视镜
8　exhaust air　废气
9　illumination　照明设备
10　layer of product　物料层
11　bed plate　分布板
12　insulation　保温层
13　hot air inlet　热空气进入
14　dry product discharge　干物料卸出
15　fluid bed dryer　流化床干燥器

3.1.17 Typical reactivation and carbon-handling system
典型再活化和活性炭处理系统

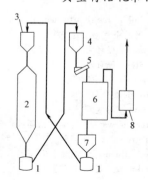

1　blowcase　吹气箱
2　adsorber　吸附器
3　reactivation carbon storage　再生活性炭储仓
4　spent carbon storage　失效活性炭储仓
5　dewatering screw　螺旋脱水器
6　reactivation furnace　再生加热炉
7　quench tank　急冷罐
8　scrubber　涤气器

3.1.18 Biophysical treatment uses activated carbon 使用活性炭的生物物理处理

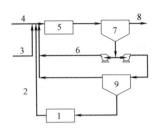

1 carbon regeneration 活性炭再生
2 regenerated carbon 再生后的活性炭
3 carbon make-up 活性炭补充
4 soluble waste 可溶性废弃物
5 aeration basin 曝气池
6 carbon-sludge recycle 活性炭-污泥循环
7 clarifier 澄清器
8 product water 生成水
9 spent carbon thickener 失效活性炭增稠器

3.1.19 Pilot system flow diagram for evaluating the feasibility of carbon adsorption 评估活性炭吸附可行性的中试装置流程

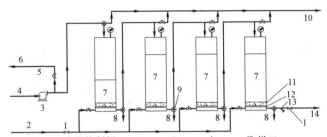

1 flow-control valve 流量控制阀
2 backwash inlet 反洗液进入
3 pump 泵
4 sample in 样品进入
5 bleedoff valve 泄压阀
6 to drain 去排放
7 carbon 活性炭
8 sample out 取样口
9 sample valve 取样阀
10 backwash (to drain) 反洗液（去排放）
11 4-in height shot gravel 4英寸高卵石层
12 filter mesh 过滤筛网
13 strainer 粗滤器
14 output 流出

3.1.20 DuPont's PACT process 杜邦 PACT 生产工艺

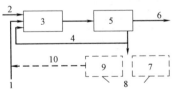

1 virgin-carbon makeup 新鲜活性炭补充
2 wastewater 污水
3 aerobic biological reactors 需氧的生物反应器
4 return sludge 污泥回流
5 secondary sludge clarifier 二次污泥澄清器
6 product water 生成水
7 other waste-sludge disposal systems 其他污水污泥处理系统
8 waste-sludge disposal choices 污水污泥处理方法选择
    (incorporates activated carbon directly into the waste stream 把活性炭直接加入到污水流中)
9 waste-sludge dewatering and carbon regeneration 污水污泥脱水和活性炭再生
10 regenerated carbon 再生后的活性炭

3.1.21 With spent solvent as a fuel, a pharmaceutical plant incorporates sludge drying as part of a total waste-treatment system 制药厂用废溶剂作燃料,把污泥干燥作为总废物处理系统的组成部分

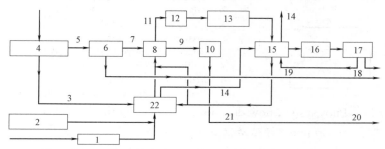

1　fresh air fan　新鲜空气鼓风机
2　solvents containing chlorine　含氯的溶剂
3　odorous exhaust air　有臭味的废气
4　sewage treatment plant　污水处理厂
5　sludge 3% TS*　污泥,含固形物 3%
6　centrifuge　离心机
7　sludge, 15% TS*　污泥,含固形物 15%
8　dryer grinder　磨碎干燥机
9　dry material, 95% TS*　干物料,含形物 95%
10　silo　筒式仓
11　vapors　蒸气
12　filter　过滤机
13　recirculation fan　循环风机
14　flue gas　烟道气
15　heat exchanger　换热器
16　induced draft fan　引风机
17　scrubber　涤气器
18　scrubber water　涤气器用水
19　concentrate　浓缩液
20　dry material　干物料
21　preheated vapors　预热后的蒸气
22　combustion chamber　燃烧炉
* TS——total solids

3.1.22　Liquid injection incineration　液体喷射焚烧过程

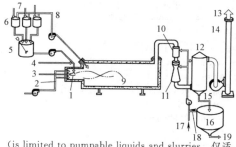

(is limited to pumpable liquids and slurries 仅适用可泵送液体和泥浆)

1　nozzle　喷嘴
2　combustion air　助燃空气
3　support gas　辅助燃气
4　support fuel if required　辅助燃料(必要时)
5　waste conditioner　废弃物调节器
6　storage　贮运器
7　liquid waste　废液
8　fumes　烟气
9　afterburner chamber　后燃烧室
10　precooler　预冷器
11　venturi scrubber　文丘里涤气器
12　cooling scrubber　冷却涤气器
13　scrubbed gases　洗涤后的气体
14　dispersion stack　弥散烟囱
15　water　水
16　water treatment　水处理
17　makeup water　补充水
18　treated water　处理后的水
19　residue　残液

3.1.23 Wet-air oxidation 湿空气氧化作用

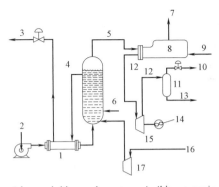

1　heat exchanger　换热器
2　raw wastewater　原污水
3　treated wastewater　处理后的废水
4　reactor　反应器
5　hot gases　热（可燃）气体
6　steam for preheating　预热用蒸汽
7　steam　蒸汽
8　steam generator　蒸汽发生器
9　boiler feed water　锅炉给水
10　CO_2 and N_2 to atmosphere　二氧化碳和氮气通大气
11　separator　分离器
12　gases　气体
13　condensate　冷凝液
14　generated power　发出的电力
15　turboexpander　透平膨胀机
16　air　空气
17　compressor　压气机

(for soluble and water-miscible organic wastes　用于可溶的及水溶混的有机废液)

3.1.24 Typical deep well used for the subsurface injection of liquid 典型的地下注入液体废物的深井

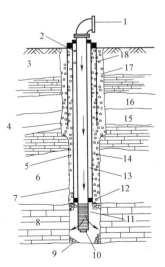

1　input　进口
2　area between inner and outer casing is sometimes packed with corrosion and biological-growth inhibitors　内、外壳之间有时填充抗腐蚀、抗生物生长的材料
3　surface water-bearing strata　地表水层
4　outer bore hole　外钻孔
5　concrete　水泥
6　shale　页岩
7　inner bore hole　内钻孔
8　disposal formation　处理系统
9　gravel pack　废料堆
10　plug　堵头
11　screened wellhead　多孔套管
12　mechanical packer　机械密垫
13　inner casing (injection tube)　内壳
14　protection casing　保护层
15　shale　页岩层
16　deep water bearing strata　深水层
17　concrete　混凝土
18　surface casing　外壳

3.1.25 Water processing flow diagram considered in the study of wastewater treatment plant in Avellino, Italy
（意大利 Avellino）废水处理厂研究中的水处理工艺

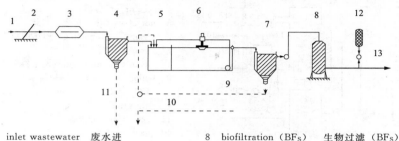

1　inlet wastewater　废水进
2　bar screens　铁栅筛
3　grit chamber　砂箱
4　primary settling tank　第一沉降池
5　denitrification　脱氮
6　oxidation/nitrification tank　氧化/硝化池
7　secondary settling tank　第二沉降池
8　biofiltration（BFs）　生物过滤（BFs）
9　return activated sludge　回流活性污泥
10　secondary activated sludge　二次活性污泥
11　primary sludge　一次污泥
12　disinfection　消毒
13　effluent　出水

3.1.26 Solid processing flow diagram considered in the study of wastewater treatment plant in Avellino, Italy
意大利 Avellino 废水处理厂研究中的固体处理工艺流程图

(a) Combined treatment (anaerobic digestion) and traditional disposal of the primary and secondary sludge　一次污泥和二次污泥传统的和厌氧消化的联合处理

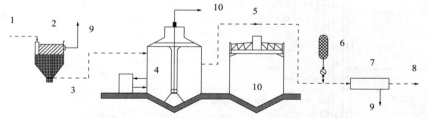

1　primary and secondary sludge　一次和二次污泥
2　gravity thickener tank　重力增稠罐
3　sludge　污泥
4　anaerobic sludge digestion tank　厌氧污泥消化池
5　stabilized sludge　稳定后的污泥
6　chemical conditioning　化学调理
7　dewatering with centrifuge and belt-filter press　用离心机和带式压滤机脱水
8　dewatered biosolids flow to disposal　脱水后生物固体流去处置
9　underflow to plant influent　浓浆至厂流入液中
10　biogas to gasometer　沼气至贮气柜

(b) Anaerobic digestion of the primary sludge and direct reuse of the secondary sludge in agriculture 一次污泥厌氧消化和二次污泥在农业中的直接再利用

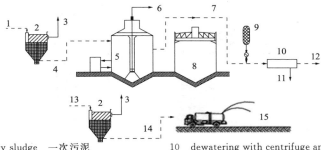

1　primary sludge　一次污泥
2　gravity thickener tank　重力增稠罐
3　underflow to plant influent　浓浆至厂进液中
4　primary thickened sludge　一次浓缩污泥
5　anaerobic sludge digestion tank　厌氧污泥消化池
6　biogas to gasometer　沼气至贮气柜
7　stabilized primary sludge　稳定后的一次污泥
8　gasometer　贮气柜
9　chemical conditioning　化学调理
10　dewatering with centrifuge and belt-filter press　用离心机和带式压滤机脱水
11　underflow to plant influent　浓浆至厂进液中
12　dewatered biosolids flow to disposal　脱水后生物固体去处置
13　secondary sludge　二次污泥
14　secondary thickened sludge　二次浓缩污泥
15　sludge for agricultural utilisation　农业可利用的污泥

3.1.27　Flow chart of CAPS (chemically assisted primary sedimentation) upgrading in the studied WWTP
废水处理厂研究中改进的化学辅助一次沉降流程图

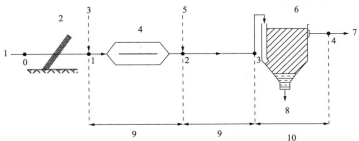

1　inlet wastewater　废水进
2　bar screens　铁栅筛
3　coagulant　凝结剂
4　grit chamber　砂箱
5　flocculant　絮凝剂
6　primary settling tank　一次沉降池
7　secondary treatment　二次处理
8　primary sludge　一次污泥
9　5 minutes　5分钟
10　120 minutes　120分钟

3.1.28 MBR (Membrane Bioreactor) process system with membrane module situated outside the bioreactor
生物反应器外设膜组件的膜生物反应器系统

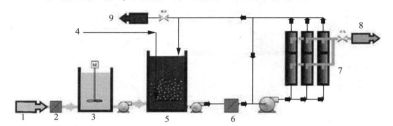

1	raw wastewater 原废水	6	security filter 安全过滤器
2	screen 筛网	7	membrane modules 膜组件
3	conditioning tank 调节池	8	treated effluent 处理后出水
4	air 空气	9	excess sludge 过剩的污泥
5	bioreactor 生物反应器		

3.1.29 Working principle of crossflow membrane filtration
错流膜过滤工作原理

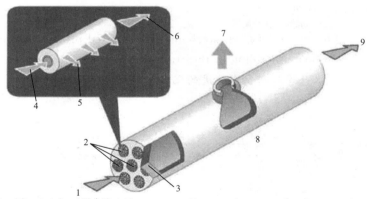

1	mixed liquor inlet 混合液入口	7	treated water outlet (permeate) 处理后的水出口（渗透）
2	bundles 过滤膜捆	8	membrane module 膜组件
3	support 支撑体	9	concentrate (retentate) 浓缩物（截留物）
4	mixed liquor 混合液		
5	treated effluent 处理后的废液		
6	retentate 截留物		

3.1.30 Comparison between MBR and equivalent traditional WWTP 膜生物反应器和等效的传统废水处理工艺比较

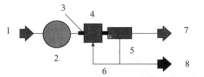

A MBR (Membrane Bioreactor)   膜生物反应器
1 raw wastewater   未经处理的废水（原废水）
2 conditioning tank   调节池
3 screen   筛网
4 bioreactor   生物反应器
5 membrane   滤膜
6 to bioreactor   去生物反应器
7 treated effluent   处理后的中水
8 excess sludge (reduced)   剩余的污泥（减少了）
9 suitable for recycle/reuse within the plant   适合在厂内循环/再利用

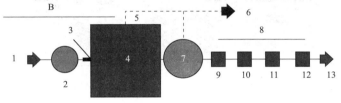

B Conventional Activated Sludge   常规的活性污泥
1 raw wastewater   原废水
2 conditioning tank   调节池
3 screen   筛网
4 bioreactor   生物反应器
5 RAS   循环活性污泥
6 WAS   废活性污泥
7 final clarifier   终端澄清器
8 additional step required   需要附加的工序
9 sand filter   砂滤器
10 GAC column   颗粒活性炭吸附塔
11 microfiltration   微孔过滤
12 disinfection   消毒
13 treated effluent   处理后的中水

3.1.31 Coffee factory WWT flow schematic 咖啡生产厂废水处理流程图

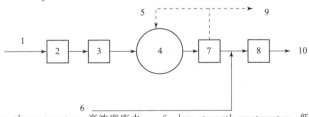

1 high-strength wastewater   高浓度废水
2 pH adjust   调节pH值
3 primary settler   主要沉降池
4 bioreactor   生物反应器
5 return biosolids   回流生物固体
6 low-strength wastewater   低浓度废水
7 membranes   滤膜
8 final control basin   终端控制池
9 excess biosolids   剩余的生物固体
10 effluent discharge   中水排放

3.1.32　The VERTREAT™ process (Courtest NORRAM Engineering and Construction Ltd.) VERTREAT (Courtesy NORRAM 工程建筑有限公司) 废水处理工艺

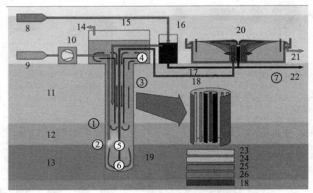

1	aeration 曝气	14	vent air 放空
2	influent injection 进水注入	15	head tank 高位槽
3	biodegradation 生物降解	16	hold tank 储液罐
4	de-gasing 脱气	17	return sludge 回流的污泥
5	polishing 精制	18	extraction line 提抽管道
6	withdrawal 提取	19	aerated shaft (300 呎) 曝气井
7	flotation 浮选	20	flotation clarifier 浮选澄清器
8	influent 进水	21	clarified effluent 澄清水排出
9	air 空气	22	waste sludge 废污泥
10	compressor 压缩机	23	reactor casing 反应器外壳
11	oxidation zone 氧化区	24	air lines 空气管道
12	mixing zone 混合区	25	influent lines 进水管道
13	saturation zone 饱和区	26	downcomer 降液管

3.1.33　Bardenpho process (Source US EPA)
Bardenpho 废水中除氮磷工艺 (取自美国环境保护局)

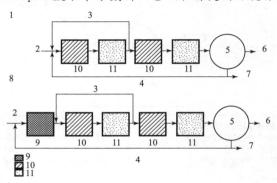

1	4-stage (Nitrogen removal) process 4级（除氮）工艺	6	effluent 出水
2	influent (Q) 进水 (Q)	7	WAS 废活性污泥
3	nitrified recycle (400%Q) 硝化循环 (400%Q)	8	5-stage (phosphorus and nitrogen removal) process 5级（脱磷和脱氮）过程
4	RAS (100%Q) 回流活性污泥 (100%Q)	9	anaerobic zone 厌氧区
5	final clarifier 终端澄清器	10	anoxic zone 缺氧区
		11	aerobic zone 需氧区

3.1.34 Process schematic for Coxsackie sewage treatment plant, New York 纽约 Coxsackie 下水道污水处理厂工艺简图

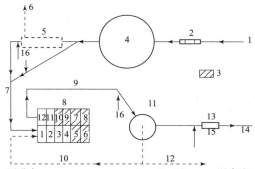

1 plant influent 厂进水
2 screen channel 滤网通道
3 anoxic cells 缺氧池
4 surge tank 缓冲罐
5 primary settler 一次沉降池
6 primary sludge 一次污泥
7 primary effluent 一次污水
8 aeration tank 曝气池
9 mixed liquor 混合液
10 return sludge 回流污泥
11 final settler 终端沉降池
12 waste sludge 废污泥
13 chlorine contact chamber 氯接触箱
14 plant effluent 厂出水
15 Cl_2 氯气
16 $FeCl_3$ 三氯化铁

3.1.35 Dual anoxic zone system 双缺氧区系统

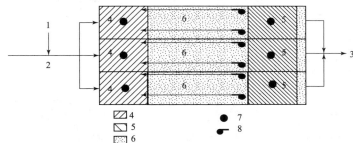

1 RAS 回流活性污泥
2 primary effluent with recycles 多次循环的一次污水
3 to secondary clarifiers 去二次澄清器
4 first anoxic zone 第一缺氧区
5 secondary anoxic zone 第二缺氧区
6 aerobic zone 需氧区
7 mixer 混合器
8 internal recycle pump 内部循环泵

3.1.36 Flow diagram of upflow fluidized-bed system
上流式流化床处理系统流程图

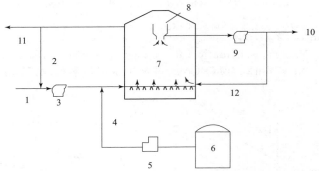

1 nitrified effluent 硝化污水
2 recycle flow 循环液流
3 fluidization pump 流态化泵
4 methanol feed 甲醇进料
5 methanol feed pump 甲醇进料泵
6 methanol storage tank 甲醇储罐
7 fluid bed reactor 流化床反应器
8 growth control device 生长控制装置
9 growth control pump 生长控制泵
10 waste biomass to solids handling 废弃生物质送去固体处理
11 denitrified effluent 脱氮的污水
12 sheared media & return biomass 切变的介质和回流生物质

3.1.37 Flow diagram of Reno-Sparks wastewater treatment plant
Reno-Sparks 废水处理厂流程图

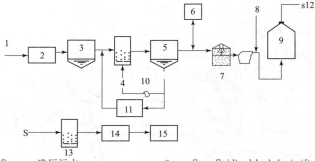

1 plant influent 进厂污水
2 preliminary treatment 初步处理
3 primary clarification 一次澄清
4 aeration (BOD removed) 曝气（生化需氧量去除）
5 secondary clarification 二次澄清
6 equalization 均压
7 nitrification 硝化
8 methanol 甲醇
9 upflow fluidized-bed denitrification 上流式流化床去硝化
10 RAS 回流活性污泥
11 phostrip 侧流除磷
12 (continued below) 转接下一行
13 post-aeration 后曝气
14 filtration 过滤
15 disinfection 消毒

3.1.38 Flow diagram of packed bed reactor system
填料床反应器系统流程图

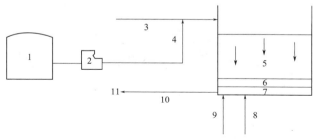

1 methanol storage tank 甲醇储罐
2 methanol feed pump 甲醇进料泵
3 nitrified influent 硝化污水进
4 methanol 甲醇
5 media 介质
6 support gravel 砾石支撑层
7 underdrain 聚水系统
8 backwash air 反洗空气
9 backwash water 反洗水
10 denitrified effluent 去硝化污水
11 to clear well 去清水池

3.1.39 PBR (Packed bed reactor) system with coarse media denitrification columns
带粗介质反硝化塔的填料床反应器

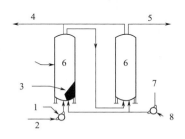

1 influent pump 进液泵
2 methanol 甲醇
3 dumped media (typical) 堆积的介质（典型的）
4 backwash water to head of plant 反洗水去厂前区
5 effluent to polishing or filtration 排水去精制或过滤
6 denitrification column 反硝化塔
7 clarified backwash water 澄清的反洗水
8 backwash pump 反洗泵

123

3.1.40 Flow diagram of Hookers Point Advanced Wastewater Treatment Plant Hookers Point 先进废水处理厂流程图

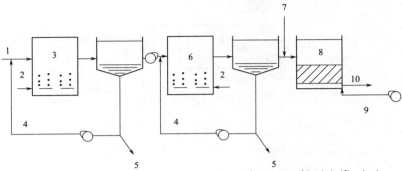

1 Q 进水
2 HPO (high purity oxygen) 高纯氧气
3 first-stage aerobic (BOD_5 removal)
   一段有氧处理 去除五天生化需氧量
4 RAS (return activated sludge) 回流活性污泥
5 WAS 废活性污泥
6 second-stage aerobic (nitrification)
   二段有氧处理（硝化）
7 methanol 甲醇
8 downflow packed-bed-system (denitrification)
   下流式填充床系统（反硝化）
9 backwash supply 反冲洗供给
10 Q 出水

3.1.41 Flow Diagram for a Biological Aerated Filter (BAF) System 生物曝气过滤系统流程图

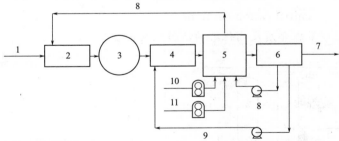

1 plant influent 厂进水
2 headworks 进水口工程
3 primary clarifier 主澄清器
4 pump station 泵站
5 BAF/biocarbone system 生物曝气过滤/生物炭系统
6 clear well 清水池
7 effluent to disinfection 流出水去消毒
8 backwash water 反洗水
9 dilution water (Optional: generally would only be needed in industrial wastewater treatment applications with strong influent concentrations) 稀释水（任选：一般只在工业废水具有很强的进水浓度处理的应用中才需要）
10 process air 工艺空气
11 air source 气源

3.1.42 Fluidized bed biological (FBB)-GAC (granular activated carbon) system 生物流动床-颗粒活性炭系统

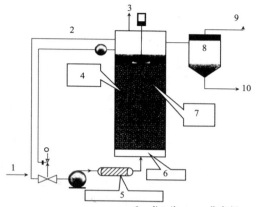

1 influent 进水
2 recycle 循环
3 vent 放空
4 fluidized absorbent media 流态化的吸收剂介质
5 froth flow contactor 泡沫流动接触器
6 distributor 分布板
7 biomass harvesting system 生物质收集系统
8 PFM 加压过滤膜
9 biomass 生物质
10 clear effluent 干净的排水

3.1.43 Powdered activated carbon activated sludge process (PACT) 粉末活性炭处理活性污泥工艺

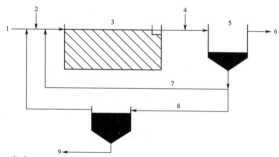

1 wastewater 废水
2 carbon addition 加炭
3 aeration tank 曝气池
4 chemical addition 化学药品添加
5 clarifier 澄清器
6 effluent 排水
7 carbon/biomass recycle 炭/生物质循环
8 waste carbon/biomass 废炭/生物质
9 regeneration or disposal 再生或处置

3.1.44 ABF (activated bio filter) process flow diagram
ABF (活化生物过滤设备) 工艺流程图

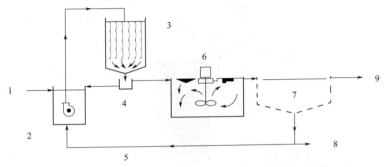

1　primary effluent feed　一次污水进
2　bio-cell lift station　生物电池提升工段
3　fixed-film bio-cell　固定膜生物电池
4　flow control & splitting　流量控制和分流
5　return sludge　回流污泥
6　aeration　曝气
7　clarifier　澄清器
8　waste sludge　废污泥
9　secondary effluent　二次出水

3.1.45 Flow diagram for covered pure oxygen activated sludge process
密闭的纯氧活化污泥工艺流程图

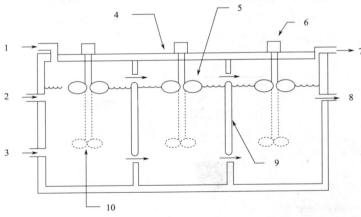

1　oxygen feed gas　氧气供气
2　raw wastewater　原废水
3　return sludge　回流污泥
4　aeration tank cover　曝气池盖
5　surface aerator　表面曝气装置
6　mixer drive　搅拌器驱动装置
7　exhaust gas　废气
8　mixed liquor to clarifier　混合液去澄清器
9　stage baffle　级间隔板
10　submerged propeller (optional)　液下搅拌叶片（任选）

3.1.46 Flow diagram for uncovered pure oxygen active sludge process
敞口的纯氧活化污泥工艺流程图

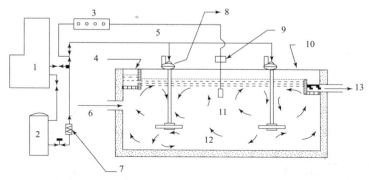

1 oxygen generator 氧气发生器
2 LOX storage (stand-by) 液氧储罐（备用）
3 control panel 控制板
4 influent raw wastewater or primary effluent 原废水流入或一次水流出
5 oxygen supply 供氧
6 return sludge 回流污泥
7 vaporizer 气化器
8 motor/gear reducer assembly 电机/齿轮减速机组件
9 D.O. analyzer 溶解氧分析仪
10 typical open basin 典型的敞口池
11 D.O. probe 溶解氧传感器
12 O_2 diffusion 氧气扩散
13 mixed liquor 混合液

3.1.47 Schematic diagram of trickling filter process
滴流式过滤工艺示意图

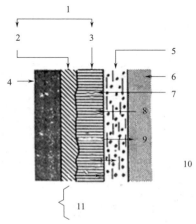

1 biological film 生物膜
2 anaerobic 厌氧的
3 aerobic 需氧的
4 stone media 石子介质
5 waste liquid 废液
6 air 空气
7 organic pollutant 有机污染物
8 O_2 氧气
9 CO_2 二氧化碳
10 metabolic products and excess cell growth 代谢产物和多余的细胞产物
11 H_2S, NH_3, organic acids 硫化氢，氨，有机酸

3.2 Biochemical wastewater treatment systems 生物化学废水处理系统

3.2.1 Typical process flow diagram for a wastewater treatment system illustrating the role of the biochemical operations
采用生化操作的废水处理系统流程

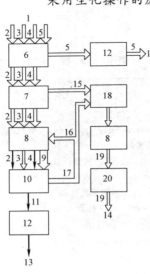

1　influent　进水
2　SOM——soluble organic matter　可溶性有机物质
3　IOM——insoluble organic matter　不溶性有机物质
4　SIM——soluble inorganic matter　可溶性无机物质
5　IIM——insoluble inorganic matter　不溶性无机物质
6　preliminary physical unit operations　初步物理单元操作
7　sedimentation　沉降
8　biochemical operation　生化操作
9　biomass　生物质
10　physical unit operation——typically sedimentation　物理单元操作——一般沉降
11　overflow　溢流
12　additional treatment　附加处理
13　effluent　出水
14　ultimate disposal　最终处置
15　underflow primary sludge　一次底流污泥
16　recycle (optional)　循环（任选的）
17　underflow secondary sludge　二次底流污泥
18　blending & thickening　掺合与增稠
19　stable residue and biomass　稳定的残渣与生物质
20　thickening & dewatering　增稠与脱水

3.2.2 Typical activated sludge process 典型活性污泥工艺

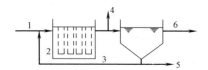

3.2.3 Selector activated sludge (SAS) process 选择器活性污泥（SAS）工艺

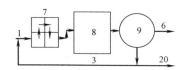

3.2.4 Single-sludge biological nutrient removal process 单一污泥生物营养素脱除工艺

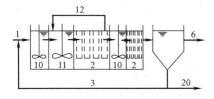

3.2.5 Anoxic/aerobic digestion 缺氧/好氧消化

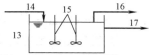

(a) intermittent feed 间歇投料

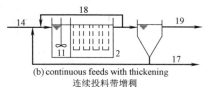

(b) continuous feeds with thickening 连续投料带增稠

1 influent 进水
2 AER——aerobic 好氧，需氧
3 RAS——return activated sludge 回流活性污泥
4 WAS——waste activated sludge (Garrett method) 废活性污泥（Garrett 法）
5 WAS (conventional) 废活性污泥（常规的）
6 effluent 出水
7 selector 选择器
8 aeration basin 曝气池
9 clarifier 澄清池
10 ANA——anaerobic 厌氧
11 ANX——anoxic 缺氧
12 MLR——mixed liquor recirculation 混合液循环
13 aeration (cycled) 曝气（循环的）
14 feed sludge 污泥加入
15 mixer (optional) 搅拌器（可有可无）
16 supernatant (optional) 上清液（可有可无）
17 digested sludge 陈化污泥
18 recirculation 循环
19 supernatant 上清液
20 WAS——waste activated sludge 废活性污泥

3.2.6 Egg shaped anaerobic digester 蛋形厌氧消化器

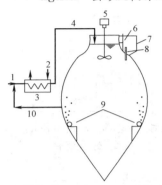

3.2.7 Low rate anaerobic process using a rectangular concrete structure 采用长方形混凝土结构的低速厌氧工艺

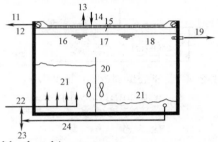

3.2.8 Upflow anaerobic sludge blanket bioreactor 厌氧污泥上流过滤层生物反应池

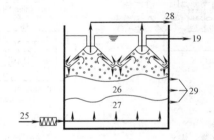

1	feed sludge 污泥加入		15	scum layer 浮渣层
2	hot water 热水		16	primary reaction zone 一次反应区
3	heat exchanger 换热器		17	secondary reaction zone 二次反应区
4	digester feed 消化器进液		18	clarification zone 澄清区
5	mixer 搅拌器		19	effluent 出水
6	scum port 浮渣出口		20	baffle 挡板
7	hopper 贮斗		21	sludge 污泥
8	supernatant withdrawal 上清液出料		22	feed 进料
9	gas spargers 气体鼓泡器		23	waste 废物
10	digesting sludge 陈化污泥		24	sludge recycle 循环污泥
11	biogas 生物气		25	influent 进水
12	neg. press. 负压		26	flocculent sludge 絮凝污泥
13	floating insulated membrane cover 绝缘膜浮动盖		27	granular sludge 颗粒污泥
			28	gas 气体
14	variable level 可变液面		29	sludge wastage 废污泥

3.2.9 Fluidized bed and expanded bed process 流化床与膨胀床工艺

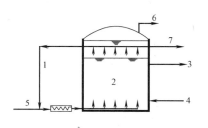

1. recycle 循环
2. fluidized/expanded media bed 介质流态化膨胀床
3. media and biomass to separation 介质和生物质去分离
4. cleaned/recycled media 净化循环介质
5. influent 进水
6. gas 气体
7. effluent 出水

3.2.10 Alternative configurations for the solids fermentation process 固体发酵工艺可供选择的各种方案

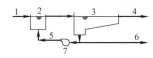 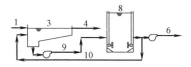

(a) activated primary tanks 一次活化槽　　(b) complete mix fermenter 全混发酵罐

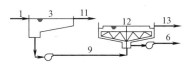 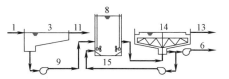

(c) single-stage fermenter/thickener 一级发酵罐/增稠器　　(d) 2-stage complete mix/thickener fermenter 二级全混/增稠器发酵罐

1. raw influent 原水
2. mixing/elutriation tank 混合/淘析槽
3. primary clarifier 一次澄清池
4. VFA-rich primary effluent to bioreactor (VFA——volatile fatty acid) 富挥发性脂肪酸一次排放物去生物反应器
5. sludge recycle 循环污泥
6. waste sludge to sludge handling 废污泥去污泥处理
7. pump 泵
8. fermenter 发酵罐
9. primary sludge 一次污泥
10. fermenter mixed liquor return 发酵罐混合液回流
11. primary effluent to bioreactor 一次排放液去生物反应器
12. fermenter/thickener 发酵罐/增稠器
13. VFA-rich fermenter supernatant to bioreactor 富挥发性脂肪酸发酵罐上清液去生物反应器
14. thickener 增稠器
15. primary sludge recycle 一次污泥循环

3.2.11 Schematic diagram of an RBC 回转生物接触器原理图

3.2.12 Schematic diagram of an FBBR 流化床生物反应器示意图

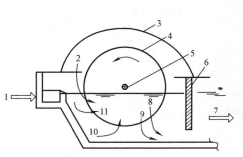
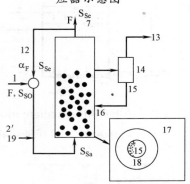

RBC rotating biological contactor 回转生物接触器

FBBR (fluidized bed bioreactor) 流化床生物反应器

1 influent 进水

2 oxygen 氧气

3 cover 顶盖

4 RBC 回转生物接触器

5 shaft 轴

6 inter stage baffle 级间挡板

7 effluent 出水

8 degradation products 降解产物

9 sludge 污泥

10 nutrient 营养素

11 food 食物

12 recirculation 循环

13 excess biomass 过量生物质

14 separator 分离器

15 carrier particles 颗粒载体

16 reactor 反应器

17 bioparticle 生物粒子

18 biofilm 生物膜；菌膜

19 chemicals (optional)
化学药品（可有可无）

3.2.13 Placement of fixed media in combined suspended and attached growth systems
固定介质悬浮与附着生长相结合的模式

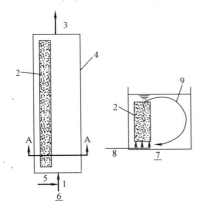

1 influent 进水
2 media 介质
3 effluent to clarifier 排放液至澄清池
4 suspended growth reactor
   悬浮生长反应器
5 RAS——return activated sludge 回流活性污泥
6 plan view 平面图

3.2.14 Configuration of combined suspended and attached growth system using freefloating media
采用自由浮动介质悬浮与附着生长相结合的模式

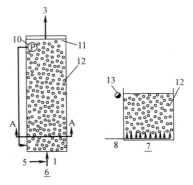

7 section A-A A-A 剖面
8 process air 工艺空气
9 mixed liquor flow pattern 混合液流型
10 airlift pump for pad recirculation 团块再循环气提泵
11 effluent screen 排液筛网
12 free-floating media 自由浮动介质
13 pad recirculation line 团块再循环轨迹

3.2.15 Schematic diagram of a circular sedimentation tank
圆形沉降槽示意图

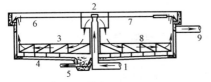

1 influent 进水
2 circular inlet baffle 圆形进口挡板
3 sludge collectors 污泥收集器
4 sludge hopper 污泥接收器
5 sludge 污泥
6 scum baffle 浮渣挡板

3.2.16 Schematic diagram of a rectangular sedimentation tank
矩形沉降槽示意图

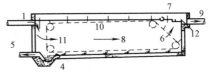

7 scum collector 浮渣收集装置
8 flow pattern 流型
9 effluent 出水
10 chain and cross flights 链及横刮板
11 inlet baffle 进口挡板
12 weir 堰

3.3 Solid waste treatment 固体废物处理

3.3.1 Thermal pyrolysis (fits into energy-recovery systems) 适合能量回收系统的热分解过程

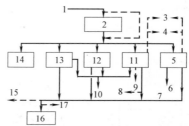

1 solid waste (refuse, biomass, sludges) 固体废物（垃圾、生物质、淤渣）
2 preprocessing 预处理
3 chemicals 化学制品
4 gaseous fuel 气体燃料
5 biochemical conversion 生物化学转化
6 soil conditioner 土壤调理剂
7 residues 残渣
8 char 炭
9 liquid fuel 液体燃料
10 steam 蒸汽
11 PTGL processes 热分解、热气体生成和熔解过程
12 incineration with steam generation 焚化伴随产生蒸汽
13 utility or industrial boiler (cofiring with coal) 公共事业或工业锅炉（与煤混烧）
14 cement kiln 水泥窑
15 to non-energy resource recovery 去非能量资源回收
16 land disposal 埋入地下
17 aggregate 堆聚

3.3.2 Fluidized-bed incinerator-system 流化床焚烧系统

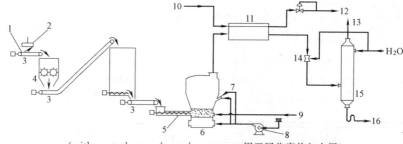

(with waste-heat and metal recovery 用于回收废热与金属)

1 waste feed 废弃物加料
2 metal collector 金属捕集器
3 conveyor 输送机
4 shredder 切碎机
5 screw feeder 螺旋加料器
6 incinerator 焚烧炉
7 preheat burner 预热燃烧器
8 fluidizing air blower 流态空气风机
9 No.2 fuel oil 2号燃料油
10 boiler feed water 锅炉给水
11 waste heat boiler 废热锅炉
12 steam 蒸汽
13 to atmosphere 通大气
14 Venturi 文丘里管
15 scrubber 气体洗涤器
16 effluent 废水

3.3.3 Multiple-hearth furnaces 多膛燃烧炉

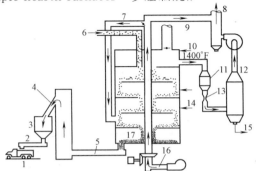

(are best suited for solids and sludges 最适用于固体和沉淀物)

1 hauling　运输车
2 ash conditioner　灰分调理器
3 ash bin　灰仓
4 bucket elevator　斗式提升机
5 screw conveyor　螺旋输送机
6 waste feed　废弃物进料
7 return air　回流空气
8 exhaust stack　排气烟囱
9 emergency bypass stack　应急旁路排风管
10 fuel burner　燃料燃烧器
11 precooler　预冷器
12 scrubber　涤气器
13 Venturi　文丘里管
14 fuel burners　燃料燃烧器
15 effluent　废水
16 cooling air for rabble arms and drive shaft
　（空心）搅拌耙臂和传动轴的冷却空气
17 furnace　燃烧炉膛

3.3.4 Fluosolids incinerator flowsheet 流化焙烧焚烧炉流程

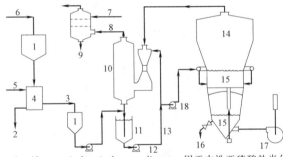

(for neutral sulfite semi-chemical waste liquor 用于中性亚硫酸盐半化学废液)

1 storage tank　贮罐
2 condensate　冷凝水
3 20% total solids　20%固含量
4 evaporator　蒸发器
5 steam　蒸汽
6 weak liquor, 10% total solids
　弱碱液，10%固含量
7 cooling water　冷却水
8 gas cooler　气体冷却器
9 hot water　热水
10 Venturi scrubber evaporator　文丘里涤气蒸发器
11 recycle tank　循环槽
12 recycle pump　循环泵
13 40% total solids　40%固含量
14 fluosolids reactor　流化反应器
15 fluid bed　流化床
16 sodium sulfate pellet product　硫酸钠颗粒产品
17 fluidizing air blower　流化空气鼓风机
18 feed pump　进料泵

3.3.5 Flowsheet for a multipurpose incinerator destroying paint residues 销毁油漆废料的多功能焚烧炉流程

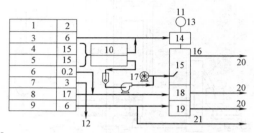

1 source 来源
2 quantity (ton/wk) 处理量（吨/星期）
3 cleaning sludge 净化的残渣
4 still bottoms 釜底残留物
5 wet flammable sludge 湿易燃残渣
6 solid organic rubbish 固体有机垃圾
7 flammable liquid waste 易燃废液
8 dry rubbish 干垃圾
9 drums 金属桶
10 separate to sludge and pumpable liquid 对残渣和可泵送液体分离
11 incinerator 焚烧炉
12 to special burner on boiler plant 去锅炉厂专用燃烧炉
13 stack 烟囱
14 dust catcher 集尘器
15 sludge hearth 泥状沉淀炉膛
16 ash 灰分
17 fan 鼓风机
18 dry rubbish grate 干垃圾格栅
19 drums and tins 金属桶和白铁罐
20 to contractor 去打包机
21 to contractor for dumping 去填埋打包机

3.3.6 Fluid-bed incinerator 流化床焚烧炉

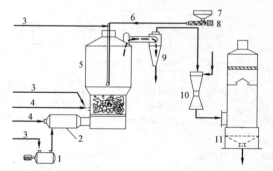

1 blower 鼓风机
2 preheater 预热器
3 atomizing air 雾化空气
4 oil 油
5 fluid bed reactor 流化床反应器
6 sludge feed 污泥进料
7 hydraulic cyclone 旋液分离器
8 feed pump 进料泵
9 cyclone separator 旋风分离器
10 Venturi scrubber 文丘里涤气器
11 separator 分离器

3.3.7 Flow diagram shows materials-handling sequence involved in the disposal of hazardous wastes 处置含危险废弃物物料的处理工序流程图

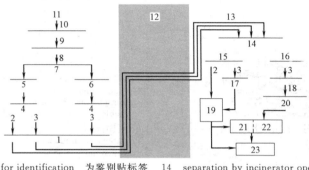

1	labeling for identification 为鉴别贴标签	14	separation by incinerator operator 由焚烧炉操作者分离
2	bulk 统装	15	pumpable 可泵送的
3	drummed 装桶	16	nonpumpable 不可泵送的
4	packaging 装箱	17	pumped from drum 从桶泵送出
5	pumpable waste 可泵送的废料	18	inspection 检查
6	nonpumpable waste 不可泵送的废料	19	burner feed tank 燃烧炉进料槽
7	segregation 分开	20	pack-and-drum feed system 箱包和桶的加料装置
8	categorization 分类	21	kiln burner 窑炉燃烧器
9	chemical identification 化学识别	22	kiln 窑炉
10	waste source 废物源	23	secondary burner 二次燃烧炉
11	manufacturing plant 制造厂		
12	transportation 输送		
13	incinerator 焚烧炉		

3.3.8 Incinerator facility 焚烧炉装置

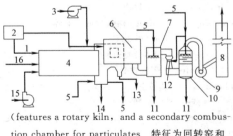

(features a rotary kiln, and a secondary combustion chamber for particulates 特征为回转窑和二次燃烧室处理粒状物)

1 liquid, pumpable 液体，可泵送的
2 burner tanks 喷燃器燃料罐
3 secondary air 二次空气
4 rotary kiln 回转窑
5 water 水
6 secondary combustion chamber 二次燃烧室
7 quench chamber 急冷室
8 stack 烟囱
9 induced draft fan 引风机
10 sieve tower (demister) 筛板塔（除雾器）
11 water ash 水灰
12 Venturi throat 文丘里管喉部
13 flash 飞灰
14 ash drums 灰筒
15 plenum air 加压空气
16 drummed, nonpumpable 装桶，不可泵送的

3.3.9 Multiple-hearth incinerator 多室焚烧炉

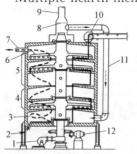

1 cooling air fan 冷却空气鼓风机
2 ash discharge 灰分排出
3 cooling zone 冷却段
4 combustion zone 焚烧段
5 drying zone 干燥段
6 rabble arm at each hearth 各层炉膛搅拌耙臂
7 flue gases out 烟道气排出
8 floating damper 浮动风门
9 cooling air discharge 冷却空气排出
10 sludge inlet 污泥进入
11 combustion-air return 已燃空气回流
12 rabble-arm drive 搅拌耙臂驱动装置

3.3.10 Section through submerged combustor 浸没式燃烧炉截面

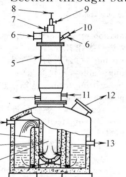

1 combustion products 燃烧产物
2 downcomer 下降管
3 weir 堰
4 diluted liquid 稀释液
5 burner incinerator 燃烧焚化炉
6 air 空气
7 auxiliary fuel gas 辅助燃料气
8 waste liquid 废液
9 atomizing air 雾化空气
10 fuel gas (for pilot burner) 燃料气(用导燃烧嘴)
11 cooling water 冷却水
12 waste gas 废气
13 concentrated (waste) liquid 浓废液

3.3.11 Molten-salt incinerator design for pesticide disposal 用于农药废弃物的熔盐焚烧炉结构

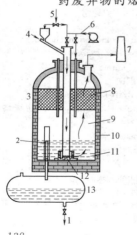

1 to salt recovery 去回收盐
2 molten-salt level control 熔盐位面控制器
3 molten-salt chamber 熔盐室
4 solid waste feed 固体废物进料
5 liquid waste feed 废液进料
6 combustion air 助燃空气
7 exhaust stack 排气烟囱
8 ignition source and molten-salt demister 点火装置及熔盐除沫器
9 secondary reaction zone 二次反应段
10 preheat burner 预热燃烧器
11 molten salt 熔盐
12 waste entrance 废物进口
13 salt quenching chamber 盐急冷室

3.3.12 Retort multiple-chamber incinerators 甑式多室焚烧炉

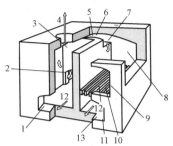

(have an upper capacity limit of about 1000 lb/h 生产能力上限约为1000磅/小时)

3.3.13 Inline multiple-chamber incinerators 直列式多室焚烧炉

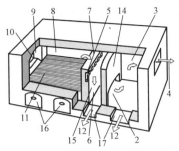

(have a low capacity limit of 750 lb/h 生产能力下限为750磅/小时)

1　cleanout door　清扫口
2　curtain wall port　幕墙通口
3　secondary combustion chamber　二次焚烧室
4　gases out　气体排出
5　secondary air port　二次空气通口
6　mixing chamber　混合室
7　flame port　火焰通口
8　ignition chamber　起燃室
9　charging door with overfire air port　兼过烧空气通道的装料门
10　waste entrance　废弃物进口
11　grates　炉条
12　ash　灰分
13　cleanout door with undergrate air port　炉条下面空气通道兼清扫口
14　curtain wall　幕墙
15　location of secondary burner　二次燃烧器的位置
16　cleanout doors with undergrate air ports　炉栅下空气通道兼（两个）清扫门
17　cleanout doors　（两个）清扫门

3.3.14 A secure burial pit　可靠的深埋坑

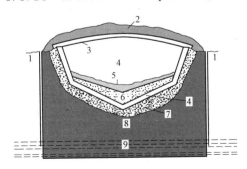

1　monitoring well　监控井
2　earth cover　土地表层
3　impervious liner　不渗透的（混凝土）衬砌
4　hazardous material　危险品
5　limestone　石灰石
6　clay　黏土
7　clay liner　黏土衬砌
8　soil　土质
9　water table　地下水位

3.3.15 Different configurations for vertical-flow reactors 不同结构的垂直流反应器

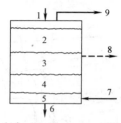

(a) directly fired, countercurrent-flow, moving-packed-bed reactor (shaft furnace) 直接燃烧,逆流,移动填充床反应器(竖式炉)

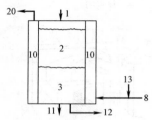

(b) indirectly fired, cocurrent-flow, moving-packed-bed reactor 间接燃烧,并流,移动填充床反应器

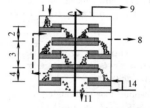

(c) directly fired, counter current-flow, moving-stirred-bed staged reactor (multiple-hearth furnaces; different flow configurations possible) 直接燃烧,逆流,移动搅拌分段式反应器(多膛炉,可能形成不同的流动形态)

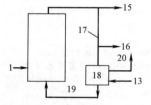

(d) indirectly fired, cocurrent-flow, entrained-bed reactor (transport reactor. mechanical recirculation of heat carrier) 间接燃烧,并流,输送床反应器(输送反应器,热载体机械循环)

1　solid waste　固体废物
2　drying and heating　干燥与加热
3　pyrolysis　热分解
4　char gasification　炭气化
5　ash or slag　灰分或炉渣
6　slag or char＋ash　炉渣或炭＋灰分
7　steam＋O_2 or air　蒸汽＋氧气或空气
8　fuel gas　燃料气
9　fuel gas (with vaporized oil and tar) 燃料气(含气化的油与焦油)
10　combustion chamber　燃烧室
11　char and ash　炭和灰分
12　fuel gas (some oil and tar) 燃料气(含一些油和焦油)
13　air　空气
14　fuel＋steam＋O_2 or air　燃料＋水蒸气＋氧气或空气
15　fuel gas and oil to recovery quench for oil 燃料气和油去油急冷回收
16　excess char＋ash　过量炭＋灰分
17　recirculating heat carrier (char and ash) 循环的热载体(炭和灰分)
18　char heater　炭加热器
19　hot char＋ash　热炭＋灰分
20　flue gas　烟道气

3.3.16 Typical flow sheet for the recovery of materials and the production of refuse-derived fuels (RDF) 回收物料及生产垃圾燃料（RDF）的典型流程图

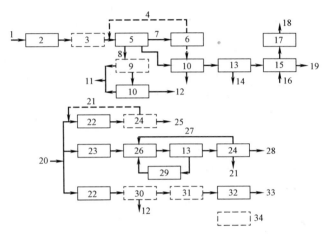

1 solid waste 固体废物
2 receiving area 接受区
3 shear shredder 撕剪机
4 option 挑选
5 trommel 转筒筛
6 shredder 撕碎机
7 large material 大物料
8 fine material 粉状物料
9 screening 筛分
10 magnetic separation 磁分离
11 to aluminum and glass recovery (if used); otherwise residue to landfill 去作铝与玻璃回收（若需要），否则残余物去掩埋
12 ferrous metal 铁金属
13 air classification 空气分类
14 residue to landfill 残余物去掩埋
15 cyclone 旋风分离器
16 air 空气
17 dust collector 集尘器
18 to atmosphere 至大气
19 mainly organic fraction (see below) 主要的有机物料（接下）
20 mainly organic fraction (see above) 主要的有机物料（接上）
21 oversize material 尺寸过大物料
22 secondary shredder 二次撕碎机
23 secondary trommel 二次转筒筛
24 screen 筛
25 fluff RDF 废物垃圾燃料
26 ball mill 球磨
27 intermediate-size material 中等尺寸物料
28 powder RDF 垃圾废料粉末
29 ball heater 球加热器
30 magnetic separator 磁分离器
31 moisture control 水分控制
32 pelletized 成粒
33 pelletized (densified) RDF 成粒（压实）垃圾废料
34 optional processing steps 供选择的加工步骤

3.3.17 Schematic diagram of anaerobic digestion system 厌氧消化系统示意图

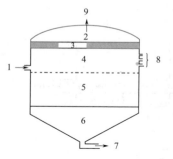

1 sludge inlet 污泥入口
2 gas 气体
3 scum layer 浮渣层
4 supernatant 上清液
5 digesting sludge 污泥降解
6 digested sludge 消化污泥
7 outlet 排泄口
8 outlet ports 出料口
9 gas exhaust 废气

3.3.18 Schematic diagram of two-stage anaerobic digestion of sludge 污泥的两级厌氧消化示意图

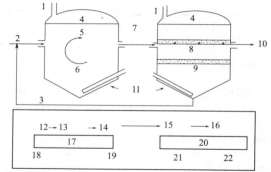

1 gas release 放气
2 sludge inlet 污泥入口
3 sludge return 污泥回流
4 gas 气体
5 zone of mixing 混合区
6 actively digesting sludge 活跃消解的污泥
7 mixed liquor 混合液
8 supernatant 上清液
9 digested sludge 消化污泥
10 supernatant removal 上清液去除
11 sludge drawoff 污泥排出管
12 suspended solids 悬浮固体
13 dissolved solids 溶解固体
14 organic acids 有机酸
15 acetate 醋酸盐
16 methane 甲烷
17 acid phase 酸相
18 liquefaction 液化
19 acidification 酸化
20 gas (methane) phase 气(甲烷)相
21 acetate formation 醋酸盐形成
22 methane formation 甲烷形成

3.3.19 Schematic of the screw press dewatering system (courtesy of Hollin Iron Works)
螺旋压榨脱水系统示意图 (Hollin 钢铁厂)

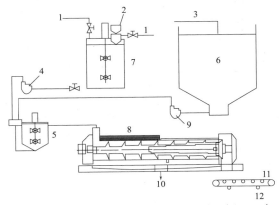

1 water 水
2 polymer powder 聚合物粉末
3 from sludge thickener 来自污泥浓缩机
4 polymer feed pump 聚合物进料泵
5 polymer and sludge mixing tank 聚合物和污泥搅拌罐
6 sludge storage tank 污泥储存罐
7 polymer dissolving tank 聚合物溶解槽
8 water shower or air spray 水喷淋或空气喷雾
9 sludge feed pump 污泥进料泵
10 filtrate 滤液
11 sludge cake 泥饼
12 cake conveyor 泥饼运输机

3.3.20 Septage odor control 腐化物除臭

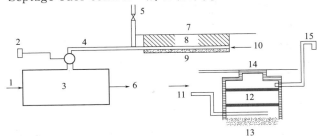

1 raw septage 原腐化物
2 controls 控制装置
3 pretreatment facilities 预处理设施
4 fan/venting system 风机/排风系统
5 bypass 旁路
6 to treatment process 去处理流程
7 soil filter 土壤过滤器
8 topsoil 表土
9 gravel 碎石
10 perforated piping 多孔管
11 air from treatment facility 来自处理设施的空气
12 Fe_2O_3/wood chip mixture 氧化铁/木屑混合物
13 iron oxide filter 氧化铁过滤器
14 manhole 检修孔
15 vent 放空

3.3.21 Schematic of a waste sludge chlorination process system (BIF) 废污泥氯化工艺系统简图（BIF 公司）

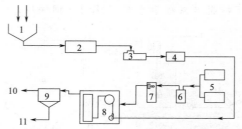

1 sludge storage 污泥储存池
2 macerator 浸解池
3 sludge feed pump 污泥进料泵
4 flow meter 流量计
5 chlorine 氯
6 vaporizer (if required) 气化器（如需要）
7 chlorinizer 氯化器
8 purifax unit 纯化装置
9 holding tank 存储槽
10 supernatant return to plant 上清液回厂
11 sludge 污泥

3.3.22 Schematic of a septage chlorination system (US EPA) 腐化物氯化系统示意图（美国 EPA）

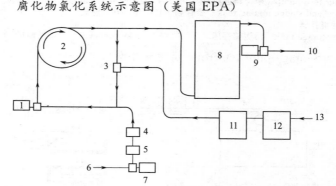

1 recirculation pump 循环泵
2 first reactor 第一反应器
3 eductor 喷射器
4 flow meter 流量计
5 disintegrator 粉碎机
6 raw septage 原腐化物
7 supply pump 供料泵
8 second reactor 第二反应器
9 pressure control pump 压力控制泵
10 conditioned sludge 调理过的污泥
11 chlorinator 氯化器
12 evaporator 蒸发器
13 chlorine supply 供应氯气

3.3.23 SilvaGas refuse gasification system flowsheet
SilvaGas 垃圾气化系统工艺流程

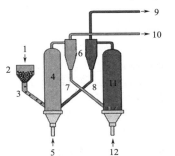

1	biomass	生物质
2	fuel storage	燃料储存器
3	feed	给料
4	gasifier	气化炉
5	steam	蒸汽
6	cyclone separators	旋风分离器
7	sand	砂子
8	sand & char	砂子和木炭
9	fuel gas	燃料气
10	medium-BTU gas	中等热值可燃气体
11	combustor	燃烧炉
12	air	空气

3.3.24 Raymond flash dryer. (Courtesy of Raymond Division, Combustion Engineering Inc.)
雷蒙德闪蒸干燥机（源自雷蒙德分部，燃烧工程公司）

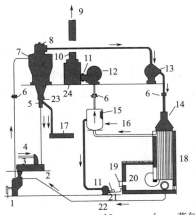

1	cage mill	笼式磨机
2	discharge spout	排液管
3	mixer	混合器
4	wet sludge conveyor	湿污泥输送机
5	manual dry divider	手动干料分流器
6	expansion joint	膨胀节
7	cyclone	旋风分离器
8	relief vent	安全放空口
9	exhaust gas	废气
10	stack	烟囱
11	automatic dampers	自动调节风门
12	induced draft fan	引风机
13	vapor fan	蒸气风机
14	remote manual dampers	遥控手动风门
15	combustion air preheater	助燃空气预热器
16	inlet air	进空气
17	dry product conveyor	干成品输送机
18	deodorizing preheater	除臭预热器
19	burners	烧嘴
20	furnace	加热炉
21	combustion air fan	助燃风机
22	hot gas duct	热气导管
23	double flap valve	双瓣阀
24	scrubber	涤气器

3.3.25 Flowsheet of direct-fired rotary dryer
直燃式回转烘干机的工艺流程

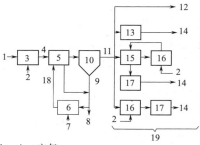

1	air 空气
2	fuel 燃料
3	furnace 加热炉
4	hot gases 热气体（1200°～1400°F）
5	rotary dryer 回转烘干机
6	blender 搅拌机
7	feed sludge 污泥进料
8	disposal 处置
9	dried sludge 干污泥
10	cyclone 旋风分离器
11	exhaust gases（300°F） 废气
12	direct discharge to atmosphere 直接排放到大气中
13	chemical scrubber 化学洗涤器
14	atmosphere 大气
15	heat exchanger 换热器
16	burner（1500°F） 燃烧炉
17	scrubber 洗涤器
18	blended sludge 搅拌过的污泥
19	alternatives available for exhaust gas deodorization and particulate removal 对于废气除臭和去除颗粒有其他方案可选

3.3.26 Detail of rabble arm of multiple hearth biosolids furnace. (Courtesy of BSP Div. Envirotech Corp.)
多层生物固体焚烧炉耙臂祥图（源于BSP环境技术公司）

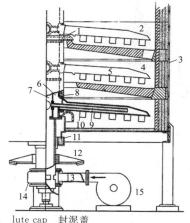

1	lute cap 封泥盖
2	out hearth 炉膛外
3	steel shell 钢外壳
4	in hearth 炉膛内
5	rabble arm 耙臂
6	hot air compartment 热风室
7	cold air tube 冷空气管
8	arm holding PIN 耙臂固定销
9	air lance 空气喷枪
10	rabble arm teeth 耙臂齿
11	sand seal 砂封
12	center shaft gear drive 中心轴齿轮传动装置
13	cooling air 冷却空气
14	air housing 空气室
15	shaft cooling air fan 主轴冷却风机

3.3.27 Cross section of fluidized bed biosolids incinerator 生物固体流化床焚烧炉截面图

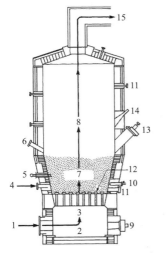

1 fluidizing air inlet 流化空气入口
2 windbox 风箱
3 refractory arch 耐火拱顶
4 sludge inlet 污泥入口
5 thermocouple 热电偶
6 sand feed 砂子进口
7 fluidized sand bed 砂子流化床
8 freeboard 分离空间
9 startup preheat burner for hot windbox 开工预热风箱的燃烧器
10 fuel gun 燃料喷枪
11 pressure tap 压力计接口
12 tuyeres 风口
13 burner 燃烧器（烧嘴）
14 sight glass 视镜
15 exhaust and ash 废气和灰分

3.3.28 Diagram of an earthworm process 蚯蚓工艺图

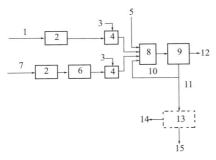

1 aerobically digested sludge 有氧消化过的污泥
2 dewatering 脱水
3 bulking agent if required 填充剂（如需要）
4 mixer 搅拌机
5 make-up earthworms 补充用的蚯蚓
6 pre-treatment（aeration, etc.）预处理（曝气等）
7 an aerobically digested sludge 有氧消化过的污泥
8 worm beds 蚯蚓床
9 earthworm harvester 蚯蚓采集器
10 Recycled earthworms, uneaten sludge particles, and (if used) bulking agent 循环的蚯蚓、吃剩的污泥颗粒和填充剂（如果使用的话）
11 surplus earthworms (if any) 过剩蚯蚓（如果有的话）
12 castings for land utilization as soil amendment 舍弃的蚯蚓可用于土地利用的土壤改良剂
13 separation of earthworms from bulking agent (if required) 从填充剂中分离出蚯蚓（如需要）
14 bulking agent for recycle 循环的填充剂
15 surplus earthworms for sale 过剩蚯蚓可出售

3.3.29 Plug-flow in-vessel composting bioreactors: (a) cylindrical; (b) rectangular; (c) tunnel 容器内活塞流复合肥生物反应器: (a) 圆筒形; (b) 矩形; (c) 隧道式

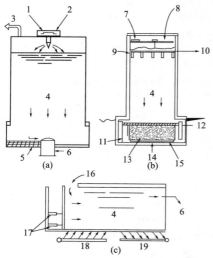

1 mixer 搅拌器
2 infeed 进料
3 CO_2 二氧化碳
4 composting mix 复合肥混合
5 air 空气
6 outfeed 出料
7 material feed conveyor 供料输送机
8 inflow traverse conveyor 流入水平往返移动输送机
9 air removal system 空气去除系统
10 off gases 废气
11 material removal system 物料去除系统
12 material extraction mechanism 物料抽提机构
13 graded stone 分层碎石
14 air 空气
15 air input system 空气输入系统
16 infeed 进料
17 hydraulic ram 液压夯锤
18 air distribution 空气分配装置
19 air removal 去除空气

3.3.30 Agitated (mixed) in-vessel composting bioreactors: (a) circular; (b) rectangular 容器内的搅拌（混合）复合肥生物反应器: (a) 圆形; (b) 矩形

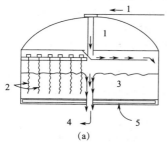
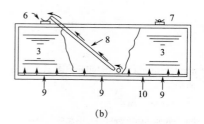

1 inflow 流入
2 augers 螺旋钻
3 composting mix 复合肥混合
4 material removal 物料卸出
5 air manifold 空气歧管
6 outfeed conveyer 出料输送机
7 infeed conveyer 进料输送机
8 extraction conveyer 抽提输送机
9 air 空气
10 plenum 通风管道

3.4 Waste gas treatment 废气处理

3.4.1 Schematic diagram of conventional trickle bed biofilter 传统的滴流床生物过滤器示意图

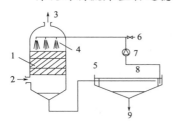

1. packed column 填料塔
2. waste gas 工业废气
3. gaseous effluent 废气排出
4. sprays 喷头
5. sludge settling tank 污泥沉淀池
6. fresh water 淡水
7. pump 泵
8. recirculating water 循环水
9. excess sludge 过剩的污泥

3.4.2 Schematic diagram of a conventional bioscubber 传统的生物涤气器示意图

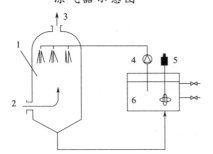

1. scrubber compartment 涤气塔空间
2. waste gas 工业废气
3. gaseous effluent 废气排出
4. pump 泵
5. stirrer 搅拌器
6. activated sludge tank 活性污泥池

3.4.3 Full-Scale biofilter design 全尺寸生物滤池设计

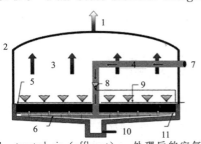

4. influent 进气
5. 4′LAVA rock (expandable) to 6′ deep LAVA rock 4英尺(呎)火山石(可膨胀)到6英尺深火山石
6. stainless steel ducting and PVC laterals 不锈钢主管道和聚氯乙烯支管
7. 40″duct 40英寸的管道
8. humidifier 增湿器
9. water or nutrient feed 水或营养物进料
10. drainage 排水
11. stainless steel grating 不锈钢格栅

1. treated air (effluent) 处理后的空气(废气)
2. 100′DIA steel tank 直径100英尺钢罐
3. treated air 处理后的空气

3.4.4 Schematic diagram of typical wet scrubbers: (a) Bubble column; (b) Packed bed/tower; (c) Spray tower; (d) Venturi scrubber 典型的湿式涤气器示意图：(a) 鼓泡塔；(b) 填充床/塔；(c) 喷雾塔；(d) 文丘里涤气器

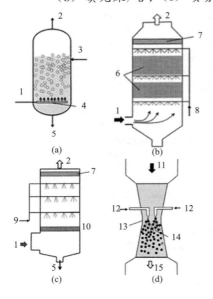

1　gas in　进气
2　gas out　出气
3　liquid supply　供液
4　gas distributor (sparger)　气体分布器（鼓泡器）
5　liquid drain　液体排放口
6　packing media　填充介质
7　mist eliminator　除沫器
8　liquid　液体
9　liquid in　液体进入
10　gas distributor plate　气体分布板
11　dirty air　脏空气
12　liquid inlet　液体入口
13　throat　喉部
14　spray or jet　喷雾或喷射
15　air-droplet mixture　空气-液滴混合物

3.4.5 VOC destruction by catalytic oxidation with recuperative heat exchanger 带同流换热器的催化氧化法分解挥发性有机化合物

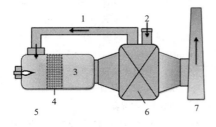

1　preheated gas　已预热的气体
2　waste gas to be oxidized (inlet)　将被氧化的工业废气（入口）
3　oxidation zone　氧化区
4　catalyst bed　催化剂床
5　incinerator　焚烧炉
6　heat exchange to preheat the waste gas stream　热交换器用来预热工业废气流
7　stack　烟囱

3.4 Waste gas treatment 废气处理

3.4.6 Simplified process diagram of an atmospheric CO_2 capture system (packed bed tower option) 大气中二氧化碳捕集系统的简化工艺流程图（填料塔方案）

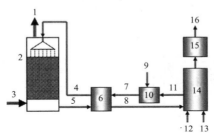

1　CO_2-depleted air　二氧化碳耗尽的空气
2　contactor　接触器
3　air　空气
4　NaOH　氢氧化钠
5　Na_2CO_3　碳酸钠
6　causticizer　苛化器
7　$Ca(OH)_2$　氢氧化钙
8　$CaCO_3$　碳酸钙
9　H_2O　水
10　slaker　消化器
11　CaO　氧化钙
12　air　空气
13　NG　天然气
14　calciner　煅烧炉
15　CO_2 capture & compression　二氧化碳捕集与压缩
16　CO_2 to storage　二氧化碳去储存

3.4.7 Wet oxidation lab equipment 湿氧化实验室设备

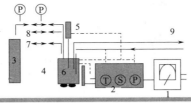

1　power supply　电源
2　digital controller (temperature, pressure and stirring speed)　数字控制器（温度，压力和搅拌速度）
3　oxygen bottle　氧气瓶
4　heater　加热器
5　stirrer　搅拌器
6　reactor　反应器
7　samples extraction　样品提取
8　gas draining　气体排放
9　cooling water　冷却水

151

3.5 Emissions measurement 排放物测定

3.5.1 Current meters are made either with vertical axis or horizontal axis 垂直轴与水平轴的流速计

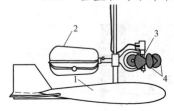
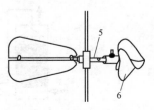

(a) vertical axis current meter
垂直轴流速计

1　weight　负荷
2　tail vane　尾部叶片
3　vertical axis　垂直轴

(b) horizontal axis current meter
水平轴流速计

4　rotating cups　旋转杯
5　horizontal axis　水平轴
6　propeller　螺旋桨

3.5.2 Velocity-measurement system　流速测量系统

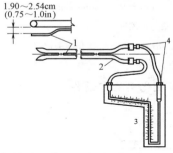

1　temperature sensor　温度传感器
2　type-S Pitot tube　S 型皮托管
3　manometer　测压计
4　leak-free connections　无泄漏联结

3.5.3 Automatic device proportions samples to flow
液体流动比例抽样自动装置

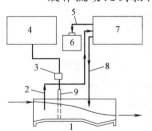

1　flume　斜坡水槽
2　sampling line　取样管线
3　transmitter　变送器
4　flow measurement, recording and proportional sampling control　流量测量、记录和比例抽样控制器
5　composite sample line　组合取样管线
6　sample bottle　取样瓶
7　continuous flow composite sampler　连续流动组合取样器
8　excess sample return line　剩余样品回流管线
9　probe　探头

3.5.4 Parshall flumes 帕里斯霍尔水槽

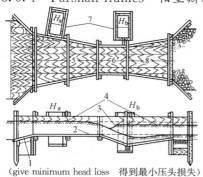

1　slope 1/4　斜面坡度 1∶4
2　level floor　水平底面
3　standing wave　驻波
4　water surface　水表面
5　converging section　收缩截面
6　throat section　喉部截面
7　stilling wells　沉降槽
8　diverging section　扩大截面

(give minimum head loss　得到最小压头损失)

3.5.5 Water-level recorders 水位记录仪

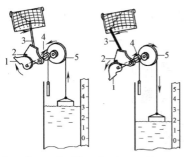

1　flow cam　滑移凸轮
2　base plate　基板
3　pen arm　笔尖杆
4　flow gears　滑移齿轮
5　float pulley　浮动滑轮

(can be used with weirs to provide a record of stream flow　利用溢流堰记录水流)

3.5.6 Particle counter monitors 粒子计数器

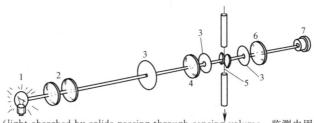

(light absorbed by solids passing through sensing volume　监测由固体通过信号传感体积所吸收的光)

1　lamp　信号灯
2　condenser lens　聚焦透镜
3　aperture　小孔
4　relay lens　中继透镜
5　sensing volume　信号传感体积
6　collection lens　聚光透镜
7　photodiode　光电二极管

3.5.7 Moisture-sample train 湿含量测定系统

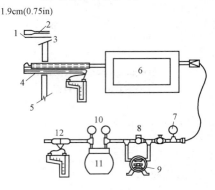

3.5.8 Grab-sample setup for molecular-weight determination 测定分子量的定时取样装置

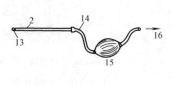

3.5.9 Integrated-sample setup for molecular-weight determination 测定分子量的积分采样装置

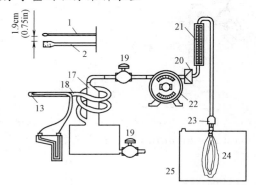

1　Pitot tube　皮托管
2　probe　探头
3　stack wall　烟囱壁
4　reverse-type Pitot tube　反向皮托管
5　Pitot manometer　皮托压力计
6　condenser-ice-bath system including silica-gel tube　包括硅胶管的冰浴冷凝器系统
7　vacuum gauge　真空计
8　bypass valve　旁通阀
9　airtight pump　气密泵
10　thermometer　温度计
11　dry-gas meter　干式气体流量计
12　orifice　孔板
13　filter (glass wool)　过滤器（玻璃棉）
14　flexible tubing　软管
15　squeeze bulb　挤球
16　to analyzer　至分析仪
17　air-cooled condenser　空冷冷凝器
18　probe-Pitot tube　皮托管探测器
19　valve　阀
20　surge tank　均压箱
21　rate meter　流量计
22　pump　泵
23　quick disconnect　快速切断
24　bag　袋
25　rigid container　刚性容器

3.5.10 Stationary sample systems 静态采样系统

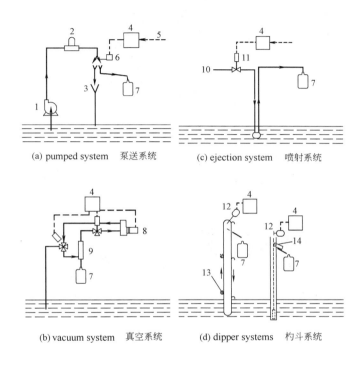

(a) pumped system 泵送系统

(c) ejection system 喷射系统

(b) vacuum system 真空系统

(d) dipper systems 杓斗系统

(provide representative composite samples for analysis
可提供有代表性的混合物分析样品)

1　sample pump　采样泵

2　pH sensor　pH 传感器

3　drain　排放

4　control　控制器

5　flow　流入

6　diverter　分流器

7　sample container　样品存储器

8　vacuum pump　真空泵

9　measuring chamber　测量室

10　air　空气

11　control valve　调节阀

12　drive　驱动装置

13　collection buckets　集料吊斗

14　dipper　杓斗

3.5.11 Particulate-sample apparatus 颗粒物采样仪器

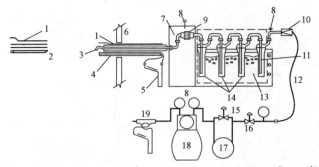

3.5.12 Continuous sample train for CO
CO 连续采样系统

3.5.13 Sampling apparatus for CO
CO 采样仪

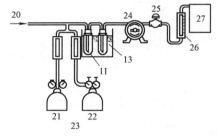

1	temperature sensor 温度传感器	17	airtight pump 气密泵
2	Pitot tube 皮托管	18	dry-gas meter 干式气体流量计
3	probe 探头	19	orifice 孔板
4	reverse-type Pitot tube 反向皮托管	20	sample 取样
5	Pitot manometer 皮托测压管	21	zero 零点
6	stack wall 烟囱壁	22	span 刻度间隔
7	heated area 加热区	23	calibration gases 标定气体
8	thermometer 温度计	24	pump 泵
9	filter holder 滤膜托	25	needle valve 针形阀
10	check valve 止逆阀	26	rate meter 流量计
11	silica gel 硅胶	27	non-dispersive infrared analyzer (NDIR) 非色散型红外线分析仪 (NDIR)
12	vacuum line 真空管线		
13	ice bath 冰浴	28	filter (glass wool) 玻璃棉
14	impingers 冲击器	29	air-cooled condenser 空冷冷凝器
15	bypass valve 旁通阀	30	to analyzer 去分析仪
16	main valve 主阀	31	valve 阀

3.5.14 Sulfur dioxide sample train 二氧化硫采样系统

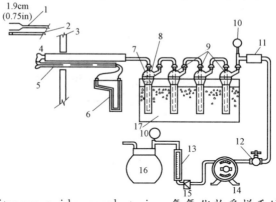

3.5.15 Nitrogen oxide sample train 氮氧化物采样系统

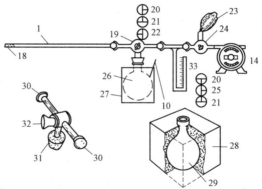

1 probe 探头
2 Pitot tube 皮托管
3 stack wall 烟囱壁
4 probe (end packed with quartz or pyrex wool) 探头（末端封填有石英或耐热玻璃棉）
5 type S Pitot tube S 型皮托管
6 Pitot manometer 皮托测压计
7 glass wool 玻璃棉
8 midget bubbler 小型鼓泡器
9 impingers 撞击器
10 thermometer 温度计
11 silica gel drying tube 硅胶干燥管
12 needle valve 针形阀
13 rate meter 流量计
14 pump 泵
15 surge tank 均压箱
16 dry gas meter 干式气体流量计
17 ice bath 冰浴
18 filter 过滤器
19 flask valve 烧瓶阀
20 evacuate 抽气
21 purge 吹扫
22 sample 采样
23 squeeze bulb 挤球
24 pump valve 泵阀
25 vent 放空
26 flask 烧瓶
27 flask shield 烧瓶屏罩
28 foam encasement 泡沫塑料外套
29 round-bottom boiling flask 圆底长颈烧瓶
30 ground glass socket 磨砂玻璃插口
31 ground glass cone 磨砂玻璃锥
32 3-way stop cock 三通旋塞
33 manometer 测压计

3.5.16 Sampling apparatus for fluoride 氟化物采样装置

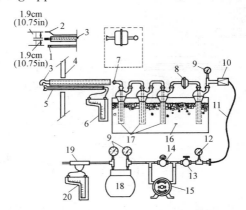

3.5.17 Sampling system for organics and hydrocarbons 有机物与烃类的采样系统

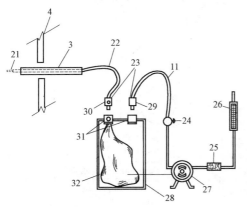

1 Pitot tube 皮托管
2 temperature sensor 温度传感器
3 probe 探头
4 stack wall 烟囱壁
5 reverse-type Pitot tube 反向皮托管
6 Pitot manometer 皮托管测压计
7 optional filter holder location 任选的过滤器夹持器配置
8 filter holder 过滤器夹持器
9 thermometer 温度计
10 check valve 单向阀
11 vacuum line 真空管线
12 vacuum gauge 真空表
13 main valve 主阀
14 bypass valve 旁通阀
15 airtight pump 气密泵
16 ice bath 冰浴
17 impingers 撞击器
18 dry gas meter 干式气体流量计
19 orifice 孔板
20 orifice manometer 孔板测压计
21 filter（glass wool） 过滤器（玻璃棉）
22 Teflon sample line 聚四氟乙烯采样管
23 male 插入式配件
24 needle valve 针形阀
25 charcoal tube 活性炭管
26 flowmeter 流量计
27 pump 泵
28 rigid leak proof container 刚性防漏容器
29 valve 阀
30 ball check valve 球形止逆阀
31 quick connects female 快装管接头
32 aluminized mylar bag 涂铝聚酯袋

3.5.18 Many options are available for outfall flow measurement
适用于排水流量测量的许多选择方案

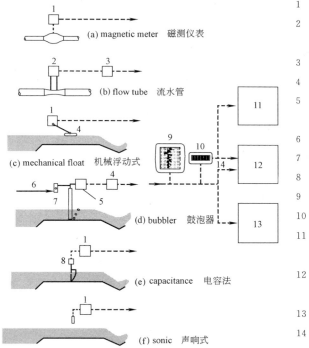

1　transmitter　变送器
2　differential-pressure cell　压差传感器
3　linearizer　线性化电路
4　float　浮子
5　pressure transmitter　压力变送器
6　air　空气
7　dip tube　沉浸管
8　probe　探头
9　record　自动记录
10　totalize　累计器
11　composite sampler　组合取样器
12　data acquisition　数据收集
13　chlorine　氯气
14　or　或者

3.5.19 Elements of pollutant monitoring system
污染物监控系统的组成单元

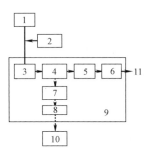

1　sample manifold　取样歧管
2　calibration unit　校正单元
3　filter and/or scrubber　过滤器和（或）涤气器
4　measurement cell　测量元件
5　flow control　流量控制
6　pump　泵
7　signal processing　信号处理
8　output　输出
9　pollutant analyzer　污染物分析器
10　data recording system　数据记录系统
11　exhaust　排出

3.5.20 Block diagram of a basic gas-chromatograph system
气相色谱系统框图

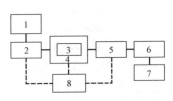

1　carrier-gas supply and control　载气供给与控制
2　injector　注射器
3　column　柱
4　oven　恒温器
5　detector　检测器
6　signal processor　信号处理系统
7　output device　输出设备
8　temperature-control electronics　温度控制电子元件

3.5.21 Cross-section of the flame-ionization detector
火焰离子化检测器截面图

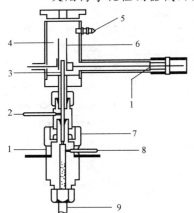

1　insulator　绝缘体
2　air　空气
3　quartz jet　石英喷嘴
4　polarizer electrode　极化器电极
5　ignitor　点火器
6　collector electrode　集电极
7　heater block　换热器单元
8　hydrogen　氢气
9　column　柱

3.5.22 Block diagram of the Hall electrolytic conductivity detector
霍尔电解质电导率检测器框图

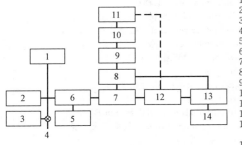

1　reaction gas　反应气体
2　gas chromatograph　气相色谱
3　vent timer　排气定时器
4　solenoid vent valve　电磁排气阀
5　temperature controller　温控器
6　reactor　反应器
7　gas/liquid contactor　气-液接触器
8　reference cell　参比池
9　ion-exchange bed　离子交换床
10　pump　泵
11　solvent reservoir　溶剂储罐
12　separator cell　分离器元件
13　differential conductivity bridge
　　　差分电导测定电桥
14　recorder　记录器

4 Heat-Transfer Equipment 传热设备

4.1 Shell-and-tube heat exchangers 管壳换热器（列管换热器）

4.1.1 Heat-exchanger component nomenclature 换热器部件名称

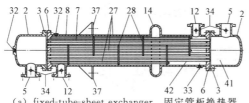
(a) fixed-tube-sheet exchanger 固定管板换热器

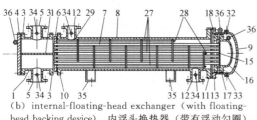
(b) internal-floating-head exchanger（with floating-head backing device） 内浮头换热器（带有浮动勾圈）

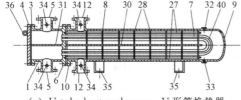
(c) U-tube heat exchanger U形管换热器

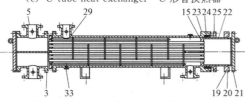
(d) outside-packed floating-head exchanger 外填料函浮头换热器

1 stationary head-channel 固定端管箱
2 stationary head-bonnet 固定端端盖
3 stationary head flange-channel [bonnet] 固定端法兰（管箱或端盖）
4 channel cover 管箱盖
5 stationary head nozzle 管箱接管
6 stationary tubesheet 固定端管板
7 tubes 管子
8 shell 壳体
9 shell cover 封头
10 stationary head end shell flange 固定端壳体法兰
11 rear head end shell flange 浮动端壳体法兰

4.1 Shell-and-tube heat exchangers 管壳换热器（列管换热器）

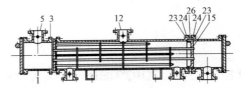

(e) exchanger with packed floating tube sheet and lantern ring
具有填料函浮动管板和套环的换热器

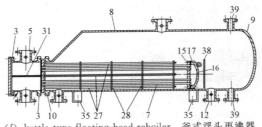

(f) kettle-type floating-head reboiler 釜式浮头再沸器

12	shell nozzle 壳体接管		28	transverse baffles or support plate 横向折流板或支持板
13	shell cover flange 封头法兰			
14	expansion joint 膨胀节		29	impingement baffle 防冲板；缓冲挡板
15	floating tubesheet 浮头管板		30	longitudinal baffle 纵向折流板
16	floating head cover 浮头盖		31	pass partition 分程隔板
17	floating head flange 浮头法兰		32	vent connection 放气接口
18	floating head backing device 浮头勾圈		33	drain connection 排液接口
19	split shear ring 对开挡圈；对开环		34	instrument connection 仪表接口
20	slip-on backing flange 活套靠背法兰		35	support saddle 鞍式支座
21	floating head cover-external 外浮头端盖		36	lifting lug 吊耳
			37	support bracket 悬挂式支座
22	floating tubesheet skirt 浮动管板裙		38	weir 堰板
23	packing box flange 填料函法兰		39	liquid level connection 液面计接口
24	packing 填料		40	tube bundle 管束
25	packing follower ring 填料压盖		41	tube（side）pass 管程
26	lantern ring 套环；灯笼环		42	shell（side）pass 壳程
27	tie rods and spacers 拉杆和定距管			

4.1.2 Front end stationary head types 前端固定封头型式

(a) channel and removable cover 管箱与可拆盖板

(b) bonnet(integral cover) 端盖(整体盖板)

(c) channel integral with tubesheet 与管板制成整体的管箱

(d) removable cover 可拆盖板, 同(c), 1∶ke (c)

(e) special high pressure closure 特殊高压密封端盖

4.1.3 Shell types 壳体型式

(f) one pass shell 单程壳体

(g) two pass shell with longitudinal baffle 双程壳体带纵向隔板

(h) split flow 分隔流动

(i) double split flow 双分隔流动

(j) divided flow 分流流动

(k) kettle type reboiler 釜式再沸器

(l) cross flow 横向流动

4.1.4 Rear end head types 后端封头型式

(m) fixed tubesheet like (a) stationary head 与固定封头(a)相似的固定管板

(n) fixed tubesheet like (b) stationary head 与固定封头(b)相似的固定管板

(o) fixed tubesheet like (d) stationary head 与固定封头(d)相似的固定管板

(p) outside packed floating head 外填料函式浮头

(q) floating head with backing device 具有背封构件的浮头

(r) pull through floating head 牵拉浮头式封头

(s) U-tube bundle U形管管束

(t) externally sealed floating tubesheet 外部密封的浮头管板

1　pass partition　分程隔板
2　heat exchange tube　换热管
3　tubesheet　管板
4　channel cover　管箱盖
5　longitudinal baffle　纵向折流板
6　weir　堰板
7　floating tubesheet skirt　浮动管板裙
8　packing box　填料函
9　internal floating head　内浮头

4.1.5 Tube arrangement 管子排列形式

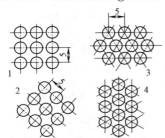

1 square pitch 正方形排列
2 rotated square pitch 转角正方形排列
3 equilateral triangle pitch 正三角形排列
4 rotated equilateral triangle pitch 转角正三角形排列
5 tube pitch 管心距

4.2 Other heat exchangers for liquids and gases 液体和气体用的其他型式换热器

4.2.1 Plate-and-frame heat exchanger 板框式换热器

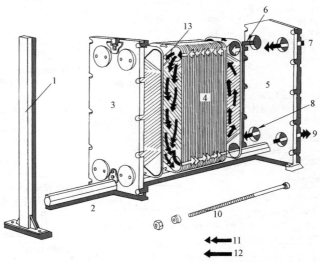

(hot fluid flows down between alternate plates, and cold fluid flows up between alternate plates 热流体在交替的板片之间向下流动,冷流体在交替的板片之间向上流动)

1 upright column 立柱
2 carrying bar 承载梁
3 moveable end cover 可移动端盖
4 heat exchange plate pack 换热板组件
5 fixed end cover 固定端盖
6 cold fluid out 冷流体出口
7 hot fluid in 热流体进入
8 cold fluid in 冷流体进入
9 hot fluid out 热流体出口
10 compression bolt 压紧螺栓
11 hot fluid 热流体
12 cold fluid 冷流体
13 seal packing 密封垫

4.2.2 Fixed type double-pipe heat exchanger 固定式套管换热器

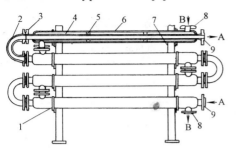

1 U-bolt U形螺栓
2 return bend 回弯头
3 flange 法兰
4 inner pipe 内管
5 support piece 支承块
6 outer pipe 外管
7 support 支柱
8 shell nozzle 壳程接管口
9 inner pipe nozzle 管程接管口

4.2.3 Spray-type coil heat exchanger 喷淋式盘管换热器

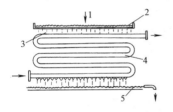

1 cooling water 冷却水
2 spray header 喷淋水管
3 uniform waterfilm 均匀水膜
4 coil pipe 盘管
5 water collecting pond 集水槽

4.2.4 Air preheater 空气预热器

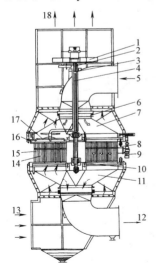

1 fan blade 风机叶片
2 top bearing 上轴承
3 seal 密封装置
4 main shaft 主轴
5 cold air inlet 冷空气入口
6 air hood collar seal 空气罩密封圈
7 upper air hood 上部空气分布罩
8 roller 滚子
9 drive motor 驱动电动机
10 lower bearing 下轴承
11 lower air hood 下部空气集气罩
12 hot gas outlet 热气体出口
13 waste gas inlet 废热气入口
14 heating element 加热元件
15 stator 定子
16 air tight equipment 气体密封装置
17 soot blower 吹灰装置
18 waste gas outlet 废气出口

4.2.5 Other heat exchangers 其他型式换热器

1. plate-type exchanger 板式换热器
2. spiral-plate exchanger 螺旋板式换热器
3. brazed-plate-fin heat exchanger 黄铜板翅式换热器
4. plate-fin cooler 散热片式冷却器
5. plate fin-and-tube type 板翅管式
6. graphite-block exchanger 石墨块体换热器
7. cascade heat exchanger 级联式换热器
8. atmospheric cooler 大气冷却器，空气冷却器
9. bayonet-tube exchanger 内插管式换热器，刺刀管换热器
10. spiral-tube exchanger 蛇管式换热器，螺旋管换热器
11. cryogenic-service spiral-tube exchanger 低温操作螺旋管换热器
12. falling-film exchanger 降膜式换热器
13. teflon heat exchanger 聚四氟乙烯换热器
14. scraped-surface exchanger 刮面式换热器
15. jacketed type exchanger 夹套式换热器
16. submerged-pipe coil exchanger 沉浸式盘管换热器
17. double-pipe exchanger 套管式换热器
18. tank coil heater 盘管式油罐加热器
19. double-pipe longitudinal finned exchanger 套管式纵向翅片换热器

4.3 Heat exchangers for solids 固体用换热器

4.3.1 Heat-transfer equipment for melting or fusion of solids 熔融或熔化固体用传热设备

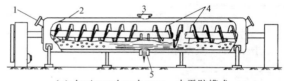

(a) horizontal-tank type 水平贮槽式

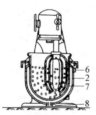   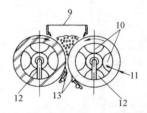

(b) agitated kettle 搅拌釜式   (c) double-drum mill 双滚筒磨式

1. solids charge opening (inlets each end) 固体颗粒加料孔（在每一端的进口）
2. liquid or steam jacket 液体或蒸汽夹套
3. vapor connection (when necessary) 蒸气接口（当必需时）
4. mixing-ribbon spirals 混合螺带螺旋
5. burden discharge 装料排出口
6. liquid or steam connection 液体或蒸汽接口
7. mixing kettle wall 混合釜壁
8. liquid or condensate connection 液体或冷凝液接口
9. dry-powdered solids hopper 干燥粉末状固体加料斗
10. steam or liquid chamber 蒸汽或液体室
11. wall 壁
12. siphon 虹吸管
13. peeling knives intermittently operated 断续操作的剥离刮刀

4.3.2 Rotating shells as indirect heat-transfer equipment 回转壳体用作间接传热设备

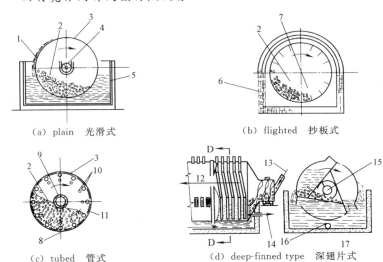

(a) plain 光滑式

(b) flighted 抄板式

(c) tubed 管式

(d) deep-finned type 深翅片式

1　water film　水膜
2　divided-solids bed　分散固体床层
3　rotating metal shell　回转的金属壳体
4　solids discharge provision　固体排料装置
5　water tank　水槽
6　brick tunnel for heated gases　砖洞道用于加热气体
7　solids discharge at bottom of open end (far end)　固体排出在底部开口端（远端）
8　divided-solids out (at far end by bottom discharge)　分散固体出口（在远端通过底部排出）
9　divided-solids in　分散固体入口
10　tubes rotate with shell and carry water, steam or hot gases　管子随壳体回转,其中通入水、水蒸气或热气体
11　stationary header gear for heat-medium supply and removal (at far end)　固定的集管装置,用于加热介质的供入与排出（在远端）
12　burden travel　装料移动
13　solids in　固体进口
14　warm-water outlet　温水出口
15　solids outlet (far end)　固体出口（远端）
16　cold-water inlet (far end)　冷水进口（远端）
17　section D-D　D-D 截面

4.3.3 Spiral conveyor adaptations (as heat-transfer equipment) 螺旋输送设备（作传热装置）

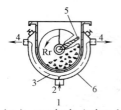
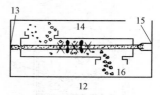

(a) standard jacketed solid flight 标准夹套固定刮板式

(b) small spiral, large shaft 小螺旋粗轴式

(c) "porcupine" medium shaft "porcupine" 中间轴式

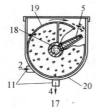
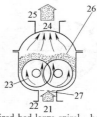

(d) large spiral, hollow flight 粗螺旋窄刮板

(e) fluidized-bed large spiral, helical flight 流态化床，粗螺旋，螺旋面刮板

1　regular-solids bed（30% to 70% fill）规则的固体床层（填充 30%～70%）
2　in　进口
3　coolant　冷却液
4　out　出口
5　section through flight　通过刮板的剖面
6　unbaffled jacket　无挡板夹套
7　annular solids bed（90% to 100% fill）环状的固体床层（填充 90%～100%）
8　jacket　夹套
9　coolant pool in shaft　轴内的冷却液池
10　coolant film　冷却液膜
11　jacket coolant　夹套冷却液
12　heavy annular solids bed　重型环状固体床层
13　heat-transfer fluid supplied in and parellel to each spindle-through shaft　供入传热流体并与每个穿过浆叶的轴平行
14　input　进料
15　drive shaft　驱动轴
16　discharge　排料
17　deep solids bed（90% to 100% fill）深的固体床层（填充 90%～100%）
18　flight coolant　刮板冷却液
19　long pitch mixing ribbon　长节距混合螺条
20　baffled jacket　带挡板的夹套
21　fluidized deep-solids bed（80% to 100% fill）流态化深的固体床层（填充 80%～100%）
22　fluidized air… hot or cold… humid or dry　流态化空气……热的或冷的……湿的或干的
23　holo-flite intermeshed screws for circulating cooling or heating fluid　中间网状螺旋用于循环冷却或加热流体
24　exhaust air removing residual water　除去残留水分的排出空气（废气）
25　low air velocity no dust carryover　低的空气速度，无粉尘带出
26　bed of material in incipient stage of fluidization　在流态化初期阶段物料床层
27　porous sintered stainless plate　烧结的多孔不锈钢板

4.3.4 Vibratory-conveyor adaptations (as indirect heat-transfer equipment) 振动输送机（作间接传热设备）

(a) heavy-duty jacketed for liquid coolant or high-pressure steam 重负荷夹套式，用于液体冷却液或高压蒸汽

(b) jacketed for coolant spraying 用冷却液喷射的夹套式

(c) light-duty jacketed construction 轻负荷夹套式结构

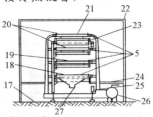

(d) jacketed for air or steam in tiered arrangement 夹套式用于层层排列的空气或蒸汽

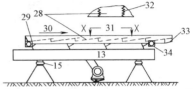

(e) jacketed for air or steam with "Mix-R-Step" surface 夹套式用于 Mix-R-Step 表面的空气或蒸汽

1 viewing port 窥视孔
2 inert-gas connection 惰性气体接口
3 pressure-tight cover 压力密封盖
4 burden chamber 装料室
5 conveying and heat-transfer surface 输送和传热表面
6 coolant, steam and condensate connections 冷却液，蒸汽和冷凝液接口
7 jacketed bottom and sides 带夹套的底部和侧面
8 deck section 台面截面
9 conveying and heat-transfer surface 输送和传热表面
10 spray chamber 喷洒室
11 water supply (under pressure) 供水（在压力下）
12 water drain (by gravity) 排水口（借助重力）
13 vibratory power unit 振动的动力装置
14 (stationary) hood-if needed (固定的) 机罩，如果需要的话
15 vibration isolators 振动隔离体
16 air-supply distribution duct (stationary) 供入空气分布导管（固定的）
17 building floor 建筑物地面
18 travel forward 向前传送
19 travel away 向后传送
20 burden travel forward 装料向前传送
21 divided solids feed (far end) 分散固体进料（远端）
22 enclosure (with venting air heaters) 机壳（带排出空气加热器）
23 "in-process" rolls or granulator 操作过程中的辊辊或造粒机
24 operating floor 操作地面
25 air distribution and recirculating ducts 空气分布与再循环风管道
26 unit heater 单独的加热器
27 granular solids discharge (near end) 粉状固体排料（近端）
28 solid-plate conveying surface 固体平板输送表面
29 divided-solids feed 分散固体进料
30 travel 传送
31 section X-X X-X 截面
32 fixed mixing vanes 固定的混合叶片
33 divided-solids discharge 分散固体排料
34 connections for air or steam (both ends) 用于空气或蒸汽的接口（两端）

4.4 Waste-heat boiler 废热锅炉

4.4.1 Reformed gas waste-heat boiler 重整气废热锅炉

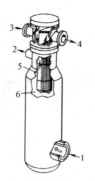

1　gas inlet　气体进口
2　gas outlet　气体出口
3　water in　水进口
4　steam/water out　蒸汽/水出口
5　metal shroud　金属套筒,金属护罩
6　refractory lining　耐火材料衬里

（arrangement of vertical U-tube water-tube boiler　属于立式 U 形管水管锅炉）

4.4.2 Reformed gas waste-heat boiler 重整气废热锅炉

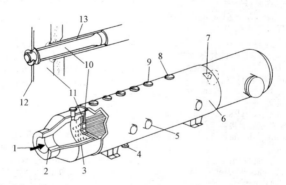

（principal features of typical natural circulation fire-tube boilers　具有典型的自然循环火管锅炉的主要特征）

1　process gas inlet　过程气进口
2　refractory concrete　耐火混凝土
3　insulating concrete　保温混凝土
4　blowdown connection　排污连接件
5　water downcomer pipes　降水管
6　external insulation　外保温层
7　process gas outlet　过程气出口
8　steam/water　蒸汽/水
9　riser pipes　上升管
10　alloy 800 ferrule　800 号合金的水管口密套
11　concrete　混凝土
12　alloy 800 production plate　800 号合金专用板
13　ferrule wrapped with insulating fibre　外包绝热纤维的水管口密套

4.5 Air-cooled heat exchangers 空冷换热器

4.5.1 Air-cooled heat exchanger 空冷换热器

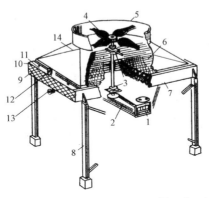

1 electric motor   电动机
2 V-belt drive   V 形带传动
3 bearing   轴承
4 axial flow fan   轴流风机
5 fan ring   风机外圈
6 finned tube bundle   翅片管束
7 channel frame   槽钢框架
8 substructure columns and braces   构架与斜撑
9 removable tube plugs   可拆管丝堵
10 header   管箱
11 vents (drains below)   放气口（排液口在下方）
12 process-fluid inlet nozzle   过程流体进口接管
13 process-fluid outlet nozzle   过程流体出口接管
14 plenum   风斗

4.5.2 Air-cooler recirculator for cold climates 用于寒冷气候空冷器循环系统

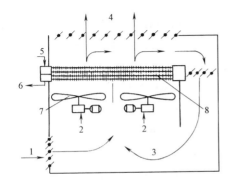

1 cold ambient air   外界环境冷空气
2 warm air   暖空气
3 recirculated hot air   再循环热空气
4 hot air out   热空气出口
5 process stream in   过程流体加入
6 process stream out   过程流体出口
7 axial flow fans   轴流风机
8 plate-fin heat exchanger   翅片式换热器

4.6 Cooling tower 凉水塔

4.6.1 Atmospheric tower 自然通风（凉水）塔

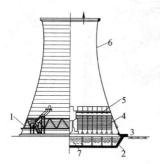

1 cold air intake 冷空气吸入口
2 cold water well 凉水池
3 G. L. (ground level) 地平面
4 plastic packing 塑料填料
5 water distribution 淋水装置
6 tower body 塔体
7 cooling water 冷却水

4.6.2 Natural draft cooling tower 自然通风式凉水塔

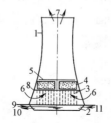

1 fan stack 风筒
2 catch basin 集水池
3 air distributor 空气分布器
4 packing 填料
5 water distributing system 配水系统
6 air 空气
7 hot air 热空气
8 air intake 进风口

4.6.3 Counter flow induced draft cooling tower 引风式逆流凉水塔

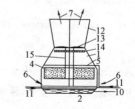

9 warm water inlet 热水进入
10 return line of cooling water 冷水返回管线
11 make-up water inlet 补充水进水管
12 diffuser 扩散器
13 exhaust fan 排风机
14 entrainment eliminator 除水器
15 column of driving device 传动装置竖筒

5 Distillation and Absorption Equipments
蒸馏与吸收设备

5.1 Type of towers 塔器类型

5.1.1 Classification of towers based on unit operation 按单元操作分类

1　distillation column　精馏塔，蒸馏塔
2　fractionating column　分馏塔，精馏塔
3　complex distillation column　复杂精馏塔
4　stripping tower　提馏塔，汽提塔
5　absorption column　吸收塔
6　desorption column　解吸塔，再生塔
7　extraction tower　萃取塔，抽提塔
8　stripping column　汽[气]提塔
9　washing tower　洗涤塔
10　cooling tower　冷却塔，凉水塔

5.1.2 Classification of towers based on internal structures 按内部结构分类

1　plate column　板式塔
2　bubble cap column　泡罩塔
3　sieve tray column　筛板塔
4　float valve tower　浮阀塔
5　packed column　填料塔
6　tube bundle column; multitubular column　多管塔
7　pulsed column　脉冲塔
8　packed column filled with liquid　充液填料塔，乳化塔
9　turbulent ball column　湍球塔
10　rotating disc column　转盘塔
11　baffle tower　挡板塔
12　bubble column　鼓泡塔
13　perforated plate tower　孔板塔，穿流筛板塔
14　wetted wall column　湿壁塔

5.2 Plate column 板式塔

5.2.1 Plate column 板式塔

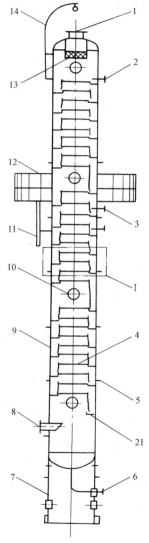
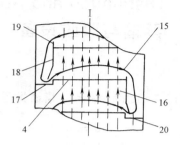

1 vapo(u)r outlet 蒸气出口
2 reflux inlet 回流口
3 feed inlet 进料口
4 tray 塔盘
5 insulating ring 保温圈
6 liquid outlet 液体出口
7 skirt support 裙座
8 reboiler return 再沸器返回口
9 tower shell 塔体
10 manhole 人孔
11 ladder 直梯
12 platform 平台
13 demister 除沫器
14 hoisting pillar 吊柱
15 liquid stream 液流
16 vapo(u)r 蒸气
17 receiving pan 受液盘
18 downcomer 降液管
19 overflow weir 溢流堰
20 weep hole 泪孔
21 liquid seal pot 液封槽

5.2.2 Single pass tray 单流塔盘

5.2.3 Two pass tray 双流塔盘

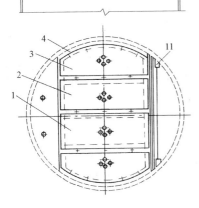
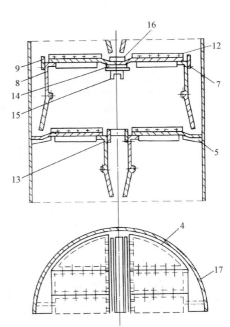

1	man way plate	通道板
2	rectangular plate	矩形板
3	segmental plate	弓形板
4	supporting ring	支持圈
5	web plate	筋板
6	seal pot	受液盘
7	support plate	支承板
8	downcomer apron	降液板
9	adjustable weir	可调堰板
10	removable downcomer	可拆降液管
11	connecting plate	连接板
12	tray sheet	塔盘板
13	center downcomer	中心降液管
14	major beam	主梁
15	support	支座
16	clamp plate	压板
17	tower shell	塔体
18	weep hole	泪孔

5.2.4 Trays with spacer tubes 带定距管塔盘

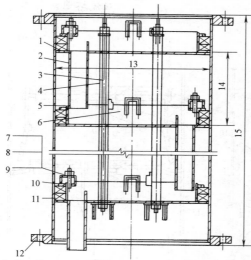

5.2.5 Superposed tray 重叠式塔板

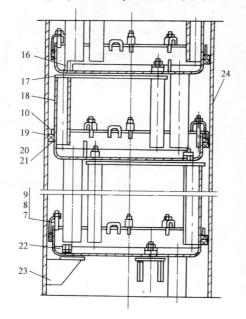

1 intact tray 整块式塔板
2 downcomer 降液管
3 tie-rod 拉杆
4 spacer tube 定距管
5 tray ring 塔盘圈
6 lifting lug 吊耳
7 bolt 螺栓
8 nut 螺母
9 clamp plate 压板
10 press ring 压圈
11 asbestos cord 石棉绳
12 vessel flange 容器法兰,设备法兰
13 tower internal diameter 塔内径
14 space of plates 板间距
15 height of sectional tower shell 塔节高度
16 adjusting bolt 调节螺栓
17 support plate 支承板
18 support leg 支柱
19 tray ring 塔板圈
20 packing 填料
21 support ring 支承圈
22 tray sheet 塔板
23 support 支座
24 tower body 塔体

5.2.6 Typical cross-flow plate(sieve) 典型的单溢流塔板（筛板）

5.2.7 Segregative tray construction 分块式塔盘结构

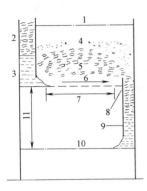

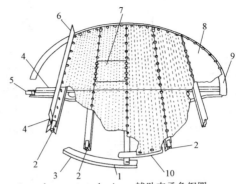

1　plate above　上一层塔板
2,5　froth　泡沫
3　clear liquid　清液
4　spray　雾沫
6　liquid flow　液流
7　active area　有效区
8　calming zone　安定区
9　downcomer apron　降液板
10　plate below　下一层塔板
11　tray spacing　板间距

1　subsupport angle ring　辅助支承角钢圈
2　minor beam support clamp　支梁固定板
3　subsupport plate ring used with angle ring
　　辅助支承环板与角钢圈一起使用
4　major beam　主梁
5　major beam clamp, welded to tower wall
　　主梁固定板与塔壁焊接
6　downcomer and weir　降液管与溢流堰
7　manway plate　通道板
8　calming area　安定区
9　plate support ring　钢板支承圈
10　peripheral ring clamps　周边圆环固定板

5.2.8 Bubble-cap tray tower 泡罩塔

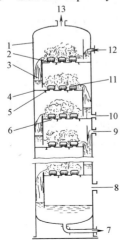

1　shell　塔外壳
2　tray　塔盘
3　down spout　降液管
4　tray support ring　塔盘支持圈
5　tray stiffer　塔盘支承
6　vapor riser　升气管
7　liquid outlet　液体出口
8　gas inlet　气体入口
9　intermediate feed　塔中部液体入口
10　side stream withdrawal　侧流抽出口
11　bubble-cap　泡罩
12　liquid feed　液体入口
13　gas outlet　气体出口

177

5.2.9 Turbulent ball column 湍球塔

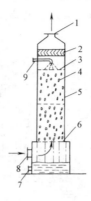

1 gas outlet　气体出口
2 entrainment separator　雾沫分离器
3 top grating　上栅板
4 ball packing　球形填料
5 tower body　塔体
6 lower grating　下栅板
7 liquid outlet　液体出口
8 gas inlet　气体入口
9 spray header　喷淋水管

5.2.10 Type of trays 塔盘类型

1　bubble cap tray　泡罩（帽）塔盘
2　Thormann tray　索曼塔盘，槽形泡罩塔盘
3　floating valve tray　浮阀塔盘
4　rectangular valve tray　条形浮阀塔盘
5　sieve tray　筛板
6　perforated tray　多孔筛板
7　large hole sieve tray　大孔筛板
8　directional sieve tray　导向筛板
9　slotted sieve tray　导向筛板
10　floating sieve tray　浮动筛板
11　vertical sieve tray　垂直筛板
12　Linde sieve tray　林德筛板
13　multiple downcomer sieve tray　多降液管筛板
14　S-shape tray　S形塔盘
15　jet tray　舌形塔盘，喷射式塔板
16　floating jet tray　浮动舌形塔盘
17　inclined hole tray　斜孔塔板
18　angle bar tray　角钢塔盘
19　floating angle tray　浮动角钢塔盘
20　perform tray　网孔塔盘
21　rotating stream tray　旋流塔盘
22　dual-flow tray　穿流塔盘
23　ripple tray　波纹（穿流）塔盘
24　single pass tray　单流塔盘
25　two pass tray　双流塔盘
26　reverse flow tray　U形流塔盘
27　intact tray　整块式塔板
28　segregative tray　分块式塔板
29　self-beam tray　自身梁式塔板

5.3 Packed column 填料塔

5.3.1 Packed column 填料塔

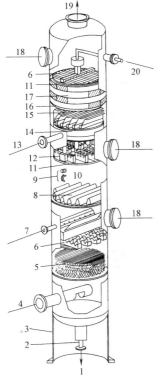

1 bottom product 塔底产品

2 circulation pipe to reboiler 循环管道再沸器

3 skirt support 裙座

4 reboiler return 再沸器返回口

5 structured packing 规整填料

6 liquid distributor 液体分布器

7 vapo(u)r feed 蒸气进口

8 support plate 支承板

9 random packing 散装填料

10 rings or saddles 拉西环或鞍形填料

11 hold-down grid 床层限制栅

12 liquid distributor/redistributor 液体分布器/再分布器

13 liquid feed 液体进料口

14 ringed channel 环形沟

15 liquid collector 液体收集器

16 support grid 支承格栅

17 structured packing 规整填料

18 manway 人孔

19 vapor outlet to condenser 蒸气出口通冷凝器

20 reflux from condenser 回流液来自冷凝器

5.3.2 Some kinds of random packings (dumped packings) 各种散装填料

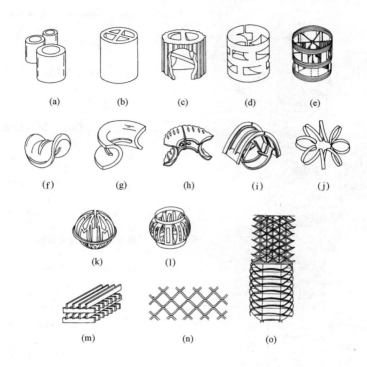

(a) Raschig ring　拉西环

(b) cross partition ring　十字格环

(c) double spiral ring　双螺旋环

(d) metal Pall ring　金属鲍尔环

(e) plastic Pall ring　塑料鲍尔环

(f) ceramic Berl saddle　瓷弧鞍填料

(g) ceramic Intalox saddle　瓷矩鞍填料

(h) plastic Intalox saddle　塑料矩鞍

(i) metal Intalox saddle　金属环矩鞍

(j) Teller Bosette packing　特勒花环填料

(k) plastic tripak　塑料球形填料

(l) metal tripak　金属球形填料

(m) wood grid　木格栅

(n) section through expanded metal packing　膨胀金属网填料

(o) section of expanded metal packings placed alternatively at right angles　膨胀金属网填料安装图

5.3.3 Demister and its applications 除沫器与应用

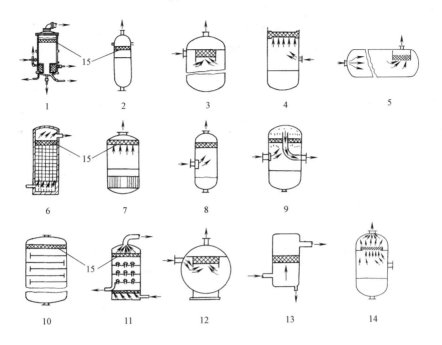

1 evaporator 蒸发器

2 batch kettle 间歇操作釜

3 oil-gas seperator 油气分离器

4 open seperator 敞开式气液分离器

5 horizontal seperator 卧式气液分离器

6 packed tower 填料塔

7 crystallizer 结晶器

8 vertical seperator 立式气液分离器

9 in-line gas scrubber 管道气液分离器

10 absorber 吸收塔

11 fractionating tower 分馏塔，精馏塔

12 spherical seperator 球形分离器

13 knock-out drum 气液分离罐，缓冲罐

14 large scale vessel 大型容器

15 demister 除沫器

5.3.4 Tower internals 塔内件

1. spray apparatus 喷淋装置
2. tubular sprayer 管式喷淋器
3. shower nozzle thrower 莲蓬头式喷淋器
4. impact spray thrower 冲击式淋洒器
5. cascade impact thrower 宝塔式淋洒器
6. mechanical spray thrower 机械式喷嘴
7. spinner thrower 漩涡式喷淋器
8. disc spray thrower 盘式淋洒器
9. liquid distributor 液体分布器
10. liquid redistributor 液体再分布器
11. perforated pipe distributor 多孔管分布器
12. orifice-type distributor 锐孔型分布器
13. trough-type distributor 槽型分布器
14. weir-riser distributor 堰-上升管分布器
15. wiper redistributor 收集型再分布器
16. bell-cap redistributor 钟罩型再分布器
17. straight side wiper 平直型塔侧收集器
18. cone side wiper 锥形塔侧收集器
19. pyramid support and redistributor 锥形支持板和再分布器
20. packing support plate 填料支承板
21. packing support 填料支承
22. welded ring support 焊接环支承
23. wire mesh support 丝网支承
24. beam-type "gas-injection" support plate for large columns 大塔的横梁式"气体喷射"支承板
25. packing press ring 填料压环
26. packing draw-off hold 填料卸出孔
27. compact set of tray 塔盘紧固件
28. wedge compact set 楔形紧固件
29. grip 卡子
30. clamp 卡子
31. support plate 支承板
32. support ring 支承圈
33. sectional tower shell 塔节
34. skirt support 裙座
35. base of stone bolt 地脚螺栓座
36. basic ring 基础环
37. hoisting pillar 吊柱
38. foam separator 除沫器
39. foam catcher 除沫器
40. foam breaker 消沫器
41. vortex breaker 防涡器
42. dumped tower packing 乱堆塔填料
43. arranged packing 规整填料
44. arranged-type packing 规整填料
45. regular packing 规则填料
46. structured packing 规整填料
47. wire-web packing 网波纹填料
48. mellapak 板波纹填料
49. grid packing 格栅填料

5.4 Absorption column 吸收塔

5.4.1 CO_2 absorber CO_2 吸收塔

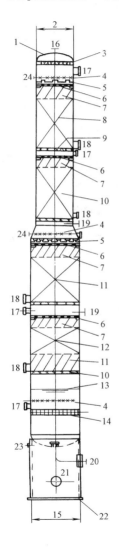

1　elliptical head　椭圆形封头
2　tower inside diameter　塔内径
3　demister　除沫器
4　spray header　喷淋头
5　liquid distributor　液体分布器
6　hold down grate　床层限制栅
7,8,10～12　packing　填料
9　packing support-gas injection plate and liquid redistributor　填料支承板-气体喷射板和液体再分布器
13　max. liquid level　最高液位
14　vortex breaker　防涡流板
15　anchor bolt circle　地脚螺栓中心圆
16　vapo(u)r out　气体出口，蒸气出口
17　manhole　人孔
18　unloading connection　填料卸出口
19　vapo(u)r sample　气体取样口
20　liquid out　液体出口
21　access opening　出入口
22　anchor bolt　地脚螺栓
23　vent pipe　通气管
24　liquid in　液体入口

183

5.4.2 Falling film type absorber 湿壁降膜吸收塔

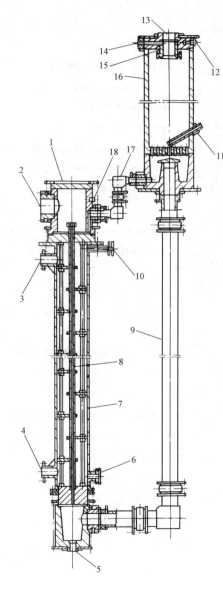

1 cooling absorber 冷却吸收器
2 HCl gas inlet HCl气体进口
3 cooling water outlet 冷却水出口
4 cooling water inlet 冷却水进口
5 hydrochloric acid outlet 盐酸出口
6 peep sight glass 窥视镜
7 rubber lined steel shell 衬橡胶钢制壳体
8 resbon tube 不透性石墨管
9 lean gas riser 贫气提升管
10 air vent 排气管
11 thermowell 热电偶套管
12 manometer tapping hole and plug U型管测压计螺纹孔与螺塞
13 gas outlet 气体出口
14 feed water inlet 进水口
15 water distributor 水分布器
16 gas scrubber 气体洗涤塔
17 acid tube 酸管
18 tapping hole and plug 螺纹孔与螺塞

5.5 Molecular distillation still 分子蒸馏釜

5.5.1 Molecular distillation and related kinds of equipment 分子蒸馏与相关的几种设备

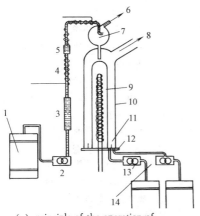

(a) principle of the operation of the falling film still 降膜式分子蒸馏设备的工作原理

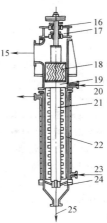

(b) thin-layer evaporator with rigid wiper blades 具有刚性刮板薄层蒸发器

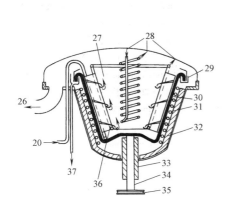

(c) centrifugal molecular still 离心式分子蒸馏釜

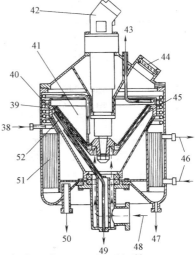

(d) the Liprotherm rotating thin film evaporator Liprotherm 旋转薄膜蒸发器

5.5 Molecular distillation still 分子蒸馏釜

1	crude distilland 待蒸馏的粗制品		26	to vacuum pumps 接真空泵
2	metering inlet pump 进料计量泵		27	distillate out 馏出液出口
3	fluid meter 流量计		28	cooling water 冷却水
4	preheater 预热器		29	gutter 出料槽
5	check valve 止逆阀		30	rotor 转子
6	to vacuum pump 接真空泵		31	electrical radiant heater 电热辐射加热器
7	degassing chamber 脱气室		32	thermal insulation 隔热层
8	to fine vacuum 接高真空		33	bearing 轴承
9	heated distilling column 加热蒸馏塔		34	rotor shaft 转轴
10	cylindrical glass condenser 圆筒形玻璃冷凝器		35	pulley 皮带轮
11	gutter separating distilland and distillate 分隔蒸馏液与馏出液的出料槽		36	condenser 冷凝器
			37	residue 釜残液
12	supporting plate 支承板		38	feed point 进料口
13	distilland withdrawal pump 蒸馏液出料泵		39	preheater 预热器
14	distillate withdrawal pump 馏出液出料泵		40	revolving vaporizer body 旋转蒸发器机体
15	product vapor 产品蒸气		41	vapor space 蒸气域
16	rotor bearing 转子轴承		42	drive 驱动装置
17	gear case 齿轮箱		43	runoff 逸出
18	centrifugal separator 离心式分离器		44	sight glass 视镜
19	distributor 分布器		45	hole for condensate 冷凝液出孔
20	feed 进料		46	coolant 冷却液
21	rotor with 3 to 8 blades 带有3～8个叶片的转子		47	distillate 馏出液
			48	heating steam 加热蒸汽
22	jacket 夹套		49	condensed heating steam and gases 冷凝加热蒸汽和气体
23	heating medium 加热介质		
24	lower rotor bearing 转子下轴承		50	vacuum 真空
25	bottom product 底产品		51	condenser 冷凝器
			52	skimmer tube 除沫管

5.5.2 Industrial wiped-film stills 工业刮膜式分子蒸馏釜

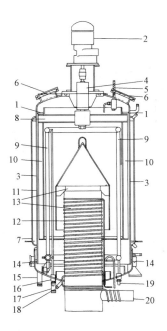

1	feed nozzle to evaporator 给蒸发器的进料管	12	diffusion pump cooling coils 扩散泵冷却盘管
2	gear head motor 电机齿轮变速箱	13	diffusion pump body 扩散泵体
3	evaporator wall 蒸发器壁	14	demountable flange connection 可拆法兰连接
4	oil-sealed stuffing box 油封填料函	15	residue outlet 釜残液出口
5	distilland feed 蒸馏液进料口	16	condenser inlet 冷凝器水进口
6	sight glass 视镜	17	distillate drain 馏出液排出口
7	jacket drain 夹套排液口	18	chamber condensate drain 室内冷凝液排出口
8	rotor 转子	19	condenser water outlet 冷凝器水出口
9	vertical tube condenser 立管冷凝器	20	to fore-pump 接前级泵
10	carbon wipers 碳石墨刮壁器		
11	entrainment separator 雾沫分离器		

5.5.3 Typical flow diagram for wiped-film still and degassers
典型的刮膜式分子蒸馏釜与脱气器的流程图

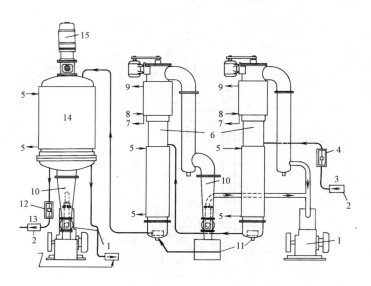

1 mechanical backing pump 前级机械泵
2 positive displacement oil-sealed liquor pump 油封式正位移式打液泵
3 feed 加料
4 feed flowmeter 加料流量计
5 diphenyl 联苯
6 rotary degassers 回转式脱气器
7 volatile tops 挥发性馏分
8 cooling water in 冷却水进口
9 cooling water out 冷却水出口
10 diffusion pump 扩散泵
11 positive displacement pump 正位移泵,容积式泵
12 product flowmeter 产品流量计
13 distillate 馏出液
14 Scott Smith molecular still Scott Smith 分子蒸馏釜
15 variable speed drive 无级变速驱动装置

6 Extraction, Ion-exchange and Adsorption Equipments 萃取、离子交换与吸附设备

6.1 Extraction equipment 萃取设备

6.1.1 Schematic diagram of extraction operation 萃取操作原理图

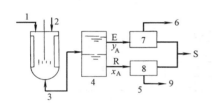

1 feed solution 原料液
2 solvent 溶剂
3 mixing tank 混合槽
4 settling and demixing 沉降分层
5 solvent removing 脱除溶剂
6 extract 萃取液
7 extractive phase 萃取相
8 raffinate phase 萃余相
9 extraction raffinate 萃余液

6.1.2 Liquid-liquid extraction of uranium 铀的液-液萃取

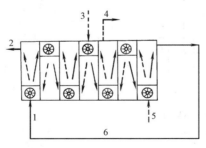

1 organic phase 有机相
2 aqueous raffinate 水相萃残液
3 aqueous feed 水相料液
4 pregnant strip solution(secondary-extract) 富集萃取液（辅助萃取液）
5 aqueous strip solution(secondary-solvent) 水相萃取液（辅助溶剂）
6 primary extraction-solvent 主级萃取溶剂

6.1.3 Solvent extraction of acetic acid from water 用溶剂从水中萃取乙酸

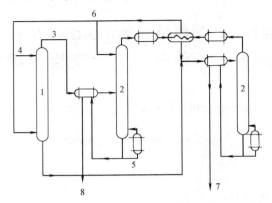

6.1.4 Udex process Udex 工艺过程

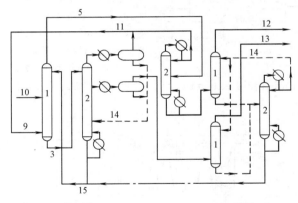

1. extraction tower 萃取塔
2. distillation tower 蒸馏塔
3. extract 萃取液
4. feed, acetic acid in water 料液，乙酸水溶液
5. raffinate 萃残液
6. recycle extraction-solvent 循环萃取溶剂
7. waste water 废水
8. acetic acid 醋酸
9. wash solvent 洗涤液
10. feed 料液
11. lights and benzene 轻组分和苯
12. aliphatics 脂肪族化合物
13. aromatics, BTX 芳香族，BTX
14. water 水
15. extraction-solvent 萃取溶剂

6.1.5 Baffled mixing vessel 内装挡板的混合容器

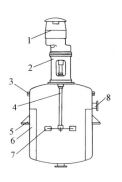

1 geared motor 带减速齿轮箱的电动机
2 frame 机架
3 vessel flange 容器法兰
4 rotor shaft 转轴
5 support 支座
6 baffles 挡板
7 impeller 桨叶
8 connection tube 接管
9 paraffin feed 烷烃原料
10 olefin feed 烯烃原料
11 fresh acid 新鲜酸
12 recycle acid 循环酸
13 spent acid 废酸

6.1.6 Mixer 混合器

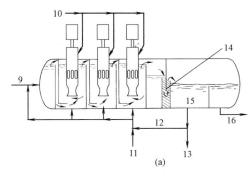

14 coalescer 聚结器
15 settler for acid 酸的澄清器
16 alkylate product 烷基化产品
17 acid 酸
18 paraffin 烷烃
19 olefin 烯烃
20 propeller 搅拌叶片

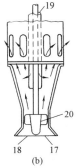

(a) cascade alkylator for sulfuric acid alkylation of paraffins and olefins, simplified 用于链烷烃和烯烃的硫酸烷烃化的多级烷化器（简化图）
(b) mixer detail 混合器的简化结构图

6.1.7 Stratco contactor Stratco 混合器

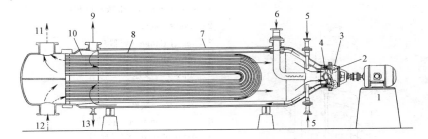

(for HF alkylation of hydrocarbons 用于碳氢化合物 HF 烷烃化)

1　motor　电动机
2　diffuser vanes　扩压器叶片
3　hydraulic head　水力头
4　impeller　桨叶
5　HC feed　HC 原料
6　acid in　进酸
7　shell　机壳
8　circulation tube　循环管
9　mix to settler　混合物去澄清器
10　tube bundle　管束
11　coolant out　冷却剂出
12　coolant in　冷却剂进
13　drain and pump out　排放和泵出

6.1.8 Combination coalescer, settler, and membrane separator 组合的聚结器、澄清器和膜分离器

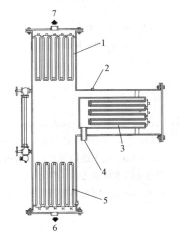

1　hydrophilic separatory membranes　亲水分离膜
2　pressure gage connection　压力表接头
3　coalescing membranes　聚结膜
4　emulsion inlet　乳化液进口
5　hydrophobic separatory membranes　憎水分离膜
6　organic phase outlet　有机相出口
7　water-phase outlet　水相出口

6.1.9 Petreco Cylectric coalescer (schematic) Petreco Cylectric 电聚结器示意图

6.1.10 Five-stage countercurrent cascade arranged for fractional extraction 用于分馏萃取的五级逆流级联

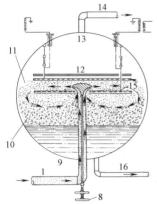

(internal circulation and electric field 内部循环流动和电场)

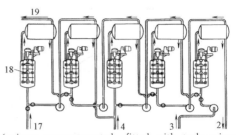

(mixers compartmented, fitted with turbo-mixer agitators 其中装有涡轮混合搅拌器，混合器互相分开)

6.1.11 Pump-Mix mixer-settler 泵混(合)式混合-澄清槽

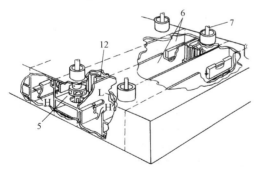

1 emulsion inlet 乳化液进口
2 fat heavy solvent 富的重溶剂
3 lean light solvent 贫的轻溶剂
4 feed 料液
5 mixer 混合器
6 settler 澄清器
7 impeller shaft 叶轮轴
8 velocity adjustor 调速器
9 distributor pipe 分配管
10 coarse coalescing weak field 粗颗粒聚结的弱电场区
11 turbulent fine particle coalescing 湍流的细颗粒聚结
12 baffle 挡板
13 quiet zone 静止区
14 oil out 油出口
15 electrodes 电极
16 water out 水出口
17 lean heavy solvent 贫的重溶剂
18 turbo-mixer agitators 涡轮混合搅拌器
19 fat light solvent 富的轻溶剂

6.1.12 Kerr-McGee multistage mixer-settler Kerr-McGee 多级混合澄清器

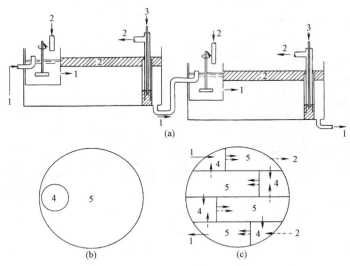

(a) and (b) for uranium, (c) for vanadium extraction
(a) 和 (b) 用于铀；(c) 用于钒萃取

6.1.13 Vitro uranium mixer-settler Vitro 混合-澄清器（用于铀萃取）

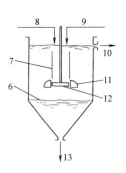

1   aqueous phase   水相
2   organic phase   有机相
3   air   空气
4   mixer   混合器
5   settler   澄清器
6   interface   界面
7   draft tube   导流筒
8   aqueous feed   水相料液
9   solvent   溶剂
10   organic effluent (extract)
     有机相流出物（萃取液）
11   shroud ring   遮环
12   impeller   叶轮
13   aqueous effluent (raffinate)
     水相流出物（萃残液）

6.1.14 Rotating-disk (RDC) extractor 转盘萃取塔（RDC）

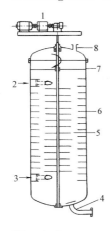

1　variable speed drive　变速的驱动装置
2　heavy liquid inlet　重液体入口
3　light liquid inlet　轻液体入口
4　heavy liquid outlet　重液体出口
5　rotor disk　转盘
6　stator ring　静止环
7　interface　界面
8　light liquid outlet　轻液体出口

6.1.15 Treybal extractor　Treybal 萃取器

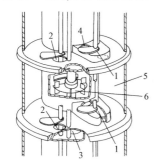

1　LL——light liquid　轻液
2　HL——heavy liquid　重液
3　LLR——light liquid recycle　轻液循环
4　HLR——heavy liquid recycle　重液循环
5　settler　澄清器
6　impeller　叶轮

6.1.16 Pulsed columns　脉冲塔

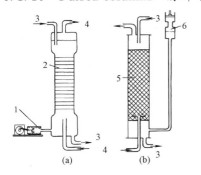

(a) perforated-plate column with pump pulse generator　带泵送脉冲发生器的筛板塔
(b) packed column with air pulser　空气脉冲填料塔

1　pump pulse generator　泵送脉冲发生器
2　perforated plates　多孔板，筛板
3　heavy liquid　重液体
4　light liquid　轻液体
5　packing　填料
6　air pulser　空气脉冲（发生）器

6.1.17 Rotocel extractor 洛特赛（Rotocel）萃取机

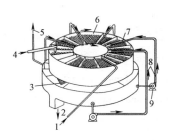

1 fresh solvent in 新鲜溶剂进口
2 leached solid discharge 浸提后的渣料排出
3 hinged screen bottom 绞链式筛网底
4 solid feed 固体加料口
5 full miscella 富液
6 rotating cells 回转隔池
7 spray 喷洒器
8 circulating mother liquor 循环母液
9 pump 泵

6.1.18 Centrifugal extractor 离心分离萃取机

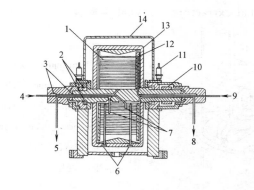

1 contacting element 接触板
2 bearing 轴承
3 mechanical seal 机械密封
4 heavy liquid inlet 重液入口
5 light liquid outlet 轻液出口
6 plug for rotor cleaning 转鼓清洗螺塞
7 heavy liquid dispersion pipe 重液分布管
8 heavy liquid outlet 重液出口
9 light liquid inlet 轻液入口
10 driving sheave 驱动轮
11 lubricating device 注油器
12 light liquid dispersion pipe 轻液分布管
13 rotor 转鼓
14 rotor cover 转鼓外罩

6.2 Ion-exchange equipment 离子交换设备

6.2.1 Typical two-bed deionizing system 典型的双床脱离子系统

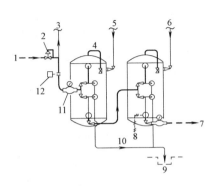

1 raw water 原料水
2 pressure regulating valve 压力调节阀
3 acid dilution and slow rinse supply 酸稀释与慢洗供应
4 air relief 空气释放
5 dilute acid and slow rinse 稀酸与慢洗
6 dilute caustic and slow rinse 稀苛性碱与慢洗
7 to storage 去贮存
8 conductivity cell 传导性电池
9 waste sump 废料坑
10 waste 废料
11 totalizing meter 总计表
12 flow indicator 流量显示器

6.2.2 Ion-exchanger regeneration 离子交换器再生

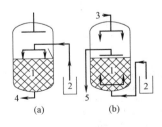

1 resin 树脂
2 acid 酸
3 air 空气
4 to drain 去排液管
5 to drain and vacuum 接排液管与真空

(a) acid is passed downflow through the cation-exchange resin bed(conventional)
酸往下流,通过阳离子交换树脂床层 (常用的)
(b) counterflow regenerant solution is introduced upflow with the resin bed held in place by a dry layer of resin 逆流,再生液引入往上流,树脂床被干层树脂保持固定

6.2.3 Internals of an upflow regenerated unit
往上流再生设备的内部结构

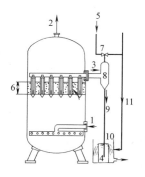

1. regenerant inlet　再生剂进口
2. vent to atmosphere　放空至大气
3. outlet air+effluent from regeneration cycle 从再生循环中排出的空气与废液
4. outlet effluent from regeneration cycle 从再生循环中排出物
5. inlet injected water　喷射用水进口
6. drained strata　排干层
7. ejector　喷射器
8. separator　分离器
9. siphon tube　虹吸管
10. water seal　水封
11. effluent from regeneration process 再生过程排出物

6.2.4 Mode of operation of the Higgins contactor
Higgins 接触器的操作方式

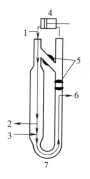 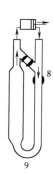 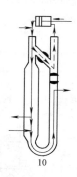

1. solution to be treated　待处理溶液
2. treated solution　经过处理的溶液
3. regenerant　再生剂
4. resin pump　树脂泵
5. resin valves　树脂阀
6. waste　废液
7. solution flow period (several minutes); resin pump stopped (or bypassed) 溶液流动期（数分钟）；停树脂泵（或旁通）
8. valves reversed　阀反向
9. resin movement period (several seconds); solution flow stopped　树脂移动期（数秒钟）；溶液停止流动
10. start of solution flow period; resin pump reversed (or bypassed); resin settles into reservoir 溶液流动期开始；树脂泵反向（或旁通）；树脂沉降至贮槽

6.2.5 Asahi countercurrent ion-exchange process
Asahi 逆流离子交换过程

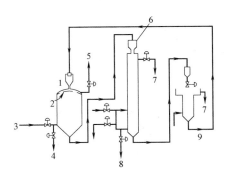

1 collector 收集器
2 adsorption tank 吸附罐
3 influent 进料
4 to waste (or recovery) 去废弃（或回收）
5 to service 去供应
6 resin hopper 树脂料斗
7 waste 废料
8 regeneration tank 再生罐
9 wash rinse tank 冲洗罐

6.2.6 Himsley continuous ion-exchange system
Himsley 连续离子交换系统

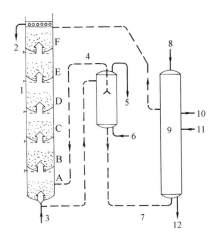

1 adsorption column 吸附柱
2 barren liquor 废液
3 feed liquor 料液
4 measuring and resin cleaning chamber 计量及树脂清洗室
5 washwater outlet 洗涤水出口
6 washwater inlet 洗涤水入口
7 resin transfer lines shown dotted 虚线表示树脂转移路线
8 rinse water inlet 冲洗水入口
9 elution column 洗提柱
10 rinse outlet 冲洗出口
11 eluate inlet 洗提剂入口
12 eluate 洗提液

6.2.7 Multicomponent ion-exchange process for chromate recovery from plating rinse water
从喷镀漂洗水中回收铬酸盐的多元离子交换过程

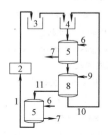

1　recovered chromic acid　回收的铬酸
2　evaporation　蒸发
3　chromate plating tank　铬酸盐喷镀槽
4　chromate rinse tank　铬酸盐冲洗槽
5　cation exchanger　阳离子交换器
6　H_2SO_4 regenerant　H_2SO_4 再生剂
7　waste regenerant　废再生液
8　anion exchanger　阴离子交换器
9　NaOH regenerant　NaOH 再生剂
10　deionized water　去离子水
11　sodium chromate　铬酸钠

6.2.8 Flow diagram of desiccant cooling cycle
干燥剂冷却循环流程图

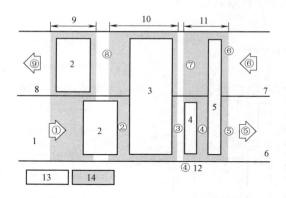

1　regeneration air　再生空气
2　humidifier　增湿器
3　heat exchanger　热交换器
4　reactivation air heater coil　再活化空气加热器盘管
5　desiccant wheel　干燥剂盘
6　to ambient　排空
7　process air　过程空气
8　supply air　供应空气
9　direct evaporative cooler　直接蒸发冷却器
10　indirect evaporative cooler　间接蒸发冷却器
11　desiccant dehumidifier　干燥剂除湿器
12　state point　状态点
13　component　部件
14　sub-assembly　分部装配

6.2.9 Principles of mixed-bed ion exchange 混合床离子交换原理

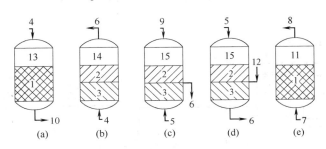

(a) service period 工作期
(b) backwash period 反洗期
(c) caustic regeneration 碱再生
(d) acid regeneration 酸再生
(e) resin mixing 树脂混合

1　mixed resin　混合树脂
2　anion　阴离子型
3　cation　阳离子型
4　raw water　原料水
5　water　水
6　drain　排放水
7　air　空气
8　air out　空气排放
9　alkali　碱
10　treated　处理过的
11　mixing　混合
12　acid　酸
13　service　工作
14　backwash　反洗
15　regeneration　再生

6.3 Adsorption equipment 吸附设备

6.3.1 Pittsburgh moving-bed system for sugar treatment 处理糖的 Pittsburgh 移动床系统

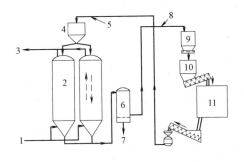

1　juice feed　糖汁进料
2　carbon columns　活性炭柱
3　treated juice return　处理过的糖汁返回
4　carbon feed tank　炭料罐
5　regenerated carbon　再生过的活性炭
6　sweeten-off tank　除甜罐
7　sweet water　甜水
8　spent carbon　用过的活性炭
9　dewatering tank　脱水罐
10　feed bin　料仓
11　regenerating furnace　再生炉

6.3.2 Flow diagram for contact filtration 接触过滤流程图

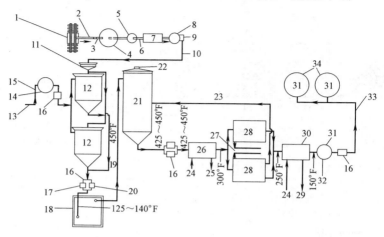

1　clay bin below R tracks　R 轨道下的黏土箱
2　belt conveyor　皮带运输机
3　bucket conveyor　翻斗运输机
4　clay storage　黏土贮存器
5　hammer mill　锤磨
6　10 mesh　10 目
7　dryer　干燥器
8　raymond mill　雷蒙磨
9　pulverized clay blower　粉碎黏土吹送机
10　pulverized clay to storage bins in filter bouse　粉碎黏土去过滤室贮斗
11　clay storage bins　黏土贮斗
12　mixing agitator　混合搅拌器
13　long residuum from batch treatment plant　来自间歇处理厂的长渣油
14　charging tank　加料罐
15　from lube pump in continuous L.O. treatment plant　来自连续润滑油处理装置的油泵
16　duplex piston pump　双活塞泵
17　operating　操作中
18　pipe still: to run continuously for 8 hours heating　管式炉：连续 8h 加热
19　oil and vapors　油与蒸气
20　standby　闲置中
21　vapor separator　蒸气分离器
22　water vapor outlet　水蒸气出口
23　circulating line(where improper filtration occurs)　循环管线（当出现不适当过滤时）
24　water in at 70°F　70°F 水进
25　water out at 200°F　200°F 水出
26　pipe precooler　管式预冷器
27　spent clay to disposal　废黏土排放
28　No. 12 sweetland filter　12 号过滤器
29　water out at 170°F　170°F 出水
30　cooling box with continuous pipe cooling coils　有连续冷却盘管的冷却箱
31　steam coils　蒸汽盘管
32　rundown tank　降温罐
33　to outside storage　至外部贮存
34　outside storage tanks　外部贮存罐

6.3.3 Sequence of operations in batch mixer-settler 间歇混合-沉降器的操作顺序

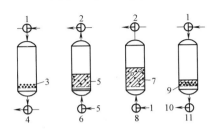

1 gas pressure 气压
2 vent 放空
3 resin particles 树脂粒子
4 empty 流空
5 solution 溶液
6 filling 注入
7 air bubbles 空气泡
8 equilibrating 平衡
9 porous support 多孔支承板
10 drain 排放
11 draining 排放

6.3.4 Two-bed TSA system with regeneration equipment 具有再生设备的双床TSA系统

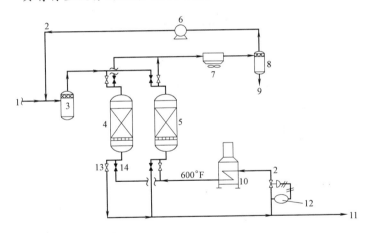

1 wet feed gas 湿气体进口
2 regeneration gas 再生气体
3 inlet separator 进气分离器
4 adsorbing 吸附
5 regenerating & cooling 再生与冷却
6 regen gas compressor 再生气体压缩机
7 regen gas cooler 再生气体冷却器
8 water knockout 气水分离器
9 water 水
10 regen gas heater 再生气体加热器
11 dry gas 干气体
12 FRC(facility remote control) 设备远距离控制
13 valve open 阀开
14 valve closed 阀闭

6.3.5 Multiple hearth furnace for carbon reactivation 活性炭再生用多层膛式炉

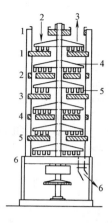

1   hearth   炉膛
2   carbon in   活性炭进入
3   gas out   气体出口
4   rabble arm   搅拌耙臂
5   rabble teeth   搅拌耙齿
6   carbon out   活性炭出口

6.3.6 Hypersorption adsorber vessel 超吸附塔

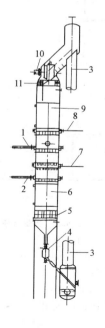

1   feed   进料
2   steam in   蒸汽进口
3   lift line   提升管
4   solids flow control valve   固体流量调节阀
5   reciprocating feed tray   往复加料盘
6   stripper   汽提段
7   bottom product   塔底产品
8   top product   塔顶产品
9   cooler   冷却段
10 lift gas return   提升气体返回口
11 cyclone   旋风分离器

6.3.7 PuraSiv HR adsorber vessel　PuraSiv HR 吸附塔

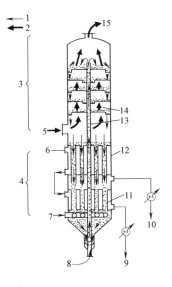

1　adsorbent flow　吸附剂流
2　gas flow　气体流
3　adsorption section　吸附段
4　desorption section　解吸段
5　raw gas　原料气体
6　steam for heating　加热用蒸汽
7　steam for desorption　解吸用蒸汽
8　adsorbent carrier gas　吸附剂载气
9　condensate　冷凝液
10　recovered solvent　回收的溶剂
11　desorption tube　解吸管
12　preheating tube　预热管
13　gas lift line　气体提升管
14　tray　塔盘
15　clean gas　净化气体

6.3.8 Annular bed for liquid separation　环形液体分离床

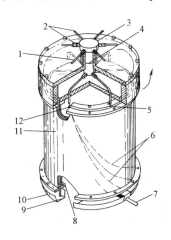

1　feed inlet　进料口
2　eluent inlet　洗脱液进口
3　stationary inlet distributor　固定式进口液分配器
4　inlet pressure seal　进口加压密封件
5　eluent streams　洗脱液流
6　separated constituents　分开的组元
7　product collection　产品收集管
8　annular bed　环形床
9　eluate exit　洗脱液出口
10　stationary eluent waste collection　固定式洗脱废液收集槽
11　rotating annular chromatograph　回转式环形色谱仪
12　continuous feed stream　连续进料流

6.3.9 Pressurized adsorber vessel 加压吸附器

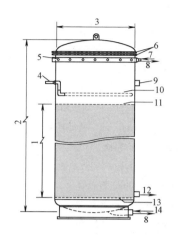

1 carbon bed height 活性炭床层高度
2 vessel height 容器高
3 vessel diameter 容器直径
4 wash water 洗涤水
5 20 1in holes 20个1in孔（1in=2.54cm）
6 bolt ring 螺栓法兰圈
7 influent 液体进入
8 backwash 反洗液
9 carbon charge 活性炭装填口
10 surface wash 表面洗涤
11 carbon bed surface 活性炭床层表面
12 carbon discharge 活性炭卸出口
13 metal screen 金属筛板
14 effluent 废液

6.3.10 Flowsketch of a process for making molecular sieve adsorbents 制备分子筛吸附剂工艺流程

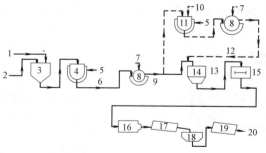

1 sodium aluminate 铝酸钠
2 sodium silicate 硅酸钠
3 makeup tank 补给罐
4 crystallization tank 结晶槽
5 steam 蒸汽
6 crystal slurry 结晶浆液
7 wash water 洗涤水
8 filter 过滤机
9 zeolite crystals 沸石晶体
10 calcium chloride 氯化钙
11 ion exchange tank 离子交换槽
12 calcium-substituted zeolite 钙取代沸石
13 clay binder 黏土黏结剂
14 weigh hopper 称量斗
15 mix muller 混合研磨机
16 pellet extruder 挤压造粒机
17 rotary dryer 回转干燥器
18 screen 筛子
19 rotary kiln 回转窑
20 activated molecular sieve pellets 活性分子筛颗粒

6.3.11 Moving bed sorber 移动床吸附器

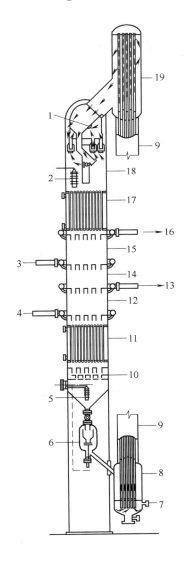

1 lift gas return 提升气体返回口
2 solid level 固体料面计
3 feed gas in 原料气进口
4 stripping steam in 解吸用蒸汽进口
5 solid level controller 固体料面调节器
6 solid flow control valve 固体流量控制阀
7 lift gas in 提升气体进口
8 lower lift drum 下部提升筒
9 lift line 提升管
10 carbon flow controller 活性炭流量调节器
11 stripper 汽提段
12 steam section 蒸汽解吸段
13 bottom product 塔底产品
14 rectifying section 提纯段
15 sorption section 吸附段
16 top product 塔顶产品
17 cooler 冷却器
18 hopper 加料段
19 upper lift drum 上部提升筒

7 Evaporators and Crystallization Equipment 蒸发器与结晶设备

7.1 Evaporators 蒸发器

7.1.1 Agitated thin-film evaporator 搅拌薄膜蒸发装置

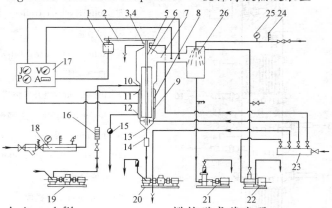

7.1.2 Agitated film evaporator 搅拌膜式蒸发器

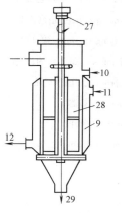

1 motor 电动机　　　　　　2 V-belt V形传动带
3 bearing 轴承　　　　　　4 shaft sealing 轴封
5 shaft 转轴　　　　　　　6 separation zone 分离段
7 vapor outlet 蒸气出口　　8 heating wall 加热面
9 heating jacket 加热夹套
10 liquor inlet 料液入口
11 heating medium inlet 热载体入口
12 drain outlet 凝液出口
13 product outlet 成品出口
14 peep hole 视孔
15 steam trap 蒸气（疏水）阀
16 rotameter 转子流量计
17 control panel 控制台
18 steam regulator 蒸汽调节器
19 feed pump 料液泵
20 discharge pump for concentrated liquor 浓缩液排出泵
21 condensate pump 冷凝液泵
22 vacuum pump 真空泵
23 water header 水总管
24 thermometer 温度计
25 pressure gauge 压力表
26 condenser 冷凝器　　　27 belt pulley 皮带轮
28 agitator 搅拌器　　　　29 concentrate 浓缩液

7.1 Evaporators 蒸发器

7.1.3 Short-tube evaporator 短管蒸发器

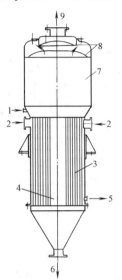

7.1.4 Long-tube evaporator 长管蒸发器

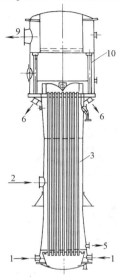

7.1.5 Falling-film evaporator 降膜蒸发器

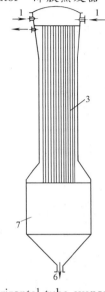

7.1.6 Evaporator with rotating brush 带回转刷的蒸发器

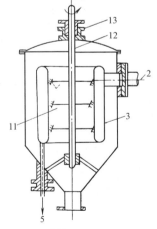

7.1.7 Horizontal-tube evaporator 水平管式蒸发器

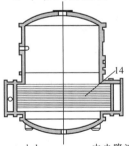

1   feed   进料口
2   steam inlet   蒸汽入口
3   heating element   加热元件
4   central downcomer   中央降液管，中央循环管
5   condensate outlet   凝液出口
6   concentrate   浓缩液
7   separator   分离室
8   demister   除沫器
9   secondary steam   二次蒸汽
10   flash chamber   蒸发室
11   rotating brush   回转刷子
12   shaft   转轴
13   packing seal   填料密封
14   horizontal tube bundle   水平管束

7.1 Evaporators 蒸发器

7.1.8 Submerged-tube F.C.(forcedcirculation)evaporator 浸没管束强制循环蒸发器

7.1.9 Upward-flow (climbing-film) evaporator 升膜蒸发器

7.1.10 Oslo-type crystallizer 奥斯陆型结晶器

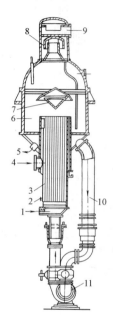
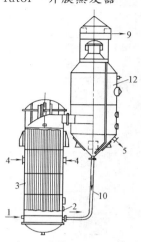
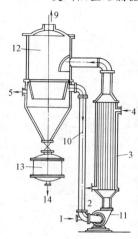

1　feed　进料
2　condensate outlet　凝液出口
3　heating tube bundle　加热管束
4　steam inlet　蒸汽进口
5　separator　分离室
6　flash chamber　闪蒸室
7　demister　除沫器

8　bell cap　钟帽式除沫器
9　secondary steam　二次蒸汽
10　circulation tube　循环管
11　circulation pump　循环泵
12　separator　分离室
13　filter　过滤器
14　filtrate outlet　滤液出口

7.1.11 Horizontal wiped-film evaporator 卧式刮壁（膜）型蒸发器

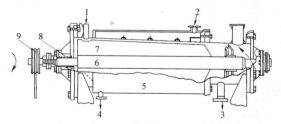

1　feed　进料
2　steam　蒸汽
3　product　产品
4　condensate　冷凝水
5　heating jacket　加热夹套
6　rotor　转子
7　blade　叶片
8　shaft seal　转轴密封装置
9　driver unit　驱动装置

7.1.12 Thermal medium evaporator 热载体蒸发器

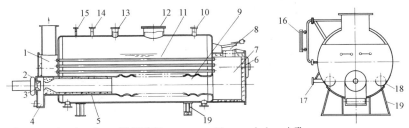

1 front smoke chamber　前烟道室
2 burner　烧嘴
3 burner tile　煤嘴耐火砖
4 wind box　风室
5 flue lining　烟道衬里
6 cleaning door　清扫门
7 rear smoke box　后烟道室
8 explosion door　防爆门
9 flue　烟道
10 nozzle for safety valve　安全阀接管
11 smoke tube　火管
12 manhole　人孔
13 main nozzle for vapor　蒸气总管
14 vent nozzle　排气管
15 nozzle for compound pressure and vacuum gauge　压力真空表接管
16 level gauge　液面计
17 thermal medium return nozzle　热载体回流口
18 cleaning hole　清灰口
19 support saddle　鞍式支座

7.1.13 Evaporator types 蒸发器型式

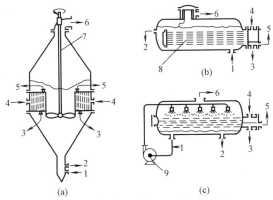

(a) propeller calandria　螺旋浆排管式 (b),(c) horizontal-tube evaporators　水平管式蒸发器

1 feed　进料
2 product　产品
3 condensate　冷凝液
4 steam　水蒸气
5 vent　放气口
6 vapor　蒸气
7 propeller agitator　螺旋桨式搅拌器
8 heating element　加热元件
9 pump　泵

211

7.1.14 Swenson single-stage recompression evaporator 单级再压缩式蒸发器(热泵蒸发)

1. feed solution　进料溶液
2. circulating pump　循环泵
3. heat exchanger　换热器
4. elutriating leg　析出段
5. vapor body　蒸发器壳体
6. purge　放出
7. mesh wash　筛网洗涤液
8. mesh separator　筛网分离器
9. water vapor　水蒸气
10. make-up steam if required　必要时，补充蒸汽
11. condensate　冷凝液
12. vapor desuperheater　蒸气降温器
13. bypass　旁路
14. vanes　导流叶片
15. single stage centrifugal vapor compressor　单级离心式蒸气压缩机
16. condensate to storage　冷凝液去贮罐
17. to centrifuge　去离心机

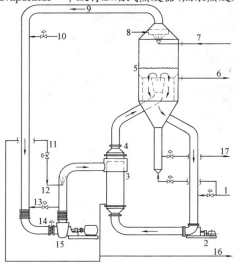

7.2 Crystallization equipment　结晶设备

7.2.1 Forced-circulation baffle surface-cooled crystallizer 强制循环有挡板表面冷却结晶器

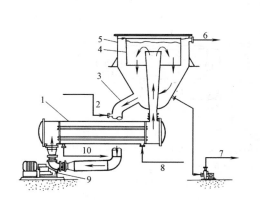

7.2.2 Direct-contact-refrigeration crystallizer (DTB type) 直接接触冷冻结晶器 (DTB型)

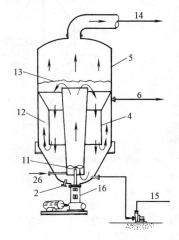

7.2 Crystallization equipment 结晶设备

7.2.3 Oslo evaporative crystallizer Oslo 蒸发结晶器

7.2.4 Oslo surface-cooled crystallizer Oslo 表面冷却结晶器

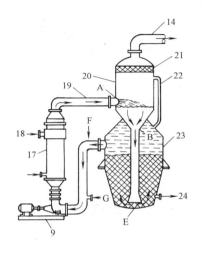
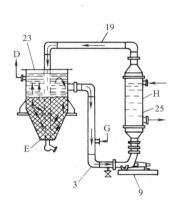

1	cooler	冷却器
2	feed inlet	进料入口
3	circulating pipe	循环管
4	skirt baffle	裙式挡板
5	body	主体
6	mother liquor outlet	母液出口
7	product crystals	产品晶体
8	coolant inlet	冷却剂入口
9	circulating pump	循环泵
10	coolant outlet	冷却剂出口
11	propeller	桨叶
12	settling zone	澄清区
13	boiling surface	沸腾表面
14	vapor outlet	蒸气出口
15	product discharge	产品排出
16	propeller drive	桨叶驱动器
17	heat exchanger	热交换器
18	steam inlet	蒸汽入口
19	recirculation pipe	再循环管
20	vaporizer	蒸发器
21	mesh separator	筛网分离器
22	vent	排气管
23	suspension chamber	悬浮室
24	product outlet	产品出口
25	cooler	冷却器
26	refrigerant inlet	冷冻剂入口

7.2.5 Forced-circulation (evaporative) crystallizer 强制循环（蒸发）结晶器

7.2.6 Draft-tube-baffle (DTB) crystallizer 导流筒-挡板（DTB）结晶器

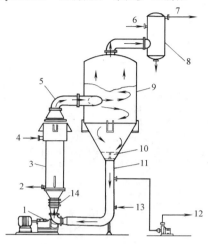

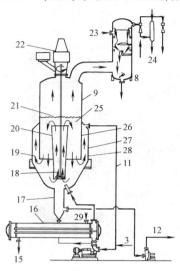

7.2.7 Conispherical magma crystallizer 锥底球形晶浆罐结晶器

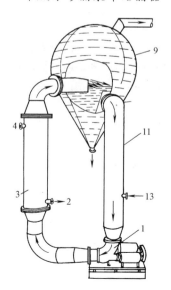

1　circulation pump　循环泵
2　condensate outlet　凝液出口
3　heat exchanger　热交换器
4　steam inlet　蒸汽入口
5　recirculation pipe　再循环管
6　cooling water inlet　冷却水入口
7　non-condensable gas outlet　不凝气出口
8　barometric condenser　大气冷凝器
9　body　主体
10　swirl breaker　破旋器
11　circulating pipe　循环管
12　product discharge　产品排出
13　feed inlet　进料口
14　expansion joint　膨胀接头
15　condensate　凝液
16　heating element　加热器
17　elutriation leg　淘析腿
18　propeller　桨叶
19　setting zone　澄清区
20　draft tube　导流筒
21　boiling surface　沸腾面
22　propeller drive　桨叶驱动器
23　cooling water　冷却水
24　air ejector　空气喷射器
25　slurry　晶浆
26　skirt baffle　裙式挡板
27　clarified M.L.　清母液
28　settler　澄清器
29　steam　蒸汽

7.2.8 Swenson reaction type DTB crystallizer 反应型 DTB 结晶器

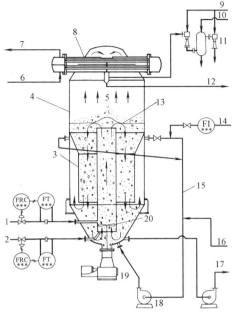

1 feed inlet gaseous NH₃ 气态 NH₃ 进料
2 98% H₂SO₄ feed inlet 98% H₂SO₄ 进料
3 internal skirt baffle 内部裙式挡板
4 vapor chamber body 蒸气室主体
5 vapor 蒸气
6 water in 水加入
7 water out 水出口
8 surface condenser 表面冷凝器
9 steam 水蒸气
10 water 水
11 vacuum equipment 真空装置
12 condensate outlet 冷凝水出口
13 boiling surface 沸腾表面
14 water addition for fines destruction 破损细粒用水
15 ML circulating pipe 母液循环管
16 ML return 母液回流
17 product slurry discharge 浆状产品排出
18 ML recycle pump 母液循环泵
19 propeller drive 螺旋浆驱动装置
20 propeller 螺旋浆式搅拌器

7.2.9 Swenson atmospheric reaction-type DTB crystallizer 常压反应型 DTB 结晶器

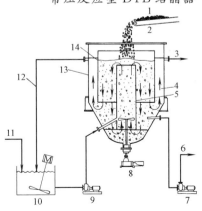

1 solids feed 固体进料
2 conveyor 输送机
3 M. L. purge by overflow level control 由溢流口液面控制母液放出
4 settler 沉降池
5 draft tube 导流筒
6 product slurry discharge 浆状产品排出
7 V-belt driven product pump V 形带传动的产品泵
8 internal circulator drive 内循环器的驱动装置
9 variable speed drive recycle pump 变速驱动循环泵
10 fines dissolving tank 细粒溶解槽
11 fines destruction water 破损细粒用水
12 fines stream 细粒流
13 body 主体
14 baffle 挡板

7.2.10 Escher-Wyss or Tsukushima DP (double propeller) crystallizer DP（双桨）型结晶器

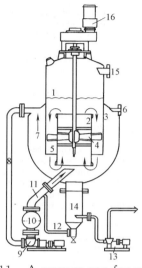

1   thickening zone   增稠区
2   draft tube   导流筒
3   evaporation chamber   蒸发室
4   double-acting circulation propeller 双作用循环搅拌叶片
5   crystal growth zone   晶体生长区
6   overflow   溢流
7   settling zone   沉降区
8   circulation pipe   循环管
9   circulation pump   循环泵
10   heater or cooler   加热器或冷却器
11   solution return   溶液回流
12   elutriation liquid feed   淘析液加入
13   slurry discharge pump   浆液排出泵
14   elutriation zone   淘析区
15   vapour outlet   蒸气出口
16   variable-speed drive   变速驱动装置

7.2.11 A vacuum pan for crystallization of sugar 糖真空结晶器

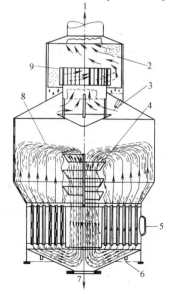

1   vapor   蒸气
2   collecting screen   捕集筛网
3   catch-all drain   除沫器排液管
4   stationary circulation louvers   固定的百叶窗式循环挡板
5   steam inlet   蒸汽加入
6   liquor inlet   液体加入
7   discharge   排出
8   strike level   抛起界面
9   centrifugal deflector   离心式导流板

7.2.12 Cooling crystallizer 冷却结晶器

7.2.13 Evaporator crystallizer 蒸发结晶器

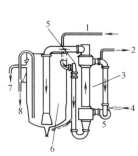

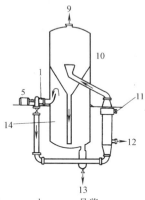

1 mother liquor　母液
2 cooling medium outlet　冷却介质出口
3 cooler　冷却器
4 cooling medium inlet　冷却介质进口
5 circulation pump　循环泵
6 crystal suspension　晶体悬浮液
7 fine salt　细盐
8 crystal paste　晶浆
9 vapor　蒸气
10 evaporator　蒸发器
11 heating steam　加热蒸汽
12 condensate　冷凝液
13 salt outlet　盐出口
14 salt suspension　盐悬浮液

7.2.14 Dense-bed center-fed column crystallizer 紧密床层中央加料式塔结晶器

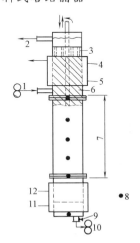

1 feed　进料
2 overflow　溢流
3 porous piston　多孔活塞
4 freezer　冻凝器
5 coolant　冷却剂
6 scraper　刮板
7 purification zone　提纯区
8 thermistor probes　热敏电阻探针
9 sample　样品
10 product　产品
11 hot water　热水
12 heater　加热器

217

8 Drying Equipment 干燥设备

8.1 Classification of dryers 干燥器分类

8.1.1 Classification of dryers by several criteria 依据不同的角度对干燥器分类

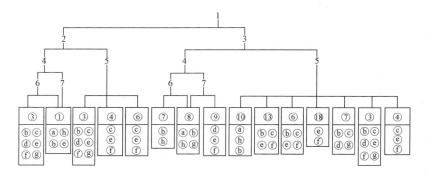

(a)

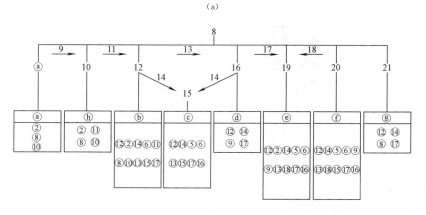

(b)

8.1 Classification of dryers 干燥器分类

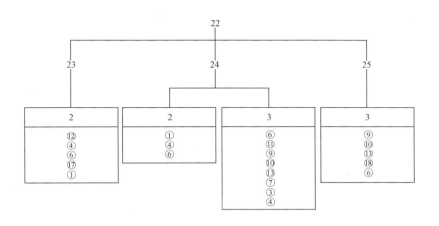

(c)

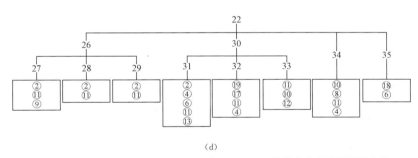

(d)

(a) classification of dryers based on method of operation 按操作方法的干燥器分类
(b) classification of dryers based on physical form of feed 按物料进入干燥器的物态对干燥器分类
(c) classification of dryers by scale of production 按生产规模的干燥器分类
(d) classification of dryers by suitability for special features 按附加特征的适应性进行干燥器分类

ⓐ liquid 液体
ⓑ paste 膏糊状
ⓒ preform 预成型物
ⓓ hard 硬（颗粒）

ⓔ granular 粒状
ⓕ fibrous 纤维状
ⓖ sheet 片状
ⓗ slurry 浆状

① agitated 搅拌式   ② agitated batch 间歇式带搅拌

219

③ tray 盘式
④ through-circulation 穿流循环
⑤ batch through-circulation 间歇、穿流循环
⑥ fluid bed 流化床
⑦ band 带式
⑧ drum 滚筒
⑨ indirect rotary 间接加热回转
⑩ spray 喷雾
⑪ vacuum band 真空带式
⑫ vacuum tray 真空盘式
⑬ pneumatic 气流
⑭ convection tray 对流盘式
⑮ convection band 对流带式
⑯ cont. through-circulation 连续穿流-循环式
⑰ cont. tray 连续盘式
⑱ direct rotary 直接加热回转
⑲ cont. band 连续带式

1 dryer 干燥器
2 batch 分批
3 continuous 连续
4 conduction 热传导
5 convection 对流
6 vacuum 真空
7 atmos.; atmosphere 常压
8 wet feed 湿物料
9 evaporate 蒸发
10 pumpable slurry or suspension 可泵送的浆状物或悬浮液
11 evaporate or back-mix 蒸发或返回混合物
12 soft paste or sludge 软的膏状物或沉淀物
13 press 压制品
14 preform 预成型物
15 preformed paste 预成型的膏状物
16 hard paste or matrix 硬的膏状物或型块
17 grind 研磨物
18 grind or preform 研磨或预成型物
19 free-flowing granular or crystalline solid 自由流动的颗粒或结晶固体
20 fibrous solid 纤维状固体
21 sheet 片状
22 (dry) process （干燥）过程
23 small scale, 20~50 kg/h 小批量, 20~50kg/h
24 medium scale, 50 to 1000kg/h 中等规模, 50~1000kg/h
25 large scale, tons/h 工业规模, t/h
26 hazard 危险品
27 dust 粉尘
28 toxic 毒品
29 flame 火焰
30 sensitive product 敏感性物料
31 temperature 温度
32 mechanical 机械
33 oxidation 氧化
34 special form of product 产品的特殊形状
35 low capital cost per unit of output 单位产量的低投资成本

8.2 Tray and compartment dryer 厢式干燥器

8.2.1 Tray dryer 厢式干燥器

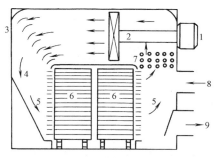

1 fan motor 风机电机
2 adjustable-pitch fan 1 to 15 hp 可调角度风机（1～15 马力）
3 turning vanes 导向叶片
4 plenum chamber 强制通风室
5 adjustable air-blast nozzles 可调式空气送风管
6 trucks and trays 车和料盘
7 fin heaters 翅片加热器
8 air-inlet duct 空气进口管
9 air-exhaust duct with damper 带调节板的空气出口管

8.2.2 Vacuum-shelf dryer 真空盘架式干燥器

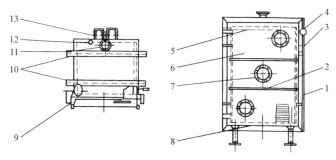

1 shell 壳体
2 trays 料盘
3 vacuum break valve 消除真空用阀
4 dial-type vacuum ga(u)ge 指针式真空表
5 top heating shelf 顶部加热架
6 steel door 钢门
7 sight glasses 视镜
8 chamber drain 室排水口
9 neoprene gasket 氯丁橡胶垫圈
10 external reinforcing ribs 外部加强圈
11 vacuum connection 真空接头
12 pressure-relief valve 减压阀
13 exterior manifolds steam or hot water 外部蒸汽或热水集管

221

8.3 Tunnel dryer and belt dryer 隧道式和带式干燥器

8.3.1 Three types of tunnel dryers 隧道式干燥器的三种类型

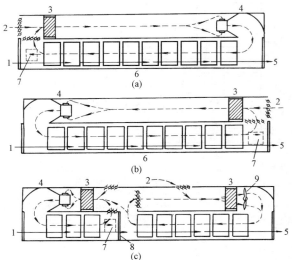

（a） countercurrent tunnel dryer 逆流隧道式干燥器
（b） parallel current tunnel dryer 并流隧道式干燥器
（c） center exhaust tunnel dryer 中间排气式隧道式干燥器
1 wet material in 湿物料入口
2 fresh-air inlet 新鲜空气入口
3 heater 加热器
4 blower 风机
5 dry material out 干物料出口
6 trucks 小车
7 exhaust-air stack 排气烟道
8 movable partition 可动隔板
9 fan 风机

8.3.2 Special conveyor dryer 特殊输送式干燥器

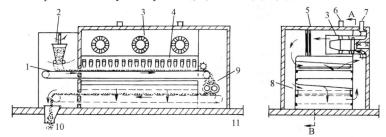

（with air jets impinging on surface of bed on first pass, dried material is crushed and passed through dryer, with air going through the now permeable bed 使空气喷射冲击到第一通道上冻层表面，被干燥的物料粉碎后再次通过干燥器与通过可渗透床层的空气接触）

1 conveyor belt 输送带
2 feed roller press 进料滚压机
3 fan 风机
4 hot-air jets 热空气喷射器
5 steam-heating unit 蒸汽加热装置
6 air-exhaust 排气口
7 air inlet 空气进口
8 air-distributor grating 空气分布器栅板
9 crusher 粉碎机
10 product 产品
11 section A-B A-B 截面

8.3.3 Section view of a continuous through-circulation conveyor dryer 连续穿流-循环输送带式干燥器截面图

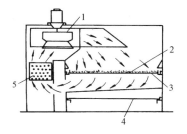

1 air circulating fan 空气循环风机
2 material 物料
3 conveyor 输送带
4 conveyor return 返回输送带
5 heater 加热器

8.3.4 Band type drying machine 带式干燥机

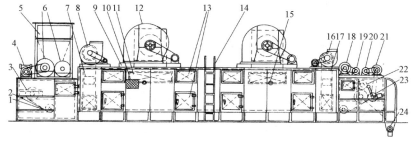

1 chain 传动链
2 perforated plate or wire net 多孔板或金属丝网
3 take-up (tray) 接料盘
4 reducer for finned drum feeder 翅片转鼓给料机用减速机
5 hopper 料斗
6 finned drum 翅片转鼓
7 exhaust damper 排气风门
8 exhaust blower 排风机
9 blower base 鼓风机底座
10 fresh air inlet 新鲜空气入口
11 heater 加热器
12 blower 鼓风机
13 testing door 检修门
14 ladder 梯子
15 thermostat 恒温器
16 exhaust fan 排风机
17 exhaust duct 排风管道
18 reducer for conveyor 输送机用减速机
19 motor for conveyor 输送机用电机
20 motor for kicker and doffer 抖动器和滚筒用电动机
21 reducer for doffer 滚筒减速机
22 kicker 抖动器
23 doffer 滚筒
24 screw conveyer 螺旋输送机

8.4 Pneumatic dryer 气流式干燥器

8.4.1 Single-stage pneumatic-conveyor dryer 单级气流输送干燥器

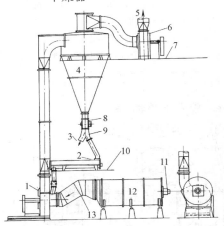

8.4.2 Strong-Scott flash dryer Strong-Scott 闪急干燥器

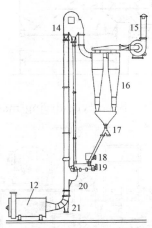

8.4.3 Air-lift pneumatic-conveyor dryer 空气提升气流输送干燥器

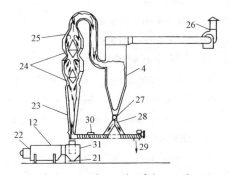

(includes partial recycle of dry product, and expanding tube and cone sections to provide longer hold-up for coarse particles 包含干燥产品的部分循环及对粗颗粒提供较长停留时间的扩大管和锥形截面)

8.4.4 Two-stage air stream and cage mill pneumatic-conveyor dryer 两级气流和笼式磨机气流输送干燥器

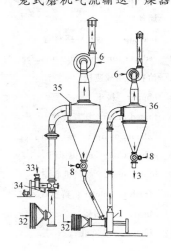

8.4 Pneumatic dryer 气流式干燥器

1. cage mill 笼式磨碎机
2. wet feed mixer 湿进料混合器
3. finished product 成品
4. cyclone 旋风分离器
5. vent to atmosphere or secondary collector 排至大气或至二级集尘器
6. vent fan 排风机
7. cyclone & vent fan support level 旋风分离器和排风机支承平面
8. airlock 空气闭锁器
9. dry divider 干料分流管
10. mixer support level 混合器支承面
11. oil or gas burner 燃油或燃气器
12. air heater 空气加热器
13. hot gas duct 热气体管道
14. classifier 分级器，料粒分选器
15. system fan 系统风机
16. multiple cyclones 并联旋风分离器
17. splitter valve 多通阀，分流阀
18. feed-all 总进料
19. turbulizer-backmixer 涡流返混混合器
20. dispersion sling 抛扬分散器
21. tramp discharge 杂质排出口
22. burner 燃烧器
23. high-velocity section 高速断面
24. expanded sections 扩大断面
25. airlift turbo drying action 空气提升涡流干燥作用区
26. fan 风机
27. rotary airlock 旋转式空气闭锁器
28. recycle control baffle 控制循环导流板
29. product 产品
30. material in 物料进口
31. air-mix chamber 空气混合室
32. steam air heater 蒸汽式空气加热器
33. wet feed 湿进料
34. wet feeder 湿加料器
35. wet-stage cyclone 湿级旋风分离器
36. dry-stage cyclone 干级旋风分离器

8.4.5 Examples of pneumatic conveying dryers 气流输送干燥器的实例

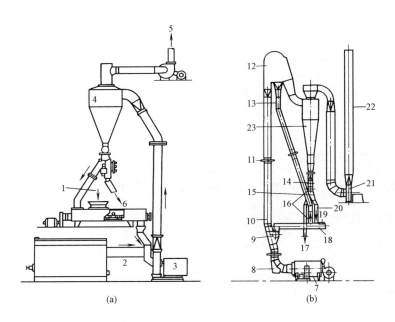

(a)

(b)

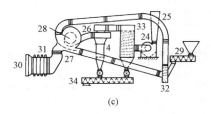

(c)

8.4 Pneumatic dryer 气流式干燥器

(a) Raymond flash dryer (with a hammer mill for disintegrating the feed and with partial recycle of product) 雷蒙闪急干燥器（用锤式粉碎机破碎物料，产品作部分循环）
(b) Buttner-Rosin pneumatic dryer Buttner-Rosin 气流干燥器
(c) Berks ring dryer (the material circulates through the ring-shaped path, product is withdrawn through the cyclone and bag filter) Berks 环形干燥器（物料在环形风管内循环，产品通过旋风分离器及袋滤器排出）

1	wet feed 湿物料加入		18	double paddle mixer 双桨混合器
2	hot gases 热气体		19	wet material 湿物料
3	hammer mill 锤式粉碎机		20	recirculated material 循环物料
4	cyclone 旋风分离器		21	exhaust fan 排风机
5	outlet 气体出口		22	stack 烟囱
6	dry product 干（成）品		23	twin primary cyclone 并联两个一级旋风分离器
7	combustion chamber 燃烧室			
8	bottom bend 底肘管		24	fan 引风机
9	distributor 分配器		25	ring duct 环形风管
10	drier duct 干燥器风管		26	manifold 歧管
11	expansion bellows 膨胀波纹管，膨胀节		27	injector 喷射器
			28	air outlet 空气出口
12	classifier 粒料分级器		29	feeder 加料器
13	recirculation duct 循环管		30	filter 过滤器
14	collecting screw 集料螺旋		31	heater 加热器
15	change over flap 换向挡板		32	disintegrator 破碎机
16	rotary valves 旋转阀		33	bag filter 袋滤器
17	dry material 干物料		34	discharge 卸料

8.4.6　P-type ring dryer　P型环形干燥器

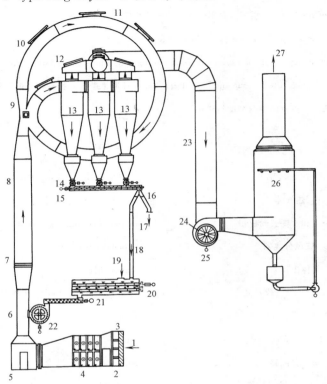

1	fresh air in　新鲜空气加入		15	discharge screw　输出螺旋
2	louvers　百叶窗		16	splitter　分流器
3	air filter　空气过滤器		17	dry product　干成品
4	air heater　空气加热器		18	recycle　循环
5	hot air box　热空气室		19	wet feed　湿物料进入
6	Venturi　文丘里管		20	double paddle mixer　双桨混合机
7	expansion joint　膨胀节		21	feed screw　进料螺旋
8	drying column　干燥塔		22	disperser　分散设备
9	manifold classifier　歧管粒料分选器		23	exhaust duct　排气管道
10	ring duct　环形管道		24	radial inlet damper　径向进口风门
11	explosion panel　防爆膜（板）		25	main fan　主风机
12	exhaust header　排气集管		26	scrubber　涤气器
13	cyclone　旋风分离器		27	to atmosphere　通大气
14	rotary valve　旋转阀			

8.5 Fluidized bed dryer 流化床干燥器

8.5.1 Multistage fluized-bed dryer 多层流化床干燥器

8.5.2 Horizontal multicompartment fluidized-bed dryer (cooler) 卧式多室流化床干燥器（冷却器）

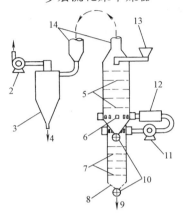
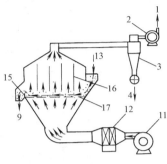

1　exhaust air　排气
2　exhaust blower　排风机
3　cyclone　旋风分离器
4　fine product　粉料成品
5　perforated plate for drying　干燥段多孔板
6　hot air blast hole　热风通入口
7　perforated plate for cooling　冷却段多孔板
8　air inlet port　进风口
9　coarse product　粗粒成品
10　rotary valve　旋转阀
11　blower　鼓风机
12　air heater　空气加热器
13　feed inlet　加料口
14　pneumatic conveying drying pipe　气流输送干燥管
15　overflow plate　溢流板
16　distance plate　隔板
17　perforated plate　多孔板

8.5.3 The flow sheet of the experimental fluidized-bed dryer with inert particles 惰性粒子流化床干燥器实验流程

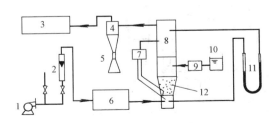

1　fan　风机
2　flowrator　转子流量计
3　bag collector　布袋捕集器
4　cyclone　旋风分离器
5　conical flask　锥形瓶
6　electric heater　电加热器
7　temperature indicator　温度显示器
8　fluidized-bed dryer　流化床干燥器
9　peristaltic pump　蠕动泵
10　feed tank　物料槽
11　U-gauge pressurer　U形管压差计
12　inert particles　惰性粒子

8.5.4 Fluidized bed dryers 流化床干燥器

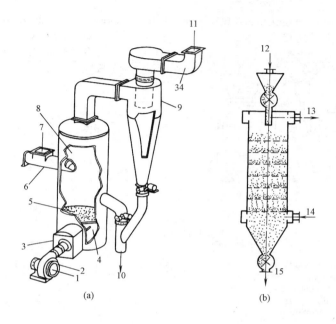

(a) basic equipment arrangement 主要设备与流程
(b) multiple bed dryer 多层流化床干燥器

8.5.5 Fluidized bed dryer 流化床干燥器

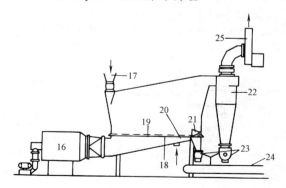

8.5.6 Fluidized-bed coal dryer 流化床煤干燥器

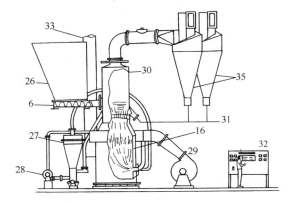

1	air inlet 空气进口		19	fluidized product 流态化物料
2	fluidizing blower 流化用的鼓风机		20	cooling zone 冷却区段
3	heat source 热源		21	weir plate 溢流堰板
4	plenum 充气室		22	dust rotor 粉末转筒,旋风分离器
5	distributor plate 分布板		23	discharge valve 排出阀
6	feeder 加料器		24	belt 输送皮带
7	wet feed 湿进料		25	fan 引风机
8	fluidizing chamber 流化室		26	wet-coal bin 湿煤仓
9	dust collector 集粉器		27	pulverizer 粉碎机
10	dry product discharge 干品出口		28	fan 通风机
11	clean gas discharge 净化气体出口		29	blower 鼓风机
12	wet material 湿物料		30	drying chamber 干燥室
13	to cyclone 去旋风分离器		31	dry product 干燥产品
14	gas 气体		32	control panel 控制盘
15	dry material 干料		33	start-up stack 开工烟囱
16	air heater 空气加热器		34	stack 烟囱
17	product feed 物料进口		35	cyclone 旋风分离器
18	gas distributor plate 气体分布板			

8.5.7 Agitated flash dryer 带搅拌的闪急干燥器

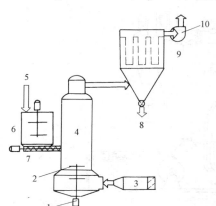

1 drive motor 驱动电机
2 agitator 搅拌装置
3 heater 加热器
4 drying chamber 干燥室
5 feed inlet 加料
6 feed tank 加料罐
7 feed dosing 定量加料器
8 product outlet 产品出口
9 bag filter 袋滤器
10 exhaust fan 排风机

8.6 Spray dryer 喷雾干燥器

8.6.1 Spray drying tower 喷雾干燥塔

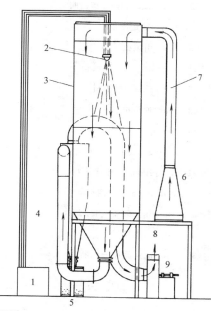

1 feed pump 进料泵
2 spraying nozzle 雾化喷嘴
3 drying column 干燥塔
4 cyclone 旋风分离器
5 product collector 成品收集器
6 furnace 热风炉
7 hot air duct 热空气导管
8 exhaust to atmosphere 废气通大气
9 fan 风机

8.6.2　Spray dryer　喷雾干燥器

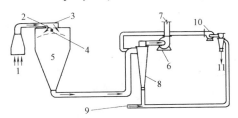

1　air heater　空气加热器
2　feed　进料
3　hot air duct　热空气导管
4　atomizer　雾化器
5　drying chamber　干燥室
6　exhaust fan　排风机
7　discharge air　排气
8　cyclone　旋风分离器
9　ambient air　环境空气
10　conveying fan　输送风机
11　product　产品

8.6.3　Gas-solids contacting methods in spray dryer 喷雾干燥塔内气固接触方式

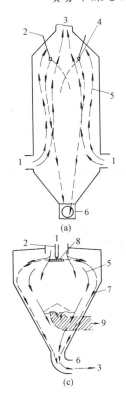
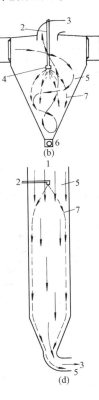

(a) countercurrent　逆流式

(b) mixed-flow　混合流式

(c) cocurrent-disk atomization
　　并流转盘雾化式

(d) cocurrent-nozzle atomization
　　并流喷嘴雾化式

1　hot air　热空气
2　feed　进料
3　exhaust　废气
4　nozzle　喷嘴
5　air flow　空气流
6　product　产品
7　spray　喷散物
8　disk atomizer　转盘雾化器
9　air outlet when using drying chamber for initial separation
　　用干燥室作为初级分离时的空气出口

8.7 Rotating drum dryer 滚筒干燥器

8.7.1 Rotating drum dryer 滚筒干燥器

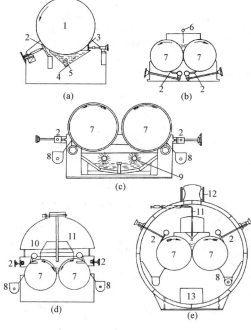

(a) single drum dryer with dip-feed 浸没布料式单滚筒干燥器
(b) top drum dryer with top-feed 顶部加料式双滚筒干燥器
(c) double drum dryer with splash-feed 飞溅布料式双滚筒干燥器
(d) double drum dryer with spray-feed 喷洒加料式双滚筒干燥器
(e) vacuum double drum dryer 双滚筒真空干燥器
1　revolving drum　转鼓，滚筒
2　knife　刮刀
3　spreader　布料器，膜厚控制器
4　drain　排放口
5　agitator and feed pan　搅拌器与料盘
6　feed pipe　进料管
7　drum　滚筒
8　conveyor　输送器
9　splashing device　溅料轮
10　vapor hood　蒸气罩
11　pendulum feed pipe　摆式进料管
12　vapor outlet　蒸气出口
13　man hole　人孔

8.7.2 The use of air impingement in drying as a secondary heat source on a double-drum dryer 干燥过程中空气冲击双鼓干燥器用作第二热源

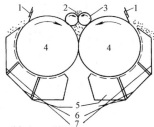
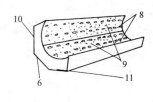

1　scraping blades　刮刀
2　feed puddle　进料槽
3　applicating rolls　从动滚
4　steam-heated drum　蒸汽加热鼓
5　high temperature air-cap　高温空气罩
6　medium temperature　中温
7　low temperature　低温
8　exhaust-air tubes　废空气管
9　high-velocity air　高速空气
10　low temperature zone　低温区
11　high temperature　高温

8.8 Rotary dryer 回转干燥器

8.8.1 Component arrangements of a countercurrent direct-heat rotary dryer 逆流直接加热式回转干燥器的部件配置图

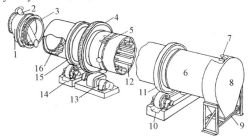

8.8.2 Alternative direct-heat rotary-dryer flight arrangement 直接加热式回转干燥器的几种抄板安置形式

8.8.3 Elevation of a 60-in.-diameter by 30-ft.-long direct-heat cocurrent rotary dryer 直径60英寸长30英尺的直接加热式并流回转干燥器

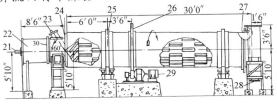

1　inlet head（counterflow only）　进料头（仅限于逆流）
2，23　feed chute　进料斜槽
3　friction seal　摩擦密封
4　girt gear　大齿轮
5　knocker　锤击器
6　shell　壳体
7　breeching seals　筒尾密封
8　breeching　烟道
9　discharge　出料
10　trunnion roll assembly　枢轴装配体
11　No. 2 riding ring　2号滚圈
12　lifting flights　提升抄板
13　drive assembly　驱动器装配体
14　trunnion and thrust roll assembly　枢轴与止推辊装配体
15　No. 1 riding ring　1号滚圈
16　spiral flights　螺旋抄板
17　radial flights　直立式（径向）抄板
18　45-deg. lip flights　45°（唇形）抄板
19　90-deg. lip flights　90°（唇形）抄板
20　shell with diaphragm section　内装隔板的筒体截面
21　burner　燃烧器
22　combustion furnace　燃烧炉
24　inlet seal　进口密封
25　riding ring　滚圈
26　gear mounting　齿轮架
27　angle seal　角封
28　discharge outlet　卸料口
29　drive　驱动器
30　60-deg.　60°

8.8.4 Rotary-dryer components 回转式干燥器零件图

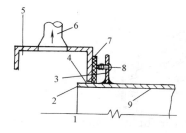

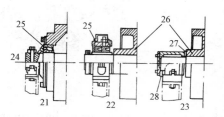

(b) alternative trunnion roll bearings
各种滚动轴承

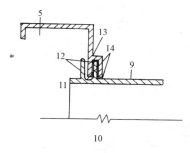

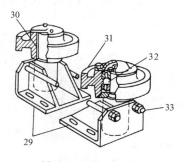

(c) thrust roll assembly
止推轮装配图

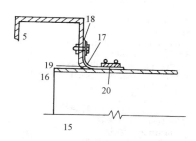

(a) alternative rotary gas seals
各种旋转气体密封

8.8 Rotary dryer 回转干燥器

1 lubricated or dry friction 润滑或干式摩擦
2 end of shell 壳体终端
3 ring welded to rotating shell 焊在回转壳体上的环
4 clearance 间隙
5 breeching or end box 筒尾或箱体端部
6 gas outlet or inlet 气体出口或进口
7 friction material (e.g., brake lining) 摩擦材料（如刹车片）
8 spring to maintain tension on rubbing surfaces 保持摩擦力的螺钉
9 dryer shell 干燥器壳体
10 labyrinth type seal where rubbing contact is to be avoided 避免接触摩擦的迷宫式密封
11 end of dryer 干燥器终端
12 rotating elements 回转件
13 stationary element 固定件
14 points of clearance 间隙点
15 flexible cloth rubbing seal 柔性织物摩擦密封
16 rotating shell 回转壳体
17 cloth or similar flexible material 织物或类似的柔性材料
18 cloth fastened to breeching 固定于烟道上的织物
19 area of friction 摩擦面
20 metal band pulled tight to effect friction seal between cloth and rotating shell 压紧金属带使织物与旋转壳体之间实现摩擦密封
21 dead shaft 静轴
22 antifriction pillow block 耐磨轴台
23 bushed-angle pillow block 角加衬的耐磨轴台
24 shaft 轴
25 antifriction bearing 耐磨轴承
26 press fit 压入配合
27 thrust washer 止推垫圈
28 bushing 轴瓦
29 thrust roll shaft and bracket-welded assembly 止推辊轴与支座焊接装配体
30 thrust roll bushing 止推辊轴瓦
31 thrust roll 止推辊
32 thrust roll bearing 止推滚动轴承
33 tie rods 系杆

8.8.5 Rotary kilns 回转窑

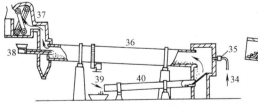 

(a) single diameter 单直径式   (b) two diameter 双直径式

34 fuel 燃料
35 burner 燃烧器
36 kiln 窑
37 waste heat boiler 废热锅炉
38 feeder 加料器
39 air 空气
40 cooler 冷却器

8.8.6 Steam-tube rotary dryer 蒸汽管式回转干燥器

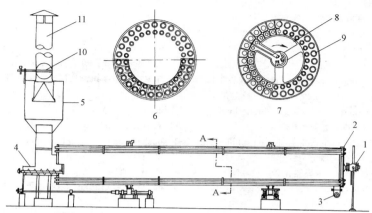

1 steam neck 蒸汽缩颈管
2 steam manifold 蒸汽分配盘
3 dried material discharge conveyor 干燥产品卸料输送器
4 wet material feed in here 湿物料入口
5 dust drum 粉尘筒
6 section at A-A A-A 剖面
7 section through steam manifold 蒸汽集管箱剖面
8 rotation 转动方向
9 steam-heated tube 蒸汽加热管
10 adjustable damper 可调挡板
11 natural-draft stack 自然抽风筒

8.8.7 Rotary steam joint for a standard steam-tube dryer 标准蒸汽管干燥器的旋转蒸汽接头

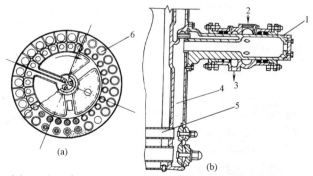

(a) section of cast steam manifold 铸造蒸汽集管箱剖面
(b) section of manifold and steam joint 集管箱与蒸汽接头剖面

1 rotary steam joint 旋转蒸汽接头
2 steam in 蒸汽进口
3 condensate out 冷凝液出口
4 steam manifold 蒸汽总管；蒸汽分配盘
5,6 steam-heated tube 蒸汽加热管

8.8.8 Gas-fired indirect-heat rotary calciner 间接加热式回转煅烧炉

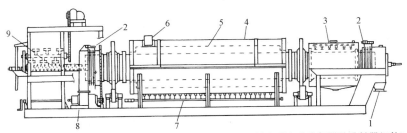

(with a water-spray extended cooler and feeder assembly 具有喷水式冷却器及进料器组件)

1　discharge chute　卸料槽
2　bellows seal　波纹管密封
3　water-spray cooler　洒水式冷却器
4　furnace　炉体
5　furnace lining　炉衬
6　stack connection　烟道连接
7　burners　燃烧器
8　motor drive　电机
9　feeder assembly　进料器组件

8.8.9 Through-flow rotary dryer 穿流回转干燥器

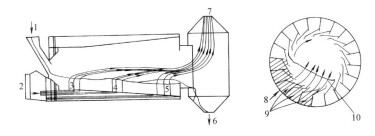

1　wet-feed inlet　湿料进口
2　hot-air inlet　热空气进口
3　wet material　湿物料
4　moist material　（半）湿物料
5　dry material　干物料
6　product outlet　产品出口
7　exhaust-gas outlet　废气出口
8　air flow through louvers and material 空气流动穿过百叶窗和物料
9　hot-air chambers　热空气室
10　material　物料

8.8 Rotary dryer 回转干燥器

8.8.10 A typical vacuum rotary dryer 典型的真空回转干燥器

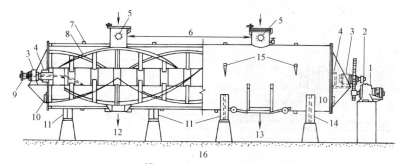

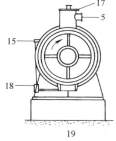

1　motor reducer drive　电机减速传动装置
2　rotary joint for steam inlet　蒸汽进口旋转接头
3　outboard bearing　外置轴承
4　stuffing box　填料函
5　vapor outlet　蒸气出口
6　vapor and charging domes　蒸气及装料筒
7　air vent　空气孔
8　double spiral agitator　双螺旋搅拌器
9　rotary joint for condensate drain from center heating tube　中心加热管冷凝液排放用的旋转接头
10　inspection and clean out door　检查与清扫门
11　supporting saddle on expansion rollers　膨胀托辊上的鞍式支座
12　discharge door　卸料门
13　discharge door and counter weighted operating levers　卸料门及平衡锤操作杆
14　fixed supporting saddle　固定鞍式支座
15　steam inlets to jacket　夹套蒸汽进口
16　elevation and partial cross section　正视与局部剖面图
17　cover　顶盖
18　door operating device　门的（自动）启闭装置
19　cross sectional view　剖面图
20　drive end view　驱动端视图

8.8.11 Rotating (double-cone) vacuum dryer 双锥型回转式真空干燥器

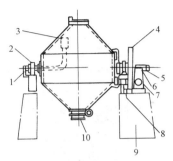
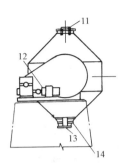

1 vacuum connection 真空接口
2 dusttight sealed roller bearings 防尘密封滚动轴承
3 dust filter 粉尘过滤器
4 oil and dusttight chain casing 链条防油防尘罩
5 steam or hotwater inlet 蒸汽或热水进口
6 condensate or hot-water return 冷凝液或热水返回口
7 variable-speed and brake motor 变速电机
8 jog switch 啮合开关
9 concrete or structural foundation 混凝土基础或结构基础
10 discharge opening 卸料口
11 charge opening 装料口
12 motor 电机
13 dust-bag sleeve 集尘袋套管
14 self-wiping vacuumtight discharge valve (worm-gear-operated) 自清洗式真空密封卸料阀（蜗轮操作）

8.8.12 Rotating tube bundle dryer 旋转管束干燥机

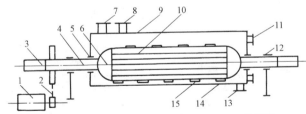

1 reducer with motor 电机减速机
2 chain-drive unit 链传动装置
3 rotary joint 旋转接头
4 hollow shaft 空心轴
5 seal 轴封
6 end cap 端盖
7 wet feed 湿料进口
8 vapor outlet 蒸气出口
9 housing cap 机箱盖
10 tube bundle 管束
11 air inlet 空气进口
12 bearing frame 轴承架
13 product outlet 产品出口
14 housing 机箱
15 scraper blade 刮板

8.9 Vertical type dryer 立式干燥器

8.9.1 Turbo-tray dryer 蜗轮转盘式干燥器

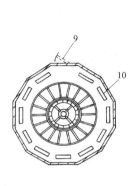
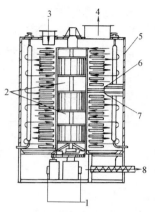

1　variable-speed driver　变速驱动器
2　turbo-fan　蜗轮风机
3　wet feed　湿进料
4　exhaust air　废气出口
5　insulated housing　绝热外壳
6　wiper　刮板
7　leveler　平料器
8　dry discharge　干料出口
9　access door　检修门
10　heaters　加热器

8.9.2 Turbo-tray dryer in closed circuit for continuous drying with solvent recovery 可回收溶剂的连续密闭循环蜗轮转盘式干燥器

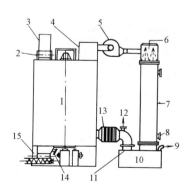

1　turbo dryer　蜗轮干燥器
2　feeder　加料器
3　feed hopper　加料斗
4　breeching　器尾
5　recirculating fan　再循环风机
6　spray head　喷头
7　condenser　冷凝器
8　cold water　冷水
9　recovered solvent　回收溶剂
10　receiver　受液槽
11　mist eliminator　除雾器
12　purge　清洗气
13　reheater　再加热器
14　inert-gas makeup　惰性气体补充口
15　air lock　空气闭锁器

8.9 Vertical type dryer 立式干燥器

8.9.3 Continuous vertical type dryer 立式连续干燥器

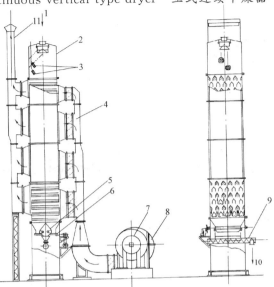

1　feed inlet　进料口
2　shell　机壳
3　level indicator　料面指示器
4　hot air duct　热空气导管
5　cut damper　节流挡板
6　shaking feeder　振动送料机
7　combustion chamber and hot air producer　燃烧炉和热风发生器
8　blower　鼓风机
9　screw conveyer　螺旋输送器
10　product outlet　成品出口
11　exhaust air　排气

8.9.4 Indirect-heated continuous plate dryer 间接加热连续操作圆盘干燥器

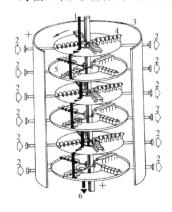

1　material　物料
2　heating or cooling medium　加热或冷却介质
3　housing　机壳
4　conveying system　输送系统
5　plate　料盘
6　product　成品

8.9 Vertical type dryer 立式干燥器

8.9.5 Bottom drive conical mixer dryer 底部驱动锥形混合干燥器

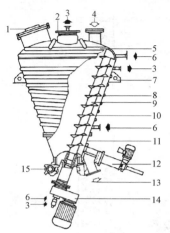

1 manhole flange 人孔法兰
2 vapor flange 蒸气出口法兰
3 heating medium outlet 加热介质出口
4 product inlet flange 物料进口法兰
5 cover heating 顶盖加热管
6 heating medium inlet 加热介质进口
7 support flange 支座法兰
8 vessel heating 容器加热管
9 mixer screw 混合器螺杆
10 vessel upper section 容器上段
11 vessel lower section 容器下段
12 product discharge valve 产品出口阀
13 product discharge 产品排出口
14 mixer screw drive unit 混合器螺旋（自转）驱动装置
15 reversing rotational drive 回转（公转）驱动装置

8.9.6 Vibrating tray dryer 振动盘式干燥器

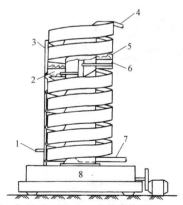

8.9.7 Spiral dryer 螺旋干燥器

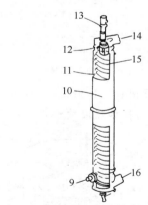

1 divided solids in 分散固体进口
2 conveying surface 输送表面
3 water drained by gravity 借助重力排水
4 solids discharge 固体排出
5 bed of solids 固体床层
6 jacket 夹套
7 hot-water supply（pressurized） 供入热水（加压下）
8 shaker 振动器

9 wet product inlet 湿物料进口
10 outer shell steam-jacket 带有蒸汽夹套的外壳
11 air deflectors 导流板
12 moisture-carrying exhaust air 携带湿分的废气
13 rotor drive 转子驱动装置
14 discharge for dried product 已干产品排出口
15 steam-heated rotor 被蒸汽加热的转子
16 carrier gas inlet 载气进口

9 Filtration Equipment, Centrifuges and Expression Equipment 过滤设备、离心机与压榨设备

9.1 Plate-and-frame type filter press 板框压滤机

9.1.1 Filling and washing flow patterns in a filter press 压滤机充料与洗涤的流动模式

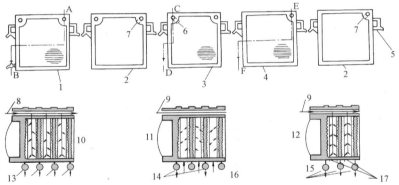

1 non-wash plate 非洗涤板
2 frame 滤框
3 wash plate 洗涤板
4 plate 滤板
5 side lug 支耳
6 wash inlet 洗水入口
7 feed inlet 料入口
8 path of mixture to filter 混合物供向过滤机的路径
9 path of wash water 洗水路径
10 section through A-B showing disposition of solids A-B剖面表示固体沉积
11 section through C-D showing thorough washing C-D剖面表示穿过式洗涤
12 section through E-F showing simple washing E-F剖面表示简单洗涤
13 filtrate exit through cocks 滤液从旋塞放出
14 wash water exit through every other cock 洗水每隔一个旋塞放出
15 wash water exit through all cocks 洗水从所有旋塞放出
16 frames filled with solids 框已全部充满固体
17 frames partly filled with solids 框已部分充满固体

245

9.1.2 Plate and frame filter press 板框压滤机

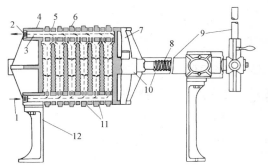

1 slurry inlet 滤浆进口
2 filtrate outlet 滤液出口
3 channel (path) 通道
4 fixed head 固定端板
5 plate 滤板
6 frame 滤框
7 movable head 活动端板
8 hand screw 手动压紧丝杠
9 closing device （自动）压紧装置
10 side rail 侧轨，横杆
11 filter cloth 滤布
12 stand leg 机架

9.1.3 Section detail of calked-gasketed-recessed filter plate 采用堵缝垫圈的凹板压滤机滤板明细截面图

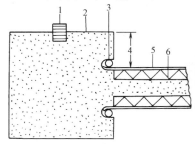

1 sealing gasket 密封垫片
2 plate joint 滤板接缝
3 calking strip 滤布嵌条
4 cake recess 滤饼槽，存滤饼的凹处
5 filter cloth 滤布
6 drainage surface of plate 滤板的排水表面
7 unfiltered liquid 未过滤的液体
8 binding 包圈
9 filter cake 滤饼
10 precoat 预覆盖层
11 chamber screen 滤室网

9.1.4 Section of precoated wire filter leaf 预覆盖的丝网滤叶截面图

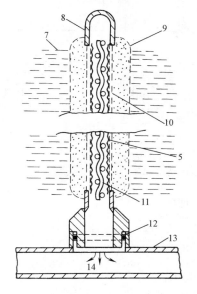

12 O-ring O形环
13 manifold 总管
14 filtered liquid 滤出液

9.2 Pressure leaf filter 加压叶滤机

9.2.1 Pressure leaf filter 加压叶滤机

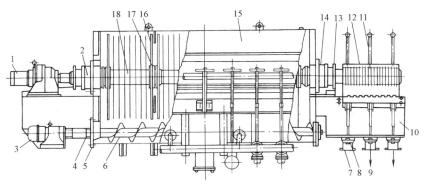

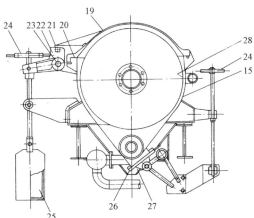

1 cycloidal planetary gear speed reducer 行星摆线针轮减速机
2 main shaft bearing 主轴轴承
3 gear motor for screw conveyer 螺旋输送器用电机减速机
4 seal cover 密封压盖
5 bearing for screw conveyer 螺旋输送器轴承
6 screw conveyer 螺旋输送器
7 valve seat 阀座
8 valve 阀
9 filtrate outlet 滤液出口
10 filtrate receiver 滤液槽
11 changer lever 切换杆
12 distributor 分配器
13 packing gland 填料压盖
14 stuffing box 填料箱
15 vessel 容器
16 filter leaf 滤叶
17 bush 轴衬
18 hollow shaft 空心轴
19 cover 顶盖
20 packing for cover 顶盖垫片
21 block 顶盖座
22 hinged lever 铰链杆
23 shaft for hinged lever 铰链轴
24 valve handle 阀门手轮
25 balance weight 平衡锤
26 packing for discharge cover 卸渣管法兰垫片
27 discharge cover 卸渣管法兰盖
28 washing nozzle 洗涤液接管

247

9.2.2 Vertical pressure leaf filter 立式叶滤机

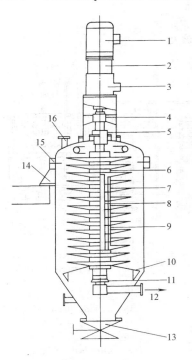

1 motor 电动机
2 gear reducer 减速机
3 coupling 联轴器
4 upper bearing 上轴承
5 stuffing box 填料函
6 filter leaf group 过滤叶片组件
7 filter leaf 过滤叶片
8 hollow shaft 空心轴
9 spacer ring 定距环
10 rake 耙子
11 lower bearing 下轴承
12 filtrate outlet 滤液出口
13 residual cake valve 排渣阀
14 support 支座
15 flange 法兰
16 feed slurry 滤浆进口

9.2.3 Schematic of a centrifugal-discharge filter 立式离心卸料加压叶滤机

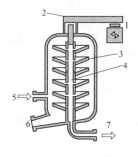 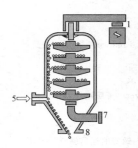

1 motor 电动机
2 driving belt 传动带
3 filter leaf 滤叶
4 hollow shaft 空心轴
5 slurry feed 加料口
6 slurried cake discharge 稀渣排放
7 filtrate 滤液
8 dry cake discharge 干渣排放口

9.2 Pressure leaf filter 加压叶滤机

9.2.4 Horizontal pressure leaf filter 卧式加压叶片过滤机

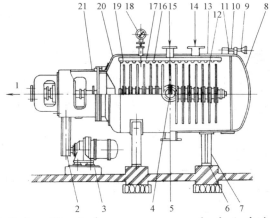

9.2.5 Vertical pressure leaf filter 立式叶滤机

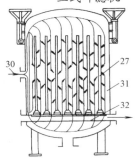

1 filtrate outlet 滤液出口
2 chain cover 传动链罩子
3 leaf driving motor 过滤叶片驱动电机
4 feed slurry inlet 滤浆入口
5 drain 排放口
6 foundation bolt 基础螺栓
7 leg 支腿
8 bottom 底封头
9 closing handle 紧固手轮
10 closing bolt 紧固螺栓
11 gasket 垫片
12 leaf 过滤叶片
13 shell 筒体
14 overflow 溢流口
15 washing water inlet 洗涤水入口
16 shower nozzle 喷头
17 shower header 喷管
18 pressure gauge 压力表
19 manifold 滤液总管
20 cover 顶盖
21 leaf driving unit 过滤叶片驱动装置
22 arm 悬臂
23 temper handle 缓启闭手轮
24 shower 喷淋管
25 clamping bolt 紧固螺栓
26 leaf support 过滤叶片支架
27 filter leaf 过滤叶片
28 leaf nozzle 过滤叶片接口
29 bottom head 底封头
30 slurry inlet 料浆进口
31 upright cylindrical pressure tank 立式圆筒形压力罐
32 filtrate manifold 滤液集管

9.2.6 Vertical type pressure leaf filter 立式加压叶片过滤机

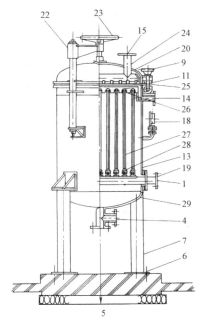

249

9.2.7 Pressure leaf filter 加压叶片过滤机

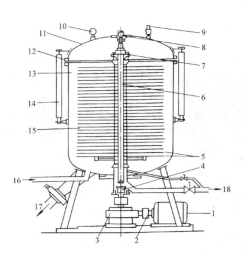

1　motor　电动机
2　coupling　联轴器
3　worm speed reducer　蜗轮减速器
4　lower bearing　下轴承
5　filter leaf for residual　残液过滤叶片
6　hollow shaft　空心轴
7　leaf fastener　叶片紧固器
8　upper bearing　上轴承
9　safety valve　安全阀
10　pressure gauge　压力表
11　cover head　顶盖
12　cover flange　封头法兰
13　vessel　容器
14　hydraulic jack　液力千斤顶
15　filter leaf　过滤叶片
16　feed slurry　滤浆进口
17　sludge outlet　滤渣出口
18　filtrate outlet　滤液出口

9.3 Rotary vacuum drum filter 转筒真空过滤机

9.3.1 Flowsheet for continuous vacuum filtration 真空连续过滤流程图

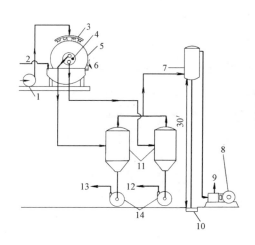

1　wash water pump　洗涤水泵
2　feed　进料
3　wash sprays　洗涤水喷头
4　air connection　空气接管
5　continuous rotary filter　连续回转过滤机
6　cake　滤饼
7　moisture trap　水分离器
8　dry vacuum pump　干式真空泵
9　air out　空气出口
10　barometric seal　水封
11　vacuum receivers　真空受液罐
12　filtrate　滤液
13　wash　洗液
14　pump　泵

9.3.2 Continuous rotary vacuum filter 真空连续回转过滤机

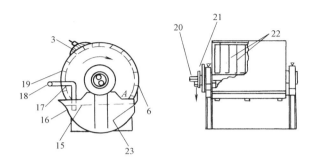

9.3.3 Rotary valve 转盘阀（分配头）

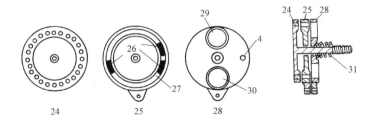

15　liquid level　液面
16　slurry trough　滤浆槽
17　inner drum　内转鼓
18　feed pipe　进料管
19　cloth-covered outer drum　铺有滤布的外转鼓
20　liquid outlets　出液口
21　rotary valve　转盘阀；分配头
22　internal pipes to various compartment　连通各室的内部管
23　doctor blade　刮刀
24　rotating plate　旋转板
25　stationary intermediate plate　中间固定板
26　division bridge　分配块
27　annulus　环形槽
28　stationary plate　固定板
29　wash vacuum connection　洗液通真空系统的接口
30　filtrate vacuum connection　滤液通真空系统的接口
31　spring　弹簧

9.3 Rotary vacuum drum filter 转筒真空过滤机

9.3.4 Oliver continuous vacuum drum filter 转筒真空过滤机

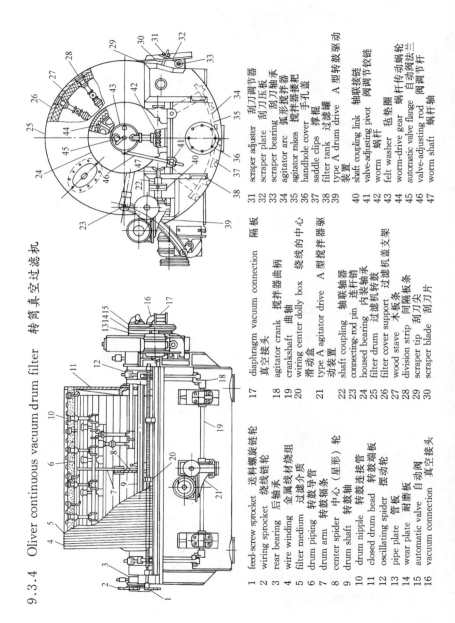

1	feed-screw sprocket	送料螺旋链轮
2	wiring sprocket	绕线链轮
3	rear bearing	后轴承
4	wire winding	金属线材绕组
5	filter medium	过滤介质
6	drum piping	转鼓导管
7	drum arm	转鼓辐条
8	center spider	中心（星形）轮
9	drum shaft	转鼓轴
10	drum nipple	转鼓连接管
11	closed drum head	转鼓端板
12	oscillating spider	摆动轮
13	pipe plate	管板
14	wear plate	耐磨板
15	automatic valve	自动阀
16	vacuum connection	真空接头
17	diaphragm vacuum connection	隔板真空接头
18	agitator crank	搅拌器曲柄
19	crankshaft	曲轴
20	wiring center dolly box	绕线的中心滑动盒
21	type A agitator drive	A型搅拌器驱动装置
22	shaft coupling	轴联轴器
23	connecting-rod pin	连杆销
24	housed bearing	内装轴承
25	filter drum	过滤机转鼓
26	filter cover support	过滤机盖支架
27	wood stave	木板条
28	division strip	间隔板条
29	scraper tip	刮刀尖
30	scraper blade	刮刀片
31	scraper adjuster	刮刀调节器
32	scraper plate	刮刀压板
33	scraper bearing	刮刀轴承
34	agitator arc	弧形搅拌器
35	agitator rakes	搅拌器搅耙
36	handhole cover	手孔盖
37	saddle clips	撑棍
38	filter tank	过滤罐
39	type A drum drive	A型转鼓驱动装置
40	shaft coupling link	轴联接链
41	valve-adjusting pivot	阀调节铰链
42	worm	蜗杆
43	felt washer	毡垫圈
44	worm-drive gear	蜗杆传动蜗轮
45	automatic valve flange	自动阀法兰
46	valve-adjusting shaft	阀调节轴
47	worm shaft	蜗杆轴

9.3.5 Diagram of string-discharge filter operation 绳索卸料过滤机操作示意图

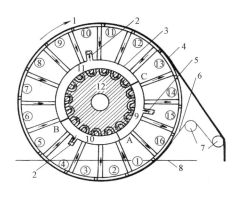

1 direction of rotation 旋转方向
2 outlet 出口
3 stationary automatic valve ring 固定的自动阀环
4 filtered cake 滤饼
5 vent to atmosphere - 出口通大气
6 strings returning to drum 绳索绕回转鼓
7 roll 滚轮
8 level of material to be filtered 过滤物料液面
9 discharging 卸料
10 filtering 过滤
11 dewatering 脱水
12 rotating element 旋转部件

9.3.6 Schematic elevation of coilfilter 绳索过滤机示意剖面图

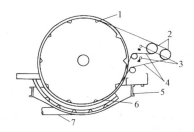

1 drainage channel 排液沟槽
2 discharge roll 卸料辊子
3 wash header 洗涤水总管
4 coil 绳索
5 frame 机架

9.3.7 Cake discharge and medium washing on an *Eimco*-belt belt-discharge filter Eimco 带过滤机的卸饼与介质洗涤

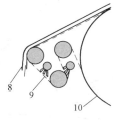

6 compartment seal 室间密封装置
7 feed and drain 进料与排放
8 cake 滤饼
9 belt 带
10 drum 转鼓

9.3.8 Operating method of vacuum precoat filter 预敷层转鼓真空过滤机操作示意图

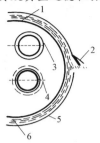

1 filter cloth 滤布
2 knife 刮刀
3 start 开始
4 finish 完成
5 thin film of solids 固体薄片
6 precoat filter medium 预敷层过滤介质

9.3.9 Drum filter 转鼓过滤机

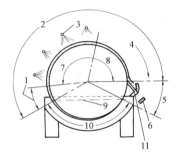

1 initial dewatering 初始脱水
2 dewatering 脱水
3 wash distributors 洗液分布器
4 rotation 旋转方向
5 discharge 卸料
6 discharged filter cake 已卸出滤饼
7 cake washing 滤饼洗涤
8 final dewatering 最后脱水
9 slurry level 浆料液面
10 filtering 过滤
11 scraper 滤饼刮刀

9.3.10 Schematic of a rotary drum vacuum filter with scraper discharge 带刮刀的转鼓真空过滤机图

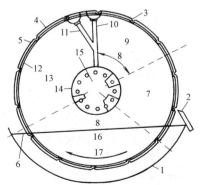

1 filter tank 滤槽
2 scraper 刮刀
3 caulking 挤缝
4 filter medium 过滤介质
5 grid support 格栅支承
6 slurry level 料浆样面
7 air blow for cake discharge 吹空气卸饼
8 vacuum applied 施真空
9 internal piping typical of all sections 内部管线各区段相同
10 lead pipe 前管
11 trail pipe 后管
12 drum shell 转鼓外壳
13 dry zone 干燥区
14 valve 阀
15 bridge blocks 过渡堵头
16 form zone 形成区
17 rotation 转向

(showing operating zones 表明各操作区)

9.4 Vacuum filter and others 真空过滤机及其他过滤机

9.4.1 Schematic illustrating the cycle of the Bird-Prayon tilting-pan filter 翻斗真空过滤机

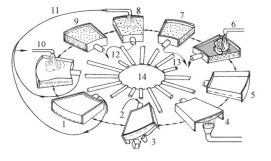

1 cake dewatering 滤饼脱水
2 air blow 空气反吹
3 cake dislodging and discharging 滤饼卸出
4 cloth cleaning 清洗滤布
5 cloth drying 滤布干燥
6 feed slurry 料浆加入
7 cake dewatering 滤饼脱水
8 cake washing 滤饼洗涤
9 cake dewatering 滤饼脱水
10 wash water 洗涤水
11 weak wash liquor 循环洗涤液
12 strong wash liquor 强制洗涤液
13 undiluted mother liquor 过滤母液
14 vacuum distributor 真空分布器

9.4.2 Continuous horizontal vacuum filter 连续水平真空过滤机

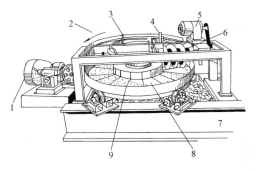

1 reduction drive-gear for filter 过滤机减速驱动装置
2 direction of rotation 旋转方向
3 support for drip-piping wash 洗涤水水管支架
4 feed inlet 进料口
5 drive motor for cake-removal screw 滤饼移出螺旋的驱动电机
6 screw 螺旋
7 cake 滤饼
8 cloths in place 铺设的滤布

9.4.3 Peterson roto-disc clarifier Peterson 转盘沉降器

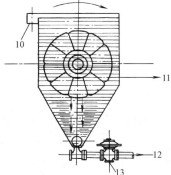

9 perforated metal cloth-support 多孔金属滤布及支架
10 feed (1% to 20% solids) 进料（1%～20%固体）
11 clear filtrate 清液
12 discharge sludge 卸渣（40%～70%固体）
13 diaphragm pump 隔膜泵

9.4 Vacuum filter and others 真空过滤机及其他过滤机

9.4.4 Schematic section of industrial horizontal tubular filter
工业用水平列管式过滤机的断面示意图

9.4.5 Top-outlet tubular filter
顶部排液的列管过滤机

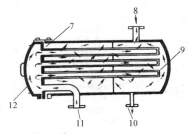

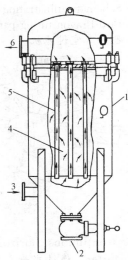

1　chamber　室
2　drain　出液
3　inlet　入口
4　flow　流向
5　tube　管
6,19　outlet　出口
7　tube sheet　管板

9.4.6 Elevation section of Sparkler horizontal plate filter Sparkler水平板式过滤机的正视断面图

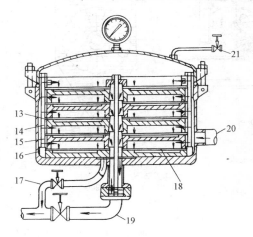

8　effluent discharge　流出物排出口
9　perforated tube with liner　带衬的多孔管
10　effluent drain　流出物放净口
11　influent inlet and drain　流入物进口与放净口
12　cover　端盖
13　filter paper, cloth or screen　滤纸、滤布或网
14　perforated plate　多孔板
15　filter plate　滤板
16　scavenger plate　清洗板
17　scavenger　清洗物
18　cake　滤饼
20　intake　进料
21　air vent　排空气

256

9.4.7 Cartridge filter 筒形过滤机

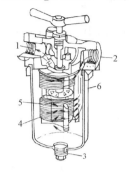

1 clarified liquid 清液
2 inlet 进料口
3 drain plug 排放口管堵
4 stationary comb 固定式梳状物组件
5 disks 圆盘板组件
6 casing 机壳

9.4.8 Belt filter 带式过滤机

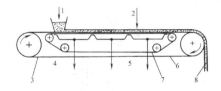

1 feed 加料
2 wash 洗涤
3 filter belt 过滤带
4 mother liquor 母液
5 wash liquor 洗液
6 filter media 过滤介质
7 support belt 支承带
8 cake 滤饼

9.4.9 Two versions of the dynamic filter 两种动态过滤机

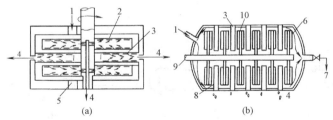

(in which cross-flow filtration is performed with rotating elements
用旋转元件实现十字流过滤)
(a) European design (only two stages shown) 欧洲型（仅表示了两级）
(b) United States design 美国型

1 feed 进料
2 rotating filter leaves 旋转滤叶
3 stationary filter plate 静止滤板
4 filtrate 滤液
5 thickened slurry 已增浓的料浆
6 wiper blade 宽刮刀
7 cake discharge 滤渣出口
8 thin cake on filter cloth 滤布上的薄层滤饼
9 solid shaft 实心轴
10 rotating disk 旋转叶轮

9.5 Centrifuge 离心机

9.5.1 Schematic diagrams of filtration centrifuge types
各种过滤离心机示意图

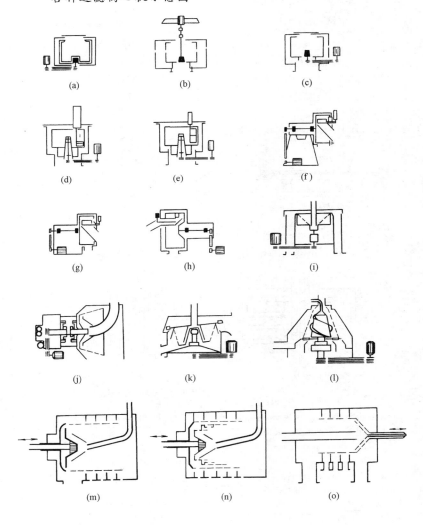

(a) bottom drive batch basket with bag 底部驱动间歇式带袋离心机
(b) top drive bottom discharge batch basket 顶部驱动底部出料间歇式离心机
(c) bottom drive bottom discharge batch basket 底部驱动底部出料间歇式离心机
(d) bottom drive automatic basket, rising knife 底部驱动,升降刮刀自动离心机
(e) bottom drive automatic basket, rotary knife 底部驱动,旋转刮刀自动离心机
(f) single-reversing knife rising knife 单一刮刀可升降、换向离心机
(g) single-speed automatic rotary knife 旋转刮刀单速自动离心机
(h) single-speed automatic traversing knife 单速横移刮刀自动离心机
(i) inclined wall self-discharge 斜壁自动卸料离心机
(j) inclined vibrating wall self-discharge 振动斜壁自动出料离心机
(k) inclined "tumbling" wall self-discharge 倾斜滚光壁自动卸料离心机
(l) inclined wall scroll discharge 倾斜壁卷轴卸料离心机
(m) traditional single-stage pusher 传统的单级推料离心机
(n) traditional multi-stage pusher 传统的多级推料离心机
(o) conical pusher with dewatering cone 锥形推料脱水离心机

9.5.2 Filtration centrifuge family tree 过滤离心机类型分类图

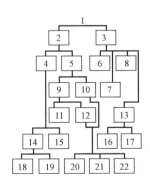

1 filtration centrifuge 过滤离心机
2 fixed bed 固定床
3 moving bed 移动床
4 batch manual 手动操作间歇式
5 cyclic automatic 自动循环式
6 scroll discharge 卷轴（螺旋）卸料
7 vibratory discharge 振动卸料
8 reciprocating push discharge 往复推料
9 vertical axis (multi speed) 垂直轴（多速）
10 horizontal axis (single speed) 水平轴（单速）
11 top drive 顶部驱动
12 bottom drive 底部驱动
13 inclined bowl discharge 倾斜转鼓卸料
14 top discharge 顶部出料
15 bottom discharge manual 底部手工出料
16 constant angle 固定角度
17 variable angle tumbler 可变角度转鼓
18 bag （料）袋
19 manual 人工的
20 rising knife 升降刮刀
21 rotary knife 旋转刮刀
22 traversing knife 横移刮刀

9.5.3 Three-column basket centrifuges 三足式转鼓离心机

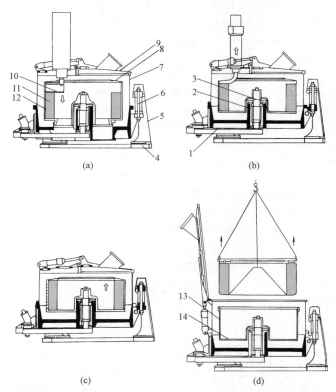

(a) with bottom discharge by scraper 刮刀下部卸料式
(b) with pneumatic top discharge 气动顶部卸料式
(c) with manual top discharge 手工顶部卸料式
(d) with top discharge by bag withdrawal 整袋顶部吊出卸料式
1　V-belt drive　V形传动带
2　bearing housing　轴承箱
3　shaft　主轴
4　bed　底座
5　supporter　支柱
6　buffer spring　缓冲弹簧
7　housing　外壳体
8　flinger ring　拦液圈
9　cover　机盖
10　scraper blade　刮刀
11　filter cake　滤饼
12　filtering medium　过滤介质
13　basket　转鼓
14　basket bottom　转鼓底

9.5.4 Basket type centrifuge or centrifugal separator 下部卸料离心机

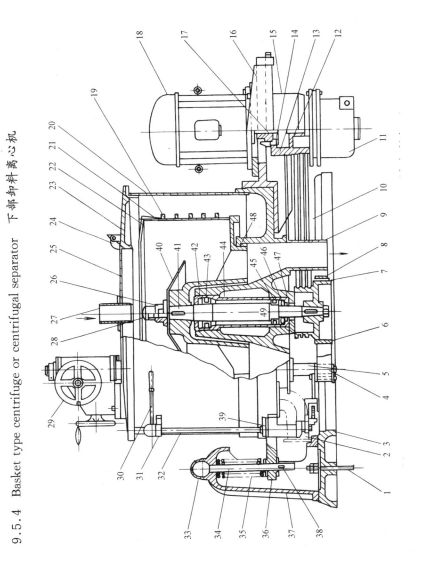

9.5 Centrifuge 离心机

1. foundation bolt 地脚螺栓
2. liquid outlet 排液口
3. bed 底座
4. arm 支杆
5. arm pin 支杆销
6. brake pulley 制动轮
7. brake lining 制动闸衬
8. brake hand 制动闸
9. solid outlet 滤渣出口
10. V-belt V形皮带
11. reducing gear 减速机
12. slip lining 滑动衬板
13. slip blade 滑动托板
14. slip blade pin 滑动托板销
15. bracket 托架
16. motor base 电机座
17. slip boss 滑动轮毂
18. main motor 主电动机
19. rein forcing ring 补强环
20. outside casing plate 外壳板
21. basket 转鼓
22. top annular plate 顶部环板
23. cover plate 盖板
24. hinge 铰链
25. cover 盖
26. grease cup 润滑脂杯（下）
27. feed liquid inlet 供液口
28. grease cup 润滑脂杯（上）
29. scraper 刮料装置
30. brake handle 制动手柄
31. upper metal 上（滑动）轴承
32. brake axle 制动轴
33. supporter cap 支柱帽
34. hanger bolt 吊挂螺栓
35. hanger spring 吊挂弹簧
36. frame 框架
37. supporter 支柱
38. cotter 开口销
39. metal (upper, lower) 上下（滑动）轴承
40. distributing plate 分布板
41. main shaft key 主轴键
42. gland 压盖
43. upper bearing 上轴承
44. pipe 管
45. lower bearing 下轴承
46. thrust bearing 止推轴承
47. bearing box 轴承座
48. basket bottom 转鼓底
49. main shaft 主轴

262

9.5.5 Cylindrical-conical helical-conveyor centrifuge
筒锥形螺旋卸料离心机

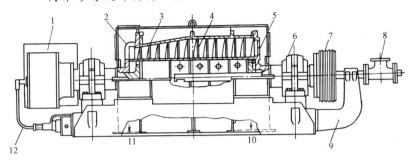

9.5.6 Top-suspended basket centrifugal
上悬式转鼓离心机

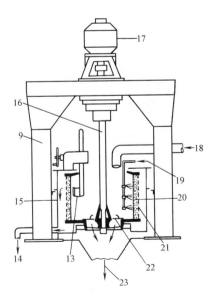

1 conveyor drive　输送器驱动装置
2 solids-discharge port　固体排出口
3 bowl　(转)鼓
4 scroll conveyor　蜗旋式输料器
5 adjustable filtrate port　可调滤液排出口
6 bearing housing　轴承箱
7 driven sheave　传动皮带轮
8 feed slurry　滤浆进料
9 frame　机架
10 filtrate discharge　滤液排出口
11 solids-discharge port　固体(滤渣)排出口
12 overload release　过载断路器
13 adjustable unloader knife　可调式刮刀
14 liquid draw off　液体排出
15 casing　机壳
16 shaft　轴
17 motor　电动机
18 feed inlet　进料口
19 wash inlet　洗液入口
20 solids cake　固体滤渣
21 perforated basket　多孔吊篮(转鼓)
22 removable valve plate　提升式(卸料)阀板
23 solids discharge　滤渣卸出口

9.5.7 Reciprocating-conveyor continuous centrifuge 活塞推料离心机

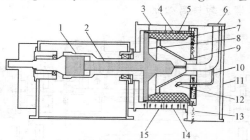

9.5.8 Automatic batch centrifugal 间歇式自动离心机

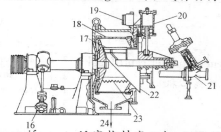

9.5.9 Pusher centrifuges 活塞推料离心机

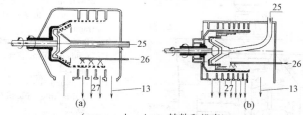

(rotor and casing 转鼓和机壳)
(a) single-stage with conical pusher screen 单级,带网板锥形推料机
(b) multistage 多级

1　servomotor　伺服电动机
2　reciprocating piston rod　往复式活塞杆
3　housing　机壳
4　basket　悬筐（转鼓）
5　screen　滤网
6　feed pipe　进料管
7　cake　滤饼
8　piston　活塞
9　feed funnel　进料斗
10　wash pipe　洗液管
11　access door　检修门
12　spray nozzle　洗液喷嘴
13　solids discharge　滤渣卸出
14　wash discharge　洗液出口
15　liquor discharge　滤液出口
16　unloader hydraulic pump　卸料装置用的液压泵
17　unloader knife　卸料刮刀
18　perforated basket　带孔转鼓
19　limit switch　极限开关
20　unloader control cylinder　卸料装置控制气缸
21　feed valve　进料阀
22　chute (solids discharge)　斜槽（固体排出）
23　crystal distributor　固体（晶体）分布器
24　liquid outlet　滤液出口
25　feed　进料
26　cake rinse　滤饼清洗液
27　liquid discharge　滤液出口

9.5.10 Schematic of typical pusher centrifuge 典型的活塞推料离心机简图

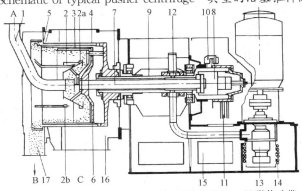

1   feed pipe   进料管
2   feed distributor   进料分布器
2a wash basket   洗涤篮筐
2b spray nozzle   喷头
3   slot screen   长眼筛板
4   basket   转鼓
5   screen retaining ring 筛板固定环
6   pusher plate   推料板
7   pusher shaft   推料轴
8   pusher control with pusher piston 带推料活塞的推送控制器
9   hollow shaft with bearings 空心轴带轴承
10 disc brake   盘式制动器
11 V-belt drive   V形传动带
12 oil feed bearing   给油轴承
13 hydraulic pump with motor 带电动机油压泵
14 oil cooler   油冷却器
15 bearing housing with oil pan and motor support 带油槽的轴承箱和电机支架
16 filtrate housing   滤液排出壳体
17 solids collecting housing with volute race discharge 带蜗壳卸料口的固体聚集箱
A   slurry   浆料
B   solids   固体
C   filtrate   滤液

9.5.11 Tubular-bowl centrifuge 管式高速离心机

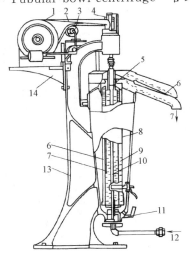

1   pulley   皮带轮
2   belt   皮带
3   idler   导轮
4   bearing   轴承
5   ring dam   环形堰板
6   light liquid   轻液
7   heavy liquid   重液
8   rotating bowl 转鼓
9   solid   固体
10 spindle   主轴
11 drag   制动器
12 liquid inlet   液体入口
13 frame   机架
14 support   支架

9.5 Centrifuge 离心机

9.5.12 Component parts of separator bowl 分离机转鼓部件

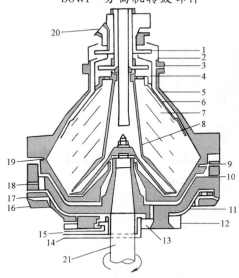

1. water paring disc 弧形水盘
2. gravity disc 重力盘
3. oil paring disc 弧形油盘
4. level ring 水平环板
5. bowl hood 转鼓外罩
6. top disc 顶层盘
7. disc stack 料盘空筒
8. distributor 分布器
9. sliding bowl bottom 滑板型转鼓底
10. bowl body 转鼓主体
11. operating slide 操作滑板
12. spring 弹簧
13. operating water paring disc 操作弧形水盘
14. opening water supply 开式供水口
15. closing and make-up water supply 定量和补充供水口
16. dosing ring 定量环
17. dosing ring chamber 定量环室
18. drain valve 排放阀
19. sludge port 污泥出口
20. liquid seal and flushing water supply 液封与冲洗水供入口
21. rotor shaft 转轴

9.5.13 Centrifugal clarifier with self-cleaning bowl 带自动卸料转鼓的澄清式离心机

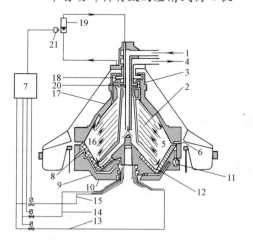

1. feed 进料
2. disc 料盘
3. centripetal pump 向心泵
4. discharge 排出口
5. sediment holding space 沉积物聚集空间
6. sediment ejection ports 沉积物排出口
7. timing unit 定时装置
8. outer closing chamber 外密闭室
9. inner closing chamber 内密闭室
10. opening chamber 敞开室
11. bowl valve 转鼓阀
12. piston 活塞
13. opening water 常开水管
14. closing water 常关水管
15. auxiliary opening water 辅助常开水管
16. sensing zone disc 传感层圆盘
17. sensing liquid clarifying disc 传感液澄清盘
18, 20 sensing liquid pump 传感液泵
19. flowmeter 流量计
21. switch 开关

9.5.14 Disk-centrifuge bowl 碟式分离机转鼓

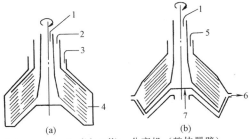

(a) separator(solid wall) 分离机（整体器壁）
(b) recycle clarifier(nozzle discharge) 再循环澄清器（喷嘴卸料）

1　feed　进料
2　light-phase effluent　轻相流
3　heavy-phase effluent　重相流
4　solids holding space　固体持留空间
5　clarified effluent　澄清液流
6　concentrated solid phase　增稠固体相
7　recycle　再循环

9.5.15 Disk centrifuge 碟式离心机

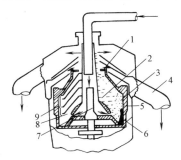

1　adjustable dam　可调堰
2　light liquid　轻液
3　heavy liquid　重液
4　discharge spout　排液管
5　rotating bowl　转鼓
6　solid　固体
7　spindle　轴
8　disk　金属碟
9　disk stack　叠装碟组

9.5.16 Cross-section of a typical nozzle centrifuge 典型的喷嘴卸料离心机截面图

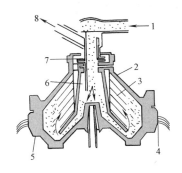

1　feed　进料
2　bowl　转鼓
3　disc set　碟组件
4　concentrate outlet　浓缩液出口
5　concentrate space　浓缩液滞留空间
6　inner bowl space　内转鼓空间
7　centripetal pump　向心泵
8　clarified liquid discharge　澄清液排出

9.6 Expression equipment 压榨设备

9.6.1 Three-roll sugar mill 三辊榨糖机

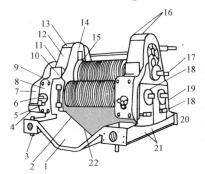

8　side cap　侧盖
9　side cap key　侧盖开关
10　side roll juice collar　侧辊糖汁环
11　crownwheel shield　冠齿轮罩
12　housing　外壳
13　top cap　上盖板
14　top cap key　上盖开关
15　top roll flange　上辊法兰
16　push-pull jacking screws　推挽式起重螺旋
17　oil pump　油泵
18　oil drip box　盛油箱
19　side roll box　侧辊箱
20　housing and bedplate cast in one piece　外壳与底座连体铸件
21　turnbeam adjusting level and screws　导向板调节杆与螺钉
22　trunnion block adjusting screw　轴颈座调节螺钉

1　juice pan　榨汁盘
2　juice shield　榨汁挡板
3　foundation bolt　基础螺栓
4　crownwheel oil pan　冠齿轮油池
5　side box pullback bolt　侧箱拉回螺栓
6　adjusting screw　调节螺钉
7　locking plate　锁紧板

9.6.2 Schematic of a belt filter press 带式压榨机简图

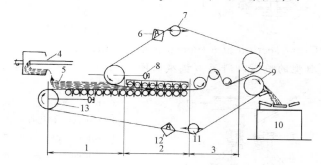

1　gravity drainage zone　重力渗滤区
2　wedge zone　挤压区
3　press zone　压榨区
4　mixing drum　混合转鼓
5　breast roll　排气辊
6　upper belt wash　上部滤带冲洗
7　upper tracking roll　上部跟踪辊
8　upper belt tensioning　上部滤带张紧器
9　drive roll　驱动辊
10　conveyor　运渣带
11　lower tracking roll　底部跟踪辊
12　lower belt wash　下部滤带冲洗
13　lower belt tensioning　下部滤带张紧器

10 Gas-solids Separation Equipments
气固分离设备

10.1 Cyclone separators 旋风分离器

10.1.1 Some commercial cyclones 几种工业旋风分离器

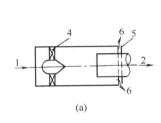

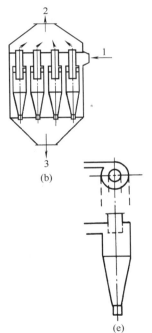

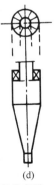

(a) uniflow cyclone 单向流动旋风分离器
(b) multiclones 多管式旋风分离器
(c) helical 螺旋面式
(d) axial 轴流式
(e) spiral 螺旋式
1 inlet 进口
2 gas 气体
3 solid 固体
4 swirl vanes 旋转叶片
5 deflector ring 折流环
6 purge gas and solid 排出气体与固体

10.1.2 Typical commercial cyclones 典型的工业旋风分离器

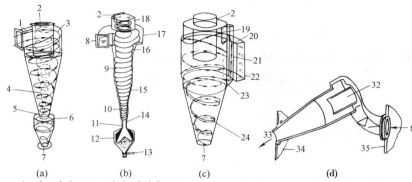

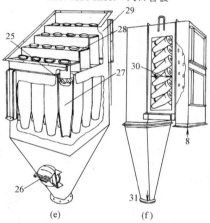

1 dust-laden-gas inlet 含尘气入口
2 clean-gas outlet 净化气出口
3 helical top 螺线顶盖
4 vortex 涡流
5 vortex shield 涡流屏
6 dust trap 集尘器
7 dust outlet 粉尘出口
8 inlet 入口
9 upper cylinder 上圆筒
10 middle cylinder 中圆筒
11 tail piece 尾管
12 receptacle 接受器
13 blast gate 风门
14 lower cone 下圆锥
15 middle cone 中圆锥
16 upper cone 上圆锥
17 body 主体
18 outlet head 出口端部
19 dust shave-off 粉尘刮面板
20 pattern of dust stream (principally the finer particles) following eddy current 随涡流而产生的粉尘流型（主要是较细颗粒）
21 shave-off-dust channel 粉尘刮沟
22 inlet for dust-laden gases 含尘气体入口
23 shave-off-reentry opening 所刮下的重返通道
24 pattern of coarser dust main stream 粗粉尘主流线
25 vane 导流叶片
26 man hole 人孔
27 cyclone tube 旋风分离管
28 tube sheet 管板
29 shell 外壳
30 gas outlet duct 气体出口管
31 dust hopper 粉尘料斗
32 gas outlet 气体出口
33 dust 粉尘
34 hopper tube sheet 料斗管板
35 inlet tube sheet 入口管板

(a) Duclone collector 杜康旋风分离器
(b) Sirocco type D collector Sirocco D 型分离器
(c) van Tongeren cyclone van Tongeren 分离器
(d) cutaway of Dustex cyclone tube Dustex 微型旋风分离器剖视图
(e) multiclone collector 多管式旋风分离器
(f) Dustex miniature collector assembly Dustex 微型（斜管）收集器组

10.1 Cyclone separators 旋风分离器

10.1.3 Two-stage multiple cyclonic separator 两级组合式旋风分离器

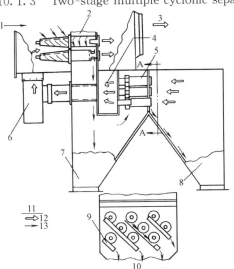

1 main flow to cyclonic separator unit 主流向旋风分离机组
2 cyclonic separator unit 旋风分离机组
3 main flow to exhauster 主流向排气装置
4 concentrator clean air plenum 经浓集器后的净化空气流
5 concentrator unit 浓集器机组
6 exhauster for concentrator 浓集器后的排气装置
7 primary dust hopper 一级集尘箱[斗]
8 concentrator dust hopper 浓集器集尘箱[斗]
9 dust chutes 微粉流道
10 view A-A A-A 剖视
11 nomenclature 命名规定
12 cleaned air 净化空气
13 dust laden air 含尘空气

10.1.4 Typical centrifugal separators 典型的离心分离器

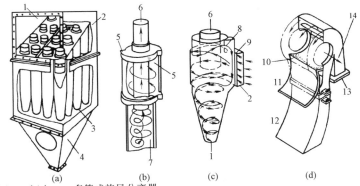

(a) multiclone 多管式旋风分离器
(b) cutaway Thermix ceramic tube Thermix 陶瓷管分离器局部剖视图
(c) van Tongeren cyclone van Tongeren 旋风分离器
(d) horizontal steam separator 卧式蒸汽分离器
1 outlet 出口
2 inlet 进口
3 cyclone tubes 旋风分离管
4 receiver 集尘器
5 gas in 气体进入
6 gas out 气体排出
7 liquid out 液体出口
8 skimmer 脱尘板
9 by-pass channel 旁路沟槽
10 skimmer edge 脱液板边缘
11 primary discharge baffle 一级排液挡板
12 secondary drain 二级排液口
13 wet steam 湿蒸汽
14 steam out 蒸汽出口

271

10.2 Bag filters 袋滤器

10.2.1 Typical cloth filters 典型的袋式过滤器

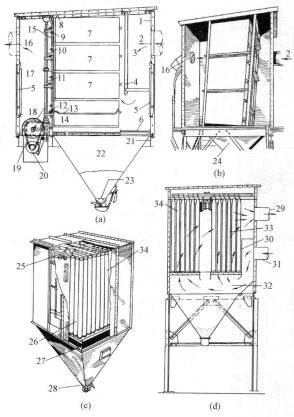

1　casing　外壳；箱体
2　air inlet　空气进口
3,32　dusty air side　含尘空气侧
4　removable baffle　可拆挡板
5　man door　人孔
6　wire mesh walkway　金属丝网通道
7　screen　筛网
8　roof　顶盖
9　clamp　夹紧装置
10　rocker arm　摇臂
11　rocker shaft　摇臂轴
12　beater connecting rod　敲打器连杆
13　screen beater　筛网击打器
14　screen with cloth removed　包有可拆滤布的筛网
15　grids　栅条
16　air outlet　空气出口
17,33　clean air side　净化空气侧
18　rapping mechanism　敲击机构
19　drive　传动装置
20　motor　电机
21　structural support　设备支座
22　hopper　集尘斗
23　hopper valve　料斗底阀
24　dust hopper　集尘斗
25　bag support and shaking mechanism　布袋支撑和振打机构
26　clean gas side　净化气侧
27　dirty gas side　脏气体侧
28　dust discharge　粉尘排出口
29　outlet pipe　出口管
30　baffle plate　挡板
31　inlet pipe　入口管
34　filter bag　滤袋

(a) screen or envelope type (sectional view)　筛网或包套型（剖视图）
(b) screen or envelope type (cutaway view)　筛网或包套型（局部剖视图）
(c) bag type (cutaway view)　袋式（局部剖视图）
(d) bag type (sectional view)　袋式（剖面图）

10.2.2 Multi-compartment vibro bag filer 多室振打袋滤器

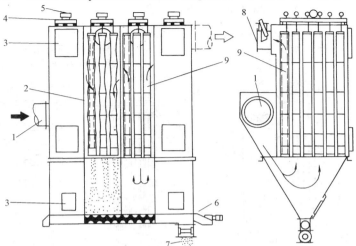

1 dust-laden gas inlet 含尘气体进口
2 compartment isolated for bag cleaning 过滤袋隔离室
3 access door 检修门
4 vibro frame 振打机架
5 vibro motor 振打电机
6 screw conveyor 螺旋输送器
7 dust discharge 粉尘排出
8 clean gas outlet 净化气体出口
9 filter bag 滤袋

10.2.3 Reverse-pulse fabric filter 逆向（反吹）脉冲纤维过滤器

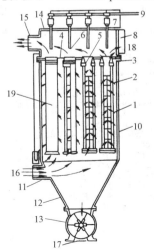

1 filter cylinder 过滤器筒体
2 wire retainer 金属丝支承圈
3 collar 卡箍
4 tube sheet 管板
5 Venturi nozzle 文丘里喷嘴
6 nozzle or orifice 喷嘴或锐孔
7 solenoid valve 电磁阀
8 timer 定时器
9 compressed air manifold 压缩空气总管
10 collector housing 收集室
11 inlet 进口
12 hopper 料斗
13 air lock 空气星形阀
14 upper plenum 上部气室
15 to exhauster 去引风机
16 dust-laden air 含尘气体
17 material discharge 下料
18 induced flow 导流器
19 filter bag 滤袋

273

10.3 Electrical precipitators 电除尘器

10.3.1 Electrostatic precipitator 静电除尘器

10.3.2 Vertical-flow heavy-duty plate precipitator 垂直流动重负载板式除尘器

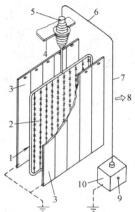
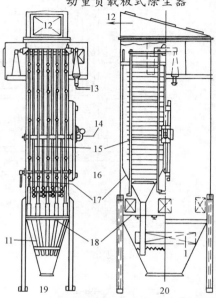

1 direction of gas flow 气体流动方向
2 discharge (negative) electrons 放电电子（负极）
3 collecting (positive) plate 集尘板（正极）
4 precipitator plate cover 除尘器平板盖
5 discharge system support insulator 支承输电系统的绝缘体
6 high voltage cable 高压电缆
7 D. C. output 直流电输出
8 clean gas outlet 净化气体出口
9 A. C. input 交流电输入

10.3.3 Two-stage electrical-precipitation principle 两级电除尘器原理图

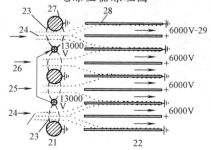

10 transformer rectifier set 整流变压器机组
11 gas inlet 气体入口
12 gas outlet 气体出口
13 H. T. inlet 高压电接入口
14 rapper bar 振打棒
15 discharge electrode 放电电极
16 H. T. rapper shaft 高压电振打器传动轴
17 collecting electrode 收集电极
18 distribution plate 分布板
19 sectional end elevation 端部剖视
20 side elevation 侧视图
21 ionizing unit 离子化单元
22 collecting unit 收集单元
23 electrostatic field 静电场
24 dust particles 粉尘粒子
25 discharge electrodes 放电电极
26 gas flow 气体流向
27 grounded receiving electrodes 接地收集电极
28 alternately grounded and charged collector plates 交替接地和放电的收尘板

10.3.4 Wet cottrell 湿式静电除尘器

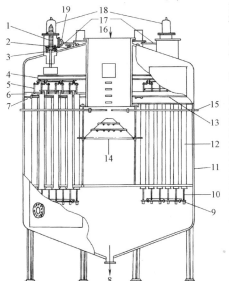

1. support insulator 支承绝缘子
2. intermittent overhead spray nozzle 间歇式顶喷嘴
3. steam coil 蒸汽盘管
4. high voltage support frames 高压电路支架
5. high voltage discharge electrode 高压放电极
6. water pond 水盘
7. adjustable weir ring 溢流堰调节环
8. drain 排液口
9. electrode weight 高压电极重锤
10. high voltage electrode 高压电极
11. shell 壳体
12. collecting electrode pipe 集尘电极管
13. header sheet 管板
14. gas deflector cone 导风锥
15. inlet flushing system for pipe and cone 管道和导风锥的冲洗装置
16. gas inlet 气体入口
17. clean gas outlet 净化气出口
18. high voltage insulator compartment 高压绝缘子室
19. hand hole cover 手孔盖

10.3.5 Pilot-scale tubular precipitator 管式电除尘器中试装置

1. natural-gas pump 天然气泵
2. direct-fired air heater 直接燃烧空气加热器
3. ash pickup air 集尘空气
4. air 空气
5. fly ash 飞灰
6. elutriator 洗提器
7. vent 排气孔
8. high-voltage power 高压电源
9. control valve 控制阀
10. dust sample line 粉尘取样管
11. rapping mechanism (tubes) 振打机构(管)
12. manhole 人孔
13. dust-lock hopper 粉尘锁闭料斗
14. electrostatic precipitator 静电除尘器
15. rapping mechanism (wires) 振打机构(导线)
16. power connection 电源接头
17. insulator 绝缘体
18. bus bar 母线条
19. emission electrode 发射电极
20. tube sheet 管板
21. collecting electrode 收集电极
22. suspension weight 悬挂重物

(for operation at 860℃. and 556kPa in. gage 操作条件：860℃，表压556kPa)

10.3 Electrical precipitators 电除尘器

10.3.6 Blast-furnace pipe precipitator 炼铁炉管式除尘器

10.3.7 Two-stage water-film pipe precipitator 两级水膜管式除尘器

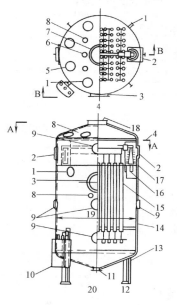
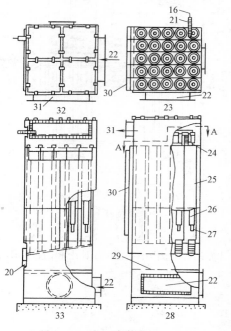

1	6″ vent 6in 排气口	18	gas outlet 气体出口
2	C. I. door(C. I＝cast iron) 铸铁门	19	13′6″ I.D. (I. D.——internal diameter)内径为13ft6in
3	gas inlet 气体入口	20	drain 排料口
4	PLAN A-A A-A 剖视	21	H. T. inlet bushing 高压电输入绝缘管
5	manholes 人孔	22	inlet 入口
6	2″～4″vents 2～4in 管口	23	section A-A A-A 剖视
7	gas outlet 气体出口	24	supporting insulators 支承绝缘体
8	spray nipple 喷雾器管接头	25	collecting electrode 收集电极
9	door(near side only) 门（只有左侧）	26	precipitating electrode 除尘器电极
10	water control box 水控制箱	27	ionizer 离子化器
11	8″ drain 8in 排料井	28	side elevation 侧视图
12	ELEVATION B-B B-B 剖视	29	distribution plate 分布板
13	4 columns 4 个支腿	30	flushing system 洗涤系统
14	⅜″外壳 ⅜英寸外壳（壁厚）	31	outlet 出口
15	90—8″diam. pipes 90 个直径为 8in 管子	32	plan 俯视图
16	H. T. line (H. T.——high tension) 高压导线	33	front elevation 正视图
17	steam coil 蒸汽盘管		

10.4 Other separators 其他分离器

10.4.1 Electrically augmented granular-bed filter 电力增强颗粒床过滤器

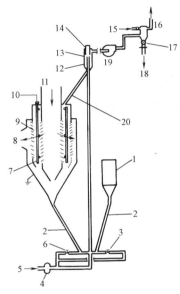

1 inventory hopper 贮料斗
2 seal leg 密封腿
3 air injector 空气注入器
4 media left air blower 介质提升气风机
5 ambient air 环境空气
6 media flow air injector 介质流动空气注入器
7 high voltage grid 高压栅
8 clean gas outlet 净化气体出口
9 media filter bed 介质过滤床层
10 high voltage 高压电
11 dirty gas inlet 脏气体进口
12 overflow vessel 溢流槽
13 air classifier 空气分级器
14 media/dust deentrainment chamber 介质/粉尘减少夹带室
15 compressed air 压缩空气
16 bag house 集尘室
17 rotary air lock 旋转气闸
18 to dust bin 去粉尘盒
19 ash conveyor fan 粉尘输送风机
20 media return 介质返回

10.4.2 H-E brink mist eliminator element 高效率纤维雾沫净除器单元

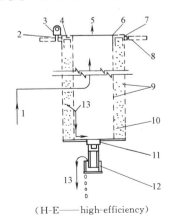

1 gas flow 气体流
2 gasket 垫片
3 two lifting lugs eq. sp. 两个等距的吊耳
4 retainer lugs 挡板凸缘
5 clean gas out 净化气体出口
6 retainer plate 挡板
7 cap screws 有帽螺钉
8 support plate 支承板
9 concentric screens 同心筛网
10 fiber packing 纤维填充物
11 half coupling 半接头
12 liquid seal pot 液封筒
13 liquid drainage 液体排泄

(H-E——high-efficiency)

10.4.3 Typical separators using impingement in addition to centrifugal force 既采用离心力又采用撞击的典型分离器

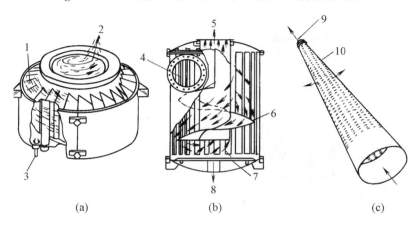

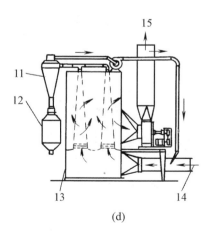
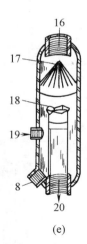

(a) Hi-eF purifier　Hi-eF 净化器
(b) flick separator　轻击分离器
(c) aerodyne tube　空气动力管
(d) aerodyne collector　空气动力捕集器
(e) type RA line separator　RA 型在线分离器

10.4 Other separators 其他分离器

1	wet steam in 湿水蒸气进入		12	receiver 集尘器
2	steam out 蒸汽出口		13	collector 捕集器
3	water out 水出口		14	intake 进气
4	vapor in 蒸气进入		15	exhaust 排气
5	vapor out 蒸气出口		16	inlet 入口
6	helical plate 螺旋板		17	helicoid tuyere 螺旋形风帽
7	perforated plate 多孔板		18	secondary vortex breaker 二次涡流破碎器
8	drain 泄液			
9	particle concentrate 粒子富集		19	vent 放气口
10	louvers 百叶窗		20	outlet 出口
11	cyclone 旋风除尘器			

10.4.4 Recirculating baffle collector 循环式挡板集尘器

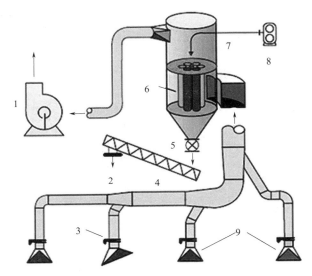

1 fan 风机
2 dust disposal 粉尘处理
3 blast gate 排气门
4 ducting 风管
5 airlock 气闸
6 bags 滤袋
7 dust collector 集尘器
8 reverse air cleaning 反向空气净化（反吹）
9 pickup hoods 抽风罩

10.5 Scrubber 涤气器

10.5.1 Impingement-plate scrubber 撞击板式涤气器

10.5.2 Scrubber equipped with vertical rotor 装有立式转子的涤气器

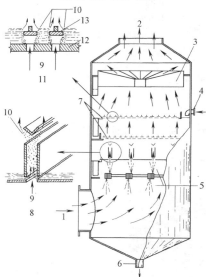

10.5.3 Spray tower 喷淋塔

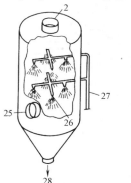

5 humidification sprays 增湿雾化器
6 dirty water outlet 脏水出口
7 impingement baffle plate stages 撞击挡板级
8 high velocity slot stage 高速沟槽级
9 gas flow 气体流向
10 water droplets atomized at edges of orifices 水滴在孔边缘雾化
11 detail of impingement plate 撞击板详图
12 sieve plate 筛孔板
13 impingement plate 撞击板
14 contaminated air inlet 脏空气入口
15 baffle 挡板
16 enclosed belt tunnel 封闭式传动皮带通道
17 exhaust fan 排风机
18 clean air outlet 净化空气出口
19 entrainment separator 雾沫分离器
20 spray zone 溅雾区
21 vertical rotor 立式转子
22 drain 排液口
23 dry precleaner 粗粉尘预沉降器
24 coarse particles out 粗粒粉尘出口
25 dirty air inlet 脏空气入口
26 spray jet 喷头
27 supply water piping 供水管
28 water and sludge drain 水和泥浆排出口

1 inlet 入口
2 clean gas outlet 净化气出口
3 entrainment separation stage 雾沫分离级
4 scrubbing water inlet 涤气水入口

10.5.4 Cyclone spray scrubber 旋风涤气器

10.5.5 Self-induced spray scrubber 自导式喷雾涤气器

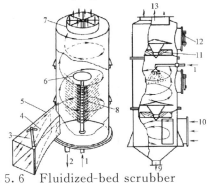

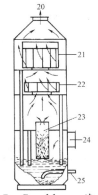

10.5.6 Fluidized-bed scrubber 游动床涤气器

10.5.7 Scrubber 涤气器

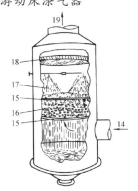

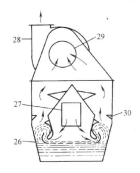

1　water inlet　水入口
2　water outlet　水出口
3　damper position indicator　调节位置指示器
4　swinging inlet damper　旋转式入口气流调节器
5　tangential gas inlet　气体切向入口
6　core buster disk　中央挡盘
7　antispin vane　消旋叶片
8　spray manifold　多头喷嘴集管
9　dirty water outlet　脏水出口
10　dust-laden air in　含尘空气入口
11　fixed vane　固定（旋流）叶片
12　inspection door　检查门
13　clean air out　净化气出口
14　dust-laden gas　含尘气体
15　retaining grid　固定栅板
16　floating bed of low-density sphere 低密度球浮动床
17　scrubbing liquor　涤气液
18　mist eliminator　脱湿器
19　cleaned gas　净化气
20　clean gas outlet　净化气体出口
21　entrainment separator plate　雾沫分离板
22　primary entrainment separator　预雾沫分离器
23　gas contacting region　气体接触区
24　gas inlet　气体入口
25　water inlet　水入口
26　liquid level　液面
27　inlet　入口
28　fan　风机
29　outlet　出口
30　baffle　挡板

10.5.8 Packed-bed scrubber 填充床涤气器

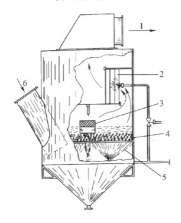

1. clean gas outlet　净化气体出口
2. zig-zag entrainment separator　曲径式雾沫分离器
3. liquid overflow pipe with screen　带筛网的溢流管
4. packed bed ¾″dia. marbles （dia.＝diameter）直径为¾in 的卵石填充床
5. spray　雾沫
6. gas inlet　气体入口

10.5.9 Mechanical scrubber 机械式涤气器

1. dust-laden gas　含尘气体
2. flooded disk　富液盘
3. erosion-resistant throat　耐磨性喉部衬环
4. to cyclone separator　去旋风分离器
5. automatic disk positioner　自动圆盘定位器
6. water　水入口

10.5.10 Fly ash scrubber 飞灰涤气器

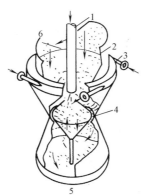

1. combined liquid inlet and cone support　液体进口与锥形机座组合体
2. clear cut liquid/gas junction　冲洗液与气体接触会合处
3. liquid inlet　液体进口
4. automatically adjustable cone for varying throat opening　可改变喉部开度的自动调节锥体
5. cleaned gas outlet　净化气体出口
6. gas inlet section　气体进入段

10.5.11 Fibrous-bed scrubber 纤维填充床涤气器

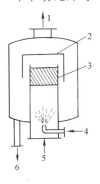

1. clean gas outlet 净化气体出口
2. cap 罩子
3. fibrous contacting element 纤维接触元件
4. liquid inlet 液体入口
5. dirty gas in 脏气体入口
6. slurry out 浆液出口

10.5.13 Venturi scrubber 文丘里涤气器

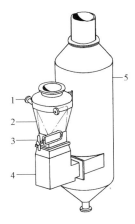

1. open shelf liquid distributor 开口搁架式液体分布器
2. flooded wall 流淌壁

10.5.12 Sprayer 喷淋[洒]器

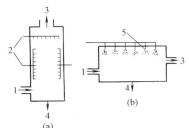

(a) spray-column cross section showing various liquid-distribution geometries 喷洒塔截面图（表示液体分布不同的几何位置）
(b) crossflow spray absorber 错流式喷淋吸收器

1. gas inlet 气体进口
2. liquid in 液体进口
3. gas outlet 气体出口
4. liquid out 液体出口
5. parallel rows of spray heads 若干喷头并联排列

10.5.14 Venturi-scrubber system 文丘里洗涤器系统

3. externally adjustable Venturi throat 外部调节的文丘里喉管
4. flooded elbow 溢流肘管
5. cyclonic separator 旋风分离器
6. odorous exhaust 有气味的废气
7. Venturi 文丘里（管）
8. scrubber water 涤气器用水
9. scrubbed exhaust 洗涤后排出气
10. separator 分离器
11. spent scrubber water 涤气器废水
12. fan 风机
13. exhaust air 废气

10.5.15 Counter-current flow scrubber 逆流洗涤塔

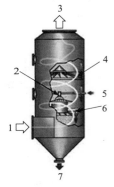

1 dirty gas inlet 脏气体进口
2 scrubbing liquid inlet 洗涤液入口
3 clean gas outlet 清洁气体出口
4 wetted fan wheel 浸湿风扇叶轮
5 spray 喷雾
6 scrubbing vanes 洗气叶片
7 slurry outlet 泥浆出口

11 Size Reduction Equipments 粉碎设备

11.1 Crushing equipment 破碎设备

11.1.1 Blake jaw crusher Blake 颚式破碎机

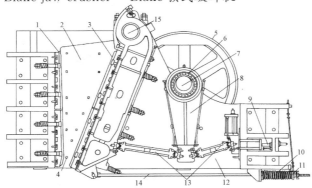

1 fixed-jaw plates 固定颚板
2 crushing chamber 破碎室
3 swing-jaw plates 摆动颚板
4 curved product outlet 曲线形产品出口
5 flywheel 飞轮
6 self-aligning roller bearing 自定心滚柱轴承
7 eccentric shaft 偏心轴
8 pitman 连杆
9 hydraulic setting control 排料口大小液压控制装置
10 tension rod spring 拉杆弹簧
11 washer (拉杆弹簧)垫圈
12 frame 框架
13 toggle plates 肘板
14 tension rod 拉杆
15 pivot 摆动中心（枢轴）

11.1.2 Fairmount single-roll crusher Fairmount 单辊破碎机

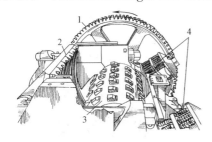

1 drive gear 主动齿轮
2 curved anvil 曲线砧座
3 sledging crushing and roll 锤碎及滚筒
4 pressure equalizing springs 压力平衡弹簧

11.1.3 Symons standard cone crusher Symons 标准圆锥破碎机

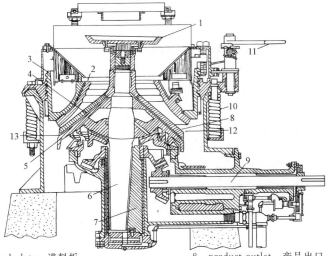

1　feed plate　进料板
2　bowl liner　圆锥筒衬里
3　bowl　圆锥筒
4　mantle　套筒
5　conical head　锥形头
6　main shaft　主轴
7　eccentric　偏心件
8　product outlet　产品出口
9　counter shaft　副轴
10　relief spring　保险弹簧
11　product size control　产品大小控制器
12　maximum opening　最大开度
13　minimum opening　最小开度

11.1.4 Gyratory crusher with spider suspension 机架悬挂的回转破碎机

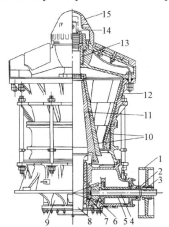

1　driving sheave　传动皮带轮
2　countershaft seal retainer　传动轴密封保持盖
3　key for driving sheave　传动皮带轮键
4　countershaft　传动轴
5　countershaft housing　传动轴套筒
6　driving pinion　传动小齿轮
7　driving gear　驱动齿轮
8　eccentric shaft　偏心轴
9　discharge　排料处
10　liners　衬里
11　crushing head　破碎头
12　bowl　圆锥筒
13　suspension bushing　悬挂轴衬
14　suspension　悬挂装置
15　spider cap　多幅架帽

11.1.5　Hammer crusher　锤式破碎机

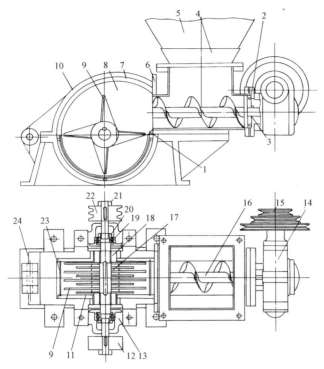

1　liner　衬板
2　connect collar　连接卡圈
3　oil seal　油封
4　hopper base　料斗支架
5　hopper　料斗
6　screw case　螺杆箱体
7　liner plate　衬板
8　side plate　侧衬板
9　knife hammer　刀头锤子
10　upper case　上盖
11　screen guide　护板定位框
12　flywheel　飞轮
13　ball bearing case　球轴承箱
14　worm speed reducer　蜗轮减速器
15　V-pulley　V形皮带轮
16　screw　螺杆
17　distance piece　定距板
18　locking nut　锁紧螺母
19　radial ball bearing　径向球轴承
20　bearing nut　轴承螺母
21　pulley nut　皮带轮螺母
22　main pulley　主动V形皮带轮
23　screen　护板
24　lower case　底座

11.1.6 Impact crusher 冲击式破碎机

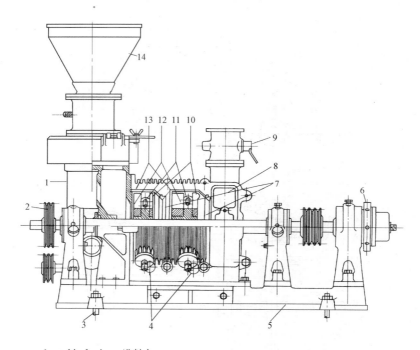

1　table feeder　进料台
2　V-pulley　V形皮带轮
3　foundation bolt　地脚螺栓
4　nozzle　喷嘴
5　commen bed　底座
6　adjusting wheel　调整轮
7　bush　衬套
8　fan　风扇
9　valve　阀门
10　No. 2 runner　2号碾碎头
11　impact pin　冲击销
12　No. 1 runner　1号碾碎头
13　liner　衬板
14　hopper　料斗

11.1.7 Double shaft hammer crusher 双轴锤式破碎机

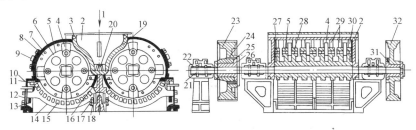

1　feed opening　进料口
2　end plate　端板
3　tie bolts　拉杆螺栓
4　heavy manganese steel hammer　重型锰钢锤
5　hammer carriers　锤盘
6　rotor　转子
7　bolts for casing side liners　壳体侧衬板螺栓
8　side liners for casing　壳体侧衬板
9　inspection door　检查门
10　flat cover　平盖
11　locking bars　锁紧棒
12　door　门
13　support bar　支承扁钢
14　grid frames　栅条架
15　grid bars　栅条
16　breaking plates　破碎板
17　cover plate　盖板
18　adjustable support bar for grid frames　栅条架的可调支承扁钢
19　grate basket　栅筐
20　anvil block　锤砧块
21　bearing oiling rings　轴承甩油环
22　bearing bushes in nalves　对开轴瓦
23　driving wheel　驱动轮
24　shearing pin bush　剪切销套
25　shearing pin　剪切销
26　cover　盖
27　end hammer carriers　端部锤盘
28　stop plate　挡板
29　bars for receiving basket　受料筐板条
30　end bars for receiving basket　受料筐端部板条
31　shearing pin　剪切销
32　flywheel　飞轮

11.1.8 Atomizer 粉碎机

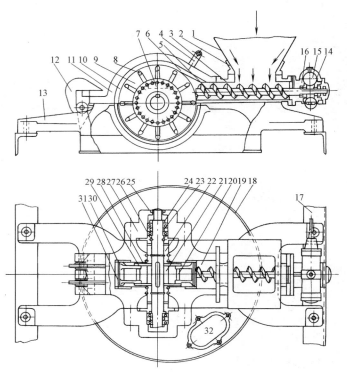

1 hopper 料斗
2 hopper base 料斗支座
3 screw case 螺杆箱体
4 grip 手柄
5 screw 螺杆
6 straight lining 直线段衬里
7 exhaust hole 排气孔
8 swing hammer 摆锤
9 side plate 侧板
10 lining plate 内衬板
11 upper case 上盖
12 hinge 铰链
13 machine body 机身
14 worm wheel 蜗轮
15 worm 蜗杆
16 worm case 蜗轮箱
17 machine supporter 机座
18 metal for product discharge 排料接口
19 feeding trough 送料槽
20 hammer pin 摆锤销轴
21 distance piece 间隔块
22 labyrinth nut 迷宫密封螺母
23 bearing distance piece 轴承间隔块
24 bearing nut 轴承螺母
25 oil seal 油封
26 radial ball bearing 径向球轴承
27 ball case 轴承箱
28 pin plate 销钉垫板
29 hammer plate 夹锤板
30 screen guide 挡板导架
31 screen 挡板
32 exhaust barrel 排气筒

11.2 Grinding equipment 研磨设备

11.2.1 Marcy grate-type continuous ball mill Marcy型栅板式连续球磨机

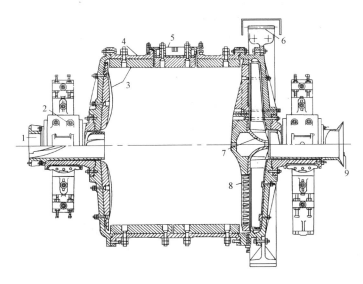

1. feed inlet 加料口
2. trunnion 耳轴
3. steel liner 钢衬里
4. shell 壳体
5. manhole cover 人孔盖
6. drive gear 传动齿轮
7. discharge cone 排料锥
8. steel grate 钢栅板
9. product outlet 产品出口

11.2.2 Types of ball-mill liners 球磨机衬里的类型

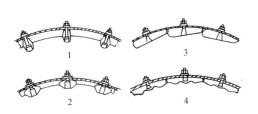

1. wedge bar liners
   楔形挡板衬里
2. Lorain liners
   Lorain型衬里
3. shiplap liners 搭叠衬里
4. corrugated liners
   波形衬里

11.2.3　Conical ball mill　锥形球磨机

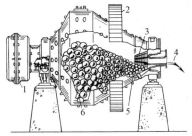

1　feed inlet　物料进口
2　drive gear　驱动齿轮
3　conical grate　锥形栅板
4　product outlet　产品出口
5　small balls　小球
6　large balls　大球

11.2.4　Rod mill　棒磨机

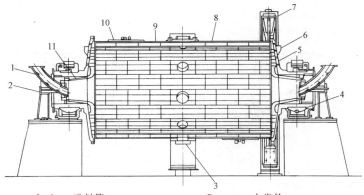

1　spout feeder　进料管
2　roding hole　磨棒装入口
3　discharge hole　成品出口
4　main bearing　主轴承
5　head　封头
6　end liner　侧衬板
7　gear　大齿轮
8　shell　筒体
9　shell liner　筒体衬板
10　man hole　人孔
11　hollow journal　空心轴颈

11.2.5　Mikro-Pulverizer hammer mill　Mikro 型锤磨机

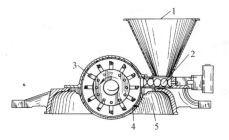

1　feed hopper　加料斗
2　screw feeder　螺旋加料器
3　T-shaped hammers　T 形锤子
4　perforated cylindrical screen discharge　多孔圆筒形排料筛
5　product outlet　产品出口

11.2.6 Mikro-pulverizer   Mikro 型粉磨机

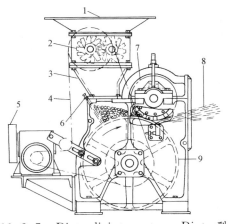

1 feed hopper   加料斗
2 double roll feed crusher
   双辊破碎给料机
3 crusher discharge chute
   破碎物料斜槽
4 chain drive for feed crusher and roll feed
   滚筒加料器与破碎给料机传动链
5 variable speed drive
   变速传动装置
6 cleanout door   清理门
7 pulverizer rotor bar hammer type
   带棒锤的粉碎机转子
8 pulverized material   磨碎物料
9 roll feeder   滚筒加料器

11.2.7 Rietz disintegrator   Rietz 型破碎机

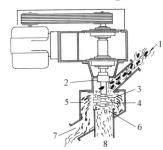

1 solid and/or liquid feed
   固体和（或）液体进料
2 seal   密封装置
3 hammers   锤子
4 screen   筛子
5 can be under pressure
   可在压力下操作
6 closure plate (optional)
   隔板（任选的）
7 primary discharge   主要排料口
8 secondary discharge (where needed or desired)   二次排料（视需要而定）

11.2.8 Whizzer air classification applied to Raymond Imp mill
具有离心空气分级的冲击型雷蒙磨碎机

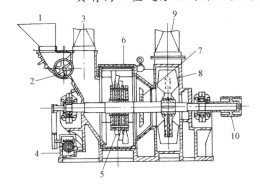

1 feed   进料
2 rotary feeder   旋转加料器
3 air inlet   空气入口
4 feeder speed reducer
   加料器减速器
5 swing hammers   摆动锤
6 replaceable liners   可更换的衬里
7 whizzer blades   离心机叶片
8 fan   风扇
9 air and material discharge
   空气与物料排出口
10 mill drive   磨碎机传动装置

11.2.9 Mikro-atomizer operating principle Mikro型粉碎机操作原理

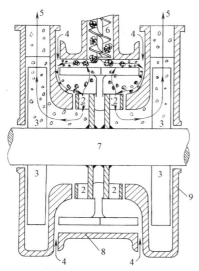

1 hammers or impact members 锤或冲击部件
2 classifier wheels 分级器叶轮
3 fan wheels 风扇叶轮
4 annular air inlets 空气环状入口
5 product outlets 产品出口
6 feed screw 进料螺旋
7 rotor shaft 旋转轴
8 anvil plate 砧面垫板
9 casing 外壳

11.2.10 Section of Mikro-ACM pulverizer Mikro-ACM型粉磨机的截面图

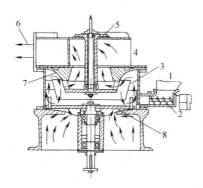

(illustrating air and material flow 图示了空气和物料的流向)

1 feed hopper 加料斗
2 feed screw 加料螺旋
3 shroud and baffle ring 箍环和挡圈
4 volute 蜗壳
5 separator shaft and bearing housing 分离器的轴和轴承箱
6 air and material out 空气和物料出口
7 air dispersion ring 空气分散环
8 pin rotor 钢栓转子

11.2.11 Raymond vertical mill 立式雷蒙磨

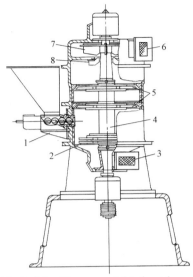

1. feeder 加料器
2. grinding hammers or element 研磨锤或研磨元件
3. air inlet 空气入口
4. main shaft 主轴
5. classifying whizzer blades 分级离心机叶片
6. air and material outlet 空气和物料出口
7. fan wheel 风扇轮
8. diffuser 扩散器

11.2.12 Raymond high-side mill with internal whizzer classifier 带有内离心分级器的高边雷蒙磨碎机

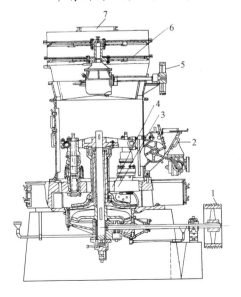

1. mill drive 磨碎机传动装置
2. feeder 加料器
3. grinding roller 研磨辊
4. grinding ring 研磨环
5. whizzer drive 离心机传动装置
6. revolving whizzers 旋转离心机
7. product outlet 产品出口

295

11.3 Non-rotary ball or bead mills 不旋转的球磨机或珠磨机

11.3.1 Vibro-energy mill 振动研磨机

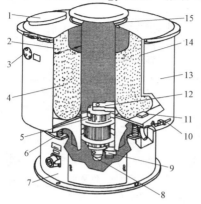

1 charge opening 装料孔
2 grinding chamber 研磨室
3 series grinding inlet 连续研磨入口
4 grinding media 研磨介质
5 motor 电动机
6 springs 弹簧
7 base 底座
8 angle lead graduated adjustment 导角逐步调节器
9 lower weight 下部重块
10 product discharge valve handle 排出产品的活门把手
11 media retainer （研磨）介质挡板
12 upper weight 上部重块
13 outer tank shell 容器外壳
14 abrasion-resistant lining 耐磨衬里
15 center column 中央圆筒

11.3.2 Vibratory ball mill 振动球磨机

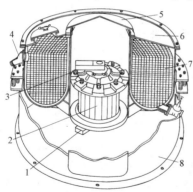

1 auxiliary motor weight 电机附加重块
2 motor 电机
3 upper motor weight 电机上端重块
4 weir type overflow 堰式溢流
5 center tube cover 中心管盖
6 mill cover 粉碎机盖
7 grinding chamber with media 存有介质的粉碎室
8 mill base 粉碎机底座

11.3.3 Batch vibrating ball mill 间歇式振动磨

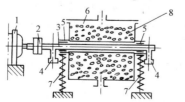

1 motor 电动机
2 coupling 联轴器
3 main shaft 主轴
4 eccentric weight 偏心重块
5 bearing 轴承
6 cylindrical shell 圆筒形机壳
7 pressure spring 压力弹簧
8 milling medium 研磨介质

11.3.4 Attrition mill 碾磨机

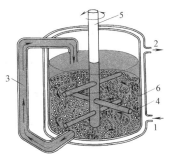

1 cooling water inlet 冷却水进口
2 cooling water outlet 冷却水出口
3 circulation system 循环系统
4 grinding medium 研磨介质
5 stirrer 搅拌器
6 radial arms 径向臂

11.4 Dispersion and colloid mills 分散磨和胶体磨

11.4.1 Dispersion mill 分散磨　　11.4.2 Colloid mill 胶体磨

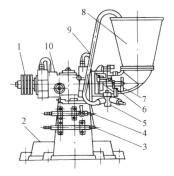

1 V-pulley   V形皮带轮
2 commen bed   底座
3 cooling water outlet   冷却水出口
4 cooling water inlet   冷却水进口
5 cock   旋塞
6 runner   碾轮
7 propeller   螺旋桨
8 hopper   料斗
9 recirculating pipe   循环管
10 main body   机体
11 cylinder   缸体
12 rotor   转盘
13 main shaft   主轴
14 liner   衬板
15 jacket   夹套
16 mechanical seal   机械密封
17 outlet   出口
18 metal   金属壳体
19 V-pulley   V形皮带轮
20 clearance adjuster   间隙调节器
21 impact pin   冲击销

11.4.3 Model M colloid mill M 型胶体磨

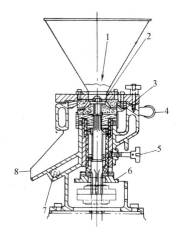

1. feed 进料
2. mixer breaks up material 混合器打碎物料
3. matched upper and lower stones 相配的上下磨石
4. water outlet(shown out of position) 水出口（未表示在应有的位置）
5. lock-screw keeps stones in adjustment 将磨石保持在调节位置的闭锁螺旋
6. adjustment wheel 调节轮
7. water inlet 进水口
8. discharge spout 排料口

11.5 Fluid-energy or jet mill 流能磨或气流粉碎机

11.5.1 Micronizer fluid-energy (jet)mill Micronizer 流能磨（气体粉碎机）

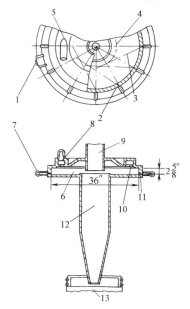

1. fluid inlet 流体入口
2. 12 drilled orifices 12 个钻孔
3. jet axes 射流轴线
4. hypothetical tangent circle 假想的相切圆
5. feed injector 射流器进料口
6. grinding chamber 研磨室
7. fluid pressure manifold 流体压力总管
8. feed manifold 进料集管
9. fluid outlet 流体出口
10. 8 feed inlets 8 个加料口
11. jet orifice 射流孔
12. concentric collector 同心收集器
13. bag or product bin 袋或成品仓

11.5 Fluid-energy or jet mill 流能磨或气流粉碎机

11.5.2 Trost jet mill Trost 型气流粉碎机

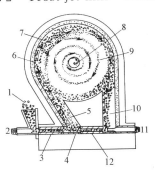

1　material input　物料加入
2　P jet　P 射流
3　P tube(variable sizes)
　　P 管（可改变尺寸）
4　impact chamber　撞击室
5　upstack　上升管
6　classification chamber　分级室
7　large particles　大颗粒
8　discharge(variable sizes)
　　排料（可改变尺寸）
9　fine particle separation　细颗粒分离
10　downstack　下降管
11　O jet　O 射流
12　O tube　O 管

11.5.3 Majac jet pulverizer Majac 型射流粉碎机

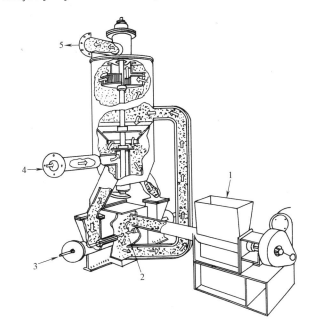

1　feed material　进料
2　pulverizing zone　粉碎区
3　compressed air, steam or gas　压缩空气，蒸汽或气体
4　fan air　风机空气
5　finished product　最终产品

11.5.4 Fluid-energy mill 流能磨

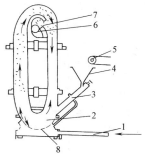

1 compressed air or superheated steam 压缩空气或过热蒸汽
2 grinding chamber 研磨室
3 injector 喷注器
4 hopper 料斗
5 feeder 加料器
6 material and spent fluid outlet 物料与废流体出口
7 classifier outlet 分级器出口
8 energizing nozzles 强化喷嘴

11.5.5 Jet mill with flat chamber 扁平室气流粉碎机

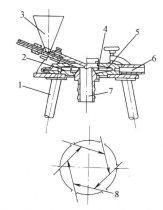

1 stand leg 机腿
2 milling chamber 粉碎室
3 feed hopper 给料斗
4 jet 喷嘴
5 jet annular disk 喷嘴环轮
6 compressed air inlet 压缩空气入口
7 discharge tube 出料管
8 jet direction of airstream 气流喷射方向

11.6 Crushing and grinding practice 破碎与研磨的实际应用

11.6.1 Open circuit grinding system 开路磨碎系统

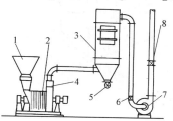

1 feed hopper 加料斗
2 turbo-mill with internal air classification 内分级蜗轮粉碎机
3 bag filter 袋滤器
4,6 butterfly valve 蝶阀
5 rotating segment feeder 星形给料器
7 exhaust blower 排风机
8 exhaust damper 排气风门

11.6.2 Simplified flow sheet of mine iron-ore concentrator 铁矿石富集器的简化流程

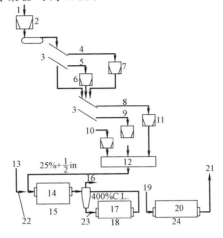

1　ore　矿石
2　48-in. gyratory crusher (6.5-in. close opening)　48in 回转破碎机 (6.5in 窄孔)
3　screens　筛子
4　+4in.　大于 4in
5　+2in.　大于 2in
6　−2in.　小于 2in
7　7-ft. cone crusher $\left(1\frac{1}{2}\text{-in. close opening}\right)$　7ft 圆锥破碎机 $\left(1\frac{1}{2}\text{in 窄孔}\right)$
8　+1in.　大于 1in
9　+1/2in.　大于 1/2in
10　$-\frac{1}{2}$in.　小于 1/2in
11　7-ft. cone shorthead $\left(\frac{3}{16}\text{-in. closed opening}\right)$　7ft 锥形短头 $\left(\frac{3}{16}\text{in 窄孔}\right)$
12　bins　料仓
13　water　水
14　rod mill 10×16ft　棒磨机 10×16ft
15　3 $\frac{1}{2}$-in. rods speed 66%～74% critical　棒长为 3 $\frac{1}{2}$in，速度为临界值的 66%～74%
16　to flotation and pelletizing　去浮选与造粒
17　ball mill 11×20ft　球磨机 11×20ft
18　1 $\frac{1}{2}$-in balls　1 $\frac{1}{2}$in 球
19　45%-325 mesh 70% solids　45%325 目 70%固体
20　regrind ball mill 11ft×20ft　再磨碎球磨机 11ft×20ft
21　to reflotation　去再浮选
22　78% solids　78%固体
23　68%～70% solids　68%～70%固体
24　$\frac{7}{8}$-in balls　$\frac{7}{8}$in 球

11.6.3 Simplified flow diagram of iron mine concentrator 铁矿石富集器简化流程图

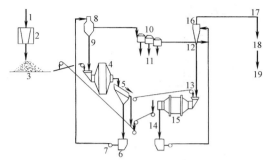

(with two autogenous wet-grinding stages
其中有两段湿式自磨机)

1　ore 25%～30% iron　矿石 25%～30%铁
2　60-in. gyratory crusher　60in 回转破碎机
3　blending bins　混合料仓
4　24×8ft. cascade mill　24×8ft 阶式磨机
5　screen　筛
6　sumps　槽
7　pump　泵
8　D. S. M. screen　D. S. M. 筛
9　recycle　再循环
10　magnetic cobbers　磁选组
11　tailings　选余物
12　feed+recycle　进料+再循环
13　$-2\frac{1}{2}$in. pebble feed　小于 $2\frac{1}{2}$in 卵石进料
14　chip return　碎粒返回
15　rock pebble mill 12ft×24ft.　12ft×24ft 卵石磨机
16　classifying cyclones　分级旋流器
17　−500mesh　小于 500 目
18　magnetic separators, filters, pelletizers　磁选分离器、过滤器、造粒机
19　63% iron　63%铁

11.6 Crushing and grinding practice　破碎与研磨的实际应用

11.6.4　Combined drying-grinding system　组合干燥-磨碎系统

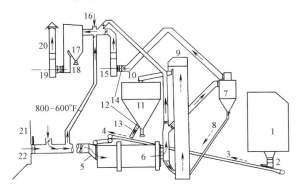

(using ball-mill and hot kiln exhaust gases　采用了球磨机和废热炉气)

1	feed bins　进料仓	12	air slide to storage 进入贮槽的气动滑阀
2	feed　进料	13	rejects　筛余粗料
3	main feed belt　主进料皮带	14	louvre damper　气门
4	cross feed belt　交叉进料皮带	15	high press. fan　高压风机
5	air lock　气闸	16	cold air inlet　冷空气入口
6	mill　磨机	17	dust filter　滤尘器
7	cyclone　旋风分离器	18	screw conveyor　螺旋输送机
8	oversize　大粒度	19	filter fan　过滤器风机
9	elevator　提升机	20	to atmosphere　通大气
10	air slide 12 in.　气动滑阀 12in	21	kiln dust chamber　炉子除尘室
11	separator　分离器	22	kiln gas　炉气

11.6.5　Hammer mill in closed circuit with air classifier
带空气分级器的锤磨机闭路循环

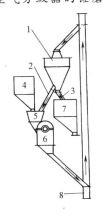

1　mechanical air classifier
　　机械式空气分级器
2　tailings　尾渣
3　fines　细粒
4　mill feed bin
　　磨碎机料仓
5　hopper　料斗
6　mill　磨机
7　finished product bin
　　最终产品仓
8　elevator　提升器

12 Mixing and Separation Equipments for Solids
固体的混合与分离设备

12.1 Solids mixing machines 固体物料混合机械

12.1.1 Rotary mixer 回转圆筒混合器

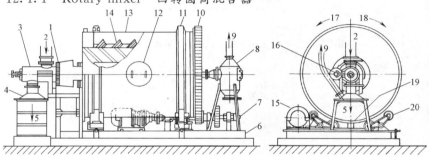

12.1.2 Vertical screw mixer 立式螺杆混合器

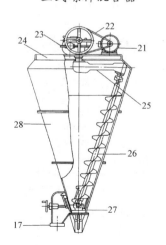

1 gear for screw feeder 螺旋送料机齿轮
2 feed 加料
3 screw feeder 螺旋送料机
4 cover of dust collector 集尘器顶盖
5 product 成品
6 common base 底座
7 drive shaft 驱动轴
8 separator 分离器
9 air 空气
10 drum gear 转鼓齿轮
11 tire 滚圈
12 man hole 人孔
13 drum 转鼓
14 mixing blade 搅拌叶片
15 geared motor 带变速齿轮箱的电动机
16 clutch 离合器
17 outlet 出料口，排出口
18 mixing 混合
19 driving rolls 驱动滚轮
20 supporting rolls 支承滚轮
21 motor 电动机
22 V-belt V形皮带
23 speed reducer 减速装置
24 inlet 加料口
25 swing arm 摆动臂
26 mixing screw 混合螺杆
27 lower bearing 下轴承
28 cone body 圆锥形壳体

304

12.1.3 Several types of solid mixing machines
固体物料混合机械的若干型式

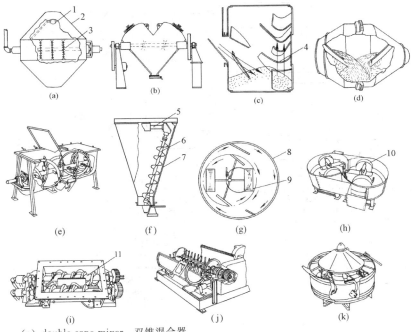

(a) double cone mixer 双锥混合器
(b) twin shell (Vee) mixer 双筒混合器（V型）
(c) horizontal drum mixer 水平转鼓混合器
(d) double-cone revolving around long axis (with baffles) 绕长轴旋转的双锥混合器（装有挡板）
(e) ribbon mixer 螺条混合器
(f) vertical screw (orbiting type) 立式螺旋混合器（环行式）
(g) batch muller 间歇式滚轮混合器
(h) continuous muller (stationary shell) 连续式滚轮混合器（固定壳）
(i) twin rotor (adapted to heat transfer-jacketed body and hollow screws) 双转子混合器（适于用机身装热交换夹套和空心螺旋）
(j) single rotor 单转子混合器
(k) turbine 蜗轮混合器

1　spray nozzle 喷嘴
2　tumbler 转鼓
3　breaking device 粉碎装置
4　baffles 挡板
5　orbiting arm （做圆周运动的）旋转臂
6　screw 螺杆
7　conical tank 锥形罐
8　pan (rotates clockwise) 槽（顺时针转动）
9　muller turret (rotates counterclockwise) 滚轮（反时针方向转动）
10　muller 滚轮
11　hollow screws 空心螺旋

12.1.4 Drum V-type and double-cone mixers 转鼓型 V 型和双锥型混合器

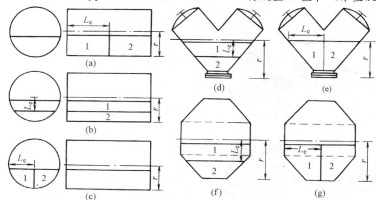

1 black particles 黑粒子   2 white particles 白粒子
(a) drum mixer-end to end loading 转鼓混合器（两端进料）
(b) drum mixer-layer-by-layer loading 转鼓混合器（分层进料）
(c) drum mixer-side-by-side loading 转鼓混合器（两侧进料）
(d) V-type mixer-layer-by-layer loading V 形混合器（分层进料）
(e) V-type mixer-side-by-side loading V 形混合器（两侧进料）
(f) double cone mixer-layer-by-layer loading 双锥形混合器（分层进料）
(g) double cone mixer-side-by-side loading 双锥形混合器（两侧进料）

12.1.5 Multiple-loop recirculation blending system 复合回路循环混合系统

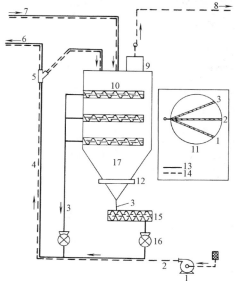

1　air blower　鼓风机
2　1500 SCFM air　每分钟 1500 标准 ft³ 空气
3　50000 lb/h solid　每小时 50000lb 固体
4　recycle 100000 lb/h solids (1500 SCFM air)　循环每小时 100000lb 固体（每分钟1500标准 ft³ 空气）
5　diverter valve　三通阀
6　final product　最终产品
7　powder stock and air　粉末原料和空气
8　air out　气体出口
9　dust collector　粉尘收集器
10　screw conveyors　螺旋输送机
11　sectional view of conveyor orientation　输送机定向剖面图
12　bin activator　料仓抖动器
13　powder　粉料
14　air　空气
15　twin-screw feeder　双螺杆加料器
16　rotary-vane feeder　回转叶片加料器
17　200000 lb capacity　200000 lb 容量

12.2 Screening machines 筛分机械
12.2.1 Motions of screens 筛子的振动方式

(a) gyrations in horizontal plane
水平面回转式

(b) gyrations in vertical plane
垂直面回转式

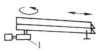

(c) gyrations at one end, shaking at other
一端回转，另一端摇动式

(d) shaking
振荡式

(e) mechanically vibrated
机械振动型

(f) electrically vibrated
电动振动型

12.2.2 Two-deck screen separator 双层筛分离器

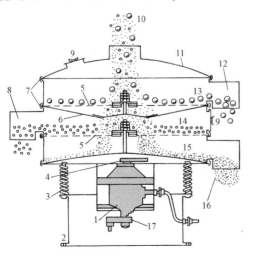

1　motor　电机
2　base　底座
3　spring assembly　弹簧组件
4　eccentric top weight　顶部偏心块
5　screen deck　筛板
6　feed tray　送料盘
7　clamp rings　夹紧环
8　middle-frame discharge　中部框式卸料口
9　inspection port　观察孔
10　material feed　进料
11　dust cover　防尘罩
12　top-frame discharge　顶部框式卸料口
13　top frame　顶部框架
14　middle frame　中部框架
15　bottom frame　底部框架
16　bottom-frame discharge　底部框式卸料口
17　eccentric bottom weight　底部偏心块

12.2.3 Gyrating screens 回转振动筛

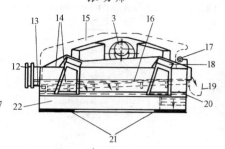

(horizontally gyrated 水平回转式)

1　eccentric　偏心轮
2　eccentric vibrator　偏心式振动器
3　vibrator　振动器
4　motor　电动机
5　feed　加料
6　upper screen　上层筛
7　flexible connection　挠性接头
8　coarse　粗料
9　fines　细料
10　lower screen　下层筛
11　ball cleaner(s)　圆形清理球
12　flexible inlet flange sealing　挠性进口法兰密封件

12.2.4 Vibrating screen 振动筛

13　inlet sealing　进口密封件
14　screen cradle　筛摇架
15　hood made of glass fibre plastics　玻璃钢罩
16　screen plate　筛板
17　shower pipe　喷淋管
18　rubber supports　橡胶垫
19　separated tailing trough　分出的粗粉料槽
20　vat　受料槽
21　damping rubber plate　减振橡胶板
22　concrete plate　混凝土板

12.2.5 Rotex-screen 转动筛分机

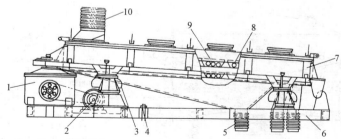

1　head motion　传动机头
2　motor　电动机
3　dust cover　防尘罩
4　foundation bolt　地脚螺栓
5, 10　flexible joint　挠性接头
6　under frame　机座
7　screen frame　筛架
8　gum ball　树胶球
9　screen　筛网

12.3 Dry classification 干式分级

12.3.1 Gayco centrifugal separator Gayco 型离心分离器

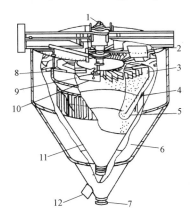

1 feed pipe 进料管
2 circulating fan 循环风机
3 centrifugal fan 离心风机
4 baffle 折流板
5 air circulation 空气循环
6 fines cone 细粉锥筒
7 fines discharge 细粉排出口
8 shutter adjustment 风门调节器
9 separating chamber 分离室
10 distributing plate 分散板
11 tailings cone 粗粉锥筒
12 tailings discharge 粗粉排出口

12.3.2 Pneumatic classifier 气力分级器

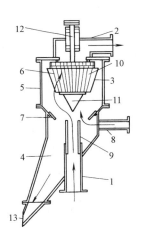

1 conveyer pipe for material 给料管
2 outlet duct for fine material 细粉排出管
3 rotor 转子
4 outlet duct for tailing 粗粉排出管
5 separator body chamber 分级器筒体
6 air way slits in rotor 转子（叶片间）狭缝空气通道
7 elutriation ring 扬析环
8 secondary air inlet pipe 二次空气入口管
9 adjustable pipe for feed flow 进料调节管
10 rotor blades 转子叶片
11 rotor guide cone 转子导流锥体
12 shaft 转轴
13 tailing outlet 粗粉排出口

12.3.3 Double-cone air classifier 双锥空气分级器

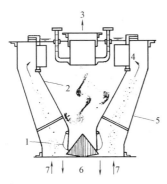

1 flap valves 片状阀
2 inner cone 内锥体
3 airborne product 气载产品
4 vanes 导流叶片
5 outer cone 外锥体
6 oversize particles 过大颗粒
7 airborne material to be classified 不同粒度的气载原料

12.3.4 Air classifier 空气分级器

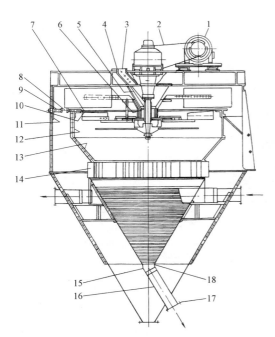

1 motor 电动机
2 Vee-belt V形带
3 feed hopper line 供料斗管线
　feed hopper liner(left) 料斗衬板(左)
4 feed hopper liner(right) 料斗衬板(右)
5 intake cone liner(upper) 进口锥体衬板(上)
6 intake cone liner(lower) 进口锥体衬板(下)
7 valve 阀板
8 valve rod 阀杆
9 handle 手柄
10 inside drum cover 内筒盖
11 outside casing liner 外壳衬板
12 inside drum liner (cylinder) 内筒衬板(圆筒)
13 inside drum liner(cone) 内筒衬板(圆锥)
14 air vane 通气格子板
15 tailing chute 粗粉溜槽
16 tailing cone elbow 弯头
17 tailing chute flange 粗粉溜槽法兰
18 tailing elbow clamp ring 粗粉溜槽夹紧环

12.3.5 Detail drawing of air classifier 空气分级器的局部详图

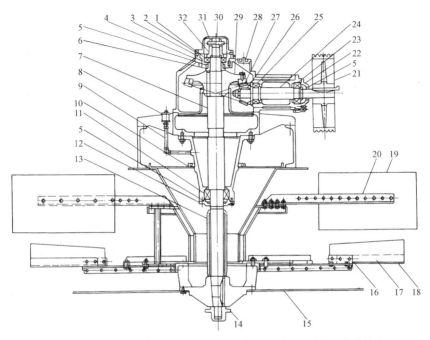

1	bearing(upper) 上轴承		17	auxiliary blade 辅助叶片
2	band 卡箍		18	auxiliary blade wear 辅助叶片防磨板
3	top bearing adjusting nut 上轴承调节螺母		19	fan blade 风机叶片
4	top bearing housing 上轴承箱		20	fan blade arm 风机叶片支架
5	oil seal 油封		21	V-belt pulley V形皮带轮
6	spacer 定距片		22	bearing cover 轴承盖
7	gear shaft 传动轴		23	pinion shaft outside bearing 小齿轮轴外侧轴承
8	glass oiler with screen 带滤网的玻璃油杯		24	pinion shaft 小齿轮轴
9	oil pipe 油管		25	inside pinion bearing 小齿轮轴内侧轴承
10	dust seal 防尘密封环		26	pinion 小齿轮
11	lower bearing housing 下轴承箱		27	gear 大齿轮
12	lower bearing housing cap 下轴承箱盖		28	oil hole cover 给油口盖
13	gear shaft liner 轴衬套		29	gear ring 大齿轮盘
14	clamping nut 锁紧螺母		30	top bearing cover 上轴承盖
15	lower distribution plate 下分配板		31	top bearing nut lock 上轴承锁紧母
16	blade retainer 叶片安装底架		32	top bearing shaft nut 上轴承螺母

12.4 Wet classifiers 湿式分级器

12.4.1 Wet classification machine 湿式分级设备

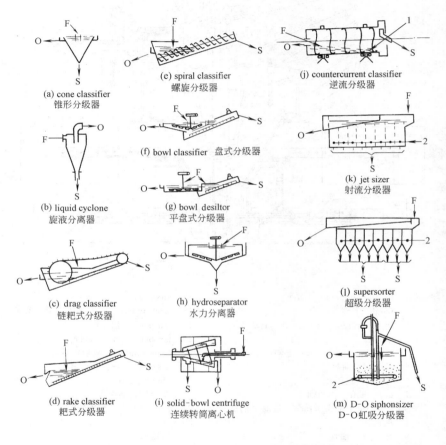

(a) cone classifier 锥形分级器
(b) liquid cyclone 旋液分离器
(c) drag classifier 链耙式分级器
(d) rake classifier 耙式分级器
(e) spiral classifier 螺旋分级器
(f) bowl classifier 盘式分级器
(g) bowl desiltor 平盘式分级器
(h) hydroseparator 水力分离器
(i) solid-bowl centrifuge 连续转筒离心机
(j) countercurrent classifier 逆流分级器
(k) jet sizer 射流分级器
(l) supersorter 超级分级器
(m) D-O siphonsizer D-O虹吸分级器

F——feed 加料; O——overflow product 溢流产品; S——sand product 砂浆产品

1　wash water　洗涤水

2　H. W.——hydraulic water　加压水

12.4.2 Crossflow wet-settling classifier 错流湿式沉降分级器

12.4.3 Rake classifier 耙式分级器

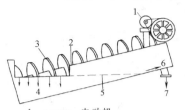

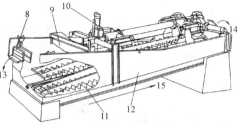

1　motor　电动机
2　feed　加料
3　helical conveyor　螺旋输送器
4　slimes overflow　泥浆溢流
5　liquid level　液面
6　settled solids　沉积固体
7　sands discharge　排砂
8　overflow weir　溢流堰
9　feed launder　进料槽

10　lifting mechanism　提升机械
11　rake　耙
12　sloping bottom tank　斜底槽
13　overflow and fines　细粒随溢流而出
14　coarse material　粗粒料
15　motion caused by rake　耙子使物料（向上）移动

12.5 Dense-media separation 稠密介质分离

12.5.1 Heavy-media separation 重介质分离

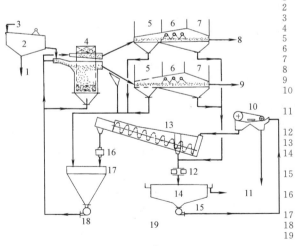

1　fines　细（颗）粒
2　pre-wet screen　预湿筛
3　feed　进料
4　separator　分离器
5　drain　排水
6　rinse　洗涤
7　dewater　脱水
8　sink　废弃
9　float　浮选
10　magnetic separator　磁力分离器
11　to clarifying system　去澄清系统
12　magnetizing block　磁铁
13　densifier　增浓器
14　medium thickener　介质沉降槽
15　contaminated medium pump　杂质泵
16　demagnetizing coil　去磁线圈
17　medium scmp　介质储槽
18　medium pump　介质泵
19　standard heavy media circuit　标准重介质分离系统

313

12.5.2 Drag-tank-type dense-media separatory vessel 拖曳槽型稠密介质分离器

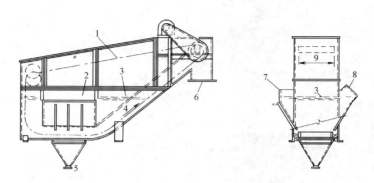

12.5.3 Revolving-drum-type dense-media separatory vessel 转筒式稠密介质分离器

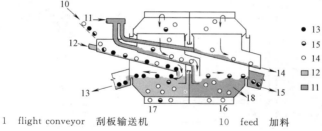

1 flight conveyor   刮板输送机
2 sluice opening   斜槽口
3 media level   介质的液面
4 travel   前进方向
5 drain outlet   排水出口
6 refuse discharge 残渣卸出口
7 outlet sluice   出口斜槽
8 inlet sluice   入口斜槽
9 flight width   刮板宽度
10 feed   加料
11 high-gravity medium   高密度介质
12 low-gravity medium   低密质介质
13 float   上浮物
14 sink   下沉物
15 middling   中间物
16 high-gravity compartment   高密度室
17 low-gravity compartment   低密度室
18 medium   介质

12.5.4 Dense-media cone vessel arrangements
锥形稠密介质分离器结构

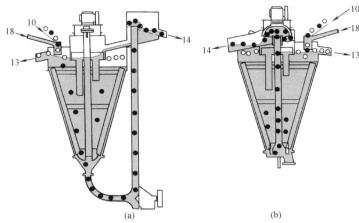

(a) single-gravity two-product system with pump sink removal
用泵排出下沉产品的单纯重力式两产品系统
(b) single-gravity two-product system with compressed-air sink removal
用压缩空气排出下沉产品的单纯重力式两产品系统

12.5.5 Typical dense-media flow sheet for coal cleaning plant
用于煤净化工厂的典型的稠密介质分离流程

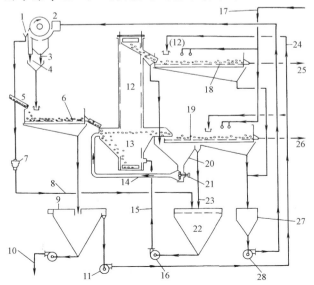

12.5 Dense-media separation 稠密介质分离

1. magnetic concentrate 磁力分离的富集物
2. magnetic separator 磁力分离器
3. tailings 尾砂
4. overflow 溢流
5. feed 加料
6. preconditioning screen 预调节筛
7. demagnetizing coil 去磁煤
8. recycle-medium concentrate 再循环的中间富集物
9. setting tank or cyclones 沉降槽或旋液分离器
10. fines to treatment for recovery or disposal 细粒去进行回收处理或排弃
11. recycle-rinsewater pump 再循环的淋洗水泵
12. elevating conveyor for sink material 下沉物料提升机
13. separating vessel 分离槽
14. heavy-medium circulation 重介质循环
15. heavy-medium return 重介质返回
16. heavy-medium pump 重介质泵
17. fresh rinse water 新鲜淋洗水
18. sink-material rinse screen 下沉物料淋洗筛
19. float-material drain and rinse screen 上浮物料排出和淋洗筛
20. circulating vessel 循环槽
21. axial flow low head circulating pump 轴流式低压头循环泵
22. heavy-medium storage sump 重介质储槽
23. heavy-medium overflow 重介质溢流
24. recycled rinse water 再循环淋洗水
25. sink product 下沉的产品
26. float product 上浮产品
27. dilute-medium sump 稀介质储槽
28. dilute-medium pump 稀介质泵

12.6 Magnetic separators 磁力分离器

12.6.1 Magnetic pulley
磁性转轮

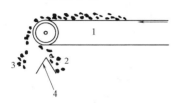

12.6.2 Magnetic separator
磁力分离器

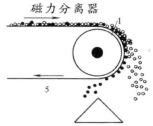

12.6.3 Arrangement of magnetic drum separators 磁性鼓式分离器的布置

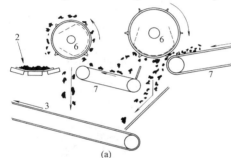

(a)　　　　　　　　　(b)

12.6.4 Multiple induced-roll magnetic separator
多级感应磁辊分离器

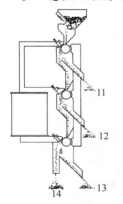

(a) magnetic drum operating as a lifting magnet　磁性鼓用作提升磁体
(b) magnetic drum operating as a pulley　磁性鼓用作转轮
1　magnetic pulley　磁性转轮
2　magnetic fraction　磁性组分
3　nonmagnetic fraction　非磁性组分
4　splitter　分流器
5　magnetic material　磁性材料
6　magnet　磁铁
7　belt conveyor　皮带输送机
8　feed distributor　加料分布器
9　stationary magnet　固定磁铁
10　revolving shell　旋转壳体
11　magnetic concentrate 1　磁性富集物1
12　magnetic concentrate 2　磁性富集物2
13　magnetic concentrate 3　磁性富集物3
14　nonmagnetic concentrate　非磁性富集物

12.6 Magnetic separators 磁力分离器

12.6.5 Concurrent wet-drum separator 并流型湿式鼓式磁力分离器

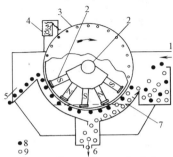

12.6.6 Counterrotation wet-drum separator 逆流转动型湿式鼓式磁力分离器

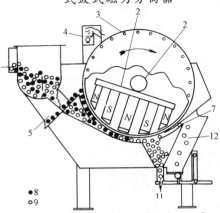

12.6.7 Operating principle of alternating-polarity Ball-Norton separator 交变极性磁力分离器操作原理

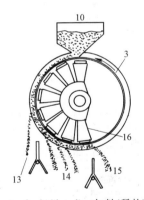

12.6.8 Operating principle of dry-fines permanent magnet separator 永磁体干细粉磁力分离器操作原理

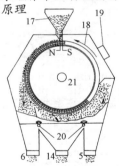

1　feed(diluted)　加料(稀的)
2　stationary-magnets position adjustable by mounting shaft　通过安装的轴使固定的磁性状态可调
3　rotating drum　转鼓
4　drum wash　转鼓洗液
5　magnetic concentrates　磁性富集物
6　non-magnetic tailings　非磁性尾砂
7　pulp level　浆状物料面
8　magnetics　磁性
9　non-magnetics　非磁性
10　feed　加料
11　non-magnetic coarse tailings　非磁性粗尾砂
12　overflow(carrying nonmagnetic fine tailings)　溢流(携带非磁性细粒尾砂)
13　tailings　尾砂
14　middling　中级品
15　concentrate　富集物
16　coils　感应线圈
17　feed hopper　加料斗
18　drum rotation　转鼓旋转方向
19　to dust collector　接集尘器
20　adjustable splitters　可调的分流器
21　stationary permanent-magnet assembly 固定永磁体组合体

12.6.9 Wet drum separator 湿磁鼓分离器

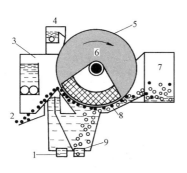

1　overflow discharge　溢流卸出
2　magnetic material　磁性材料
3　repulping box　二次浆化箱
4　drum wash　磁鼓洗涤
5　rotating drum　旋转磁鼓
6　magnetic drum　磁鼓
7　feed　进料
8　stationary magnet assembly　固定磁性装置
9　nonmagnetic tailings discharge　非磁性渣料卸出

12.6.10 Dry drum separator 干磁鼓分离器

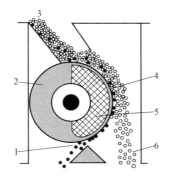

1　automatic magnetic discharge　磁性体自动卸料
2　revolving cylinder　旋转圆筒体
3　mixed material　混合物料
4　working face　工作表面
5　stationary Indox V magnet assembly　固定的 V 型 Indox 永久磁铁组件
6　nonmagnetic material　非磁性材料

12.6.11 Induced-roll separator 磁感应分离器

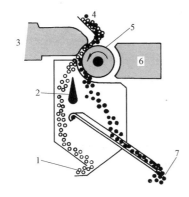

1　nonmagnetic material　非磁性材料
2　adjustable splitter　可调分流器
3　primary magnet pole　主要磁极
4　feed　进料
5　magnetic roll　磁鼓
6　magnet　磁铁
7　magnetic material　磁性材料

12.7 Electrostatic separator and optical separator 静电分离器和光学分离器

12.7.1 Electrostatic separator 静电分离器

12.7.2 Separation by ionic bombardment 利用离子辐射分离

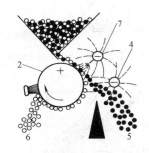

12.7.3 Electrostatic sieve for fine metal powders 用于细金属粉的静电筛

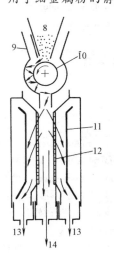

1 active electrode 激活电极
2 earthed [grounded] rotor 接地转子
3 splitter 分流器
4 static electrode 静电极
5 conductors 导体介质
6 dielectrics 绝缘材料
7 ionic electrode 离子电极
8 feed 加料
9 attracting plate(−) 吸引板(−)
10 electrostatic feeder 静电加料器(+)
11 attracting grid(−) 吸引栅(−)
12 325 or finer mesh screen(+) 325目或更细目的筛网
13 undersize 筛下物料
14 oversize 筛上物料

12.7.4 Sortex 711M optical separator　Sortex 711M 光学分离器

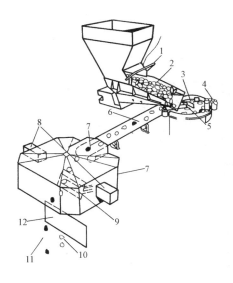

1　water spray(wet machine only)
　　喷水器(只指湿式机器)
2　vibrating feeder　振动加料器
3　aligning disc　调准盘
4　belt drive motor　皮带的驱动电机
5　aligning belts　调准皮带
6　conveyor belt　输送带
7　optical chamber　光室
8　cameras　照相机
9　high speed air valves　高速空气阀
10　main stream　主物流
11　deflected stream　偏离的物流
12　dividing edge　分流板

12.7.5 Diamond separation process　金刚石分离工艺

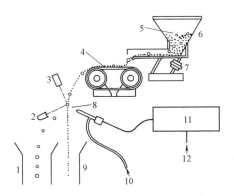

1　diamond bin　金刚石料仓
2　photomultiplier tube
　　光电倍增管
3　X-ray source　X 射线源
4　feed belt　加料传送带
5　diamond concentrate
　　金刚石精矿
6　feed hopper　料仓
7　vibrator　振动器
8　air jet　空气喷嘴
9　waste bin　废料仓
10　compressed air　压缩空气
11　electronic amplifier and ejector
　　control unit　电子放大器和喷射
　　器控制设备
12　from photomultiplier tube　源自
　　光电倍增管

(uses X-rays to detect diamonds in ore　采用 X 射线从矿石中)

321

12.8 Flotation 浮选

12.8.1 Flotation 浮选

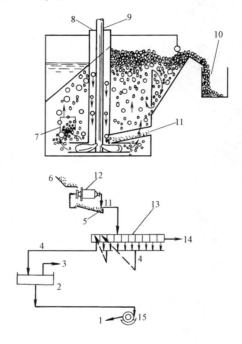

(a popular method for separating solids 分离固体的一种普遍使用方法)

1. lead concentrate 浓铅
2. thickener 浓缩器
3. overflow(waste) 溢流(废水)
4. froth 泡沫
5. classifier 分级器
6. fine ore feeder (Pb-Zn) 细矿进料器(铅-锌)
7. tailing outlet 尾砂出口
8. air 空气
9. impeller shaft 旋转混合器轴
10. concentrate 精选矿
11. feed 进料
12. ball mill 球磨机
13. flotation machine (lead) 浮选机械(铅)
14. to zinc flotation 去锌的浮选
15. filter 过滤机

12.8.2 Flow sheet of a simple flotation plant
简单的浮选工厂流程图

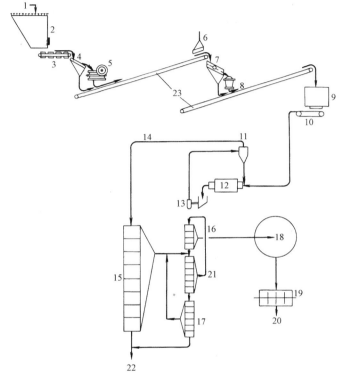

1	ore 矿石		13	pump 泵
2	coarse-ore bin 粗矿石储槽		14	flotation feed 为浮选加料
3	apron feeder 板式加料器		15	rougher flotation 粗器浮选
4	grizzly 栅筛		16	recleaner flotation 再净化器浮选
5	jaw crusher 颚式破碎机		17	cleaner-scavenger flotation 净化器排出物浮选
6	magnet 磁铁		18	thickener 增稠器
7	vibrating screen 振动筛		19	filter 过滤器
8	secondary cone crusher 二级圆锥轧碎机		20	final concentrate 最终富集物
9	fine ore bin 细矿石储槽		21	cleaner flotation 净化器浮选
10	feeder 加料器		22	final tailing 最终的尾砂
11	cyclone 旋液分离器		23	conveyors 输送机
12	ball mill 球磨机			

13 Granulation Equipments 造粒设备

13.1 Prilling tower 造粒塔

13.1.1 Prilling tower for ammonium nitrate 硝酸铵造粒塔

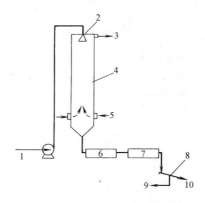

(product size range 0.4~2.0mm, the dryer is not needed if the moisture content of the melt is less than about 0.5% 成品相当粒径为 0.4~2.0mm，若熔融物含湿量低于 0.5%，则无需干燥器)

    1 ammonium nitrate melt　熔融硝酸铵
    2 atomizing device　雾化器
    3 air outlet　空气出口
    4 prilling tower　造粒塔
    5 cooling air　冷却空气
    6 dryer　干燥器
    7 cooler　冷却器
    8 screens　筛子
    9 fines　细粉
    10 ammonium nitrate pills to coating and bagging　硝酸铵颗粒去包衣（涂膜）和装袋

13.1.2 Prill spray assembly 造粒喷头组件

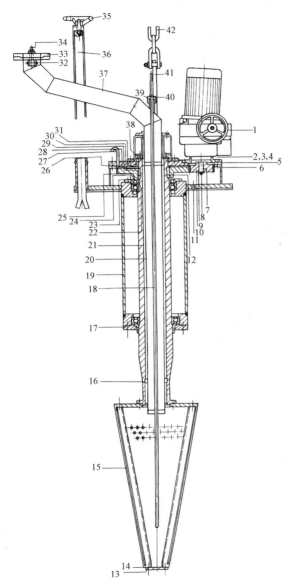

1. motor reducer variator 电机减速机
2. stud end 双头螺栓
3. washer 垫圈
4. nut 螺母
5. cover 盖
6. screw 螺钉
7, 12 key 键
8. bolt 螺栓
9. ring 环
10, 28 tooth 齿轮
11. plug 丝堵
13. cover 底盖
14, 26 gasket 垫片
15. prill bucket 造粒筒
16. bearing 轴承
17. cover 轴承底盖
18. level indication pipe 液面指示管
19. support prill bucket 造粒筒支承套
20. charging funnel 进料筒
21. external shaft 外轴
22, 31 outer safety ring 安全外挡圈
23. cover 轴承上盖
24, 30, 39 O-ring O形密封圈
25. oil sealed ring 油封环
27. air breather 透气管
29. distance ring 间隔环
32. pin 销子
33. flange 法兰
34. clamping bolt 夹紧螺栓
35. handwheel 手轮
36. column 立柱
37. turn tube 弯管
38. ring 环
40. gland 压盖
41. lifting yoke 吊钩
42. chain 链

325

13.2 Extrusion pelleting equipment 挤压造粒设备

13.2.1 Two types of extrusion pelleting equipment 两种挤压造粒设备

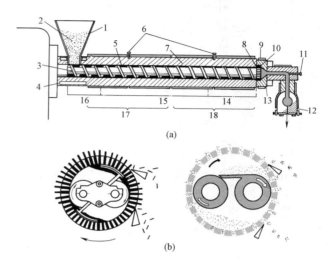

(a) Screw-type extruder for molten plastics 用于熔融塑料的螺杆式挤压机
(b) Ring extruders 圆环挤压机

1　hopper　料斗
2　resin　树脂
3　screw　螺杆
4　hopper cooling jacket　料斗冷却夹套
5　hardened liner　淬硬套管
6　thermocouples　热电偶
7　barrel　机筒
8　screen pack　筛组
9　adapter heater　管接头加热器
10　breaker plate　缓冲衬板
11　melt thermocouple　熔融物料热电偶
12　die　模具
13　adapter　管接头
14　metering section　计量段
15　compression section　压缩段
16　feed section　加料段
17　back heat zone　后加热区
18　front heat zone　前加热区

13.2.2 Equipment for compacting, briquetting, and pelleting 压实、团块与造粒设备

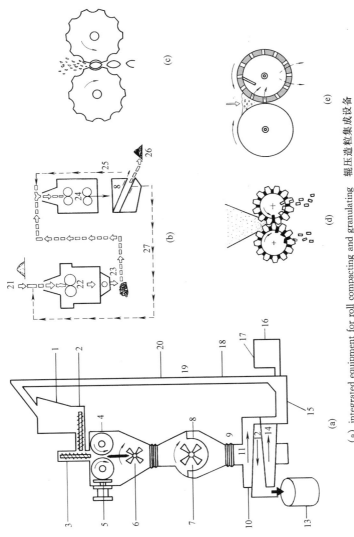

(a) integrated equipment for roll compacting and granulating 辊压造粒集成设备
(b) flowsketch of a process for compacting fine powders, then granulating the mass 先压实细粉末后造粒工艺流程
(c) briquetting rolls 压块辊子 (d) gear pelleter 齿轮造粒机 (e) double roll extruder 双辊挤压机

13.2 Extrusion pelleting equipment 挤压造粒设备

1　upper feed hopper (collects virgin and recycle product)
　　上方加料斗（聚集原始料与循环料）
2　horizontal feed screw (controlled feed)　水平螺旋加料（可控加料量）
3　vertical screw (pre-compresses and deaerates/product)
　　垂直螺杆（对产品预压缩与脱气）
4　compaction rolls　挤压双辊
5　pressure applied, air to hydraulic actuator regulates pressure exerted on rolls
　　施加压力（空气对液压传动装置调控对辊子施加的压力）
6　prebreak (breaks sheets into chips and flakes)　预破碎（将大片打碎成鳞片状小片）
7　granulator (sizes compacted material to desired particle size)
　　造粒（将物料压实为所设定的粒径）
8　screen　筛板
9　screener　筛选器
10　overs (oversized granules recycled)　筛上粗料（筛上粗粒循环回去）
11　overs　筛上粗料
12　product　产品
13　finished product (granular free flowing dustless controlled particle size)
　　成品（颗粒自由流动无粉尘粒径合格）
14　fines　细粉末
15　recycle　循环（物料）管道
16　lower hopper　下方料斗
17　original powder feed　原始粉体进料
18　recycle system　循环装置
19　oversize and fines for reprocessing　重新再加工的筛上粗料与筛下细末
20　feed and recycle　（加入的）新料与循环料
21　fine material　细粉料
22　compacting mill　挤压机
23　flake breaker　切片机
24　granulating mill　成粒机
25　oversize returned　粒度过大的返回
26　granular product　颗粒成品
27　undersize returned　粒度过小的返回

13.2.3 Single-screw extruder 单螺杆挤压机

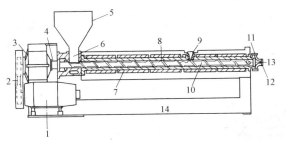

1 drive motor 驱动电机
2 drive belt 传动皮带
3 gear reducer 齿轮减速器
4 thrust bearing 止推轴承
5 hopper 料斗
6 feed throat 加料喉管
7 heater/cooler 加热器/冷却器
8 barrel 机筒
9 vent 放气口
10 screw 螺杆
11 clamp 夹紧装置
12 die 模具
13 filtration plate and screens 过滤板和筛子
14 base 机座

13.2.4 Pelletizing plant 造粒设备

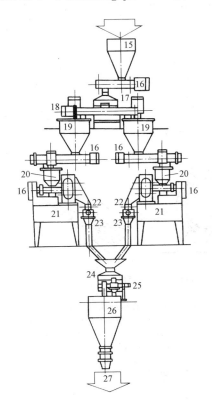

15 bin 料斗
16 screw feeder 螺旋加[送]料器
17 screening unit 筛选机组
18 distributor with helical blades 带螺旋形叶片的分配管
19 deaerating tank 脱气槽
20 stirred vessel 搅拌容器
21 sheeting machine 压片机
22 breaking plant 粉碎设备
23 primary shaper 一次整形机
24 screening unit for finished product 成品筛
25 secondary shaper 二次整形机
26 finished product tank 成品罐
27 finished stock 成品

13.3 Rotating dish granulators 转盘造粒机

13.3.1 Sketch of a rolling drum granulator 滚筒式造粒机示意图

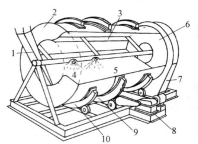

1　solid feed chute　固体进料斜槽
2　inlet dam ring　进口挡料圈
3　scraper bar　刮刀杆
4　sprays　雾滴
5　granule bed　颗粒床
6　exit dam ring　出口挡料圈
7　exit chute　出口斜槽
8　drive assembly　驱动装置组件
9　girt gear　齿圈
10　riding ring　滚圈

13.3.2 Flow charts for tumbling agglomerators 滚动团聚器流程图

 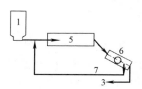

1　bin　料仓
2　disc　转盘
3　product　产品
4　disc or pan　转盘或盆
5　balling drum　成球转鼓
6　screen　筛网
7　recycle　循环（返回）

13.3.3 Rotating dish granulator 转盘造粒机

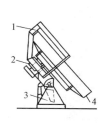 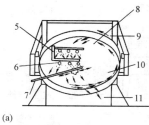 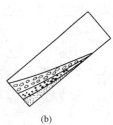

(a)　　　　　　　　　　　　　　　(b)

(a) edge and face view of a dish granulator (diameters to 25ft).
　　盘式造粒机（直径达 25ft）的侧视图与正视图
(b) stratification of particle sizes during rotation　回转过程粗细粒子分层现象

1　frame　框架
2　drive　驱动装置
3　base support　机座
4　collar　套圈
5　concentrated solution　浓溶液
6　solution sprays　溶液雾滴
7　reciprocating scraper　往复式刮刀
8　recycle fines　循环返回的细粉末
9　rotation　回转
10　undersize　细粒
11　product　产品

13.3.4 Pan granulator 盆式造粒机

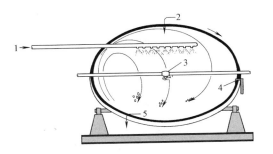

1 feed melt 熔融物料加入
2 recycle 循环物料
3 reciprocating floor scraper 往复式平台刮刀
4 stationary sidewall scraper 固定的侧壁刮刀
5 granules 粒料

(sprays melt over stirred mass of seed particles 熔融物料喷洒在被搅拌的晶种粒子上)

13.3.5 Pan granulation 盘式固化

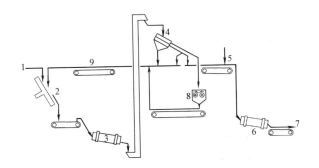

(solidifies and cools in separate operations 造粒与冷却分步进行)

1 feed melt 熔融物料加入
2 pan granulator 盘式造粒机
3 precooler 预冷却器
4 screens 筛分
5 other trains 其他辅料
6 product cooler 产品冷却器
7 product 成品
8 crusher 压碎机
9 recycle 循环物料

13.4 Fluidized bed and spouted bed granulators 流化床与喷动床造粒机

13.4.1 Fluidized bed and spouted bed granulators 流化床与喷动床造粒机

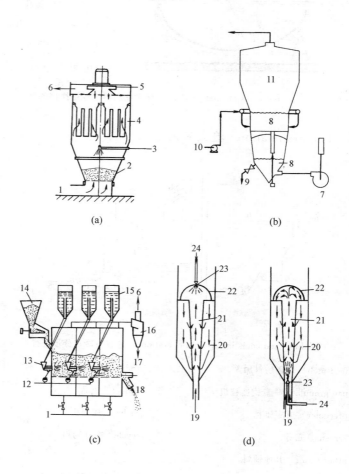

13.4 Fluidized bed and spouted bed granulators 流化床与喷动床造粒机

(a) batch fluidized bed granulator (used in the pharmaceutical industry) 间歇式流化床造粒机（用于制药工业）

(b) part of a fluidized bed incineration process for paper mill waste recovering sodium sulfate pellets 流化床中焚烧纸浆废液回收硫酸钠造粒工艺（设备之一）

(c) three-stage fluidized bed granulator (for more complete control of process conditions and more nearly uniform size distribution) 三级流化床造粒机（可以更精确控制工艺条件，使颗粒大小更均匀）

(d) two mode of injection of spray to spouted beds (into the body on the left and at the top on the right) 喷动床造粒中料液喷入有两种模式（一个由底往上喷，另一个在顶往下喷）

1 fluidizing air 流化用气
2 batch charge 分批投料
3，15 liquid supply 供入液体
4 filter 过滤器
5 fan 引风机
6 exhaust 废气
7 fluidizing air blower 流化用鼓风机
8 fluid bed 流化床
9 sodium sulfate pilled product 硫酸钠颗粒成品
10 feed pump (40% total solids) 进料泵（含固形物40%）
11 fluosolids roasting 流态化焙烧
12 compressed air 压缩空气
13 spray nozzles 喷嘴
14 solids feeder 固体加料器
16 collector 捕集器
17 fines 细粉
18 product 产品
19 spouting gas 喷动用气
20 annulus 环形空间
21 spout 喷发
22 particle fountain 颗粒喷泉
23 atomizing nozzle 雾化喷嘴
24 feed granulating liquid 成粒液加入

13.5 Pelletizing by solidification method 固化法造粒

13.5.1 Flaker 结片机

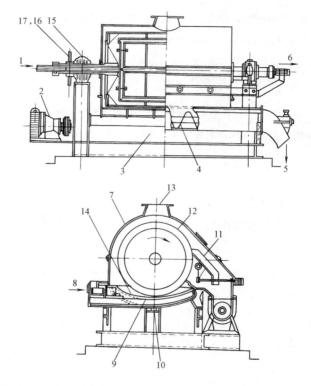

1 cooling water inlet 冷却水进口
2 variable speed reduction motor 电机减速机
3 supportor 支架
4 screw conveyer 螺旋输送机
5 products outlet 产品出口
6 cooling water outlet 冷却水出口
7 cover 外罩
8 feed inlet 进料口
9 jacket 夹套
10 drain outlet 排放口
11 scraper 刮刀
12 drum 转鼓
13 vapor outlet 排气口
14 melt 熔融物料
15 bearing 轴承
16 chain wheel 链轮
17 chain 传动链条

13.5.2 Heat-transfer equipment for batch solidification 批料凝固用传热设备

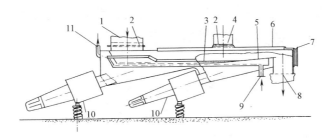

(vibrating-conveyor type 振动输送式)

1. melted burden inlet (liquid state) 熔融物料进口（液态）
2. gastight connections 密闭连接
3. solidifying burden 凝固物料
4. inert-gas supply to burden chamber 惰性气体供入装料室
5. liquid dam 液体堰
6. gastight cover 密封盖
7. viewing and lighting windows 窥视和采光孔
8. burden outlet (solids state) gastight connection 粒料出口（固态）密封连接
9. cooling-fluid inlet (to jacket) 冷却流体进口（去夹套）
10. vibratory power units 振动的动力装置
11. cooling-fluid outlet (from jacket) 冷却流体出口（来自夹套）

13.5.3 Heat-transfer equipment for solidification 凝固用传热设备

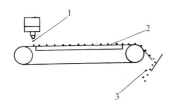

(belt type for the operation of pastillization 带式，用于制锭操作)

1. feed (drops of molten chemical) 进料（熔融化学品的液滴）
2. water-cooled steel belt as heat-transfer surface 水冷却的钢带作为传热表面
3. discharge (solid pastilles as end-product) 卸料（固体颗粒作为最终产品）

14 Fluidized-bed Systems and Reactors 流化床系统和反应器

14.1 Design of fluidized-bed systems 流化床系统的设计

14.1.1 Fluidization phase diagram for a fine powder 细颗粒流态化相图

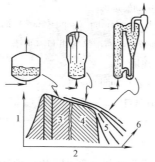

1　log pressure gradient　对数压降
2　log velocity　对数速度
3　bubbling　鼓泡床
4　turbulent　湍流床
5　fast　快速床
6　higher solid rate　固体速率增大

(showing schematic diagrams of equipment for use in bubbling, turbulent, and fast fluidization regimes 表示了在鼓泡、湍流和快速流态化工况下所用装置的简图)

14.1.2 Influence of the fluid velocity on the pressure gradient 流体速度对压力降的影响

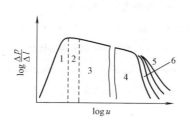

1　fixed bed　固定床
2　expanded bed　膨胀床
3　bubbling bed　鼓泡床
4　turbulent bed　湍流床
5　fast bed　快速床
6　higher fine powder rate　提高细颗粒速率

14.1.3 Fluidization 流态化

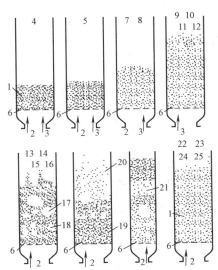

14.1.4 Fluidized-bed systems 流化床系统

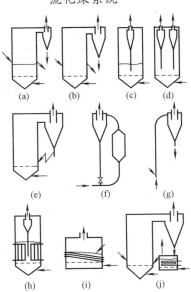

1 fine particle 细颗粒
2 gas 气体
3 liquid 液体
4 fixed bed 固定床
5 expanded bed 膨胀床
6 distributor 分布器
7 incipiently fluidized bed 初始流化床
8 minimum fluidization 临界流态化
9 particulately fluidized bed 散式流化床
10 homogeneously fluidized bed 均一流化床
11 smoothly fluidized bed 平稳流化床
12 liquid fluidized bed 液体流化床
13 aggregative fluidized bed 聚式流化床
14 heterogeneously fluidized bed 非均一流化床
15 bubbling fluidized bed 鼓泡流化床
16 gas fluidized bed 气体流化床
17 bubbling 鼓泡
18 channeling 沟流
19 dense-phase fluidized bed 密相流化床
20 dilute-phase fluidized bed 稀相流化床
21 slugging 腾涌
22 dispersed-phase fluidized bed 稀相流化床
23 lean-phase fluidized bed 稀相流化床
24 broad fluidization 广义流态化
25 moving bed 移动床

(a) bubbling bed, external cyclone
    鼓泡床，外部旋风分离器
(b) turbulent bed, external cyclone
    湍流床，外部旋风分离器
(c) bubbling bed, internal cyclone
    鼓泡床，内部旋风分离器
(d) turbulent bed, internal cyclone
    湍流床，内部旋风分离器
(e) circulating (fast) bed, external cyclones 循环（快速）床，外部旋风分离器
(f) circulating bed 循环床
(g) transport 输送床
(h) bubbling or turbulent bed with internal heat transfer 具有内部热交换的鼓泡床或湍流床
(i) bubbling or turbulent bed with internal heat transfer 具有内部热交换的鼓泡床或湍流床
(j) circulating bed with external heat transfer 具有外部热交换的循环床

14.1.5 Fluidized-bed cyclone arrangements 流化床旋风分离器布置方式

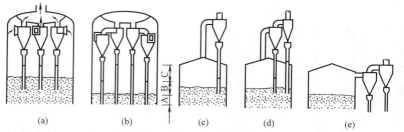

(a) single-stage internal cyclone 单级内旋风分离器
(b) two-stage internal cyclone 双级内旋风分离器
(c) single-stage external cyclone; dust returned to bed 单级外旋风分离器，粉尘返回床内
(d) two-stage external cyclone; dust returned to bed 双级外旋风分离器，粉尘返回床内
(e) two-stage external cyclone; dust collected externally 双级外旋风分离器，外部收集粉尘

14.1.6 Solids-flow-control devices 固体流量控制装置

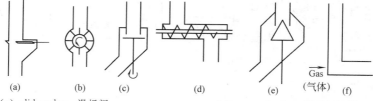

(a) slide valve 滑板阀
(b) star valve 星形阀
(c) table feeder 盘式加料器
(d) screw feeder 螺旋加料器
(e) cone valve 锥形阀
(f) L valve L形阀

14.1.7 Cyclone solids-return seals 旋风分离器固体回流密封装置

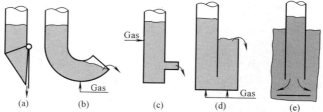

(a) Flapper valve   Flapper 阀
(b) J valve   J 阀
(c) ICI valve   ICI 阀
(d) fluid-seal pot   流体密封槽
(e) "Dollar" plate   "Dollar" 板

(a), (b), (c) and (d) may be used above the bed; (a) and (e) are used below the bed
(a)、(b)、(c)、(d) 可用于床层上面，而 (a) 和 (e) 则用于床面之下

14.1.8 Gas distributor for gases containing solids
气体含有固体时所用的气体分布器

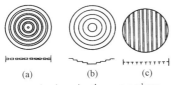

(a) concentric rings in the same plane, with the annuli open 环隙开口的在同一平面上的同心圆环
(b) concentric rings in the form of a cone 锥形同心环
(c) grids of T bars or other structural shapes T形或其他结构型式的栅板
(d) flat metal perforated plates 多孔金属平板
(e), (f) dished and perforated plates concave both upward and downward 上凹和下凹的碟形多孔板

14.1.9 Fluo seal
Fluo 密封装置

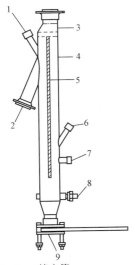

1 fill pipe 填充管
2 overflow 溢流
3 eccentric reducer 偏心变径接头
4 body 主体
5 partition 隔板
6 pressure-tap nozzle 测压管口
7 secondary air inlet 二次空气入口
8 primary air inlet 主空气入口
9 dumping valve 卸料阀

14.1.10 Fluidized-bed seal leg
流化床密封腿

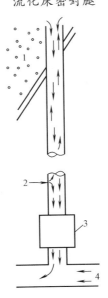

1 fluidized bed 流化床
2 seal and stripping gas 密封和脱除气体
3 solids valve 颗粒阀
4 carrier gas 载气

14.1.11 Quench tank for overflow or cyclone solids discharge
用于溢流或旋风分离器固体卸料的急冷槽

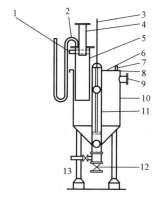

1　quench-water inlet　急冷水进口
2　safety seal　安全封
3　air inlet　空气进口
4　inlet pipe　进口管
5　seal pipe　密封管
6　inspection cover　观察罩
7　gas vent　出气孔
8　tank cover　槽盖
9　outlet pipe　出口管
10　quench tank　急冷槽
11　air-lift pipe　气提管
12　drain valve　排放阀
13　flush water　冲洗水

14.1.12 Non-catalytic fluidized bed system
非催化流化床系统

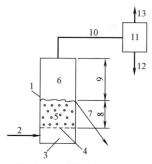

14.1.13 Catalytic fluidized-bed system
催化流化床系统

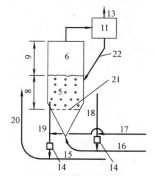

1　solids feed　固体进料
2　gas in　气体进口
3　windbox or plenum chamber　风箱或气室
4　gas distributor or constriction plate　气体分布板或收缩板
5　fluid bed　流化床
6　disengaging space　分离空间
7　solids discharge　固体排出
8　bed depth　床层深度
9　freeboard　自由空间
10　gas and entrained solids　气体和所夹带固体
11　dust separator　粉末分离器
12　dust　粉末
13　gas　气体
14　solids flow-control valves　固体流量控制阀
15　regenerating gas　再生气
16　feed gas　原料气
17　seal and stripping gas　密封和汽提气
18　regenerated catalyst　再生催化剂
19　spent catalyst　用过的催化剂
20　to regenerator　去再生器
21　distributor　分布板
22　dust return　粉末回流

14.2 Uses of fluidized-bed 流化床的应用

14.2.1 Circulating-fluid-bed calciner 循环流化床煅烧炉

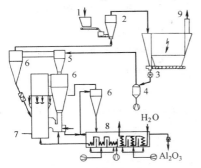

1　Al(OH)$_3$(moist)　Al(OH)$_3$（湿）
2　dryer　干燥器
3　electrostatic precipitator　静电除尘器
4　airlift　气力提升器
5　preheater　预热器
6　cyclone　旋风分离器
7　calcining furnace fuel　煅烧炉燃料
8　cooler　冷却器
9　stack　烟囱

14.2.2 Fluidized-bed steam generator 流化床蒸汽发生器

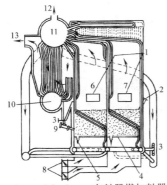

1　spreader coal feeders　布料器煤加料器
2　limestone feed pipe　石灰石加料管
3　downcomer　下水管
4　air distribution grid　空气分布器炉栅
5　bed material drain　床层排料口
6　bed A　床 A
7　bed B　床 B
8　air inlet　空气进口
9　fly ash reinjector　飞灰再注器
10　mud drum　泥浆包
11　steam drum　汽包
12　steam outlet　蒸汽出口
13　gas outlet　气体出口

14.2.3 Fluidized bed for gas fractionation 用于气体分级的流化床

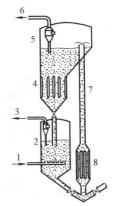

1　feed gas　进料气体
2　absorber　吸收器
3　lean gas　贫气
4　cooler　冷却器
5　gas-solids separator　气固分离器
6　heavy product　饱和产物
7　lift line　提升管
8　heater desorber　加热解吸器

341

14.2.4 Dorrco Fluosolids reactor Dorrco Fluosolids 反应器

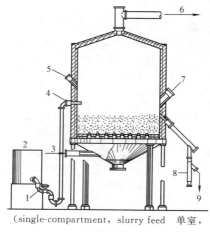

1 slurry pump 淤浆泵
2 slurry storage 淤浆储槽
3 air 空气
4 slurry feed gun 淤浆进料枪
5 water spray 水喷头
6 gas outlet 气体出口
7 startup burner 开工燃料烧嘴
8 seal 密封
9 product 产品

(single-compartment, slurry feed 单室，浆态进料)

14.2.5 Fluosolids lime kiln Fluosolids 石灰窑

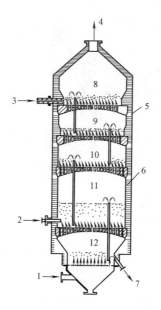

1 air 空气
2 fuel 燃料
3 feed 进料
4 stack gas to cyclones 烟气去旋风分离器
5 steel 钢板
6 firebrick 耐火砖
7 product 产品
8 preheating compartment No.1 第1预热室
9 preheating compartment No.2 第2预热室
10 preheating compartment No.3 第3预热室
11 calcining compartment 煅烧室
12 cooling compartment 冷却室

14.2.6 Fluidization processes in petroleum refinery 石油炼制中的流化过程

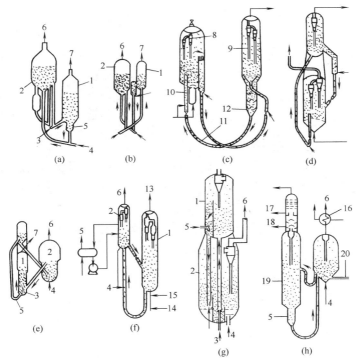

(a) SOD model Ⅱ catalytic cracking unit
    SOD Ⅱ 型催化裂化装置
(b) SOD model Ⅲ catalytic cracking unit
    SOD Ⅲ 型催化裂化装置
(c) SOD model Ⅳ catalytic cracking unit
    SOD Ⅳ 型催化裂化装置
(d) UOP stacked unit UOP 烟囱式装置
(e) Shell unit 壳牌装置
(f) SOD hydroformer SOD 临氢重整装置
(g) Kellogg orthoflow unit Kellogg 正流式装置
(h) fluidized coking unit 流化焦化装置
1 reactor 反应器
2 regenerator 再生器
3 feed 进料
4 air 空气
5 steam 蒸汽
6 flue gas 烟道气
7 product 产物
8 cyclone 旋风分离器
9 dip leg 料腿
10 stand pipe 立管
11 cat circulation riser 循环催化剂提升管
12 stripping steam 汽提蒸汽
13 reformed gas 重整气
14 hydrogen-rich recycle gas 富氢循环气
15 naphtha feed 石脑油进料
16 waste-heat boiler 废热锅炉
17 gas oil 瓦斯油
18 slurry recycle 浆料循环
19 pitch feed 沥青进料
20 quench water 急冷水

343

14.2.7 Model Ⅰ catalytic-cracking unit Ⅰ型催化裂化装置

1. catalyst coolers 催化剂冷却器
2. feed 给水
3. regenerated catalyst hopper 再生催化剂料斗
4. catalyst makeup 催化剂补充
5. regenerator 再生器
6. catalyst charging 催化剂进料
7. air 空气
8. feed 进料
9. reactor 反应器
10. waste-gas cooler 废气冷却器
11. cottrell precipitator 电除尘器
12. cracked gas 裂解气
13. spent-catalyst hopper 用过的催化剂料斗
14. fresh-oil feed 新鲜油进料
15. flue gas 烟道气
16. waste-heat boiler 废热锅炉
17. multicyclones 多管式旋风分离器
18. to fractionation 去精馏段
19. steam to stripping section 蒸汽去汽提段
20. recycle 循环物料
21. main air blower 主空气风机
22. startup heater 开工预热器
23. stripping steam 汽提蒸汽
24. cyclone stripper vessel 旋风汽提器
25. transfer line reactor 传输线反应器

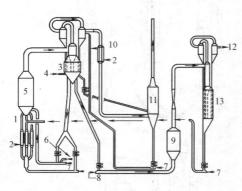

14.2.8 Downflow model Ⅱ catalytic-cracking unit 下流式Ⅱ型催化裂化装置

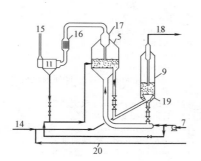

14.2.9 Exxon transfer-line catalytic-cracking unit Exxon 输送管催化裂化装置

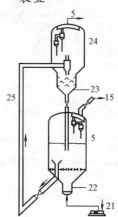

14.3 Reactors 反应器

14.3.1 Examples of reactors for specific liquid-gas processes
特定气-液过程反应器举例

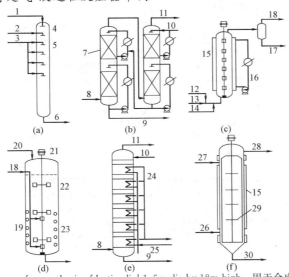

(a) trickle reactor for synthesis of butinediol 1.5m dia by 18m high 用于合成丁炔二醇的滴流床反应器, 直径 1.5m, 高 18m
(b) nitrogen oxide absorption in packed columns 填料塔内进行氧化氮吸收
(c) continuous hydrogenation of fats 油脂的连续加氢反应
(d) stirred tank reactor for batch hydrogenation of fats 油脂间歇加氢的搅拌釜式反应器
(e) nitrogen oxide absorption in a plate column 板式塔内进行氧化氮吸收
(f) a thin film reactor for making dodecylbenzene sulfonate with SO_3 用 SO_3 生产十二烷基苯磺酸盐的薄膜反应器

1 aqueous formaldehyde 含水甲醛
2 acetylene feed 乙炔进料
3 cold acetylene 冷乙炔
4 liquid distributor 液体分布器
5 gas distributors 气体分布器
6 butinediol 丁炔二醇
7 packed towers 填料塔
8 air with 10vol% $NO+NO_2$ 空气中含 $(NO+NO_2)$ 10% (体积)
9 nitric acid 硝酸
10 water 水
11 offgases 尾气
12 catalyst 催化剂
13 fat 油脂
14 hydrogen 氢
15 cooling jacket 冷却夹套
16 recycle cooling 循环冷却
17 product 产品
18 hydrogen 氢
19 heat transfer surface 传热表面
20 charge 加料
21 75~150RPM 75~150r/min (每分钟转数)
22 turbine impellers 涡轮式搅拌器
23 gas distributor 气体分布器
24 cooling coils 冷却盘管
25 water or liquid ammonia 水或液氨
26 air with 7%~9vol% SO_3 空气中含 SO_3 7%~9% (体积)
27 dodecylbenzene 十二烷基苯
28 vent gas 排出气
29 spacing plates 定距板
30 dodecylbenzene sulfonate 十二烷基苯磺酸盐

345

14.3.2 Types of reactors for synthetic fuels 合成燃料的反应器类型

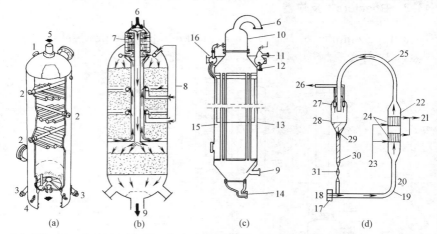

(a) ICI methanol reactor (showing internal distributors) ICI 甲醇反应器（展示内部分布器）
(b) ICI methanol reactor (with internal heat exchange and cold shots) ICI 甲醇反应器（具有内部换热器和冷态粒子）
(c) fixed bed reactor for gasoline from coal synthesis gas; dimensions 10×42ft, 2000 2-in. dia tubes packed with promoted iron catalyst, production rate 5 tons/day per reactor 由煤合成气生产汽油的固定床反应器，每个反应器生产能力为 5 吨/日，外廓尺寸为10×42ft，内有 2000 根 2in 管，管内填充激励铁催化剂
(d) Synthol fluidized bed continuous reactor system (for gasoline from coal synthesis gas) Synthol 流化床合成燃料连续反应器系统（由煤的合成气生产汽油）

1	catalyst input 催化剂入口		18	fresh feed and recycle 新鲜料与循环料加入
2	cold shot nozzles 冷态粒子接管		19	gas and catalyst mixture 气体与催化剂混合物
3	catalyst dropout 催化剂排放口			
4	thermocouple 热电偶		20	riser 提升管
5	inlet 进口		21	cooling-oil outlet 冷却油出口
6	gas inlet 气体进口		22	reactor 反应器
7	internal heat exchanger 内部换热器		23	cooling-oil inlet 冷却油进口
8	cold shots 冷态粒子		24	cooler groups 冷却器组
9	gas outlet 气体出口		25	gooseneck 鹅颈管（大弯管）
10	steam heater 蒸汽加热器		26	tail gas 尾气
11	steam outlet 蒸汽出口		27	cyclones 旋风分离器
12	feedwater inlet 供水口		28	catalyst-settling hopper 催化剂沉积料斗
13	inner shell 内壳层			
14	wax outlet 蜡出口		29	catalyst 催化剂
15	tube bundle 管束		30	stand pipe 立管
16	steam collector 蒸汽包		31	slide valves 滑阀
17	feed preheater 加料预热器			

14.3.3 Reactors with moving beds of catalyst or solids for heat supply 带供热固体或催化剂的移动床反应器

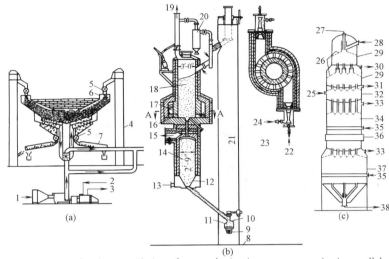

(a) pebble reactor for direct oxidation of atmospheric nitrogen; two units in parallel, one being heated with combustion gases and the other used as the reactor 常压氮气直接氧化的卵石反应器, 两个装置并联, 其中一个以燃烧气体加热, 另一个用作反应器
(b) pebble heater which has been used for making ethylene from heavier hydrocarbons 卵石加热器, 可作为用高沸点碳氢化合物生产乙烯的装置
(c) moving bed catalytic cracker and regenerator (for 20000 bps. d the reactor is 16ft dia, catalyst circulation rate 2~7lbs/lb oil, attrition rate of catalyst 0.1~0.5lb/ton circulated, pressure drop across air lift line is about 2 psi.) 移动床催化裂化装置与再生器 (生产能力为日产 20000 桶的反应器, 直径为 16ft, 催化剂循环率为 2~7lb/lb 油, 催化剂的磨耗率为 0.1~0.5lb/t 循环量, 空气提升压力损失约为 $2lb/in^2$)

1 air 空气
2 steam supply 供入蒸汽
3 exhaust steam to NH_3 generator 乏汽通 NH_3 发生器
4 pebble elevator 卵石提升机
5 valve 阀
6 pebble hopper 卵石料斗
7 trommel screen 转筒筛
8 base line of elevator 提升机基线
9 door for removing pebbles 卵石移出门
10 door for adding pebbles 卵石加入门
11 pebble feeder 卵石加料器
12 screen 网屏
13 vapor inlet 蒸气进口
14 lower pebble bed 下层卵石床
15 vapor outlet 蒸气出口
16 hot gas inlet 热气入口
17 combustion chamber 燃烧室
18 upper pebble bed 上层卵石床
19 flue gas outlet 烟道气 (废气) 出口
20 dust separator 粉尘分离器
21 continuous bucket elevator 连续型斗式提升机
22 fuel gas 燃料气体
23 section A-A A-A 剖面
24 combustion air 助燃空气
25 purge steam 冲洗蒸汽
26 catalyst level 催化剂层面
27 fresh vapor inlet 新鲜蒸气进口
28 catalyst inlet 催化剂进口
29 reaction section 反应段
30 cracked vapor outlet 裂化蒸气出口
31 purge vapors 冲洗蒸气
32 catalyst purge section 催化剂冲洗段
33 flue gas 烟道气
34 first regeneration zone 第一再生层段
35 air inlet 空气进口
36 steam generator 蒸气发生器
37 second regeneration zone 第二再生层段
38 catalyst recycle 催化剂循环

14.3.4 Fluidized bed reactor processes for the conversion of petroleum fractions 用于石油馏分转化的流化床反应器工艺

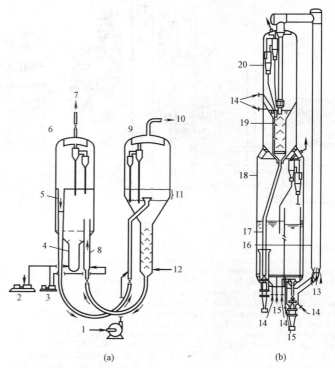

(a) Exxon Model Ⅳ fluid catalytic cracking (FCC) unit sketch Exxon Ⅳ 型催化裂化 (FCC) 装置设计简图

(b) a modern FCC unit utilizing active zeolite catalysts; the reaction occurs primarily in the riser which can be as high as 45m 利用活性沸石催化剂的现代催化裂化装置，其反应主要发生在高达 45m 的提升管内

(c) fluidized bed hydroformer (in which straight chain moiecuies are converted into branched ones in the presence of hydrogen at a pressure of 1500atm. The process has been largely superseded by fixed bed units employing precious metal catalysts) 流化床临氢重整装置（其中在 1500 大气压氢气存在的条件下，直链分子转化为支链分子。这一工艺大部分被固定床装置所取代，其中采用了贵金属催化剂）

(d) a fluidized bed coking process; units have been built with capacities of 400～12000 tons/day 流化床焦化工艺，装置生产能力达 400～12000 吨/日

14.3 Reactors 反应器

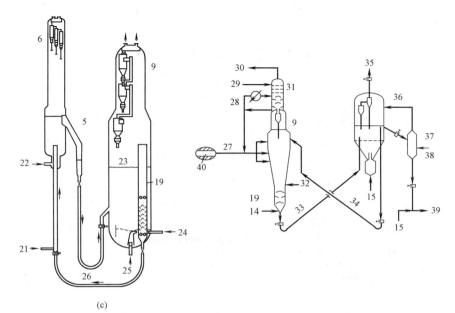

(c)

1	raw oil charge 原料油加入		21	control air inlet 调节空气入口
2	main air blower 主空气风机		22	main air inlet 主空气入口
3	control blower 调节风机		23	normal level 正常液面
4	startup heater 开工预热器		24	recycle gas 循环气
5	overflow well 溢流管		25	naphtha feed 石脑油进口
6	regenerator 再生器		26	U bends U形弯管
7	flue gas 烟道气		27	residuum feed 渣油进料
8	cat circulation control riser 催化剂循环控制提升管		28	slurry recycle 浆液循环
9	reactor 反应器		29	reflux 回流
10	product 产品		30	reactor products to fractionator 反应器产品通精馏塔
11	variable level 可变液位		31	scrubber 涤气器
12	stripping steam 汽提蒸汽		32	attrites 磨料
13	oil feed 油供入		33	cold coke 冷焦炭
14	steam 蒸汽		34	hot coke 热焦炭
15	air 空气		35	stack or CO boiler 烟囱或CO锅炉
16	riser reactor 提升管反应器		36	burner 燃烧器
17	stand pipe 立管		37	quench elutriator 急冷淘析器
18	two stage regenerator 两级再生器		38	water 水
19	stripper 汽提段		39	product coke to storage 焦炭产品去贮槽
20	disengager 分离段		40	start 开工

14.4　Agitated reactors　搅拌式反应器

14.4.1　Agitated reactor　搅拌式反应釜

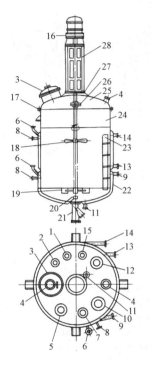

1　reflux return nozzle　回流管口
2　spare nozzle　备用管口
3　man hole　人孔
4　peep hole　视孔
5　gas inlet　气体入口
6　medium discharge　介质排出口
7　raw material inlet　原料入口
8　nozzle for compound pressure and vacuum gauge　压力真空表接管口
9　lower inlet for medium　介质下部入口管
10　air vent nozzle　排气管口
11　lower return nozzle for medium　介质下部回流管口
12　nozzle for reflux condenser　回流冷凝器接管口
13　upper return nozzle for medium　介质上部回流管口
14　upper inlet for medium　介质上部入口
15　nozzle for thermometer　温度计接管口
16　speed reducer　减速机
17　cover flange　封头法兰
18　propeller　搅拌桨
19　turbine agitator　蜗轮搅拌器
20　vibration-stopper　消振器
21　bottom discharge valve　底部排料阀
22　lower jacket　下部夹套
23　baffle　挡板
24　kettle　釜体
25　cover head　顶封头
26　coupling　联轴器
27　base for speed reducer　减速机底座
28　frame　机架

14.4.2　Low-speed agitators　低速搅拌器

(a) paddle　平桨

(b) anchor　锚（式）

(c) helical ribbon　螺带（式）

14.4.3 Electromagnetic stirring type superhigh-pressure autoclave 电磁搅拌式超高压反应釜

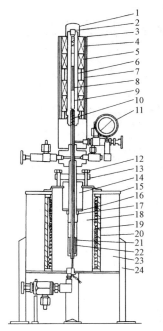

1   cap   罩
2   blank cap   盖帽
3   covering plate   盖板
4   magnet cover   磁铁罩
5   magnet   磁铁
6   shaft guide   轴导套
7   magnet receiver   磁铁座
8   upper shaft   上段轴
9   magnetic induction part   磁感应元件
10 spring   弹簧
11 pressure gauge   压力表
12 setting bolt   固定螺栓
13 gland   填料压盖
14 packing box   填料函
15 covering plate   盖板
16 packing gland   填料压盖
17 seal ring   密封环
18 autoclave body   釜体
19 heater   加热器
20 porcelain   陶瓷筒
21 low shaft   下段轴
22 agitator   搅拌器
23 heat insulator   保温材料
24 autoclave legs   釜支腿

14.4.4 Basic impeller types 搅拌器的主要类型

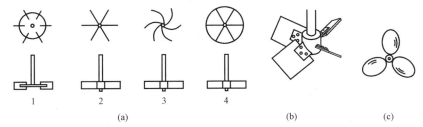

(a) turbine impeller   涡轮式搅拌器
(b) pitched bladed turbine   折叶开启涡轮式
(c) marine propeller   船舶螺旋桨，推进式
1 disc-mounted flatblade turbine   平直叶圆盘涡轮式
2 hub-mounted flatblade turbine   平直叶开启涡轮式
3 hub-mounted curved-blade turbine   弯叶开启涡轮式
4 shrouded turbine impeller   蔽式涡轮搅拌器

351

15 Miscellaneous Processes 其他化工过程

15.1 Sublimation 升华

15.1.1 Generalized schematic for simple sublimation
简单升华的综合性示意图

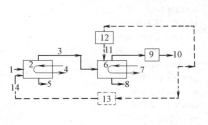

1　feed solids　固体进料
2　sublimer　升华器
3　vapor and entrainer　蒸气与夹带剂
4　heating medium　加热介质
5　residue　残渣
6　condenser　凝聚器
7　cooling medium　冷却介质
8　solid product　固体产品
9　cold trap　冷阱
10　vacuum source or blower　真空源或引风机
11　quench gas　淬冷气体
12　cooler　冷却器
13　heater　加热器
14　entrainer gas　夹带剂气体

15.2 Membrane processes 膜分离过程

15.2.1 Schematic of hollow-fiber module Permasep permeator
中空纤维型 Permasep 反渗透器

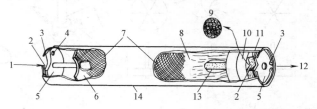

1　feed　加料
2　O-ring seal　O 形密封圈
3　snap ring　弹性挡环，止动环
4　concentrate outlet　浓缩液出口
5　end plate　端板
6, 8　hollow fiber　中空纤维
7　flow screen　流动网
9　open ends of fibers　中空纤维的开放末端
10　epoxy tube sheet　环氧树脂管板
11　porous backup disk　多孔状支撑片
12　permeate　渗透液
13　porous-feed distributor tube　多孔状进料分配管
14　shell　外壳

15.2.2 Membrane permeation module designs 膜渗透单元的设计

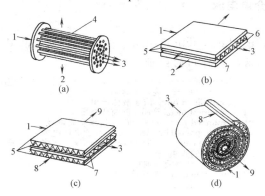

(a) tube bundle 管束型
(b) stack 积层块型
(c) bi-flow stack 双流向积层块型
(d) spiral 卷筒型

1 feed in 加料
2 permeate out 渗透液流出
3 retentate out 渗余液流出
4 hollow, thin-walled plastic tube 薄壁中空塑料管
5 membrane 膜
6 porous sheet 多孔片
7 corrugated spacer 瓦楞状隔板
8 carrier in 载体加入
9 permeate and carrier out 渗透液及载体流出

15.2.3 Plate-and-frame reverse osmosis system 板框型反渗透器

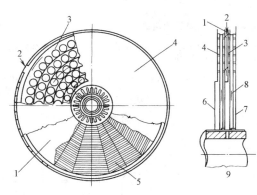

1 substrate 底基
2 feed water 加料水
3 spiral baffle 螺旋挡板
4 membrane 膜
5 support plate 支持板
6 plate No.1 第一块板
7 plate No.2 第二块板
8 product water 成品水
9 exploded cross section at center with collector shaft shown 显示收集轴的中部横断面（上半部）

15.2.4 Tubular membrane ultrafiltration 管式膜超滤

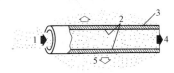

1 waste water feed 废水投入
2 membrane 膜
3 fibreglass-reinforced epoxy support tube 环氧玻璃纤维增强支持管
4 concentrate 浓缩液
5 permeate 渗透液

15.3 Dielectrophoresis 介电电泳

15.3.1 Diagram illustrating the function of an electrostatic liquid cleaner 液体静电洁净器作用的图示

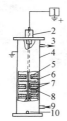

1 high-voltage dc power supply 高压直流动力供应器
2 HV insulator 高压绝缘器
3 clean liquid out 洁净液体流出
4 cylindrical metal shell 圆筒状金属壳
5 grounded metal electrode 接地金属电极
6 needle-point HV electrode 针状高压电极
7 particle collection in porous matrix 在多孔基质上粒子的收集
8 liquid path 液体通道
9 dirty liquid in 脏液体入口
10 drain 排放口

15.4 Diffusional separation processes 扩散分离过程

15.4.1 Mass-diffusion column 质量扩散塔

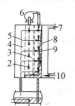

1 sweep vapor boiler 扫掠蒸气锅炉
2 process gas (countercurrent) flow 加工气（逆流）流
3 mass diffusion screen 质量扩散筛
4 porous wall 多孔壁
5 sweep vapor 扫掠蒸气
6 process gas pump 加工气体泵
7 cooling water out 冷却水出口
8 condenser wall 冷凝器壁
9 falling film of condensed sweep vapor 冷凝扫掠蒸汽的降膜
10 cooling water in 冷却水入口

15.4.2 Separation nozzle 分离喷嘴

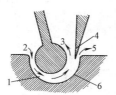

15.4.3 A commercial separation-element tube 一种商用分离元件管

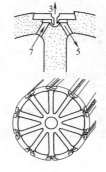

1 nozzle 喷嘴
2 feed gas 5% UF_6, 95% He 进料气体 5% UF_6, 95% He
3 light fraction 轻部分（成分）
4 knife edge 刀刃
5 heavy fraction 重部分（成分）
6 curved wall 弯曲壁
7 feed gas 进料气体

15.5 Coalescence processes 聚并过程

15.5.1 Vertical tube coalescing separator 垂直管聚并分离器

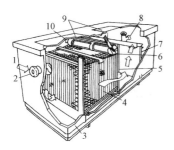

1 influent 流入液体
2 integral fittings 整体接头
3 moulded corrosion-resistant fibreglass envelope 模制耐腐蚀玻璃纤维容器
4 oil attracting tubes (removable) 吸油管（可拆的）
5 oil retention baffle 截油挡板
6 rotary pipe skimmer 旋转管式撇油器
7 weir 堰
8 T-pipe outlet for effluent 废液T形出口管
9 cover (2) 顶盖（2个）
10 oil skimmer outlet 撇油器出口接头

15.5.2 High capacity coalescer filters 大容量聚并滤清器

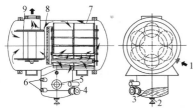

1 inlet 进口
2 drain cock 排污旋塞
3 stripped water 溶脱过的水
4 air actuated water dump valve 气动排水阀
5 sight glass 视镜
6 water detection probes 水检测探头
7 coalescer elements 聚并器元件
8 stripper elements 溶脱器元件
9 outlet 出口

(have horizontal element stacks whereas in smaller units the cartridges are installed vertically 内有水平层叠的元件，而小设备中的许多滤芯是垂直安装的)

15.5.3 A bath feed separator 选矿池进料分离器

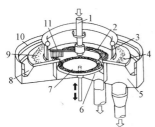

1 slurry feed pipe 浆液进料管
2 heavy mineral ripple 重的矿石粒子
3 wash-water curtain 洗涤水水幕（水帘）
4 outer annular surface 外侧环形表面
5 wash-water outflow and light mineral collection tank 洗涤水流出，轻质矿石收集池
6 heavy mineral collection tank 重质矿石收集池
7 central disc 中央圆盘
8 light fraction collection launder 轻成分聚集洗矿槽
9 light mineral ripple 轻的矿石粒子
10 adjustable weir 可调堰
11 rotating rake 转动耙

(standing wave principle applied to 应用驻波原理)

15.5.4 Vertical tube coalescing principle of operation 垂直管聚并操作原理

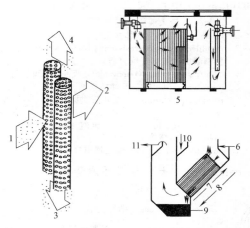

1. influent of oils water and solids 流入物有油、水、固体
2. clean effluent 净化的流出液
3. settleable solids 沉淀的固体
4. free oils 游离油
5. vertical tube (perforated) coalescing separator flow diagram 垂直管（多孔管）聚并分离器流向图
6. influent 流入物
7. oil 油
8. waste water 废水
9. heavy components 重组分
10. light components 轻组分
11. treated water 净化水

Examples of coalescing separators 聚并分离器实例

15.5.5 Cutaway of filter coalescer 过滤聚结器的局部剖视图

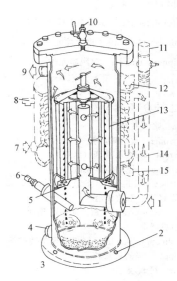

1. dirty oil inlet when heater is not used 脏油进口（不用加热器时）
2. fixing holes 地脚螺栓孔，固定孔
3. dotted outline only refers to modules 虚线轮廓线仅表示那些组件
4. drain valve (automatic control optional) 放泄阀（自动控制可有可无）
5. division plate 隔板
6. water level detector 水位探测器
7. dirty oil inlet when heater is used 脏油进口（采用加热器时）
8. relief valve 减压阀
9. clean and dry oil 干净油
10. air vent 放气口
11. thermostatically controlled steam valve 蒸汽恒温控制阀
12. steam inlet 蒸汽进入
13. PTFE coated metal mesh 涂覆聚四氟乙烯的金属筛网
14. dirty oil outlet from heater to filter 脏油从加热器排出至过滤器
15. steam outlet 蒸汽出口

15.6 Sedimentation operations 沉降操作

15.6.1 Unit thickener with bridge-supported mechanism 带桥式支撑机构的单室增稠器

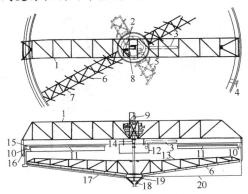

15.6.2 Rectangular clarifier 矩形澄清器

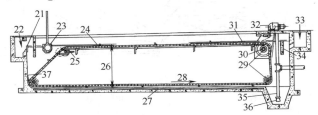

1　superstructure　上层构架
2　drive motor　驱动马达
3　feed pipe　加料管
4　overflow launder outlet　溢流溜槽出口
5　short arms　短臂
6　blades　刮板
7　long arms　长臂
8　drive control with load indicator　带负荷指示的驱动控制器
9　motorized lifting device　动力升举机构
10　overflow launder　溢流溜槽
11　weir　堰
12　feed well　加料井
13　arm　臂
14　guide bearing　导引轴承
15　liquid level　液面
16　off-tank support　槽外支撑
17　concrete tank　水泥槽
18　discharge cone　卸料锥
19　cone scraper　锥形刮板
20　steel tank　钢槽
21　adjustable weir　可调堰
22　effluent　流出
23　revolving scum skimmer　转动除沫器
24　rigid flights　刚性刮片
25　take-up　松紧装置
26　water depth variable　可变水深
27　tee rail　T形轨
28　collector travel　收集器的移动
29　self-aligning bearings　自位轴承
30　head shaft　主轴
31　water level　液面
32　drive unit　驱动装置
33　influent　流入
34　baffle　挡板
35　sludge pipe　泥浆管
36　sludge hopper　泥浆料斗
37　pivoted flight　枢轴刮片

15.6.3 Reactor-clarifier of the high-rate, solids-contact type 高速率固体接触式反应——澄清器

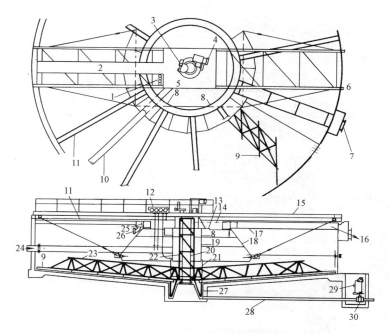

1　chemical pipes　试剂管
2　walkway　人行道
3　rake drive motor　耙子驱动马达
4　turbine drive motor　蜗轮驱动马达
5　operating platform　操作平台
6　outlet radial　径向出口
7　outlet pocket　卸出槽
8　baffles　挡板
9　blades　刮片（刀）
10　inlet pipe　进料管
11　radial launders　径向溜槽
12　chemical feed piping　试剂加入管
13　turbine　蜗轮
14　maximum water level　最高水平面
15　superstructure　上层构架
16　outlet　出口
17　outlet launder　出液溜槽
18　reaction well　反应井
19　recirculation drum　循环筒
20　center column　中央塔
21　rotating drum　转筒
22　cage　笼架
23　rake arm　耙臂
24　inlet　入口
25　annular collection launder　环形收集溜槽
26　auto launder drain　溜槽自动放液管
27　sludge sump scraper　泥浆坑刮片
28　sludge pipe　泥浆管
29　sludge sump　泥浆坑
30　de-sludging　除渣

15.6.4 Coagulator 凝聚器

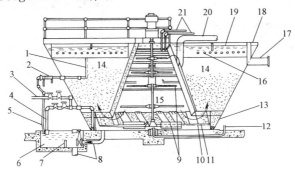

1 rotary type sampling tube   旋转式采样管
2 indicator dial   指示盘
3 washing pipe   清液管
4 automatic sludge discharge pipe   自动排污管
5 sampling tube   采样管
6 sump   污水槽
7 cut-off wall   堰板
8 drain   排污管
9 mixing blade   搅拌器
10 feed water distribution orifice   给水分布孔
11 baffle   挡板
12 pressure type waterpro of lubricating device   压力式防水注油器
13 sludge collector   污泥沉积室
14 blanket   过滤层
15 mixing chamber   混合室
16 water collecting orifice   集水孔
17 treated water   清水
18 treated water conduit   清水导管
19 water level   水平面
20 feed inlet   供液口
21 coagulant feed pipe   凝结剂加入管

15.6.5 Sludge collector 集泥机

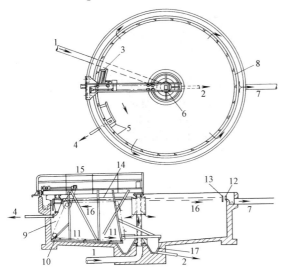

1 feed inlet   供液口
2 sludge pipe   污泥管
3 scum collector   浮渣收集装置
4 scum pipe   浮渣导管
5 scum box and baffle   浮渣室和挡板
6 perforated baffle   多孔挡板
7 outlet   出口
8 overflow plate   溢流堰
9 flight chain conveyer (double lined)   双列刮板输送机
10 sludge collecting board   集泥板
11 sludge   污泥
12 overflow   溢流
13 scum baffle   浮渣挡板
14 skimmer   撇沫板
15 walkway   走道
16 water direction   水流方向
17 sludge hopper   污泥贮斗

15.7 Forming machine for rubber and plastics, mixing machine 橡胶塑料加工成型及混合机械

15.7.1 Banbury mixer 生胶混炼机

15.7.2 Extruder 挤压机

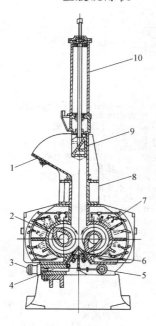

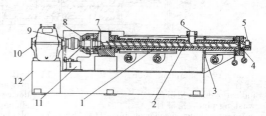

1　barrel cooler　筒体冷却器
2　screw　螺杆
　　feed section　给料段
　　compression section　压缩段
　　metering section　计量段
3　barrel　筒体
4　breaker plate　分流板
5　pressure-adjusting valve　压力调节阀
6　vent　放气口
7　hopper　加料斗
8　heavy-load thrust bearing　重型止推轴承
9　flexible coupling　挠性联轴器
10　herring bone gearbox　人字齿轮箱
11　forced oil circulation pump　强制循环油泵
12　machine-base　机座

1　feed hopper　料斗
2　spiral-shaped rotor　蜗壳形转子
3　latch cylinder　排料启闭器气缸
4　door latch　排料启闭器
5　door support　排料门底座
6　discharge door　排料门
7　spray pipe　喷水管
8　extend neck　接筒
9　floating weight　移动式重锤
10　weight cylinder　重锤导向筒

15.7.3 Injection moulding machine 注射成型机

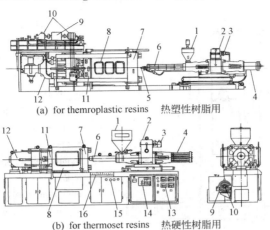

(a) for thermoplastic resins 热塑性树脂用

(b) for thermoset resins 热硬性树脂用

1	hopper 料斗	10	hydraulic pump 液压泵
2	reducer 减速机	11	movable die plate 可拆模板
3	hydraulic motor 液压马达	12	die clamping cylinder 装夹模具用气缸
4	injection cylinder 注塑缸	13	thermometer 温度计
5	nozzle 喷嘴	14	automatic temperature controller 自动温度控制器
6	heating barrel 加热筒	15	oil pressure gauge 油压表
7	stationary die plate 固定模板	16	control panel 操作盘
8	safety guard 安全护罩		
9	motor 电动机		

15.7.4 Mixing & grinding machine 混合研磨机

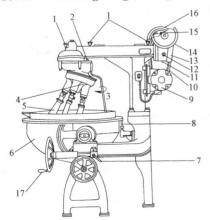

1. oil cup 油杯
2. oil inlet 注油口
3. scraper 刮棒
4. oil draw out 排油口
5. pestle 捣棒
6. pan 锅
7. lever for pan position 固定锅体手把
8. adjusting screw for pan position 锅体水平调节螺杆
9. motor 电动机
10. V-belt V形皮带
11. oil outlet for reducer 减速机出油口
12. oil cup (L type) L形油杯
13. reduction gear 减速箱
14. flywheel 飞轮
15. oil filling port 注油口
16. V-pulley V形带轮
17. handle for pan position 确定锅位置的手轮

361

15.7.5 Inflation film manufacturing machine 吹膜机

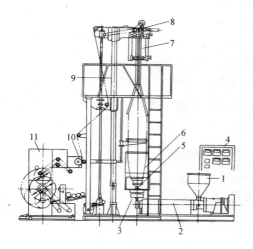

1　hopper　料斗
2　extruder　挤出机
3　die turning gear　塑模旋转装置
4　control panel　控制盘
5　die　模具
6　cooling ring　冷却圈
7　water cooling stabilizer panel　水冷稳定板
8　nipping roll (upper)　上部压紧胶辊
9　take up unit　松紧装置
10　nipping roll (lower)　下部压紧胶辊
11　wind-up unit　卷扬机

15.7.6 Homogenizer 匀化器

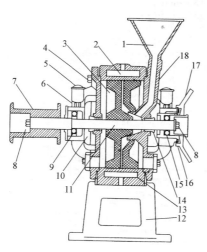

1　funnel　漏斗
2　jacket　夹套
3　disk　轮盘
4　nut　螺母
5　oil cup　油杯
6　cover　机盖
7　pulley　皮带轮
8　rotary shaft set screw　转轴固定螺钉
9　ball bearing　滚珠轴承
10　gland nut　压紧螺母
11　shaft　轴
12　base　机座
13　stator　定子
14　seal　密封
15　ball bearing　滚珠轴承
16　gap adjuster　间隙调节器
17　adjust handle　调节手轮
18　funneled feed duct　带漏斗的进料导管

16 Pumps 泵

16.1 Classification of pumps 泵分类

16.1.1 Classification of pumps Ⅰ 泵分类Ⅰ

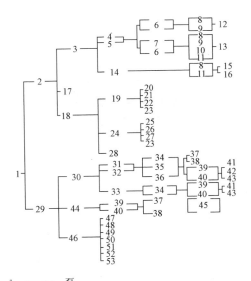

1 pumps 泵
2 positive displacement 容积（式）（正位移式）
3 reciprocating 往复（式）
4 piston 活塞（式）
5 plunger 柱塞（式）
6 double acting 双作用
7 single acting 单作用
8 simplex 单缸
9 duplex 双列，双缸
10 triplex 三列，三缸
11 multiplex 多列，多段
12 steam 蒸汽
13 power 电力，动力
14 diaphragm 隔膜（式）
15 fluid operated 流体驱动的
16 mech. operated 机械驱动的
17 blow case 酸蛋
18 rotary 回转（式）
19 single rotor 单转子
20 vane 叶片（式）
21 piston 活塞（式）
22 flexible member 挠性构件
23 screw 螺杆
24 multiple rotor 多转子
25 gear 齿轮
26 lobe 凸轮（式）；凸叶（式）
27 circumferential piston 圆环多活塞
28 liquid-ring 液环（式）
29 kinetic 动力（式）
30 centrifugal 离心（式）
31 radial flow 径流（式）
32 mixed flow 混流（式）
33 axial flow 轴流（式）
34 single suction 单吸
35 double suction 双吸
36 double valves 双阀
37 self-priming 自吸式
38 non-self-priming 非自吸式
39 single stage 单级
40 multistage 多级
41 open impeller 开式叶轮
42 semi-open impeller 半开式叶轮
43 closed impeller 闭式叶轮
44 peripheral 圆周（式）
45 regenerative 再生（泵）；漩涡（泵）
46 special 特种
47 peristaltic 蠕动（泵）
48 jet（ejector-boosted） 喷射式（喷射增压式）
49 gas lift 气升（式）
50 hydraulic ram 液压滑块（式）
51 electromagnetic 电磁（式）
52 screw centrifugal 螺杆离心（式）
53 rotating casing 旋转壳体（泵）

363

16.1.2　Classification of pumps Ⅱ　泵分类 Ⅱ

16.1.2.1　Operating characteristic of pump　泵的操作特性

1　standard pump　标准泵
2　low pressure pump　低压泵
3　medium pressure pump　中压泵
4　high pressure pump　高压泵
5　high lift pump　高扬程泵，高压泵
6　very high pressure pump　超高压泵
7　booster pump　增压泵
8　vacuum pump　真空泵
9　high vacuum pump　高真空泵
10　cryogenic pump　低温泵，深冷泵
11　high viscosity triple screw pump　高黏度三螺杆泵

16.1.2.2　Design feature of pumps　泵的结构特点

1　positive displacement pump　容积泵（正位移泵）
2　reciprocating pump　往复泵
3　plunger pump　柱塞泵
4　piston pump　活塞泵
5　diaphragm pump　隔膜泵
6　screw pump　螺杆泵
7　screw though pump　槽式螺旋泵
8　gear pump　齿轮泵
9　swinging vane pump　转叶泵
10　lobe pump　凸轮泵
11　rotary vane type pump　旋转式叶片泵
12　sliding vane pump　滑片泵
13　radial pump　径流泵
14　centrifugal pump　离心泵
15　multistage centrifugal pump　多级离心泵
16　high speed centrifugal pump　高速离心泵
17　turbine pump　蜗轮泵
18　vortex pump　漩涡泵
19　peripheral pump　漩涡泵
20　regenerative pump　再生泵
21　high-specific speed pump　高比转数泵
22　helical impeller pump　螺旋叶轮泵
23　propeller pump　螺旋桨泵
24　axial flow pump　轴流泵
25　mixed flow pump　混流泵
26　back-to-back impeller pump　背靠背叶轮泵
27　volute casing pump　蜗壳泵
28　tubular casing pump　管壳泵
29　elbow casing pump　弯管壳体泵
30　horizontal pump　卧式泵
31　barrel pump　筒形泵
32　canned motor pump　屏蔽泵
33　vertical can-type pump　立式屏蔽泵
34　armoured pump　铠装泵
35　submerged pump　液下泵
36　underwater pump　水下泵
37　underwater motor pump　水下电机泵
38　submersible motor pump　潜水（电机）泵
39　deep well pump　深井泵
40　borehole pump　深井泵
41　vertical spindle well pump　立轴式井泵
42　sliding-vane vacuum pump　滑板式真空泵
43　water ring pump　水环泵
44　liquid ring pump　液环泵
45　jet pump　喷射泵
46　ejector pump　喷射泵
47　diffusion pump　扩散泵
48　oil diffusion pump　油扩散泵
49　mercury diffusion pump　水银扩散泵
50　air pump　空气泵
51　air-lift pump　空气提升泵
52　acid egg　酸蛋
53　Nash pump　纳氏泵
54　swash plate pump　斜盘式泵，甩板泵
55　roller-type pump　滚柱式泵
56　insert pump　插入式泵
57　priming pump　灌液泵
58　self-priming pump　自吸泵
59　free-flow pump　自由流动泵
60　multiflow pump　多流式泵
61　ring-section pump　环段泵
62　cam-and-piston pump　凸轮-活塞泵
63　cam pump　凸轮泵
64　squeegee pump　挤压泵，胶管泵
65　peristaltic pump　蠕动泵
66　shuttle block pump　滑块泵
67　universal-joint pump　万向接头
68　Roots pump　罗茨泵

16.1.2.3 Service condition of pumps 泵的使用情况

1. clean water pump 清水泵
2. water supply pump 供水泵
3. feed pump 给水泵
4. boiler feed pump 锅炉给水泵
5. boiler circulating pump 锅炉循环泵
6. waterworks pumps 水厂泵
7. irrigation pump 灌溉用泵
8. drainage station pump 排灌站水泵
9. cellar drainage pump 地下室排水泵
10. drainage pump 疏水泵
11. hot water pump 热水泵
12. central heating circulating pump 集中供热循环泵
13. sea water pump 海水泵
14. wet-pit pump 排水泵
15. dewater pump 排水泵
16. faeces pump 排污泵
17. dirty water pump 脏水泵
18. sewage pump 污水泵
19. fire-fighting pump 消防泵
20. sprinkler pump 喷淋泵
21. cooling water pump 冷却水泵
22. coolant pump 冷却剂泵
23. condensate pump 冷凝液泵
24. oil pump 油泵
25. lubricating oil pump 润滑油泵
26. cargo oil pump 货船油泵
27. charge pump 加料泵
28. acid pump 耐酸泵
29. pulp pump 纸浆泵
30. sump pump 底料泵
31. sludge pump 污泥泵
32. mercury pump 水银泵
33. liquefied gas pump 液化气泵
34. standard chemical pump 标准化工泵
35. chemical pump 化工用泵
36. process pump 化工工艺用泵；流程泵
37. refinery pump 炼厂泵
38. reactor pump 反应堆泵
39. circulating pump 循环泵
40. primary circulating pump 主循环泵
41. heat pump 热泵
42. heat transfer media pump 热载体泵
43. tank farm pump 罐区泵
44. tank farm submersible pump 罐区潜水泵
45. constant rate pump 计量泵
46. metering pump 计量泵
47. diaphragm type metering pump 膜式计量泵
48. plunger type metering pump 柱塞式计量泵
49. pressure test pump 试压泵
50. marine pump 船用泵
51. dock pump 船坞泵
52. pipeline pump 管道泵
53. in-line pump 管线泵
54. piping pump 管式泵
55. slurry pump 淤泥泵
56. sand pump 砂泵

16.1.2.4 Drive type of pump 泵驱动型式

1. geared pump 齿轮传动泵
2. turbine driven pump 透平驱动泵
3. water jet pump 水喷射泵
4. steam jet 蒸汽喷射泵
5. hand pump 手动泵，手摇泵
6. direct acting steam pump 直动泵
7. hermetically sealed magnetic drive pump 封闭式电磁泵
8. duplex acting steam-driven reciprocating pump 双作用蒸汽往复泵
9. high pressure manual pump 手摇高压泵

16.1.2.5 Constructional materials of pump 泵的构造材料

1. plastic pump 塑料泵
2. glass pump 玻璃泵
3. rubber-lined pump 橡胶衬里泵
4. concrete casing pump 水泥壳泵
5. ceramic process pump 陶瓷化工泵
6. silicon iron pump 硅铁泵
7. high-silicon cast iron pump 高硅铸铁泵
8. graphite pump 石墨泵
9. corrosion resistant pump 耐腐蚀泵

16.2 Reciprocating pump 往复泵

16.2.1 Schematic showing direction of forces on pistons 活塞受力简图

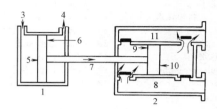

16.2.2 Schematic of a double-acting liquid end 双作用泵缸示意图

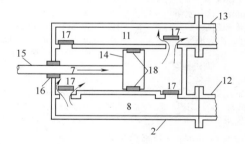

1	steam cylinder 蒸汽缸		10	discharge pressure 排出压力
2	liquid cylinder 泵缸		11	discharge manifold 排出腔
3	steam in 蒸汽进入		12	suction pipe 吸入管
4	steam exhaust 乏汽出口		13	discharge pipe 排出管
5	steam pressure 蒸汽压力		14	piston 活塞
6	back pressure 背压力,反压		15	piston rod 活塞杆
7	motion （往复）运动		16	piston-rod packing 活塞杆填料
8	suction manifold 吸入腔		17	valve 阀
9	suction pressure 吸入压力		18	piston packing 活塞密封环

16.2.3 Duplex acting steam-driven reciprocating pump 双作用蒸汽往复泵

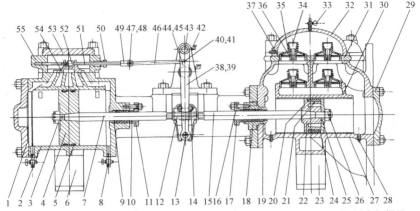

1	steam-cylinder head	蒸汽缸端盖
2	steam-cylinder with cradle	带托架的汽缸
3	steam piston nut	蒸汽活塞螺母
4	steam piston	蒸汽活塞
5	steam piston rings	蒸汽活塞环
6	steam-cylinder foot	汽缸支腿
7	steam piston rod	蒸汽活塞杆
8	drain valve for steam end	汽缸端的放液阀
9	steam piston-rod stuffing box bushing	蒸汽活塞杆填料函衬套
10	steam piston-rod stuffing box gland lining	蒸汽活塞杆填料函压盖衬套
11	steam piston-rod stuffing box gland	蒸汽活塞杆填料函压盖
12	steam piston spool	蒸汽活塞杆接头
13	connecting rod pin	连杆销
14	piston-rod spool bolt	活塞杆接头螺栓
15	liquid piston rod	液缸活塞杆
16	liquid piston-rod stuffing box gland lining	泵缸活塞杆填料函压盖衬套
17	liquid piston-rod stuffing box gland	泵缸活塞杆填料函压盖
18	liquid piston-rod stuffing box	泵缸活塞杆填料函
19	liquid piston-rod stuffing box bushing	泵缸活塞杆填料函衬套
20	liquid piston snap rings	泵缸活塞弹力环
21	liquid piston bull rings	泵缸活塞耐磨环
22	liquid piston body	泵缸活塞
23	liquid-cylinder foot	泵缸支腿
24	liquid piston fibrous packing rings	液缸活塞纤维填料环
25	liquid piston follower	泵缸活塞随动板
26	liquid piston-rod nut	泵缸活塞杆螺母
27	liquid cylinder	泵缸
28	drain plug for liquid end	泵缸端的放液螺塞
29	liquid-cylinder head	泵缸端盖
30	liquid cylinder lining	泵缸衬套
31	valve plate	阀片
32	force chamber	压力室
33	air cock	放气阀
34	valve guard	阀挡
35	valve spring	阀弹簧
36	discharge valve	排出阀
37	valve seat	阀座
38	short connecting rod	短连杆
39	long connecting rod	长连杆
40	crankpin nut	曲柄销螺母
41	crankpin	曲柄销
42	cross stand	十字支架
43	connecting rod key	连杆键
44	lower rock shaft	下（面）摇臂轴
45	long crank upper rock shaft	长曲柄上（面）摇臂轴
46	valve-rod link	阀杆拉杆
47	valve-rod head-pin nut	阀杆接头销螺母
48	valve-rod head pin	阀杆接头销
49	valve-rod head	阀杆接头
50	valve-rod stuffing box gland	阀杆填料函压盖
51	valve rod	阀杆
52	valve-rod nut	阀杆螺母
53	steam-chest cover	蒸汽室顶盖
54	slide valve	滑阀
55	steam chest	进汽室

16.2.4 Simplex plunger pump with forged steel cylinder 锻钢缸体的单缸柱塞泵

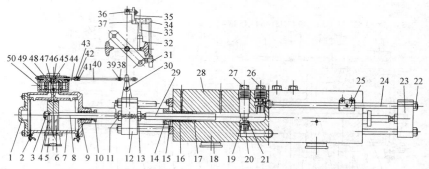

1	steam cylinder head 蒸汽缸端盖	26	discharge-valve plug 排出阀螺塞
2	steam cylinder drain valve 汽缸放液阀	27	suction-valve plug 吸入阀螺塞
3	steam piston-rod nut 蒸汽活塞杆螺母	28	liquid cylinder 泵缸
4	steam piston 蒸汽活塞	29	plunger 柱塞
5	steam-cylinder foot 汽缸支座	30	fulcrum pin 支轴销
6	steam piston rings 蒸汽活塞环	31	cross-head pin 十字头销
7	steam cylinder with cradle 带托架的汽缸	32	lever 杠杆
8	steam piston rod 蒸汽活塞杆	33	cross stand 十字支架
9	piston-rod stuffing box bushing 活塞杆填料函衬套	34	lever key 杠杆键
10	steam piston-rod stuffing box gland 蒸汽活塞杆填料函压盖	35	crank 曲柄
11	steam piston-rod jam nut 蒸汽活塞杆锁紧螺母	36	tappet 挺杆
12	front cross head of plunger 柱塞前十字头	37	crank key 曲柄键
13	plunger nut 柱塞螺母	38	lost-motion locknut 空转锁定螺母
14	plunger gland flange 柱塞填料压盖法兰	39	lost-motion adjusting nut 空转调节螺母
15	plunger gland lining 柱塞填料压盖衬套	40	valve-rod link 阀杆连接杆
16	lantern gland 润滑环密封垫	41	valve-rod link head 阀杆连接接头
17	plunger lining 衬套	42	valve-rod head pin 阀杆接头销
18	liquid cylinder foot 液缸底座	43	valve-rod head 阀杆接头
19	valve seat 阀座	44	valve rod 阀杆
20	metal valve 金属阀	45	slide valve 滑阀
21	valve spring 阀弹簧	46	steam-chest cover 蒸汽室顶盖
22	side-rod nut 边杆螺母	47	piston valve 活塞阀
23	rear cross head of plunger 柱塞后十字头	48	steam chest 蒸汽室
24	side rod 侧杆	49	valve-rod stuffing box gland 阀杆填料函压盖
25	side-rod guide 侧杆导轨	50	steam-chest head 蒸汽室端盖

16.2 Reciprocating pump 往复泵

16.2.5 Plunger type metering pump 柱塞式计量泵

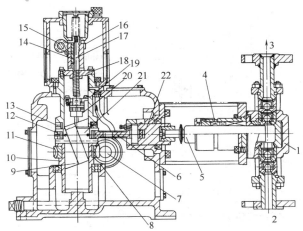

1	pump cylinder 泵缸	12	needle roller bearing 滚针轴承
2	intake 进口	13	thrust bearing 推力轴承
3	outlet 出口	14	slide-key B 滑键 B
4	stuffing box 填料函	15	adjusting worm 调节蜗杆
5	plunger 柱塞	16	adjusting wormgear 调节蜗轮
6	connecting rod 连杆	17	adjusting screw 调节螺杆
7	worm 蜗杆	18	adjusting seat 调节座
8	flat key 平键	19	upper sleeve 上套筒
9	slide-key A 滑键 A	20	eccentric slide block 偏心滑块
10	lower sleeve 下套筒	21	eccentric sleeve 偏心套筒
11	worm gear 蜗轮	22	crosshead 十字头

16.2.6 Plunger pump 柱塞泵

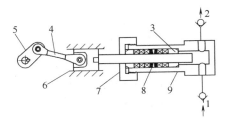

1. suction 吸入
2. discharge 排出
3. plunger 柱塞
4. connecting rod 连杆
5. crank 曲柄
6. crosshead guide 十字头滑道
7. packing take-up device 填料压紧装置
8. packing 填料
9. cylinder 气缸，泵体

16.2.7 Construction details of a diaphragm-type pump 隔膜泵结构详图

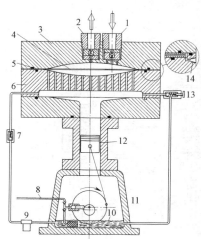

1 inlet check valve 进口止逆阀
2 discharge check valve 排出止逆阀
3 upper head 上盖
4 diaphragm 隔膜
5 O ring seal O形密封环
6 lower head 下盖
7 check valve 止逆阀
8 injection pump assembly 喷射泵组件
9 filter 过滤器
10 oil 油
11 base crankcase 机座曲轴箱
12 pulsing piston 脉冲活塞
13 relief valve 安全阀
14 leak detection port 泄漏检测孔

16.2.8 Mechanically actuated diaphragm 机械作用的隔膜

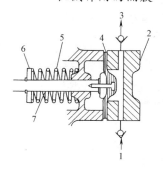

1 suction 吸入
2 liquid head 泵缸盖
3 discharge 排出
4 diaphragm 隔膜
5 return spring 复位弹簧
6 spring retainer 弹簧座

16.2.9 Flat circular diaphragm 圆形平膜

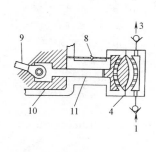

7 diaphragm push rod 隔膜推杆
8 adjustable bypass valve 可调旁通阀
9 connecting rod 连杆
10 hydraulic fluid 液压用液体
11 plunger 柱塞

16.3 Rotating pump 回转泵

16.3.1 Gear pump 齿轮泵

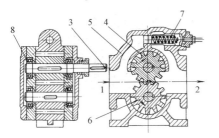

1 inlet 入口
2 outlet 出口
3 pump shaft 泵轴
4 driving gear 主动齿轮
5 pump casing 泵壳
6 driven gear 从动齿轮
7 relief valve 安全阀
8 bearing 轴承

16.3.2 Two-screw pump 双螺杆泵

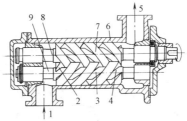

16.3.3 Three-screw pump 三螺杆泵

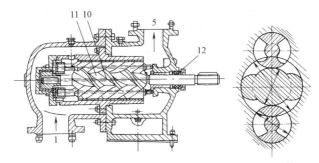

1 suction inlet 吸入口
2 through to suction chamber 与吸入腔相通
3 seal cavity 密封腔
4 through to discharge chamber 与排出腔相通
5 discharge outlet 排出口
6 pump casing 泵体
7 screw 螺杆
8 driving shaft 主动轴
9 driven shaft 从动轴
10 driving screw 主动螺杆
11 driven screw 从动螺杆
12 seal assembly 密封组件

16.3.4 Flexible rotor pump 挠性转子泵

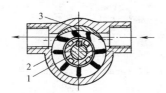

1　neoprene vane　橡胶叶轮
2　pump casing　泵壳
3　crescent　月牙形块

16.3.5 Sliding-vane pump 滑片泵

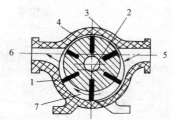

1　sliding vane　滑片
2　eccentric　偏心
3　pump casing　泵壳
4　rotor　转子
5　inlet　入口
6　outlet　出口
7　direction of rotation　旋转方向

16.3.6 Rotary lobe pump 旋转凸轮泵

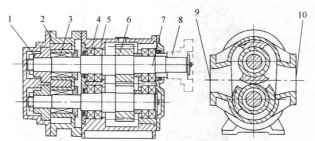

1　lobe　凸轮
2　pump body　泵体
3　seal by special rubber ring　特殊橡胶圈密封
4　bearing　轴承
5　bearing pedestal　轴承座
6　synchronizing gear　同步传动齿轮
7　drive shaft　传动轴
8　coupling　联轴器
9　discharge side　排出端
10　suction side　吸入端

16.3.7 Roots pump 罗茨泵

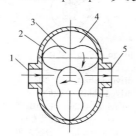

1　suction port　吸入口
2　pump body　泵壳
3　rotor　转子
4　working space　工作室
5　discharge port　排出口

16.4 Centrifugal pump 离心泵
16.4.1 Single stage centrifugal pump 单级离心泵

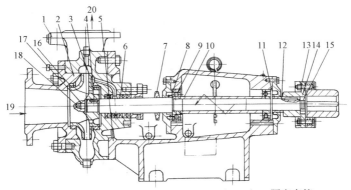

1　pump casing　泵壳
2　suction cover　吸入端泵盖
3　impeller　叶轮
4　shaft　轴
5　bracket　托架
6　shaft seal　轴封
7　flinger ring　挡液环
8,11　oil thrower　挡油圈
9　ball bearing　滚珠轴承
10　spacing tube　隔离套筒
12　spacer sleeve　挡套
13　coupling　联轴器
14,18　thrust washer　止推垫圈
15　round nut　圆螺母
16　sealing ring　密封环
17　impeller nut　叶轮螺母
19　pump suction nozzle　泵吸入口
20　pump discharge　泵出口

16.4.2 Horizontal single-stage single suction centrifugal pump 卧式单级单吸离心泵

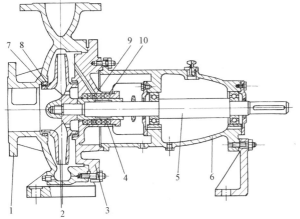

1　pump body　泵体
2　impeller　叶轮
3　pump bonnet　泵盖
4　seal assembly　密封组件
5　pump shaft　泵轴
6　bracket support　悬架支承部件
7　impeller ring　叶轮口环
8　pump body wearing ring　泵体耐磨环
9　lantern ring　封液环
10　packing ring　填料环

373

16.4.3 Typical single-stage end-suction volute pump 典型的单级侧面进口蜗壳泵

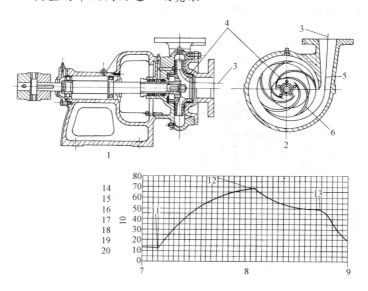

1	typical pump section 典型的泵剖面图	14	200 GPM 200 加仑/分
2	flow through impeller and volute 流经叶轮与蜗壳	15	166 FT TOTAL HEAD 总压头166英尺
3	flow line 流线	16	3500 RPM 3500r/min
4	point of entrance to impeller vanes 叶轮叶片的入口点	17	$2\frac{1}{2}$ IN SUCTION DIAM 吸入口直径 $2\frac{1}{2}$ 英寸
5	volute 蜗壳		
6	impeller 叶轮	18	2IN DISCHARGE DIAM 排出口直径2英寸
7	suction flange 吸入口法兰		
8	developed path 液流途径	19	$6\frac{3}{4}$ IN IMPELLER DIAM 叶轮直径 $6\frac{3}{4}$ 英寸
9	discharge flange 排出口法兰		
10	absolute velocity, in feet per second 绝对速度, ft/s	20	$\frac{5}{8}$ IN IMPELLER WIDTH 叶轮宽度 $\frac{5}{8}$ 英寸
11	suction vane tip 吸入叶尖		
12	discharge vane tip 排出叶尖		
13	volute throat 蜗壳喉部		

16.4.4 Typical pump used for recycling coke quench water 典型炼焦用循环冷却水泵

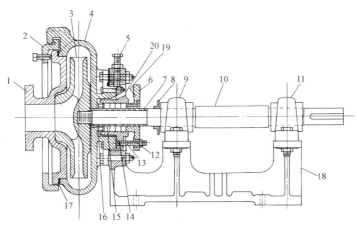

1	plate 法兰板		12	gland nut 压盖螺母
2	ring 圆环		13	gland stud 填料压盖双头螺柱
3	impeller 叶轮		14	stuffing box 填料函
4	casing 泵壳		15	packing 填料
5	bolt 螺栓		16	lantern ring 液封环
6	gland follower 填料压盖随动件		17	gaskets 垫片
7	shaft sleeve 轴套		18	bearing stand 轴承架
8	flinger 抛油环		19	grease to gland 润滑脂通入压盖
9	thrust bearing 推力轴承		20	liquid seal to lantern ring 密封液通入液封环
10	shaft 轴			
11	radial bearing 径向轴承			

16.4.5 Types of impellers 叶轮类型

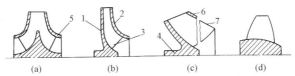

(a) centrifugal impeller, double-suction, enclosed design 离心式叶轮，双吸，闭式结构
(b) centrifugal impeller, single-suction, enclosed design 离心式叶轮，单吸，闭式结构
(c) mixed-flow impeller, single-suction, enclosed & open design 混流式叶轮，单吸，闭式与开式结构
(d) propeller or axial-flow impeller, single-suction, open design 螺旋桨或轴流式叶轮，单吸，开式结构

1	disk 轮盘		3-3	forward curved vane 前弯叶片
2	shroud 轮盖		4	impeller hub 叶轮轮毂
3	impeller vane 叶轮叶片		5	impeller ring 叶轮口环
3-1	backward-curved vane 后弯叶片		6	enclosed 闭式
3-2	radial vane 径向叶片		7	open 开式

16.4.6 Parts of a double-suction impeller 双吸叶轮部件

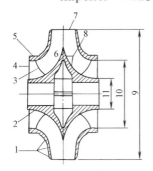

16.4.7 Typical diffuser-type pump 典型的扩压型泵

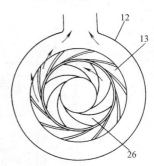

16.4.8 Single labyrinth of intermeshing type 交叉型单一迷宫式密封

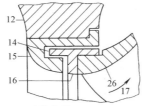

1　shroud　叶轮轮盖
2　hub　轮毂
3　suction vane edge or tip　吸入叶片之前缘
4　suction eye　吸入口
5　wearing ring　耐磨环
6　vane　叶片
7　discharge vane edge or tip　排出叶片之边缘
8　center dividing wall of double suction impeller　双吸叶轮之中心分隔壁
9　impeller diameter　叶轮直径
10　eye diameter　吸入口直径
11　hub diameter　轮毂直径
12　casing　泵壳
13　diffuser　扩压器
14　relief chamber　节流降压室
15　casing ring　泵壳口环
16　impeller ring　叶轮口环
17　flow　液流方向
18　balancing disk　平衡盘
19　discharge pressure　排出压力
20　balancing disk head　平衡盘上部
21　to suction　接吸入口
22　restricting orifice　节流降压装置
23　balancing chamber　平衡室
24　axial clearance　轴向间隙
25　back pressure　背压力
26　impeller　叶轮
27　shaft　轴
28　gland　填料压盖
29　shaft nut　轴螺母
30　set screw　止动螺钉
31　sleeve　轴套
32　impeller nut　叶轮螺母
33　key　键
34　liner ring　衬环

16.4 Centrifugal pump 离心泵

16.4.9 Simple balancing disk． 普通平衡盘

16.4.10 Sleeve construction 轴套结构图

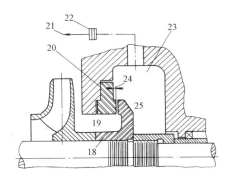

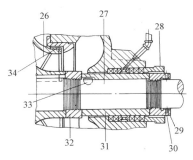

with internal impeller nut, external shaft-sleeve nut, and separate key for sleeve
内有叶轮螺母，外有轴套螺母，还带键

16.4.11 Vertical split centrifugal pump 垂直剖分式离心泵

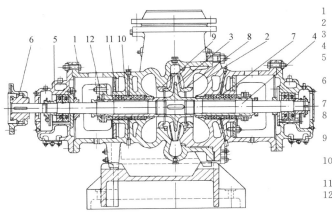

1 pump body 泵体
2 pump cover 泵盖
3 impeller 叶轮
4 pump shaft 泵轴
5 bearing pedestal 轴承座
6 gear coupling 齿轮联轴器
7 shaft sleeve 轴套
8 impeller ring 叶轮口环
9 casing ring 泵壳口环
10 lantern ring 填料封液环
11 packing 填料
12 water cooled gland 水冷式填料压盖

377

16.4.12 Double suction centrifugal pump 双吸离心泵

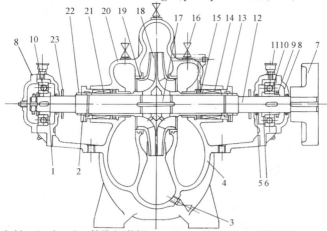

1 bearing holder (outboard) 轴承座(外侧)
2 packing gland 填料压盖
3 drain outlet 排放口
4 lower casing 下壳体
5 oil seal 油封圈
6 bearing holder (inboard) 轴承座（内侧）
7 coupling 联轴器
8 bearing cover 轴承箱盖
9 bearing nut 轴承锁紧螺母
10 ball bearing 滚珠轴承
11 grease cup 油脂杯
12 main shaft 主轴
13 packing ring 填料隔圈
14 packing 填料
15 lantern ring 液封环
16 air cock 排气旋塞
17 impeller key 叶轮轴键
18 impeller 叶轮
19 casing wear ring 泵壳耐磨环
20 upper casing 上壳体
21 shaft sleeve 轴套
22 sleeve nut 轴套螺母
23 flinger 挡液环

16.4.13 Horizontal single-stage double-suction volute pump 卧式单级双吸蜗壳泵

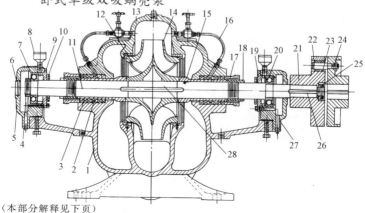

（本部分解释见下页）

16.4.14 Sectional view of a vertical-shaft end-suction pump with a double-volute casing 具有双蜗壳的立式端吸离心泵剖视图

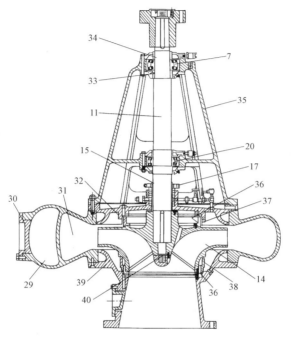

1　casing (lower half)　泵壳（下半部）
2　seal cage　封液环
3　packing　填料
4　bearing lock nut　轴承锁紧螺母
5　bearing housing (outboard)　轴承箱（外侧）
6　bearing end cover　轴承箱端盖
7　bearing (outboard)　轴承（外侧）
8　grease (oil) cup　润滑脂（油）杯
9　bearing cover (outboard)　轴承盖（外侧）
10　deflector　挡板
11　pump shaft　泵轴
12　casing ring　泵壳口环，阻漏环
13　casing (upper half)　泵壳（上半部）
14　impeller　叶轮
15　shaft sleeve　轴套
16　seal piping (tubing)　密封液管
17　gland　填料压盖
18　shaft sleeve nut　轴套螺母
19　bearing cover (inboard)　轴承箱端盖（内侧）
20　bearing (inboard)　轴承（内侧）
21　coupling (pump half)　联轴器（泵侧）
22　coupling bushing　联轴器销衬套
23　coupling pin　联轴器柱销
24　coupling (driver half)　联轴器（驱动装置侧）
25　coupling lock nut　联轴器锁母
26　coupling key　联轴器键
27　bearing housing (inboard)　轴承箱（内侧）
28　impeller key　叶轮键
29　casing　泵壳
30　handhole cover　手孔盖
31　double volute　双蜗壳
32　stuffing box cover　填料函盖
33　bearing spacer　轴承定距环
34　shaft collar　轴迷宫圈
35　frame　（泵）机架
36　impeller ring　叶轮口环
37　stuffing box cover ring　填料函盖圈
38　suction head ring　吸入接头套圈
39　suction cover　吸入口盖
40　impeller nut　叶轮螺母

16.4.15 Peripheral pump 漩涡泵（再生泵）

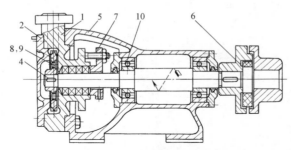

16.4.16 Regenerative turbine pump schematic of operation 再生泵操作示意图

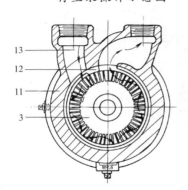

1 pump body 泵体
2 pump casing cover 泵盖
3 impeller 叶轮
4 pump shaft 泵轴
5 bracket 托架
6 coupling 联轴器
7 stuffing box gland 填料压盖
8 balancing hole 平衡孔
9 bolthole for dismounting 拆卸用螺孔
10 bearing 轴承
11 casing 泵壳
12 stripper 限流环
13 fluid particles 流体质点

16.4.17 Regenerative pump construction 再生泵（漩涡泵）结构图

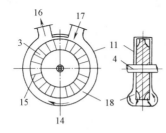 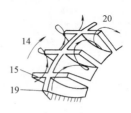

14 direction of rotation 转向
15 vane 叶片
16 outlet 出口
17 inlet 进口
18 flow passage 流道
19 disk 轮盘
20 longitudinal vortex 纵向漩涡

16.4.18 Multi-stage centrifugal pump Ⅰ 多级离心泵（一）

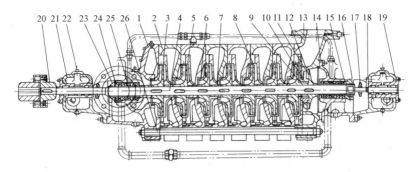

1 inlet portion 进口段
2 intermediate portion 中段
3 impeller 叶轮
4 shaft 轴
5 guide plate 导向隔板
6 wearing ring 承磨环
7 distance collar 定距轴环
8 guide vane 导向叶片
9 balance hole 平衡孔
10 balance sleeve 平衡套
11 balance ring 平衡环
12 guide plate of outlet portion 排出段的导向隔板
13 outlet portion 出口段
14 rear cover 后盖
15 shaft sleeve A 轴承 A
16 sleeve nut 牵紧螺母
17 flinger ring 挡液环
18 pointer of balance disk 平衡盘指针
19 bearing assembly B 轴承部件 B
20 coupling 联轴器
21 bearing assembly A 轴承部件 A
22 oil cup 油杯
23 shaft sleeve B 轴套 B
24 packing gland 填料压盖
25 lantern ring 液封环
26 staybolt 拉紧螺栓

16.4.19 Multi-stage centrifugal pump Ⅱ 多级离心泵（二）

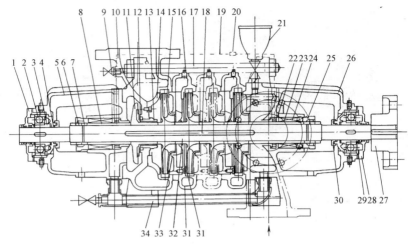

1	discharge side bearing cover	出口侧轴承压盖
2	radial ball bearing	径向滚珠轴承
3	bearing holder	轴承座
4	bearing sheet	轴承垫片
5	bearing bracket	轴承架
6	shaft nut	轴套定位螺母
7	packing ring	填料环
8	discharge side sleeve	出口侧轴套
9	air valve	排气阀
10	discharge casing	排出口壳体
11	balance disk	平衡盘
12	balance seal	平衡盘密封
13	balance bush	平衡盘套筒
14	discharge ring	出口环
15	impeller (last stage)	末级叶轮
16	plug for air vent	放气螺塞
17	clamping bolt	固定螺栓
18	impeller key	轴键
19	sealing water pipe	水封管
20	sealing water pipe connection	水封管接头
21	priming cup	灌液漏斗
22	seal cage	液封环
23	suction side sleeve	进口侧轴套
24	gland packing	填料函
25	packing gland	填料压盖
26	suction casing	进口壳体
27	coupling	联轴器
28	suction side bearing cover	吸入侧轴承压盖
29	retaining ring	轴定位环
30	flinger	挡液环
31	middle casing	中间壳体
32	main shaft	主轴
33	liner ring	衬环
34	balance pipe	平衡管

16.4.20 Inline pump 管道泵

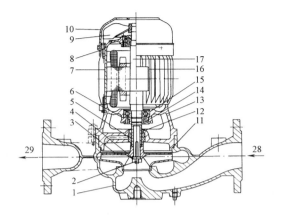

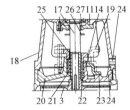

1 volute casing 蜗壳
2 impeller nut 叶轮螺母
3 impeller 叶轮
4 lock washer 锁紧垫圈
5 mechanical seal 机械密封
6 labyrinth seal 迷宫密封
7 motor casing 电机壳体
8 casing stud 壳体双头螺栓
9 fan impeller 风扇叶轮
10 cowl 风扇罩
11 flat gasket 平垫片
12 shaft protecting sleeve 轴套
13 bearing bracket 轴承架
14 O-ring O形环
15 deep groove ball bearing 深沟球轴承
16 terminal box 接线盒
17 rotor 转子
18 soft-packed stuffing box 软填料填函箱
19 cooling water outlet and inlet 冷却水进出口
20 discharge cover 排出端泵盖
21 stuffing box packing 填料
22 throat ring 喉部垫环，填料衬环
23 cooling compartment cover 冷却室盖
24 leakage fluid drain 排液口
25 stuffing box gland 填料压盖
26 thrower 甩油环
27 bearing cover 轴承盖
28 inlet 进口
29 outlet 出口

16.4.21 Vertical pump with flexible shafting 具有挠性轴系的立式泵

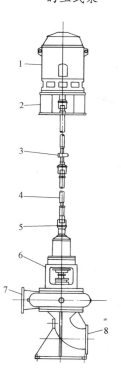

1 motor   电动机
2 motor stand   电机机架
3 guide bearing   导向轴承
4 tubular shaft   管式轴，空心轴
5 universal joint   万向接头
6 pump bearing frame   泵轴承架
7 discharge   排出口
8 suction   吸入口

16.4.22 Submerged pump 液下泵

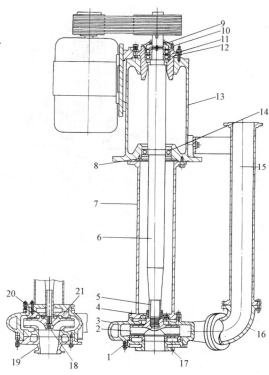

1 pump cover   泵盖
2 impeller of pump   泵叶轮
3 pump body   泵体
4 throttle ring   节流圈
5 spindle sleeve   轴套
6 spindle   主轴
7 support column   支承管
8, 9 oil seal ring   油封圈
10 bearing cover   轴承压盖
11 bearing box   轴承箱
12 thrust bearing   推力轴承
13 bearing pedestal   轴承架
14 radial bearing   径向轴承
15 discharge spout   排液管
16 discharge elbow   排液弯管
17, 19 suction cover   吸入端盖
18 impeller nut   叶轮螺母
20 impeller key   叶轮键
21 wearing ring   耐磨环

16.4.23 Double-suction wet-pit pump 双吸排水泵

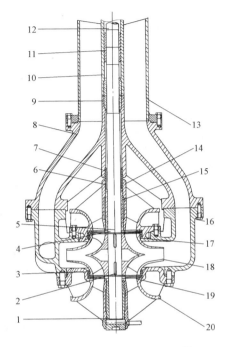

1. rotating shaft　转轴
2. impeller key　叶轮键
3. casing ring　泵壳口环
4. protecting collar　防护环
5. seal　密封件
6. seal ring　密封环
7. bearing sleeve　轴承座套
8. transition diffuser　过渡段扩压器
9. connector bearing　接头轴承
10. shaft enclosed tube　轴的密封套管
11. shaft coupling　轴联接件
12. line shaft　中间轴
13. column pipe　圆筒管
14. suction bell（upper）吸水喇叭口（上部）
15. suction bell bearing　吸水喇叭口轴承
16. casing　泵壳
17. impeller ring　叶轮口环
18. impeller　叶轮
19. locating ring　定位环
20. suction bell（lower）吸水喇叭口（下部）

16.4.24 Vertical wet-pit diffuser pump bowl 立式扩压型排水泵

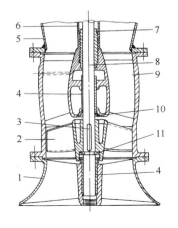

1. suction bowl　吸入筒
2. propeller　叶轮，螺旋桨
3. propeller key　叶轮键，螺旋桨键
4. bearing bushing　轴承衬套
5. column pipe　圆筒管
6. shaft enclosing tube　泵轴套管
7. connector bearing　接头轴承
8. pump shaft　泵轴
9. discharge bowl　排出筒
10. seal　密封件
11. thrust collar　止推环

16.4.25 Section of vertical turbine pump 立式涡轮泵剖面图

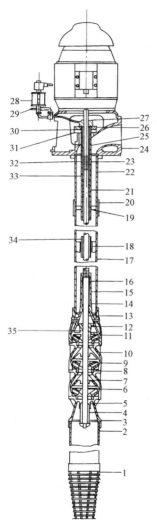

1. strainer 过滤器
2. suction pipe 吸入管
3. suction head 吸入口接头
4. suction head bearing 吸入口接头轴承
5. bearing cap 轴承盖
6. impeller bushing 叶轮衬套
7. bowl bearing 圆筒支架
8. wearing ring 耐磨环
9. impeller 叶轮
10. bowl 圆筒
11. top bowl 上部圆筒
12. seal ring 密封环
13. top bowl bearing 上部圆筒支架
14. seal ring spacer 密封环隔圈
15. top bowl connector bearing 上部圆筒连接器支座
16. impeller shaft 叶轮轴
17. column pipe 管状圆筒
18. shaft tube stabilizer 轴管固定件
19. column pipe spacer 管状圆筒定距块
20. column coupling 圆筒连接器
21. enclosed line shaft bearing 密闭的中间轴轴承
22. top column pipe 上部圆筒管
23. shaft coupling 联轴器
24. discharge head 排出口接头
25. top shaft 上部轴
26. stuffing box 填料函
27. shaft tube tension bearing 轴套管压力支承块
28. solenoid oiler 圆筒形油杯
29. sight feed valve 可视给油阀
30. packing follower 填料压盖
31. packing 填料
32. top shaft tube 上部轴套管
33. line shaft 中间轴
34. shaft tube 轴套管
35. retainer plate 挡板

(with closed impellers and enclosed line shafting, oil lubrication 具有闭式叶轮和密闭的中间轴,油润滑)

16.4.26 Section of bowls of a vertical turbine pump (with closed impellers) 立式涡轮泵泵体剖面图（具有闭式叶轮）

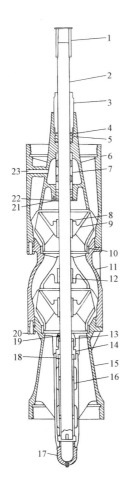

1 impeller shaft coupling 叶轮轴联轴器
2 impeller shaft 叶轮轴
3 connector bearing 接头轴瓦
4 seal ring 密封环
5 seal spacer 密封环隔圈
6 top bowl 上筒
7 top bowl bearing 上筒轴瓦
8 impeller bushing 叶轮衬套
9 impeller 叶轮
10 wearing ring 耐磨环
11 bowl 外筒
12 bowl bearing 外筒轴瓦
13 bearing cap 轴承盖
14 suction head cap 吸入接头盖帽
15 suction head 吸入接头
16 suction head bearing 吸入接头支承面
17 suction head plug 吸入接头管塞
18 clipper seal 钳压密封
19 bearing cap set screw 轴承盖止动螺钉
20 hex head bolt 六角头螺栓
21 retainer plate screw 挡板螺钉
22 retainer plate 挡板
23 relief port 放气口

16.4.27 Vertical multistage centrifugal pump with barrel casing 具有筒形泵壳的立式多级离心泵

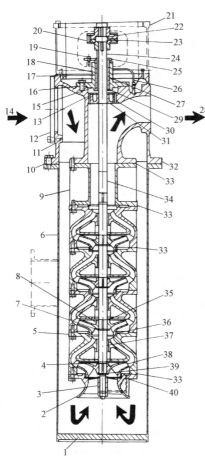

1 base plate 底板
2 sectional bell 喇叭形吸入口
3 lower sleeve bearing 下套筒轴承
4 first stage snap ring 第一级开口环
5 diffuser 扩压器
6 tank 筒体
7 first stage impeller retaining collar 第一级叶轮挡圈
8 second stage impeller and above 第二级叶轮（往上类推）
9 spacer column 隔离竖筒
10 intermediate gasket 中间垫片
11 blind flange gasket 盲板法兰垫片
12 blind flange 盲板法兰
13 lower shaft sleeve 下轴套
14 inlet 进口
15 stuffing box bushing 填料函衬套
16 upper gasket 上垫片
17 intermediate shaft sleeve 中轴套
18 seal cage 密封隔环，液封环
19 gland bolt 填料压盖螺栓
20 complete coupling 联轴器部件
21 motor support columu 电动机支架
22 motor half coupling lock nut 电机侧联轴器锁紧螺母
23 pump half coupling lock nut 泵侧联轴器锁母
24 upper shaft sleeve 上轴套
25 gland 填料压盖
26 complete piping 连通管
27 packing 填料
28 discharge 排出口
29 stuffing box 填料函
30 nozzle head bushing 出口接头衬套
31 balance disc 平衡盘
32 nozzle head 出口接头
33 lower gasket 下垫片
34 shaft with key 带键的轴
35 second stage snap ring and above 第二级开口环（往上类推）
36 diffuser ring 扩压器口环
37 upper sleeve ring 上套筒环
38 first stage impeller 第一级叶轮
39 first stage impeller retaining ring 第一级叶轮挡圈
40 suction bell ring 喇叭形吸入罩

16.4.28 Submersible motor pump 潜水泵

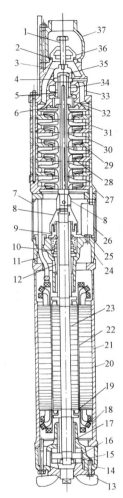

1	check valve spindle	止逆阀顶丝
2	O-ring	O形环
3	shaft cap	轴盖
4	nut	螺母
5	shaft sleeve	轴套
6	balance bush	平衡衬套
7	strainer	滤网
8	coupling	联轴器
9	upper end cover	上端盖
10	mechanical seal	机械密封
11	lead wire	引出线
12	sleeve	轴套
13	bottom end cover	下端盖
14	bottom bracket	下托架
15	thrust metal	推力轴承衬瓦
16	thrust runner	推力轴承滑道
17	sleeve bearing	轴瓦
18	stator coil	定子线圈
19	end ring	终端环
20	motor frame	电机机架
21	stator core	定子芯
22	rotor core	转子芯
23	motor shaft	电机轴
24	bracket	上托架
25	feed water plug	注水口旋塞
26	protector for water swelling 防水胀保护器	
27	bottom case	底泵壳
28	liner ring	衬环
29	impeller	叶轮
30	guide vane	导向叶片
31	middle case	中间泵壳
32	balance piston	平衡活塞
33	discharge case	出口泵壳
34	bearing (upper)	上轴承
35	pump shaft	泵轴
36	check valve	止逆阀
37	check valve case	止逆阀罩

16.4.29 Horizontal mixed flow pump 卧式混流泵

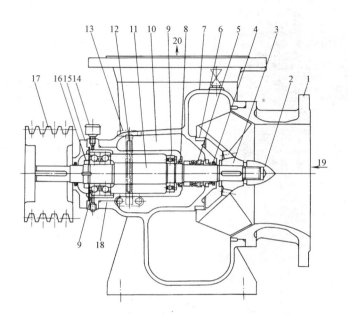

1	suction cover 吸入口壳体	11	main shaft 主轴
2	impeller nut 叶轮紧固螺母	12	oil gauge 油标
3	impeller 叶轮	13	oil cap 油杯盖
4	oil seal 油封环	14	grease cup 油脂杯
5	mechanical seal 机械密封	15	bearing cover 轴承压盖
6	liner ring 阻漏环，衬环	16	bearing nut 轴承紧固螺母
7	casing 泵壳体	17	V-pulley V形皮带轮
8	oil seal 油封圈	18	bearing seat 轴承座
9	ball bearing 滚珠轴承	19	suction 吸入口
10	oil reservoir (sump) 油腔	20	discharge 排出口

16.4.30 Vertical axial flow pump 立式轴流泵

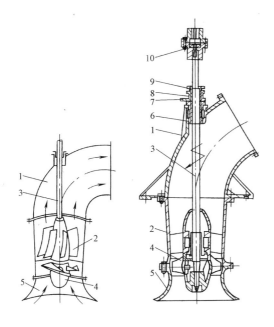

1 elbow for discharge 出水弯管
2 guide vane 导叶
3 pump shaft 泵轴
4 impeller 叶轮
5 suction casing 进口壳体
6 bearing 轴承
7 packing box 填料盒
8 packing 填料
9 packing gland 填料压盖
10 coupling 联轴器

16.5 Special pump 特种泵

16.5.1 Gas-lift pump 气升泵

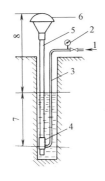

1 compressed air 压缩空气
2 pressure gage 压力表
3 air pipe 空气管
4 mixer 混合器
5 air and liquid mixed tube 气液混合管
6 separator 分离器
7 submerged length 沉浸深度
8 delivery lift 提升高度

16.5.2 Siphon 虹吸管

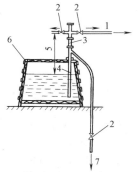

1　to vacuum pump　接真空泵
2　valve　阀
3　sight indicator　观察罩
4　siphon pipe　虹吸管
5　delivery lift　扬水高度
6　storage sump　贮槽
7　outlet　液体出口

16.5.3 Rotating casing pump　旋转壳体泵

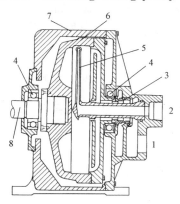

1　suction　吸入口
2　discharge　排出口
3　mechanical seal　机械密封
4　bearing　轴承
5　Pitot tube　毕托管
6　rotor　转子
7　rotor housing　转子外壳
8　shaft　轴

16.5.4 Schematic diagram of d.c. electromagnetic pump 直流电操作的电磁泵示意图

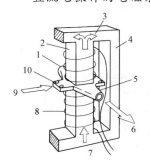

1　inlet　进口
2　pole face　极面
3　core　铁芯
4　magnetic-return circuit　磁回路
5　fluid duct　流体导管
6　flow　流动方向
7　field　磁场
8　exciting winding　激磁绕组
9　current　电流
10　electrode　电极

16.5　Special pump　特种泵

16.5.5　Hermetically sealed magnetic drive pump
　　　　密封式磁力驱动泵

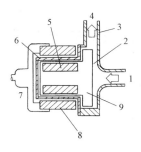

1　inlet　进口
2　fluid pumped　泵送的流体
3　front casing　前壳体
4　outlet　出口
5　impeller magnet　叶轮磁体
6　rear casing　后壳体
7　motor　电机
8　driving magnet　驱动磁铁
9　impeller　叶轮

16.5.6　Schematic diagram of a radial piston pump　径向活塞泵示意图

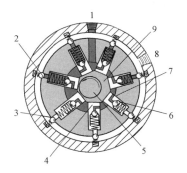

1　in　进口
2　ball check　球形止逆阀
3　outlet passage　排出通道
4　piston foot　活塞底座
5　pump shaft　泵轴
6　skirt　套筒
7　expanding lip piston
　　能伸缩唇形活塞
8　out　出口
9　eccentric　偏心轮

393

17 Compressors and Fans 压缩机与风机

17.1 Type of compressor 压缩机类型

17.1.1 Type of compressor 压缩机类型

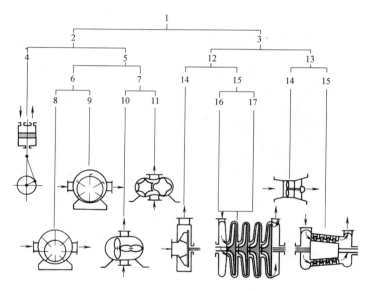

1　compressors　压缩机
2　positive displacement
　　容积式（正位移）
3　dynamic　动力式
4　reciprocating　往复式
5　rotary　回转式
6　one rotor　单转子
7　two rotor　双转子
8　sliding vane　滑片式
9　liquid ring　液环式
10　Roots　罗茨式
11　screw　螺杆式
12　centrifugal　离心式
13　axial flow　轴流式
14　one stage (fan)　单级（排风机）
15　multistage　多级
16　blower　鼓风机
17　compressor　压缩机

17.1.2 Classification of compressors based on compressed gas 压缩机按压送的气体分类

1. ammonia compressor 氨压缩机
2. ammonia compression refrigerating machine 氨压缩冷冻机
3. chlorine gas compressor 氯气压缩机
4. nitrogen compressor 氮气压缩机
5. hydrogen gas compressor 氢气压缩机
6. oxygen compressor 氧气压缩机
7. air compressor 空气压缩机
8. carbon dioxide compressor 二氧化碳压缩机
9. ethylene compressor 乙烯压缩机

17.2 Reciprocating compressor 往复式压缩机

17.2.1 Various types of cylinder arrangement 气缸布置型式

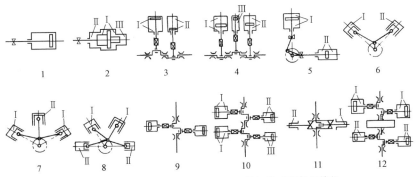

1. simplex horizontal reciprocating compressor 单缸卧式往复压缩机
2. straight-line reciprocating compressor 直线型往复压缩机
3. duplex vertical reciprocating compressor 双缸立式往复压缩机
4. triplex vertical compressor 三缸立式压缩机
5. L-type reciprocating compressor L型往复压缩机
6. V-type reciprocating compressor V型往复压缩机
7. W-type compressor W型压缩机
8. semiradial reciprocating compressor 扇型往复压缩机
9. double row balanced opposed compressor 双列对称平衡压缩机
10. M-type balanced opposed reciprocating compressor M型对称平衡式压缩机
11. horizontal opposed reciprocating compressor 卧式对置式往复压缩机
12. H-type reciprocating compressor H型往复压缩机

17.2.2 Single acting cylinder 单作用气缸

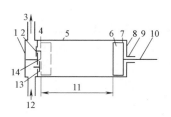

1　cylinder head end　气缸盖侧
2　discharge valve　排气阀
3　high pressure stream　高压气流
4　end of stroke　（行程末端）外止点
5　compressor cylinder　压缩机气缸
6　piston　活塞
7　inner dead point　内止点（死点）
8　crank end　曲轴侧
9　packing　填料函
10　piston rod　活塞杆
11　stroke length　行程长度
12　low pressure stream　低压气流
13　suction valve　吸气阀
14　clearance pocket　余隙容积

17.2.3 Four-corner four-stage compressor (plan view) H型四级压缩机（俯视图）

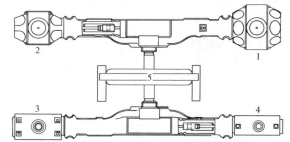

1　low pressure cylinder　低压一级缸
2　second stage cylinder　二级缸
3　third stage cylinder　三级缸
4　fourth stage cylinder　四级缸
5　motor　电动机

17.2.4 Two-stage single-acting opposed piston in a single-step-type cylinder 级差式气缸中两级对置的单作用活塞

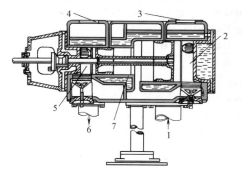

1　intake　进气口
2　first stage　第一级
3　to intercooler　通级间冷却器
4　from intercooler　来自级间冷却器
5　second stage　第二级
6　discharge　排气口
7　step-type cylinder　阶梯式气缸

17.2.5 Typical single-stage, double-acting water-cooled compressor 典型水冷式单级双作用压缩机

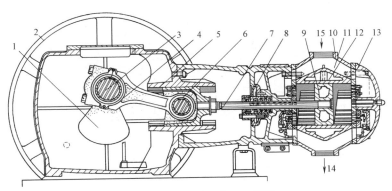

1 counterweights 平衡重
2 flywheel 飞轮
3 crank shaft 曲轴
4 connecting rod 连杆
5 crosshead guide 十字头滑道
6 crosshead 十字头
7 piston rod 活塞杆
8 packing 填料（函）
9 piston 活塞
10 compressor cylinder 气缸
11 air passage 空气通道
12 water jacket for cooling 冷却水夹套
13 cylinder head 气缸盖
14 gas outlet 气体出口
15 gas inlet port 气体进口

17.2.6 Typical double-acting piston and compressor cylinder 典型的双作用压缩机气缸与活塞

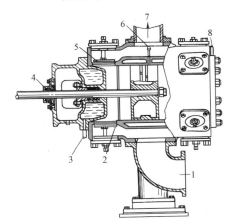

1 air intake 空气进口
2 inlet valve 进气阀
3 cooling water inlet 冷却水进口
4 oil wiper rings 刮油环
5 discharge valve 排气阀
6 force feed lubricator connection 强制润滑供油接头
7 air discharge 排气
8 cooling water outlet 冷却水出口

17.2.7 Two-stage double-acting compressor cylinders with intercooler 带级间冷却器的两级双作用压气机气缸

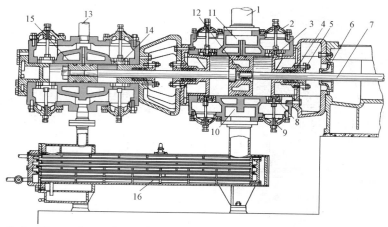

17.2.8 Forged-steel single-acting high-pressure cylinder 锻钢制单作用高压气缸

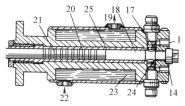

1 inlet 气体进口
2 valve catcher case 进气阀压罩
3 first stage suction valve 第一级吸气阀
4 packing 填料（函）
5 gland 填料压盖
6 wiper ring 刮油环
7 piston rod 活塞杆
8 first stage discharge valve 第一级排气阀
9 catcher screw 压紧器螺钉
10 discharge passage 排气通道
11 inlet passage 进气通道
12 first stage piston 第一级活塞
13 discharge 排气口
14 discharge valve of second stage 第二级排气阀
15 second stage piston 第二级活塞
16 intercooler 级间冷却器
17 suction valve 吸气阀
18 cylinder 气缸
19 cooling water outlet 冷却水出口
20 piston 活塞
21 piston ring 活塞环
22 cooling water inlet 冷却水进口
23 cooling water 冷却水
24 discharge valve 排气阀
25 cylinder liner 气缸衬套

17.2.9 Duplex reciprocating compressor 双缸往复式压缩机

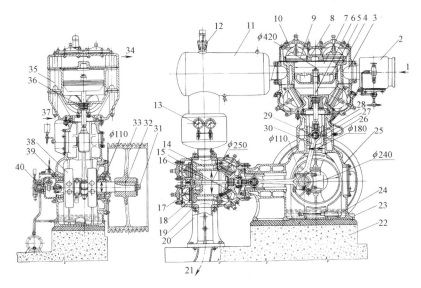

1 air inlet to compressor 压缩机空气进口
2 unloader valve 减荷阀
3 air passage 空气通道
4 piston ring 活塞环
5 first stage piston 第一级活塞
6 first stage piston rod 第一级活塞杆
7 first stage cylinder 第一级气缸
8 cylinder head 气缸盖
9 first stage discharge valve 第一级排气阀
10 pressing device （阀门）压紧装置
11 intercooler 级间冷却器
12 safety valve 安全阀
13 pressure gage 压力表
14 second stage cylinder 第二级气缸
15 second stage suction valve
　　第二级吸入阀
16 second stage piston rod 第二级活塞杆
17 second stage piston 第二级活塞
18 second stage discharge valve
　　第二级排气阀
19 washer 垫片
20 cylinder support 缸体支承
21 pressurizing air delivery 压缩空气出口
22 foundation 基础
23 foundation bolt 地脚螺栓
24 lubrication oil 润滑油
25 machine frame 机身
26 crosshead guide 十字头滑道
27 crosshead liner 十字头衬瓦
28 crosshead 十字头
29 crosshead pin 十字头销
30 connecting rod 连杆
31 shaft end nut 轴端螺母
32 key 键
33 V-pulley V形皮带轮
34 cooling water outlet 冷却水出口
35 cylinder jacket for cooling water
36 seal ring set 密封环组
37 cooling water inlet 冷却水进口
38 counterweights 平衡重
39 bearing 轴承
40 gear oil pump 齿轮油泵

17.2.10 Piston for a non-lubricated cylinder 无油润滑气缸活塞

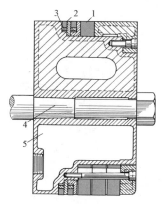

1　wearing ring　耐磨环
2　carbon piston ring　石墨活塞环
3　elastic ring　弹力环
4　piston rod　活塞杆
5　hollow piston　空心活塞

(equipped with carbon piston and wearing rings　配有石墨活塞环和承磨导向环)

17.2.11 Valve system for positive displacement compressor 容积式压缩机的阀门组件

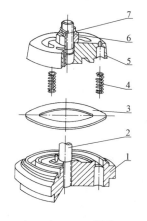
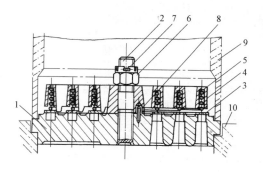

1　valve seat　阀座
2　coupling bolt　连结螺栓
3　valve plate　阀片
4　valve spring　阀弹簧
5　lift stop of a valve　阀升程限制器
6　nut　螺母
7　cotter　开口销
8　dowel pin　定位销
9　pressing device　压紧装置
10　compressor cylinder　压缩机气缸

17.2.12 Inlet valve unloader （顶开）吸气阀卸荷器

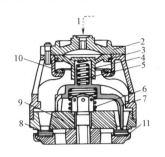

1 pressure from control device 来自控制装置的压力
2 valve head 阀盖
3 seal ring 密封环
4 spring seat 弹簧座
5 return spring 复位弹簧
6 pressure spring 压力弹簧
7 crab claw 蟹爪式阀罩
8 valve plate 阀板
9 stop plate 升程限制器
10 rubber ring 橡胶圈
11 inlet valve 吸气阀

17.2.13 Clearance control cylinder 具有余隙调节的气缸

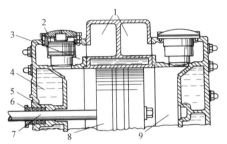

1 clearance pockets 余隙腔
2 clearance valve 余隙阀
3 gas passage 气体通道
4 cooling water 冷却水
5 jacket 夹套
6 packing 填料压盖
7 piston rod 活塞杆
8 piston 活塞
9 cylinder 气缸

17.2.14 High-pressure, low-capacity compressor having a hydraulically actuated diaphragm 液压传动的高压小流量隔膜压气机

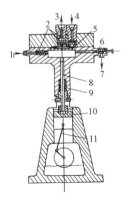

1 liquid inlet 进液
2 check valves 单向阀
3 gas out 出气
4 gas in 进气
5 diaphragm 隔膜
6 adjustable relief valve 可调减压阀
7 liquid out 出液
8 plunger 柱塞
9 seal 密封装置
10 crosshead 十字头
11 connecting rod 连杆

17.2.15 Very high pressure compressor for polyethylene process 聚乙烯超高压压缩机

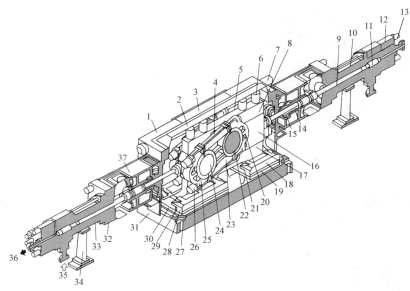

1   frame   机身
2   frame top cover   机身顶盖
3   blind cover   盲盖
4   complete connecting rod with bearing and bolts   连同轴承和螺栓的连杆组件
5   complete crosshead frame with slipper   十字头组件（带滑块）
6   crosshead bolts with nuts   十字头螺栓螺母
7   frame tie bolts   机身紧固螺栓
8   upper crosshead frame   十字头上体
9   tungsten carbide plunger   碳化钨柱塞
10 cylinder with shrunk liner   带热压缸套的气缸
11 central valve   中心型组合阀
12 cylinder head   气缸盖
13 cylinder assembling bolt   气缸装配螺栓
14 intermediate guide   中体导向装置
15 intermediate rod   中体导杆
16 indermediate crosshead frame   中体十字头框
17 crosshead slipper with lower frame   十字头滑块板
18 hind counter guide   （后）副十字头导轨
19 hind guiding plate   副十字头导板
20 big end bearing   大头轴承盖
21 hind crosshead metal   副十字头（导轨）镶条
22 crankpin   曲柄销
23 fork connecting rod   叉式连杆
24 crosshead pin   十字头销
25 small end bearing   小头轴承盖
26 connecting rod bolts with nuts   连杆螺栓螺母
27 foundation bolt hole   基础螺栓孔
28 front counter guide   （前)主十字头导轨
29 front crosshead metal   主十字头（导轨）镶条
30 front guiding plate   主十字头导板
31 base   底座
32 cylinder bottom   气缸座
33 complete metallic packing   金属填料组件
34 cylinder support   气缸支承
35 suction port   进气口
36 delivery port   排气口
37 distance piece   隔离室

17.3 Rotary blower 回转式鼓风机

17.3.1 Some kinds of rotary compressors 各种回转压缩机

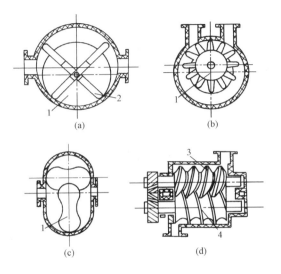

(a) sliding vane compressor
滑片式压缩机
(b) liquid ring compressor
液环鼓风机
(c) Roots blower
罗茨鼓风机
(d) double screw compressor
双螺杆压缩机

17.3.2 Screw compressor constructional detail
螺杆式压缩机结构图

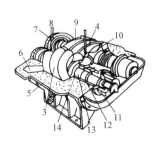

1　rotor　转子
2　sliding vane　滑片
3　male screw　阳螺杆
4　female screw　阴螺杆
5　cylinder　气缸
6　suction port　吸入口
7　synchromesh gear　同步齿轮
8　suction　吸入
9　discharge　排出
10　discharge port　排出口
11　thrust bearing　推力轴承
12　bearing　轴承
13　oil retainer　挡油环
14　oil seal ring　轴封环

403

17.4 Centrifugal compressor 离心压缩机

17.3.3 Sliding-vane type of rotary compressor 回转式滑片压气机

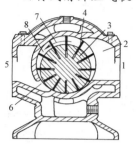

17.3.4 Liquid-piston type of rotary compressor 回转式液环压气机

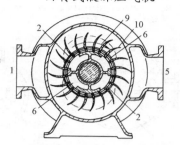

17.3.5 Two-impeller type of rotary positive-displacement blower 双叶型旋转正位移式（容积式）鼓风机（罗茨鼓风机）

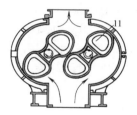

1　inlet　进气
2　inlet port　进气孔
3　rotor　转子
4　sliding vane　滑片
5　discharge　排气
6　discharge port　排出孔
7　counter clockwise (C.C.W.)　逆时针（旋转）
8　annular cell　环形气室
9　clockwise (C.W)　顺时针（旋转）
10　forward curved blade　前弯式叶片
11　straight-lobe　直叶瓣，直凸轮

17.4 Centrifugal compressor 离心压缩机

17.4.1 Section through of first stage to second stage 第一级至第二级的通流截面

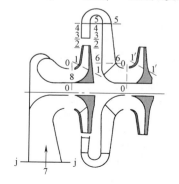

1　1—1 section through blade inlet 叶片进口截面
2　2—2 section through blade outlet 叶片出口截面
3　3—3 diffuser inlet　扩压器进口
4　4—4 diffuser outlet　扩压器出口
5　5—5 return passage inlet　回流器进口
6　6—6 return passage outlet　回流器出口
7　j—j section through stage inlet 通过级进口的截面
8　0—0 section through impeller inlet 叶轮进口截面

17.4.2 Centrifugal compressor 离心式压缩机

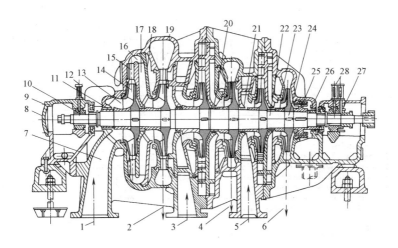

1	first section inlet 第一段进口		16	return bend 回流弯道
2	first section outlet 第一段出口		17	return passage 回流器
3	second section inlet 第二段进口		18	diaphragm 隔板
4	second section outlet 第二段出口		19	volute 蜗壳
5	third section inlet 第三段进口		20	interstage labyrinth seal 级间迷宫密封
6	third section outlet 第三段出口		21	impeller spacer 叶轮定距环
7	suction 吸入口		22	rotor assembly 转子组件
8	end cover 端盖		23	wearing ring 耐磨环
9	end cover bolt 端盖螺栓		24	shaft 轴
10	sleeve bearing 滑动轴承		25	balancing piston 平衡活塞
11	front labyrinth seal 前迷宫密封		26	rear labyrinth seal 后迷宫密封
12	intake end thermometer 进气端温度计		27	thrust bearing 推力轴承
13	inlet guide vane 进口导流叶片		28	discharge end thermometer 出口端温度计
14	impeller 叶轮			
15	diffuser 扩压器			

17.4.3 Various configurations of centrifugal compressors
离心压缩机的多种构型

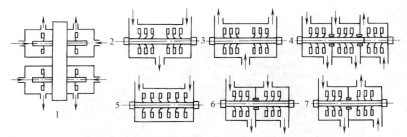

1 four post, three cooling points 分成四组，经三次冷却
2 parallel flow, suction in ends 两端进气，并流
3 parallel flow, suction in center 中央进气，并流
4 series flow, two cooling points 各级串联，其中经二次冷却
5 series flow (basic compressor) 各级串联（基本型压缩机）
6 series flow, one cooling point 各级串联，其中经一次冷却
7 series flow, one cooling point suction on ends, cool end
   两端进气，其中经一次冷却，各级串联

17.5 Fan 风机

17.5.1 Centrifugal fan 离心式风机

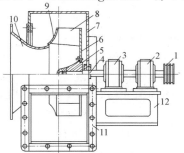

1 V-pulley V形皮带轮
2 bearing box (outboard) 轴承座（外侧）
3 bearing box (inboard) 轴承座（内侧）
4 mainshaft 主轴
5 shaft-cup 轴盖
6 disk 轮盘
7 volute 蜗壳
8 blade 叶片
8.1 straight-blade 直叶片
8.2 forward-curved-blade 前弯叶片
8.3 backward-curved-blade 后弯叶片
9 shroud 轮盖
10 air intake port 进风口
11 delivery port 出风口
12 underframe 底架

17.5.2 Designation of direction of rotation and discharge
旋转方向与排风口方向的标志

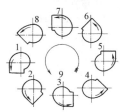

1 down blast 自上往下送风
2 bottom angular down 下倾角出风
3 bottom horizontal 下水平出风
4 bottom angular up 下仰角出风
5 up blast 自下往上送风
6 top angular up 上仰角出风
7 top horizontal 上水平出风
8 top angular down 上倾角出风
9 counter-clockwise 逆时针（方向）

17.6 Compressed air installation 压缩空气站

17.6.1 A typical compressed air installation 典型的压缩空气站

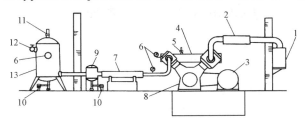

1	air intake filter 空气进口过滤器	8	two-stage double acting air compressor 两级双作用空气压缩机
2	air silencer 空气消声器	9	moisture separator 去湿器，水分分离器
3	motor 电动机	10	automatic drain valve 自动放泄阀
4	intercooler 级间冷却器，中间冷却器	11	safety valve 安全阀
5	relief valve 安全阀，减压阀	12	stop valve 截止阀
6	pressure gauges 压力表	13	air receiver 贮气罐
7	aftercooler 后置冷却器		

17.6.2 Correct air line installation 正确的空气管路装置

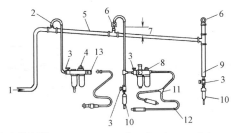

1	from receiver 来自贮气罐	8	plug in type lubro-control unit 组合式润滑油控制装置
2	wide pattern return bend 大半径180°弯头	9	trap or water leg 集水器或水竖管
3	shut-off valve 切断阀	10	drip leg drain 竖的引管排水口
4	plug in type vitalizer unit 插入式重要部件	11	wye connector 三通接头
5	main 总管	12	flexible hose 挠性软管
6	branch main 支管	13	quick release joint 快速松脱接头
7	pitch with flow 流动坡度		

18 Vacuum Equipments 真空设备

18.1 Rotary oil sealed mechanical pump 油封式旋转机械真空泵

18.1.1 Slide valve vacuum pump 滑阀式真空泵

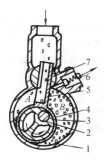
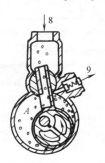

1 pump body 泵体
2 shaft 轴
3 eccentric 偏心轮
4 slide valve ring 滑阀环
5 slide valve spindle 滑阀杆
6 discharge valve 排气阀
7 sliding way 导轨
8 suction nozzle 吸气口
9 discharge nozzle 排气口

18.1.2 Single stage rotating blade vacuum pump 单级旋片式真空泵

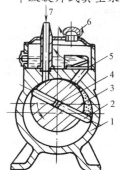

1 pump body 泵体
2 blade 旋片
3 rotor 转子
4 spring 弹簧
5 discharge valve 排气阀

18.1.3 Double stage rotating blade vacuum pump 双级旋片式真空泵

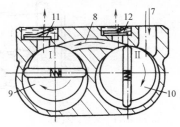

6 discharge nozzle 出气口
7 suction nozzle 吸气口
8 passage 通道
9 backing pump 前级真空泵
10 secondary pump 后级真空泵
11 backing stage discharge valve 前级排气阀
12 secondary discharge valve 后级排气阀

18.1.4 Rotary moving blade vacuum pump 旋转刮板真空泵

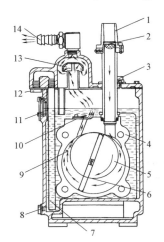

1 vacuum connection 真空接头
2 filter 过滤器
3 oil filler plug 滤油器丝堵
4 stator 定子
5 rotor 转子
6 blade 旋片
7 pressure drain tube 加压排液管
8 gravity drain plug 自流排放旋塞
9 air ballast inlet hole 气镇入口孔
10 discharge valve 排出阀
11 oil level sight glass 油位视镜
12 pressure drain plug 加压排液螺塞
13 oil spray arrester 除油雾装置
14 discharge nozzle 排气接管

18.2 Diffusion pump 扩散泵

18.2.1 A typical diffusion pump 典型的扩散泵

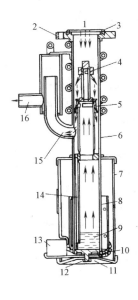

1 gas from system 来自系统的气体
2 pump flange 泵法兰
3 O ring seal O形密封圈
4 upper jet 上部喷嘴
5 lower jet 下部喷嘴
6 pump body 泵体
7 radiation shield 辐射防护屏
8 heater 加热器
9 fluid charge 注入液体
10 asbestos washer 石棉垫圈
11 bottom cover 底盖
12 bottom cover securing nut 底盖锁紧螺母
13 terminal block 接线盒
14 oil return tube 回油管
15 backing tube 通前级管
16 backing connection 前级（泵）接头

18.2.2　Structural representation of oil diffusion pump　油扩散泵的结构示意图

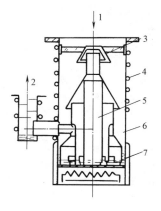

1　suction inlet　进气口
2　discharge outlet　排气口
3　jet　喷嘴
4　cooling water　冷却水
5　guide flow cylinder　导流筒
6　pump body　泵体
7　heater　加热器

18.2.3　Functional diagram of oil booster pump　油增压泵工作原理图

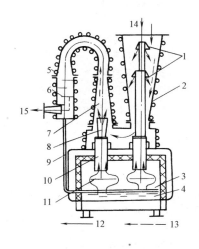

1　bevel jet　伞形喷嘴
2　pump body　泵体
3　boiler　锅炉
4　working fluid　工作液
5　cooling device　冷却设备
6　oil separator　油分离器
7　diffuser　扩压管
8　Laval nozzle　拉瓦尔喷管
9　oil return pipe　回油管
10　guide cylinder　导流筒
11　baffle plate　挡片
12　vapor stream　蒸气流
13　extracted gas　被抽气体
14　inlet　入口
15　outlet　出口

18.2.4 Components of diffusion pump 扩散泵的构件

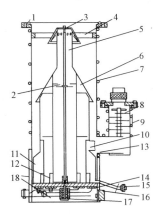

1 O-ring sealed joints with metal-to-metal contact
    金属与金属接触的 O 型环密封接头
2 integral splash baffle 整体式防喷溅挡板
3 sealed top jet cap 上喷嘴密封帽
4 guard ring to reduce backstreaming in larger pump sizes
    保护圈，在大型泵可减小回流
5 removable jet assembly 可拆的喷嘴组合件
6 interior bright-plated to reduce radiation loss 为减小热辐射损失，里侧镀出亮泽
7 copper cooling-coils 铜的冷却盘管
8 backing line condenser 前级管线冷凝器
9 mounting position for thermal cut-out 热熔断路器安装位置
10 ejector stage 喷射器级
11 double interior wall for thermal insulation of returning pump fluid
    内部双层壁，可防止泵内回流液散热
12，13 fluid fractionating system 流体分级系统
14 heaters 加热器
15 oil drain 排油口
16 drip shield 液滴防护屏
17 heater retaining stud 加热器固定（双头）螺栓
18 heat reflectors 热反射器

18.3 Ejector pump 喷射泵

18.3.1 Typical steam-jet ejector 典型的蒸汽喷射泵

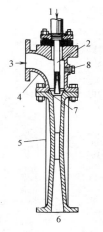

1 operating steam 工作蒸汽
2 steam nozzle 蒸汽喷嘴
3 inlet 吸气
4 suction chamber 吸气室
5 diffuser body 扩压器机体
6 discharge 排气
7 self-centering flange 自动对中法兰
8 vacuum ga (u) ge connection 真空表接口

18.3.2 Booster ejector with multiple steam nozzles 多蒸汽喷嘴增压式喷射泵

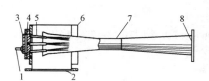

1 operating steam inlet 工作蒸汽入口
2 vapor inlet 蒸汽入口
3 steam dome 蒸汽室
4 nozzle plate 喷嘴固定板
5 steam nozzle 蒸汽喷嘴
6 booster body 增压泵机体
7 throat 喉部
8 discharge 排出口

18.3.3 Single-stage ejector 单级喷射泵

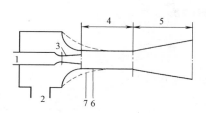

1 motive gas "a" 动力气体"a"
2 suction gas "b" 吸入气体"b"
3 nozzle 喷嘴
4 mixing section 混合段
5 diffuser 扩压器
6 constant area mixing section
   等截面混合段
7 constant pressure mixing section
   恒压混合段

18.3.4 Arrangements of two-stage ejectors with condensers 带冷凝器的两级喷射泵布置图

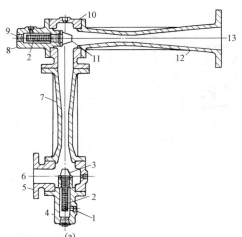

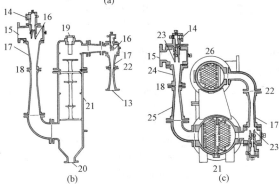

(a) main parts of a two-stage ejector 两级喷射泵的主要部件
(b) a two-stage ejector with interstage barometric condenser 两级喷射泵，级间带大气冷凝器
(c) a two-stage ejector with surface condensers interstage and terminal 两级喷射泵，级间及末端带表面冷凝器

1. high-pressure steam inlet, first stage 第一级高压蒸汽进口
2. steam strainer 滤汽器［网］
3. first stage steam nozzle 第一级蒸汽喷嘴
4. first stage steam chest 第一级蒸汽室
5. first stage suction head 第一级吸气接头
6. suction 吸气口
7. first stage combining throat 第一级混合喉管
8. second stage steam chest 第二级蒸汽室
9. high-pressure steam inlet, second stage 第二级高压蒸汽进口
10. second stage suction head 第二级吸气接头
11. second stage steam nozzle 第二级蒸汽喷嘴
12. second stage combining throat 第二级混合喉管
13. discharge 排气
14. steam inlet 蒸汽进口
15. vapor inlet 蒸气进口
16. nozzle 喷嘴
17. diffuser 扩压器
18. first stage 第一级
19. water inlet 水进口
20. water discharge 水出口
21. intercondenser 级间冷凝器
22. second stage 第二级
23. steam nozzle 蒸汽喷嘴
24. diffuser inlet 扩压器进口
25. diffuser discharge 扩压器出口
26. after condenser 后冷凝器

18.3.5 Five-stage steam ejector 五级蒸汽喷射泵

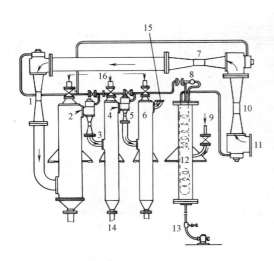

1. tertiary augmentor
   三次扩压器
2. main condenser 主冷凝器
3. first ejector 一级喷射泵
4. auxiliary condenser
   辅助冷凝器
5. second ejector 二级喷射泵
6. after condenser or silencer
   后冷凝器或消音器
7. secondary augmentor
   二次扩压器
8. needle valve 针阀
9. main steam supply
   主供蒸汽
10. primary augmentor
    一次扩压器
11. suction branch 吸入支管
12. steam dryer 蒸汽干燥器
13. trap 阱
14. barometric legs to seal pot
    大气腿伸至液封槽
15. steam 蒸汽
16. water 水

18.3.6 Barometric condenser for steam-jet system
用于蒸汽喷射系统的大气冷凝器

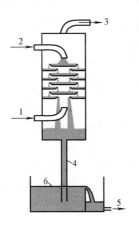

1. steam from booster ejector
   来自增压喷射泵
2. condenser water inlet
   冷凝器水进口
3. to air (secondary) ejectors
   通空气（二级）喷射泵
4. tail pipe 尾管
5. condenser and condensed water outlet
   冷凝器和冷凝水出口
6. hotwell 热水池

18.4 Water-ring vacuum pump 水环真空泵

18.4.1 Operation process chart of water-ring pump 水环泵工作流程图

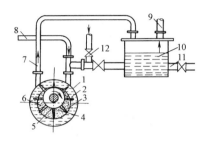

1 pump body 泵体
2 rotor 转子
3 liquid-ring 液体环
4 inlet-hole 进气孔
5 operating chamber 工作室
6 discharge hole 排气孔
7,9 discharge pipe 排气管
8 inlet pipe 进气管
10 waterbox 水箱
11 tubing 管道
12 control valve 控制阀

18.4.2 Series operation flowchart of water-ring pump and ejector pump 水环-喷射泵串联操作流程图

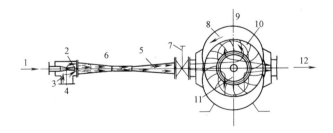

1 operating gas (air) 工作气体（空气）
2 operating nozzle 工作喷嘴
3 mixing chamber 混合室
4 extracted gas 被抽气体
5 diffuser 扩压器
6 jet ejector 喷射泵
7 gate valve 闸阀
8 pump body 泵体
9 water-ring vacuum pump 水环真空泵
10 vane wheel 叶轮
11 distributor 分配器
12 exhaust to silencer (or gas-water separator)
排气，接消音器（或气水分离器）

18.5 Vacuum system 真空系统

18.5.1 High vacuum distilling apparatus 高真空蒸馏装置

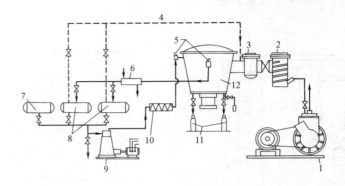

1 mechanical backing pump 前级机械泵
2 diffusion pump 扩散泵
3 refrigerated trap 冷阱
4 vacuum balancing line 真空平衡管线
5 feed & residue thermocouples 进料液与残液热电偶
6 residue heat exchanges 残液换热器
7 feed tank 进料液槽
8 residue recycling tanks 残液循环槽
9 feed pump 加料泵
10 feed preheater 料液预热器
11 distillate outlets 馏出液出口
12 molecular still 分子蒸馏器

19 Sealing of Rotating Shafts 旋转轴密封

19.1 Mechanical seal 机械密封

19.1.1 Mechanical seal components 机械密封部件

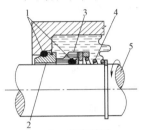

1 static seals 静密封圈
2 stationary seal ring 密封静环
3 rotating seal ring 密封动环
4 spring 弹簧
5 rotary shaft 回转轴
6 drive seat 传动座

19.1.2 External mechanical seal 外装式机械密封

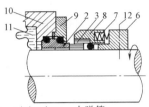

7 minisprings 小弹簧
8 thrust ring 推环
9 gland 密封压盖
10 pump bonnet 泵盖
11 operating fluid 工作液体
12 set screw 定位螺钉

19.1.3 Internal mechanical seal 内装式机械密封

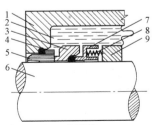

1 sealing cavity 密封腔
2 operating liquid 工作液体
3 rotating seal ring 动环
4 seal ring 密封圈
5 stationary seal ring 静环

19.1.4 Balanced internal mechanical seal 平衡型内装式机械密封

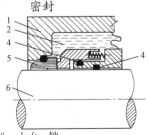

6 shaft 轴
7 spring seat 弹簧座
8 spring 弹簧
9 drive seat 传动座

19.1.5　External mechanical seal for slurry or sewage service
用于浆料和污水的外装式机械密封

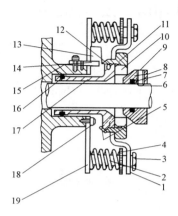

1　yoke plate　托架板
2　adjustable spring　可调式弹簧
3　spring adjustment bolt
　　弹簧调整螺栓
4　spring adjustment nut
　　弹簧调整螺母
5　flush connection port
　　冲洗液连通孔
6　shaft to be "dimpled"
　　被打有微坑的轴
7　cone "O" ring
　　圆锥体内 O 形密封圈
8　cone　圆锥体
9　set screw　紧定螺钉
10　split end seal　对开式端面密封
11　gland nut　压紧螺母
12　nylon insert　尼龙衬垫
13　guide lug　导耳（导缘）
14　anti-rotation stud or bar　防转短轴或棒
15　stuffing box area　密封盒外表面
16　gland sleeve "O" ring
　　密封套 "O" 形圈
17　gland sleeve　密封套
18　abutment plate mounting screws
　　支承板固定螺钉
19　abutment plate　支承面板
20　throttle bushing　节流衬套
21　flushing inlet　冲洗液进口

19.1.6　External mechanical seal with throttle bushing
带节流衬套的外装式机械密封

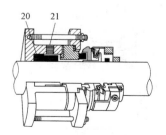

19.1.7 Mechanical contact shaft seal 机械接触式轴封

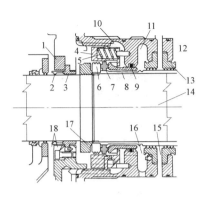

1 bypass orifice 旁通小孔
2 seal wiper ring 封闭刮油环
3 floating babbitt-faced steel ring 钢表面衬巴氏合金的浮动环
4 spring retainer 弹簧导座
5 spring 弹簧
6 stationary seal ring 密封静环
7 rotating carbon ring 碳石墨动环
8 stationary sleeve 固定轴套
9 seal ring 密封圈
10 seal oil 密封用油
11 suction gas chamber 气体吸入腔
12 process gas 过程气体
13 labyrinth seal 迷宫式密封
14 shaft 轴
15 buffer gas injection port 缓冲气体注入孔
16 gas and contaminated oil drain 气体与污染油排放口
17 shaft nut 轴螺母
18 seal oil drain line 密封油排放管

19.1.8 Double mechanical seal 双端面密封

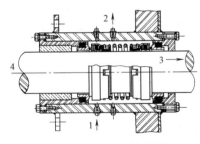

1 flushing fluid in 冲洗液入口
2 flushing fluid outlet 冲洗液出口
3 impeller end 叶轮端
4 motor end 电机端

19.2 Stuffing box 填料箱

19.2.1 A typical high-pressure stuffing box using metallic packing 典型的金属填料高压填料函

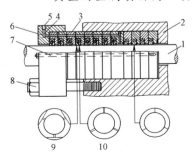

1 reciprocating shaft 往复轴
2 sealing box 密封盒
3 packing case 填料盒
4 oil filler pipe 注油管
5 oil filler point 注油孔
6 gland 压盖
7 tie-rod 长螺栓
8 gland bolt 压盖螺栓
9 metallic packing of crosshead side 十字头侧金属填料
10 metallic packing of pressure side 压力侧金属填料

19.2.2 Soft-packed stuffing box 软填料填料箱

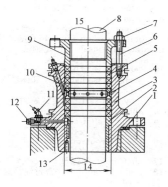

1 gasket 垫片
2 throttle bushing 节流衬套
3 stuffing box 填料箱
4 lantern ring or seal cage 封液环或隔离环
5 soft packing 填料
6 stud 双头螺栓
7 nut 螺母
8 rotating shaft 转轴
9 gland 压盖
10 oil-hole 油孔
11 oil cup 油杯
12 plug 螺塞
13 bush set screw 衬套止动螺钉
14 impeller end 叶轮端
15 motor end 电机端

19.2.3 Type of packing 填料类型

1 square braided packing 方形编织填料
2 plaited packing 折叠编织填料
3 twisted packing 麻花填料
4 interbraid packing 交叉编织填料
5 laminated cloth packing 叠层织物填料
6 folded cloth packing 折叠织物填料
7 rolled cloth packing 卷压织物填料
8 fibrous compression packing 压紧纤维填料
9 metallic packing 金属填料
10 metal foil spiral wrapped packing 金属箔螺旋卷制填料
11 metal foil crinkled and twisted packing 绉状金属箔卷制填料
12 metal core packing 金属芯填料
13 flexible graphite packing 柔性石墨填料
14 lubricated plastic packing 润滑塑料填料
15 asbestos rubber packing 石棉橡胶盘根（填料）
16 asbestos packing 石棉填料

19.3 Gas seal 气体密封

19.3.1 Floating-type restrictive ring seal 浮环节流型密封

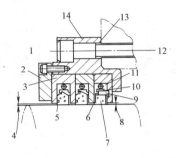
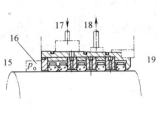

1　atmosphere (outlet) p_o
　　大气压（出口处）p_o
2　end plate　侧板
3　metal chamber　金属小室
4　moderate clearance　适度间隙
5　circumferential groove　环向开槽
6　radial vent　径向通气
7　segmented carbon seal
　　扇形的石墨密封块
8　small radial clearance　小的径向间隙
9　key　键
10　segmented metal retainer　扇形的金属
　　定位块
11　extension spring　拉伸弹簧
12　process gas (inlet)，p_i
　　工艺过程气体（进口）p_i
13　flange gasket　法兰垫片
14　seal housing　密封盒
15　bearing　轴承
16　outlet　出口
17　buffer inlet p_a
　　缓冲液进口（压力），p_a
18　sensing pressure　p_s　传感压力，p_s
19　process p_p　工艺过程（压力）p_p

19.3.2 Floating-ring seal construction drawing 浮环密封结构图

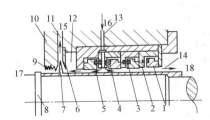

1　floating ring　浮环
2　L type fixed ring　L型固定环
3　pin　销钉
4　spring　弹簧
5　shaft sheave　轴套
6　oil retainer　挡油环
7　oil slinger　甩油环
8　shaft　轴
9　seal labyrinth of high pressure side
　　高压侧的密封梳齿
10　comb　梳齿座
11　return opening of high pressure side
　　高压侧回流孔
12　cavity　空腔
13　oil intake port　进油孔
14　oil return cavity of low pressure side
　　低压侧回油空腔
15　to oil separator　去油气分离器
16　oil from elevated tank　高位槽来油
17　high pressure gas　高压气体
18　low pressure gas　低压气体

19.3.3 Various configurations of labyrinth seals 各种结构的迷宫式密封

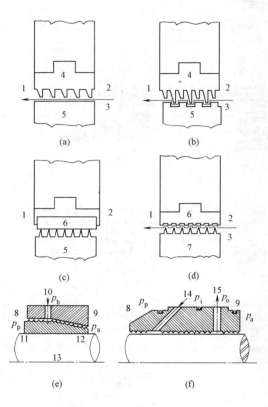

1　low pressure　低压
2　high pressure　高压
3　leakage path
　　泄漏途径
4　stationary　静止的
5　rotating surface (steel)
　　旋转表面（钢）
6　stationary sleeve
　　固定衬套
7　rotating surface
　　转动表面
8　process，p_p
　　过程压力 p_p
9　atmosphere，p_a
　　大气压力 p_a
10　buffered (or barrier) inlet
　　缓冲液（或阻挡液）进口
11．straight labyrith
　　平直迷宫
12　stepped labyrith
　　阶梯形迷宫
13　shaft　轴
14　inlet　进口
15　outlet　出口

(a) simplest design (labyrinth materials: aluminum, bronze, babbitt or steel)
　　最简单结构（迷宫材料：铝、青铜、巴氏合金或钢）
(b) more difficult to manufacture but produces a tighter seal (same material as in a)
　　制造较困难，但密封效果较好，[材料同 a]
(c) rotating labyrinth type (sleeve material: babbitt, aluminum, nonmetallic or other soft materials)
　　迷宫旋转型（衬套材料：巴氏合金、铝、非金属或其他软材料）
(d) rotating labyrinth, after operation (radial and axial movement of rotor cuts grooves in sleeve material to simulate staggered type shown in b)
　　旋转的迷宫（运转之后，转子的径向和轴向运动会在衬套上切割出槽，如同 b 所示的交错形）
(e) buffered combination labyrinth　注入缓冲液的组合式迷宫
(f) buffered-vented straight labyrinth　缓冲液注入又导出的平直迷宫

19.4 Vacuum seal 真空密封

19.4.1 Rotating shaft seal for liquid metal 液态金属转动轴密封

19.4.2 High-speed rotating shaft seal for liquid metal with very high vacuum 超高真空液体金属密封高速转轴

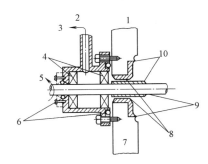

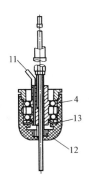

19.4.3 Rotating shaft seal for very high vacuum 超高真空转轴密封

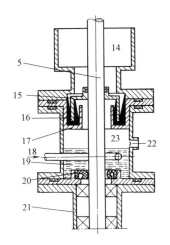

1　vacuum　真空
2　atmosphere　大气
3　to mechanical vacuum pump　到机械真空泵
4　ball bearing　滚珠轴承
5　rotating shaft　旋转轴
6　rubber seal ring　橡胶密封圈
7　wall　壁
8　liquid metal (film thickness 0.13mm)
　　液态金属（膜厚 0.13mm）
9　argon arc welding　氩弧焊接
10　stainless-clad steel　钢表面覆盖不锈钢
11　to vacuum pump　接真空泵
12　liquid metal with low-vapor pressure
　　低蒸气压的液态金属
13　Wilson seal　威尔逊密封
14　very high vacuum　超高真空
15　rotating seal ring　密封动环
16　liquid tin-base alloy　液态锡基合金
17　stationary seal ring　密封静环
18　cooling water　冷却水
19　lubricating-oil　润滑油
20　JO type polytetrafluoroethylene seal ring
　　JO 型聚四氟乙烯密封环
21　bearing pedestal　轴承座
22　mechanical vacuum pump　机械（真空）泵
23　low vacuum　低真空

20 Refrigeration 制冷

20.1 Refrigeration system 制冷系统
20.1.1 Basic refrigeration systems 基本制冷系统

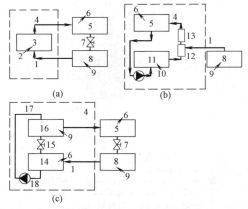

(a) mechanical refrigeration 机械压缩制冷
(b) ejector (steam-jet) refrigeration 喷射（蒸汽喷射）式制冷
(c) absorption refrigeration 吸收制冷

1 low-pressure vapor 低压蒸气
2 work 作功
3 compressor 压缩机
4 high-pressure vapor 高压蒸气
5 condenser 冷凝器
6 removing heat 移出热量
7 expansion valve 膨胀阀
8 evaporator 蒸发器
9 absorbing heat 吸热
10 applying heat 加热
11 boiler 锅炉
12 ejector 喷射器
13 diffuser 扩压器
14 absorber 吸收器
15 throttling valve 节流阀
16 generator 发生器
17 solution 溶液
18 pump 泵

20.1.2 Methods of transforming low-pressure vapor into high-pressure vapor in refrigeration systems 制冷系统中低压蒸气变换为高压蒸气三种方法

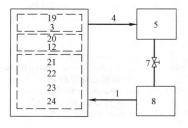

19 vapor compression 蒸气压缩
20 ejector (steam jet) 喷射器（蒸汽喷射）
21 absorption 吸收
22 absorb vapor in liquid while removing heat 移出热量的同时，液体吸收蒸气
23 elevate pressure of liquid with pump 用泵使液体压力升高
24 release vapor by applying heat 通过加热释放出蒸气

20.1 Refrigeration system 制冷系统

20.1.3 Basic refrigeration cycle 基本制冷循环

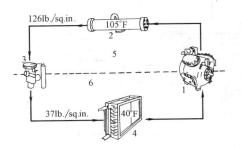

20.1.4 Centrifugal refrigeration system 离心制冷系统

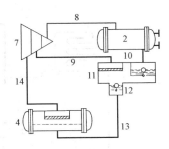

20.1.5 Basic two-stage cascade system 基本的两级逐级系统

20.1.6 Diagram of multistage centrifugal system 多级离心压缩制冷系统简图

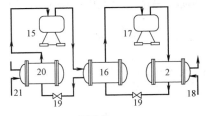

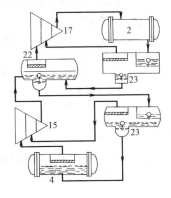

1 compressor   压缩机
2 condenser   冷凝器
3 expansion valve 膨胀阀
4 evaporator   蒸发器
5 high-pressure side 高压侧
6 low-pressure side   低压侧
7 centrifugal compressor 离心式压缩机
8 discharge gas   排出（高压）气体
9 flash gas   闪蒸气
10 liquid   液体
11 float valve   浮子阀
12 economizer   节能器
13 intermediate temperature liquid
   中间温度的液体
14 suction gas   吸入气体
15 low-stage compressor 低压级压缩机
16 cascade condenser 逐级冷凝器
17 high-stage compressor 高压级压缩机
18 water   水
19 expansion device 膨胀机械
20 low-stage evaporator 低压级蒸发器
21 product   物品
22 intercooler desuperheater
   中间冷却器（脱过热器）
23 intercooler   中间冷却器

425

20.1.7 Two-stage cascade system 两级逐级循环

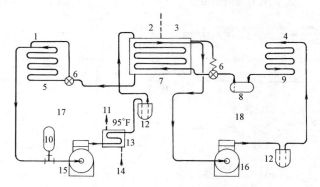

[to produce $-110°F$ refrigerant 产生$-110°F$（$-78.9℃$）制冷剂]

1 press. 168.1 lb./sq. in. abs. temp., $-110°F$ 绝压 168.1 lb./(in.)², 温度$-110°F$
2 cond. press., 150.1 lb./sq. in. abs. cond. temp., $-10°F$ 冷凝压力（绝）150.1 lb/(in)², 冷凝温度$-10°F$
3 evap. press., 19.7 lb./sq. in. abs. evap. temp., $-30°F$ 蒸发压力（绝）19.7 lb/(in)², 蒸发温度$-30°F$
4 press. 198.4 lb./sq. in. abs., temp. $+95°F$ 绝压 198.4 lb/in², 温度$+95°F$
5 evaporator 蒸发器
6 expansion valve 膨胀阀
7 cascade condenser 逐级冷凝器
8 receiver 受液槽
9 high-stage condenser 高压级冷凝器
10 expansion tank 膨胀槽
11 water out 水出口
12 oil separator 油分离器
13 desuperheater 脱过热器
14 water in 水进口
15 low-stage compressor R-13 低压级压缩机（R-13）
16 high-stage compressor R-22 高压级压缩机（R-22）
17 F-13 system F-13 系统
18 F-22 system F-22 系统

R-13——chlorotrifluoromethane 一氯三氟甲烷
R-22——dichlorodifluoromethane 二氯二氟甲烷

20.1.8 Lithium bromide absorption cycle 溴化锂吸收循环

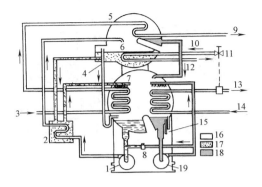

1	hermetic-solution pump 密封溶液泵	11	capacity-control valve 容量控制阀
2	heat exchanger 热交换器	12	evaporator 蒸发器
3	condensing water in 冷凝器用水进口	13	chilled water out 冷水出口
4	generator overflow tube 发生器溢流管	14	chilled water in 冷水入口
5	condenser 冷凝器	15	reed switches 簧片开关
6	generator 发生器	16	weak solution 稀溶液
7	absorber 吸收器	17	strong solution 浓溶液
8	reclaim valve 回收阀	18	refrigerant 制冷剂
9	condensing water out 冷凝用水出口	19	hermetic-refrigerant pump 密封制冷剂泵
10	steam 蒸汽		

20.1.9 Simplified ammonia-water absorption cycle 简单的氨-水吸收循环

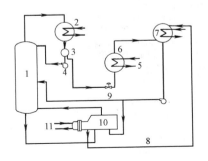

1 distillation column 蒸馏塔
2 condenser 冷凝器
3 receiver 受液器
4 pump 泵
5 to process 去过程
6 evaporator 蒸发器
7 absorber 吸收器
8 weak aqua 稀液
9 strong aqua 浓液
10 generator 发生器
11 waste heat 废热

20.1.10 Steam-jet refrigeration cycle 蒸汽喷射制冷循环

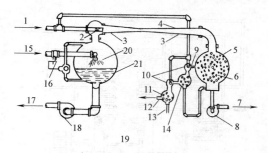

20.1.11 Ejector(steam-jet)refrigeration cycle 喷射(蒸汽喷射)式制冷循环

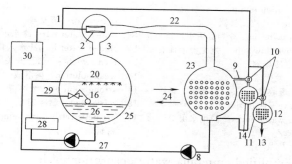

(with surface-type condenser 采用表面型冷凝器)

1	high pressure steam 高压蒸汽		17	chilled water to air conditioning equipment 去空调设备的冷水
2	steam nozzle 蒸汽喷嘴		18	circulating pump 循环泵
3	primary ejector 主喷射器		19	steam jet water cooling unit 蒸汽喷射冷却单元
4	diffuser 扩压器		20	spray nozzles 喷嘴
5	primary condenser 主冷凝器		21	flash tank 闪蒸槽
6	water tubes 水管		22	booster ejector 升压喷射器
7	condensate to boiler 冷凝液去锅炉		23	primary condenser surface type shown 主冷凝器(表面型)
8	condensate pump 冷凝液泵			
9	air take-off 脱空气		24	cooling water 冷却水
10	secondary ejectors 二次喷射器		25	flash tank-evaporator 闪蒸槽蒸发器
11	aftercondenser 后置冷凝器		26	chilled water 急冷水
12	air vent 放空		27	cold water circulating pump 冷水循环泵
13	drain 排液		28	heat load 热负载
14	intercondenser 中间冷凝器		29	make-up water 补充水
15	return water from air conditioning equipment 来自空调设备的回水		30	boiler 锅炉
16	float valve 浮子阀			

20.2 Refrigerator 制冷机

20.2.1 Turbo refrigerator 透平制冷压缩机

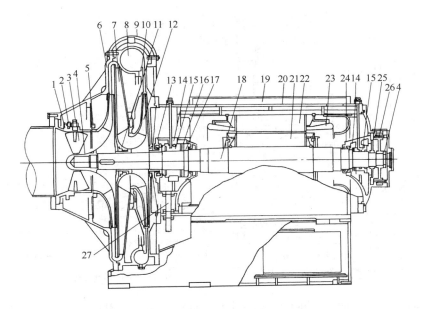

1	suction port 压缩机吸入口	15	bearing 轴承
2	suction vane gear 入口叶片齿轮	16	thrust pad 止推轴承调节垫
3	suction vane 入口导流叶片	17	thrust collar 止推轴承定位环
4	shaft nut 主轴螺母	18	shaft 主轴
5	impeller labyrinth 叶轮迷宫密封	19	cooling duct 冷却管
6	suction casing 吸入口机壳	20	motor frame 电机座
7	return guide vane 回流导叶	21	rotor 转子
8	volute casing 蜗形机壳	22	stator 定子
9	exhaust port 压缩机出口	23	coil 线圈
10	diffuser plate 扩压板	24	bearing labyrinth 轴承迷宫密封
11	shaft labyrinth 主轴迷宫密封	25	balance cylinder 平衡室
12	impeller 叶轮	26	balance runner 平衡盘
13	bearing labyrinth 轴承迷宫密封	27	oil exit pipe 排油管
14	bearing box 轴承箱		

20.2.2 Typical central-station air-conditioning unit and control system 典型的中央空调机组与控制系统

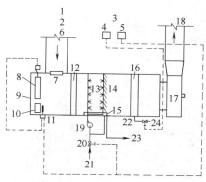

1 N. O.——normally open 常开的
2 N. C.——normally closed 常闭的
3 room 室内
4 dry-bulb controller 干球温度控制器
5 wet-bulb or humidity controller 湿球温度（或湿度）控制器
6 return air 返回空气（回风）
7 R. A damper（N. O.）D_2 回风调节风门（常开）D_2
8 maximum O. A. damper（N. C.）D_1 最大室外空气调节风门（常闭）D_1
9 outside air 外界空气
10 minimum O. A. damper（manual）（N. O.） 最小室外空气手动调节风门（常开）
11 outdoor air W. B. limit thermostat T_1 室外空气湿球恒温极限值 T_1
12 filters 过滤器
13 sprays 喷水器
14 eliminators 除沫器
15 spray pan 雾水盆
16 reheater 再加热器
17 fan 风机
18 supply air 供出空气
19 spray pump 喷水用泵
20 chilled water valve V_1（N. C.） 冷水阀 V_1（常闭）
21 chilled water supply 冷水供入
22 reheat coil 再加热盘管
23 return water to refrigeration machine 回水至制冷机
24 steam valve V_2（N. O.） 蒸汽阀 V_2（常开）
   O. A——outdoor air 室外空气

20.2.3 Turbo refrigeration machine（closed type） 封闭式透平制冷机

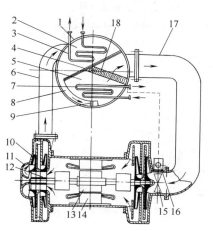

1 con.-cooler shell 冷凝-冷却器壳体
2 condenser 冷凝器
3 condenser tube 冷凝管
4 partition wall 隔板
5 cooler shell 冷却器
6 delivery pipe 排出管
7 cooler tube 冷却器管
8 refrigerant return pipe 制冷剂回流管
9 refrigerant reducing orifice 制冷剂节流孔板
10 compressor 压缩机
11 return channel 回流室
12 axial flow fan 轴流送风机
13 motor rotor 电动机转子
14 motor stator 电动机定子
15 suction vane 入口导流叶片
16 vane control motor 叶片控制电机
17 suction duct 吸入管
18 eliminator 除沫器

20.2.4 Typical p-H diagram for R-12 R-12 的典型 p-H 图

20.2.5 p-H diagram for vapor-compression cycle 蒸气压缩循环的 p-H 图

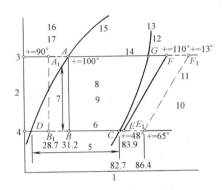
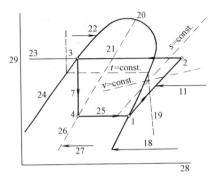

1 heat content, H, (enthalpy) Btu/lb. 热容 H（焓） Btu/磅

2 pressure, p 压力, p

3 131.6 lb/sq.in.abs. 131.6 磅/英寸²（绝压）

4 51.6 lb/sq.in.abs. 51.6 磅/英寸²（绝压）

5 latent heat 潜热

6 evaporating temp. 40℃, 蒸发温度 40℃

7 expansion 膨胀

8 wet zone 湿区

9 liquid and vapor 液体和蒸气

10 superheat zone 过热区

11 compression 压缩

12 saturated vapor curve 饱和蒸气曲线

13 simple refrigeration cycle (R-12) 简单制冷循环 (R-12)

14 condensing temp., 100°F 冷凝温度, 100°F

15 saturated liquid curve 饱和液体曲线

16 liquid zone 液相区

17 liquid subcooling 液态过冷

18 saturated vapor line 饱和蒸气线

19 superheated vapor 过热蒸汽

20 critical point 临界点

21 condensation 冷凝

22 saturated liquid line 饱和液体线

23 condensing pressure 冷凝压力

24 subcooled liquid 过冷液体

25 evaporation 蒸发

26 wet vapor-liquid and vapor 湿蒸气-液体和蒸气

27 X-quality X-质量

28 H——enthalpy, kJ/kg H——焓, kJ/kg

29 pressure, Pa 压力, Pa

R-12—— fluoromethane 一氯二氟甲烷

21 Pressure Vessel and Attachment
   压力容器及附件

21.1 Pressure vessel 压力容器

21.1.1 Construction details of pressure vessel 压力容器结构详图

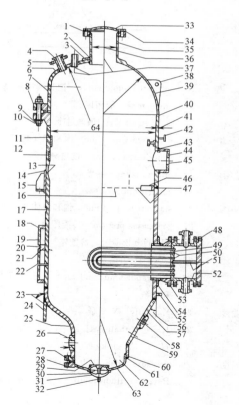

1　full face gasket　全平面垫片
2　welded connection　焊接连接
3　reinforcement pad　加强圈
4　code termination of vessel, lap jt. stub end　容器规范范围,搭接短管端
5　loose type flange　活套法兰
6　ellipsoidal head　椭圆形封头
　　int. pressure ellipsoidal head　受内压椭圆形封头
7　head skirt　封头直边段
8　optional type flanges　任意型式法兰
9　nuts & washers　螺母和垫圈
10　studs & bolts　螺柱和螺栓
11　applied linings　敷设的衬里
12　integrally clad plate　整体复合板
13　corrosion　腐蚀
14　stiffener plate　加强垫板
15　support lugs　耳式支座
16　longitudinal joints　纵缝接头
17　tell tale holes　指示孔
18　attachment of jacket　夹套附件

21.1 Pressure vessel 压力容器

19. jacketed vessels 夹套容器
20. plug welds 塞焊
21. bars & structural shapes used for stays, stayed surfaces 支撑用的杆和棒，支撑表面
22. stay bolts 支撑螺栓
23. 1/2 apex angle 半顶角
24. support skirt 裙座，裙式支座
25. toriconical head 折边锥形封头
    ext. pressure toriconical head 受外压折边锥形封头
26. studded connections 螺柱连接
27. optional type flange 任意型式法兰
28. bolted flange, spherical cover 螺栓法兰，球面封头
29. manhole cover plate 人孔盖板
30. flued openings 翻边开孔
31. yoke 支架
32. studs, nuts, washers 螺柱，螺母，垫圈
33. spherically dished covers 球面碟形盖
34. flat face flange 平面法兰
35. welded connection 焊接连接
36. opening 开孔
37. multiple openings 多孔
38. non pressure parts 非受压件
39. hemispherical head 半球形封头
    pressure hemispherical head 受压半球形封头
    int. pressure hemispherical head 受内压半球形封头
40. unequal thickness 不等厚度
41. shell thickness 壳壁厚度
42. stiffening rings 加强圈
43. welded connection 焊接连接
44. flat head 平封头
45. opening, flat heads 开孔，平板盖
46. backing strip 焊接垫板
47. circumferential joints 环焊缝接头
48. flat head 平板盖
49. tube sheet 管板
    no code 无规范
50. tubes 管子
    ext. pressure tube 外压管
    int. pressure tube 内压管
51. baffle 隔板
52. channel section 管箱；分配室
53. integral type flange 整体式法兰
54. reinforcement pad 加强圈
55. compression ring 承压环
56. 1/2 apex angle 半顶角
57. conical head 锥形封头
    int. pressure conical head 受内压锥形封头
58. small welded fittings 小型焊接管件
59. threaded openings 套扣开孔（螺纹孔）
60. head attachment 封头附件
61. fillet welds 填角焊缝
62. knuckle radius 过渡半径
63. torispherical head 碟形封头
    int. pressure torispherical head 受内压碟形封头
    ext. pressure torispherical head 受外压碟形封头
64. inside diameter 内径
    outer diameter 外径

21.1.2 General notes of pressure vessel 压力容器一般注释

1. heat treatment 热处理
2. inspection 检查
3. joint efficiency 焊缝系数
4. lethal substance 致死物质
5. loadings 负载
6. low temperature 低温
7. meterials 材料
   - carbon steel 碳钢
   - cast steel 铸钢
   - cast iron 铸铁
   - forging 锻件
   - weld const. 焊接结构
   - ductile cast iron 球墨铸铁
8. max. allowable working pressure 最大许用操作压力
9. design temperature 设计温度
   - design pressure 设计压力
10. pressure vessels subject to direct firing 用于直接明火的受压容器
11. radiographic exam. 射线照相检验
    - spot exam. of welded joint 焊缝接头局部检验
12. no radiographic exam. 不需要射线检验
    - visual exam. 外观检验
12. relief devices 泄压装置
13. repairs 维修
14. max. allow. stress value 最大许用应力值
15. hydrostatic testing 流体静压试验
    - pneumatic test 气压试验
    - pressure testing 压力试验
    - non-destructive testing 无损检验
    - mag. particle exam. 磁粉探伤
    - liq. penetration exam. 液体渗透探伤
    - ultrasonic exam. 超声波探伤
    - impact testing 冲击试验
16. stamping and data 标志与日期
17. unfired steam boilers 非火蒸汽锅炉
18. appendix 附录
19. telerance 允差（公差）
20. ASME code 美国机械工程师协会规范
21. pressure vessel code 压力容器规范

21.2 Storage and process vessels 贮罐及工艺容器

21.2.1 Some types of atmospheric storage tanks 一些型式的常压贮罐

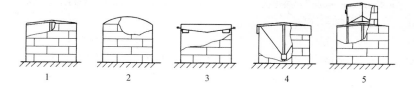

1. cone-roof (tank-supported roof) 锥顶罐（支撑顶）
2. dome-roof (tank-self-supported roof) 拱顶罐（自撑顶）
3. floating-roof (tank-Wiggins-Hidek type) 浮顶罐（W-H 型）
4. lifter-roof (tank-Wiggins dry-seal type) 升降顶贮罐（干封型）
5. variable-vapor-space (tank-Wiggins dry-seal type) 可变蒸气空间贮罐（干封型）

21.2.2 Refrigerated storage tank 低温贮罐

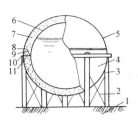

1. concrete foundation 混凝土基础
2. tie rod 拉杆
3. support 支柱
4. spherical double-wall tank 双壳球罐
5. outer shell 外壳体
6. inner shell 内壳体
7. pearlite 珍珠岩
8. compression ring 承压圈
9. hanger rod 吊杆
10. stiffener ring 加强圈
11. sway bar 稳定杆

21.2.3 Spherical storage tank 球形贮罐

1. lightening rod 避雷针
2. relief valve 安全阀
3. platform 平台
4. top crown 上极板
5. north temperate zone plate 北温带板
6. equator zone plate 赤道带板
7. support column 支柱
8. south temperate zone plate 南温带板
9. bottom crown 下极板
10. base plate 底板
11. tie rod 拉杆
12. stairway 盘梯

21.2.4 Horizontal cylindrical vessel 卧式圆筒形容器

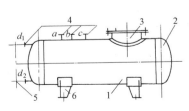

1. cylindrical shell 筒体
2. head 封头
3. manhole 人孔
4. connecting pipe 接管
5. liquid level gauge 液面计
6. supporting block 支座

435

21.2.5 Calculation of partially filled horizontal tank 卧式贮罐部分充填容积的计算

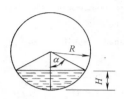

$$V = LR^2 \left(\frac{\alpha}{57.30} - \sin\alpha \cdot \cos\alpha \right)$$

H——depth of liquid 液体深度
R——radius of cylinder 筒体半径
L——length of cylinder 筒体长度
α——half the included angle 夹角的一半
V——liquid volume of partially filled horizontal cylinder 卧式筒体内部分充填液体的容积

21.3 Support 支座

21.3.1 Skirt support 裙式支座

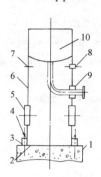

1 base 基础
2 base ring 基础环板
3 seat of bolt 螺栓座
4 anchor bolt 地脚螺栓
5 manhole 人孔
6 body 座体
7 vent hole 通气孔
8 vent nozzle 排气管
9 port for outgoing pipe 引出管孔
10 columns 塔器

21.3.2 Saddle support 鞍式支座

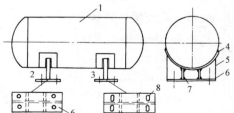

1 horizontal vessel 卧式容器
2 fixed saddle 固定鞍座
3 sliding saddle 滑动鞍座
4 stiffening pad 加强垫板
5 web plate 腹板
6 base plate 底板
7 rib 肋板
8 slotted hole 长圆孔

21.3.3 Lug support 耳式支座　21.3.4 Column support 柱式支架

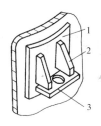

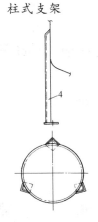

1　pad plate　垫板
2　web　筋板
3　base plate　基板
4　leg　支腿

21.4 Head 封头

21.4.1 Some types of heads 封头类型

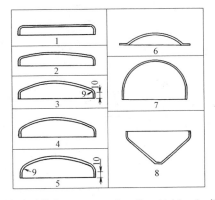

1　flanging-only head　折边平封头
2　flanging shallow dished head　带折边的浅碟形封头
3　flanging standard dished head　折边标准碟形封头
4　torispherical head　准球形封头
5　ellipsoidal head　椭圆形封头
6　spherically dished cover　球面碟形盖
7　hemispherical head　半球形封头
8　toriconical head　折边锥形封头
9　inside-corner radius　内转角半径
10　head skirt　封头直边段

22　Valves and Fittings　阀门与管件

22.1　Valves　阀

22.1.1　Types of valves　各种阀门

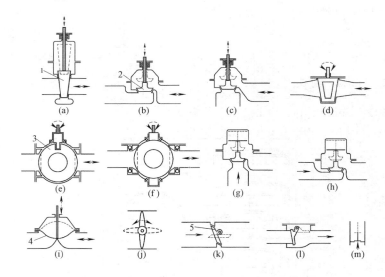

(a) gate valve　闸阀
(b) globe valve　截止阀；球心阀
(c) angle valve　角接阀
(d) plug cock　旋塞
(e) free ball valve　自由式球阀
(f) fixed ball valve　固定式球阀
(g) angle lift check valve　角接式升降止逆阀
(h) globe lift check valve　截止式升降止逆阀
(i) diaphragm valve　隔膜阀
(j) butterfly valve　蝶阀
(k) tilting-disk check valve　斜盘式止逆阀
(l) swing check valve　旋启式止逆阀
(m) vartical lift check valve　直立升降式止逆阀

1　wedge gate　楔形闸板
2　globe type disc　球心形阀盘
3　ball　球体
4　diaphragm　隔膜
5　disc　阀盘

22.1.2 Gate valve (slide valve) 闸阀（滑阀）

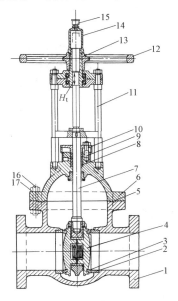

1　body　阀体
2　body rings　阀体密封圈
3　renewable seal ring　可更换的密封圈
4　gate disc　闸板
5　bonnet gasket　阀盖垫片
6　bonnet　阀盖
7　valve plug stem　阀杆
8　packing　填料
9　packing flange stud　填料压盖双头螺柱
10　packing flange　填料压盖
11　flange stud bolt　凸缘双头螺柱
12　hand wheel　手轮
13　stem sleeve　阀杆衬套
14　cap　护罩
15　lubricator　油杯
16　bonnet bolts　阀盖螺栓
17　bonnet nuts　阀盖螺母

22.1.3 Globe valve 截止阀（球心阀）

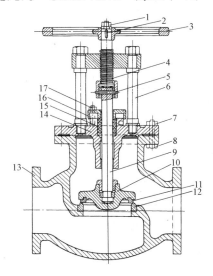

1　wheel nut　手轮螺母
2　key　键
3　valve handle　阀手轮
4　lifting screw　升降螺杆
5　pin　销子
6　yoke bolts　支架螺栓
7　bonnet nuts　阀盖螺母
8　bonnet bolts　阀盖螺栓
9　valve rod　阀杆
10　valve disc　阀盘
11　valve ring　阀环
12　valve seat ring　阀座圈
13　valve body　阀体
14　bonnet　阀盖
15　stem packing　阀杆填料
16　gland　压盖
17　gland nut　压盖螺母

22.1 Valves 阀

22.1.4 Diaphragm valve 隔膜阀

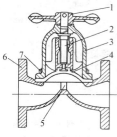

22.1.5 Butterfly valve 蝶形阀

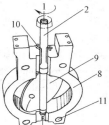

22.1.6 Plug valve 旋塞阀

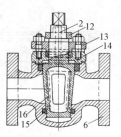

1 hand wheel 手轮
2 stem 阀杆
3 yoke 轭架
4 flexible diaphragm 挠性膜片
5 weir 堰
6 valve housing 阀体
7 yoke bolting 轭架螺栓
8 wafer type body 薄片型阀体

9 butterfly plate 蝶形板
10 seal 密封
11 hole for connection 连接孔
12 lock ring 防松环
13 packing gland 填料压盖
14 stuffing box 填料盒
15 seat ring 阀座圈
16 valve plug 阀芯

22.1.7 Ball valve 球阀

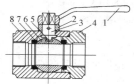

1 handle 手柄
2 lock ring 防松环
3 stem 阀杆
4 body cap 阀体盖
5 stem gasket 阀杆垫片
6 ball 球体
7 ball seat 阀座
8 body 阀体

22.1.8 Spring safety-relief valve 弹簧安全泄压阀

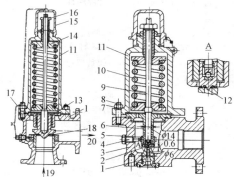

1 valve seat 阀座
2 valve clack 阀瓣
3 valve body 阀体
4 dowel pin 定位销
5 adjusting ring 调节环
6 guide bushing 导向套
7 bonnet stud nut 阀盖螺母
8 bonnet stud bolt 阀盖双头螺栓
9 valve spindle 阀杆
10 spring 弹簧
11 bonnet 阀罩
12 nozzle 喷嘴
13 guide 导向盘
14 spring retainer 弹簧座
15 cap 护罩
16 compression screw 压紧螺丝
17 lead seal 铅封
18 surfacing hard alloy 堆焊硬质合金
19 from pressure vessel 来自压力容器
20 to atmosphere 通大气

22.1.9 Control valve 调节阀

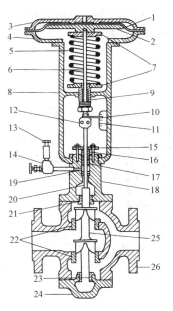

1　diaphragm　膜片
2　diaphragm disk　膜盘
3　diaphragm cap　膜头盖
4　diaphragm capsule　膜盒
5　actuator spring　执行机构弹簧
6　actuator stem　传动杆
7　spring seat　弹簧座
8　yoke　轭架
9　spring adjustor　弹簧调整装置
10　travel indicator　行程指针
11　travel indicator scale　行程标尺
12　stem connector　阀杆接头
13　packing lubricator assembly　填料润滑装置部件
14　isolating valve　隔断阀
15　packing flange　填料压盖
16　packing follower　填料压环
17　packing　填料
18　bonnet　阀盖
19　lantern ring　液封环
20　valve plug stem　阀杆
21　guide bushing　导向套
22　seat ring　阀座环
23　guide ring　导向环
24　bottom flange　底法兰
25　valve plug　阀芯
26　valve body　阀体

22.1.10 Non-return valve, check valve, hinged disc type 铰链盘式单向阀，止逆阀

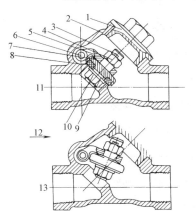

1　bonnet cover　阀盖
2　valve body　阀体
3　split pin　开口销
4　disc nut　阀盘螺母
5　hinged disc　铰接盘
6　hinge pin　铰链销
7　valve disc　阀盘
8　seal ring　密封环
9　spring washer　弹簧垫圈
10　bolt　螺栓
11　valve closed　阀关闭
12　flow　流向
13　valve fully open　阀全开

22.2 Traps 疏水阀

22.2.1 Thermodynamic trap 热动力式疏水阀

22.2.2 Disk type steam trap 盘式疏水阀

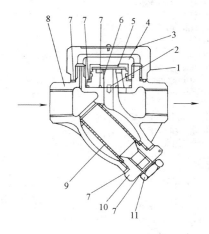
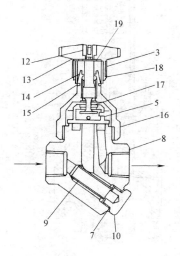

1　outer cover　外盖

2　knock pin　定位销

3　name plate　铭牌

4　inner cover　内盖

5　dick valve　盘形阀

6　valve seat　阀座

7　gasket　垫片

8　body　阀体

9　screen　滤网

10　screen holder　滤网架

11　drain plug　排污螺塞

12　nut　螺母

13　valve handle　阀柄

14　O-ring　O型环

15　snap-ring　开口环

16　bonnet　阀盖

17　spindle　阀杆

18　name plate set　铭牌位置

19　washer　垫圈

22.2 Traps 疏水阀

22.2.3 Open bucket steam trap 浮桶式蒸汽疏水阀

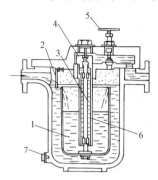

22.2.4 Floating steam trap 浮球式冷凝水排除器

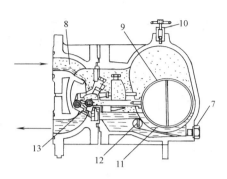

22.2.5 Inverted bucket steam trap 倒吊桶式蒸汽疏水阀

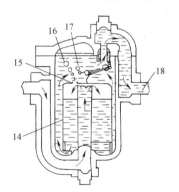

22.2.6 Thermostatic trap 恒温式疏水阀

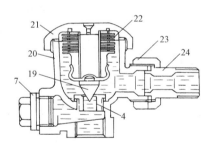

1	bucket	浮桶
2	baffle	挡板
3	needle	阀针
4	valve seat	阀座
5	air vent valve	放气阀
6	center pipe	中心管
7	drain plug	排放丝堵
8	condensate inlet	凝液入口
9	floating ball	浮球
10	air vent	排气阀
11	lever	杠杆
12	crank shaft	曲轴
13	valve disc	阀盘
14	inverted bucket	倒吊桶
15	vent hole	排气孔
16	steam bubble	蒸汽泡
17	air bubble	空气泡
18	water out	水出口
19	valve plug	阀芯
20	body	壳体
21	bonnet	压盖
22	bellows	波纹管
23	coupling	联接器
24	union nipple	接管接头

22.3 Fittings 管件

22.3.1 Flanged fittings 法兰管件

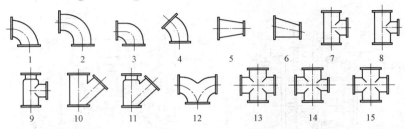

1	90° flanged elbow	90°带法兰弯头
2	90° long radius elbow	90°长半径弯头
3	90° reducing elbow	90°异径弯头
4	45° flanged elbow	45°带法兰弯头
5	flanged reducer	异径管，带法兰大小头
6	eccentric reducer	偏心异径管
7	flanged tee (straight size)	带法兰等径三通
8	reducing tee on outlet	出口为异径的三通
9	reducing tee on one run	主管一端为异径的三通
10	reducing lateral on branch	支管为异径的斜三通
11	reducing lateral on one run	主管一端为异径的斜三通
12	double branch elbow	双支管弯头
13	cross straight size	等径四通
14	reducing cross on one outlet	出口一端为异径的四通
15	reducing cross on both outlet	出口两端均为异径的四通

22.3.2 Threaded fittings 螺纹管件

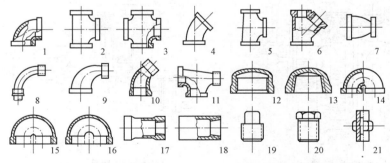

1	90° threaded elbow	90°螺纹弯头
2	threaded tee	带螺纹三通
3	threaded cross	带螺纹四通
4	45° threaded elbow	45°螺纹弯头
5	threaded reducing tee	带螺纹异径三通接头
6	45° Y-branches (straight size)	45°Y形三通（等径）
7	reducing coupling	缩径管接头
8	street elbow	带内外螺纹的弯管接头
9	reducing threaded elbow	异径螺纹弯头
10	45° street elbow	45°异径弯头
11	street tee	异径三通
12	cap with recess	带退刀槽的管帽
13	cap without recess	不带退刀槽的管帽
14	open pattern return bend	敞开式180°回弯头
15	close pattern return bend	封闭式180°回弯头
16	medium pattern return bend	普通型180°回弯头

17 coupling with band   带箍管接头
18 coupling without band   不带箍管接头
19 square head plug   方头丝堵
20 hex head plug   六角头丝堵
21 union   活接头

22.4 Flanges 法兰

22.4.1 Flanged connection 法兰联接

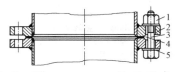

1 nut   螺母
2 flange   法兰
3 bolt   螺栓
4 gasket   垫片
5 washer   垫圈

22.4.2 Flange type 法兰型式

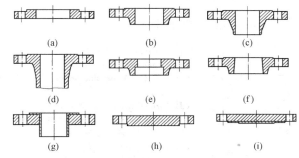

(a) plate-type flat welding flange 板式平焊法兰
(b) slip-on weld hubbed flange 带颈平焊法兰
(c) welding neck flange 带颈对焊法兰
(d) integral type flange 整体法兰
(e) socket-welding flange 承插焊接法兰
(f) threaded flange 螺纹法兰
(g) lap joint flange 松套法兰
(h) blind flange 法兰盖
(i) lined blind flange 衬里法兰盖

22.4.3 Type of flange facing 法兰密封面型式

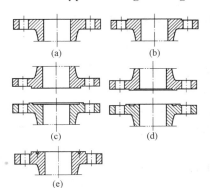

(a) flat face 全平面密封面
(b) raised face 突面
(c) male-female seal contaced face 凹凸密封面
    male flange 凸面法兰
    female flange 凹面法兰
(d) tongue-groove seal contace face 榫槽密封面
    tongued flange 榫面法兰
    grooved flange 槽面法兰
(e) ring type joint face 环槽密封面

22.4.4 Method of connecting glass pipeline to metal systems 玻璃管路与金属管路连接方法

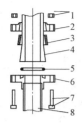

1 nuts 螺母
2 backing flange 背衬法兰
3 insert 衬垫
4 glass pipe 玻璃管
5 O-ring O形环
6 metal flange 金属法兰
7 bolts 螺栓
8 steel pipe 钢管

22.5 Expansion joints 膨胀节

22.5.1 Expansion joints 膨胀节

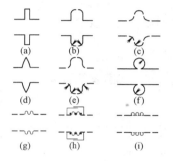

(a) flat plates 平板式
(b) flanged only heads 法兰接头
(c) flared shell or pipe segments 扩口壳体或弓形管段
(d) formed heads 成型接头
(e) flanged and flued heads 法兰和喇叭形接头
(f) toroidal 环形接头
(g) bellows type 波纹管式
(h) bellows type with reinforcing rings and insulation cover 具有增强环和绝热罩的波纹管
(i) high pressure toroidal bellows 高压环形波纹管

22.5.2 Action of expansion bellows under various movements 波纹管膨胀的各种移动

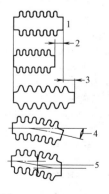

1　undeflected position 未变形位置
2　Δ —— axial compression
　　Δ ——轴向压缩
3　Δ —— axial extension
　　Δ ——轴向伸长
4　θ —— angular rotation（radians）
　　θ ——转角（弧度）
5　h —— offset
　　h ——偏置距离

22.5.3 Hinged expansion joint 铰接膨胀节

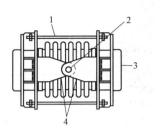

1 limit rods 限位杆
2 hinge pin provided with fitting for service lubrication 铰接销装有润滑操作的管件
3 welding end construction shown (if flanged, hinges are usually bolted to the flanges) 所示为焊接端结构（若用法兰连接，铰接件通常是用螺栓连接到法兰上）
4 hinges 铰接

22.5.4 Constrained-bellows expansion joints 限位波形膨胀节

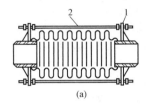 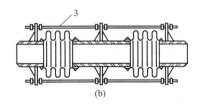

(a) tied expansion joint 带拉杆膨胀节
(b) universal-type expansion joint 通用型膨胀节

1 swivel or hinged connection 活节或铰接连接
2 tie rods 拉杆
3 limit rods 限位杆

22.6 Piping system 管系

22.6.1 Flexibility classification for piping systems 管系的柔度分类

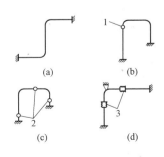

(a) stiff 刚性
(b) semirigid 半刚性
(c) non-rigid 非刚性
(d) free movement 自由移动
1 hinge 铰接
2 hinges 铰接
3 expandable joints 膨胀节

22.6.2 Pipe-systems materials 管系材料

1. metals 金属
① cast iron 铸铁
② malleable cast-iron 可锻铸铁
③ high silicon cast iron 高硅铸铁
④ carbon steel 碳钢
⑤ intermediate-alloy steel 中合金钢
⑥ low-alloy steel 低合金钢
⑦ plain nickel steel 普通镍钢
⑧ carbon-manganese steel 锰钢
⑨ manganese-vanadium steel 锰钒钢
⑩ carbon-silicon steel 硅钢
⑪ carbon-molybdenum steel 钼钢
⑫ manganese-molybdenum-vanadium steel 锰钼钒钢
⑬ chromium-vanadium steel 铬钒钢
⑭ silicon-killed carbon steel 硅镇静碳钢
⑮ high-alloy (stainless) steels 高合金钢（不锈钢）
⑯ austenitic stainless steels 奥氏体不锈钢
⑰ ferritic stainless steel 铁素体不锈钢
⑱ nickel-base alloys 镍基合金
⑲ nickel-copper alloys 镍铜合金
⑳ aluminum alloys 铝合金
㉑ copper alloys 铜合金
㉒ bronze 青铜
㉓ brass 黄铜
㉔ titanium and titanium alloys 钛和钛合金
㉕ zirconium and zirconium alloys 锆和锆合金

2. nonmetals 非金属
① thermoplastics 热塑性塑料
② ABS (acrylonitrile-butadiene-styrene) 丙烯腈-丁二烯-苯乙烯
③ polyethylene 聚乙烯
④ polypropylene 聚丙烯
⑤ nylon 尼龙
⑥ reinforced thermosetting resins 增强热固性树脂

3. nonmetallic pipe and lined pipe systems 非金属管系和非金属衬里管系
① asbestos cement 石棉水泥
② impervious graphite 不透性石墨
③ cement-lined steel 钢衬水泥
④ chemical ware 化学陶瓷管
⑤ vitrified-clay sewer pipe 釉面陶土排污管
⑥ concrete 混凝土
⑦ glass pipe and tubing 玻璃管
⑧ glass-lined steel pipe 衬玻璃钢管
⑨ chemical-porcelain pipe 化学瓷管
⑩ fused silica or fused quartz 熔融硅石或熔融石英
⑪ wood and wood-lined steel pipe 木质和衬木钢管
⑫ plastic lined steel pipe 衬塑料钢管
⑬ rubber lined steel pipe 衬橡胶钢管
⑭ plastic pipe 塑料管
⑮ rubber pipe 橡胶管

23 Conveying and Weighing Equipments
输送与称量设备

23.1 Pneumatic conveying systems 气流输送系统

23.1.1 Basic equipment arrangements of pneumatic conveying systems 气流输送系统主要设备配置图

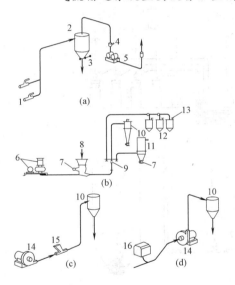

(a) vacuum system with several sources and one destination, multiple pickup 真空系统，若干处料源，多头吸料，一个目的地

(b) pressure system with rotary valve feeder, one source and several destinations, multiple discharge 压送系统，一处料源，由旋转阀加料，若干个目的地，多头卸料

(c) pressure system with Venturi feed for friable materials 用于易碎物料由文丘里进料的压送系统

(d) pull-push system in which the fan both picks up the solids and delivers them 拉-推系统，吸取物料及输送物料均由风机完成

1	pickup device 吸取装置	9	switch station 切换装置
2	gravity-flow hopper 自流式料斗	10	collector 收集器
3	counter-weighted door 平衡动作门	11	filter receiver 过滤接收器
4	3-way valve 三通阀	12	bins 料仓
5	blower 引风机	13	vent 放气口
6	blower and motor 鼓风机与电机	14	fan 鼓风机
7	rotary valve 旋转阀	15	Venturi 文丘里管
8	material in 进料	16	process machine 工艺过程机械

23.1 Pneumatic conveying systems 气流输送系统

23.1.2 Types of air-conveying systems 空气输送系统的各种型式

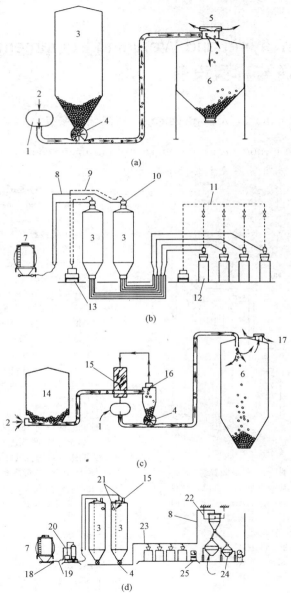

23.1 Pneumatic conveying systems 气流输送系统

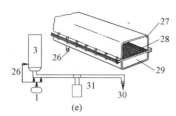

(e)　　　　　　　　(f)

(a) pressure　压力系统
(b) vacuum　真空系统
(c) pressure-vacuum　压力-真空系统
(d) pressure-vacuum unloading and transfer
　　压力-真空卸料与输送系统
(e) fluidizing system　流化系统
(f) blow tank　风送槽
1　blower　鼓风机
2　air in　空气进口
3　silo　料库
4　rotary air lock　旋转气塞阀
5　filter air out　滤后空气出口
6　material destination　物料收集器
7　rail car　铁路货车
8　material line　物料管线
9　vacuum line　真空管线
10　vacuum hopper　真空罐
11　sequencing valve　顺序阀
12　vacuum sequencing system　真空顺序系统
13　power unit　动力装置
14　material source　物料仓
15　filter　过滤器
16　cyclone separator　旋风分离器
17　air out　空气出口
18　car connection　货车联结点
19　flexible tube　软管
20　vacuum-pressure & material receiver
　　真空-压力和物料接收器
21　level sensors　物位传感器
22　weigh beam　称量杆
23　diverter valve　分流阀
24　blender　搅拌器
25　system controls　系统控制装置
26　air　空气
27　fluidized powder compartment
　　粉粒体流化室
28　porous medium　多孔介质
29　air compartment　空气室
30　loading spout　装料嘴
31　side discharge　侧卸料
32　compressed air　压缩空气
33　feed inlet with airtight valve　带气密阀的加料口
34　outlet valve　出口阀
35　conveying line　输送管线

23.1.3　Mass-flow bin
密相流料仓

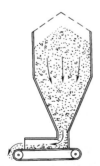

(material does not channel on discharge
仓内物料卸出时无沟流)

23.1.4　Funnel-flow bin
漏斗流料仓

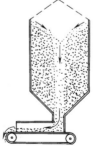

(material segregates and develops rat holes
料仓内物料离析并产生鼠洞)

451

23.1.5 Positive-pressure conveying 正压输送

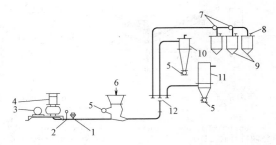

(is best suited to transport from one pickup location to various points
最适宜于从一处取料输送到若干处)

1	controls 控制系统		7	diverter valve 分流阀
2	gage 压力表		8	vent 通气口
3	motor 电机		9	bins 料仓
4	blower 鼓风机		10	collector 收集器
5	rotary valve 旋转阀		11	filter receiver 过滤接收器
6	material in 进料		12	switch station 切换装置

23.1.6 Vacuum system assists feed entry 物料吸入真空系统

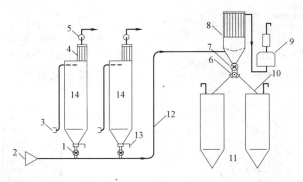

(is useful for delivering material to one receiver 可用于将物料输送到一个接收器中)

1	airlock feeder 旋转阀进料器		8	filter receiver 过滤收集器
2	air inlet 进气口		9	centrifugal fan 离心式风机
3	material inlet 进料口		10	gravity spouts 重力下料管
4	dust filter 粉尘过滤器		11	process bins 工艺过程料仓
5	vent fan 排气风机		12	conveying-line 输送管线
6	two-way diverter 双向分流器		13	slide gate 滑动闸阀
7	airlock discharger 旋转卸料阀		14	silo 储仓

23.1.7 Air lift 空气提升器

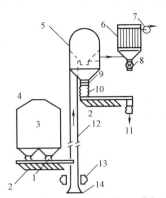

1 flexible connection 挠性联接
2 natural frequency conveyor 固有频率输送器
3 railcar 机动有轨车
4 gravity discharge 重力卸料,自流出料
5 disengaging chamber 分离室
6 dust filter 粉尘过滤器
7 exhaust fan 排气引风机
8 rotary airlock 气密旋转出料器
9 double-gate airlock 双闸进料器
10 bellows 波纹管
11 to process 去下道工序
12 air lift 空气提升机
13 photocell dropout alarm 光电物料下落报警设备
14 bellmouth nozzle 喇叭形进气口

(for friable material handles solids sensitive to impact at pipe bends 用于脆性物料输送固体在管路拐弯处易感受撞击)

23.1.8 Positive pressure systems 正压输送系统

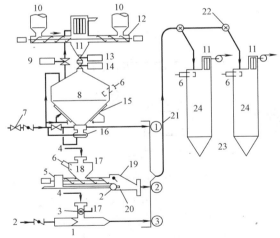

1 injector tee T形喷射器
2 air inlet 空气进口
3 rotary airlock 气密旋转加料器
4 material 物料
5 screw pump 螺杆泵
6 Hi-level 料面高度计
7 air inlet valve 进气阀
8 blow tank 放空罐
9 vent valve 排气阀
10 silo 筒仓
11 vent filter 排气过滤器
12 screw conveyor 螺旋输送机
13 batching valve 定量调节阀
14 sealing valve 密封阀
15 aeration pad 充气缓冲器
16 proportioning valve 比例调节阀
17 vent 放气口
18 hopper 料斗
19 mixing chamber 混合室
24 bin 料仓
① air-into-material 空气进入物料中
② air-mixing 与空气混合
③ material-into-air 物料进入空气中

(fed by a variety of modes and serving several receivers 有各种进料方式有若干个接收装置)

20 air nozzle 空气喷嘴
21 conveying line 输送管线
22 diverter valve 换向阀
23 process bins 生产过程料仓

23.1.9 A two-stage cyclone receiver 两级旋风分离接收器

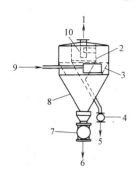

1 air 空气
2 secondary cyclone 第二级旋风分离器
3 inner skirt 内裙罩
4 secondary discharge lock 第二级卸料阀
5 dust 粉尘
6 material 物料
7 primary discharge lock 第一级卸料阀
8 primary cyclone 第一级旋风分离器
9 air and material 空气与物料
10 thimble 套管

23.2 Belt-conveyor systems 带式输送系统

23.2.1 Schematic of belt-conveyor system shows major components 带式输送机系统截面图及主要零部件

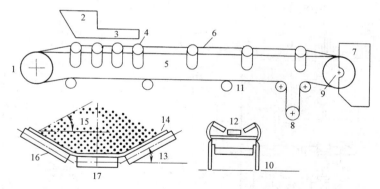

1 tail pulley 尾轮
2 feed chute 加料斜槽
3 loading skirts 加料裙罩
4 closely spaced idlers at loading point 靠近承载点托辊
5 troughing carrying-idlers 槽形承载托辊
6 troughed conveyor belt 槽式输送机皮带
7 discharge chute 卸料槽
8 vertical gravity takeup 垂直重力张紧装置
9 head pulley and drive 前轮和驱动装置
10 ground 地面
11 return idlers （回程）下托辊
12 simply supported belt conveyor 简单支承的皮带输送机
13 troughing angle 槽形角度
14 belt 皮带
15 angle of surcharge 堆积角
16 idler roll 惰辊
17 typical cross-section of troughed-belt conveyor 典型槽式皮带输送机截面图

23.2.2 Belt feeders 带式给料机

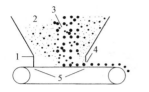

(a) belt feeder——poorly designed feed hopper
带式给料机——拙劣设计的加料斗

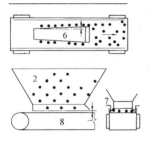

(b) belt feeder——well-designed feed hopper
带式给料机——良好设计的加料斗

1   poor-flowing or non-flowing region   流动不良或死区
2   feed hopper   加料斗
3   solids-flow channel (may shift position during operation)   固体流通道（操作时可能变动方位）
4   compaction and jamming as solids leave hopper   离开料斗时，固体被挤压与阻塞
5   frictional drag on belt   作用在皮带上的摩擦曳力
6   5 deg divergence   扩张角 5°
7   skirts   裙罩
8   1 to 3 deg slope   倾斜角 1°～3°
(can transport almost any solid at high loadings 在高负荷下几乎可以输送任何固体)

23.2.3 Bucket-wheel reclaimer 戽斗链轮再装料器

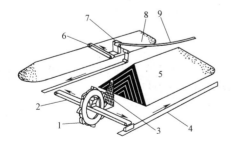

(digging buckets mounted on wheel discharge on belt conveyor for material transfer 挖斗安装在链轮上，并卸料于输送物料的带式运输机上)

1   bridge-mounted rotary bucket wheel reclaimer   跨接旋转漏斗链轮再堆放器
2   reclaim conveyor   再堆放运送器
3   harrow   耙
4   outgoing conveyor   外运输送器
5   material bed (reclaiming)   物料层（再堆放）
6   slewing boom   旋转尾端运送器
7   tripper stacker   自动转换堆放器
8   material bed (filling)   物料床(装料)
9   distributing conveyor   分配送料器

23.2.4 Belt-conveyor discharge arrangements 带式输送机卸料方式

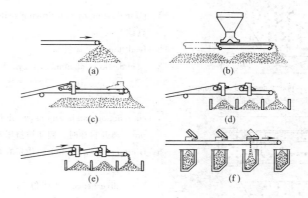

(a) discharge over end pulley forms conical pile at end of belt 越过末端皮带轮卸料，在皮带末端处堆成锥形料堆

(b) discharge over either end pulley to distribute lengthwise by reversible shuttle conveyor 利用可换向的振荡输送机越过两个末端皮带轮卸料堆成长堆

(c) discharge through travelling tripper, with or without cross conveyor, to distribute material to one or both sides of conveyor for entire distance of tripper travel 利用移动式卸料器（横向送料器可有可无），在卸料器可移动范围内，将物料卸在输送机的一边或两边

(d) discharge through fixed trippers, with or without cross conveyor to one or both sides of belt, to fixed bin openings or pile locations 利用固定倾料器，依有无横向送料器，把物料卸在输送带的一边或两边上的固定料仓开口处或堆料处

(e) discharge from multiple conveyors through fixed discharge chutes, with or without cross conveyor to one or both sides of belt, to fixed bin openings or pile locations 从多路输送器通过固定料槽卸料，依有无横向送料器，把物料卸在皮带的一边或两边，卸在固定料仓开口处或堆料处

(f) discharge by hinged plows to one or more fixed locations along one or both sides of conveyor 利用铰链刮料器把物料卸在输送机一边或两边的一个或多个固定位置上

23.2.5 Belt-conveyor storage and reclaiming in a flat-floor building 在平底料仓内,用带式输送机进行储料与再装料操作

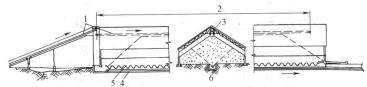

1 30-in. belt conveyor 200tons/hr
   30 英寸带式输送机,200t/hr
2 350ft 350 英尺
3 belt tripper 带式倾料器
4 42-in. belt conveyor 700tons/hr
   42 英寸带式输送机,700t/hr
5 gated floor openings discharge to conveyor 带有闸板的底板孔,卸料至输送机
6 section through storage building 储料仓断面图

23.3 Vibrating conveyor and miscellaceous 振动输送机及其他

23.3.1 Oscillation of a vibratory conveyor causes particles to move forward 振动输送机的振动引发颗粒向前运动

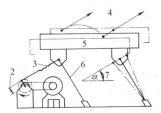

1 motor 电机
2 eccentric 偏心轮
3 flexible connection 柔性联接结构
4 particle trajectory 颗粒轨迹
5 trough 输送槽
6 spring 板簧
7 throw angle 抛射角

23.3.2 Vibrating-conveyor classification 振动式输送机分类

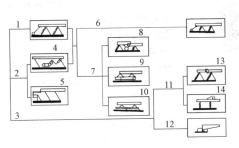

1 positive drive 正位移驱动装置
2 semipositive drive 半正位移驱动装置
3 non-positive drive 非正位移驱动装置
4 mechanical dash pot 机械减振器
5 hydraulic and pneumatic 液压和气动式
6 unbalanced 非平衡式
7 balanced 平衡式
8 split pan 分离盘式
9 double deck 双台式
10 counterweight 配重式
11 mechanical 机械式
12 electrical 电力式
13 directional 定向式
14 non-directional 非定向式

(modern materials handling 现代物料输送方法)

23.3.3 Typical arrangements and applications for continuous-flow conveyors 连续流动输送机典型构造与应用情况

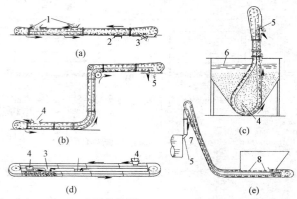

(a) horizontal conveyor 水平输送机
(b) Z-type conveyor-elevator Z形输送式提升机
(c) loop-feed elevator used for dewatering 用于脱水的环形加料提升机
(d) side-pull horizontal recirculating conveyor 侧边拖曳水平再循环输送机
(e) horizontal-inclined conveyor-elevator 水平倾斜输送式提升机

1 loading points 装料口
2 slide gate 滑板闸门
3 discharge points 卸料口
4 feed point 加料口
5 discharge point 卸料口
6 water level 水平面
7 kiln 窑
8 measured feed 计量后进料

23.3.4 Cross-section of properly withdrawn storage bin with bin activator 带松动器顺利卸料的料仓剖面图

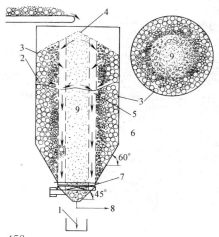

1 open-drop discharge-arbitrary operation of bin activator 敞口卸料-料仓松动器非定时操作
2 typical proper with drawl bin level 料仓正常卸料时的料面高度
3 aggregate 团粒
4 original fill line 初始装填线
5 annular flow 环形下流
6 note-remix 再混合
7 bin stream-deflector 料仓弧形导流器 bin activator 料仓松动器
8 feeder at outlet "cycle-flow" required 加料器出口需形成循环流
9 fines 细颗粒

23.3.5 Chain conveyors perform well on an incline 链式输送机在斜面上运行良好

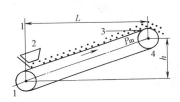

1 takeup 松紧装置
2 material loading 装料
3 drive socket 传动装置轴座
4 material discharge 卸料

23.3.6 Types of skip hoists and skip-hoist paths 倒卸式起重机型式及运行路径

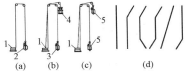

(a) unterweighted 无平衡块
(b) couterweighted 带平衡块
(c) balanced 平衡型
(d) skip-hoist paths 倒卸式起重机运行路径
1 winding machine 卷扬机
2 skip 翻斗
3 bucket 抓斗
4 counterweight 平衡块
5 skip bucket 翻斗式抓斗

23.3.7 Drag-type enclosed conveyor-elevator (Redler Design) 密闭型刮板式输送提升机

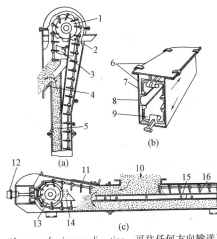

(b) carrying and return runs 运载和回程
(c) loading end 装料端
1 drive sprocket 主动链轮
2 cleaning finger and stripper 清理爪与刮料板
3 push out plate 推料板
4 discharge point 卸料口
5 solid column of material 物料竖筒
6 steel casing 钢质框架
7 empty return run 空回程
8 steel track 钢轨
9 material in carrying run 载运的物料
10 self-feeding 自动加料
11 inspection door 检查门
12 knob operates take-up 旋柄控制拉紧
13 tail wheel 尾轮
14 stripper 刮料板
15 feed plate 送料板
16 flight 刮板

(for transfer in any direction 可往任何方向输送)

(a) head and discharge end of elevator 提升机的头部与卸料端

23.3.8 Fluidizing outlets for hopper cars 底卸式货车的流态化卸料口

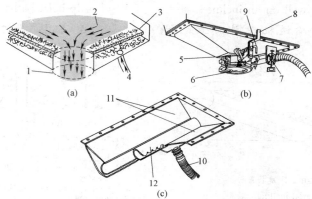

(a) air introduced through fluidizing pad makes powder flow toward opening 空气通过流化气垫，使粉粒体流向出口
(b) ACF center-flow butterfly outlet controls discharge of fluidized bulk powders ACF 中心流动货车的蝶阀式出口控制流态化粉粒体卸货
(c) another type of fluidizing butterfly outlet 另一种流态化蝶形阀式出口
1　discharge opening　卸料口
2　lading slides to discharge opening on pad of air　物料在空气垫上滑到卸料口
3　fluidizing pad　流态化气垫
4　air injected into area around discharge opening　空气喷射进入卸料口四周区域
5　operating handle locking bolt　操作手柄固定销
6　sanitary shield　防尘罩
7　fluidizing connection　流态化接管
8　operating handle　操作手柄
9　position retaining handle　固位手柄
10　fluidizing air line　流态化空气管线
11　permeable stainless steel slope sheets　可穿过的不锈钢斜板
12　control valve　控制阀

23.3.9 Ship-loading system with trimmer and telescoping chute 装备有堆料器和套叠式斜槽的轮船装料系统

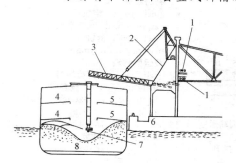

1　belt conveyor　带式输送机
2　travelling gantry　移动式吊架
3　gantry belt conveyor　吊架式带式输送机
4　tween　两甲板间的空间
5　decks　甲板
6　pier　码头
7　centrifugal trimmer　离心堆料器
8　lower hold　底货仓

23.4 Packaging and handling of solid and liquid products 固体和液体产品的包装与运送

23.4.1 Typical high-precision liquid filling and weighing system 典型高精度液体装料和称量系统

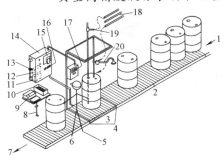

3 drum filling 正装料圆桶
4 load cell scale platform & roller conveyor section 负载传感秤,平台和辊式输送机段
5 drumming operator 圆桶操作人员
6 drum closing 封闭圆桶
7 to shipping 发运
8 torque wrench 扭力扳手
9 label printer 说明打印机
10 bung dust caps 桶口灰尘罩
11 intrinsic barrier 内阻层
12 solid-state weight controller 固态电路重量控制器
13 purge fittings & devices 吹扫配件和组件
14 remote-control panel 遥控板
15 printer recessed in panel 遥控板内打印机
16 data-entry thumb wheels 数据进口压轮
17 digital weight selector & display 数字式重量选定机和显示器
18 existing supply lines 现有供应线
19 retractable filling lance valve 伸缩式装料长杆阀
20 top-fill valve & nozzle 顶部装料阀和接管

(for packaging 208L (55-gal) steel drums and similar smaller containers 208L (55gal) 钢圆桶和类似较小容器包装用)
1 from empty drum storage 从空桶储存仓来
2 drum waiting to be filled 待装料空桶

23.4.2 Typical four-tube force flow valve-bag packer 典型4管强制流阀式袋包装机

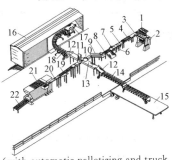

(with automatic palletizing and truck and railcar-loading facilities 装备有自动码垛和货车以及机动有轨车的装货设备)
1 4-tube force flow bag packer 4管强制流袋包装机
2 packer conveyor 包装机输送器
3 filled bag take-away conveyor 满袋移去输送器
4 accumulator conveyor 累积器输送器
5 metering conveyor 计量输送器
6 feed conveyer 喂送输送器
7 check-weigh conveyor 校核称量输送器
8 reject conveyor 不合格品输送器
9 elevating flattener 提升整平机
10 diverter 转向器
11 bridge 桥
12 chute 斜槽
13 reversible conveyor 可逆输送器
14 rigid truck loader 刚性滚轴搬运器
15 flat bed trailer 平板拖车
16 railroad boxcar 火车闷罐车
17 stacker 码垛机
18 flexible car loader 挠性装车机
19 inventory conveyor 库存料输送器
20 accelerator conveyor 加速器输送器
21 automatic palletizer 自动码垛机
22 filled-pallet gravity roll-conveyor 满货板重力辊子输送器

23.4.3 Typical pinch-bottom system with automatic bag hanging and semiautomatic palletizing 自动装袋吊挂及半自动码垛的典型粘压底部系统

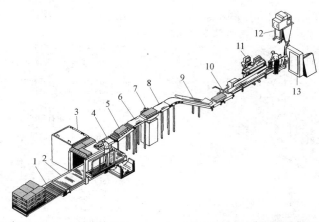

1　gravity conveyor　重力输送器
2　filled-pallet take-away conveyor　满货板移出输送器
3　air-floation palletizer　气浮码垛器
4　feed conveyor　喂送输送机
5　accumulator conveyor　累积器，输送机
6　reject conveyor　不合格品输送机
7　check-weigher　校核称量秤
8　metering conveyor　计量输送机
9　elevating flattener　提升整平机
10　bag turner　袋旋转器
11　pinch-bottom field closer　压紧底部现场封口机
12　preweigh scale　预称量秤
13　automatic bag placer　自动袋放置器

23.5 Weighing of bulk solids　粉粒体的称量

23.5.1 Mass-flow hopper for free-flowing products 自由流动产品的密相流料斗

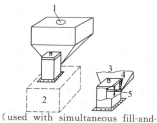

(used with simultaneous fill-and-weigh scales　它与装料机和磅秤配合使用)

1　material from process or bulk storage　从过程或散装仓来的物料
2　weigh scale　秤
3　detail of flow-adjustment feature　流量调节部件详图
4　gate clamp screws　阀门压紧螺钉
5　adjustable gate　调节阀

24 Industrial Furnaces 工业炉

24.1 Industrial furnace 工业炉

24.1.1 Thermal medium heater 热载体加热炉

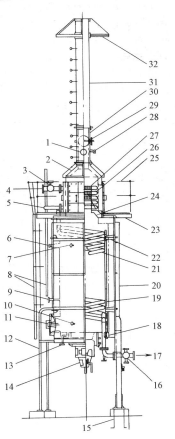

1. nozzle for thermometer of exhaust gas　排气温度计接管
2. shell of convection chamber　对流段外壳
3. header of thermal medium inlet　热载体入口总管
4. medium inlet　热载体入口
5. connection tube　接管
6. sight hole　视孔
7. spare nozzle for thermometer for gas in furnace　测炉内气温用温度计备用接管
8. ladder and cage　梯子和罩
9. nozzle for thermometer for gas in furnace　测炉内气温用温度计接管
10. spare nozzle for furnace inspection　炉内检查备用孔
11. manhole with sight glass　带视镜人孔
12. ladder to manhole　人孔梯子
13. steam inlet for snuffing　熄火蒸汽吹入管
14. gas jet burner　煤气喷燃器
15. foundation bolt　地脚螺栓
16. header of thermal medium outlet　热载体出口总管
17. medium outlet　热载体出口
18. seat for heating tube support　炉管支承架底座
19. heating tube support　炉管支承架
20. shell of radiation chamber　辐射段外壳
21. heating tube in radiant section　辐射段炉管
22. vibration-stopper　减振装置
23. stage for operation　操作台
24. tube sheet support　管板支承板
25. hand rail　栏杆
26. heating tube in convection section　对流段炉管
27. tube sheet　管板
28. inspection hole for exhaust gas pressure　排气压力检查孔
29. damper with remote control device　遥控风门机构
30. connected stack　烟囱连接段
31. stack　烟囱
32. beam　吊车梁

463

24.1.2 Heating furnace (tubular furnace) 加热炉(管式炉)

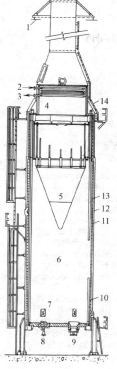

1 monorail beam 单轨梁
2 inlet 入口
3 outlet 出口
4 auxiliary convection section 辅助对流段
5 reradiating cone 辐射锥
6 combustion chamber 燃烧室
7 burner 烧嘴
8 gas 煤气
9 oil and gas 油和煤气
10 heating tube 加热管
11 steel shell 钢制炉体
12 insulation board 保温板
13 refractory brick 耐火砖
14 header door 顶盖门

24.1.3 Cylindrical-furnace tubular heater 圆筒形管式加热炉

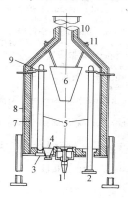

1 oil or gas burner 油或气体燃烧器
2 process fluid inlet and outlet 工艺过程流体进、出口
3 coil support 盘管支座
4 observation port 观察孔
5 tube coils 盘管
6 metal radiating cone 金属辐射锥
7 refractory 耐火材料
8 cylindrical shell 圆筒形壳体
9 coil bracket 盘管托架
10 metal stack 金属烟囱
11 cone support 辐射锥支架

24.1.4 Small industrial or commercial incinerator without fly-ash control 小型工业焚烧炉，未设飞灰控制器

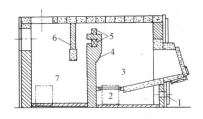

1 air port 空气入口
2 ash pit 灰坑
3 primary combustion chamber 主要燃烧室
4 bridge wall 火墙
5 air ports 空气入口
6 drop arch 凹形拱墙
7 settling chamber 沉降室

24.1.5 Mechanically fired industrial incinerator with reciprocating grates 机械添煤并有往复炉排的工业煅烧炉

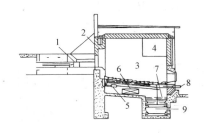

1 feed ram 进料推料机
2 adjustable feed gate 可调加料口
3 furnace 炉子
4 gas exit 气体出口
5 grate drive cylinder 炉排驱动缸
6 air inlet 进气口
7 water level 水位
8 dump grate operating cylinder 煤渣炉排操作气缸
9 ash drag 炉灰拖运机

24.1.6 Submerged-exhaust burner 沉浸式排气燃烧炉

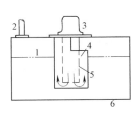

1 liquid level 液面
2 vent pipe 排气管
3 burner 燃烧器
4 weir 堰
5 open-end downcomer from burner 上接燃烧器下端开口的下降管
6 tank 槽

24.1.7 Pit furnace 槽式炉

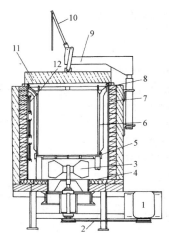

1. motor 电机
2. V-belts V形传动皮带
3. alloy fan 合金风机
4. fire brick 耐火砖
5. alloy grid plate 合金栅板
6. alloy basket 合金框
7. refractory insulating brick 耐火绝热砖
8. insulating block 绝热砌块
9. cover supporting bracket 顶盖支架
10. cover lift lever 顶盖升降控制杆
11. insulated cover 绝热顶盖
12. electric heating elements 电热元件

24.1.8 Car-bottom furnace 车底炉

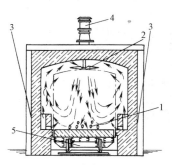

1. material 物料
2. fan 风机
3. burner 燃烧器
4. driver 驱动装置
5. car 小车

24.1.9 Rotary-hearth furnace 旋转膛式炉

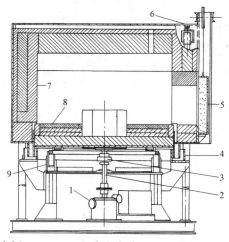

1. vari-speed drive 变速驱动装置
2. driving shaft 传动轴
3. coupling 联轴器
4. seal trough 密封槽
5. door 门
6. hearth clearance adjustment 炉床间隙调节器
7. furnace 炉
8. rotary hearth 旋转炉床
9. rollers 辊子

24.1.10 Small muffle furnace 小型隔焰炉

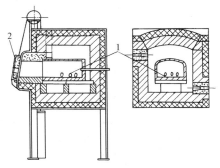

1 material 物料
2 inspection door 检查门

24.1.11 Mannheim-type mechanical hydrochloric acid furnace Mannheim 型机械盐酸炉

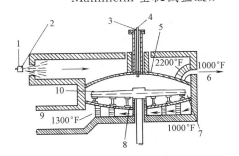

1 oil 油
2 burner 烧嘴
3 salt 盐
4 sulfuric acid 硫酸
5 carborundum arch 碳化硅炉拱
6 HCl gases to absorption system HCl 气体去吸收系统
7 salt cake 盐饼
8 pot with hearth 炉膛底
9 combustion gases to stack 燃烧气去烟道
10 rotating arms with ploughs 装有搅拌耙的旋转臂

24.1.12 Shaft furnace for lime production 竖式石灰窑

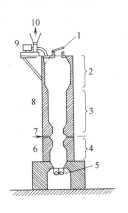

1 charge opening 加料口
2 preheating 预热段
3 calcining and post-heating 煅烧及其后加热段
4 cooling 冷却段
5 discharge gates 卸料阀
6 cooler 冷却器
7 oil burners 油燃烧器
8 kiln 窑
9 fan 风机
10 exhaust gases 废气

467

24.1.13 Pacific multiple-hearth furnace Pacific 多层床炉

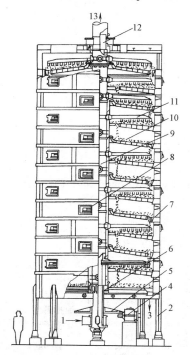

1 cooling air in   冷却空气入口
2 supporting construction   支承结构
3 discharge port   卸料口
4 motor with educer   电机减速机
5 bevel gear   伞齿轮
6 lower bearing housing   下轴承箱
7 shell   壳体
8 observation window   观察窗
9 hollow shaft   空心轴
10 materials   物料
11 agitator blade   搅拌耙
12 top bearing housing   上轴承箱
13 heated air out   热气出口

24.2 Burner 燃烧器

24.2.1 Circular burner for pulverized coal, oil or gas 燃烧煤粉、油或燃气的圆形燃烧器

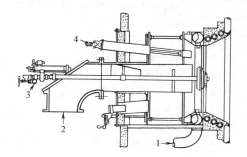

1 gas   气
2 coal   煤
3 oil   油
4 gas-fired lighter   气体点火器

24.2.2 Burner and furnace configurations for pulverized-coal firing 粉煤燃烧用燃烧器与炉子的配置

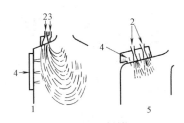

(a) vertical firing 立式燃烧

(b) tangential firing 切向燃烧

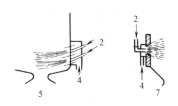

(c) horizontal firing 水平燃烧

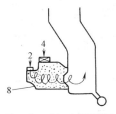

(d) cyclone firing 旋风燃烧

(e) opposed-inclined firing 对置倾角燃烧

1　fantail　扇尾式
2　primary air and coal　一次空气与煤
3　tertiary air　三次空气
4　secondary air　二次空气
5　multiple intertube　复式内管
6　plan view of furnace　炉的俯视图
7　circular　圆形
8　cyclone　旋风器

24.2.3 Cyclone furnace 旋风炉

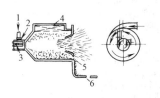

1　coal　煤
2　primary air　一次空气
3　tertiary air　三次空气
4　tangential secondary air　切向二次空气
5　cyclone slag-tap hole　旋风器出渣孔
6　primary furnace slag-tap hole　主炉膛出渣孔

24.2.4 Types of grates used with spreader stokers 布煤机——炉排使用的炉栅类型

(a) stationary 固定式

(b) dumping 翻斗式

(c) oscillating 振动式

(d) traveling 移动式

24.2.5 Principal types of oil burners 油燃烧器的主要类型

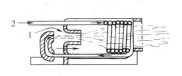

(a) pressure-type vaporizing burner
压力式气化燃烧器

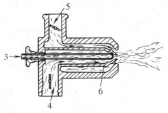

(d) low-pressure air-atomizing burner
低压空气雾化燃烧器

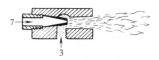

(b) high-pressure steam or air-atomizing burner
高压蒸汽或空气雾化燃烧器

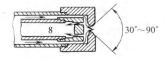

(e) mechanical or oil-pressure atomizing burner
机械或油压雾化燃烧器

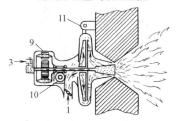

(c) horizontal rotary-cup atomizing oil burner
水平旋杯式雾化油燃烧器

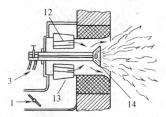

(f) complete mechanical or oil-pressure atomizing
burner unit 全机械或油压雾化燃烧器装置

1　air　空气
2　fuel　燃料
3　oil　油
4　primary air　一次空气
5　secondary air　二次空气
6　primary whirling vanes　一次旋流叶片
7　steam or air　水蒸气或空气
8　returned oil　回油
9　motor　电动机
10　oil pump　油泵
11　mounting hinge　支架
12　adjustable registers　可调配风器
13　wind box　风箱
14　air deflectors　空气导流装置

24.2.6 Atmospheric industrial gas-burner 常压工业气体燃烧器

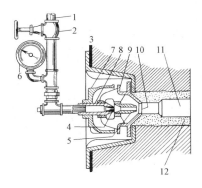

1. gas line 燃气管线
2. gas valve 燃气阀
3. wall casting 墙固定板
4. spud 锥体
5. burner throat 燃烧器喉管
6. pressure gage 压力表
7. shutter 空气调节器
8. spud holder 锥固定器
9. burner body 燃烧器本体
10. nozzle 喷嘴
11. tunnel 通道
12. burner cement 燃烧器胶结水泥

(installation with individual air control to each burner 每一燃烧器安设有单独的空气控制)

24.2.7 Premix burner 预混燃烧器

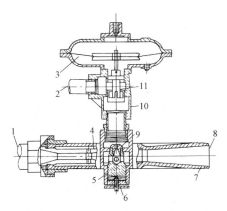

1. air inlet 空气进口
2. gas inlet 燃气进口
3. governor diaphram 调节器隔膜
4. insert 衬芯
5. spud holder 锥固定器
6. plug 堵头
7. inspirator body 引射器本体
8. mixture outlet 混合器出口
9. spud 锥体
10. governor body 调节器本体
11. governor valve 调节阀

(in which a proportional mixer uses air velocity to draw in a measured amount of gas 带有比例混合器，利用空气速度引入一定量的燃气)

24.2.8 High-velocity burner 高速燃烧器

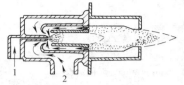

1 fuel 燃料
2 air 空气

(high heat release produces combustion results similar to hydroxylation 放热产生的燃烧效果类似于羟基化作用)

24.2.9 Various ways of mixing gas and air at the burner 燃烧器中空气与燃气混合的几种方法

1 gas supply 燃气供入
2 ignition tube 点火管
3 air register 空气调节器
4 gas-outlet holes 燃气出口孔
5 gas ring 燃气环形管
6 air 空气
7 spider vane 辐射形叶片
8 integral fan 整体式风机

(a) blunt pipe 圆头管　(b) small ring 小环管

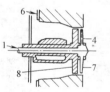

(c) turbine 涡轮式　(d) large ring 大环管

24.2.10 Pressurized fluidized-bed combustor 加压流化床燃烧室

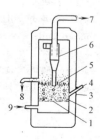

1 air distributor 空气分布器
2 heat-absorption tubes 吸热管
3 pressure vessel 压力容器
4 coal in 煤入口
5 fluidized bed of coal and ash 煤与灰的流化床
6 cyclone dust collector 旋风集尘器
7 combustion products 燃烧产物
8 residue 煤渣
9 air 6 atm 6个大气压的空气

24.3 Coal gasifier 煤气化炉

24.3.1 Lurgi gasifier
Lurgi 气化炉

24.3.2 Winkler gasifier
Winkler 气化炉

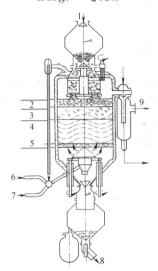

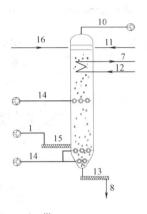

24.3.3 Koppers-Totzek gasifier
Koppers-Totzek 气化炉

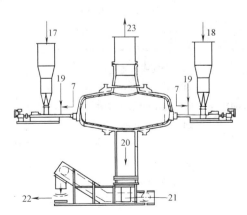

1　coal　煤
2　drying　干燥段
3　carbonization　碳化段
4　gasification　气化段
5　combustion　燃烧段
6, 19　oxygen (O_2)　氧气
7　steam　水蒸气
8　ash　炉灰
9　crude gas　粗制煤气
10　crude product gas　粗制产品煤气
11　process steam　工艺过程蒸汽
12　feed water　给水
13　ash screw　炉灰螺旋输送器
14　steam and oxygen　水蒸气和氧气
15　screw feeder　螺旋加料器
16　condensate　冷凝液
17　coal dust　粉煤
18　crude dust　粗粉煤
20　liquid slag　液态炉渣
21　slag quench tank　炉渣猝灭槽
22　granulated slag　粒状炉渣
23　crud gas　粗制杂煤气

25 Power Plant 动力装置

25.1 Cogeneration systems 联合生产系统

25.1.1 Alternative configurations for cogenerating heat engines 联合生产热机的各种组合

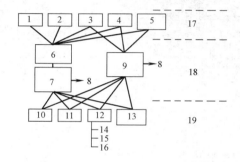

1　solar energy　太阳能
2　nuclear energy　核能
3　biomass fuels　生物质燃料
4　fossil fuels　化石燃料
5　coal or shale derived synfuels
　　煤或页岩衍生的合成燃料
6　combustor/heater（for gas or liquid）
　　燃烧器/加热器（气体或液体用）
7　external-combustion heat engine（1）
　　外燃式热机（1）
8　electric power　电力
9　internal-combustion heat engine（2）
　　外燃式热机（2）
10　district heating　地区加热
11　space heat　空间加热
12　process heat　过程加热
13　atmosphere or water bodies
　　大气或水体
14　industrial processes　工业过程
15　agriculture　农业
16　desalination　淡化
17　heat-source alternatives　各种热源
18　power-conversion-system alternatives
　　各种电力转换系统
19　heat-rejection alternatives
　　各种热力排放

25.1.2 Extraction-condensing-steam-turbine cogeneration cycle 抽汽-凝汽-蒸汽透平的联合生产循环

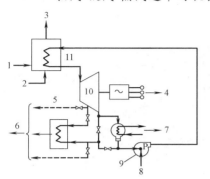

1　fuel　燃料
2　air　空气
3　stack gas　烟道气
4　electric power　电力
5　direct steam use　直接使用蒸汽
6　heat load　热负荷
7　cooling load　冷却负荷
8　pump power　泵动力
9　pump　泵
10　turbine　透平
11　combustor　锅炉

25.2　Nuclear reactor　核反应堆

25.2.1　Fast-breeder reactor　快中子增殖（反应）堆

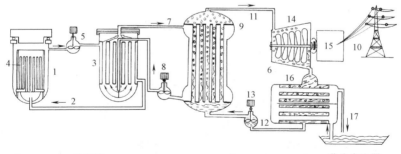

1　reactor　反应堆
2　primary circuit (primary loop)　一次电路（回路）
3　heat exchanger　热交换器
4　fuel rods; fuel pins　燃料棒
5　primary sodium pump　一次钠泵
6　cooling water flow circuit　冷却水（流动）回路
7　secondary circuit (secondary loop)　二次电路（回路）
8　secondary sodium pump　二次钠泵
9　steam generator　蒸汽发生器
10　transmission line　输电线路
11　steam line　蒸汽管线，蒸汽管路
12　feedwater line　给水管线，给水管路
13　feed pump　给水泵
14　steam turbine　汽轮机，蒸汽涡轮机
15　generator　发电机
16　condenser　冷凝器，凝结器
17　cooling water　冷却水

25.2.2 Nuclear reactor 核反应堆

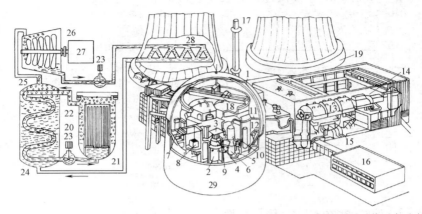

(a pressurized-water reactor, nuclear power plant, atomic power plant 一种压水反应堆，核动力站，原子能动力站)

1. steel containment (steel shell) with air extraction vent 带抽气孔的钢（安全）壳
2. reactor pressure vessel 反应堆压力容器
3. control rod drive 控制棒驱动装置
4. control rods 控制棒
5. primary coolant pump 一次冷却剂泵
6. steam generator 蒸汽发生器
7. fuel-handling hoists 燃料装卸吊车（起重机）
8. fuel storage 燃料仓库
9. coolant flow passage 冷却剂流通管
10. feedwater line 给水管线
11. prime steam line 主蒸汽管道
12. manway 人员专用道，走道
13. turbogenerator set 汽轮发电机组
14. turbogenerator 汽轮发电机
15. condenser 冷凝器
16. service building 服务大楼
17. exhaust gas stack 废气排放管
18. polar crane 极性起重机
19. cooling tower (a dry cooling tower) 冷却塔（一种干式冷却塔）
20. pressurized-water system 压水系统
21. reactor 反应堆
22. primary circuit (primary loop) 一次电路［回路］
23. circulation pump (recirculation pump) 循环泵（再循环泵）
24. heat exchanger (steam generator) 热交换器（蒸汽发生器）
25. secondary circuit (secondary loop) 二次电路［回路］
26. steam turbine 汽轮机，蒸汽涡轮机
27. generator 发电机
28. cooling system 冷却系统
29. concrete shield (reactor building) 混凝土屏蔽层（反应堆建筑物）

25.2.3 Boiling water system 沸水系统

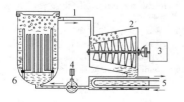

1. steam and recirculation water flow paths 蒸汽和再循环水流径
2. steam turbine 汽轮机，蒸汽涡轮机
3. generator 发电机
4. circulation pump (recirculation pump) 循环泵（再循环泵）
5. coolant system (cooling with water from river) 冷却系统（用河水冷却）
6. reactor 反应堆

25.2.4 Radioactive waste storage in salt mine 放射性废料存放在盐矿中

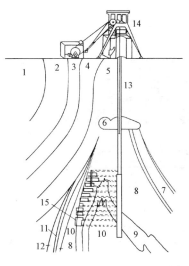

1~12 geological structure of abandoned salt mine converted for disposal of radioactive waste 放置放射性废料的废盐矿的地质构造
1 Lower Keuper 下［早］考依波（泥灰岩及砂岩）
2 Upper Muschelkalk 上［晚］壳灰岩层
3 Middle Muschelkalk 中壳灰岩层
4 Lower Muschelkalk 下［早］壳灰岩层
5 Bunter downthrow 班砂岩降侧，班特下落断层块
6 residue of leached Zechstein 淋溶镁灰残留物，沥滤的蔡希斯坦残渣
7 Aller rock salt 阿勒岩盐
8 Leine rock salt 莱茵岩盐
9 Stassfurt seam (potash salt seam, potash salt bed) 斯塔斯弗矿层（钾盐层，钾盐床）
10 Stassfurt salt 斯塔斯弗盐
11 grenzanhydrite 限界流体，越界流体
12 Zechstein shale 蔡希斯坦页岩，镁灰油页岩
13 shaft 矿井，立井
14 minehead buildings 井口建筑物
15 storage chamber 贮存室

25.2.5 Storage of medium-active waste in salt mine 中放射性废料存放在盐矿中

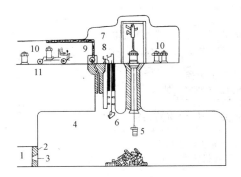

1 511m level 511米深的水平坑道
2 protective screen (anti-radiation screen) 保护壁（抗辐射屏障）
3 lead glass window 铅玻璃窗
4 storage chamber 贮存室
5 drum containing radioactive waste 装放射性废料的金属桶
6 television camera 电视摄像机
7 charging chamber 进［装］料室
8 control desk (control panel) 控制台［盘］
9 upward ventilator 向上通风装置
10 shielded container 屏蔽好的容器
11 490m level 490米深的水平坑道

477

25.3 Gas-turbine 燃气轮机

25.3.1 Gas-turbine cross section 燃气轮机

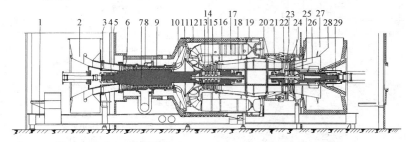

1 metallic foundation　金属底座
2 inlet air casing　空气进口壳体
3 compressor inlet guide vanes　压缩机进口导叶
4 compressor radial——axial bearing　压缩机径向——止推轴承
5 conical inlet casing with bearing support　带轴承座的锥形进口套管
6 compressor stator blades　压缩机静叶片
7 gas-generator turbine rotor　燃气发生器透平转子
8 compressor rotor blades　压缩机转子叶片
9 compressor stator　压缩机定子
10 intermediate conical casing with diffusor　带扩压段的中间锥形外壳
11 hot gas casing　热燃气套管
12 combustion-chamber main casing　燃烧室主壳体
13 gas-generator turbine stator blades　燃气发生器透平定子叶片
14 gas-generator turbine stator blade carrier　燃气发生器透平定子叶片托架
15 gas-generator turbine rotor blades　燃气发生器透平转子叶片
16 intermediate casing　中间壳体
17 nine incorporated combustion chambers　九个结合的燃烧室
18 compressor radial bearing　压缩机径向轴承
19 central casing　中心套管
20 conical intermediate piece　锥形的中间零件
21 power turbine radial bearing　动力涡轮机径向轴承
22 power turbine rotor　动力涡轮机转子
23 power turbine adjustable rotor blades　动力涡轮可调转子叶片
24 power turbine stator blade carrier　动力涡轮机定子叶片托架
25 power turbine stator blades　动力涡轮机定子叶片
26 power turbine rotor blades　动力涡轮机转子叶片
27 outlet casing　出口外壳
28 outlet diffusor　出口扩压器
29 power turbine radial-axial bearing　动力涡轮机径向——止推轴承

25.3.2 Gas-turbine cycles 燃气轮机循环

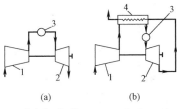
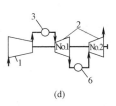

(a) basic Brayton or Joule cycle 基本 Brayton 或 Joule 循环
(b) gas turbine with regeneration 有再生回热的燃气轮机
(c) gas turbine with intercooling 有中间冷却的燃气轮机
(d) gas turbine with reheating 有中间再热的燃气轮机

25.3.3 Regenerative-cycle gas turbine 采用再生回热循环的燃气轮机

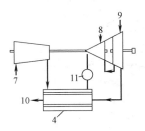

(two-shaft arrangement with separate power turbines in series 双轴布置，并有串联型单独的功率燃气轮机)

25.3.4 Closed-cycle gas turbine 闭式循环燃气轮机

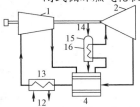

25.3.5 Semiclosed, internally fired gas-turbine cycle 半闭式内部燃烧型燃气轮机循环

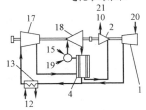

1　compressor　压气机
2　turbine　汽轮机
3　combustion chamber　燃烧室
4　regenerator　回热器
5　intercooler　中间冷却器
6　reheating combustion chamber　再热燃烧室
7　inlet　进气口
8　high-pressure turbine　高压汽轮机
9　low-pressure turbine　低压汽轮机
10　exhaust　排气口
11　fuel combustor　燃烧室
12　cooling medium　冷却工质
13　precooler　预冷器
14　combustion air　燃烧用空气
15　fuel　燃料
16　air heater　空气加热器
17　main compressor　主压气机
18　main turbine　主汽轮机
19　combustor　燃烧室
20　air intake　空气进口
21　pump up or charging set　泵送或给料设备

25.4 Electric parts and power plant 电工器件与发电厂

25.4.1 Electric parts 电工器件

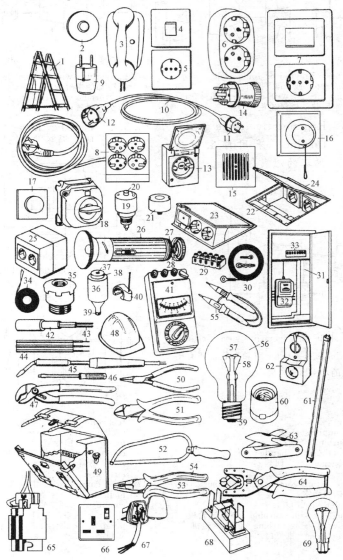

25.4 Electric parts and power plant 电工器件与发电厂

1. ladder 梯子
2. bell push (for low-voltage safety current) 电铃按钮（使用低压安全电流）
3. house telephone with call button 带呼唤按钮的户内电话
4. flush-mounted rocker switch 嵌装式摇杆开关
5. flush-mounted earthed socket 嵌装式接地插座
6. earthed double socket 接地双插座
7. switched socket 带开关的插座
8. four-socket (four-way) adapter 四连插座，四插座转接器
9. earthed plug 接地插头
10. extension lead 延长线，接长用引线
11. extension plug 延长线插头
12. extension socket 延长线插座
13. surface-mounted three-pole earthed socket with neutral conductor 带有中性导体的表面安装式三眼接地插座
14. three-phase plug 三相插头
15. electric bell (electric buzzer) 电铃（蜂鸣器）
16. pull-switch 拉线开关
17. strip of sheet metal 薄金属片
18. drill-cast rotary switch 旋转式定位开关
19. miniature circuit breaker (screw-in circuit breaker, fuse) 小型断路器（熔断器）
20. resetting button 复位［重接］按钮
21. set screw 定位螺钉
22. underfloor sockets 地板下的插座
23. hinged floor socket for power lines and communication lines 电源线和通信线用的铰链式地面插座
24. sunken floor socket with hinged lid (snap lid) 带铰链盖的凹入地面的插座
25. surface-mounted socket outlet box 表面安装式插座引出线盒
26. pocket torch （袖珍）手电筒
27. dry cell battery 干电池
28. contact spring 接触弹簧
29. strip of thermoplastic connectors 热塑连接器的插座板
30. steel draw-in wire with threading key, and ring attached 带插入键和金属环的钢拉线
31. electricity meter cupboard 电表柜
32. electricity meter 电表
33. miniature circuit breakers 小型断路器
34. insulating tape 绝缘（胶）带
35. fuse holder 熔断器座
36. circuit breaker, a fuse cartridge with fusible element 断路器（熔断器），带易熔元件的保险线盒
37. colour indicator 颜色指示器
38, 39 contact maker 断续器，接通器
40. cable clip （塑料）电缆夹（头）
41. universal test meter (multiple meter for measuring current and voltage) 多用途电表，万用表（测量电流、电压的多用电表）
42. thermoplastic moisture-proof cable 热塑防潮电线
43. copper conductor 铜导体，铜芯线
44. three-core cable 三芯电线
45. electric soldering iron 电烙铁
46. screwdriver 螺丝起子
47. pipe wrench 管子扳手
48. shock-resisting safety helmet 防震安全帽
49. tool case 工具箱
50. round-nose pliers 圆嘴钳
51. cutting pliers 剪钳
52. junior hacksaw 轻型钢锯
53. combination cutting pliers 钢丝钳
54. insulated handle 绝缘柄
55. continuity tester 线路通断测试器
56. electric light bulb 电灯泡
57. glass bulb 玻璃灯泡
58. coiled-coil filament 绕线式灯丝
59. screw base 螺旋灯头
60. lampholder 灯座
61. fluorescent tube 日光［荧光］灯管
62. bracket for fluorescent tubes 日光灯管座
63. electrician's knife 电工刀
64. wire strippers 剥线钳
65. bayonet fitting 卡口灯座
66. three-pin socket with switch 带开关的三眼插座
67. three-pin plug 三脚插头
68. fuse carrier with fuse wire 带保险丝的熔断架
69. light bulb with bayonet fitting 卡口灯泡

481

25.4.2 Control room 控制室

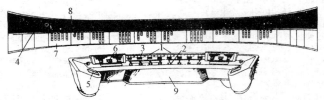

1 control board (control panel) for the alternators 交流发电机控制板
2 master switch 主［总］开关
3 signal light 信号灯
4 feeder panel 馈电屏
5 controls 监控盘
6 monitoring controls for the switching systems 开并系统监控盘
7 revertive signal panel 返回信号盘
8 matrix mimic board 电路模拟板
9 control console (control desk) 控制台

25.4.3 Transformer 变压器

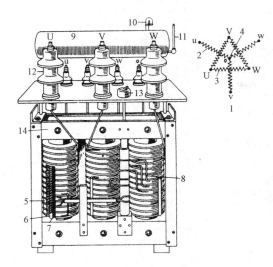

1 transformer connection 变压器连接方式
2 star connection (star network, Y-connection) 星形连接（Y连接）
3 delta connection (mesh connection) 三角连接（网形连接）
4 neutral point 中（性）点
5 primary winding 初级绕阻（高压绕阻）
6 secondary winding (low-voltage winding) 次级绕阻（低压绕阻）
7 core 铁芯
8 tap (tapping) 分接头
9 oil conservator 储油柜，油箱
10 breather 呼吸器
11 oil gauge 油位计
12 feed-through terminal 馈电端子（瓷套管）
13 on-load tap changer 有载换接器
14 yoke 铁轭

25.4.4 Steam turbine 汽轮机

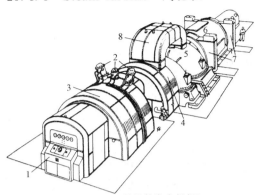

1. turbine monitoring panel with measuring instruments 带测量仪表的汽轮机监控盘
2. jet nozzle 喷射嘴
3. high-pressure cylinder 高压汽缸
4. medium-pressure cylinder 中压汽缸
5. low-pressure cylinder 低压汽缸
6. three-phase generator 三相发电机
7. hydrogen cooler 氢冷却器
8. leakage steam path 漏泄蒸汽通道

(turbogenerator unit 一种涡轮发电机组)

25.4.5 Automatic voltage regulator 自动电压调节器

25.4.6 Synchro 同步器

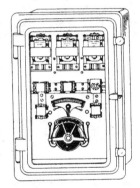

25.4.7 Cable box 电缆（接线）盒

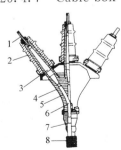

1. conductor （导电）芯线，导线
2. feed-through insulator 馈电端子（瓷套管）
3. core 电缆轴心，线芯
4. casing 电缆套［外套］
5. filling compound 封填绝缘膏
6. lead sheath 铅包皮，铅护套
7. lead-in tube 铅封管
8. cable 电缆

25.4.8 High voltage cable (for three-phase current) （三相）高压电缆

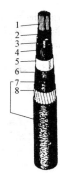

1. conductor （导电）芯线
2. metallic paper (metallized paper) 金属化纸（镀金箔纸）
3. tracer (tracer element) 描绘器（描绘器元件）
4. varnished-cambric tape 漆布带，黄蜡布带
5. lead sheath 铅包皮，铅护套
6. asphaltum paper 沥青纸
7. jute serving 黄麻护层
8. steel tape or steel wire armour 钢带或钢丝护套

25.4.9 Air-blast circuit breaker 空气吹弧断路器

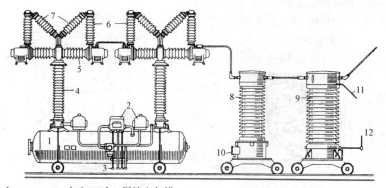

1. compressed-air tank 压缩空气罐
2. control valve (main operating valve) 控制阀（主操作阀）
3. compressed-air inlet 压缩空气入口
4. support insulator, a hollow porcelain supporting insulator 支架绝缘子，一种空心的瓷支架绝缘子
5. interrupter 断续器
6. resistor 电阻器
7. auxiliary contacts 辅助接点
8. current transformer 变流器
9. voltage transformer (potential transformer) 变压器
10. operating mechanism housing 操作机械盒
11. arcing horn 消弧角，角形避雷器
12. spark gap 火花（放电）间隙

25.5 Steam-generation system 蒸汽发生系统

25.5.1 Atmospheric-fluidized-bed-combustor arrangements 常压流化床燃烧锅炉的类型

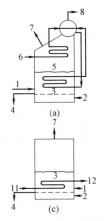
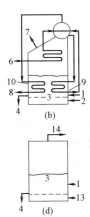

(a) saturated-steam boiler　饱和蒸汽锅炉
(b) superheater-steam generator　过热蒸汽发生器
(c) indirect-process heater　间接过程加热器
(d) direct-process heater　直接过程加热器
1　fuel and sorbent inlet　燃料与吸收剂入口
2　fluidizing-air inlet　流化空气入口
3　fluidized bed　流化床
4　ash disposal　排灰
5　economizer　省煤器
6　feed water inlet　水加入口
7　off gas　排出气体
8　saturated-steam outlet　饱和蒸汽出口
9　boiling　沸腾
10　superheating　过热
11　working fluid inlet　工作液入口
12　hot working fluid for process　热工作液供过程用
13　air inlet (fluidizing and process)　空气入口（流化与过程）
14　hot gases for process needs　热气体供过程用

25.5.2 Pressurized-fluidized-bed-combustion concepts 加压流化床燃烧的概念流程

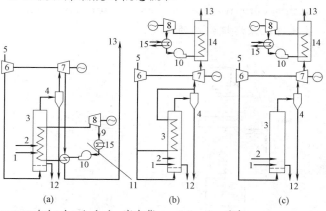

(a) steam-cooled tubes in bed 床内蒸汽冷却管
(b) air-cooled tubes in bed 床内空气冷却管
(c) bed cooled by excess air (300 percent); no in-bed tubes 床内用过剩（300％）空气冷却，床内无冷却管
1　coal　煤
2　sorbent　吸收剂
3　pressurized combustor　加压燃烧室
4　particulate removal　除微粒
5　air　空气
6　compressor　压缩机
7　gas turbine　气体透平
8　steam turbine　蒸汽透平
9　water　水
10　boiler feed water　锅炉给水
11　condenser heat recovery　冷凝器回收热
12　ash disposal　排灰
13　stack　烟囱
14　waste-heat boiler　废热锅炉
15　condenser　冷凝器

25.5.3 Operational schematic of a fluidized-bed steam-generation system 流化床蒸汽发生系统的操作流程

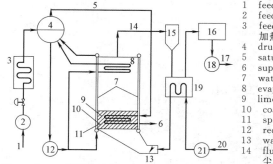

1　feedwater　进水
2　feedpump　进料泵
3　feedwater heater economizer　进水加热器省煤器
4　drum　汽包
5　saturated steam　饱和蒸汽
6　superheated steam　过热蒸汽
7　waterwalls　水汽壁
8　evaporator　蒸发器
9　limestone　石灰石
10　coal　煤
11　spent bed　废料床
12　recirculation pump　循环泵
13　warm air to furnace　入炉热空气
14　flue gas and dust particles　烟气与尘粒
15　cyclone　旋风分离器
16　bag filter　袋滤器
17　stack gas　烟囱气体
18　ID fan (ID——induced draft)　引风机
19　air preheater　空气预热器
20　atmospheric air　空气
21　FD fan (FD——forced-draft)　鼓风机

25.6 Heat transport 热的输送

25.6.1 Dowtherm vaporizer 导热姆蒸发器

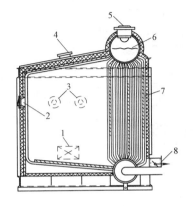

1 access door 检修门
2 inspection port 观察孔
3 burners 燃烧器
4 explosion port 安全防爆孔
5 vapor outlet 蒸汽出口
6 liquid level 液面
7 unheated downcomers 不加热下降管
8 gas outlet to stack 气体出口去烟道

25.6.2 Dowtherm gravity-return system 导热姆的重力回路系统

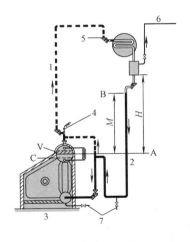

V, C——maximum and minimum liquid levels in vaporizer drum 蒸发器鼓内最高与最低液面

M, H——range of liquid heights in condensate leg between A and B to cause flow 造成系统流动的 A 与 B 之间冷凝液柱高度范围

1 Dotted lines show vapor path 虚线表示蒸气路线
2 solid lines the liquid path 实线表示液体路线
3 Dowtherm vaporizer 导热姆蒸发器
4 temperature controller 温度控制器
5 throttle valve 节流阀
6 vent line to vent condenser 放空线，至放空冷凝器
7 drains to storage tank 排污至储罐

25.6.3 Dowtherm pumped-return system 导热姆的泵回路系统

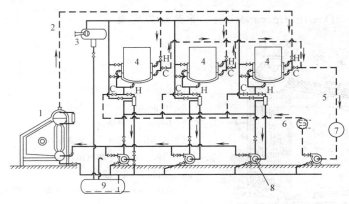

1 Dowtherm vaporizer 导热姆蒸发器
2 heating circuit 加热管线
3 vacuum unit 真空装置
4 users 用热设备
5 cooling circuit 冷却管线
6 cooler 冷却器
7 receiver 接收器
8 pumps 泵
9 storage tank 储罐

25.6.4 Combination vapor and liquid heating with Dowtherm 导热姆蒸气与液体的联合加热

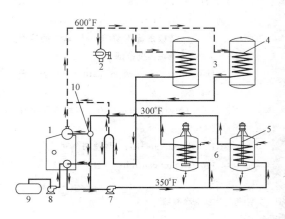

1 Dowtherm vaporizer 导热姆蒸发器
2 vent condenser 放空冷凝器
3 high-temperature vessels 高温罐
4 heating coil 加热盘管
5 agitator 搅拌器
6 low-temperature vessels 低温罐
7 circulating pump 循环泵
8 charge pump 加料泵
9 storage tank 储罐
10 thermostat-controlled valve 定温控制阀

25.6.5 Fire-tube salt-bath heater 火管式盐浴加热器

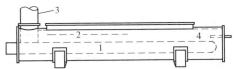

1 firetube 火管
2 salt level 盐液面
3 stack 烟道
4 flow coil 流体盘管

25.6.6 Circulating salt system for heating and cooling operation 加热与冷却操作两用的循环盐系统

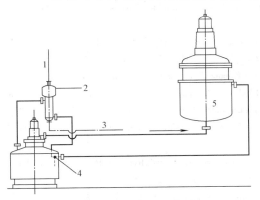

1 steamline
  蒸汽管线
2 steam generator
  蒸汽发生器
3 water line 水管线
4 three-way valve 三通阀
5 reactor 反应器

25.6.7 Typical hot-oil circulating system 典型的热油循环系统

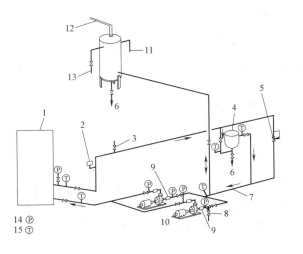

1 heater 加热器
2 over-temperature cutout
  超温切断
3 vent 放空
4 user 用热设备
5 bypass pressure-relief
  valve 旁路调节阀
6 drain 排空
7 return 回流
8 filling connection
  注入接头
9 strainer 滤网
10 circulating pumps 循
   环泵
11 overflow 溢流
12 vent line 放空管线
13 2in. fill line 2英寸
   加料管线
14 ⓟ ——pressure
   gage 压力表
15 ⓣ ——thermometer
   温度计

主要参考文献

[1] 化工名词审定委员会. 化学工程名词. 北京：科学出版社，1995.
[2] Perry RH. Chemical Engineers' Handbook，6th ed. New York：McGraw-Hill，1984.
[3] Perry RH. Chemical Engineers' Handbook 7th ed. New York：McGraw-Hill，1997.
[4] McCabe WL. Unit Operation of Chemical Engineering. New York：McGraw-Hill，1993.
[5] Stanley MW. Chemical Process Equipment，selection and design. London：Butterworth，1988.
[6] McCabe，Smith. Unit Operations in Chemical Engineering，sixth edition. New York：McGraw-Hill，2001.
[7] Greankopkis CJ. Transfer Process，2nd ed. New York：McGraw-Hill，1980.
[8] King CJ. Separation Process，2nd ed. New York：McGraw-Hill，1980.
[9] Christopher D. Filters and Filtration Handbook 3rd ed. Elsevier Science Publishers. 1992.
[10] Arun S Mujumdar. Handbook of Industrial Drying. New York：Marcel Dekker，Inc.，1987.
[11] Terry L. Henshaw. Reciprocating Pump. New York：van Nostrand Reinhold Co.，1987.
[12] James G Speight. Chemical and Process Design Handbook. New York：McGraw-Hill，2002.
[13] Warring RH. Pumping Manual，7th ed. London：Trade & Technical Press Limited，1984.
[14] Igor J Karassik. Pump Handbook. New York：McGraw-Hill，1986.
[15] Robert E McCabe. Metering Pump Handbook. Industrial Press，1984.
[16] Heinz P Bloch. Compressors and Expanders. New York：Marcel Dekker，Inc，1982.
[17] Sadik K，Hongtan Liu. Heat Exchangers. New York：CRC Press，1997.
[18] Coulson JM，Richardson JF. Chemical Engineering，sixth edition. Butterworth Heineman，1999.
[19] Leslie Grady CP. Biological Wastewater Treatment，second edition. New York：Marcel Dekker，Inc.，1999.
[20] 余国琮. 化学工程辞典. 北京：化学工业出版社，2003.
[21] 陈国桓. 机械基础. 北京：化学工业出版社，2001.
[22] 李家珍. 染料、染色工业废水处理. 北京：化学工业出版社，1997.
[23] The Oxford-Duden Pictorial English-Chinese Dictionary. 北京：商务印书馆、牛津大学出版社，1998.
[24] 西德杜登出版社、英国牛津大学出版社字典编辑部编. 牛津-杜登英汉图解词典. 卜纯英等译. 北京：化学工业出版社．1984.
[25] Coulson JM，Richardson JF. Chemical Engineering，Vol. 6 Design. New York：Pergamon Press，1983.
[26] Coulson JM，Richardson JF. Chemical Engineering，Vol. 1. New York：Pergamon

Press, 1964.
[27] Liu K, Song C, Subramani V. Hydrogen and Syngas Production and Purification Technologies. American Institute of Chemical Engineers, 2009.
[28] Bell DA, Towler BF, Fan M. Coal Gasification and its Applications. 北京：科学出版社, 2011.
[29] 丁云杰, 等. 煤制乙醇技术. 北京：化学工业出版社, 2014.
[30] Parkash S. Refining Processes Handbook. Oxford：Gulf Professional Publishing, 2003.
[31] 吴秀章. 煤制低碳烯烃工艺与工程. 北京：化学工业出版社, 2014.
[32] Muradov N. Lecture Notes in Energy, Liberating Energy from Carbon：Introduction to Decarbonization. New York：Springer Science+Business Media, 2014.
[33] Prodan VD, Klyushenkova MI, Borodacheva EI. Low-Tonnage Methanol Production, Chemical and Petroleum Engineering. New York：Springer Science + Business Media, 2013.
[34] Bertau M, Offermanns H, Plass L, et al. Methanol：The Basic Chemical and Energy Feedstock of the Future. Berlin Heidelberg：Springer-Verlag, 2014.
[35] Ray C, Jain R. Strategies for Sustainability, Drinking Water Treatment Focusing on Appropriate Technology and Sustainability. Netherlands：Springer, 2011.
[36] Cai H, Jia CQ. Mercury removal from aqueous solution by coke-derived sulphur-impregnated activated carbon. Industrial and Engineering Chemistry Research, 2010.
[37] Bahadori A, Clark M, Boyd B. Essentials of Water Systems Design in the Oil, Gas, and Chemical Processing Industries. Springer Briefs in Applied Sciences and Technology, 2013.
[38] Lofrano G. Removal Green Technologies for Wastewater Treatment. Springer Briefs in Molecular Science, Energy Recovery and Emerging Compounds, 2012.
[39] Wang LK, Shammas NK, Hung Y. Advanced Biological Treatment Processes in Handbook of Environmental Engineering. LLC：Springer Science+Business Media, 2009.
[40] Bahadori A. Pollution Control in Oil, Gas and Chemical Plants. Switzerland：Springer International Publishing, 2014.
[41] Wang LK, Pereira NC, Hung Y. Biological Treatment Processes in Handbook of Environmental Engineering. LLC：Humana Press, a part of Springer Science+Business Media, 2009.
[42] Tan Z. From Basic Concepts to Engineering Applications for Air Emission Control, Air Pollution and Greenhouse Gases, Green Energy and Technology. Singapore：Springer Science+Business Media, 2014.
[43] Muradov N. Lecture Notes in Energy, Liberating Energy from Carbon：Introduction to Decarbonization. New York：Springer Science+Business Media, 2014.

英汉对照词汇部分

A

abandoned salt mine 废盐矿 25.2.4.1~12
ABF(activated bio filter) process flow diagram
ABF(活化生物过滤设备)工艺流程图 3.1.44
abrasion-resistant lining 耐磨衬里 11.3.1.14
absolute alcohol product 无水乙醇产品 2.2.59.6
absolute velocity 绝对速度 16.4.3.10
absorbed methanol 吸收甲醇 2.2.55.4
absorbent regenerator 吸收剂再生器 2.2.106.33
absorber 吸收塔 2.2.27.6;2.2.61.8;2.2.63.11;2.2.102.23;2.3.15.3
吸收器 20.1.2.14
absorbing heat 吸热 20.1.2.9
absorbing section 吸收工段 2.2.56.8
absorbing tower 吸收塔 2.2.70.6
absorb vapor in liquid 液体吸收蒸气 20.1.2.22
absorption 吸收 1
absorption column 吸收塔 5.4
absorption cycle 吸收循环 20.1.8
absorption refrigeration 吸收制冷 20.1.(c)
absorption tower 吸收塔 2.2.49.3
abutment plate 支承面板 19.1.5.19
abutment plate mounting screws 支承板固定螺钉 19.1.5.18
accelerator 加速器 9.5.18.5
accelerator conveyor 加速器输送器 23.4.2.20
access door 检修门 8.9.1.9
access opening 出入口 5.4.1.21
accumulator 收集器 2.2.5.20
accumulator conveyor 累积器输送器 23.4.2.4
accumulator conveyor 累积器,输送机 23.4.3.5
AC electricity 交流电 2.2.113.2
acetate 乙酸酯 2.2.40.6

acetate 醋酸盐 3.3.18.15
acetate formation 醋酸盐形成 3.3.18.22
acetic acid 乙酸 2.2.1.19;2.2.40.5;2.2.54.14;2.2.57.12;2.2.59.2
acetic acid esterification 乙酸酯化 2.2.40.7
acetonitrile 乙腈 2.3.15.16
acetonitrile purification column 乙腈净化塔 2.3.15.7
acetylene 乙炔 2.2.1.9
acetylene feed 乙炔进料 14.3.1.2
acetylene saturator 乙炔饱和塔 2.2.68.17
ACF center-flow ACF中心流动货车 23.3.8.(b)
acid 酸 2.3.6.3
acid phase 酸相 3.3.18.17
acid dilution 酸稀释 6.2.1.3
acid egg 酸蛋 16.1.2.2.52
acid gas 酸性气 2.2.67.8
acid-gas absorber 酸性气体吸收器 2.2.106.37
acid gas removal 脱除酸性气体 2.2.101.12
acidification 酸化 3.3.18.19
acid in 酸进 6.1.7.6
acid pump 耐酸泵 16.1.2.3.28
acid regeneration 酸再生 6.2.9.(d)
acid tube 酸管 5.4.2.17
A.C.input 交流电输入 10.3.1.9
AcOEt,MeOH 乙酸乙酯,甲醇 2.2.41.11
A combination of oxy-combustion and IGCC with CCS (carbon capture and seguestration) 有氧燃烧与带有碳捕集与封存的整体煤气化联合循环的组合工艺 2.2.112
acration (BOD removal) 曝气(生化需氧量去除) 3.1.37.4
acrylonitrile purification columns 丙烯净化塔 2.3.15.8
activated molecular sieve pellets 活性分子筛颗粒 6.3.10.20
activated primary tanks 一次活化槽 3.2.10.(a)
activated sludge 活性污泥 1
activated sludge process 活性污泥工艺 3.

493

activated

2.2
activated-sludge system 活性污泥系统 3.1.8
activated sludge tank 活性污泥池 3.4.2.6
active area 有效区 5.2.6.7
active electrode 激活电极 12.7.2.1
actively digesting sludge 活跃消解的污泥 3.3.18.6
active zeolite catalyst 活性沸石催化剂 14.3.4.(b)
activity coefficient 活度系数 1
actuator spring 执行机构弹簧 22.1.9.5
actuator stem 传动杆 22.1.9.6
adapter 管接头 13.2.1.13
adapter heater 管接头加热器 13.2.1.9
additional step required 需要附加的工序 3.1.30.B.8
additional treatment 附加处理 3.2.1.12
additive 添加剂 2.2.35.15
additive pump 添加剂泵 2.2.35.8
additive tank 添加剂槽 2.2.35.7
adiabatic process 绝热过程 1
adjustable air-blast nozzles 可调式空气送风管 8.2.1.5
adjustable bypass valve 可调旁通阀 16.2.8.8
adjustable dam 可调堰 9.5.15.1
adjustable damper 可调挡板 8.8.6.10
adjustable feed gate 可调加料口 24.1.5.2
adjustable filtrate ports 可调滤液排出口 9.5.5.5
adjustable gate 调节阀 23.5.1.5
adjustable inlet weir 可调进口堰 3.1.4.10
adjustable outlet weir 可调出口堰 3.1.4.6
adjustable pipe for feed flow 进料调节管 12.3.2.9
adjustable-pitch fan 可调角度风机 8.2.1.2
adjustable registers 可调配风器 24.2.4.12
adjustable relief valve 可调减压阀 17.2.14.6
adjustable splitters 可调的分流器 12.6.4.20
adjustable spring 可调式弹簧 19.1.5.2
adjustable support bar 可调支承扁钢 11.1.7.18
adjustable unloader knife 可调式刮刀 9.5.5.13
adjustable weir 可调堰(板) 15.5.3.10
adjustable weir ring 溢流堰调节环 10.3.4.7
adjust handle 调节手轮 15.7.6.17
adjust H_2/CO ratio 调节 H_2/CO 比值 2.2.36.5
adjusting bolt 调节螺栓 5.2.4.16

adjusting ring 调节环 22.1.8.5
adjusting screw 调节螺杆 16.2.5.17
adjusting screw for pan position 锅体水平调节螺杆 15.7.4.8
adjusting seat 调节座 16.2.5.18
adjusting the clearance between the knives 调整刀片间隙 2.3.14.74
adjusting wheel 调整轮 11.1.6.6
adjusting worm 调节蜗杆 16.2.5.15
adjusting wormgear 调节蜗轮 16.2.5.16
adjustment wheel 调节轮 11.4.3.6
adsorbent carrier gas 吸附剂载气 6.3.7.8
adsorbent flow 吸附剂流 6.3.7.1
adsorber 吸附器 3.1.17.2
adsorbing 吸附 6.3.3.4
adsorption 吸附 1
adsorption column 吸附柱 6.2.6.1
adsorption equipment 吸附设备 6
adsorption section 吸附段 6.3.7.3
adsorption tank 吸附罐 6.2.5.2
AER(aerobic) 好氧,需氧 2.2.2.2
aerated lagoon 曝气池 3.1.1.6
aerated recycle water 经曝气循环水 3.1.14.17
aerated shaft 曝气井 3.1.32.19
aeration 曝气 3.1.2.10;3.1.32.1;3.1.44.6
aeration basin 曝气池 2.2.2.8
aeration pad 充气缓冲器 23.1.8.15
aeration recycle 曝气循环 3.1.8.6
aeration tank 曝气罐 3.1.14.16
曝气池 3.1.43.3;3.1.34.8
aeration tank cover 曝气池盖 3.1.45.4
aerobic 需氧的 3.1.47.3
aerobically digested sludge 有氧消化过的污泥 3.3.28.1
aerobic bacteria 好氧细菌 1
aerobic biological reactors 需氧的生物反应器 3.1.20.3
aerobic zone 需氧区 3.1.33.11;3.1.35.6
aerodyne collector 空气动力捕集器 10.4.3.(d)
aerodyne tube 空气动力管 10.4.3.(c)
affinity 亲和势 1
afterburner chamber 后燃烧室 3.1.22.9
after condenser 后(置)冷凝器 20.1.10.11
after cooler 后置冷却器 17.6.1.7
aggregate 团粒 23.3.4.3
堆聚 3.3.11.7
aggregative fluidized bed 聚式流化床 14.1.3.13
agitated 搅拌式 8.1.1.①

agitated batch 间歇式带搅拌 8.1.1.②
agitated film evaporator 搅拌膜式蒸发器 7.1.2
agitated flash dryer 带搅拌的闪急干燥器 8.5.7
agitated kettle 搅拌釜式 4.3.1.(b)
Agitated (mixed) in-vessel composting bioreactors: (a) circular; (b) rectangular 容器内搅拌(混合)复合肥生物反应器: (a) 圆形; (b) 矩形 3.3.30
agitated reactor 搅拌式反应釜[器] 14.4.1
agitated thin-film evaporator 搅拌薄膜蒸发装置 7.1.1
agitator 搅拌器 2.3.10.47
agitator arc 弧形搅拌器 9.3.3.34
agitator blade 搅拌耙 25.1.13.11
agitator crank 搅拌器曲柄 9.3.3.18
agitator rakes 搅拌器搂耙 9.3.3.35
agriculture 农业 25.1.1.15
air 空气 2.2.33.5; 2.2.45.6; 2.2.64.14; 2.2.67.1; 2.2.68.2; 2.2.69.3; 2.2.71.13; 2.2.72.1; 2.2.80.1; 2.2.81.1; 2.2.110.12; 2.2.111.3; 2.2.113.5; 2.2.114.6; 2.2.115.1; 2.2.116.1; 2.2.117.6; 2.2.118.1; 2.2.119.1; 2.3.15.11; 2.3.17.3; 3.1.28.4; 3.1.32.9; 3.1.47.6; 3.3.23.12; 3.3.25.1; 3.3.29.5; 3.3.29.14; 3.3.30.9; 3.4.6.3; 3.4.6.12
air actuated water dump valve 气动排水阀 15.5.2.4
air and liquid mixed tube 气液混合管 16.5.1.5
air and material 空气与物料 23.1.13.9
air and material discharge 空气与物料排出口 11.2.8.9
air and material out 空气和物料出口 11.2.10.6
air and material outlet 空气和物料出口 11.2.11.6
air ballast inlet hole 气镇入口孔 18.1.4.9
air-blast circuit breaker 空气吹弧断路器 25.4.9
air blow 空气反吹 9.4.1.2
air blower 鼓风机 12.1.5.1; 2.3.16.8
air blow for cake discharge 吹空气卸饼 9.3.10.7
air-blown firebox 鼓风式燃烧器 2.2.88.16
air booster compressor 空气增压压缩机 2.3.4.7
airborne material 气载原料 12.3.3.7
airborne product 气载产品 3.3.3
air breather 透气管 13.1.2.27

air bubble 空气泡 22.2.3.17
air circulation 空气循环 12.3.1.5
air classification 空气分类 3.3.16.13
air classifier 空气分级器 12.3.4
air cock 放气阀 16.2.3.33
　　 排气旋塞 16.4.12.16
air compressor 空气压缩机 2.3.3.5
air conditioning equipment 空调设备 20.1.10.15
air connection 空气接管 9.3.1.4
air control 空气控制 24.2.6
air-cooled condenser 空冷冷凝器 3.4.7.17
air-cooled heat exchanger 空冷换热器 4.5.1
Air-cooled heat exchangers 空冷换热器 4.5
air-cooled tubes in bed 床内空气冷却管 25.5.2.(b)
air cooler 空气冷却器 2.2.9.21; 2.2.44.10
air-cooler recirculator 空冷器循环系统 4.5.2
air cylinder 气动罐
air deflectors 空气导流装置 24.2.4.14
　　 导流板 8.9.6.11
air discharge 排气 17.2.6.7
air dispersion ring 空气分散环 11.2.10.7
air distribution 空气分布 4.3.4.25
　　 空气分配装置 3.3.29.18
air distribution grid 空气分布器炉栅 14.2.4
air distributor 空气分布器 24.2.10.1
air-distributor grating 空气分布器栅板 8.3.2.8
air-droplet mixture 空气一液滴混合物 3.4.4.15
air ejector 空气喷射器 7.2.5.24
air-exhaust 排气口 8.3.2.6
air-exhaust duct 空气出口管 8.2.1.9
air filter 空气过滤器 2.2.53.5; 8.4.6.3
air-flotation palletizer 气浮码垛器 23.4.3.3
air flow 空气流 2.3.16.2; 8.6.3.5
air flow through louvers and material 空气流穿过百叶窗和物料 8.8.9.8
air from treatment facility 来自处理设施的空气 3.3.20.11
air heater 空气加热器 8.6.2.1
air hood collar seal 空气罩密封圈 4.2.4.6
air housing 空气室 3.3.26.14
air impingement 空气冲击 8.7.2
air in 空气进口 23.1.2.2
air injector 空气注入器 10.4.1.3
air inlet 空气入[进]口 11.2.11.3
　　 进气口 23.1.6.2

air-inlet duct 空气进口管 8.2.1.8
air inlet port 进风口 8.5.1.8
air inlet to compressor 压缩机空气进口 17.2.9.1
air inlet valve 进气阀 23.1.8.7
air input system 空气输入系统 3.3.29.15
air intake 空气进口 17.2.6.1
  进风口 4.6.2.8
air intake filter 空气进口过滤器 17.6.1.1
air intake port 进风口 17.5.1.10
air-into-material 空气进入物料 23.1.8.①
air jet 空气喷嘴 12.7.5.8
air jets impinging 空气喷射冲击 8.3.2
air lance 空气喷枪 3.3.26.9
air lift 气力提升器 14.2.1.4
  空气提升机 23.1.7.12
air-lift pipe 气提管 14.1.1.11
air-lift pneumatic-conveyor dryer 空气提升气流输送干燥器 8.4.3
air-lift pump 空气提升泵 16.1.2.2.51
airlift pump for pad recirculation 团块再循环汽提泵 3.2.13.10
airlift turbo drying action 空气提升涡流干燥作用区 8.4.1.25
air line installation 空气管路装置 17.6.2
air lines 空气管道 3.1.32.24
airlock 空气闭锁器 8.9.2.15
  气闸 10.4.4.5;11.6.4.5
  阻气门 11.6.9.4
airlock discharger 旋转卸料阀 23.1.6.7
airlock feeder 旋转阀进料器 23.1.6.1
air manifold 空气歧管 3.3.30.5
air-mix chamber 空气混合室 8.4.1.31
air-mixing 与空气混合 23.1.8.②
air nozzle 空气喷嘴 23.1.8.20
air,O₂,steam 空气,氧气,蒸汽 2.2.24.1
air out 空气出口[排放] 23.1.2.17
  气体出口 12.1.5.8
air passage 空气通道 17.2.5.11
air pipe 空气管 16.5.1.3
air port 空气入口 24.1.4.1
air preheater 空气预热器 2.3.16.7;4.2.4
air pulser 空气脉冲(发生)器 6.1.16.6
air pump 空气泵 16.1.2.2.50
air receiver 贮气罐 2.2.53.3;17.6.1.13
air register 空气调节器 24.2.9.3
air relief 空气释放 6.2.1.4
air removal 去除空气 3.3.29.19
air removal system 空气去除系统 3.3.29.9
air(secondary)ejectors 空气(二级)喷射泵 18.3.6.3

air separation 空气分离 2.2.67.2;2.2.78.4
air separation plant 空分装置 2.3.1.27
air separation unit 空分装置 2.2.33.6;2.2.36.1;2.2.115.3;2.2.117.5;2.2.118.2
air silencer 空气消声器 17.6.1.2
air sink removal 压缩空气排出下沉产品 12.5.4.(b)
air slide 气动滑阀 11.6.4.12
air slide 12in 气动滑阀12in 11.6.4.10
air source 气源 3.1.41.11
air-supply distribution duct 供入空气分布导管 4.3.4.16
air take-off 脱空气 20.1.10.9
air tight equipment 气体密封装置 4.2.4.16
airtight pump 气密泵 3.4.7.9
airtight valve 气密阀 23.1.2.33
air to recycle 空气去循环 2.2.106.22
air valve 排气阀 16.4.19.9
air vane 通气格子板 12.3.4.14
air vent 放气口 15.5.5.10
  放空 20.1.10.12
air vent 排气管 5.4.2.10
air vent nozzle 排气管口 14.4.1.10
air vent valve 放气阀 22.2.3.5
air way slits in rotor 转子(叶片间)狭缝空气通道 12.3.2.6
aligning belt 调准皮带 12.7.4.5
aligning disc 调准盘 12.7.4.3
aliphatics 脂族化合物 6.1.3.12
alkali 碱 2.2.71.16;6.2.9.9
alkali-cellulose crumbs 碱性纤维素碎片 2.3.7.7
alkali-cellulosein 碱纤维素 2.3.7.9
alkylate product 烷基化产品 6.1.5.16
alkylation 烷基化 2.1.1.72;2.2.75.10
Aller rock salt 阿勒岩盐 25.2.4.7
alloy basket 合金框 24.1.7.6
alloy fan 合金风扇 24.1.7.3
alloy grid plate 合金栅板 24.1.7.5
alloy 800 production plate 800号合金专用板 4.4.2.12
alternately grounded and charged collector plates 交替接地和放电的收尘板 10.3.1.28
alternatives available for exhaust gas deodorization and particulate removal 对于废气除臭和去除颗粒有其他方案可选 3.3.25.19
alternator 交流发电机 2.2.81.14
aluminized mylar bag 涂铝聚酯袋 3.5.17.32
aluminum 铝 19.3.3.(a)

A process

aluminum alloys　铝合金　22.6.2.②
aluminum recovery　铝回收　3.3.16.11
ambient air　环境空气　8.6.2.9
American Petroleum Institute　美国石油学会　3.1.3
amine scrubber　脱胺塔　2.1.9.44
ammonia　氨　2.3.15.10
ammonia compression refrigerating machine　氨压缩冷冻机　17.1.2.2
ammonia compressor　氨压缩机　17.1.2.1
ammonia nitrogen fertillzers　氨,氮肥　2.2.3.18
ammonia or electricity or heat　氨或电或热　2.2.78.14
ammonia plant　合成氨厂　2.3.2
　合成氨装置　2.3.3
ammonia refrigeration　氨冷冻　2.2.38.14
ammonia scrubber　氨气洗涤塔　2.3.14.23
ammonia stripper　合成气溶液　2.3.6.13
ammonia synthesis　氨合成　2.2.38.12;2.3.1.14
　合成氨　2.2.33.13;2.3.2.15
ammonia synthesis loop, showing a quench reactor　氨合成回路,显示了急冷反应器　2.2.37
ammonia synthesis process　合成氨工艺流程　2.2.38
ammonia-water absorption cycle　氨-水吸收循环　20.1.9
ammonium nitrate melt　熔融硝酸铵　13.1.1.1
ammonium nitrate pills　硝酸铵颗粒　13.1.1.10
ammonium sulphate　硫酸铵　2.3.14.36
ammonium sulphate solution　硫酸铵溶液　2.3.8.26
ANA（anaerobic）　厌氧　3.2.2.10
anaerobic　厌氧的　3.1.47.2
anaerobic bacteria　厌氧细菌　1
an aerobically digested sludge　有氧消化过的污泥　3.3.28.7
Anaerobic digestion of the primary sludge and direct reuse of secondary sludge agriculture　一次污泥厌氧消化和二次污泥在农业中的直接再利用　3.1.26.9
anaerobic sludge digestion tank　厌氧污泥消化池　3.1.26.(a).4;3.1.26.(b).5
anaerobic zone　厌氧区　3.1.33.9
anchor bolt　地脚螺栓　5.4.1.22
anchor bolt circle　地脚螺栓中心圆　5.4.1.15
anergy　烌,无效能　1
angle bar tray　角钢塔盘　5.2.10.18

angle lead graduated adjustment　导角逐步调节器　11.3.1.8
angle lift check valve　角接式升降止逆阀　22.1.(g)
angle of surcharge　堆积角　23.2.1.15
angle ring　角钢圈　5.2.7.3
angle seal　角封　8.8.1.27
angle valve　角接阀　22.1.1.(c)
angular rotation　转角　22.5.2.4
anion　阴离子型　6.2.9.2
anion exchanger　阴离子交换器　6.2.7.8
annealing lehr　退火窑　2.3.9.18
annular air inlets　空气环状入口　11.2.9.4
annular bed　环形床　6.3.8.8
annular bed for liquid separation　环形液体分离床　6.3.8.8
annular cell　环形气室　17.3.3.8
annular collection launder　环形收集溜槽　15.6.3.25
annular flow　环形下流　23.3.4.5
annular solids bed　环状的固体床层　4.3.3.7
annuli open　环隙开口　14.1.8.(a)
annulus　环形空间　13.4.1.20
annulus　环形槽　9.3.2.27
anhydrous ammonia　无水氨　2.3.1.15
anode coke　氧化焦炭　2.2.91.11
An overview on water treatment process　水处理过程概述　2.4.5
anoxic/aerobic digestion　缺氧/好氧消化　3.2.5
anoxic cells　缺氧池　3.1.34.3
anoxic zone　缺氧区　3.1.33.10
antifriction pillow block　耐磨轴台　8.8.5.22
anti-radiation screen　抗辐射屏障　25.2.5.2
anti-rotation stud [bar]　防转短轴[棒]　19.1.5.14
antispin vane　消旋叶片　10.5.4.7
anvil block　锤砧块　11.1.7.20
anvil plate　砧面垫板　11.2.9.8
ANX（anoxic）　缺氧　3.2.2.11
aperture　小孔　3.5.6.3
1/2 apex angle　半顶角　21.1.1.23
API separator　美国石油学会标准分离器　2.1.9.6
apparent density　视密度;表观密度　1
applicating rolls　从动滚　8.7.2.3
applied　敷设的　21.1.1.11
applying heat　加热　20.1.2.10
A process flow diagram to manufacture vinyl chloride monomer　氯乙烯单体生产工艺流程　2.3.18

A process flow diagram to produce ethylene oxide 环氧乙烷生产工艺 2.3.19
apron feeder 板式加料器 12.8.2.3
aqueous 水相 6.1.12.1
aqueous effluent 水相流出物 6.1.12.13
aqueous feed 水相料液 6.1.2.3
aqueous formaldehyde 含水甲醛 14.3.1.1
aqueous layer 水层 2.2.60.3
aqueous raffinate 水相萃残液 6.1.2.2
aqueous strip solution 水相萃取液 6.1.2.5
arbitrary operation 非定时操作 23.3.4.1
arcing horn 消弧角,角形避雷器 25.4.9.11
area of friction 摩擦面 8.8.5.17
argon arc welding 氩弧焊接 19.4.1.9
ari compartment 空气室 23.1.2.29
arithmetic mean temperature difference 算术平均温差 1
arm (悬)臂 9.2.5.22
　　支杆 9.5.4.4
arm base pin 悬臂支承销 13.3.5.20
arm holding pin 耙臂固定销 3.3.26.8
armoured pump 铠装泵 16.1.2.2.34
arm pin 支杆销 9.5.4.5
aromatics 芳香族 6.1.3.13
arranged(type) packing 规整填料 5.3.4.44
arrangement 布置 12.6.3
arrgt(arrangement) (1)排列;安排;整理
artificial intelligence 人工智能 1
Asahi countercurrent ion-exchange process Asahi 逆流离子交换过程 6.2.5
asbestos cement 石棉水泥 22.6.2.①
asbestos cord 石棉绳 5.2.4.11
asbestos packing 石棉填料 3.3.2.3.16
asbestos rubber packing 石棉橡胶盘根(填料) 19.2.3.15
asbestos washer 石棉垫圈 18.2.1.10
ash 煤渣 2.3.1.18
　　灰(分) 2.2.17.3;2.3.1.21
　　炉灰 2.2.118.9;24.3.1.8
　　煤灰 2.2.115.7;2.2.117.10
ash and residue 灰渣 2.2.94.16
ash bin 灰仓 3.3.3.3
ash conditioner 灰分调理器 3.3.3.2
ash conveyor fan 粉尘输送风机 10.4.1.19
ash depressurization 灰分卸压 2.2.24.5
ash dewatering 灰分脱水 2.3.1.31
ash discharge 灰分排出 3.3.9.2
ash disposal 排灰 2.2.89.16
ash drag 炉灰拖运机 24.1.5.9
ash drums 灰筒 3.3.8.14
ash heat exchanger 炉灰热交换器 2.2.88.3

ash lock 闭锁式灰斗 2.2.12.11
　　灰锁斗 2.2.13.6
ash lock hopper 闭锁式煤渣斗 2.2.11.11
ash pickup air 集尘空气 10.3.5.3
ash pit 灰坑 24.1.4.2
ash removal 灰分去除 2.2.19.9
ash residue 灰渣 2.2.97.14
ash screw 炉灰螺旋输送器 24.3.1.13
ash separator 炉灰分离器 2.2.88.9
ASME code 美国机械工程师协会规范 21.1.2.20
asphalt 沥青 2.1.1.86
asphalted paper 沥青纸 25.4.8.6
assembly 装配图 8.8.4.(c)
　　装配体 8.8.5.29
ASU 空分装置 2.2.119.2
atm 大气 2.3.17.10
atmosphere 大气 3.3.25.14;19.4.1.2
　　放空 2.2.5.7
atmosphere(atmos) 常压 8.1.1.7
atmospheric air 空气 25.5.3.20
atmospheric and vacuum distillation 常减压蒸馏 2.1.1.4
atmospheric cooler 大气冷却器,空气冷却器 4.2.5.8
atmospheric-fluidized-bed-combustor 常压流化床燃烧锅炉 25.5.1
atmospheric industrial gas-burner 常压工业气体燃烧器 24.2.6
atmospheric nitrogen 常压氮气 14.3.3.(a)
atmospheric storage tanks 常压贮罐 21.2.1
atmospheric tower 常压塔 2.1.3.1
atomic power plant 原子能动力站 25.2.2
atomized 雾化 10.5.1.10
atomizer 粉碎机 11.1.8
atomizer 雾化器 8.6.2.4
atomizing air 雾化空气 3.3.6.3
atomizing device 雾化器 13.1.1.2
atomizing nozzle 雾化喷嘴 13.4.1.23
atmospheric methanol distillation column 常压甲醇蒸馏塔 2.2.43.14
atmospheric tower 自然通风(凉水)塔 4.6.1
attachment 附件 21
attachment of jacket 夹套附件 21.1.1.18
attracting grid 吸引栅 12.7.2.11
attracting plate 吸引板 12.7.2.9
attriters 磨料 14.3.4.32
attrition mill 碾磨机 11.3.4
attrition rate 磨耗率 14.3.3.(c)
A typical Integrated Gasification Combined Cy-

cle plant 典型的整体煤气化联合循环工厂 2.2.110
augers 螺旋钻 3.3.30.2
augmentor 扩压器 18.2.5.1
austenitic stainless steels 奥氏体不锈钢 22.6.2.⑯
autoclave body 釜体 14.4.3.18
autoclave legs 釜支腿 14.4.3.24
auto launder drain 溜槽自动放液管 15.6.3.26
automatically adjustable cone 自动调节锥体 10.5.10.4
automatic bag hanging 自动装袋吊挂 23.4.3
automatic bag placer 自动袋放置器 23.4.3.13
automatic batch centrifugal 歇式自动离心机 9.5.8
automatic control 自动控制 15.5.5.4
automatic cutter 自动截切机 2.3.9.19
automatic dampers 自动调节风门 3.3.24.11
automatic device proportions samples to flow 液体流动比例抽样自动装置 3.5.3
automatic disk positioner 自动圆盘定位器 10.5.9.5
automatic drain valve 自动放泄阀 17.6.1.10
automatic magnetic discharge 磁性体自动卸料 12.6.10.1
automatic palletizer 自动码垛机 23.4.2.21
automatic palletizing 自动码垛 23.4.2
automatic sludge discharge pipe 自动排污管 15.6.6.4
automatic temperature controller 自动温度控制器 15.7.3.14
automatic valve 自动阀 9.3.3.15
automatic valve flange 自动阀法兰 9.3.3.45
automatic valve ring 自动阀环 9.3.5.3
automatic voltage regulator 自动电压调节器 25.4.4
auxiliary blade 辅助叶片 12.3.5.17
auxiliary blade wear 辅助叶片防磨板 12.3.5.18
auxiliary condenser 辅助冷凝器 18.3.5.4
auxiliary contact 辅助接点 25.4.9.7
auxiliary convection section 辅助对流段 25.1.2.4
auxiliary fuel gas 辅助燃料气 3.3.10.7
auxiliary motor weight 电机附加重块 11.3.2.1
auxiliary opening water 辅助常开水管 9.5.13.15
availability 㶲;有效能 1
axial 轴流式 10.1.1.(d)
axial clearance 轴向间隙 16.4.6.24
axial compression 轴向压缩 22.5.2.2
axial extension 轴向伸长 22.5.2.3
axial flow 轴流(式) 16.1.1.33
axial flow impeller 轴流式叶轮 16.4.5.(d)
axial flow pump 轴流泵 16.1.2.2.24
axial flow low head circulating pump 轴流式低压头循环水泵 12.5.5.21
azeotrope 共沸物;恒沸物 1
azeotropic distillation 共沸蒸馏,恒沸蒸馏 1
azeotropic distillation column 共沸蒸馏塔

B

babbitt 巴氏合金 19.3.3.(a)
back heat zone 后加热区 13.2.1.17
backing connection 前级(泵)接头 18.2.1.16
backing flange 背衬法兰 22.4.4.2
backing line condenser 前级管线冷凝器 18.2.4.8
backing pump 前级真空泵 18.1.2.9
backing stage discharge valve 前级排气阀 18.1.2.11
backing strip 焊接垫板 21.1.1.46
backing tube 通前级管 18.2.1.15
back-mix 返回混合物 8.1.1.11
back mixing 返混 1
back pressure 背压力,反压 16.2.1.6
backpressure valve 止回阀 3.1.14.18
backstreaming 回流 18.2.4.4
back-to-back impeller pump 背靠背叶轮泵 16.1.2.2.26
back to gasification 回造气 2.2.50.1
back to the absorption tower 回吸收塔 2.2.50.2
backward-curved-blade 后弯叶片 17.5.1.8.3
backward-curved vane 后弯叶片 16.4.5.3.1
backwash 反洗 3.1.10.12
反洗液 3.1.19.10
backwash air 反洗空气 3.1.38.8
backwash period 反洗期 6.2.9.(b)
backwash pump 反洗泵 3.1.39.8
backwash supply 反冲洗供给 3.1.40.9
backwash water 反洗水 3.1.38.9;3.1.41.8
backwash water to head of plant 反洗水去厂

499

前区 3.1.39.4
BAF/biocarbon system 生物曝气过滤/生物炭系统 3.1.41.5
baffle 挡板 10.5.1.15
折流板 12.3.1.4
隔板 21.1.1.51
baffled jacket 带挡板的夹套 4.3.3.20
baffled mixing vessel 内装挡板的混合容器 6.1.5
baffle plate 挡板[片] 10.2.1.30
baffle ring 挡圈 11.2.10.3
baffle ring centrifuge 环形挡板离心机 9.5.20
baffle ring insert 嵌入环形挡板 9.5.20.4
baffle tower 挡板塔 5.1.2.11
bag (料)袋 9.5.2.18
bag collector 布袋捕集器 8.5.6.3
bag filter 袋滤器 10.2
袋式过滤器 2.3.16.14；2.3.17.8
bagging 装袋 13.1.1.10
bag house 集尘室 10.4.1.16
bags 滤袋 10.4.4.6
bag support 布袋支撑 10.2.1.25
bag turner 袋旋转器 23.4.3.10
bag type 袋式 10.2.1.(c)
balance bush 平衡盘套筒 16.4.19.13
平衡衬套 16.4.28.6
balance cylinder 平衡分 20.2.1.25
balanced 平衡式(型) 23.3.2.7
balanced internal mechanical seal 平衡型内装式机械密封 19.1.4
balance disc[disk] 平衡盘 16.4.27.31
balanced opposed reciprocating compressor 对称平衡式压缩机 17.2.1.10
balance for measuring basis weight 测重秤 2.3.11.9
balance hole 平衡孔 16.4.18.9
balance pipe 平衡管 16.4.19.34
balance piston 平衡活塞 16.4.28.32
balance ring 平衡环 16.4.18.11
balance runner 平衡盘 20.2.1.26
balance seal 平衡盘密封 16.4.19.12
balance sleeve 平衡套 16.4.2.1.10
balance weight 平衡锤 9.2.1.25
balancing chamber 平衡室 16.4.6.23
balancing disk 平衡盘 16.4.6.18
balancing disk head 平衡盘上部 16.4.6.20
balancing hole 平衡孔 16.4.15.8
balancing piston 平衡活塞 16.4.2.25
bale of polyamide staple 聚酰胺纤维包 2.3.8.62

baling press (压力)打包机 2.3.7.33
ball 球体 22.1.1.3
ball bearing 滚珠轴承 15.7.6.15
ball bearing case 球轴承箱 11.1.5.13
ball case 轴承箱 11.1.8.27
ball check 球形止逆阀 16.5.6.2
ball check valve 球形止逆阀 3.5.16.30
ball cleaner(s) 圆形清理球 12.2.3.11
ball cover 轴承上盖 13.3.5.2
ball elevator 小球提升机 2.2.5.9
ball heater (小)球加热器 2.2.5.14
balling drum 成球转鼓 13.3.1.5
ball mill 球磨(机) 11.6.2.17
ball mill liners 球磨机衬里 11.2.2
Ball Norton separator 交变极性磁力分离器 12.6.7
ball packing 球形填料 5.2.9.4
ball seat 阀座 22.1.7.7
ball valve 球阀 22.1.7
Banbury mixer 生胶混炼机 15.7.1
band 带式 8.1.1.⑦
band type drying machine 带式干燥机 8.3.4
bar 巴 2.2.66.7
bar hammer 棒锤 11.2.6.7
Bardenpho process (Source US EPA) Bardenpho 废水中除氮磷工艺(取自美国环境保护局) 3.1.33
barley elevator 大麦提升机 2.3.12.33
barley hopper 大麦料斗 2.3.12.2
barley reception 大麦的接收 2.3.12.41
barley silo 大麦筒仓 2.3.12.31
barometric condenser 大气冷凝器 18.3.4 (b)
barometric leg 大气腿 18.3.5.14
barometric seal 水封 9.3.1.10
barrel 机筒 13.2.1.7
筒体 15.7.2.3
barrel casing 筒形泵壳 16.4.27
barrel cooler 筒体冷却器 15.7.2.1
barrel pump 筒形泵 16.1.2.2.31
barrel store 酒桶仓库 2.3.13.16
barren liquor 废液 6.2.6.2
barrier 阻液 19.3.3.10
bars 杆 21.1.1.21
bar screens 铁栅筛 3.1.25.2；3.1.27.2
bars for receiving basket 受料筐板条 11.1.7.29
base 底座 11.3.1.7
机座 12.2.7.13
基础 21.3.1.1

base crankcase 机座曲轴箱 16.2.7.11
base for speed reducer 减速机底座 14.4.1.27
base line of elevator 提升机基线 14.3.3.8
base of stone bolt 地脚螺栓座 5.3.4.35
base plate 底板 16.4.27.1
　支座板 13.3.5.19
　基板 3.5.5.2
base ring 基础环板 21.3.1.2
base support 机座 13.3.3.3
BASF process of acetic acid synthesis from methanol carbonylation BASF 公司甲醇羰基化制乙酸工艺流程 2.2.54
basic Brayton or Joule cycle 基本 Brayton 或 Joule 循环 25.3.2.(a)
basic compressor 基本型压缩机 17.5.3.5
basic equipment arrangement 主要设备与流程 8.5.1.(a)
　主要设备配置图 23.1.1
basic material 基本原料 2.3.7.1
basic refrigeration cycle 基本制冷循环 20.3
basic refrigeration system 基本制冷系统 20.1
basic ring 基础环 5.3.4.36
basic SNG process 基本合成天然气工艺 2.2.103
basic system 基本系统 3.1.2.30
basket 转鼓 9.5.10.4
basket 悬筐 9.5.6.4
　催化剂筐 2.2.39.4
basket bottom 转鼓底 9.5.4.48
basket hub 转鼓毂 9.5.17.37
basket lip 转鼓唇缘 9.5.17.24
basket skirt 转鼓裙 9.5.17.25
batch 分批 8.1.1.2
batch charge 分批投料 13.4.1.2
batch distillation 间歇蒸馏,分批蒸馏 1
batch feeder 原料进给装置 2.3.9.13
batch fluidezed bed granulator 间歇式流化床造粒机 13.4.1(a)
batch hydrogenation of fats 油脂间歇加氢 14.3.1.(d)
batching valve 定量调节阀 23.1.8.13
batch kettle 间歇操作釜 5.3.3.2
batch manual 手动操作间歇式 9.5.2.4
batch mixer-settler 间歇混合沉降器 6.3.3
batch muller 间歇式滚轮混合器 12.1.3.(g)
batch solidification 批料凝固 13.5.2
batch still 间歇蒸馏釜 3.1.6.18
batch-still system 间歇蒸馏系统 3.1.6

batch through-circulation 间歇穿流循环 8.1.1.⑤
batch treatment plant 间歇处理厂 6.3.2.13
batch vibrating ball mill 间歇式振动磨 11.3.3
bath feed separator 选矿池进料分离器 15.5.3
battery of coke ovens 炼焦炉组 2.3.14.8
bayonet fitting 卡口灯座 25.4.1.65
bayonet-tube exchanger 内插管式换热器,刺刀管换热器 4.2.5.9
BDO 丁二醇 2.2.1.10
bead mills 珠磨机 11.3
beam 吊车梁 24.1.1.32
beam-type"gas-injection"support plate 横梁式"气体喷射"支承板 5.3.4.24
bearing assembly A 轴承部件 A 16.4.18.21
bearing assembly B 轴承部件 B 16.4.18.19
bearing box 轴承箱[座] 16.4.22.11
bearing bushes in halves 对开轴瓦 11.1.7.22
bearing bushing 轴承衬套 16.4.24.4
bearing cap 轴承盖 16.4.25.5
bearing cap set screw 轴承盖止动螺钉 16.4.26.19
bearing collar 轴承定位环 13.3.5.9
　轴承(压)盖 16.4.22.10
bearing distance piece 轴承间隔块 11.1.8.23
bearing end cover 轴承(箱)端盖 16.4.19.6
bearing for screw conveyer 螺旋输送器轴承 9.2.1.5
bearing frame 轴承架 8.8.12.12
bearing holder 轴承座 16.4.12.1
bearing housing 轴承箱[座] 16.4.19.27
bearing housing with oil pan 带油槽的轴承箱 9.5.10.15
bearing labyrinth 轴承迷宫密封 20.2.1.13
bearing lock nut 轴承锁紧螺母 16.4.9.4
bearing nut 轴承(锁紧)螺母 13.3.5.6
bearing oiling rings 轴承甩油环 11.1.7.21
bearing outboard 轴承外侧 16.4.19.7
bearing pedestal 轴承座[架] 16.3.6.5
bearing seat 轴承座 16.4.29.18
bearing sheet 轴承垫片 16.4.19.4
bearing sleeve 轴承座套 16.4.23.7
bearing spacer 轴承定距环 16.4.19.33
bearing stand 轴承架 16.4.4.18
bearing support 轴承座 25.3.1.5
bearing(upper) 上轴承 12.3.5.1
beater connecting rod 敲打器连杆 10.2.1.12
bed 底座 9.5.4.3
bed A 床 A 14.2.2.6

501

bed B　床B　14.2.2.7
bed cooled by excess air　床内用过剩空气冷却　25.5.2.(c)
bed depth　床(层)深(度)　3.1.2.7
bed material drain　床层排料口　14.2.2.5
bed of material　物料床层　4.3.3.26
bed of solids　固体床层　8.9.6.5
bed plate　分布板　3.1.16.11
beer barrel　啤酒桶　2.3.13.17
beer bottle　瓶装啤酒　2.3.13.26
beer can　罐装啤酒　2.3.13.25
beer filter　啤酒滤器　2.3.13.15
beer glass　啤酒杯　2.3.13.30
bell cap　钟帽式除沫器　7.1.8.8
bell cap redistributor　钟罩型再分布器　5.3.4.16
bellmouth nozzle　喇叭形进气口　23.1.7.14
bellows type　波纹管式　22.5.1.(g)
bell push　电铃按钮　25.4.1.2
belt　皮带　23.2.1.14
belt conveyor　带式输送机　23.3.9.1
belt conveyor discharge arrangements　带式输送机卸料方式　23.2.3
belt conveyor system　带式输送系统　23.2
belt drive motor　皮带的驱动电机　12.7.4.4
belt dryer　带式干燥器　8.3
belt filter　带式过滤机　9.4.8
belt filter press　带式压榨机　9.6.2
belt pulley　皮带轮　7.1.2.27
belt tripper　带式倾料器　23.2.4.3
Benfield process　本菲尔德脱碳工艺　2.2.116.13
benzene　苯　2.1.1.51
benzene chlorination　苯氯化　2.3.8.8
benzene scrubber　苯洗涤塔　2.3.14.24
Berks ring dryer　Berks环形干燥器　8.4.5.(c)
Bernoulli equation　伯努利方程　1
bevel gear　伞齿轮　13.3.5.14
bevel jet　伞形喷嘴　7.2.3.1
big end bearing　大头轴承盖　17.2.15.20
bin　料斗　13.2.4.15
　　料仓　23.1.8.24
bin activator　料仓松动器　23.3.4.7
bin activator　料仓抖动器　12.1.5.12
binding　包圈　9.1.3.8
bin level　料面高度　23.3.4.2
bin stream-deflector　料仓弧形导流器　23.3.4.7
biocatalytic reaction　生物催化反应　1
biocell　生物电池　1

bio-cell lift station　生物电池提升工段　3.1.44.2
biochemical conversion　生物化学转化　3.3.1.5
biochemical engineering　生化工程　1
biochemical operation　生化操作　3.2.1.8
biochemical oxygen demand　生化需氧量　1
biochemical separation　生化分离　1
biochemical wastewater treatment system　生物化学废水处理系统　3.2
biodegradation　生物降解　3.1.32.3
bioengineering　生物工程　1
biofilm　生物膜;菌膜　3.2.11.18
biofiltration(BFs)　生物过滤(BFs)　3.1.25.8
biogas　生物气　3.2.6.11
biogas to gasometer　沼气至贮气柜　3.1.26(a).10;3.1.26(b).6
biological agent　生物制剂　1
biological film　生物膜　3.1.47.1
biological-growth inhibitors　抗生物生长材料　3.1.24.2
biomass　生物质　2.2.3.7;2.2.31.1;3.1.42.9;3.2.1.9;3.3.23.1
biomass fuels　生物质燃料　25.1.1.3
biomass harvesting system　生物质收集系统　3.1.42.7
bioparticle　生物粒子　3.2.11.17
biophysical treatment uses activated carbon　活性炭生物物理处理　3.1.18
bioreactor　生物反应器　3.1.28.5;3.1.30.A.4;3.1.30.B.4;3.1.31.4;3.2.10.4
biotechnology　生物技术　1
biotransformation　生物转化　1
Bird-Prayon tilting-pan filter　翻斗真空过滤机　9.4.1
bitumen　沥青　2.1.9.21
black liquor filter　黑液过滤器　2.3.10.28
black liquor storage tank　黑液贮槽　2.3.10.29
black particle　黑粒子　12.1.4.1
blade　叶片　17.5.1.8
blade　旋片　18.1.2.2
blade coating machine　刮刀涂布机　2.3.11.29~35
blade retainer　叶片安装底架　12.3.5.16
blades　刮板　15.6.1.6
　　刮片(刀)　15.6.3.9
　　叶片　5.5.1.21
Blake jaw crusher　Blake颚式破碎机　11.1.1
blank cap　盖帽　14.4.3.2

blanket 过滤层 15.6.4.14
blast gate 风门 10.1.2.13
  排气门 10.4.4.3
bleaching 漂白 2.3.7.21
bleed 排放 10.5.18.12
bleed off valve 泄压阀 3.1.19.5
blend 调和 2.1.1.85
blended gasoline 调和汽油 2.1.1.60
blended sludge 搅拌过的污泥 3.3.25.18
blender 搅拌器 3.3.25.6;23.1.2.24
blending 共混 2.2.75.12
  掺和 3.2.1.18
blending bin 混合料仓 11.6.3.3
blending tank 混合槽 3.1.8.2
blending & thickening 掺和 3.2.1.18
blind cover 盲盖 17.2.15.3
blind flange 盲板法兰 16.4.27.12
blind flange gasket 盲板法兰垫片 16.4.27.11
block 顶盖座 9.2.1.21
block diagram (方)框图 2.3.3
blow-and-blow process 吹制过程 2.3.9.22
blow case 酸蛋 16.1.1.17
  吹气箱 3.1.17.1
blowdown connection 排污连接件 4.4.2.4
blower 鼓风机 2.3.10.4
  风机 8.3.1.4
blower base 鼓风机底座 8.3.4.9
blowing 吹制 2.3.9.28
blowing iron 吹玻璃用的铁管 2.3.9.39
blowing processes 吹制法 2.3.9.22-37
blow mould 吹制模 2.3.9.26
blow pit 喷放池 2.3.10.13
blow tank 喷放罐 2.3.10.11
  风送槽 23.1.2.(f)
  放空罐 23.1.8.8
blow valve 喷放阀 2.3.11.12
blunt pipe 圆头管 25.2.9.(a)
body 主体 10.1.2.17
  座体 21.3.1.6
  阀体 22.1.2.1
  壳体 22.2.3.20
body cap 阀体盖 22.1.7.4
body rings 阀体密封圈 22.1.2.2
boiler 锅炉 18.2.3.3
boiler ash 炉灰 2.3.1.26
boiler circulating pump 锅炉循环泵 16.1.2.3
boiler feed pump 锅炉给水泵 16.1.2.3.4
boiler feed water 锅炉给水 2.2.25.6;2.2.30.8;2.2.42.5;2.2.43.1;2.2.44.5;2.2.45.7;2.2.46.12;2.2.47.11;2.2.111.17;2.2.116.3;3.1.23.9
boiler-feed water blowdown 锅炉给水排污 2.3.6.30
boiling 沸腾 25.5.1.9
boiling surface 沸腾表面 7.2.1.13
boiling water system 沸水系统 25.2.3
block 顶盖座 9.2.1.21
bolt 螺栓 5.2.4.7
bolted flange 螺栓法兰 21.1.1.28
bolt for casing side liners 壳体侧衬板螺栓 11.1.7.7
bolthole for dismounting 拆卸用螺孔 16.4.15.9
bolt ring 螺栓法兰圈 6.3.9.6
bonnet 阀盖 22.1.2.6
  阀罩 22.1.8.11
  压盖 22.2.3.21
  端盖 4.1.2.(b)
bonnet bolt 阀盖螺栓 22.1.2.16
bonnet cover 阀盖 22.1.10.1
bonnet gasket 阀盖垫片 22.1.2.5
bonnet nut 阀盖螺母 22.1.2.17
bonnet stud bolt 阀盖双头螺栓 22.1.8.8
bonnet stud nut 阀盖螺母 22.1.8.7
booster body 增压泵机体 18.3.1.6
booster ejector 升[增]压喷射器 20.1.10.22
booster ejector with multiple steam nozzles 多蒸汽喷嘴增压式喷射泵 18.3.2
booster pump 增压泵 16.1.2.1.7
borehole pump 深井泵 16.1.2.2.40
bot. (bottom) (1)底(2)底板(商)
both ends 两端 4.3.4.34
bottle-making machine 制瓶机 2.3.9.21
bottle-washing machine 洗瓶机 2.3.13.19
bottle-washing plant 洗瓶车间 2.3.13.18
bottling 装瓶 2.3.13.22
bottom angular down 下倾角出风 17.5.2.2
bottom angular up 下仰角出风 17.5.2.4
bottom ash 炉底灰 2.2.18.12;2.2.24.4
bottom bend 底肘管 8.4.5.8
bottom bracket 下托架 16.4.28.14
bottom case 底泵壳 16.4.28.27
bottom cover 底盖 18.2.1.11
bottom cover securing nut 底盖锁紧螺母 18.2.1.12
bottom crown 下极板 21.2.3.9
bottom discharge manual 底部手工出料 9.5.2.15
bottom discharge valve 底部排料阀 14.4.

503

1.21
bottom drive 底部驱动 9.5.2.12
bottom drive automatic basket, rising knife 底部驱动,升降刮刀自动离心机 9.5.1.(d)
bottom drive automatic basket, rotary knife 底部驱动,旋转刮刀自动离心机 9.5.1.(e)
bottom drive batch basket with bag 底部驱动间歇式带袋离心机 9.5.1.(a)
bottom drive bottom discharge batch basket 底部驱动底部出料间歇式离心机 9.5.1.(c)
bottom drive conical mixer dryer 底部驱动锥形混合干燥器 8.9.5
bottom end cover 下端盖 16.4.28.13
bottom flange 底法兰 22.1.9.24
bottom frame 底部框架 12.2.1.15
bottom-frame discharge 底部框式卸料口 12.2.1.16
bottom head 底封头 9.2.5.29
bottom horizontal 下水平出风 17.5.2.3
bottom product 塔底产品 5.3.1.1
bottoms 塔底残留物 2.1.4.28
bottom scoop 底部勺 9.5.21.4
boundary condition 边界条件 1
bowl 圆锥筒 11.1.3.3
圆筒 16.4.25.10
泵体 16.4.26
外筒 16.4.26.11
转鼓 9.5.16.2
bowl bearing 圆筒支架 16.4.25.7
外筒轴瓦 16.4.26.12
bowl body 转鼓主体 9.5.12.10
bowl hood 转鼓外罩 9.5.12.5
bowl liner 圆锥筒衬里 11.1.3.2
bowl valve 转鼓阀 9.5.12.11
brace 斜撑 4.5.1.8
brace angle 斜撑 5.6.1.13
bracket 托架 16.4.1.5
bracket for fluorescent tubes 日光灯管座 25.4.1.62
bracket support 悬架支撑部件 16.4.2.6
bracket-welded assembly 支座焊接装配体 8.8.5.29
brake axle 制动轴 9.5.4.32
brake hand 制动闸 9.5.4.8
brake handle 制动手柄 9.5.4.30
brake lining 刹车片 8.8.4.7
制动闸衬 9.5.4.7
branch circuit 支线 2.1.7.2
branched (molecules) 支链(分子) 14.3.4.(c)

branch main 支管 17.6.2.6
brass 黄铜 22.6.2.㉓
brazed-plate-fin heat exchanger 黄铜板翅式换热器 4.2.5.3
breaker plate 缓冲衬板 13.2.1.10
分流板 15.7.2.4
breaking device 粉碎装置 12.1.3.3
breaking plant 粉碎设备 13.2.4.22
breaking plate 破碎板 11.1.7.16
breast box of a paper machine 造纸机料箱 2.3.11.11
breast roll 排气辊 9.6.2.5
breather 呼吸器 25.4.3.10
breeching 烟道 8.8.1.8
器尾 8.9.2.4
筒尾 8.8.5.5
breeching seal 筒尾密封 8.8.1.7
brewer 啤酒酿造者 2.3.12.49
brewhouse 啤酒厂 2.3.13
brewing 啤酒酿造 2.3.12
brick tunnel for heated gases 砖洞道用于加热气体 4.3.2.6
bridge 桥 23.4.2.11
bridge blocks 过渡堵头 9.3.10.15
bridge-mounted rotary bucket wheel reclaimer 跨接旋转漏斗链轮再堆放器 23.2.5.1
bridge-supported mechanism 桥式支撑机构 15.6.1
bridge wall 火墙 24.1.4.4
brine 盐水 2.4.3.3
brine cooler 盐水冷却器 20.12.10
brine discharge 盐水排放 2.4.2.9
brink mist eliminator element 雾沫净除器单元 10.4.2
brine to other stages 盐水去其他级 2.4.3.4
British Gas Corp 英国天然气协会 2.2.14.1
broaching tap 取汁龙头 2.3.13.14
broad fluidization 广义流态化 14.1.3.24
bronze 青铜 19.3.3.(a)
brz(bronze) 青铜 2
bubble-cap 泡罩 5.2.8.11
bubble cap column 泡罩塔 5.1.2.2
bubble cap tray 泡罩(帽)塔盘 5.2.10.1
bubble cap tray tower 泡罩塔 5.2.8
bubble column 鼓泡塔 5.1.2.12
bubble point 泡点 1
bubbler 鼓泡器 3.5.18.(d)
bubbling 鼓泡 14.1.3.17
bubbling bed 鼓泡床 14.1.2.3
bubbling fluidized bed 鼓泡流化床 2.2.18.10;14.1.3.15

bucket 浮桶 22.2.3.1
抓斗 23.3.6.3
bucket conveyor 翻斗运输机 6.3.2.3
bucket elevator 斗式提升机 2.3.16.17;3.3.3.4
bucket-wheel reclaimer 斗链轮再装料器 23.2.5
buffer 缓冲 2.2.56.7
buffered 缓冲液 19.3.3.10
buffered combination labyrinth 注入缓冲液的组合式迷宫 19.3.3.(e)
buffer gas 缓冲气体 19.1.7.15
buffer inlet 缓冲液进口 19.3.1.17
buffer spring 缓冲弹簧 9.5.3.6
building floor 建筑物地面 4.3.4.17
bulk 统装 3.3.7.2
bulk density 堆密度 1
bulking agent for recycle 循环的填充剂 3.3.28.14
bulking agent if required 填充剂（如需要） 3.3.28.3
bulk STG 散装储存货架 2.3.17.14
bulk storage 散装仓 23.5.1.1
bulk transporting vehicle 散装运输车辆 2.3.16.15
bung dust caps 桶口灰尘罩 23.4.1.10
Bunsen burner 本生灯 2.3.11.5
Bunter downthrow 班砂岩降侧,班特下落断层块 25.2.4.6
burden chamber 装料室 4.3.4.4
burden discharge 装料排出口 4.3.1.5
burden outlet (solids state) gastight connection 粒料出口（固态）密封连接 13.5.2.8
burden travel 装料移动 4.3.2.12
burden travel forward 装料向前传送 4.3.4.20
bundles 过滤膜捆 3.1.29.2
burial pit 深埋坑 3.3.14
burned-out zone 燃尽段 2.2.85.12
burner 燃烧器 3.3.27.13;14.3.4.36
烧嘴 2.2.26.3;3.3.24.19;24.1.11.2
燃烧炉 2.2.20.3;3.3.25.16
burner and furnace configurations 燃烧器与炉的配置 24.2.2
burner body 燃烧器本体 24.2.6.9
burner box 箱式燃烧器 2.2.111.2
burner cement 燃烧器胶结水泥 24.2.6.12
burner feed tank 燃烧炉进料槽 3.3.7.19
burner incinerator 燃烧器焚化炉 3.3.10.5
burner tanks 喷燃器燃料罐 3.3.8.2
burner throat 燃烧器喉管 24.2.6.7
burner ventilator 燃烧送风机 2.3.12.17

bus bar 母线条 10.3.5.18
bush 衬套 11.1.6.7
轴衬 9.2.1.17
bushed-angle pillow block 角加衬的耐磨轴台 8.8.5.23
bushing 轴瓦 8.8.5.28
bushing containing 套筒 2.3.9.50
bushing tips 套筒端 2.3.9.51
bush set screw 衬套止动螺钉 19.2.3.13
butinediol 丁炔二醇 14.3.1.(a)
n-butene 正丁烯 2.1.1.69
butterfly outlet 蝶阀式出口 23.3.8.(b)
butterfly plate 蝶形板 22.1.4.9
Butterfly valve 蝶(形)阀 22.1.5
Buttner-Rosin pneumatic dryer Buttner-Rosin气流干燥器 8.4.5.(b)
by-pass 旁路 2.2.88.14;3.3.20.5
by-pass channel 旁路沟槽 10.1.4.9
by-passed 旁通 6.2.4.7
by-pass orifice 旁通小孔 19.1.7.1
by-pass pressure-relief valve 旁路调节阀 25.6.7.5
by-pass valve 旁路阀 2.1.7.21
旁通阀 3.5.11.15
by-product 副产物［品］ 1,2.2.25;2.2.54.15
by-product gases (fuel for drier and power plant) 副产品气体(作为烘干机或电厂燃料) 2.3.16.20
by product steam 副产蒸汽 2.2.73.13

C

C_2 oxygen containing compounds C_2 含氧化合物 2.2.40.9
$C_{3,4}$ 碳3碳4 2.2.75.14
$C_2^=$ 乙烯 2.2.69.19
C_2^0 乙烷 2.2.69.24
C_2- 2个碳(含C_2)以下的烃类 2.2.77.15
$C_3^=$ 丙烯 2.2.69.20
C_3^0 丙烷 2.2.69.23
C_{4+} 4个碳(含C_4)以上的烃类 2.2.71.21
C_{5+} 5个碳(含C_5)以上的烃类 2.2.69.22;2.2.75.15
>C_{30}(solid wax) >C_{30}(固体蜡) 2.2.73.11
$C_{13} \sim C_{19}$ (diesel fuel) $C_{13} \sim C_{19}$(柴油) 2.2.73.9
$C_5 \sim C_{12}$ (naphtha and gasoline) $C_5 \sim C_{12}$(石

脑油及汽油） 2.2.73.8
$C_2 \sim C_4$ (olefin hydrocarbon and liquefied petroleum gas) $C_2 \sim C_4$（烯烃及液化气） 2.2.73.7
$C_{20} \sim C_{30}$ (paraffin level heavy oil) $C_{20} \sim C_{30}$（石蜡级重油） 2.2.73.10
cable 电缆 25.4.5.8
cable box 电缆(接线)盒 25.4.7
cable clip （塑料）电缆夹(头) 25.4.1.40
cage 笼架 15.6.3.22
cage mill 笼式磨碎机 3.3.24.1；8.4.1.1
cake 滤饼 9.3.1.6
cake conveyor 泥饼运输机 3.3.19.12
cake dewatering 滤饼脱水 9.4.1.1
cake discharge 滤渣出口 9.4.9.7
cake dislodging and discharging 滤饼卸出 9.4.1.3
cake drwatering 滤饼脱水 9.4.1.7
cake recess 滤饼槽；存滤饼的凹处 9.1.3.4
cake rinse 滤饼清洗液 9.5.6.26
cake washing 滤饼洗涤 9.3.9.7
calcination 煅烧 2.1.1.45
calciner 煅烧炉 3.4.6.14
calcining 煅烧 24.1.12.3
calcining compartment 煅烧室 14.2.5.11
calcining furnace fuel 煅烧炉燃料 14.2.1.7
calcium carbide 电石 2.2.1.6
calcium chloride 氯化钙 6.3.10.10
calcium-substituted zeolite 钙取代沸石 6.3.10.12
calender 压光机 2.3.11.36
calender roll 压光辊 2.3.11.38
calibration gases 标定气体 3.4.11.23
calibration unit 校正单元 3.4.19.2
calked-gasketed-recessed filter plate 堵缝垫圈的凹板压滤机滤板 9.1.3
calking strip 滤布嵌条 9.1.3.3
call button 呼唤按钮 25.4.1.3
calming area 安定区 5.2.7.8
calming zone 安定区 5.2.6.8
cam-and-piston pump 凸轮-活塞泵 16.1.2.2.62
cameras 照相机 12.7.4.8
cam pump 凸轮泵 16.1.2.2.63
can be under pressure 可在压力下操作 11.2.7.5
candle filter 烛型过滤器 2.2.22.6；2.2.23.9
canned motor pump 屏蔽泵 16.1.2.2.32
cap （护）罩 22.1.2.14
capacitance 电容法 3.4.18.(e)

capacity 容量 12.1.5.17
capacity-control valve 容量控制阀 20.1.8.11
caprolactam 己内酰胺 2.3.8.28
caprolactam oil 己内酰胺油 2.3.8.25
cap screws 有帽螺钉 10.4.2.7
cap without recess 不带退刀槽的管帽 22.3.3.13
cap with recess 带退刀槽的管帽 22.3.2.12
car 小车 25.1.8.5
carbon （活性）炭 3.1.19.7
carbon addition 加炭 3.1.43.2
carbon bed hight 活性炭床层高度 6.3.9.1
carbon bed surface 活性炭床层表面 6.3.9.11
carbon/biomass recycle 炭/生物质循环 3.1.43.7
carbon-black 炭黑原料 2.2.91.9
carbon black flow 炭黑流 2.3.16.1
carbon black process diagram 炭黑工艺流程图 2.3.16
carbon charge 活性炭装填口 6.3.9.9
carbon column 活性炭柱 6.3.1.2
carbon dioxide 二氧化碳 2.2.6.14
carbon dioxide compressor 二氧化碳压缩机 17.1.2.8
carbon discharge 活性炭卸出口 6.3.9.12
carbon disulphide 二硫化碳 2.3.7.8
carbon feed tank 炭料罐 6.3.1.4
carbon flow controller 活性炭流量调节器 6.3.11.10
carbon in 活性炭进入 6.3.5.2
carbonization 碳化段 24.3.1.3
carbonizer 碳化塔 2.2.108.13
carbon make-up 活性炭补充 3.1.18.3
carbon-manganese steel 锰钢 22.6.2.⑧
carbon-molybdenum steel 钼钢 22.6.2.⑪
carbon monoxide 一氧化碳 2.2.54.11
carbon out 活性炭出口 6.3.5.6
carbon piston ring 石墨活塞环 17.2.10.2
carbon regeneration 活性炭再生 3.1.18.1
carbon sieve container 碳分子筛容器 2.2.53.7
carbon-silicon steel 硅钢 22.6.2.⑩
carbon-sludge recycle 活性炭-污泥循环 3.1.18.6
carbon steel 碳钢 21.1.2.7
carbon wipers 碳石墨刮壁器 5.5.2.10
carborundum arch 碳化硅炉拱 24.1.11.5
car-bottom furnace 车底炉 24.1.8
car connection 货车联结点 23.1.2.18
cargo oil pump 货船油泵 16.1.2.3.26

Carnot cycle 卡诺循环 1
carrier 载体 1
carrier gas inlet 载气进口 8.9.6.16
carrier-gas supply and control 载气供给与控制 3.4.20.1
carrier in 载体加入 15.2.2.8
carrier out 载体流出 15.2.2.9
carrier particles 颗粒载体 3.2.11.15
carrying 运载 23.3.7.(b)
carrying bar 承载梁 4.2.1.2
cartridge 滤芯 15.5.2
cartridge filter 筒形过滤机 9.4.7
cascade alkylator 多级烷化器 6.1.1.6.(a)
cascade condenser 逐级冷凝器 20.3.16
cascade heat exchanger 级联式换热器 4.2.5.7
cascade impact thrower 宝塔式淋洒器 5.3.4.5
cascade mill 阶式磨机 11.6.3.4
cascade system 逐级系统 20.1.5
casing 外[机,泵]壳;箱体 10.2.1.1 套管 2.1.2.18
casing ring 泵壳口环,阻漏环 16.4.19.12
castings for land utilization as soil amendment 舍弃的蚯蚓可用于土地利用的土壤改良剂 3.3.28.12
casing stud 壳体双头螺栓 16.4.20.8
casing wear ring 泵壳耐磨环 16.4.12.19
cast iron 铸铁 21.1.2.7
cast steam manifold 铸造蒸汽集管箱 8.8.7.(a)
cast steel 铸钢 21.1.2.7
catalyst 催化剂 2.2.39.5;2.2.64.13;14.3.1.12
catalyst addition 催化剂加入 2.1.10.1
catalyst bed 催化剂床 3.4.5.4
catalyst charging 催化剂进料 14.2.7.6
catalyst circulation rate 催化剂循环率 14.3.3.(c)
catalyst cooler 催化剂冷却器 14.2.7.1
catalyst dropout 催化剂排放口 14.3.2.3
catalyst inlet 催化剂进口 14.3.3.28
catalyst input 催化剂入口 14.3.2.1
catalyst level 催化剂(层)面 2.2.98.9
catalyst makeup 催化剂补充 14.2.7.4
catalyst purge section 催化剂冲洗段 14.3.32
catalyst recirculation 催化剂循环 2.3.5.14
catalyst recycle 催化剂循环 14.3.3.38
catalyst separator column 催化剂分离塔 2.2.54.8
catalyst-settling hopper 催化剂沉积料斗 14.3.2.28
catalyst stripper 催化汽提塔 2.1.5.4
catalyst withdrawal 催化剂出料 2.1.10.4
catalytic 催化 2.2.97.3
catalytic cracked gasoline 催化裂化汽油 2.1.1.26
catalytic cracking 催化裂化 2.1.1.24
catalytic cracking unit 催化裂化装置 2.1.5
catalytic diesel oil 催化裂化柴油 2.1.1.27
catalytic distillation column 催化蒸馏塔 2.2.65.b
catalytic fluidized-bed system 催化流化床系统 14.1.13
catalytic(H-coal)liquefaction 催化(H-coal)液化 2.2.90.5
catalytic hydrogenation 催化加氢 2.2.2.30
catalytic reforming 催化重整 2.1.1.47
catch-all drain 除沫器排液管 7.2.11.3
catch basin 集水池 4.6.2.2
catcher screw 压紧器螺钉 17.2.7.9
cat circulation control riser 催化剂循环控制提升管 14.3.4.8
cat circulation riser 循环催化剂提升管 14.2.6.11
categorization 分类 3.3.7.8
cation 阳离子型 6.2.9.3
cation exchanger 阳离子交换器 6.2.7.5
cation-exchange resin bed 阳离子交换树脂床层 6.2.2.(a)
caulking 挤缝 9.3.10.3
caustic 氢氧化物 2.3.6.4
caustic tower 烧碱塔 2.3.18.19
causticizer 苛化器 3.4.6.6
causticizing agitators 苛化搅拌器 2.3.10.49
caustic regeneration 碱再生 6.2.9.(c)
caustic scrubber 碱洗塔 2.2.69.9;2.3.18.8
caustic soda 苛性钠;烧碱;氢氧化钠 2.3.7.3
caustic soda solution 苛性钠溶液 2.3.8.10
caustic solution 碱液 2.1.9.4
caustic washing tower 碱洗塔 2.2.71.9
cavitation 气蚀 1
cavity 空腔 19.3.2.12
C_4 conversion reactor C_4 转化反应器 2.2.69.6
cell culture 细胞培养 1
cellar drainage pump 地下室排水泵 16.1.2.3.9
cellulose xanthate 纤维素黄酸酯 2.3.7.9
cem.(cement) 水泥

cement kiln 水泥窑 3.3.1.14
cement-lined steel 钢衬水泥 22.6.2.③
center column 中央圆筒 11.3.1.15
 中央塔 15.6.3.20
center dividing wall 中心分隔壁 16.4.6.8
center downcomer 中心降液管 5.2.2.13
center exhaust tunnel dryer 中间排气式隧道式干燥器 8.3.1.(c)
center heating tube 中心加热管 8.8.10.9
center pipe 中心管 22.2.3.6
center shaft gear drive 中心轴齿轮传动装置 3.3.26.12
center spider 中心(星形)轮 9.3.3.8
center tube cover 中心管盖 11.3.2.5
central casing 中心套管 25.3.1.19
central control room 中央控制室 2.3.12.25
central disc 中央圆盘 15.5.3.7
central downcomer 中央降液管,中央循环管 7.1.3.4
central heating circulating pump 集中供热循环泵 16.1.2.3.12
centralized separator 集中分离器 2.3.10.15
central-station air-conditioning unit 中央空调机组与控制系统 20.2.2
central-station air-conditioning unit and control system 中央空调机组与控制系统 20.2.2
central valve 中心型组合阀 17.2.15.11
centrifugal 离心(式) 16.1.1.30
centrifugal clarifier 澄清式离心机 9.5.13
centrifugal clarifier with self-cleaning bowl 带自动卸料转鼓的澄清式离心机 9.5.13
centrifugal cleaners 造纸机料箱 2.3.11.11
centrifugal compressor 离心(式)压缩机 17.4.2
centrifugal decanter 离心或滗析器 3.1.16.5
centrifugal deflector 离心式导流板 7.2.11.9
centrifugal-discharge filter 立式离心卸料加压叶滤机 9.2.3
centrifugal extractor 离心分离萃取机 6.1.17
centrifugal fan 离心(式)风机 17.5.1
centrifugal impeller, double-suction, enclosed design 离心式叶轮,双吸,闭式结构 16.4.5.(a)
centrifugal molecular still 离心式分子蒸馏釜 5.5.1.(c)
centrifugal pumps 离心泵 16.4
centrifugal refrigeration system 离心制冷系统 20.1.4
centrifugal screen 离心筛 2.3.10.22
centrifugal separator 离心式分离器 10.1.4
centrifugal trimmer 离心堆料器 23.3.9.7

centrifugate distillation 离心液精馏 2.2.64.9
centrifuge 离心机 2.2.64.8;9.5
centripetal pump 向心泵 9.5.13.3
ceramic Berl saddle 瓷弧鞍填料 5.3.2.(f)
ceramic Intalox saddle 瓷矩鞍填料 5.3.2.(g)
ceramic process pump 陶瓷化工泵 16.1.2.5.5
C_2^1 fractionator C_2 分馏塔 2.2.69.14
C_3^1 fractionator C_3 分馏塔 2.2.69.16
CH_4 甲烷 2.2.71.23
chain 链,传动链条 13.1.2.42
chain conveyor 链式输送机 23.3.5
chain cover 传动链罩子 9.2.5.2
chain drive 传动链 11.2.6.4
chain-drive unit 链传动装置 8.8.12.2
chain wheel 链轮 13.5.1.16
chamber 室 9.4.4.1
chamber condensate drain 室内冷凝液排出口 5.5.2.18
chamber drain 室排水口 8.2.2.8
chamber screen 滤室网 9.1.3.11
change over flap 换向挡板 8.4.5.15
changer lever 切换杆 9.2.1.11
channel 沟流 23.1.2
channel and removable cover 管箱与可拆盖板 4.1.2.(a)
channel cover 管箱盖 4.1.1.4
channel frame 槽钢框架 4.5.1.7
channeling 沟流 14.1.3.18
channel integral with tubesheet 与管板制成整体的管箱 4.1.2.(c)
channel 通道 9.1.2.3
channel section 管箱;分配室 21.1.1.52
char 炭 2.2.106.3
 灰渣 2.2.22.7;2.2.23.7
char cake 炭滤饼 2.2.106.6
charcoal tube 活性炭管 3.5.16.25
char gasification 炭气化 3.3.15.4
charge 加料 14.3.1.20
charge bin 加料料仓 2.2.20.5
charged 放电 10.3.1.28
charge opening 装[加]料孔[口] 11.3.1.1
charge pump 加料泵 16.1.2.3.27
charging chamber 进[装]料室 25.2.5.7
charging dome 装料筒 8.8.10.6
charging door 装料门 3.3.12.9
charging funnel 进料筒 13.1.2.20
charging tank 加料罐 6.3.2.14
charging zone 装料区 2.2.84.12
char heater 炭加热器 3.3.15.18

508

circulating

char product 产品炭 2.2.99.12
char slurry tank 炭料浆罐 2.2.106.18
CH_3COOH final product 乙酸成品 2.2.56.10
check valve 止逆阀,单向阀 16.2.7.7
check valve case 止逆阀罩 16.4.28.37
check valve spindle 止逆阀顶丝 16.4.28.1
check-weigh conveyor 校核称量输送器 23.4.2.7
check-weigher 校核称量秤 23.4.3.7
chemical absorption 化学吸收 1
chemical addition 化学药品添加 3.1.43.4
chemical conditioning 化学调理 3.1.26.(a).6;3.1.26.(b).9
chemical engineering 化学工程 1
chemical engineering science 化学工程学 1
chemical engineering thermodynamics 化工热力学 1
chemical equilibrium 化学平衡 1
chemical feed piping 试剂加入管 15.6.3.12
chemical feedstocks 化学原料 2.2.93.10
chemical identification 化学识别 3.3.7.9
chemical industry 化学工业 1
Chemical Industry and Engineering Society of China 中国化工学会 1
chemical machinery 化工机械 1
chemical oxygen demand(COD) 化学需氧量,化学耗氧量 1
chemical pipes 试剂管 15.6.3.1
chemical-porcelain pipe 化学瓷管 22.6.2.⑨
chemical process 化工工艺 2.3
chemical processing of coal 煤化工 2.2
chemical pump 化工用泵 16.1.2.3.35
chemical reaction engineering 化学反应工程 1
chemicals 化学(制)品 2.2.89.11
chemical scrubber 化学洗涤器 3.3.25.13
chemicals from petroleum 石油化工产品 1
chemical systems engineering 化工系统工程 1
chemical towers 化工塔类 1
chemical vapor deposition(CVD) 化学气相沉积 1
chemical ware 化学陶瓷管 22.6.2.④
chemisorption 化学吸附 1
CHEM SYSTEMS Process 化工系统过程 2.2.9
chilled methanol 急冷甲醇 2.2.28.15
chilled water (急)冷水 20.1.10.26
chilled water in 冷水入口 20.1.8.14
chilled water out 冷水出口 20.1.8.13
chilled water supply 冷水供入 20.2.2.21

chilled water valve 冷水阀 20.2.2.20
chip container 碎片容器 2.3.8.40
chip packer 木片装料机 2.3.10.3
chipper 削片机 2.3.10.1
chip return 碎粒返回 11.6.3.14
chlorination 加氯消毒 2.4.5.8
chlorine 氯(气) 2.3.8.7;2.3.18.2;3.3.21.5
chlorine contact chamber 氯接触箱 3.1.34.13
chlorine gas compressor 氯气压缩机 17.1.2.3
chlorinizer 氯化器 3.3.21.7;3.3.22.11
chlorine supply 供应氯气 3.3.22.13
CH_3OH 甲醇 2.2.56.13;2.2.61.4;2.2.63.2
CH_3OH 甲醇,水 2.2.63.17
CH_3OH recycle 甲醇循环 2.2.63.9
chromate plating tank 铬酸盐喷镀槽 6.2.7.3
chromate rinse tank 铬酸盐冲洗槽 6.2.7.4
chromium-vanadium steel 铬钒钢 22.6.2.⑬
CH_3SH 甲硫醇 2.2.63.6
$CH_3SH,CH_3OH,(CH_3)_2S,H_2O$ 甲硫醇,甲醇,二甲硫醚,水 2.2.63.14
$CH_3SH,(CH_3)_2S$ 甲硫醇,二甲硫醚 2.2.63.15
CH_3SH product 甲硫醇产品 2.2.63.18
$(CH_3)_2S$ product or recycle 二甲硫醚产品或循环 2.2.63.19
chute 斜槽,(颗粒)出口通道 23.4.2.12
C_4-hydrocarbons 碳 4-烃 2.2.65.1
circuit breaker 断路器;熔断器 25.4.1.36
circular 圆形 24.2.2.7
circular burner 圆形燃烧器 24.2.1
circular inlet baffle 圆形进口挡板 3.2.15.2
circular sedimentation tank 圆形沉降槽 3.2.15
circulated water 循环水 2.2.66.15
circulating bed 循环床 14.1.4.(f)
circulating cooling 循环冷却 4.3.3.23
circulating fan 循环风机 12.3.1.2
circulating(fast)bed 循环(快速)床 14.1.4.(e)
circulating-fluid-bed calciner 循环流化床煅烧炉 14.2.1
circulating line 循环管线 6.3.2.23
circulating loop 循环回流管 2.1.7
circulating mother liquid 循环母液 2.2.55.3
circulating mother liquor 循环母液 6.1.17.8
circulating pipe 循环管 7.2.5.11
circulating pump 循环泵 16.1.2.3.39
circulating quench-water 循环急冷水 2.2.

509

14.2
circulating salt system 循环盐系统 25.6.6
circulating vessel 循环槽 12.5.5.20
circulation 环流 1
circulation louvers 百叶窗式循环挡板 7.2.14.4
circulation pipe 循环管 7.2.10.8
circulation pipe to reboiler 循环管通再沸器 5.3.1.2
circulation pump 循环泵 25.2.3.4
circulation system 循环系统 11.3.4.3
circulation tube 循环管 6.1.7.8
circulator 循环机 2.2.46.6;2.2.47.6
circumferential groove 环向开槽 19.3.1.5
circumferential joints 环焊缝接头 21.1.1.47
circumferential piston 圆环多活塞 16.1.1.27
Cl_2 氯气 3.1.34.15
clamp 夹紧装置 10.2.1.9
卡子 5.3.4.30
clamping bolt 夹紧螺栓,固定螺栓,紧固螺栓 13.1.2.34
clamping nut 锁紧螺母 12.3.5.14
clamp plate 压板 5.2.2.16
clamp ring 夹紧环 12.2.1.7
clappers 拍板 2.3.9.42
clarification zone 澄清区 3.2.6.18
clarified backwash water 澄清的反洗水 3.1.39.7
clarified effluent 澄清流液 9.5.14.5
澄清水排出 3.1.32.21
clarified liquid 清液 9.4.7.1
clarified M.L. 清母液 7.2.5.27
clarifier 澄清器[池] 2.2.29.16;2.2.30.15;2.3.6.8;3.1.43.5;3.1.44.7
clarified liquid discharge 澄清液排出 9.5.16.8
clarifying system 澄清系统 12.5.1.11
classical fluidization 经典流态化 1
classifier 分级器 12.8.1.5
classification 分类 16.1
classification chamber 分级室 11.5.2.6
classified 粒度 12.3.3.7
classifier 分级器,粒料分选[级]器 8.4.1.14
分选机 2.3.16.18
classifier outlet 分级器出口 11.5.4.7
classifier wheel 分级器叶轮 11.2.9.2
classifying cyclone 分级旋流器 11.6.3.16
classifying whizzer blades 分级离心机叶片 11.2.11.5
Claus converter 克劳斯转化器 2.1.9.46
Clause process 克劳斯(二段脱硫)工艺 2.117.15

Claus sulfur plant 克劳斯制硫装置 2.3.1.32
clay 黏土 3.3.12.6
clay bin 黏土箱 6.3.2.1
clay binder 黏土黏结剂 6.3.10.13
clay liner 黏土衬砌 3.3.12.7
clay storage 黏土贮存器 6.3.2.4
clay storage bins 黏土贮斗 6.3.2.11
clean air out 净化空气出口 10.5.4.13
clean air outlet 净化空气出口 10.5.1.18
clean air side 净化空气侧 10.2.1.33
clean and dry oil 干净油 15.5.5.9
clean-coke 洁净焦炭 2.2.93
clean-coke demonstration plant 洁净焦炭实验厂 2.2.93.2
clean gas 净化煤气 2.2.19.6
cleaned air 净化空气 10.1.3.12
cleaned bottle 已洗净的瓶子 2.3.13.21
cleaned gas 净化气 10.5.4.19
cleaned gas outlet 净化气体出口 10.5.10.5
cleaned/recycled media 净化循环介质 3.2.9.4
clean effluent 净化的流出液 15.5.4.2
cleaner flotation 净化器浮选 12.8.2.21
cleaner-scavenger flotation 净化器排出物浮选 12.8.2.17
clean gas 净化气(体) 6.3.7.15
净化煤气 2.2.19.6
clean gas discharge 净化气体出口 8.5.6.11
clean gaseous fuels 洁净气体燃料 2.2.2.13
clean gas out 净化气(体)出口 10.4.2.5
clean gas outlet 净化气(体)出口 10.2.2.8
clean gas outlet 净化气体出口 10.4.1.8
清洁气体出口 10.5.5.3
clean gas side 净化气侧 10.2.1.26
cleaning finger 清理爪 23.3.7.2
cleaning machine 清洗机 2.3.12.36
cleaning sludge 净化的残渣 3.3.5.3
clean liquid fuels 洁净液体燃料 2.2.2.27
clean liquid out 洁净液体流出 15.3.1.3
cleanout 清扫 3.1.6.13
cleanout door 清理门 11.2.6.6
清扫口 3.3.12.1
clean solid fuels 洁净固体燃料 2.2.2.37
clean SRC 净化的溶剂精制煤 2.2.94.20
clean syngas 洁净合成气 2.2.36.7
clean water 净化水 3.1.14.6
clean-water outlet channel 净化水流出通道 3.1.4.4
clean water pump 清水泵 16.1.2.3.1

clear air side 净化空气侧 10.2.1.17
clearance 间隙 8.8.5.4
clearance adjuster 间隙调节器 11.4.1.20
clearance control 余隙调节 17.2.13
clearance control cylinder 具有余隙调节的气缸 17.2.13
clearance pocket 余隙容积 17.2.2.14 余隙腔 17.2.13.1
clearance valve 余隙阀 17.2.13.2
clear cut liquid 冲洗液 10.5.10.2
clear cut liquid/gas junction 冲洗液与气体接触会合处 10.5.10.2
clear effluent 干净的排水 3.1.42.10
clear filtrate 清液 9.4.2.11
clear liquid 清液 5.2.6.3
clear well 清水池 2.4.5.9;3.1.41.6
climbing-film evaporator 升膜蒸发器 7.1.9
clipper seal 钳压密封 16.4.26.18
clockwise(CW) 顺时针(旋转) 17.3.3.9
clone 克隆 1
closed circuit 闭路循环 11.6.5
closed-cycle gas turbine 闭式循环燃气轮机 25.3.4
closed drum head 转鼓端板 9.3.3.11
closed impeller 闭式叶轮 16.1.1.43
close opening 窄孔 11.6.2.2
close pattern return bend 封闭式180°回弯头 22.3.3.15
closing bolt 紧固螺栓 9.2.5.10
closing device (自动)压紧装置 9.1.2.9
closing handle 紧固手轮 9.2.5.9
closing water 常关水管 9.5.13.14
closure plate 隔板 11.2.7.6
cloth 织物 8.8.5.19
滤布 9.3.2.19
cloth cleaning 清洗滤布 9.4.1.4
cloth drying 滤布干燥 9.4.1.5
cloth fastened to breeching 固定于烟道上的织物 8.8.5.18
cloth filters 袋式过滤器 10.2.1
cloth in place 铺设的滤布 9.4.2.8
clutch 离合器 12.1.1.16
CO 一氧化碳 2.2.56.12;2.2.62.3
CO_2 二氧化碳 2.2.32.7;2.2.33.17;2.2.67.9;2.2.68.9;2.2.70.8;2.2.71.22;2.2.80.12;2.2.109.13;2.2.114.14;2.2.116.14;2.2.117.7;2.2.118.15;2.2.119.8;2.3.19.10;3.1.47.9;3.3.29.3
CO_2 absorber CO_2 吸收塔 2.2.9.15
CO_2 absorber 二氧化碳吸收塔 2.2.118.12;2.3.19.8

coagulant 絮凝剂 3.1.12.5
coagulant 凝固剂 2.4.5.2
凝结剂 3.1.27.3
coagulant feed pipe 凝结剂加入管 15.6.4.21
coagulating basin 凝固槽 2.3.7.15
coagulator 凝聚器 15.6.4
coal 煤 2.2.2.1;2.2.13.1;2.2.18.3;2.2.29.2;2.2.30.3;2.2.36.2;2.2.38.1;2.2.78.1;2.2.79.2;2.2.109.2;2.2.110.1;2.2.115.5
煤炭 2.2.3.5;2.2.25.2;2.2.117.2;2.2.118.6
coal,air,steam 煤,空气,蒸汽 2.2.23.1
coal-based syngas 煤荃合成气 2.2.73.2
coal bed 煤床 2.2.17
coal bunker 煤仓 2.2.13.2;2.2.18.1
coal cleaning plant 煤净化工厂 12.5.5
coal-derived liquid 煤衍生液体 2.2.2.28
coal distributor 煤分布器 2.2.11.8
coal distributor/stirrer 煤分布器/搅拌器 2.2.14.9
coal dust 粉煤 24.3.1.17
coal energy and chemical industrial chain 煤炭能源化工产业链 2.2.1.1
coalescence 聚并,凝并 1
coalescence processes 聚并过程 15.5
coalescer 聚结器 6.1.5.14
coalescer element 聚并器元件 15.5.2.7
coalescer filter 聚并滤清器 15.5.2
coalescing membrane 聚结膜 6.1.8.3
coalescing separator 聚并分离器 15.5.4
coal from lock hoppers 从煤锁斗来的煤 2.2.24.6
coal from minemouth 来自矿井的煤 2.2.101.3
coal gasification 煤气化 2.2.1.11;2.2.36.3;2.2.40.1
coal gasifier 煤气化炉 24.3
coal-handling and preparation 煤的加工制备 2.3.2.1
coal in 煤入口 24.2.10.4
coal liquefaction 煤液化 2.2.1.20
coal lock 闭锁式煤斗 2.2.12.2
coal lock-hopper 密封煤斗 2.2.14.7
coal,oxygen,steam in 煤,氧气,蒸汽进入 2.2.16.1
coal prep 煤的准备 2.2.100.1
coal preparation 煤预处理[加工] 2.2.98.1
煤的制备 2.3.1.4
coal pulverizer 煤粉碎 2.2.92.5
磨煤机 2.2.110.2
coal seam 煤层 2.2.81.17

coal slurry 煤浆 2.2.21.1;2.2.102.5
coal slurry/O_2 煤浆/氧气 2.2.21.2
coal slurry pump 煤浆泵 2.2.35.14
coal slurry tank 煤浆槽 2.2.35.13
coal storage hopper 储煤斗 2.2.35.1
coal slurry discharge tank 煤浆出料槽 2.2.35.10
coal gravimetric feeder 煤称量给料机 2.2.35.2
coal/limestone slurry 石灰石煤浆 2.2.116.4
coal storage 储煤仓 2.2.89.2
coal synthesis gas 煤合成气 14.3.2.(c)
coal tar 煤焦油 2.2.1.4
coal tar extraction 焦油的提取 2.3.14.18
coal tower 煤塔 2.3.4.5
coal tower conveyor 煤塔输送机 2.3.14.4
coal/water slurry 水煤浆 2.2.29.8
coal/water slurry, oxygen 水煤浆,氧气 2.2.17.1
coarse 粗料[粒] 12.2.3.8
coarse coalescing 粗颗粒聚结 6.1.9.10
coarse material 粗粒料 12.4.2.14
coarse-ore bin 粗矿石储槽 12.8.2.2
coarse particles 粗粒 8.4.3
coarse particles out 粗粒粉尘出口 10.5.1.24
coarse product 粗粒成品 8.5.1.9
coater for the top side 顶面涂布器 2.3.11.31
coater for the underside 底面涂布器 2.3.11.34
coating 去包衣(涂膜) 13.1.1.10
CO_2 by product CO_2 副产品 2.3.3.12
CO_2 capture & compression 二氧化碳捕集与压缩 3.4.6.15
CO_2 capture 二氧化碳捕集 2.2.109.10
cock 旋塞 11.4.1.5
CO conversion 一氧化碳变换 2.2.67.6
cocurrent-flow 并流 3.3.15.(b)
CO cycle 一氧化碳循环 2.2.66.8
CO_2-depleted air 二氧化碳耗尽的空气 3.4.6.1
CO_2 desorber 二氧化碳解吸塔 2.3.19.9
code termination of vessel 容器规范范围 21.1.1.4
COED coal pyrolysis COED煤热解过程 2.2.99
COED gas COED气体 2.2.99.6
COED oil COED油 2.2.99.9
coefficient of performance(COP) 性能系数 1
CO_2-free flue gas 无二氧化碳烟气 2.2.116.11
coffee factory WWT flow schematic 咖啡生产厂废水处理流程 3.1.31
CO_2 - free flue gas to atmosphere 无CO_2烟气排放到大气 2.2.27
cogenerating 联合生产 25.1.1
cogeneration systems 联合生产系统 25.1
$CO+H_2$ 一氧化碳+氢气 2.2.67.12
coil 线圈 20.2.1.23
  绳索 9.3.7.4
  感应线圈 12.6.4.16
coil bracket 盘管托架 24.1.3.9
coiled-coil filament 绕线式灯丝 25.4.1.58
coil filter 绳索过滤机 9.3.6
coil pipe 盘管 4.2.3.4
coil support 盘管支座 25.1.3.3
coke 焦炭 2.1.1.44;2.2.1.3
coke drum 焦炭罐 2.1.6.2
coke for aluminum anodes 铝氧化用焦炭 2.2.93.7
coke guide 焦炭导槽 2.3.14.9
coke loading 焦炭装载 2.3.14.15
coke loaking bay 凉焦台 2.3.14.12
coke oven gas 焦炉气 2.2.1.5
coke-oven gas processing 炼焦炉煤气的加工 2.3.14.16~45
coker distillate 焦化塔馏出物 2.2.93.9
coker naphtha 焦化石脑油 2.1.6.7
coke side bench 放焦台 2.3.14.13
coke to gsaification 焦炭去气化 2.2.79.16
coke waste 炼焦废液 3.1.5
coking 煤焦化 2.2.1.2
coking diesel oil 焦化柴油 2.1.1.42
coking gasoling 焦化汽油 2.1.1.41
coking plant 焦化厂 2.3.14
coking plant for dry coal distillation 煤干馏用的焦化设备 2.3.8.2
coking waxy oil 焦化蜡油 2.1.1.43
cold acetylene 冷乙炔 14.3.1.3
cold air inlet 冷空气入口 11.6.4.16
cold air intake 冷空气吸入力 4.6.1.1
cold air tube 冷空气管 3.3.26.7
cold ambient air 外界环境冷空气 4.5.2.1
cold coke 冷焦炭 14.3.4.33
cold coke 冷焦炭 2.1.9.30
cold-flow model experiment 冷模试验 1
cold fluid 冷流体 4.2.1.12
cold fluid in 冷流体进入 4.2.1.8
cold fluid out 冷流体出口 4.2.1.6
cold product gas 冷成品气 2.2.28.3
cold shot nozzles 冷态粒子接管 14.3.2.2
cold shots 冷态粒子 14.3.2.(b)
cold trap 冷阱 15.1.1.9

cold water 冷水 8.9.2.8
cold water circulating pump 冷水循环泵 20.1.10.27
cold-water inlet 冷水进口 4.3.2.16
cold-water makeup 冷却水补充 3.1.6.12
cold-water recycle pump 冷水循环泵 3.1.6.15
cold-water reservoir 冷水贮罐 3.1.6.14
cold water tank 冷水箱 2.3.12.9
cold water well 凉水池 4.6.1.2
collar 卡箍 10.2.3.3
套圈 13.3.3.4
collecting electrode 收集电极 10.3.1.17
collecting electrode-pipe 集尘电极管 10.3.4.12
collecting(positive)plate 集尘板（正极） 10.3.1.3
collecting screen 捕集筛网 7.2.11.2
collecting screw 集料螺旋 8.4.5.14
collecting tank for backwater 回水收集槽 2.3.10.26
collecting unit 收集单元 10.3.1.22
collection bucket 集料吊斗 3.5.10.13
collection chamber 聚集室 3.1.10.1
collection efficiency 捕集效率 1
collection lens 聚光透镜 3.4.6.6
collector 捕集器 10.4.3.13
收集器 23.1.1.10
collector electrode 集电极 3.5.21.6
collector housing 收集室 10.2.3.10
collector plates 收尘板 10.3.1.28
collector shaft 收集轴 15.2.3.9
collector travel 收尘极的移动 15.6.1.28
CO_2-loaded solution 会 CO_2 溶液 2.2.27.5
CO_2 loaded solvent 含有 CO_2 的溶剂 2.2.32.3
colloid mill 胶体磨 11.4.2
colour indicator 颜色指示器 25.4.1.37
column （立）柱 13.1.2.36
column coupling 圆筒连接器 16.4.25.20
column of driving device 传动装置竖筒 4.6.3.15
column pipe 圆筒管 16.4.23.13
管状圆筒 16.4.25.17
column pipe spacer 管状圆筒定距块 16.4.25.19
columns 支腿 13.3.6.25
塔器 21.3.1.10
column support 柱式支架 21.3.4
column washer 洗涤塔 1
comb 梳齿座 19.3.2.10

combination labyrinth 组合式迷宫 19.3.3.(e)
combination vapor and liquid heating with Dowtherm 导热姆蒸气与液体的联合加热 25.6.4
combination drying-grinding system 组合干燥-磨碎系统 11.6.4
combined liquid inlet and cone support 液体进口与锥形机架组合体 10.5.10.1
Combined treatment (anaerobic digestion) and traditional disposal of the primary and secondary sludge 一次消化和二次污泥传统的和厌氧消化的联合处理 3.1.26(a)
combining throat 混合喉管 18.3.4.7
combustion 已燃;燃烧 2.2.85.23
combustion air 助燃空气 14.3.3.24
combustion air fan 助燃风机 3.3.24.21
combustion air preheater 助燃空气预热器 3.3.24.15
combustion-air return 已燃空气回流 3.3.9.11
combustion chamber 燃烧室 2.2.81.6
combustion-chamber main casing 燃烧室主壳体 25.3.1.12
combustion chamber 燃烧炉 8.9.3.7
combustion furnace 燃烧炉 8.8.1.22
combustion gas 燃烧气体 14.3.3.(a)
combustion products 燃烧产物 3.3.10.7
combustion result 燃烧效果 24.2.8
combustion zone 燃烧段 2.2.82.7
焚烧段 3.3.9.4
combustor 燃烧器[室] 25.1.1.6
锅炉 25.1.2.11
燃烧炉 2.2.113.6;2.2.119.10;3.3.23.11
commercial cyclone 工业旋风分离器 10.1.1
commercial incinerator 工业焚烧炉 24.1.4
commercial separation-element tube 商用分离元件管 15.4.3
common base 公共底座 12.1.1.6
common bed 底座 11.1.6.5
communication lines 通信线 24.4.1.23
compaction 挤压 23.2.2.4
compact set of tray 塔盘紧固件 5.3.4.27
comparison between MBR and equivalent traditional WWTP 膜生物反应器和等效的传统废水处理工艺比较 3.1.70
compartment isolated for bag cleaning 过滤袋隔离室 10.2.2.2
compartment seal 室间密封装置 9.3.7.6
complete 全部的 2.2.104.9
complete condenser 全凝器 2.2.57.6

complete connecting rod 连杆组件 17.2.
15.4
complete connecting rod with bearing and bolts
连同轴承和螺栓的连杆组件 17.2.15.4
complete coupling 联轴器部件 16.4.27.20
complete crosshead frame 十字头组件 17.
2.15.5
completed vessel 成品容器 2.3.9.37
complete metallic packing 金属填料组件
17.2.15.33
complete mix fermenter 全混发酵罐 3.2.
10.(b)
complete piping 连通管 16.4.27.26
complex distillation column 复杂精馏塔 5.
1.1.3
component 部件,构件,零件 6.2.8.13
component arrangements 部件配置图 8.8.1
component parts of separator bowl 分离机转
鼓部件 9.5.12
component solvent 混合溶剂 2.1.1.78
composite flow model 组合流动模型 1
composite membrane 复合膜 1
composite sample 复合样品 3.1.2.27
composite sample line 组合取样管线 3.5.
3.5
composite sampler 组合取样器 3.4.18.12
composting mix 复合肥混合 3.3.29.4;3.3.
30.3
compressed air 压缩空气 2.2.81.3
compressed air injection well 压缩空气注入
井 2.2.85.20
compressed air inlet 压缩空气入口 11.5.5.6
compressed air installation 压缩空气站 17.
6
compressed air manifold 压缩空气总管 10.
2.3.9
compressed-air tank 压缩空气罐 25.4.9.1
compressed air washing unit 压缩空气洗涤
装置 2.3.12.3
compressed gas 压送的气体 17.1.2
compressibility factor 压缩因子 1
compression 压缩 2.2.101.17
compression bolt 压紧螺栓 4.2.1.10
compression process 压缩工序 2.2.45.12
compression refrigeration 压缩制冷 1
compression ring 承压环 21.1.1.55
compression ring 承压环 21.2.8
compression screw 压紧螺丝 22.1.8.16
compression section 压缩段 13.2.1.15
compressor 压缩机 2.2.20.1;2.2.31.13;2.
2.63.4;2.2.68.10;2.2.76.15;2.2.77.16;2.2.
81.2;2.2.111.4;2.2.115.16;2.2.116.2;3.1.
32.10;17.1.1.17
压气机 3.1.23.17
compressor cylinder 压缩机气缸 17.2.2.5
compressor cylinder 气缸 17.2.5.10
compressor/gas turbine/generator 压缩机/
燃气轮机/发电机 2.2.112.3
compressor inlet guide vanes 压缩机进口导
叶 25.3.1.7
compressor radial-axial bearing 压缩机径向
止推轴承 25.3.1.4
compressor radial bearing 压缩机径向轴承
25.3.1.18
compressor rotor blades 压缩机转子叶片
25.3.1.8
compressors 压缩机 17.1.1.1
compressor stator 压缩机定子 25.3.1.9
compressor stator blades 压缩机静叶片
25.3.1.6
computer aided process design(CAPD) 计算
机辅助过程设计 1
CO_2,N_2 二氧化碳,氮气 2.2.68.5
concentrate 富集物 12.6.4.15
精选矿 12.8.1.10
浓缩液 3.1.21.19
concentrate (retentate) 浓缩物(截留物)
3.1.29.9
concentrated liquid 浓废液 3.3.10.13
concentrated solid phase 增稠固体相 9.5.
14.6
concentrated solution 浓溶液 13.3.3.5
concentrate outlet 浓缩液出口 15.2.1.4
concentrate space 浓缩液滞留空间 9.5.
16.5
concentration profile 浓度[分布]剖面[图] 1
concentration tower 提浓塔 2.2.58.11
concentrator 浓集器 10.1.3.6
浓缩器 2.3.10.24
concentrator clean air plenum 经浓集器后的
净化空气流 10.1.3.4
concentrator dust hopper 浓集器集尘箱[斗]
10.1.3.8
concentrator for the weak wash liquor 稀洗液
浓缩器 2.3.10.44
concentrator unit 浓集器机组 10.1.3.5
concentric ring 同心圆环 14.1.8.(a)
concentric rings in the form of a cone 锥形同
心环 14.1.8.(b)
concentric screens 同心筛网 10.4.2.9
concentric collector 同心收集器 11.5.1.12
con.-cooler shell 冷凝-冷却器壳体 20.2.3.1

concrete 混凝土 22.6.2.⑥
concrete 混凝土 3.1.24.17
水泥 3.1.24.5
concrete casing pump 水泥壳泵 16.1.2.5.4
concrete foundation 混凝土基础 21.2.2.1
concrete plate 混凝土板 12.2.3.22
concrete shield 混凝土屏蔽层 25.2.2.29
concrete tank 水泥槽 15.6.1.17
concurrent-disk atomization 并流转盘雾化式 8.6.3.(c)
concurrent-nozzle atomization 并流喷嘴雾化式 8.6.3.(d)
concurrent wet-drum separator 并流型湿式鼓式磁力分离器 12.6.5
condensate 冷凝[物,水] 2.2.9.14;2.4.3.2
condensate drain 冷凝液排放 8.8.10.9
condensate inlet 凝液入口 22.2.3.8
condensate out 冷凝液出口 8.8.7.3
冷凝水出 2.4.2.2
condensate outlet (冷)凝液[水]出口 7.1.3.5
condensate pump 冷凝液泵 16.1.2.3.23
condensate return flow 冷凝液回流 4.8.1.8
condensate water 凝结水 2.2.45.6
condensation 冷凝 20.2.4.21
condensed heating steam and gases 冷凝加热蒸汽和气体 5.5.1.49
condensed sweep vapor 冷凝扫掠蒸汽 15.4.1.9
condensed water outlet 冷凝水出口 18.3.6.5
condenser 凝聚[结]器 15.1.1.6
冷凝器 2.2.27.12;2.2.60.1;2.2.76.14;2.2.81.16;2.2.111.19;2.2.113.10;2.2.114.13
condenser heat recovery 冷凝器回收热 25.5.2.11
condenser-ice-bath system 冰浴冷凝器系统 3.5.7.6
condenser inlet 冷凝器水进口 5.5.2.16
condenser len 聚焦透镜 3.5.6.2
condenser tube 冷凝管 20.2.3.3
condenser wall 冷却壁 4.4.1.8
condenser water inlet 冷凝器水进口 18.3.6.2
condenser water outlet 冷凝器水出口 5.5.2.19
condensing pressure 冷凝压力 20.2.4.23
condensing temp. 冷凝温度 20.2.4.14
condensing water 冷凝用水 20.2.4.8.9
condensing water in 冷凝器用水进口 20.1.

8.3
conditioned sludge 调理过的污泥 3.3.22.10
conditioning agent 调节剂 3.1.11.4
conditioning tank 调节池 3.1.28.3;3.1.30.2; A.2;3.1.30.B.2
cond.press. 冷凝压力 20.1.7.2
conduction 热传导 8.1.1.4
conductivity cell 传导性电池 6.2.1.8
conductor (导电)导线,导线 25.4.5.1
导体介质 12.7.2.5
cone 导风锥 10.3.4.15
cone 圆锥(体) 19.1.5.8
cone"O"ring 圆锥体内O形密封圈 19.1.5.7
cone body 圆锥形壳体 12.1.1.28
cone breaker 锥形碎纸机 2.3.10.82
cone crusher 圆锥破碎机 11.6.2.7
cone form 锥形丝筒 2.3.7.25
cone-roof 锥顶罐 21.2.1.1
cone scraper 锥形刮板 15.6.1.19
cone section 锥形截面 8.4.3
cone shorthead 锥形短头 11.6.2.11
cone side wiper 锥型塔侧收集器 5.3.4.18
cone support 锥形机座 10.5.10.1
辐射锥支架 24.1.3.11
cone valve 锥形阀 14.1.6.(e)
cone-winding machine 锥形绕线机 2.3.7.26
configuration of combined suspended and attached growth system using freefloating media 采用自由浮动介质悬浮与附着生长相结合的模式 3.2.14
configurations 各种组合 25.1.1
CO_2,N_2,H_2,CO 二氧化碳,氮气,氢气,一氧化碳 2.2.38.7
conical ball mill 锥形球磨机 11.2.3
conical grate 锥形栅板 11.2.3.3
conical head 锥形(封)头 21.1.1.57
conical inlet casing 锥形进口套管 25.3.1.5
conical intermediate piece 锥形的中间零件 25.3.1.20
conical pile 锥形料堆 23.2.3.(a)
conical refiner 锥形磨浆机 2.3.10.27
conical tank 锥形槽 11.1.3.7
conispherical magma crystallizer 锥底球形晶浆罐结晶器 7.2.7
coolant pump 冷却剂泵 16.1.2.3.22
connect collar 连接卡圈 11.1.5.2
connected stack 烟囱连接段 24.1.1.30
couecting electrode 收集电极 10.3.5.21
connecting glass pipeline to metal systems 玻璃管路与金属管路连接 22.4.4
connecting pipe 接管 21.2.4.4

515

connecting plate 连接板 5.2.2.11
connecting rod bolts with nuts 连杆螺栓螺母 17.2.12.26
connecting rod key 连杆键 16.2.3.43
connecting rod pin 连杆销 16.2.3.13
connection 接口 9.3.2.29
connection tube 接管 24.1.1.5
connector bearing 接头轴承 16.4.23.9
Conoco Phillips gasifier 康菲气化炉 2.2.21
Conoco Phillips E-Gas gasifier 康菲 E-Gas 气化炉 2.2.22
constant angle 固定角度 9.5.2.16
constant area mixing section 等截面混合段 18.3.3.6
constant pressure mixing section 恒压混合段 18.3.3.7
constant rate pump 计量泵 16.1.2.3.45
constrained-bellows expansion joints 限位波形膨胀节 22.5.4
constriction plate 收缩板 14.1.12.4
constructional materials of pump 泵的构造材料 16.1.2.5
construction details 结构详图 21.1.1
contact filtration 接触过滤 6.3.2
contacting element 接触板 6.1.18.1
contact maker 断续器,接触接通器 25.4.1.38,39
contactor 接触器 3.4.6.2
contact spring 接触弹簧 25.4.1.28
contaminated air inlet 脏空气入口 10.5.1.14
contaminated medium pump 杂质泵 12.5.1.15
contaminated oil drain 污染油排放口 19.1.7.16
continued below 转接下一行 3.1.37.12
continuity tester 线路通断测试器 25.4.1.55
continuous 连 8.1.1.3
continuous band 连续带式 8.1.1.⑲
continuous bucket elevator 连续型斗式提升机 14.3.3.21
continuous chain grinder 连续链式磨木机 2.3.10.53
continuous feed stream 连续进料流 6.3.8.12
continuous feeds with thickening 连续投料带增稠 3.2.5.b
continuous filament process 连续拉丝 2.3.9.48
continuous flow composite sampler 连续流组合采样器 3.5.3.7
continuous-flow conveyors 连续流动输送机 23.3.3

continuous horizontal vacuum filter 连续水平真空过滤机 9.4.2
continuous hydrogenation of fats 油脂的连续加氢反应 14.3.1.(c)
continuous L.O.treatment plant 连续润滑油处理工厂 6.3.2.15
continuous muller 连续式滚轮混合器 12.1.3.(h)
continuous phase 连续相 6.1.18.5
continuous pipe cooling coils 连续冷却盘管 6.3.2.30
continuous rotary filter 连续回转过滤机 9.3.1.5
continuous rotary vacuum filter 真空连续回转过滤机 9.3.2
continuous sample train for COCO 连续采样系统 3.5.12
continuous through-circulation 连续穿流-循环式 8.1.1.⑯
continuous through-circulation conveyor dryer 连续穿流-循环输送带式干燥器 8.3.3
continuous tray 连续盘式 8.1.1.⑰
continuous vacuum filtration 真空连续过滤 9.3.1
continuous vertical type dryer 立式连续干燥器 8.9.3
contractor 打包机 3.3.5.20
contractor for dumping 填埋打包机 3.3.5.21
control 控制器 3.5.10.4
control air inlet 调节空气入口 14.3.4.21
control blower 调节风机 14.3.4.3
control board[panel] 控制盘 25.4.2.1
control console[desk] 控制台 25.4.2.9
control desk 操纵台 2.3.14.1
控制台[盘] 25.2.5.8
control device 控制装置 17.2.12.1
control panel 操作盘 15.7.3.16
控制板 3.1.46.3
控制盘[台] 15.7.5.4
control rod 控制棒 25.2.2.4
control rod drive 控制棒驱动装置 25.2.2.3
control room 控制室 25.4.2
controls 控制装置 3.3.20.2
control system 控制系统 20.2.2
control tap 控制龙头 2.3.10.9
control valve 控制阀 10.3.5.7
control valve 控制阀,调节阀 18.4.1.12;22.1.9
convection 对流 8.1.1.5
convection band 对流带式 8.1.1.⑮

convection tray 对流盘式 8.1.1.⑭
conventional 常用的 6.2.2.(a)
conventional activated sludge 常规的活性污泥 3.1.30.B
converging section 收缩截面 3.5.4.5
conversion 转化率,转化 1 变换 2.2.33.8;2.2.36.4;2.2.38.5;2.2.78.6
conversion of petroleum fractions 石油馏分转化 14.3.4
conversion reactor 转化釜 2.2.55.7
conversion reactors 转化反应器 2.2.75.5;2.3.4.5
conveyer pipe for material 给料管 12.3.2.1
conveying and heat-transfer surface 输送和传热表面 4.3.4.5
conveying (equipment) 输送设备 23
conveying fan 输送风机 8.6.2.10
conveying line 输送管线 23.1.2.35
conveying surface 输送表面 8.9.6.2
conveying system 输送系统 8.9.4.4
conveyor 输送机[带] 3.3.2.3 运渣带 9.6.2.10
conveyor belt 输送带 12.7.4.6
conveyor drive 输送器驱动装置 9.5.5.1
conveyor-elevator 输送式提升机 23.3.3.(b)
conveyor return 返回输送带 8.3.3.4
cool 冷却 2.2.115.14
coolant-collecting plant 冷却液收集装置 2.3.12.7
coolant film 冷却液膜 4.3.3.10
coolant flow passage 冷却剂流通管 25.2.2.9
coolant pool in shaft 轴内的冷却液池 4.3.3.9
coolant system 冷却系统 25.2.3.5
cooled inert hydrocarbon liquid 冷却的惰性烃液体 2.2.39.b
cooled stripping gas as concentrated water back to refining 汽提气冷却后作为浓缩水回炼 2.2.72.13
cooler 冷却器 2.2.5.2;2.2.27.9
cooler groups 冷却器组 14.3.2.24
coolers 冷却器 2.2.54.2;2.2.74.9
cooler shell 冷却器 20.2.3.5
cooler tube 冷却器管 20.2.3.7
cool flue gas 冷烟道气 2.2.111.20
cool gas 冷气体 2.2.6.15
cooling 冷却 2.2.53.2;2.2.53.6
cooling absorber 冷却吸收器 5.4.2.1
cooling air 冷却空气 3.3.26.13;13.1.1.5

cooling air discharge 冷却空气排出 3.3.9.9
cooling air fan 冷却空气鼓风机 3.3.9.1
cooling air in 冷却空气入口 24.1.13.1
cooling box 冷却箱 6.3.2.30
cooling box with continuous pipe cooling coils 有连续冷却盘管的冷却箱 6.3.2.30
cooling circuit 冷却管线 25.6.3.5
cooling compartment 冷却室 14.2.5.12
cooling compartment cover 冷却室盖 16.4.20.23
cooling compressor 冷冻压缩机 2.3.12.22
cooling crystallizer 冷却结晶器 7.2.12
cooling cylinder 冷却滚筒 2.3.8.27
cooling device 冷却设备 3.1.3.5
cooling duct 冷却管 20.2.1.19
cooling-fluid 冷却流体 13.5.2.9
cooling jacket 冷却夹套 2.2.26.8;14.3.1.15
cooling load 冷却负荷 25.1.2.7
cooling medium 冷却介[工]质 15.1.1.7
cooling of the polyamide 聚酰胺的冷却 2.3.8.29
cooling-oil inlet 冷却油进口 14.3.2.23
cooling-oil outlet 冷却油出口 14.3.2.21
cooling ring 冷却圈 15.7.5.6
cooling roll 冷却滚轮 2.3.11.25
cooling screen 冷却帘栅 2.2.26.5
cooling scrubber 冷却涤气器 3.1.22.12
cooling system 冷却系统 2.2.80.10;25.2.28
cooling tank 冷却池 2.3.9.15
cooling tower 凉水塔 4.6 冷却塔 3.1.6.11
cooling-tower feed surge pond 冷却塔进料缓冲池 2.3.6.9
cooling unit 冷却器 2.2.61.5;2.2.63.16;2.2.76.10 冷却装置 2.2.62.4
cooling unit for catalyst 催化剂的冷却器 2.2.76.12
cooling wall 冷却壁 2.2.16.4
cooling water 冷却水 2.2.15.4;2.2.34.23;2.2.37.9;2.2.45.13;2.2.48.8;2.2.110.15;3.4.23;3.4.7.9
cooling water flow circuit 冷却水(流动)回路 25.2.1.6
cooling water in 冷却水入[进]口 15.4.1.10
cooling water inlet 冷却水进口 17.2.6.3
cooling water out 冷却水出口 11.3.4.2
cooling water outlet 冷却水出口 11.3.4.2
cooling water outlet and inlet 冷却水进出口 16.4.20.19

cooling water pump 冷却水泵 16.1.2.3.21
cooling-water return 冷却水返回 2.3.6.18
cooling water to heat incinerators 冷却水去热焚烧炉 2.3.6.20
cooling with water from river 用河水冷却 25.2.3.5
cooling zone 冷却段 3.3.9.3
cooling zone 冷却区段 8.5.6.20
cool lock hopper 闭锁式煤斗 2.2.11.2
cool recycle gas 冷循环气 2.2.87.14
copper alloys 铜合金 22.6.2.㉑
copper conductor 铜导体,铜芯线 25.4.1.43
copper cooling-coils 铜的冷却盘管 18.2.4.7
core 铁芯 16.5.4.3
电缆轴心,线芯 25.4.5.3
core buster disk 中央挡盘 10.5.4.6
CO_2 removal 脱二氧化碳 2.2.3.15;2.2.68.8;2.3.2.13
corn removal suction 吸除麦芽装置 2.3.12.38
CO_2 removal tower 脱CO_2塔 2.2.70.9
corrosion inhibitors 抗腐蚀材料 3.1.24.2
corrosion resistant pump 耐腐蚀泵 16.1.2.5.9
corrugated liners 波形衬里 11.2.2.4
corrugated parallel plates 波纹平行板 3.1.4.14
corrugated spacer 瓦楞状隔板 15.2.2.7
corrugated wire gauze packing 网波纹填料 1
corrugate-plate interceptor 波纹板油水分离器 3.1.4
corrosion 腐蚀 21.1.1.13
CO-shift reactor 一氧化碳变换反应炉 2.2.109.8
CO_2/steam turbine 二氧化碳/汽轮机 2.2.114.12
CO_2 storage 二氧化碳存储 2.2.117.13;2.2.118.17
CO_2 stripper 二氧化碳汽提塔 2.2.118.16
CO_2 to compression CO_2 去压缩 2.2.27.13
CO_2 to compression & sequestration 二氧化碳去压缩和储存 2.2.112.6
CO_2 to recycle CO_2 去循环 2.2.106.32
cotter 开口销 17.2.11.7
cottrell precipitator 电除尘器 14.2.7.11
coucher 伏(辊)工 2.3.11.49
counter ball case 副轴滚珠轴承箱 13.3.5.1
counter clockwise(C.C.W.) 逆时针(旋转) 17.3.3.7
counter-clockwise 逆时针(方向) 17.5.2.9

countercurrent 逆流式 8.6.3.(a)
countercurrent direct-heat rotary dryer 逆流直接加热式回转干燥器 8.8.1
countercurrent-flow 逆流 3.3.15.(a)
counter-current flow scrubber 逆流洗涤塔 10.5.15
countercurrent tunnel dryer 逆流隧道式干燥器 8.3.1.(a)
counterflow 逆流 6.2.2.(b)
counterrotation wet-drum separator 逆流转动型湿式鼓式磁力分离器 12.6.6
counter shaft 副轴 11.1.3.9
传动轴 11.1.4.4
countershaft housing 传动轴套筒 11.1.4.5
countershaft seal retainer 传动轴密封保持盖 11.1.4.2
counterweight 配重式 23.3.2.10
counter-weighted door 平衡重动作门 23.1.1.3
counter weighted operating levers 平衡锤操作杆 8.8.10.13
coupling 联轴器 9.2.2.3
coupling bolt 连接螺栓 17.2.11.2
coupling bushing 联轴器销衬套 16.4.9.22
coupling key 联轴器键 16.4.9.26
coupling lock nut 联轴器锁母 16.4.9.25
coupling pin 联轴器柱销 16.4.9.23
coupling with band 带箍管接头 22.3.2.17
coupling without band 不带箍管接头 22.3.18
couterweighted 带平衡块 23.3.6.(b)
coven lift lever 顶盖升降控制杆 24.1.7.10
cover 盖 11.1.7.26
cover 底盖 13.1.2.13
cover 轴承底盖 13.1.2.17
cover 轴承上盖 13.1.2.23
cover 盖 13.1.2.5
外罩 13.5.1.7
顶盖 2.2.39.2;15.5.1.9
covered glasshouse pot 有盖的熔化玻璃坩埚 2.3.9.46
cover flange 封头法兰 9.2.7.12
cover head 顶封头 14.4.1.25
cover head 顶盖 9.2.7.11
cover heating 顶盖加热管 8.9.5.5
covering plate 盖板 14.4.3.15
cover of dust collector 集尘器顶盖 12.1.1.4
cover plate 盖板 11.1.7.25
cover supporting bracket 顶盖支架 24.1.7.9
cow1 风扇罩 16.4.20.10
C_{4+} product 4 个碳(含 C_4)以上的烃类产品

518

2.2.68.15
crab claw 蟹爪式阀罩 17.2.12.7
cracked gas 裂解气 14.2.7.12
cracked vapor outlet 裂化蒸气出口 14.3.3.30
crank 曲柄 16.2.4.35
crank end 曲轴侧 17.2.2.8
crank key 曲柄键 16.2.4.24
crankpin 曲柄销 16.2.3.41
crankpin nut 曲柄销螺母 16.2.3.40
crank shaft 曲轴 17.2.5.3
crescent 月牙形块 16.3.4.3
cresols 混合甲酚 2.2.91.9
crimping of tow 丝束的起皱 2.3.8.59
critical exponent 临界指数 1
critical point 临界点 20.2.4.20
cross conveyor 横向送料器 23.2.3.(c)
cross feed belt 交叉进料皮带 11.6.4.4
cross flights 横刮板 3.2.15.10
cross flow 横向流动 4.1.2.(l)
cross-flow filtration 十字流过滤 9.4.9
cross-flow plate 单溢流塔板 5.2.6
crossflow spray absorber 错流式喷淋吸收器 10.5.11.(b)
crossflow wet-settling classifier 错流湿式沉降分级器 12.4.2
crosshead 十字头 16.2.5.22
crosshead bolts with nuts 十字头螺栓螺母 17.2.15.6
crosshead guide 十字头滑道 16.2.6.6
crosshead liner 十字头衬瓦 17.2.9.27
crosshead pin 十字头销 16.2.4.31
crosshead slipper with lower frame 十字头滑块板 17.2.15.17
cross partition ring 十字格环 5.3.2.(b)
cross sectional view 剖面图 8.8.10.19
cross section of fluidized bed biosolids incinerator 生物固体流化床焚烧炉截面图 3.3.27
cross stand 十字支架 16.2.3.42
cross straight size 等径四通 22.3.1.13
crown blocks 顶滑车,天车 2.1.2.4
crown cork closure 冠状软木瓶盖 2.3.13.27
crown safety platform 架顶安全平台 2.1.2.3
crownwheel oil pan 冠齿轮油池 9.6.1.4
crownwheel shield 冠齿轮罩 9.6.1.11
crude 原油 2.1.4.4
crude acetic acid 粗乙酸 2.2.55.11
crude and reboiler 原油与再沸器 2.1.4.6
crude atmospheric tower 原油常压塔 2.1.4
crude benzol refining 粗苯精制 2.2.1.7
crude benzol tank 粗苯罐 2.3.14.42
crude charge 原油加入 2.1.4.3

crude distilland 待蒸馏的粗制品 5.5.1.1
crude dust 粗粉煤 24.3.1.18
crude ethanol from hydrogenation process 粗乙醇来自加氢工序 2.2.59.4
crude ethyl acetate to hydrogenation process 粗乙酸乙酯去加氢工序 2.2.59.3
crude gas 粗制煤气 24.3.1.9
crude gas from cold gasification 来自煤气化装置的粗煤气 2.2.34.21
crude gasoline stock 粗汽油料 2.2.91.8
crude methanol 粗甲醇 2.2.42.4;2.2.45.17;2.2.46.18;2.2.47.16;2.2.48.10;2.2.75.3;2.3.4.2
crude methanol to rectification 粗甲醇去耗馏 2.2.44.17
crude methanol to stabilization tower 粗甲醇至稳定塔 2.2.44.18
crude methyl formate 粗甲酸甲酯 2.2.61.9
crude oil 原油 2.1.1.3
crude oil drilling 石油钻探 2.1.2
crude-oil-topping petroleum refinery 原油初馏炼厂 3.1.1
crude phenol 粗酚 3.1.5.8
crude phenol tank 粗酚罐 2.3.14.40
crude product gas 粗制品煤气 24.3.1.10
crude tar tank 粗焦油罐 2.3.14.32
crude unit 原油蒸馏装置 2.1.4
crud gas 粗制杂煤气 24.3.1.23
crushed coal 粉碎后的煤 2.2.99.1
crusher 压碎机 13.3.5.8
crusher 粉碎机 8.3.2.9
crusher discharge chute 破碎物料斜槽 11.2.6
crushing 破碎 11.6
crushing chamber 破碎室 11.1.1.2
crushing equipment 破碎设备 11.1
crushing head 破碎头 11.1.4.11
cryogenic liquid ammonia washing 低温液氨洗 2.2.38.10
cryogenic pump 低温泵,深冷泵 16.1.2.1.10
cryogenic separation 低温分离 2.2.78.15;2.2.104.11
cryogenic separation tank 低温分离罐 2.2.74.10
cryogenic-service spiral-tube exchanger 低温操作螺旋管换热器 4.2.5.11
crystal 晶体 9.5.19.4
crystal distributor 固体(晶体)分布器 9.5.6.23
crystal growth zone 晶体生长区 7.2.10.5

crystalline solid 结晶固体 8.1.1.19
crystallisation 结晶 2.2.64.7
crystallization equipment 结晶设备 7.2
crystallization tank 结晶槽 6.3.10.4
crystallizer 结晶器 5.3.3.7
crystal paste 晶浆 7.2.12.8
crystal slurry 结晶浆液 6.3.10.6
crystal suspension 晶体悬浮液 7.2.12.6
CSTR reactor 连续搅拌釜式反应器 2.3.15.1
Cu-Chromite 铜铬铁矿 2.2.66.13
culture medium 培养基 1
curing floor 熟化层 2.3.12.18
curing kiln 熟化窑 2.3.12.30
current 电流 16.5.4.9
current meter 流速计 3.5.1
current transformer 变流器 25.4.9.8
curtain wall 幕墙 3.3.12.14
curtain wall port 幕墙通口 3.3.12.2
curved anvil 曲线砧座 11.1.2.2
curved baffle 曲面挡板 9.5.17.19
curved product outlet 曲线形产品出口 11.1.1.4
curved wall 弯曲壁 15.4.2.6
cutaway view 局剖视图 10.2.1.(b)
cut damper 节流挡板 8.9.3.5
cut-groove type fitting 切削槽式管件 22.3.1.23
cut-off wall 堰板 15.6.4.7
cutter 切纸机 2.3.11.44
cutter opening 切割(取样)器 11.7.1.8
cutting plier 剪钳 25.4.1.51
cutting tip 截割喷嘴 26.1.45
CVD(chemical vapor deposition) 化学气相淀积 附录 2
CW(cooling water) 冷却水 2.1.4.29;2.2.46.10;2.2.47.9
C.W(cooling water) 冷却水 2.2.9.7
CWR 冷却水回路 2.2.34.24
cyclic automatic 自动循环式 9.5.2.5
cycle-flow 循环流 23.3.4.8
cyclohexanone 环己酮 2.3.8.20
cyclohexanoxime 环己酮肟 2.3.8.22
cycloidal planetary gear speed reducer 行星摆线针轮减速机 9.2.1.1
cyclone 旋风分离器 2.2.19.7;2.2.76.2;3.3.24.7;3.3.25.10
cyclone dust collector 旋风集尘器 24.2.10.6
cyclone furnace 旋风炉 24.2.3
cyclone separators 旋风分离器 3.3.23.6;10.1

cyclone slag-tap hole 旋风器出渣孔 24.2.3.5
cyclone solids-return seals 旋风分离器固体回流密封装置 14.1.7
cyclone spray scrubbers 旋风涤气器 10.5.4
cyclone stripper vessel 旋风汽提器 14.2.7.24
cyclone tube 旋风分离管 10.1.2.27
cyclonic separator unit 旋风分离机组 10.1.3.2
cyclonic separator 旋风分离器 10.5.13.5
cyl.(cylinder)·(1)圆筒(2)量筒(3)钢瓶(4)汽缸 附录2
cylinder 缸体,圆筒,气缸,泵体 11.4.1.11
cylinder arrangement 气缸布置 17.2.1
cylinder assembling bolt 气缸装配螺栓 17.2.15.13
cylinder bottom 气缸座 17.2.15.32
cylinder head 气缸盖 17.2.15.12
cylinder head end 气缸盖侧 17.2.2.1
cylinder jacket for cooling water 17.2.9.35
cylinder liner 气缸衬套 17.2.7.25
cylinder support 气缸支承 17.2.15.34
cylinder with shrunk liner 带热压缸套的气缸 17.2.15.10
cylindrical-conical helical-conveyor centrifuge 筒锥形螺旋卸料离心机 9.5.5
cylindrical-furnace tubular heater 圆筒形管式加热炉 24.1.3
cylindrical gear reducer 圆柱齿轮减速机 26.2.5.6
cylindrical glass condenser 圆筒形玻璃冷凝器 5.5.1.10
cylindrical metal shell 圆筒状金属壳 15.3.1.4
cylindrical shell 圆筒形机壳 11.3.3.6
cylindrical shell 筒体 21.2.4.1
cylindrical shell 圆筒形壳体 24.1.3.8
cylindrical vibration plate 圆筒形振动板 12.2.7.2

D

damper 调节板 8.2.1.9
damper position indicator 调节位置指示器 10.5.4.3
damper with remote control device 遥控风门机构 24.1.1.29
damping rubber plate 减振橡胶板 12.2.3.21
data 日期 21.1.2.16
data acquisition 数据收集 3.5.18.13

data-entry thumb wheels 数据进口压轮 23.4.1.16
data recording system 数据记录系统 3.4.19.10
Davy methanol synthesis process 英国Davy公司甲醇合成工艺流程 2.2.44
d.c.electromagnetic pump 直流电操作的电磁泵 16.5.4
D.C.(output) 直流电输出 10.3.1.7
deactivation 失活 1
dead shaft 静轴 8.8.5.21
deaerating tank 脱气槽 13.2.4.19
debenzoling 脱苯 2.3.14.27
debutanizer 脱丁烷塔 2.2.75.9
decantation 倾析 1
decantation tank 滗析槽 3.1.8.13
decanter 倾析器 2.2.9.10
  滗析器 2.2.106.25
decarbonizing 脱碳 2.3.3.11
decarbonization 脱碳 2.2.38.8
decarbonization tower 脱碳塔 2.2.50.6
decay of activity 活性衰减 1
de-C_3 column 脱C_3塔 2.2.69.15
de-C_4 column 脱C_4塔 2.2.69.17
deck 甲板 23.3.9.5
deck section 台面截面 4.3.4.8
declamping tank 分层罐 2.2.62.12
decontamination factor(DF) 去污指数;净化指数 1
de-C_1 tower 脱C_1塔 2.2.69.13
de-C_2 tower 脱C_2塔 2.2.69.11
deep-pinned type 深翅片式 4.3.2.(d)
deep solids bed 深的固体床层 4.3.3.17
deep water bearing strata 深水层 3.1.24.16
deep well pump 深井泵 16.1.2.2.39
deep well used for the subsurface injection of liquid 地下注入液体废物的深井 3.1.24
deethanizer 脱乙烷塔 2.2.68.12;2.2.71.8
deflected stream 偏离的物流 12.7.4.11
deflector 挡板 16.4.9.10
deflector ring 折流环 10.1.1.5
degassers 脱气器 5.5.3
de-gassing 脱气 3.1.32.4
degassing chamber 脱气室 5.5.1.7
degassing tower 脱气塔 2.2.54.6
5 deg divergence 扩张角5° 23.2.2.6
45-deg.lip flights 45°(唇形)抄板 8.8.1.18
90-deg.lip flights 90°(唇形)抄板 8.8.1.19
degradation products 降解产物 3.2.11.8
dehydration 脱水 2.2.68.11
dehydration tower 脱水塔 2.2.55.12;2.2.56.4
dehydrator 脱水机 2.3.18.c;9.5.19
dehydrogenation 脱氢 2.1.1.57
deionized water 去离子水 6.2.7.10
delayed coker 延迟焦化装置 2.2.93.8
delayed coking 延迟焦化 2.1.1.40
delayed-coking unit 延迟焦化装置 2.1.6
delivery 输送 2.3.9.37
delivery lift 提升[扬水]高度 16.5.1.8
delivery of the completed vessel 成品容器的输送 2.3.9.29
delivery pipe 排出管 20.2.3.6
delivery port 排气口 17.2.15.36
  出风口 17.5.1.11
delivery reel 卷取纸轴 2.3.11.28
delta connection 三角连接 25.4.3.3
demagnetizing coil 去磁线圈 12.5.1.16
demagnetizing coil 去磁煤 12.5.5.7
demister 除沫器 2.2.74.8
demethanizer 脱甲烷塔 2.2.68.18;2.2.71.5
demountable flange connexion 可拆法兰连接 5.5.2.14
denitrification 脱氮 3.1.25.5
denitrification column 反硝化塔 3.1.39.6
denitrified effluent 脱氮的污水 3.1.36.11
  去硝化污水 3.1.38.10
dense-bed center-fed column crystallizer 紧密床层中央加料式塔结晶器 7.2.14
dense-media cone vessel arrangements 锥形稠密介质分离器结构 12.5.4
dense-media flow sheet 稠密介质分离流程 12.5.5
dense-media Separation 稠密介质分离 12.5
dense phase 密相,浓相 1
dense-phase fluidized bed 密相流化床 14.1.3.19
densifier 增浓器 12.5.1.13
deodorizing preheater 除臭预热器 3.3.24.18
deoxyribonucleic acid(DNA) 脱氧核糖核酸 1
dephenolized effluent 脱酚后液体排出 3.1.5.7
depolymerizer 解聚装置 2.2.102.13
depropanizer 脱丙烷塔 2.2.68.13;2.2.71.6
depth of liquid 液体深度 21.2.5
derived 衍生的 25.1.1.5
E_{12}-desalinated water heater E_{12}-脱盐水加热器 2.2.34.11
desalination 淡化 25.1.1.16
desalter 脱盐设备 2.1.4.7
desiccant cooling cycle 干燥剂冷却循环 6.2.8

desiccant dehumidifier 干燥剂除湿器 6.2.8.11
desiccant wheel 干燥剂盘 6.2.8.5
design feature of pumps 泵的结构特点 16.1.2.2
design of catalytic reactor 催化反应器的设计 2.2.39
design of fluidized-bed systems 流化床系统的设计 14.1
design pressure 设计压力 21.1.2.9
design temperature 设计温度 21.1.2.9
de-sludging 除渣 15.6.3.30
desorption column 解吸塔,再生塔 5.1.1.6
desorption section 解吸段 6.3.7.4
desorption tube 解吸管 6.3.7.11
destroying 销毁 3.3.5
desulfuration 脱硫 2.2.38.6;2.3.3.6
desulfurization 脱硫 2.1.9.49
desulfurization and decarbonzation 脱硫,脱碳 2.2.36.6
desulfurization tower 脱硫塔 2.2.45.3
desulphurization 脱硫 2.3.14.28
desulphurizing 脱硫 2.3.7.20
desuperheater 脱过热器 20.1.7.13
detail 详图 10.5.1.11
detail drawing 局部详图 12.3.5
detarring asphalt 脱油沥青 2.1.1.39
developed path 液流途径 16.4.3.8
device (pasteurizer) 设备(巴氏消毒器) 2.4.1.8
dewater 脱水 12.5.1.7
dewatered biosolids flow to disposal 脱水后生物固体流去处理 3.1.26.(a).8;3.1.26.(b).12
dewatering 脱水 3.1.2.21;3.3.28.2
dewatering screw 螺旋脱水器 3.1.17.5
dewatering tank 脱水罐 6.3.1.9
dewatering with centrifuge and belt-filter press 用离心机及带式压滤机脱水 3.1.26.(a).7;3.1.26.(b).10
dewater pump 排水泵 16.1.2.3.15
dew point 露点 1
diagrammatic flow sheet 流程图 2.1.1
Diagram of an earthworm process 蚯蚓工艺图 3.3.28
dial-type vacuum ga(u)ge 指针式真空表 8.2.2.4
dialysis 渗析;透析 1
diameter of cylinder 筒体直径 21.2.5
diameters 直径 13.3.3(a)
diamond bin 金刚石料仓 12.7.5.1

diamond concentrate 金刚石精矿 12.7.5.5
diamond separation 采用X射线在矿石中分离出金刚石工艺 12.7.5
diaphragm 隔膜(式) 16.1.1.14 膜片 22.1.9.1
diaphragm cap 膜头盖 22.1.9.3
diaphragm capsule 膜盒 22.1.9.4
diaphragm disk 膜片 22.1.9.2
diaphragm pump 隔膜泵 16.1.2.2.5
diaphragm push rod 隔膜推杆 16.2.8.7
diaphragm relief valve 膜式泄压阀 2.1.7.20
diaphragm type metering pump 膜式计量泵 16.1.2.3.47
diaphragm-type pump 隔膜泵 16.2.7
diaphragm vacuum connection 隔板真空接头 9.3.3.17
diaphragm valve 隔膜阀 22.1.4
100′DIA steel tank 直径100英尺钢罐 3.4.3.2
dick valve 盘形阀 22.2.1.5
die 模具 13.2.1.12
die clamping cylinder 装夹模具用气缸 15.7.3.12
dielectrics 绝缘材料 12.7.2.6
dielectrophoresis 介电电泳 15.3
diesel 柴油 2.1.4.12
die turning gear 塑模旋转装置 15.7.5.3
differential conductivity bridge 差分电导测定电桥 3.4.22.13
differential-pressure cell 压差传感器 3.5.18.3
diffuser 扩散器 11.2.11.8 扩压器 16.4.27.5
diffuser body 扩压器机体 18.3.1.5
diffuser cone 锥形扩散器 3.1.14.10
diffuser discharge 扩压器出口 18.3.4.25
diffuser inlet 扩压器进口 18.3.4.24
diffuser outlet 扩压器出口 17.4.1.7
diffuser plate 扩压板 20.2.1.10
diffuser ring 扩压器口环 16.4.27.36
diffuser-type pump 扩压型泵 16.4.7
diffuser vane 扩压器叶片 6.1.7.2
diffusional separation processes 扩散分离过程 15.4
diffusion coefficient 扩散系数 1
diffusion device 扩散装置 3.1.3.7
diffusion pump 16.1.2.2.7
diffusion pump body 扩散泵体 5.5.2.13
diffusion pump cooling coils 扩散泵冷却盘管 5.5.2.12
digested sludge 陈化污泥 3.2.2.17

消化污泥　3.3.17.6;3.3.18.9
digester　蒸解锅　2.3.10.7
老化器　3.1.2.14
digester feed　消化器进液　3.2.6.4
digesting sludge　陈化污泥　3.2.6.10
污泥降解　3.3.17.5
digging buckets　挖斗　23.2.5
digital controller（temperature, pressure and stirring speed）　数字控制器（温度,压力和搅拌速度）　3.4.7.2
digital weight selector & display　数字式质量选定机和显示器　23.4.1.17
dilatant fluid　胀塑性流体　1
dilute acid　稀酸　6.2.1.5
dilute caustic　稀苛性碱　6.2.1.7
diluted　稀的　12.6.4.1
diluted froth　稀释过的泡沫　2.1.9.18
diluted H_2SO_4　稀释的硫酸　2.2.66.2
diluted liquid　稀释液　3.3.10.4
dilute-medium pump　稀介质泵　12.5.5.28
dilute-medium sump　稀介质储槽　12.5.5.27
dilute phase　稀相　1
dilute-phase fluidized bed　稀相流化床　14.1.3.20
dilution tank　稀释罐　2.1.9.17
dilution water（Optional: generally would only be needed in industrial wastewater treatment applications with strong influent concentrations）　稀释水（任选：一般只在工业废水具有很强的进水浓度处理的应用中才需要）　3.1.41.9
dimensional analysis　量纲分析;因次分析　1
dimensionless group　无量纲数群　1
dimethyl benzene　二甲苯　2.1.1.53
dimethylether　二甲醚　2.2.1.18;2.2.49.10;2.2.50.9;2.2.54.16
dimethyl terephthalate（DMT）production through p-xylene oxidation　对二甲苯氧化制对苯二甲酸二甲酯　2.2.64
diphenyl　联苯　5.5.3.5
dip leg　料腿　14.2.6.9
dipper　构斗　3.4.10.14
dipper systems　构斗系统　3.5.10.(d)
dip tube　沉浸管　3.5.18.8
direct acting steam pump　直动泵　16.1.2.4.6
direct chlorination　直接氯化　2.3.18.A
direct-coal-liquefaction　煤直接液化　2.2.90
direct contact cooler　直接接触式冷却器　2.2.27.3
direct-contact-refrigeration crystallizer（DTB type）　直接接触冷冻结晶器（DTB型）　7.2.2
direct desulfurization　直接脱硫　2.2.2.38
direct discharge to atmosphere　直接排放到大气中　3.3.25.12
direct evaporative cooler　直接蒸发冷却器　6.2.8.9
direct-fired air heater　直接燃烧空气加热器　10.3.5.2
direct firing　直接明火　21.1.2.10
direct-heat cocurrent rotary dryer　直接加热式并流回转干燥器　8.8.3
direct-heat rotary-dryer　直接加热式回转干燥器　8.8.2
directional　定向式　23.3.2.13
directional sieve tray　导向筛板　5.2.10.8
direction of forces on pistons　活塞受力　16.2.1
direction of gas flow　气体流动方向　10.3.1.1
direction of rotation　旋转方向　16.3.5.7
direct liquefaction　直接液化　2.2.89.12
directly fired　直接燃烧　3.3.15.(a)
direct oxidation　直接氧化　14.3.3.(a)
direct-process heater　直接过程加热器　25.5.1.(c)
direct rotary　直接加热回转　8.1.1.⑱
direct steam use　直接使用蒸汽　25.1.2.5
dirty air　脏空气　3.4.4.11
dirty air inlet　脏空气入口　10.5.1.25
dirty gas in　脏气体入口　10.5.11.5
dirty gas inlet　脏气体进口　10.4.1.11;10.5.15.1
dirty gas side　脏气体侧　10.2.1.27
dirty liquid in　脏液体入口　15.3.1.9
dirty oil　脏油　15.5.5.14
dirty oil inlet　脏油进口　15.5.5.1
dirty water outlet　脏水出口　10.5.1.6
dirty water pump　脏水泵　16.1.2.3.17
disc（转）盘　13.3.1.2
disc attrition mill　圆盘磨　11.2.15
disc brake　盘式制动器　9.5.10.10
discharge cone　卸料锥　15.6.1.18
discharge　排气口　17.2.4.6
discharge air　排气　8.6.2.7
discharge arrangements　卸料方式　23.2.3
discharge bowl　排出筒　16.4.24.9
discharge case　出口泵壳　16.4.28.33
discharge casing　排出壳体　16.4.19.10
discharge chamber　排出腔　16.3.2.4
discharge channel　分离液出口　9.5.19.15

discharge check valve 排出止逆阀 16.2.7.2
discharge chute 卸料槽 23.2.1.7
discharge coefficient 流量系数;孔流系数 1
discharge cone 排料锥,卸料锥 11.2.1.7
discharge cover 排出端泵盖 16.4.20.20
　卸渣管法兰盖 9.2.1.27
discharged filter cake 已卸出滤饼 9.3.9.6
discharge dia. 排出口直径 16.4.3.18
discharge door 排料门,卸料门 15.7.1.6
discharge elbow 排液弯管 16.4.22.16
discharge electrode 放电电极 10.3.1.15
discharge end 卸料端 23.3.7.(a)
discharge end thermometer 出口端温度计 17.4.2.28
discharge flange 排出口法兰 16.4.3.9
discharge for dried product 已干产品排出口 8.9.6.14
discharge gas 排出(高压)气体 20.1.3.8
discharge gates 卸料阀 24.1.12.5
discharge grate 出料格栅 2.2.87.3
discharge head 出口接头 16.4.25.24
discharge hole 成品出口 11.2.4.3
　排气孔 18.4.1.6
discharge lock 卸料阀 23.1.13.4
discharge manifold 排出腔 16.2.1.11
discharge mechanism 卸料机械装置 2.2.13.6
discharge(negative) electrons 放电电子(负极) 10.3.1.2
discharge nozzle 排[出]气口 18.1.1.9
　排气接管 18.1.4.14
discharge of gas from the coke ovens 炼焦炉煤气排出(口) 2.3.14.16
discharge opening 排出口 23.3.8.1
discharge outlet 排出[气]口 16.3.2.5
　卸料口 8.8.1.28
discharge passage 排气通道 17.2.7.10
discharge pipe 排出[气]管 16.2.1.13
discharge point 卸料口 23.3.3.5
discharge port 排出口[孔] 16.3.7.5
　卸料口 24.1.13.3
discharge pressure 排出压力 16.2.1.10
discharge pump for concentrated liquor 浓缩液排出泵 7.1.2.20
discharge ring 出口环 16.4.19.14
discharge roll 卸料辊子 3.9.7.2
discharge screw 输出螺旋 8.4.6.15
discharge side 排出端 16.3.6.9
discharge side bearing cover 出口侧轴承压盖 16.4.19.1
discharge side sleeve 出口侧轴套 16.4.19.8

discharge sludge 卸渣 9.4.2.12
discharge spout 排料口 16.4.22.15
　排液管 3.3.24.2;11.4.3.8
discharge system 输电系统 10.3.1.5
discharge system support insulator 支承输电系统的绝缘体 10.3.1.5
discharge tube 出料管 11.5.5.7
discharge valve 排气阀 16.2.3.36
　排气阀 17.2.2.2
discharge valve of second stage 第二级排气阀 17.2.7.14
discharge-valve plug 排出阀螺塞 16.2.4.26
discharge vane edge or tip 排出叶片之边缘 16.4.6.7
discharge vane tip 排出叶尖 16.4.3.12
discharge water 排放水 2.2.49.12;2.2.50.11
discharging 卸料 9.3.5.9
disc-mounted flatblade turbine 平直叶圆盘涡轮式 14.4.4.1
disc nut 阀盘螺母 22.1.1.10.4
disc set 碟组件 9.5.16.3
disc spray thrower 盘式淋洒器 5.3.4.8
disc stack 料盘空筒 9.5.12.7
disengager 分离段 14.3.4.20
disengager 释放器 2.3.5.7
disengaging chamber 分离室 23.1.7.5
disengaging space 分离空间 14.1.12.6
dished and perforated plates concave both upward and downward 上凸和下凹的碟形多孔板 14.1.8.(e),(f)
dish granulator 盘式造粒机 13.3.3(a)
disinfection 消毒 3.1.25.12;3.1.30.B.12;3.1.37.15
disintegrating the feed 破碎物料 8.4.5.(a)
disintegration 破碎 1
disintegrator 气体洗涤机 2.1.8.10
　粉碎机 2.3.10.5;3.3.22.5
　破碎机 8.4.5.32
disk 轮盘 15.7.6.3
　金属盘 9.5.15.8
disk atomizer 转盘雾化器 8.6.3.8
disk centrifuge 碟式离心机 1
disk-centrifuge bowl 碟式分离机转鼓 9.5.14
disk stack 叠装碟组 9.5.15.9
disk-type centrifuge 盘式离心机 2.1.9.23
disk type steam trap 盘式疏水阀 22.2.2
dispatch 运送 2.3.8.6
dispersed-phase fluidized bed 稀相流化床 14.1.3.22

disperser 分散设备 8.4.6.22
disperser breaks 分散器 3.1.13.3
dispersion mill 分散磨 11.4.1
dispersion sling 抛扬分散器 8.4.1.20
dispersion stack 弥散烟囱 3.1.22.14
disposable pack 一次性包装 2.3.13.28
disposal 排弃 12.5.5.10
　　处置 2.2.5.23;3.3.25.8
disposal formation 处理系统 3.1.24.8
disposition of solids 固体沉积 9.1.1.10
dissolution in solvent 溶剂溶解 2.2.2.31
dissolve 溶解 2.2.90.4
dissolved-air flotation 溶气浮选 3.1.1.5
dissolved-air flotation system 溶气浮选系统 3.1.14
dissolved solids 溶解固体 3.3.18.13
dissolver 溶解器 2.2.94.6
dissolving tank 溶解槽 2.3.10.37
distance collar 定距轴环 16.4.18.7
distance piece 定距板 11.1.5.17
　　间隔块 11.1.8.21
　　隔离室 17.2.15.37
distance plate 隔板 8.5.1.16
distance ring 间隔环 13.1.2.29
distilland 蒸馏液
distilland feed 蒸馏液进料口 5.5.2.5
distilland withdrawal pump 蒸馏液出料泵 5.5.1.13
distillate 馏出液 2.2.79.11;5.5.1.47
distillate drain 馏出液排出口 5.5.2.17
distillate out 馏出液出口 5.5.1.27
distillate outlets 馏出液出口 18.5.1.11
distillate pipe 馏出液管 5.5.4.25
distillate withdrawal pump 馏出液出料泵 5.5.1.14
distillation 蒸馏 2.2.97.8
distillation column 精馏塔,蒸馏塔 2.2.61.12;2.2.62.8;2.2.70.11;5.1.1.1
distillation equipments 蒸馏设备 5
distillation process 蒸馏工序 2.2.45.15
distillation tower 蒸馏塔 6.1.3.2
distillation with chemical reaction 反应蒸馏 1
distilled gasoline 直馏汽油 2.1.1.7
distributing conveyor 分配送料器 23.2.5.9
distributing plate 分散板 12.3.1.10
　　分布板 9.5.4.40
distribution plate 分布板 10.3.1.18
distributor 分配管[器] 13.2.4.18
　　分布板[器] 3.1.42.6;14.1.3.6
distributor 分配器 18.4.2.11
distributor 布料器 2.2.12.7

distributor grid plate 分布器栅板 2.1.10.9
distributor with helical blades 带螺旋形叶片的分配管 13.2.4.18
district heating 地区加热 25.1.1.10
divided-solids inlet 分散固体入口 4.3.2.9
diverging section 扩大截面 3.5.4.8
diverter 转向器 23.4.2.10
　　分流器 3.5.10.6
diverter valve 三通阀 12.1.5.5
diverter valve 分流阀 23.1.2.23
diverter valve 换向阀 23.1.8.22
divided flow 分流流动 4.1.2.(j)
divided-solids bed 分散固体床层 4.3.2.2
divided-solids discharge 分散固体排料 4.3.4.33
divided solids feed 分散固体进料 4.3.4.21
divided solids in 分散固体进口 8.9.6.1
divided-solids out 分散固体出口 4.3.2.8
dividing edge 分流板 12.7.4.12
division bridge 分配块 9.3.2.26
division plate 隔板 15.5.5.5
division strip 间隔板条 9.3.3.28
DME 二甲醚 2.2.51.9;2.2.70.14
DME agueous solution 二甲醚水溶液 2.2.70.10
DME column 二甲醚塔 2.2.52.3
DME+H_2O 二甲醚+水 2.2.77.4
DME product 二甲醚产品 2.2.52.7
DME(dimethyl ether)reactor 二甲醚反应器 2.2.51.6;2.3.4.1
DME reactor feed/product HX 二甲醚反应器进料/产物氢交换 2.2.51.4
DME tower 二甲醚塔 2.2.49.8;2.2.50.7;2.2.51.3
DMT 对苯二甲酸二甲酯 2.2.64.16
DMT column 对苯二甲酸二甲酯塔 2.2.64.10
Do(1)(dissolved oxygen) 溶解氧量 3.1.2.8
D.O. analyzer 溶解氧分析仪 3.1.46.9
dock pump 船坞泵 16.1.2.3.51
doctor blade 刮刀(片) 9.3.2.23
dodecylbenzene 十二烷基苯 14.3.1.27
dodecylbenzene sulfonate 十二烷基苯磺酸盐 14.3.1.(f)
doffer 滚筒 8.3.4.23
"Dollar"plate "Dollar"板 14.1.7.(e)
dome of the tun 桶顶 2.3.12.45
dome-roof 拱顶罐 21.2.1.2
door 门 11.1.7.12
door for adding pebbles 卵石加入门 14.3.3.10

door for removing pebbles 卵石移出门 14.3.3.9
door latch 排料启闭器 15.7.1.4
door operating device 门的(自动)启闭装置 8.8.10.18
door support 排料门底座 15.7.1.5
D. O. probe 溶解氧传感器 3.1.46.11
Dorrco Fluosolids reactor Dorrco Fluosolids 反应器 14.2.4
dosing 定量 9.5.12.15
dosing ring 定量环 9.5.12.16
dosing ring chamber 定量环室 9.5.12.17
dotted outline 虚线轮廓线 15.5.5.3
double acting 双作用 16.1.1.6
double-acting circulation propeller 双作用循环搅拌叶片 7.2.10.4
double-acting liquid end 双作用泵缸 16.2.2
double-acting piston and compressor cylinder 双作用压缩机气缸与活塞 17.2.6
double-bowl vacuum centrifuge 双鼓真空离心过滤机 9.5.17
double branch elbow 双支管弯头 22.3.1.12
double-cone air classifier 双锥空气分级器 12.3.3
double cone mixer 双锥(形)混合器 12.1.4(f)
double-cone revolving around long axis 绕长轴旋转的双锥混合器 12.1.3.(d)
double deck 双台式 23.3.2.9
double-drum dryer 双鼓干燥器 8.7.2
double drum dryer with splash-feed 飞溅布料式双滚筒干燥器 8.7.1.(c)
double drum dryer with spray-feed 喷洒加料式双滚筒干燥器 8.7.1.(d)
double drum dryer with top-feed 顶部加料式双滚筒干燥器 8.7.1.(b)
double-drum mill 双滚筒磨式 4.3.1.(c)
double flap valve 双瓣阀 3.3.24.23
double-gate airlock 双闸进料器 23.1.7.9
double interior wall 内部双层壁 18.2.4.11
double mechanical seal 双端面密封 19.1.8
double paddle mixer 双桨混合器[机] 8.4.5.18
double-pipe exchanger 套管换热器 4.2.5.17
double-pipe longitudinal finned exchanger 套管式纵向翅片换热器 4.2.5.19
double roll feed crusher 双辊破碎给料机 11.2.6.2
double row balanced opposed compressor 双列对称平衡压缩机 17.2.1.9
double screw compressor 双螺杆压缩机 17.3.1.(d)

double shaft hammer crusher 双轴锤式破碎机 11.1.7
double spiral agitator 双螺旋搅拌器 8.8.10.8
double spiral ring 双螺旋环 5.3.2.(c)
double split flow 双分隔流动 4.1.2.(i)
double stage rotating blade vacuum pump 双级旋片式真空泵 18.1.3
double suction 双吸 16.1.1.35
double suction centrifugal pump 双吸离心泵 16.4.12
double-suction impeller 双吸叶轮 16.4.6
double-suction wet-pit pump 双吸排水泵 16.4.23
double valve 双阀 16.1.1.36
double volute 双蜗壳 16.4.9.31
double-volute casing 双蜗壳 16.4.14
dowel pin 定位销 22.1.8.4
down blast 自上往下送风 17.5.2.1
downcomer 下水管 14.2.2.3
　下降管 3.3.10.2
　降液管 3.1.32.26;5.2.1.18
downcomer apron 降液板 5.2.2.8
downflow model Ⅱ catalytic-cracking unit 下流式Ⅱ型催化裂化装置 14.2.8
downflow packed-bed-system (denitrification) 下流或填充床系统(反硝化) 3.1.40.8
down spout 降液管 5.2.8.3
downstack 下降管 11.5.2.10
downstream processing 下游处理;后处理 1
Dowtherm gravity-return system 导热姆的重力回路系统 25.6.2
Dowtherm pumped-return system 导热姆的泵回路系统 25.6.3
Dowtherm vaporizer 导热姆蒸发器 25.6.1
draft tube 导流筒 6.1.12.7
draft-tube-baffle crystallizer 导流筒-挡板结晶器 7.2.6
drag-type enclosed conveyor-elevator 密闭型刮板式输送提升机 23.3.7
drain 排水 12.5.1.5
drainage 排水 3.4.3.10
drainage channel 排液沟槽 9.3.7.1
drainage deck 排液层面 9.4.11.6
drainage pump 疏水泵 16.1.2.3.10
drainage station pump 排灌站水泵 16.1.2.3
drainage surface 排水表面 9.1.3.6
drain cock 排污阻塞 15.5.2.2
drain connection 排液接口 4.1.1.33
drained straum 排干层 6.2.3.6

draining 排放 6.3.3.11
drain outlet 排水出口 12.5.2.5
　　排放口 13.5.1.10
drain plug 排污螺塞 22.2.1.11
　　排放丝堵 22.2.3.7
drain plug for liquid end 泵缸端的放液螺塞 16.2.3.28
drain valve 排放阀 14.1.11.12
　　放泄阀 15.5.5.4
drain valve for steam end 汽缸端的放液阀 16.2.3.8
drang-tank-type dense-media separatory vessel 拖曳槽型稠密介质分离器 12.5.2
drawing machines 拉制机 2.3.9.7
drawn upwards 引上 2.3.9.10
draw work 绞车 2.1.2.11
dray 制动器 9.5.11.11
dried material discharge conveyor 干燥产品卸料输送器 8.8.6.3
dried sludge 干污泥 3.3.25.9
drier 干燥机 2.3.8.38
　　干燥器 2.2.115.17
　　烘干机 2.3.16.13
drier duct 干燥器风管 8.4.5.10
drill-cast rotary switch 旋转式定位开关 25.4.1.18
drilled orifices 钻孔 11.5.1.2
drilling bit 钻头 2.1.2.21
drilling cable 钻油井用的钢丝绳 2.1.2.7
drilling pipe 钻杆 2.1.2.19
drilling rig 钻井机械 2.1.2.1
drill pipes 钻杆 2.1.2.6
drip leg drain 竖的引管排水口 17.6.2.10
drip shield 液滴防护屏 18.2.4.16
drive 传动装置 2.2.12.3
　　驱动装置 3.5.10.12
　　驱动器 8.8.1.29
drive assembly 驱动装置组件 13.3.1.8
　　驱动器装配体 8.8.1.13
drive belt 传动皮带 13.2.3.2
drive control 驱动控制器 15.6.1.8
drive end view 驱动端视图 8.8.10.20
drive gear 主动齿轮 11.1.2.1
　　传动齿轮 11.2.1.6
　　驱动齿轮 11.2.3.2
drive motor 驱动马达 15.6.1.2
drive motor for cake-removal screw 滤饼移出螺旋的驱动电机 9.4.2.5
driven gear 从动齿轮 11.1.3.6
driven screw 从动螺杆 13.6.2.11
driven shaft 从动轴 12.6.2.9

driven sheave 传动皮带轮 9.5.7.7
driver 驱动装置 24.1.8.4
driver half 驱动装置侧 16.4.9.24
drive roll 驱动辊 9.6.2.9
driver unit 驱动装置 7.1.11.9
drive seat 传动座 19.1.1.6
drive shaft 驱动轴 12.1.1.7
　　传动轴 16.3.6.7
drive socket 传动装置轴座 23.3.5.3
drive sprocket 主动链轮 23.3.7.1
drive type of pump 泵驱动型式 16.1.2.4
drive unit 驱动装置 15.6.1.32
drive zone 推进区 2.2.85.24
driving belt 传动带 9.2.3.2
driving gear 驱动齿轮 11.1.4.7
　　主动齿轮 16.3.1.4
driving magnet 驱动磁铁 16.5.5.8
driving pinion 传动小齿轮 11.1.4.6
driving roll 驱动滚轮 12.1.1.19
driving screw 主动螺杆 16.3.2.10
driving shaft 主动轴 16.3.2.8
　　传动轴 24.1.9.2
driving sheave 传动皮带轮 11.1.4.1
　　驱动轮 6.1.18.10
driving wheel 驱动轮 11.1.7.23
drop 液滴 13.5.3.1
drop arch 凹形拱墙 24.1.4.6
drop separator 液滴分离器 2.1.8.11
drum 转鼓 12.1.1.13
drum 汽包 25.5.3.4
　　滚筒 8.1.1.⑧
　　转鼓 9.3.7.10
drum arm 转鼓辐条 9.3.3.7
drum closing 封闭圆桶 23.4.1.6
drum filling 正装料圆桶 23.4.1.3
drum filter 转鼓过滤机 9.3.9
drum gear 转鼓齿轮 12.1.1.10
drummed 装桶 3.3.7.3
drumming operator 圆桶操作人员 23.4.1.5
drum mixer 转鼓混合器 12.1.4.(b)
drum nipple 转鼓连接管 9.3.3.10
drum piping 转鼓导管 9.3.3.6
drum rotation 转鼓旋转方向 12.6.4.18
drum shaft 转鼓轴 9.3.3.9
drum shell 转鼓外壳 9.3.10.12
drum waiting to be filled 待装料空桶 23.4.1.2
drum wash 转鼓洗液 12.6.9.2
　　磁鼓洗涤 12.6.9.4
dry 干燥 2.2.90.2
dry-bulb controller 干球温度控制器 20.2.

2.4
dry cake discharge 干渣排放口 9.2.3.8
dry cell battery 干电池 25.4.1.27
dry classification 干式分级 12.3
dry cooling tower 干式冷却塔 25.2.2.19
dry discharge 干料出口 8.9.1.8
dry divider 干料分流管 8.4.1.9
dry drum separator 干磁鼓分离器 12.6.10
dryer 干燥器 2.2.71.2;8.1
　干燥机 2.2.92.5
　烘干机 2.3.17.13
dryer grinder 磨碎干燥机 3.1.21.8
dryer shell 干燥器壳体 8.8.5.9
dry felt 干毛布 2.3.11.23
dry-fines permanent magnet separator 永磁体干细粉磁力分离器 12.6.8
dry gas 干气(体) 2.2.69.18;6.3.3.11
dry-gas meter 干式气体流量计 3.4.7.11
drying 干燥 2.2.101.16
drying chamber 干燥室 2.3.8.49
drying column 干燥塔 8.4.6.8
drying cylinder 烘缸 2.3.11.22
drying equipment 干燥设备 8
drying floor 干燥层 2.3.12.16
drying in heated room 烘房内烘干 2.3.7.24
drying kiln 烘干窑 2.3.12.29
drying of tow 丝束的烘干 2.3.8.58
drying tower 干燥塔 2.2.69.10
drying zone 干燥段 3.3.9.5
dry material 干物料 3.1.21.20
dry nitrogen 干燥氮气 2.2.53.9
dry oil-shale separator 干的油页岩分离器 2.2.88.11
dry polyamide chips 干燥的聚酰胺(碎)片 2.3.8.39
dry-powdered solids hopper 干燥粉末状固体加料斗 4.3.1.9
dry precleaner 粗粉尘预况降器 10.5.1.23
dry process 干燥过程 8.1.1.22
dry product 干成品 8.4.5.6
　干燥产物 8.5.6.31
dry product conveyor 干成品输送机 3.3.24.17
dry product discharge 干物料卸出 3.1.16.14
　干品出口 8.5.6.10
dry pulverized coal 干煤粉 2.2.105.1
dry rubbish 干垃圾 3.3.5.8
dry rubbish grate 干垃圾格栅 3.3.5.18
dry solid removal 干固体去除 2.2.25.13
dry solids removal 干固体去除器 2.2.31.17
dry-stage cyclone 干级旋风分离器 8.

1.36
dry vacuum pump 干式真空泵 9.3.1.8
dry zone 干燥区 9.3.10.13
D.S.M.screen D.S.M.筛 11.6.3.8
Dual anoxic zone system 双缺氧区系统 3.1.35
dual-flow tray 穿流塔盘 5.2.10.22
Duclone collector 杜康旋风分离器 10.1.2.(a)
40″duct 40英寸的管道 3.4.3.7
ductile cast iron 球墨铸铁 21.1.2.7
ducting 风管 10.4.4.4
dump chest 卸料池 2.3.10.81
dumped media (typical) 堆积的介质(典型的) 3.1.39.3
dumped packing 散装填料;乱堆填料 1;5.3.2
dumped tower packing 乱堆塔填料 5.3.4.42
dump grate operating cylinder 煤渣炉排操作气缸 24.1.5.8
dumping of coking coal 焦煤卸料场 2.3.10.1
dumping valve 卸料阀 14.1.9.9
duplex 双列，双缸 16.1.1.9
duplex acting steam-driven reciprocating pump 双作用蒸汽往复泵 16.2.3
duplex oil filter 复式滤油器 2.1.7.16
duplex piston pump 双活塞泵 6.3.2.16
duplex reciprocating compressor 双缸往复压缩机 17.2.9
duplex vertical reciprocating compressor 双缸立式往复压缩机 17.2.1.3
Dupont's ethylene glycol process 杜邦公司乙二醇生产工艺 2.2.66
Dupont's PACT process 杜邦PACT生产工艺 3.1.20
durene 均四甲苯 2.1.1.62
during rotation 回转过程 13.3.3(b)
dust 粉尘 10.1.2.33
dust air side 含尘空气侧 10.2.1.3
dust and tar 粉尘与焦油 2.2.11.7
dust-bag sleeve 集尘袋套管 8.8.11.13
dust bin 粉尘斗 10.4.1.18
dust catcher 集尘器 3.3.5.14
dust chute 微粉出口通道 13.3.5.18
　微粉流道 10.1.3.9
dust collected externally 外部收集粉尘 14.1.5.(e)
dust collector 粉尘收集器 12.1.5.9
　集尘[粉]器 10.4.4.7;12.6.4.19
dust contained gas 含尘气体 10.5.15.2
dust cover 防尘罩 12.2.1.11

dust discharge 粉尘排出(口) 10.2.1.28
dust disposal 粉尘处理 10.4.4.2
dust drum 粉尘筒 8.8.6.5
Dustex cyclone tube Dustex 微型旋风分离器 10.1.2.(d)
Dustex miniature collector assembly Dustex 微型(斜管)收集器组 10.1.2.(f)
dust extractor 吸尘装置 2.3.10.1
除尘器 2.3.12.40
dust filter 滤尘器 11.6.4.17
粉尘过滤器 23.1.6.4
dust hopper 粉尘料斗 10.1.2.31
集尘斗 10.2.1.24
dust laden air 含尘空气[气体] 10.1.3.13
dust-laden air in 含尘空气入口 10.5.4.10
dust-laden gas 含尘气(体) 10.5.4.14
dust-laden gas inlet 含尘气体进口 10.2.2.1
dust particles 粉尘粒子 10.3.1.24
dust-removal chamber 除尘室 2.2.88.18
dust remover 除尘 2.2.108.16
dust return 粉末回流 14.1.12.22
dust rotor 粉末转筒,旋风分离器 8.5.6.22
dust sample line 粉尘取样管 10.3.5.10
dust seal 防尘密封环 12.3.5.10
dust separator 粉末分离器 14.1.12.11
dust-settling chamber 灰尘沉淀室 2.3.10.6
dust shave-off 粉尘刮面板 10.1.2.19
dust system 除尘系统 23.1.9.13
dusttight sealed roller bearings 防尘密封滚动轴承 8.8.11.2
dust-tock hopper 粉尘锁闭料斗 10.3.5.13
dust trap 集尘器 10.1.2.6
dusty air side 含尘空气侧 10.2.1.32
dynamic 动力式 17.1.1.3
dynamic filter 动态过滤机 9.4.9
dynamic simulation 动态模拟 1
dynamic viscosity 动力黏度 1

E

early-mode 早期式样 9.5.21
early warning 事先警告 3.1.2.2
earth cover 土地表层 3.3.12.2
earthed double socket 接地双插座 25.4.1.6
earthed plug 接地插头 25.4.1.9
earthed rotor 接地转子 12.7.2.2
earthworm harvester 蚯蚓采集器 3.3.28.9
ebullition pump 沸腾泵 2.1.10.11
ebullating bed hydrotreater 沸腾床加氢装置 2.2.79.9

ebullient-bed catalytic reactor 沸腾床催化反应器 2.2.98.8
eccentric 偏心,偏心轮[套] 16.3.5.2
eccentric bottom weight 底部偏心块 12.2.1.17
eccentric reducer 偏心变径接头 14.1.9.3
eccentric reducer 偏心异径管 22.3.1.6
eccentric shaft 偏心轴 11.1.1.7
eccentric sleeve 偏心套(筒) 16.2.5.21
eccentric slide block 偏心滑块 16.2.5.20
eccentric top weight 顶部偏心块 12.2.1.4
eccentric vibrator 偏心式振动器 12.2.3.2
eccentric weight 偏心重块 11.3.3.4
economizer 省煤器,废气预热器 2.2.81.13
节能器 20.1.3.12
EDC (ethylene dichloride) 二氯乙烷 2.3.18.11
EDC NO.1 column EDC 一塔 2.3.18.12
EDC NO.2 column EDC 二塔 2.3.18.13
EDC cracker EDC 裂解装置 2.3.18.E
EDC purification EDC 净化 2.3.18.D
EDC quencher EDC 急冷塔 2.3.18.16
EDC recovery column EDC 回收塔 2.3.18.15
eddy current 涡流 10.1.2.20
eddy diffusion 涡流扩散 1
eddy flow 涡流 1
edges of orifices 孔边缘 10.5.1.10
edge(view) 侧视图 13.3.3(a)
eductor 喷射器 3.3.22.3
eductor mixer 喷射混合器 2.2.105.3
effectiveness factor 有效因子 1
effluent 流出(液,物) 3.1.1.8
废水(液) 3.1.15.5
出水 3.1.25.13;3.1.33.6;3.2.1.13
排水 3.1.43.6
effluent air cooler 流出液空冷器 2.3.4.16
effluent cooler 反应器流出液冷却器 2.3.4.19
effluent discharge 流出物排出口 9.4.4.8
中水排放 3.1.31.10
effluent drain 流出物放净口 9.4.4.10
effluent flume 废水槽 3.1.3.17
effluent screen 排液筛网 3.2.13.11
effluent sewer 废水管 3.1.3.16
effluent to ash conveyor 排放物去炉灰输送机 2.3.6.29
effluent to clarifier 排放液至澄清池 3.2.13.3
effluent to disinfection 流出水去消毒 3.1.41.7
effluent to polishing or filtration 排水去精制或过滤 3.1.39.5
effluent weir and wall 废水溢流堰板 3.1.3.15

egg shaped anaerobic digester 蛋形厌氧消化器 3.2.6
Eimco belt belt-discharge filter Eimco 带过滤机 9.3.7
ejection system 喷射系统 3.4.10.(c)
ejector-boosted 喷射增压式 16.1.1.48
ejector control unit 喷射器控制设备 12.7.5.11
ejector pump 喷射泵 18.3;16.1.2.2.46
ejector stage 喷射器级 18.2.4.10
ejector(steam-jet)refrigeration 喷射(蒸汽喷射)式制冷 20.1.1.(b)
ejector(steam-jet)refrigeration cycle 喷射(蒸汽喷射)式制冷循环 20.1.11
elastic ring 弹力环 17.2.10.3
elbow casing pump 弯管壳体泵 16.1.2.2.29
elbow for discharge 出水弯管 16.4.30.1
electrical 电力式 23.3.2.12
electrically augmented granular-bed filter 电力增强颗粒床过滤器 10.4.1
electrical precipitators 电除尘器 10.3
electrical radiant heater 电热辐射加热器 5.5.1.31
electric bell 电铃 25.4.1.15
electric buzzer 蜂鸣器 25.4.1.15
electric generator 发电机 2.2.111.7
electric heater 电(加)热器 8.5.3.6
electric heating elements 电热元件 24.1.7.12
electrician's knife 电工刀 25.4.1.63
electricity 电 2.2.108.4
电力 2.2.80.8;2.2.109.15;2.2.110.23;2.2.119.13
electricity grid 电网 2.2.81.15
electricity meter 电表 25.4.1.32
electricity meter cupboard 电表柜 25.4.1.31
electric light bulb 电灯泡 25.4.1.56
electric motor 电动机 4.5.1.1
electric parts 电工器件 25.4.1
electric power 电力 25.1.1.8
electric soldering iron 电烙铁 25.4.1.45
electric steam generator 电热蒸汽发生器 3.1.6.3
electrode 电极 16.5.4.10
electrode weight 高压电极重锤 10.3.4.9
electrodialysis 电渗析 1
electromagnetic 电磁(式) 16.1.1.51
electromagnetic stirring type superhigh-pressure autoclave 电磁搅拌式超高压反应釜 14.4.3
electronic amplifier 电子放大器 12.7.5.11
electrophoresis 电泳 1

electrostatic feeder 静电加料器(十) 12.7.2.10
electrostatic field 静电场 10.3.1.23
electrostatic liquid cleaner 液体静电洁净器 15.3.1
electrostatic precipitator 静电除[集]尘器 2.2.6.7
electrostatic separator 静电分离器 12.7.1
electrostatic sieve 静电筛 12.7.3
elevated tank 高位槽 2.2.57.3
elevating conveyor for sink material 下沉物料提升机 12.5.5.12
elevating flattener 提升整平机 23.4.3.9
elevation 正视 8.8.10.16
elevator 提升机 11.6.4.9
提叶器 11.6.5.8
eliminator 除沫器 20.2.3.18
elliptical head 椭圆形封头 5.4.1.1
eluant 洗提剂[液] 6.2.6.11
eluate exit 洗脱液出口 6.3.8.9
eluent inlet 洗脱液进口 6.3.8.2
eluent stream 洗脱液流 6.3.8.5
elutriating leg 析出段 7.1.14.4
elutriation 扬析 1
elutriation leg 淘析腿 7.2.5.17
elutriation liquid feed 淘析液加入 7.2.10.12
elutriation ring 扬析环 12.3.2.7
elutriation zone 淘析区 7.2.10.14
elutriator 洗提器 10.3.5.6
emergency bypass stack 应急旁路排风管 3.3.3.9
emission electrode 发射电极 10.3.5.19
emissions measurement 排放物测定 3.5
empirical model 经验模型 1
empty 流空 6.3.3.4
empty drum storage 空桶储存仓 23.4.1.1
empty return run 空回程 23.3.7.7
emulsion inlet 乳化液进口 6.1.8.4
encapsulation 胶囊化 1
enclosed 闭式 16.4.5.6
enclosed belt tunnel 封闭式传动度带通道 10.5.1.16
enclosed design 闭式结构 16.4.5.(b)
enclosed line shaft bearing 密闭的中间轴轴承 16.4.25.21
enclosed line shafting 密闭的中间轴 16.4.25
enclosure 机壳 4.3.4.22
end bars for receiving basket 受料筐端部板条 11.1.7.30
end cap 端盖 13.3.5.28
end cover 端盖 17.4.2.8

end cover bolt 端盖螺栓 17.4.2.9
end hammer carriers 端部锤盘 11.1.7.27
end liner 侧衬板 11.2.4.6
end of dryer 干燥器终端 8.8.5.11
end of shell 壳体终端 8.8.5.2
end of stroke 外止点(行程末端) 17.2.2.4
endothermic reaction 吸热反应 1
end plate 端板 11.1.7.2
end-product 最终产品 13.5.3.3
end pulley 末端皮带轮 23.2.3.(b)
end ring 终端环 16.4.28.19
end to end loading 两端进料 12.1.4.(a)
energizing nozzles 强化喷嘴 11.5.4.8
energy balance 能量衡算;能量平衡 1
energy-recovery systems 能量回收系统 3.3.1
engine 发动机 2.1.3.12
enrichment 富集 1
enthalpy 焓
enthalpy-entropy diagram 焓熵图 1
entrained-bed reactor 输送床反应器 3.3.15.(d)
entrained solids 所夹带固体 14.1.12.10
entrainer 夹带剂 15.1.1.3
entrainer gas 夹带剂气体 15.1.1.14
entrainment 雾沫夹带
entrainment eliminator 除水器 4.6.2.14
entrainment separation stage 雾沫分离级 10.5.1.3
entrainment separator plate 雾沫分离板 10.5.4.21
entrance of feed liquid 料液入口 9.5.19.10
entrance of washing water 洗涤水入口 9.5.19.8
entropy generation 熵产生 1
envelope type 包套型 10.2.1.(b)
enzymatic hydrolysis 酶法水解 1
enzymatic reaction kinetics 酶反应动力学 1
enzyme membrane 酶膜 1
EO desorber 环氧乙烷解吸塔 2.3.19.11
EO scrubber 环氧乙烷洗涤塔 2.3.19.6
epoxy tube sheet 环氧树脂管板 15.2.1.10
eq.sp.(equip space) 等距 10.4.2.3
equality constraint 等式约束 1
equalization 均压 3.1.37.6
equalization basin 平衡水池 3.1.1.4
equation of state(EOS) 状态方程 1
equator zone plate 赤道带板 21.2.3.6
equilateral triangle pitch 正三角形排列 4.1.5.3
equilibrating 平衡 6.3.3.8

equilibrium constant 平衡常数 1
equilibrium still 平衡釜 1
erlenmeyer flask 锥形(烧)瓶 2.3.11.2
erosion-resistant 耐磨性 10.5.9.3
erosion-resistant throat 耐磨性喉部衬环 10.5.9.3
Escher-Wyss or Tsukushima DP (double propeller) crystallizer DP(双桨)型结晶器 7.2.10
Essential features of a flow-through pasteurizer 流през式巴氏消毒器的基本特征 2.4.1
ester column 酯塔 2.2.64.5
ester distilled tower 酯蒸出塔 2.2.57.7
ester drying tower 酯干燥塔 2.2.57.10
esterification 酯化 2.2.66.11
esterification reaction tower 酯化反应器 2.2.57.4
esterification reactor 酯化釜 2.2.60.5
酯化反应器 2.2.64.2
esterification tower 酯化塔 2.2.58.10;2.2.59.14
ethane 乙烷 2.2.71.18
ethanol 乙醇 2.2.40.11;2.2.57.14;2.2.59.1
95% EtOH 95%乙醇 2.2.41.13
ethanol,acetic acid 乙醇,乙酸 2.2.58.1
ethanol tower 乙醇塔 2.2.59.16
ethyl acetate and ethanol 乙酸乙酯与乙醇 2.2.59.5
Ethyl acetate production process 乙酸乙酯生产的工艺流程 2.2.58
ethyl acetate products 乙酸乙酯产品 2.2.58.3
ethyl acetate tower 乙酸乙酯塔 2.2.59.15
ethylbenzene 乙苯 2.1.1.54
ethylene 乙烯 2.2.67.22;2.2.68.20;2.2.71.17;2.3.18.1;2.3.19.5;14.3.5.31
ethylene compressor 乙烯压缩机 17.1.2.9
ethylene fractionator 乙烯耗馏塔 2.2.71.10
ethylene glycol 乙二醇 2.2.66.14
Ethylene oxide 环氧乙烷 2.3.19.15
Ethylene oxide with water 含水的环氧乙烷 2.3.19.17
European design 欧洲型 9.4.9.(a)
exhaust fan 排气引风机 23.1.7.7
evacuate 抽气 3.5.14.20
evacuation chamber 排气室 2.2.84.9
evaluating 评估 3.1.19
evaporation 蒸发 2.3.6.19
evaporate 蒸发 8.1.1.9
evaporating temp. 蒸发温度 20.2.4.6

evaporation 蒸发 1
evaporation chamber 蒸发室 7.2.10.3
evaporation of chlorobenzene and caustic soda 单氯苯和苛性钠的蒸发 2.3.8.11
evaporator 蒸发器 2.2.55.8;2.2.61.3;2.2.76.8;2.3.8.11;3.3.22.12;7.1;20.2.8
evaporator crystalizer 蒸发结晶器 7.2.13
evaporator-feed surge pond 蒸发器-进料缓冲池 2.3.6.23
evaporator or feed surge tank 蒸发器或进料缓冲槽 2.3.6.21
evaporator types 蒸发器型式 7.1.13
evaporator wall 蒸发器壁 5.5.2.3
evaporator with rotating brush 带回转刷的蒸发器 7.1.6
evap.press. 蒸发压力 20.1.7.3
evap.temp. 蒸发温度 20.1.7.3
evolutionary operation(EVOP) 调优操作 1
Evonik catalytic distillation methyl t-butyl ether (MTBE) process Evonik 催化蒸馏制甲基叔丁基醚(MTBE)工艺 2.2.65
excess air 过剩空气 3.1.14.15
excess biomass 过量生物质 3.2.11.13
excess biosolids 剩余的生物固体 3.1.31.9
excess char 过量炭 3.3.15.16
excess condensate to deep well 剩余冷凝液去深井 2.3.6.31
excess enthalpy 超额焓;过量焓 1
excess sample return line 剩余样品回流管线 3.5.3.8
excess sludge 过剩的污泥 3.1.28.9;3.4.1.9
excess sludge (reduced) 过剩的污泥(减少了) 3.1.30.A.8
exchanger with packed floating tube sheet and lantern ring 具有填料函浮动管板和套环的换热器 4.1.1.(e)
exciting winding 激磁绕组 16.5.4.8
exergy 㶲;有效能;可用能 1
exhaust 废气 2.2.113.12;13.4.1.6
排气(口) 25.3.2.10
排出 3.5.19.11
exhaust air 废气 10.5.13.13
排(出空)气 4.3.3.24
exhaust-air stack 排气烟道 8.3.1.7
exhaust-air tubes 废空气管 8.7.2.8
exhaust and ash 废气和灰分 3.3.27.15
exhaust barrel 排气筒 11.1.8.32
exhaust blower 排风机 11.6.1.7
exhaust chamber 排气风门 11.6.1.8
exhaust duct 排风[气]管道 8.3.4.17
exhauster 排气装置 10.1.3.6

引风机 10.2.3.15
exhaust fan 排风[气]机 2.3.12.15
exhaust gas 废气 3.1.45.7;3.3.24.9;3.3.25.11;25.1.12.10
exhaust-gas outlet 废气出口 8.8.9.7
exhaust gas stack 废气排放管 25.2.2.17
exhaust header 排气集管 8.4.6.12
exhaust hole 排气孔 11.1.8.7
exhaust stack 排气烟囱 3.3.11.7
exhaust steam 乏汽 14.3.3.3
废蒸汽 2.2.110.18
exhaust to atmosphere 废气通大气 8.6.1.8
existing plant 现有装置 2.3.2.11
existing supply lines 现有供应线 23.4.1.8
exit chute 出口斜槽 13.3.1.7
exit damring 出口斜料圈 13.3.1.6
exothermic reaction 放热反应 1
expandable joints 膨胀节 22.6.1.3
expanded bed 膨胀床 14.1.2.2
expanded level 膨胀液面 21.1.10.7
expanded metal packing 膨胀金属网填料 5.3.2.(n)
expanded metal packings placed alternatively at right angles 膨胀金属网填料 5.3.2.(o)
expanded sections 扩大断面 8.4.1.24
expanding lip piston 能伸缩唇形活塞 16.5.6.7
expanding tube 扩大管 8.4.3
expansion 膨胀 20.2.4.7
expansion bellows 膨胀波纹管,膨胀节 8.4.5.11
expansion device 膨胀机械 20.1.3.19
expansion engine 膨胀机 20.2.5.10
expansion joint 膨胀节,膨胀接头 3.3.24.6;4.1.1.14
expansion rollers 膨胀托辊 8.8.10.11
expansion tank 膨胀槽 20.1.7.10
膨胀罐 2.2.43.13
expansion valve 膨胀阀 20.2.5.9
expert system(ES) 专家系统 1
exploded cross section at center 中部横断面 15.2.3.9
explosion panel 防爆膜(板) 8.4.6.11
explosion port 安全防爆孔 25.6.1.4
expression equipment 压榨设备 9.6
expression rate 压榨速率 1
extend neck 接筒 15.7.1.8
extend aeration 延时曝气 1
extension lead 延长线,接长用引线 25.4.1.10

extension plug 延长线插头 25.4.1.11
extension socket 延长线插座 25.4.1.12
extension spring 拉伸弹簧 19.3.1.11
extent of reaction 反应进度,反应程度 1
exterior 外部 8.2.2.13
external-combustion heat engine 外燃式热机 25.1.1.7
external cyclone 外部旋风分离器 14.1.4.(b)
external diffusion 外扩散 1
external heat transfer 外部热交换 14.1.4.(j)
external insulation 外保温层 4.4.2.6
external mechanical seal 外装式机械密封 19.1.2
externally adjustable 外部调节的 10.5.13.3
externally sealed floating tubesheet 外部密封的浮头管板 4.1.4.(t)
external mechanical seal 外装式机械密封 19.1.5
external reformer 外重整炉 2.2.113.4
external reinforcing ribs 加部加强圈 8.2.2.10
external shaft 外轴 13.1.2.21
ext.pressure toriconical head 受外压折边锥形封头 21.1.1.25
ext.pressure torispherical head 受外压碟形封头 21.1.1.63
ext.pressure tube 外压管 21.1.1.50
extract 萃取液,提取液 6.1.1.6
extracted gas 被抽气体 18.2.3.13
extraction 萃取 6
  提取 2.3.8.3
extraction column 萃取塔 3.1.5.3
extraction-condensing-steam-turbine cogeneration cycle 抽汽-凝汽-蒸汽透平的联合生产循环 25.1.2
extraction conveyor 抽提输送机 3.3.30.8
extraction equipment 萃取设备 6.1
extraction line 抽提管道 3.1.32.18
extraction operation 萃取操作 6.1.1
extraction raffinate 萃余液 6.1.1.9
extraction-solvent 萃取溶剂 6.1.3.15
extraction tower 萃取塔,抽提塔 5.1.1.7
extractive distillation 萃取蒸馏 1
extractive distillation column 萃取精馏塔 2.3.15.6
extractive oil 抽出油 2.1.1.80
extractive phase 萃取相 6.1.1.7
extractor 萃取器 2.3.8.37
extruder 挤压机 15.7.2
extrusion pelleting equipment 挤压造粒设备 13.2
Exxon donor solvent process Exxon 供氢溶剂过程 2.2.97
Exxon Donor Solvent process for direct hydrogenation of coal to produce liquid fuels Exxon Donor 煤直接加氢制液体燃料的溶剂生产工艺 2.2.79
Exxon Mobil MTG (methanol to gasoline) process. The feedstock is a mixture of dimethyl ether (DME) and water produced from methanol 埃克森美孚公司 MTG(甲醇制汽油)工艺流程,原料是由甲醇制得的二甲醚(DME)和水的混合物 2.2.77
Exxon Model Ⅳ fluid catalytic cracking (FCC) unit Exxon Ⅳ 型催化裂化(FCC)装置 14.3.4.(a)
Exxon transfer-line catalytic-cracking unit Exxon 输送管催化裂化装置 14.2.9
eye diameter 吸入口直径 16.4.6.10

F

F(1)(feed) ①加料,进料 ②供给 12.4.1
face view 正视图 13.3.3(a)
faeces pump 排污泵 16.1.2.3.16
failure diagnosis 故障诊断 1
Fairmount single-roll crusher Fairmount 单辊破碎机 11.1.2
falling film 降膜 15.4.1.9
falling-film evaporator 降膜蒸发器 7.1.5
falling-film exchanger 降膜式换热器 4.2.5.12
falling film type absorber 湿壁降膜吸收塔 5.4.2
fan 风机 2.2.112.5;2.2.115.9;10.4.4.1;10.5.13.12
  风扇 11.1.6.8
  引风机 13.4.1.5
  排风机 17.1.1.14
  鼓风机 23.1.1.14
fan air 风机空气 11.5.3.4
fan blade 风机叶片 4.2.4.1;12.3.5.19
fan blade arm 风机叶片支架 12.3.5.20
fan motor 风机电机 8.2.1.1
fan ring 风机外圈 4.5.1.5
fan stack 风筒 4.6.2.1
fantail 扇尾式 25.2.2.1
fan/venting system 风机/排风系统 3.3.20.4
fan wheel 风扇轮 11.2.11.7
far end 远端 4.3.2.7

fast 快速床 14.1.1.5
fast bed 快速床 14.1.2.5
fast-breeder reactor 快中子增殖(反应)堆 25.2.1
fast fluidization 快速流态化 14.1.1
fat 油脂 14.3.1.13
fat heavy solvent 富的重溶剂 6.1.9.2
fat light solvent 富的轻溶剂 6.1.9.19
FBBR(fluidized bed bioreactor) 流化床生物反应器 FBBR 3.2.12
FeCl₃ 三氯化铁 3.1.34.16
FD fan 鼓风机 25.5.3.21
feed 进料,加料 11.2.8.1
　　给料 3.3.23.3
feed air compressor 进料空气压缩机 2.2.53.1
feed-all 总进料 8.4.1.18
feed and drain 进料与排放 9.3.7.7
feedback control 反馈控制 1
feed bin 进料仓 11.6.4.1
feed-bottom exchanger 料液-塔底液换热器 3.1.7.2
feed box 进料槽 15.6.5.2
feed chain 给料链;进料链 2.3.10.67
feed coal 原料煤 2.2.14.8
　　喂送输送
feed compressor 进料压缩机 2.2.37.2
　　供料压缩机 2.2.48.1
feed conveyer[conveyor] 喂送输送器[机] 23.4.2.6
feed crusher 破碎给料机 11.2.6.4
feed distributor 进[加]料分布器 9.5.10.2
feed dosing 定量加料器 8.5.7.7
feeder 加[给]料器 11.2.11.1
feeder assembly 进料器组件 8.8.8.9
feeder panel 馈电屏 25.4.2.4
feeder speed reducer 加料器减速器 11.2.8.4
feed flowmeter 加料流量计 5.5.3.4
feed funnel 进料斗 9.5.6.9
feed gas 进料气体 14.2.3.1
　　原料气 2.2.7.1;2.2.37.6;2.2.48.5
feed gas compressor 原料气压缩机 2.3.3.4
feed granulating liquid 成粒液加入 13.4.1.24
feed hopper 加料斗 11.2.10.1
feed hopper line 供料斗管线 12.3.4.3
feed hopper liner 料斗衬板 12.3.4.4
feed in 加料 15.2.2.1
　　进料 2.3.11.13
feeding in the batch 分批进料 2.3.9.2
feeding trough 送料槽 11.1.8.19

feed injector 射流器进料口 11.5.1.5
feed inlet 加料口 11.2.1.1
feed inlet 供液口 15.6.4.20
feed inlet gaseous 气态进料 7.2.8.1
feed inlet with airtight valve 带气密阀的加料口 23.1.2.33
feed launder 进料槽 12.4.2.9
feed liquid inlet 供液口 9.5.4.27
feed liquor 料液 6.2.6.3
feed manifold 进料集管 11.5.1.8
feed material 进料 11.5.3.1
feed melt 熔融物料加入 13.3.4.1
feed nozzle to evaporator 给蒸发器的进料管 5.5.2.1
feed opening 进料口 11.1.7.1
feed pan 料盘 8.7.1.5
feed plate 进料板 11.1.3.1
feedpoint 进[加]料口 5.5.1.38
feed preheater 料液[进料]预热器 18.5.1.10
feed preparation 进料预处理 2.3.5.3
feed puddle 进料槽 8.7.2.2
feed pump 进料泵 13.4.1.10
　　给水泵 16.1.2.3.3
　　加料泵 18.5.1.9
feed ram 进料推料机 24.1.5.1
feed & residue thermocouples 进料液与残液热电偶 18.5.1.5
feed roller press 进料滚压机 8.3.2.2
feed screw 进[进]料螺旋 11.2.9.6
feed-screw sprocket 送料螺旋;链轮 9.3.3.1
feed section 加[给]料段 13.2.1.16
feed sludge 污泥加入 3.2.2.14
　　污泥进料 3.3.25.7
feed slurry 滤浆进口 9.2.2.16
feed slurry inlet 滤浆入口 9.2.5.4
feed solids 固体进料 15.1.1.1
feed solution 进料溶液 7.1.14.1
feedstock 原料 1
feedstock coal 原料煤 2.2.33.1;2.2.35.17
feedstock oil 原料油 2.3.16.3
feedstocks 原料 2.2.3.1
feed tank 进料(液)槽[罐] 18.5.1.7
feed throat 加料喉管 13.2.3.6
feed-through insulator[terminal] 馈电端子 25.4.5.2
feed tray 送料盘 12.2.1.6
feed valve 进料阀 9.5.6.21
feedwater 24.3.1.12
　　给水 2.2.110.8
feedwater 进水;给水;供水 25.5.3.1

feedwater distribution orifice　给水分布孔　15.6.4.10
feedwater heater economizer　进水加热器省煤器　25.5.3.3
E 02-feed water heater for intermediate pressure boiler　E 02-中压锅炉给水加热器　2.2.34.2
E 17-feed water heater for intermediate pressure boiler　E 17-中压锅炉给水加热器　2.2.34.13
E 11-feed water heater for low pressure boiler　E 11-低压锅炉给水加热器　2.2.34.10
feedwater inlet　供水口　14.3.2.12
feedwater inlet　水加入口　5.5.1.6
feedwater inlet　进水口　5.4.2.14
feedwater line　给水管线；给水管路　25.2.1.12
feedwater line　给水管线　25.2.2.10
feedwater plug　注水口旋塞　16.4.28.25
feed well　加料井　15.6.1.12
felt　毛布，(毛)毡　2.3.11.51
felt washer　毡垫圈　9.3.3.43
female flange　凹面法兰　22.4.3.
female screw　阴螺杆　17.3.2.4
Fe₂O₃/wood chip mixture　氧化铁/木屑混合物　3.3.20.12
fermentation　发酵　1
fermentation thermometer　发酵温度计　2.3.13.9
fermentation vessel　发酵槽　2.3.13.8
fermenter　发酵罐　3.2.10.8
fermenter mixed liquor return　发酵罐混合液回流　3.2.10.10
fermenter/thickener　发酵罐/增稠器　3.2.10.12
fermenting cellar　发酵室　2.3.13.7
ferritic stainless steel　铁素体不锈钢　22.6.3.⑰
ferrous metal　铁金属　3.3.16.12
ferrule wrapped with insulating fibre　外包绝热纤维的水管口密套　4.4.2.13
fertilizer　肥料　2.2.89.12
fiber packing　纤维填充物　10.4.2.10
fibreglass-reinforced epoxy support tube　环氧玻璃纤维增强支持管　15.2.4.3
fibrous　纤维状　8.1.1
fibrous-bed scrubber　纤维填充床涤气器　10.5.11
fibrous compression packing　压紧纤维填料　19.2.3.8
fibrous contacting element　纤维接触元件　10.5.11.3
fibrous solid　纤维状固体　8.1.1.20
field　磁场　16.5.4.7
filament tow　拉出来的丝束　2.3.7.28
filled bag take-away conveyor　满袋移去输送器　23.4.2.3
filled-pallet gravity roll-conveyor　满货板重力辊子输送器　23.4.2.22
filled-pallet take-away conveyor　满货板移出输送器　23.4.3.2
fillet welds　填角焊缝　21.1.1.61
filling　装料　23.2.5.8
　　注入　6.3.3.6
　　充料　9.1.1
filling compound　封填绝缘膏　25.4.5.5
filling connection　注入接头　25.6.7.8
filling end　加料端　2.3.9.2
fill line　加料管线　25.6.7.13
fill pipe　填充管　14.1.9.1
　　加入管　2.1.7.13
film heat transfer coefficient　传热膜系数　1
filtrate connection　液连接件　9.5.18.8
filter　过滤器　2.2.20.4;2.2.117.8;2.2.118.10
filter air out　滤后空气出口　23.1.2.5
filter bag　滤袋　10.2.1.34
filter bag　滤袋　2.3.16.10;10.2.2.9
filter belt　过滤带　9.4.8.3
filter bouse　过滤室贮斗　6.3.2.10
filter cake　滤饼　9.1.3.9
filter cloth　滤布　9.1.2.11
filter coalescer　过滤聚结器　15.5.5
filter compartment　过滤层段　3.1.10.4
filter cover support　过滤机盖支架　9.3.3.26
filter cylinder　过滤器筒体　10.2.3.1
filter drum　过滤机转鼓　9.3.3.25
filtered cake　滤饼　9.3.5.4
filtered liquid　滤出液　9.1.3.14
filter fan　过滤器风机　11.6.4.19
filter holder　滤膜托　3.4.11.9
　　过滤器夹持器　3.4.16.8
filter holder location　过滤器夹持器配置　3.5.16.7
filtering　过滤　9.3.5.10
filtering medium　过滤介质　9.5.3.12
filtering off the wort　过滤麦芽汁　2.3.12.50
filter leaf　滤叶　9.2.1.16
　　(过)滤(叶)片　9.2.2.7
filter leaf for residual　残液过滤叶片　9.2.7.5
filter leaf group　过滤叶片组件　9.2.2.6
filter medium[wedːo]　过滤介质　9.3.3.5
filter mesh　过滤筛网　3.1.19.12

filter receiver 过滤接收[收集]器 23.1.5.11
filter tank （过)滤罐 9.3.3.38
filtrate 滤液 3.1.9.11;3.3.19.10
filtrate discharge 滤液排出口 9.5.5.10
filtrate housing 滤液排出壳体 9.5.10.16
filtrate manifold 滤液集管 9.2.5.32
filtrate outlet 滤液出口 7.1.8.14
filtrate outlet 滤液出口 9.2.2.12
filtrate pump 滤液泵 3.1.9.10
filtrate receiver 滤液槽 9.2.1.10
filtrate vacuum connection 滤液通真空系统的接口 9.3.2.30
filtration 过滤 2.2.2.32;2.4.5.7;3.1.37.14
filtration centrifuge 过滤离心机 9.5.2.1
filtration equipment 过滤设备 9
filtration membrane 滤膜 1
filtration plate 过滤板 13.2.3.13
final clarifier 终端澄清器 3.1.30.B.7;3.1.33.5
final concentrate 最终富集物 12.8.2.20
final control basin 终端控制池 3.1.31.8
final dewatering 最后脱水 9.3.9.8
final methanation reactor 最终甲烷化反应器 2.2.14.9
final product 最终产品 12.1.5.6
final settler 终端沉降池 3.1.34.11
final shaping 最后成形 2.3.9.36
final stretching 最后拉伸 2.3.8.47
final tailing 最终的尾砂 12.8.2.22
fine material 粉状物料 3.3.16.8
fine metal powder 细金属粉 12.7.3
fine ore bin 细矿石储槽 12.8.2.9
fine ore feeder 细矿进料器 12.8.1.6
fine particle 细颗粒 14.1.3.1
fine particle coalescing 细颗粒聚结 6.1.9.11
fine particle separation 细颗粒分离 11.5.2.9
fine product 粉料成品 8.5.1.4
fine purification 精制 2.1.1.83
finer mesh screen 更细目的筛网 12.7.2.12
finer particles 较细颗粒 10.1.2.20
fines 细粒[料,粉] 11.6.5.3
fine salt 细盐 7.2.12.7
fines cone 细粉锥筒 12.3.1.6
fines destruction water 破损细粒用水 7.2.9.11
fines discard 细颗粒抛弃 2.2.82.16
废矿粉 2.2.82.3
fines discharge 细颗粒排出口 12.3.1.7
fines dissolving tank 细粒溶解槽 7.2.9.10
fines return 细粉回流 2.2.18.9
fines stream 细粒流 7.2.9.12

fine vacuum 接高真空 5.5.1.8
fin heaters 翅片加热器 8.2.1.7
finish 完成 9.3.8.4
finished malt collecting hopper 麦芽成品收集漏头 2.3.12.20
finished product 最终产品 11.5.3.5
成品 8.4.1.3
finished product bin 最终产品仓 11.6.5.7
finished product tank 成品罐 13.2.4.26
finished stock 成品 13.2.4.27
finished tower 成品塔 2.2.55.13
finishing of tow for spinning 纺纱用丝束的整理 2.3.8.57
finite element method 有限元法 1
finned drum 翅片转鼓 8.3.4.6
finned drum feeder 翅片转鼓给料机 8.3.4.4
finned tube bundle 翅片管束 4.5.1.6
firebrick 耐火砖 14.2.5.6
fire brick 耐火砖 25.1.7.4
fired preheater 火焰预热器 2.2.98.3
fire-fighting pump 消防泵 16.1.2.3.19
firetube 火管 25.6.5.1
fire-tube salt-bath heater 火管式盐浴加热器 25.6.5
first anoxic zone 第一缺氧区 3.1.35.4
first blowing 第一次吹制 2.3.9.24
first ejector 一级喷射泵 18.3.5.3
first law of thermodynamics 热力学第一定律 1
first pass 第一通道 8.3.2
first press 第一道压榨 2.3.11.19
first reactor 第一反应器 3.3.22.2
first regeneration zone 第一再生层段 14.3.3.34
first stage 第一级 2.4.2.3;17.2.4.2
一段 2.2.21.4
first-stage aerobic（BOD$_5$ removal） 一段有氧处理(去除五天生化需氧量) 3.1.40.3
first stage combining throat 第一级混合喉管 18.3.4.7
first stage cylinder 第一级气缸 17.2.9.7
first stage discharge valve 第二级排气阀 17.2.7.8
first stage discharge valve 第一级排气阀 17.2.9.9
first stage impeller 第一级叶轮 16.4.27.38
first stage impeller retaining collar 第一级叶轮挡圈 16.4.27.6
first stage impeller retaining ring 第一级叶轮挡圈 16.4.27.39
first stage piston 第一级活塞 17.2.7.12
first stage piston 第一级活塞 17.2.9.5

first stage piston rod 第一级活塞杆 17.2.9.6
first stage snap ring 第一级开口环 16.4.27.4
first stage steam chest 第一级蒸汽室 18.3.4.4
first stage steam nozzle 第一级蒸汽喷嘴 18.3.4.3
first stage suction head 第一级吸气接头 18.3.4.5
first stage suction valve 第一级吸气阀 17.2.7.3
first synthesis column 第一合成塔 2.2.44.4
first tower 1 塔 2.2.41.7
first wet felt 第一道湿毛布 2.3.11.17
Fischer-Tropsch low temperature slurry bed oil synthesis process 费-托低温浆态床合成油工艺流程 2.2.74
Fischer-Tropsch synthesis 费-托合成 2.2.2.18
fittings 管件 22.3
five-stage countercurrent cascade 五级逆流级联 6.1.10
five-stage steam ejector 五级蒸汽喷射泵 18.3.5
fixed ball valve 固定式球阀 22.1.1.(f)
fixed bed 固定床 9.5.2.2
fixed-bed hydrotreatment 固定床加氢处理 2.2.99.10
fixed bed reactor 固定床反应器 14.3.2.(c)
fixed bed synthesis 固定床合成 2.2.100.5
fixed bed unit 固定床装置 14.3.4.(c)
fixed discharge chute 固定料槽 23.2.3.(e)
fixed end cover 固定端盖 4.2.1.5
fixed film bio-cell 固定膜生物电池 3.1.44.3
fixed head 固定端板 9.1.2.4
fixed-jaw plates 固定颚板 11.1.1.1
fixed location 固定位置 23.2.3.(f)
fixed mixing vanes 固定的混合叶片 4.3.4.32
fixed ring 固定环 19.3.2.2
fixed saddle 固定鞍座 21.3.1.2
fixed supporting saddle 固定鞍式支座 8.8.10.14
fixed trippers 固定倾料器 23.2.3.(d)
fixed tubesheet 固定管板 4.1.4.(m)
fixed-tube-sheet exchanger 固定管板换热器 4.1.1.(a)
fixed type double-pipe heat exchanger 固定式套管换热器 4.2.2
fixed vane 固定(旋流)叶片 10.5.4.11
fixing holes 地脚螺栓孔,固定孔 15.5.5.2
flame 火焰 8.1.1.29
flame-ionization detector 火焰离子化检测器 3.5.21
flame port 火焰通口 3.3.12.7
flammable liquid waste 易燃废液 3.3.5.7
flange 法兰 22.4
flanged connection 法兰连接 22.4.1
45° flanged elbow 45°带法兰弯头 22.3.1.4
90° flanged elbow 90°带法兰弯头 22.3.1.1
flanged fittings 法兰管件 22.1.2
flanged only heads 法兰封头 22.5.1.(b)
flanged reducer 异径管,带法兰大小头 22.3.1.5
flanged tee 带法兰等径三通 22.3.1.7
flange gasket 法兰垫片 19.3.1.13
flange stud bolt 凸缘双头螺柱 22.1.2.11
flange type 法兰型式 22.4.2
flanging 折边 21.4.1.3
flanging-only head 折边平封头 21.4.1.1
flanging shallow dished head 带折边的浅碟形封头 21.4.1.2
flanging standard dishes head 折边标准碟形封头 21.4.1.3
Flapper valve Flapper 阀 14.1.7.(a)
flap valve 片状阀 12.3.3.1
flared shell 扩口壳体 22.5.1.(c)
flash 闪蒸 2.2.92.12;2.2.115.11
flash chamber 闪蒸室 7.1.8.6
flash drum 闪蒸罐 2.1.4.8
flash dryer 闪急干燥器 8.5.7
flash drying mill 干燥粉碎机 2.2.106.11
flashed vapour 闪蒸气 2.2.55.9
flash evaporation 闪蒸 1
flash gas 闪蒸气 20.1.3.9
flash gas compressor 闪蒸气体压缩机 2.2.28.10
flash separator 闪蒸分离器 2.2.98.20
flash steam 闪蒸汽 2.2.44.16;2.2.46.17;2.2.47.5
flash tank 闪蒸槽 2.2.44.13;2.2.46.9;2.2.47.8;20.1.10.21
闪蒸罐 2.2.56.2
flash tank-evaporator 闪蒸槽蒸发器 20.1.10.25
flask 烧瓶 3.5.14.26
flask shield 烧瓶屏罩 3.4.14.27
flask valve 烧瓶阀 3.4.14.19
flat bed trailer 平板拖车 23.4.2.15
flat circular diaphragm 圆形平膜 16.2.9
flat cover 平盖 11.1.7.10
flat face 全平面密封面 22.4.3.(a)
flat face flange 平面法兰 21.1.1.34
flat-floor building 平底料仓 23.2.4

flat gasket 平垫片 16.4.20.11
flat head 平封头,平板盖 21.1.1.44
flat key 平键 16.2.5.8
flat metal perforated plates 多孔金属平板 14.1.8.(d)
flat plates 平板式 22.5.1.(a)
flesh syngas 新鲜气 2.2.44.1
flexibility （操作)弹性；柔性；适应性 1
   柔度 22.6.1
flexible car roader 挠性装车机 23.4.2.18
flexible cloth rubbing seal 柔性织物摩擦密封 8.8.5.15
flexible connection 挠性接头[连接] 12.2.3.7
   柔性连接结构 23.3.1.3
flexible coupling 挠性联轴器 15.7.2.9
flexible diaphragm 挠性膜片 22.1.4.4
flexible graphite packing 柔性石墨填料 19.2.3.13
flexible hose 挠性软管 17.6.2.12
flexible inlet flange sealing 挠性进口法兰密封件 12.2.3.12
flexible joint 挠性接头 12.2.5.5,10
flexible materials 柔性材料 8.8.5.19
flexible member 挠性构件 16.1.1.22
flexible rotor pump 挠性转子泵 16.3.4
flexible shafting 挠性轴系 16.4.21
flexible tube 软管 23.1.2.19
flexible tubing 软管 4.2.7.14
flexicoker 灵活焦化器 2.2.79.15
flexicoking 灵活焦化 2.2.97.12
flick separator 轻击分离器 10.4.3.(b)
flight 刮板 23.3.7.16
flight chain conveyer (double lined) 双列刮板输送机 15.6.5.9
flight conveyer 刮板输送机 12.5.2.1
flight coolant 刮板冷却液 4.3.3.18
flighted 抄板式 4.3.2.(b)
flight scraper chain 刮泥板链条 3.1.3.11
flight scraper chain sprocket 刮泥板链轮 3.1.3.9
flight width 刮板宽度 12.5.2.9
flinger 挡液环 16.4.12.23
   抛油环 16.4.4.8
flinger ring 挡液环 16.4.1.7
   拦液圈 9.5.3.8
float 浮选(物) 3.1.15.6
   上浮物 12.5.2.13
float bath 浮槽 2.3.9.16
float glass process 浮法玻璃制造法 2.3.9.12

floating angle tray 浮动角钢塔盘 5.2.10.19
floating babbitt-faced steel ring 钢表面衬巴氏合金的浮动环 19.1.7.3
floating ball 浮球 22.2.3.9
floating bed of low-density sphere 低密度球浮动床 10.5.4.16
floating damper 浮动风门 3.3.9.8
floating head backing device 浮头勾圈 4.1.1.18
floating head cover 浮头盖 4.1.1.16
floating head cover-external 外浮头端盖 4.1.1.21
floating head flange 浮头法兰 4.1.1.17
floating head heat exchanger 浮头换热器 1
floating head with backing device 具有背封构件的浮头 4.1.4.(q)
floating insulated membrane cover 绝缘膜浮动盖 3.2.6.13
floating jet tray 浮动舌形塔盘 5.2.10.16
floating ring 浮环 19.3.2.1
floating ring seal 浮环密封 19.3.2
floating-roof 浮顶罐 21.2.1.3
floating sieve tray 浮动筛板 5.2.10.10
Floating steam trap 浮球式冷凝水排除器 22.2.4
floating tubesheet 浮头管板 4.1.1.15
floating tubesheet skirt 浮动管板裙 4.1.1.22
floating-type restrictive ring seal 浮环节流型密封 19.3.1
floating valve tray 浮阀塔盘 5.2.10.3
floating weight 移动式重锤 15.7.1.9
float-material drain 上浮物料排出 12.5.5.19
float product 上浮产品 12.5.5.26
float pulley 浮动滑轮 3.4.5.5
float valve 浮子阀 20.1.10.16
float valve tower 浮阀塔 5.1.2.4
flocculant 絮凝剂 3.1.27.5
flocculant preparation 絮凝剂制剂 3.1.16.4
flocculation 絮凝(作用) 2.4.4.1;3.1.12.9
flocculent sludge 絮凝污泥 3.2.6.26
flooded disk 富液盘 10.5.9.2
flooded elbow 溢流肘管 10.5.13.4
flooded wall 流淌壁 10.5.13.2
flooding point 泛点 1
floor openings 底板孔 23.2.4.4
floor wax 地蜡 2.1.1.84
flotation 浮选 3.1.32.7;12.8
flotation clarifier 浮选澄清器 3.1.32.20
flotation feed 浮选加料 12.8.2.14
flotation machine 浮选机械 12.8.1.13
flotation plant 浮选工厂 12.8.2

flotation unit 浮选（装置，单元） 3.1.15.4
flow 液流方向 16.4.6.17
　流（动方）向 16.5.4.6
flow-adjustment feature 流量调节部件 23.5.1.3
flow cam 滑移凸轮 3.5.5.1
flow chart 流程图 13.3.1
flow chart for two-stage synthesis of methanol 双级合成甲醇工艺流程示意 2.2.47
Flow chart of CAPS（chemically assisted primary sedimentation）upgrading in the studied WWTP 废水处理厂研究中改进的化学辅助一次沉降流程图 3.1.27
flow coil 流体盘管 25.6.5.4
flow control 流量控制 3.4.19.5
flow control & splitting 流量控制和分流 3.1.44.4
flow-control valve 流量控制阀 3.1.19.1
flow diagram 流程图 1
　流向图 15.5.4.5
Flow diagram for a Biological Aerated Filter（BAF）system 生物曝气过滤系统流程图 3.1.41
Flow diagram for covered pure oxygen activated sludge process 密闭的纯氧活化污泥工艺流程图 3.1.45
Flow diagram for uncovered pure oxygen active sludge process 敞口的纯活化污泥工艺流程图 3.1.46
flow diagram of chemical process 化工工艺流程图 2
Flow diagram of Hookers Point advanced wastewater treatment plant. Hookers Point 先进废水处理厂流程图 3.1.40
Flow diagram of packed bed reactor system 填料床反应器系统流程图 3.1.38
Flow diagram of Reno-Sparks wastewater treatment plant Reno-Sparks 废水处理厂流程图 3.1.37
Flow diagram of the Sohio process for the manufacturing of acrylonitrile 美孚公司丙烯腈生产工艺流程图 2.3.15
Flow diagram of upflow fluidized-bed system 上流式流化床处理系统流程图 3.1.36
flow gears 滑移齿轮 3.4.5.4
flow indicator 流量显示器 6.2.1.12
flow line 流线 16.4.3.3
flow meter 流量计 3.3.21.4;3.3.22.4
flow passage 流道 16.4.15.18
flow pattern 流型 3.2.15.8
　流动模式 9.1.1

flowrator 转子流量计 8.5.3.2
flow screen 流动网 15.2.1.7
flow sheet 流程图 1
Flow sheet of direct-fired rotary dryer 直燃式回转烘干机的工艺流程 3.3.25
flow sheet of the experimental 实验流程 8.5.6
flow through impeller and volute 流经叶轮与蜗壳 16.4.3.2
flow tube 流水管 3.4.18.(b)
flued heads 喇叭形接头 22.5.1.(e)
flued openings 翻边开孔 21.1.1.30
flue gas 烟道气 2.2.27.1;2.2.71.15;2.2.87.11;2.2.116.5;2.2.118.7
flue gas and dust particles 烟气与尘粒 25.5.3.14
flue gas desulfurization 烟气脱硫 2.2.115.20
flue gases out 烟道气排出 3.3.9.7
flue gas outlet 烟道气(废气)出口 14.3.3.19
flue gas recycle 烟气循环 2.2.115.8
flue gas to electrostatic precipitator 烟道气去静电除尘器 2.2.88.15
flue gas to heat recovery system 烟气至热量回收系统 2.2.72.8
flue gas to sequestration 烟气去储存待用 2.2.115.13
fluegas to stack 烟道气去烟囱 2.3.6.32
fluegas treating 加工烟道气 2.3.1.30
fluff RDF 废物垃圾燃料 3.3.16.25
fluid bed 流化床 14
fluid bed dryer 流化床干燥器 3.1.16.15
fluid-bed incinerator 流化床焚烧炉 3.3.6
fluid bed reactor 流化床反应器 2.2.71.1;3.1.36.7;3.3.6.5
fluid bed synthesis 流化床合成 2.2.100.9
fluid catalyst bed 催化流化床 2.2.9.12
fluid charge 注入液体 18.2.1.9
fluid duct 流体导管 16.5.4.5
fluid dynamics 流体动力学 1
fluid-energy mill 流能磨 11.5.4
fluid-energy or jet mill 流能磨 11.5
fluid fractionating system 流体分级系统 18.2.4.12,13
fluidised bed dryer 流化床干燥器 8.5.5
fluidization 流态化 14.1.3
fluidization phase diagram for a fine powder 细颗粒流态化相图 14.1.1
fluidization processes 流化过程 14.2.6
fluidization pump 流态化泵 3.1.36.3
fluidized absorbent media 流态化的吸收剂介质 3.1.42.4

539

fluidized air 流态化空气 4.3.3.22
fluidized bed 流化床 2.2.70.13;14.1.10.1
液态床 4.3.3.(e)
fluidized bed and expanded bed process 流化床与膨胀床工艺 3.2.9
Fluidized bed biological (FBB)-GAC (granular activated carbon) system 生物流动床-颗粒活性炭系统 3.1.42
fluidized-bed coal dryer 流化床煤干燥器 8.5.6
fluidized-bed coker 流化床焦化塔 2.1.9.31
fluidized bed coking process 流化床焦化工艺 14.3.4.(d)
fluidized-bed cyclone arrangements 流化床旋风分离器布置方式 14.1.5
fluidized bed dryer 流化床干燥器 2.2.107.7
fluidized bed dryer with inert particles 惰性粒子流化床干燥器 8.5.3
fluidized bed hydroformer 流化床临氢重整装置 14.3.4.(c)
fluidized-bed incinerator-system 流化床焚烧炉 3.3.2
fluidized bed of coal and ash 煤与灰的流化床 24.2.10.5
fluidized bed reactor processes 流化床反应器工艺 14.3.4
fluidized-bed scrubber 游动床涤气器 10.5.5
fluidized-bed seal leg 流化床密封腿 14.1.10
fluidized-bed steam-generation system 流化床蒸汽发生系统 25.5.3
fluidized-bed steam generator 流化床蒸汽发生器 14.2.2
fluidized-bed system 流化床系统 14
fluidized bulk powders 流态化粉粒体 23.3.8.(b)
fluidized coking unit 流化焦化装置 14.2.6.(h)
fluidized deep-solids bed 流态化深的固体床层 4.3.3.21
fluidized/expanded media bed 介质流态化膨胀床 3.2.9.2
fluidized powder compartment 粉粒体流化室 23.1.2.27
fluidized product 流态化物料 8.5.6.19
fluidized sand bed 砂子流化床 3.3.27.7
fluidizing air 流态化用气 13.4.1.1
fluidizing air blower 流化用鼓风机 13.4.1.7
fluidizing air blower 流化空气鼓风机 3.3.2.8
fluidizing-air inlet 流化空气入口 3.3.27.1;25.5.1.2
fluidizing air line 流态化空气管线 23.3.8.10
fluidizing blower 流化用的鼓风机 8.5.6.2

fluidizing butterfly outlet 流态化蝶形阀式出口 23.3.8.(c)
fluidizing chamber 流化室 8.5.6.8
fluidizing connection 流态化接管 23.3.8.7
fluidizing gas 流化载气 2.2.99.2
fluidizing outlet 流态化卸料口 23.3.8
fluidizing pad 流态化气垫 23.3.8.3
fluidizing system 流化系统 23.1.2.(e)
fluid meter 流量计 5.5.1.3
fluid operated 流体驱动的 16.1.1.15
fluid outlet 流体出口 11.5.1.9
fluid particles 流体质点 16.4.15.13
fluid pressure manifold 流体压力总管 11.5.1.7
fluid pumped 泵送的流体 16.5.5.2
fluid-seal pot 流体密封槽 14.1.7.(d)
fluid velocity 流体速度 14.1.2
flume 斜坡水槽 3.5.3.1
fluorescent tube 日光[荧光]灯管 25.4.1.61
fluorides incinerator 流化焙烧焚化炉 3.3.4
Fluo seal Fluo 密封装置 14.1.9
Fluosolids lime kiln Fluosolids 石灰窑 14.2.5
fluosolids reactor 流化反应器 3.3.4.14
fluosolids roasting 流态化焙烧 13.4.1.11
flush connection port 冲洗液连通孔 19.1.5.5
flushing fluid in 冲洗液入口 19.1.8.1
flushing fluid outlet 冲洗液出口 19.1.8.2
flushing inlet 冲洗液进口 19.1.5.21
flush-mounted earthed socket 嵌装式接地插座 25.4.1.5
flush-mounted rocker switch 嵌装式摇杆开关 25.4.1.4
flush water 冲洗水 14.1.11.13
fluter feeder 抖动式给料机 12.2.7.3
flux 通量 1
fluid inlet 流体入口 11.5.1.1
fly ash 飞灰 3.3.8.13;10.3.5.5
粉灰 2.2.31.19
粉煤灰 2.2.24.14;2.2.25.3;2.2.116.12
fly-ashcontrol 飞灰控制器 24.1.4
fly ash (recycled) 粉煤灰(循环的) 2.2.25.15
fly ash reinjector 飞灰再注器 14.2.2.9
fly ash scrubber 飞灰涤气器 10.5.10
fly ash silo 粉灰筒仓 2.2.31.19
flywheel 飞轮 11.1.1.5
foam breaker 消沫器 5.3.4.40
foam catcher 除沫器 5.3.4.39
foam encasement 泡沫塑料外套 3.4.14.28
foam separator 除沫器 5.3.4.38

folded cloth packing 折叠织物填料 19.2.3.6
food 食物 3.2.11.11
force chamber 压力室 16.2.3.32
forced-circulation baffle surface-cooled crystallizer 强制循环有挡板表面冷却结晶器 7.2.1
forced-circulation (evaporative) crystallizer 强制循环(蒸发)结晶器 7.2.1
forced oil circulation pump 强制循环油泵 15.7.2.11
force feed lubricator connection 强制润滑供油接头 17.2.6.6
forebay 前池 3.1.3.5
fore-pump 前级泵 5.5.2.20
FOR-fuel oxidation reactor 燃料氧化反应炉 2.2.80.4
forged steel cylinder 锻钢缸体 16.2.4
forged-steel single-acting high-pressure cylinder 锻钢制单作用高压气缸 17.2.8
forging 锻件 21.1.2.7
fork connecting rod 叉式连杆 17.2.15.23
forklift truck 叉车 2.3.13.23
formaldehyde + water vapor from methanol-oxidation 甲醇氧化生成甲醛和水蒸气 2.2.66.3
formed heads 成型接头 22.5.1.(d)
form gone 形成区 3.3.10.16
forming machine for rubber and plastics 橡胶塑料加工成型机械 15.7
for overhead tanks only 压力罐用 2.1.7.6
forward curved blade 前弯(式)叶片 17.3.3.10
forward curved vane 前弯叶片 16.4.5.3.3
fossil fuels 化石燃料 25.1.1.4
fouling 污垢;结垢 1
foundation 基础 17.2.9.22
foundation bolt 地脚[基础]螺栓 11.1.6.3
foundation bolt hole 基础螺栓孔 17.2.15.27
Fourcault glass-drawing machine 弗克法玻璃拉制机 2.3.9.8
four-corner four-stage compressor H 型四级压缩机 17.2.3
four-socket (four-way) adapter 四连插座,四插座转接器 25.4.1.8
four-stage compressor 四级压缩机 17.2.3
fourth stage cylinder 四级缸 17.2.3.4
four-tube force flow valve-bag packer 4 管强制流阀式袋包装机 23.4.2
fractional extraction 分馏萃取 6.1.10
fractionating column 分馏塔,精馏塔 5.1.1.2
fractionating tower 分馏塔,精馏塔 5.3.3.11

fractionator 精馏塔,分馏塔 2.1.5.12
分馏器 2.2.60.2
fractured shale zone 已破碎的页岩区 2.2.85.19
frame 框架 11.1.1.12
(泵)机架 16.4.19.35
滤框 9.1.1.2
frame tie bolts 机身紧固螺栓 17.2.15.7
frame top cover 机身顶盖 17.2.15.2
FRC(1) (facility remote control) 设备远距控制 6.3.3.12
free ball valve 自由式球阀 22.1.1.(e)
freeboard 自由空间 2.2.18.5;14.1.12.9
分离空间 3.3.27.8
free-floating media 自由浮动介质 3.2.13.12
free-flowing granular 自由流动的颗粒 8.1.1.19
free-flowing products 自由流动产品 23.5.1
free-flow pump 自由流动泵 16.1.2.2.59
free movement 自由地移动 22.6.1.(d)
free oil 浮油 3.1.3
游离油 15.5.4.4
free sedimentation 自由沉降 1
freeze drying 冷冻干燥 1
freezer 冻凝器 7.2.14.4
fresh acid 新鲜酸 6.1.5.11
fresh-air 新鲜空气 8.3.1.2
fresh air fan 新鲜空气鼓风机 3.1.21.1
fresh air in 新鲜空气加入 8.4.6.1
fresh air inlet 新鲜空气入口 8.3.4.10
fresh carbon oxide 新鲜的碳氧化物 2.2.66.9
fresh feed 新鲜料 14.3.2.17
进料 2.1.6.4
fresh feed gas 新鲜的原料气 2.2.37.1;2.2.48.1
fresh liquor 新鲜溶液 2.2.7.13
fresh methanol 新鲜甲醇 2.2.55.17
fresh-oil feed 新鲜油进料 14.2.7.14
fresh rinse water 新鲜淋洗水 15.5.5.17
fresh solvent in 新鲜溶剂进口 6.1.17.1
fresh vapor inlet 新鲜蒸气进口 14.3.3.27
fresh water 新鲜水 2.2.35.16
淡水 2.2.66.16;3.4.1.6
friable material 脆性物料 23.1.7
易碎物料 23.1.1.(c)
frictional drag 摩擦曳力 23.2.2.5
friction material 摩擦材料 8.8.5.7
friction seal 摩擦密封 8.8.1.3
from aeration 来自曝气池 3.1.8.7
from heat exchanger 来自换热器 2.2.110.15

from intercooler 来自级间冷却器 17.2.4.4
from pressure vessel 来自压力容器 22.1.8.19
from sludge thickener 来自污泥浓缩机 3.3.19.3
front casing 前壳体 16.5.5.3
front counter guide （前）主十字头导轨 17.2.15.28
front crosshead metal 主十字头（导轨）镶条 17.2.15.29
front cross head of plunger 柱塞前十字头 16.2.4.12
front end stationary head types 前端固定封头型式 4.1.3
front guiding plate 主十字头导板 17.2.15.30
front heat zone 前加热区 13.2.1.18
front labyrinth seal 前迷宫密封 17.4.2.11
front movement 朝前移动 2.2.85.22
froth 泡沫 12.8.1.4
froth flow contactor 泡沫流动接触器 3.1.42.5
F-13 system F-13 系统 20.1.7.17
F-22 system F-22 系统 20.1.7.18
F-T synthesis 费-托法合成 2.2.78.9
fuel 燃料 2.1.6.5；2.2.26.1；2.2.80.5；2.2.111.1；2.2.119.6；3.3.25.2
fuel and sorbent inlet 燃料与吸收剂入口 25.5.1.1
fuel burner 燃料燃烧器 3.3.3.10
fuel combustor 燃烧室 25.3.2.11
fuel gas 燃料气（体） 2.2.43.8；2.2.45.9；2.2.106.2；3.3.23.9
　　　　燃气 2.2.119.12
fuel gun 燃料喷枪 3.3.27.10
fuel-handling hoists 燃料装卸吊车（起重机） 25.2.2.7
fuel oil 燃料油 2.2.79.12；2.2.91.5；2.3.16.5
fuel rod[pin] 燃料棒 25.2.1.4
fuel storage 燃料仓库 25.2.2.8
　　　　燃料储存器 3.3.23.2
fugacity coefficient 逸度系数 1
fulcrum pin 支轴销 16.2.4.30
full face gasket 全平面垫片 21.1.1.1
full miscella 富液 6.1.17.5
Full-scale biofilter design 全尺寸生物滤池设计 3.4.3
fumes 烟气 3.1.22.8
functional diagram 工作原理图 18.2.3
funnel 漏斗 15.7.6.1
funneled feed duct 带漏斗的进料导管 15.7.6.18
funnel-flow bin 漏斗流料仓 23.1.3
furnace 燃烧炉 2.2.9.19；2.2.117.17；2.2.118.5
　　　　加热炉 2.2.61.1；2.2.87.12；3.3.24.20；3.3.25.3
　　　　炉 24.1.9.7
　　　　热风炉 8.6.1.6
furnace lining 炉衬 8.8.8.5
Furnace process（for carbon black production） 炉式工艺制炭黑流程 2.3.17
fuse carrier with fuse wire 带保险丝的熔断架 25.4.1.68
fuse cartridge 易熔元件 25.4.1.36
fused silica or fused quartz 熔融硅石或熔融石英 22.6.2.⑩
fuse holder 熔断器座 25.4.1.35
fusible element 保险线盒 25.4.1.36

G

GAC column 颗粒活性炭吸附塔 3.1.30.B.10
gage 压力表 23.1.5.2
Galoter oil-shale retorting system 加劳特油页岩干馏装置 2.2.88
gantry belt conveyor 吊架式带式输送机 23.3.9.3
gap adjuster 间隙调节器 15.7.6.16
gas 气（体） 2.1.8.5；2.2.77.10；3.3.17.2；3.3.18.4
　　　　煤气 2.2.5.10；2.2.15.6；2.2.76.19
gas and catalyst mixture 气体与催化剂混合物 14.3.2.19
gas burner 煤气燃烧器 2.2.84.5
gas centrifuge 气体离心机 9.5.21
gas chromatograph 气相色谱 3.4.22.2
gas-chromatograph system 气相色谱系统 3.4.20
gas cleaning system 气体净化系统 2.2.107.11
gas cleaning unit 气体净化装置 2.2.104.4；2.2.110.10
gas cleanup 气体净化 2.2.100.4；2.2.109.9
gas-collecting main 焦气主管 2.3.14.17
gas compressor 煤气压缩机 2.3.14.26
gas contacting region 气体接触区 10.5.4.23
gas cooler 煤气冷却器 2.3.14.19
　　　　气体冷却器 3.3.4.8
gas-cooled reactor 气冷反应器 2.2.42.6
gas-cooled synthesis reactor 气冷式合成反

应器 2.2.43.10
gas cooling 气体冷却 2.3.1.9
煤气冷却 2.3.14.29
gas deflector cone 导风锥 10.3.4.14
gas desulfurization 气体脱硫 2.3.2.4
gas-disengaging plenum 气体释放空间 2.2.87.5
gas distributor 气体分布器[板] 14.3.1.23
gas distributor plate 气体分布板 3.4.4.10; 8.5.6.18
gas distributor (sparger) 气体分布板(鼓泡器) 3.4.4.4
gas draining 气体排放 3.4.7.8
gas drying 气体干燥 2.2.103.8
煤气干燥 2.3.10.30
gaseous effluent 废气排出 3.4.1.3; 3.4.2.3
gaseous fuel 气体燃料 3.3.1.4
gaseous methanol 气相甲醇 2.2.72.4
gaseous waste 废气 2.2.4.11
gases out 气体排出 3.3.12.4
gas exhaust 废气 3.3.17.9
gas exit 气体出口 24.1.5.4
gas extractor 煤气抽出器 2.3.14.21
gas-fired indirect-heat rotary calciner 间接加热式回转煅烧炉 8.8.8
gas-fired lighter 气体点火器 25.2.1.4
gas flow 气体流(向,量) 10.5.1.9
gas fluidized bed 气体流化床 14.1.3.16
gas fractionation 气体分级 14.2.3
gas generator 煤气发生器 2.2.108.11
gas-generator turbine rotor 燃气发生器透平转子 25.3.1.7
gas-generator turbine rotor blades 燃气发生器透平转子叶片 25.3.1.15
gas-generator turbine stator blade carrier 燃气发生器透平定子叶片托架 25.3.1.14
gas-generator turbine stator blades 燃气发生器透平定子叶片 25.3.1.13
gas holder 贮气罐 2.3.4.25
gasifer 气化炉 2.2.13.8; 2.2.19.1; 2.2.20.6; 2.2.22.9; 2.2.25.11; 2.2.29.10; 2.2.30.7; 2.2.31.7; 2.2.107.4; 2.2.109.5; 2.2.110.5; 2.2.117.1; 3.3.23.4
gasification 气化 2.2.3.2; 2.2.78.3
(煤)气化 2.3.1.7; 2.2.67.4
气化作用 2.2.101.7
造气 2.2.38.3
gasification agent 气化剂 2.2.38.2
gasification and related technologies 气化和相关技术 2.2.3
gasification reactor 气化反应器 2.2.100.2

gasification unit 造气装置 2.3.2.2
gasifier 气化炉[器] 2.1.8.4
gasifier body 气化炉主体 2.2.23.2
gas in 气体进口[入] 14.1.12.2
进气 3.4.4.1; 17.2.14.4
gas injection plate 气体喷injection板 5.4.1.9
gas injector 进气喷射器 2.2.87.4
gas inlet 气体入[进]口 10.3.1.11
gas inlet port 气体进口 17.2.5.15
gas inlet section 气体进入段 10.5.10.6
gas jet burner 煤气喷燃器 24.1.1.14
gasket 垫片 10.4.2.2
gas lift 气升(式) 16.1.1.49
gas lift line 气体提升管 6.3.7.13
gas-lift pump 气升泵 16.5.1
gas line 燃气管线 24.2.6.1
gas/liquid contactor 气-液接触器 3.5.22.7
gas-liquid equilibrium(GLE) 气液平衡 1
gas/liquid product to separators 气体/液体产品去分离器 2.1.10.3
gas/liquid removal 气体/液体出口 9.4.11.5
gas-liquid separator 气液分离器 2.3.4.13
gas (methane) phase 气(甲烷)相 3.3.18.20
gas oil 瓦斯油 14.2.6.17
gas oil hydrotreater 瓦斯油加氢处理装置 2.1.9.40
gasoline 汽油 2.1.1.74; 2.2.75.13
gasoline alkylate 烷基化汽油 2.1.1.73
gasoline and diesel oil 汽油和柴油 2.2.1.21
Gasoline mode of the methanol-to-olefins gasoline plant at Wesseling, Germany (© Uhbe ExxonMobil) 德国韦瑟灵公司甲醇制烯烃汽油厂的汽油模式(© Uhbe, 埃克森美孚) 2.2.76
Gasoline synthesis process 合成油工艺流程 2.2.78
gasometer 贮气柜 3.1.26.(b).8
gas out 气体排出 10.1.4.6
出气 17.2.14.3; 3.4.4.2
气体出口 5.5.5.3
gas outlet 气体出口 10.1.2.32
gas outlet 排气口 10.5.16.16
煤气出口 2.2.14.5; 2.2.26.9
gas outlet duct 气体出口管 10.1.2.30
gas-outlet hole 燃气出口孔 24.2.9.4
gas pressure 气压 6.3.3.1
gas processing 煤气加工 2.2.108.19
gas product 气体产品 2.2.4.12
gas purification 气体精制 2.2.94.13
气体净化 2.2.95.13
gas quench 煤气急冷 2.2.14.6

543

gas recycle compressor 气体循环压缩机 2.3.4.20
gas reforming and conditioning 气体转化与调理 2.2.103.1
gas release 放气 3.3.18.1
gas ring 燃气环形管 24.2.9.5
gas scrubber 气体洗涤塔 2.2.29.15;5.4.2.16
gas scrubbing and processing 气体洗涤与加工 2.2.99.7
gas seal 气体密封 19.3
gas separating 气体分离 2.1.1.63
gas separation 气体分离 2.2.94.13
gas separation unit 气体分离装置 2.2.119.9
gas-solid contacting methods 气固接触方式 8.6.3
gas-solid separation equipment 气固分离设备 10
gas-solid separator 气固分离器 14.2.3.5
gas spager 气体鼓泡器 3.2.6.9
gas supply 燃气供入 24.2.9.1
gastight connections 封闭连接 13.5.2.2
gastight cover 密封盖 13.5.2.6
gas to user 煤气去用户 2.2.108.20
gas-turbine 燃气轮机 2.2.110.13;2.2.113.7;2.2.114.8;2.2.117.18;25.3 气体透平 25.5.2.7
gas-turbine cross section 燃气轮机 25.3.1
gas-turbine cycle 燃气轮机循环 25.3.2
gas turbine with intercooling 有中间冷却的燃气轮机 25.3.2.(c)
gas turbine with regeneration 有再生回热的燃气轮机 25.3.2.(b)
gas turbine with reheating 有中间再热的燃气轮机 25.3.2.(d)
gas valve 燃气阀 24.2.6.2
gas vent 出气孔 14.1.11.7
gas-water separator 气水分离器 18.4.2.12
gate clamp screws 阀门压紧螺钉 23.5.1.4
gated floor opening 带有闸板的底板孔 23.2.4.4
gate disc 闸板 22.1.2.4
gate valve 闸阀 18.4.2.7
gateway pier 门框墙墩 3.1.3.6
gauge well 表孔 2.1.7.7
Gayco centrifugal separator Gayco型离心分离器 12.3.1
gear 大齿轮 11.2.4.7 齿轮(式) 16.1.1.25
gear box 齿轮箱 13.3.5.11
gear case 齿轮箱 5.5.1.17
gear coupling 齿轮联轴器 16.4.11.6

geared motor 带变速齿轮箱的电动机 12.1.1.15
geared pump 齿轮传动泵 16.1.2.4.1
gear for screw feeder 螺旋送料机齿轮 12.1.1.1
gear head motor 电机齿轮变速箱 5.5.2.2
gear motor 电机减速机 9.2.1.3
gear mounting 齿轮架 8.8.1.26
gear oil pump 齿轮油泵 17.2.9.40
gear pump 齿轮泵 16.3.1
gear reducer 齿轮减速器 13.2.3.3
gear ring 大齿轮盘 12.3.5.29
gear shaft 传动轴 12.3.5.7
gear shaft liner 轴衬套 12.3.5.13
generalized flow diagram 一般流程 2.2.90
generalized fluidization 广义流态化 1
generalized schematic 综合性示意图 15.1.1
general note 一般注释 21.1.2
generated power 发出的电力 3.1.23.14
generator 发生器 20.1.2.16 发电机 2.2.110.14;2.2.113.8;2.2.114.11;2.2.116.9;2.2.117.19;2.2.118.12;25.2.1.15
generator overflow tube 发生器溢流管 20.1.8.4
genetic engineering 基因工程;遗传工程 1
geological structure 地质构造 25.2.4.1～12
geometries 几何位置 10.5.11.(a)
girt gear 齿圈 13.3.1.9 大齿轮 8.8.1.4
G.L.(1)(ground level) 地平面 4.6.1.3
gland 压盖 13.1.2.40
gland bolt 压盖螺栓 19.2.1.8
gland follower 填料压盖随动件 16.4.4.6
gland nut 压紧螺母 15.7.6.10
gland nut 压紧螺母 19.1.5.11
gland nut 压盖螺母 22.1.3.17
gland packing 填料函 16.4.19.24
gland sleeve 密封套 19.1.5.17
gland sleeve"O"ring 密封套"O"形圈 19.1.5.16
gland stud 填料压盖双头螺柱 16.4.4.13
glass bulb 玻璃灯泡 25.4.1.57
glass fibre products 玻璃纤维制品 2.3.9.56-58
glass filaments 玻璃纤维丝 2.3.9.52
glass furnace 玻璃熔炉 2.3.9.49
glass furnace for the Fourcault process 弗克法玻璃窑 2.3.9.1
glasshouse pot 熔化玻璃坩埚 2.3.9.46
glass-lined steel pipe 衬玻璃钢管 22.6.2.⑧

glassmaker 吹玻璃工人 2.3.9.38
glassmaker's chair 玻璃工人坐椅 2.3.9.45
glassmaking 玻璃制造 2.3.9.38-47
glass oiler 玻璃油杯 12.3.5.8
glass pipe 玻璃管 22.4.4.4
glass pipe and tubing 玻璃管 22.6.2.⑦
glass pipeline 玻璃管路 22.4.4
glass production 玻璃制造 2.3.9
glass pump 玻璃泵 16.1.2.5.2
glass recovery 玻璃回收 3.3.16.11
glass ribbon 玻璃板 2.3.9.10
glass wool 玻璃棉 3.5.14.7
glass yarn 玻璃丝 2.3.9.56
globe lift check valve 截止式升降止逆阀 22.1.1.(h)
globe type disc 球心型阀盘 22.1.1.2
Globe valve 截止阀,球心阀 22.1.3
glycol dryer 乙二醇干燥器 2.2.15.20
glycolic acid + H_2O 乙醇酸+水 2.2.66.1
goblet 高脚杯 2.3.9.41
gob of molten glass 熔融的玻璃料滴 2.3.9.23
god 熔解的玻璃原料 2.3.9.40
godet wheel 导丝轮 2.3.7.16
golden section method 黄金分割法 1
gooseneck 鹅颈管(大弯管) 14.3.2.25
governor body 调节器本体 24.2.7.10
governor diaphram 调节器隔膜 24.2.7.3
governor valve 调节阀 24.2.7.11
grab-sample setup 定时取样装置 3.4.8
graded stone 分层碎石 3.3.29.13
gradient 梯度 1
gradual distillation 分馏 2.3.8.4
grain size analyzer 粒度分析仪 1
granular 粒状 8.1.1
granular-media filter operates on 粒状介质过滤操作 3.1.10
granular sludge 颗粒污泥 3.2.6.27
granular solids discharge 粉状固体排料 4.3.4.27
granulated active carbon 粒状活性炭 2.4.3
granulated slag 粒状炉渣 2.2.26.11;24.3.1.22
granulate-water-mixture 颗粒-水混合物 9.5.20.1
granulation equipment 造粒设备 13
granulation water 造粒用水 3.3.16.11
granulator 造粒机 4.3.4.23
granule 粒料 13.3.4.5
granule bed 颗粒床 13.3.1.5

graphical method 图解法 1
graphite-block exchanger 石墨块体换热器 4.2.5.6
graphite pump 石墨泵 16.1.2.5.8
grate 栅板 2.2.18.8
炉栅 2.2.13.5
grate basket 栅筐 11.1.7.19
grate drive 转栅驱动器 2.2.12.9
grate drive cylinder 炉排驱动缸 24.1.5.5
grates 炉条 3.3.12.11
grates used with spreader stokers 布煤机炉排使用的炉栅 24.2.4
gravel 碎石 2.2.20.9
gravel pack 废料堆 3.1.24.9
gravity conveyor 重力输送器 23.4.3.1
gravity disc 重力盘 9.5.12.2
gravity discharge 重力卸料,自流出料 23.1.7.4
gravity drainage zone 重力渗滤区 9.6.2.1
gravity drain plug 自流排放旋塞 18.1.4.8
gravity filter 重力过滤器 3.1.2.17
gravity-flow hopper 自流式料斗 23.1.1.2
gravity spout 重力下料管 23.1.6.10
gravity thickener tank 重力增稠罐 3.1.26.(a).2;3.1.26.(b).2
grease cup 油脂杯 16.4.12.11
润滑脂杯 9.5.3.26
grease(oil)cup 润滑脂(油)杯 16.4.9.8
green liquor preheater 绿液预热器 2.3.10.43
green malt 绿麦芽 2.3.12.23
grid 栅条 10.2.1.15
grid bar 栅条 11.1.7.15
grid frame 栅条架 11.1.7.14
grid of T bars T型栅板 14.1.8.(c)
grid packing 格栅填料 5.3.4.49
grid support 格栅支承 9.3.10.5
grind 磨碎,研磨 2.2.90.2
研磨物 8.1.1.17
grinder pump with water 研磨水泵 2.2.35.5
grinding 研磨 11.6
grinding chamber 研磨室 11.3.1.2
粉碎室 11.3.2.7
grinding,drying,feeding 研磨,干燥,进料 2.2.25.4
grinding element 研磨元件 11.2.11.2
grinding equipment 研磨设备 11.2
grinding hammer 研磨锤 11.2.11.2
grinding medium 研磨介质 11.3.4.4
grinding recycled dry product 磨细的循环干物料 3.1.16.2

grinding ring 研磨环 11.2.12.4
grinding roller 研磨辊 11.2.12.3
grinding stone 磨石 2.3.10.71
grip 手柄 11.1.8.4
卡子 5.3.4.29
grit chamber 砂箱 3.1.25.3;3.1.27.4
grizzly 栅筛 12.8.2.4
grooved flange 槽面法兰 22.4.3.(d)
gross error 过失误差 1
ground 地面 23.2.1.10
ground coal 碾碎的煤 2.2.19.3
grounded 接地 10.3.1.28
grounded metal electrode 接地金属电极 15.3.1.5
grounded receiving electrodes 接地收集电极 10.3.1.27
grounded rotor 接地转子 12.7.2.2
ground glass cone 磨砂玻璃锥 3.4.14.31
ground glass socket 磨砂玻璃插口 3.5.14.30
ground level 地平 2.1.7.10
ground surface 地表面 2.2.85.11
groundwood 磨木浆 2.3.10.68
groundwood mill 磨木浆厂 2.3.10.53-65
groundwood pulp 磨木纸浆 2.3.10.77
group activity coefficient 基团活度系数 1
growth control device 生长控制装置 3.1.36.8
growth control pump 生长控制泵 3.1.36.9
growth factor 生长因子 1
guard ring 保护圈 18.2.4.4
guide 导向盘 22.1.8.13
guide bearing 导引轴承 15.6.1.14
guide bushing 导向套 22.1.8.6
guide cylinder 导流筒 18.2.3.10
guide flow cylinder 导流筒 18.2.2.5
guide lug 导耳(导缘) 19.1.5.13
guide plate 导向隔板 16.4.18.12
guide ring 导向环 22.1.9.23
guide vane 导(向)叶(片) 16.4.18.8
gum ball 树胶球 12.2.5.8
gutter 出料槽 5.5.1.11
gypsum 石膏 2.2.115.19
gyrating screens 回转振动筛 12.2.3
gyratory crusher 回转破碎机 11.6.3.2

H

H_2 氢气 2.2.33.9;2.2.43.17;2.2.61.7;2.2.67.18;2.2.68.16;2.2.109.14;2.2.117.12
H_2,CO,CO_2,H_2S,N_2, etc 氢气,一氧化碳,二氧化碳,硫化氢,氮气等 2.2.38.4
half coupling 半接头 10.4.2.11
half the included angle 夹角之一半 21.2.5
Hall electrolytic conductivity detector 霍尔电解质电导率检测器 3.5.22
hammer carrier 锤盘 11.1.7.5
hammer crusher 锤式破碎机 11.1.5
hammer mill 锤磨(机) 11.6.5
锤式粉碎机 8.4.5.(a)
hammer pin 摆锤销轴 11.1.8.20
hammer plate 夹锤板 11.1.8.29
hammers or impact members 锤或冲击部件 11.2.9.1
hand-blown 人工吹制 2.3.9.41
hand-blown goblet 人工吹制的高脚杯 2.3.9.41
handhole cover 手孔盖 9.3.3.36
handle 手柄 12.3.4.9
手轮 15.7.4.17
handle for pan position 确定锅位置的手轮 15.7.4.17
handling 运送 23.4
hand pump 手动泵,手摇泵 16.1.2.4.5
handrail 栏杆 25.1.1.25
hand screw 手动压紧丝杠 9.1.2.8
hanger bolt 吊挂螺栓 9.5.4.34
hanger rod 吊杆 21.2.2.9
hanger spring 吊挂弹簧 9.5.4.35
hard 硬 8.1.1.7
hard wax 硬蜡 2.2.74.15
hardened liner 淬硬套管 13.2.1.5
hard paste 硬的膏状物 8.1.1.16
harrow 耙 23.2.5.3
hauling 运输车 3.3.3.1
hazard 危险品 8.1.1.26
hazardous material 危险品 3.3.12.4
hazardous waste 危险废弃物 3.3.7
HC feed HC原料 6.1.7.5
HC gas 气体烃 2.2.95.11
HCl column 盐酸塔 2.3.18.17
HCN 氰化氢 2.3.15.15
HCN column 氰化氢塔 2.3.15.5
H-coal process 氢-煤过程 2.2.98
$HCOOCH_3$ 甲酸甲酯 2.2.61.11
head 扬程 1
封头 21.4
head attachment 封头附件 21.1.1.60
header 管箱 4.5.1.10
header door 顶盖门 24.1.2.14
header of thermal medium inlet 热载体入口总管 24.1.1.3

heat

header of thermal medium outlet 热载体出口总管 24.1.1.16
header sheet 管板 10.3.4.13
head motion 传动机头 12.2.5.1
head pulley and drive 前轮和驱动装置 23.2.1.9
head shaft 主轴 15.6.1.30
head skirt 封头直边段 21.4.1.10
head tank 高位槽 3.1.32.15
head works 进水口工程 3.1.41.2
hearth 炉膛 6.3.5.1
hearth clearance adjustment 炉床间隙调节器 24.1.9.6
heat 热 2.2.114.5
heat-absorption tubes 吸热管 24.2.10.2
heat carrier 载热体 2.2.70.2;2.2.88
heat-carrier preparation chamber 热载体制备室 2.2.84.10
heat-carrier separator 热载体分离器 2.2.88.13
heat content 热容 20.2.4.1
heated air out 热气出口 24.1.13.13
heated area 加热区 3.5.11.7
heated distilling column 加热蒸馏塔 5.5.1.9
heated drying cylinder 加热烘缸 2.3.11.33
heat engine 热机 25.1.1
heater 加热器 2.1.6.1;2.2.6.6;2.2.63.3;3.4.7.4
heater block 换热器单元 3.4.21.7
heater/cooler 加热器/冷却器 13.2.3.7
heater desorber 加热解吸器 14.2.3.8
heater retaining stud 加热器固定(双头)螺栓 18.2.4.17
heat exchange 热交换 2.2.117.20
heat exchange plate pack 换热板组件 4.2.1.4
heat exchange process 换热过程 2.1.4.5
heat exchanger 换热器 2.2.46.3;2.2.47.3;2.2.108.7;2.2.110.7;3.3.25.15
热交换器 2.2.76.13;2.3.10.19;2.3.19.3;2.4.1.4
heat-exchanger component 换热器部件 4.1.1
heat exchangers for solid 固体用换热器 4.3
heat exchange to preheat the waste gas stream 热交换器用来预热工业废气流 3.4.5.6
heat exchange tube 换热管 4.1.4.2
heat flux 热通量 1
heating barrel 加热筒 15.7.3.6
heating circuit 加热管线 25.6.3.2
heating coil 加热盘管 25.6.4.4

heating element 加热元件 4.2.4.14
heating fluid 加热流体 4.3.3.23
heating furnace 加热炉 24.1.2
heating jacket 加热夹套 7.1.11.5
heating medium 载热体 1
加热介质 15.1.1.4
heating medium inlet 加热介质进口 8.9.5.6
heating or cooling medium 加热或冷却介质 8.9.4.2
heating steam 加热蒸汽 5.5.1.48
heating tube 加热管 24.1.2.10
heating tube bundle 加热管束 7.1.8.3
heating tube in convection section 对流段炉管 24.1.1.26
heating tube in radiant section 辐射段炉管 24.1.1.21
heating tube support 炉管支承架 24.1.1.19
heating unit 加热器 2.2.76.5
heating wall 加热面 7.1.2.8
heat insulator 保温材料 14.4.3.23
heat load 热负载[荷] 20.1.10.28
heat loss 热损失 2.4.1.9
heat-medium 加热介质 4.3.2.11
heat of absorption 吸收热 1
heat of formation 生成热 1
heat pipe 热管 4.8
heat pump 热泵 16.1.2.3.41
heat recovery 热回收 2.4.3.5
heat recovery stages 热回收级 2.4.2.4
heat-recovery unit 热回收装置 2.2.4.10
heat reflector 热反射器 18.2.4.18
heat rejection stage 热耗损级 2.4.2.6
heat rejection 热耗损 2.4.3.6
heat-rejection alternatives 各种热力排放 25.1.1.19
heat release 放热 24.2.8
heat remover 取热器 2.2.69.1
heat source 热源 8.5.6.3
heat supply 供热固体 14.3.3
heat transfer 传热,热量传递 1
heat-transfer equipment 传热装置 4.3.3
heat-transfer equipment for melting or fusion of solids 熔融或熔化固体用传热设备 4.3.1
heat transfer-jacketed body 热交换夹套 12.1.3.(i)
heat transfer media pump 热载体泵 16.1.2.3.42
heat transfer surface 传热表面 14.3.1.19
heat transport 热的输送 25.6
heat treatment 热处理 21.1.2.1

547

heavies　重馏分　2.2.69.25
　　重组分　2.3.18.22
heavier hydrocarbons　高沸点碳氢化合物　14.3.3.(b)
heavy annular solids bed　重型环状固体床层　4.3.3.12
heavy aromatic　重芳烃　2.1.1.55
heavy aromatic separation　重芳烃分离　2.1.1.59
heavy bottom slurry　塔底重油浆　2.2.97.10
heavy catalytic gas oil　重催化裂化粗柴油　2.1.5.7
heavy components　重组分　15.5.4.9
heavy distillate　重馏出物　2.1.4.26
　　重馏分　2.2.95.25
heavy-duty jacketed　重负荷夹套式　4.3.4.(a)
heavy fraction　重成分　15.4.2.5
heavy gas oil　重柴油　2.1.4.13
heavy gasoline　重质汽油　2.2.77.13
heavy liquid　重液(体)　6.1.14.2
　　重质液体　2.2.90.11
heavy liquid dispersion pipe　重液分布管　6.1.18.7
heavy liquid inlet　重液入口　6.1.18.4
heavy liquor pump　重液泵　2.3.10.34
heavy-load thrust bearing　重型止推轴承　15.7.2.8
heavy manganese steel hammer　重型锰钢锤　11.1.7.4
heavy-medium circulation　重介质循环　12.5.5.14
heavy-medium overflow　重介质溢流　12.5.5.23
heavy-medium pump　重介质泵　12.5.5.16
heavy-medium return　重介质返回　12.5.5.15
heavy-medium storage sump　重介质储槽　12.5.5.22
heavy mineral collection tank　重质矿石收集池　15.5.3.6
heavy mineral ripple　重矿石粒子　15.5.3.2
heavy naphtha　重石脑油　2.1.1.17
heavy oil　重油　2.2.5.20
heavy-phase effluent　重相流　9.5.14.3
heavy product　饱和产物　14.2.3.6
heavy shale oil　重质页岩油　2.2.6.13
H-E(high-efficiency)brink mist eliminator element　高效率纤维雾沫净除器单元　10.4.2
height　扬程　1
height equivalent of a theoretical plate(HETP)　等[理论]板高度;理论板当量高度　1

height of a[heat]transfer unit　传热单元高度　1
height of overall transfer unit　总传质单元高度　1
height of sectional tower shell　塔节高度　5.2.4.15
height shot gravel　高卵石层　3.1.19.11
helical　螺旋面式　10.1.1.(c)
helical blades　螺旋形叶片　13.2.4.18
helical conveyor　螺旋输送器　12.4.2.3
helical flight　螺旋面刮板　4.3.3.(e)
helical impeller pump　螺旋叶轮泵　16.1.2.2.22
helical plate　螺旋板　10.4.3.6
helical top　螺线顶盖　10.1.2.3
helicoid tuyere　螺旋形风帽　10.4.3.17
hemispherical head　半球形封头　21.1.1.39
hermetically sealed magnetic drive pump　封闭式电磁泵　16.1.2.4.7
　　密封式磁力驱动泵　16.5.5
hermetic-refrigerant pump　密封制冷剂泵　20.1.8.19
hermetic-solution pump　密封溶液泵　20.1.8.5
herring bone gearbox　人字齿轮箱　15.7.2.10
heterogeneously fluidized bed　非均一流化床　14.1.3.14
heterogeneous reaction　非均相反应　1
hex head bolt　六角头螺栓　16.4.26.20
hex head plug　六角头丝堵　22.3.2.20
HGT unit　重质汽油处理器　2.2.75.17
HGT(heavy gasoline treating)reactor　重质汽油处理反应器　2.2.77.14
HHV(high heat value)　高热值　2.3.1.①
Hi-eF purifier　Hi-eF 净化器　10.4.3.(a)
Higgins contactor　Higgins 接触器　6.2.4
high-alloy(stainless)steels　高合金钢(不锈钢)　22.6.2.⑮
high-Btu　高热值　2.2.2.19
high Btu fuel gas　高热值燃料气　2.2.106.43
high-Btu gas product　高热值气体产品　2.2.87.15
high-Btu pipeline gas　高热值管道气　2.2.89.13
high capacity coalescer filter　大容量聚并滤清器　15.5.2
higher fine powder rate　提高细颗粒速率　14.1.2.6
higher solid rate　固体速率增大　14.1.1.6
high-gravity compartment　高密室　12.5.2.16
high-gravity medium　高密度介质　12.5.2.11

high lift pump 高扬程泵,高压泵 16.1.2.1.5
high-precision liquid filling 高精度液体装料 23.4.1
high press.fan 高压风机 11.6.4.15
high pressure 高压 19.3.3.2
high pressure absorbing tower 高压吸收塔 2.2.55.16
high-pressure cylinder 高压汽缸 25.4.4.3
high pressure gas 高压气体 19.3.2.17
high-pressure gas main 高压煤气总管 2.3.14.45
high-pressure, low-capacity compressor having a hydraulically actuated diaphragm 液压传动的高压小流量隔膜压气机 17.2.14
high pressure manual pump 手摇高压泵 16.1.2.4.9
high pressure pump 高压泵 16.1.2.1.4
high pressure side 高压侧 20.1.3.5
high pressure steam 高压蒸汽 2.2.42.8; 2.2.43.11; 2.2.45.11; 2.2.111.8; 20.1.10.1 高压气流 17.2.2.3
high pressure steam inlet, 2nd stage 第二级高压蒸汽进口 18.3.4.9
high pressure steam inlet, 1st stage 第一级高压蒸汽进口 18.3.4.1
high pressure steam turbine & generator 高压蒸汽涡轮机和发电机 2.2.111.9
high pressure stuffing box using metallic packing 金属填料高压填函 19.2.1
high pressure supply line 高压供油线 2.1.7.19
high pressure toroidal bellows 高压环形波纹管 22.5.1.(i)
high pressure turbine 高压气轮机 25.3.2.8
high pressure underground gasification 地下高压气化 2.2.81
high pressure vapor 高压蒸气 20.1.2.4
high pressure water 高压水 15.6.5.5
high silicon cast iron 高硅铸铁 22.6.2.③
high silicon cast iron pump 高硅铸铁泵 16.1.2.5.7
high-specific speed pump 高比转数泵 16.1.2.2.21
high speed air valves 高速空气阀 12.7.4.9
high speed centrifugal pump 高速离心泵 16.1.2.2.16
high speed rotating shaft seal for liquid metal with very high vacuum 超高真空液体金属密封高速转轴 19.4.2
high-stage compressor 高压级压缩机 20.1.3.17

high-stage condenser 高压级冷凝器 20.1.7.9
high-strength wastewater 高浓度废水 3.1.31.1
high temperature 高温 8.7.2.11
high temperature air-cap 高温空气罩 8.7.2.5
high-temperature nuclear reactor 高温核反应堆 2.2.107.18
high temperature shift 高温变换 2.3.3.9
high-temperature vessel 高温罐 25.6.4.3
high tensile strength 高抗拉强度 2.3.8.46
high vacuum distilling apparatus 高真空蒸馏装置 18.5.1
high vacuum pump 高真空泵 16.1.2.1.9
high-velocity air 高速空气 8.7.2.9
high-velocity burner 高速燃烧器 25.2.8
high-velocity section 高速断面 8.4.1.23
high-velocity slot stage 高速沟槽级 10.5.1.8
high viscosity triple screw pump 高黏度三螺杆泵 16.1.2.1.11
high voltage 高压电 10.4.1.10
high voltage cable 高压电缆 10.3.1.6
high voltage cable for three-phase current (三相)高压电缆 25.4.8
high-voltage dc power supply 高压直流动力供应 15.3.1.1
high-voltage discharge electrode 高压放电极 10.3.4.5
high-voltage electrode 高压电极 10.3.4.10
high-voltage grid 高压栅 10.4.1.7
high-voltage insulator compartment 高压绝缘子室 10.3.4.18
high-voltage power 高压电源 10.3.5.8
high-voltage support frames 高压电路支架 10.3.4.4
Hi-level 料面高度计 23.1.8.6
Himsley continuous ion-exchange system Himsley 连续离子交换系统 6.2.6
hind counter guide (后)副十字头导轨 17.2.15.18
hind crosshead metal 副十字头导轨镶条 17.2.15.21
hind guiding plate 副十字头导板 17.2.15.19
hinge 铰链 11.1.8.12
铰接 22.5.3.4
hinged connection 铰接连接 22.5.4.1
hinged disc 铰接盘 22.1.10.5
hinged disc type check valve 铰链盘式单向阀 22.1.10

hinged door 带铰链端盖 3.1.6.13
Hinged expansion joint 绞接膨胀节 22.5.3
hinged floor socket 铰链式地面插座 25.4.1.23
hinged lever 铰链杆 9.2.1.22
hinged lid 铰链盖 25.4.1.24
hinged plow 铰链刮料器 23.2.3.(f)
hinged screen bottom 绞链式筛网底 6.1.17.3
hinge pin 铰链销 22.1.10.6
HL(1)(heavy liquid) 重液 6.1.15.2
HLR(heavy liquid recycle) 重液循环 6.1.15.4
H_2O 水 2.2.33.2;2.2.63.10;2.2.65.4;2.2.70.5;2.2.71.11;2.2.114.15;3.4.6.9
H_2O,AcOH 水,乙酸 2.2.41.10
hoisting pillar 吊柱 5.2.1.14
hold down grate 床层限制栅 5.4.1.6
hold-down grid 床层限制栅 5.3.1.11
holding equalization 稳定水池 3.1.2.1
holding tank 存储槽 3.3.21.9
hold tank 储液槽 3.1.32.16
hold-up 停留时间 8.4.3
hole 孔 6.3.9.5
hole for condensate 冷凝液出孔 5.5.1.45
hole for connection 连接孔 22.1.4.11
hollow fiber 中空纤维 1.6,8
hollow-fiber module 中空纤维组件 1 中空纤维型 15.2.1
hollow-fiber module Permasep permeator 中空纤维型 Permasep 反渗透器 15.2.1
hollow flight 粗螺旋窄刮板 4.3.3.(d)
hollow journal 空心轴颈 11.2.4.11
hollow piston 空心活塞 17.2.10.5
hollow porcelain supporting insulator 空心的瓷支架绝缘子 25.4.9.4
hollow screws 空心螺旋 12.1.3.11
hollow shaft 空心轴 8.8.12.4
hollow shaft with bearings 空心轴带轴承 9.5.10.9
hollow thin-walled plastic tube 薄壁中空塑料管 15.2.2.4
holo-flite intermeshed screws 中间网状螺旋 4.3.3.23
homogeneously fluidized bed 均一流化床 14.1.3.10
homogenization 匀化 1
homogenizer 匀化器 15.7.6
hood made of glass fibre plastics 玻璃钢罩 12.2.3.15
hook 大钩 2.1.2.9

hop boiler for boiling the wort 蒸煮麦芽汁用的锅炉 2.3.12.52
hopper 集尘斗 10.2.1.22
料斗 2.2.31.5;2.2.110.3;10.2.3.12
加料斗 15.7.2.7
贮液器 2.3.5.13
hopper base 料斗支架[座] 11.1.5.4
hopper cars 底卸式货车 23.3.8
hopper cooling jacket 料斗冷却夹套 13.2.1.4
hopper tube sheet 料斗管板 10.1.2.34
hopper valve 料斗底阀 10.2.1.23
H_2O return to absorbing tower H_2O 返回吸收塔 2.2.70.7
horizontal axis 水平轴 3.5.1.5
horizontal axis current meter 水平轴流速计 3.5.1.(b)
horizontal axis(single speed) 水平轴(单速) 9.5.2.10
horizontal conveyor 水平输送机 23.3.3.(a)
horizontal cylindrical vessel 卧式圆筒形容器 21.2.4
horizontal drum mixer 水平转鼓混合器 12.1.3.(c)
horizontal element stacks 水平层叠的元件 15.5.2
Horizontal-inclined conveyor-elevator 水平倾斜输送式提升机 23.3.3.(e)
horizontally gyrated 水平回转式 12.2.3
horizontal mixed flow pump 卧式混流泵 16.4.29
horizontal multi compartment fluidizedbed dryer (cooler) 卧式多室流化床干燥器(冷却器) 8.5.2
horizontal opposed reciprocating compressor 卧式对置式往复压缩机 17.2.1.11
horizontal pressure leaf filter 卧式加压叶片过滤机 9.2.4
horizontal pump 卧式泵 16.1.2.2.30
horizontal seperator 卧式气液分离器 5.3.5
horizontal single-stage double-suction volute pump 卧式单级双吸蜗壳泵 16.4.13
horizontal single-stage single suction centrifugal pump 卧式单级单吸离心泵 16.4.2
horizontal steam separator 卧式蒸汽分离器 10.1.4.(d)
horizontal tank 卧式贮罐 21.2.5
horizontal-tank type 水平贮槽式 4.3.1.(a)

horizontal tube bundle 水平管束 7.1.3.14
horizontal-tube evaporator 水平管式蒸发器 7.1.7
horizontal-tube evaporators 水平管式蒸发器 7.1.12.(b),(c)
horizontal vessel 卧式容器 21.3.1.1
horizontal wiped-film evaporator 卧式刮壁(膜)型蒸发器 7.1.11
hospital 医院 17.6.3.34
hot air 热空气 2.2.106.12
hot air blast hole 热风通入口 8.5.1.6
hot air box 热空气室 8.4.6.5
hot-air chambers 热空气室 8.8.9.9
hot air compartment 热风室 3.3.26.6
hot air duct 热空气导管 8.6.1.7
hot air inlet 热空气进入 3.1.16.13
hot-air jet 热空气喷射器 8.3.2.4
hot air producer 热风发生器 8.9.3.7
hot ball 热球 2.2.5.15
hot bitumen 热沥青 2.1.9.26
hot-bitumen storage 热沥青贮罐 2.1.9.25
hot break 高温凝固物 2.3.13.2
hot char 热炭 3.3.15.19
hot coke 热焦炭 2.1.9.27
hot flue gas 热烟道气 2.2.5.1;2.2.111.6
hot fluid 热流体 4.2.1.11
hot fluid in 热流体进入 4.2.1.7
hot fluid out 热流体出口 4.2.1.9
hot gas 热气体 2.2.6.14;3.3.25.4
热(可燃)气体 3.1.23.5
hot gas casing 热燃气套管 25.3.1.11
hot gas duct 热气体管道 8.4.1.13
热气导管 3.3.24.22
hot gas inlet 热气入口 14.3.3.16
hot gas outlet 热气体出口 4.2.4.12
hot gas recycle methanator 热气循环甲烷转化器 2.2.106.42
hot kiln exhaust gases 废热炉气 11.6.4
hot-oil circulating system 热油循环系统 25.6.7
hot separator tank 高温分离罐 2.2.74.11
hot solid 热固体颗粒 2.2.4.6
hot spent shale 热的废页岩 2.2.5.22
hot syngas 热合成气 2.2.112.2
hot water 热水 2.1.9.4
hot water pump 热水泵 16.1.2.3.11
hot-water supply 供入热水 8.9.6.7
hot water tank 热水槽[箱] 2.3.10.18
hot working fluid 热工作液 25.5.1.12
housed bearing 内装轴承 9.3.3.24
house telephone 户内电话 25.4.1.3

house telephone with call button 带呼唤按钮的户内电话 25.4.1.3
housing 机箱 8.8.12.14
housing and bedplate cast in one piece 外壳与底座连体铸件 9.6.1.20
housing cap 机箱盖 8.8.12.9
HPO (high purity oxygen) 高纯氧气 3.1.40.2
HRSG 余热锅炉 2.2.110.16
HRSG (heat recovery steam generator) 热回收蒸汽发生器 2.2.111.14;2.2.112.8;2.2.114.9
H_2S 硫化氢 2.2.32.8;2.2.33.16;2.2.63.1;2.2.117.14
H_2S,CH_3SH 硫化氢,甲硫醇 2.2.63.7
H_2S,CO_2 separation 硫化氢,二氧化碳分离 2.2.117.16
H_2S loaded solvent 含有硫化氢的溶剂 2.2.32.9
H_2S,NH_3,Organic acids 硫化氢,氨,有机酸 3.1.47.11
H_2SO_4 硫酸 2.3.15.12
H_2S recycle 硫化氢循环 2.2.63.5
HSR naphtha HSR 石脑油 2.1.4.14
H_2S stripper 硫化氢汽提塔 2.2.32.10
HTGR (high-temperature gas-cooled reactor) 气冷式高温反应堆 2.2.108.6
H.T.inlet 高压电接入口 10.3.1.13
H.T.rapper shaft 高压电振打器传动轴 10.3.1.16
H-type reciprocating compressor H 型往复压缩机 17.2.1.12
hub 轮毂 16.4.6.2
hub diameter 轮毂直径 16.4.6.11
hub-mounted curved-blade turbine 弯叶开启涡轮式 14.4.4.3
hub-mounted flatblate turbine 平直叶开启涡轮式 14.4.4.2
humidification sprays 增湿雾化器 10.5.1.5
humidifier 增湿器 6.2.8.2;3.4.3.8
humidity controller 湿度控制器 20.2.2.5
HVGO(heavy vacuum gas oil) 重减压粗柴油 2.1.4.18
HV insulator 高压绝缘器 15.3.1.2
H.W.(hydraulic water) 加压水 12.4.1.2
hydraulic and pneumatic 液压和气动式 23.3.2.5
hydraulic cyclone 旋液分离器 3.3.6.7
hydraulic fluid 液压用液体 16.2.8.10
hydraulic head 水力头 6.1.7.3
hydraulic jack 液力千斤顶 9.2.7.14

551

hydraulic motor 液压马达 15.7.3.3
hydraulic plant 液压设备 2.3.12.13
hydraulic pump 液压泵 15.7.3.10
hydraulic pumping unit 液力泵送装置 9.4.12.1
hydraulic pump with motor 带电动机油压泵 9.5.10.13
hydraulic radius 水力半径 1
hydraulic ram 液压滑块(式) 16.1.1.50 液压夯锤 3.3.29.17
hydraulic setting control 排料口大小液压控制装置 11.1.1.9
hydraulic system 液压系统 2.3.11.37
hydro carbon 2.2.2.21
hydrocarbon gas 气体烃 2.2.98.18 碳氢化合物气体 2.2.94.11
hydrocarbon liquids 液烃 2.2.77.11
hydrocarbons 碳氢化合物,烃类 6.1.7
hydrocarbylation 烃基化 2.1.1.56
hydrochloric acid outlet 盐酸出口 5.4.2.5
hydroclone 水力旋流器 2.2.98.25
hydrocracking 加氢裂化 2.1.1.14
hydrocracking aviation fuel 加氢裂化航空燃料 2.1.1.18
hydrocracking diesel oil 加氢裂化柴油 2.1.1.19
hydrocy clone 旋液分离器 3.1.8.11 水力旋流分离器 2.2.18.8
hydrofining 加氢精制 2.1.1.13
hydro-gasification 加氢气化 2.2.2.15
hydrogasifier 加氢气化器 2.2.6.16
hydrogen 氢(气) 2.2.8.1;2.2.41.6
hydrogen and carbon dioxide 氢气和二氧化碳 2.2.39.c
hydrogenated diesel fuel 加氢柴油 2.1.1.75
hydrogenated gasoline 加氢汽油 2.1.1.75
hydrogenated kerosene 加氢煤油 2.1.1.75
hydrogenating 加氢反应 2.2.66.13
hydrogenation of phenol 酚氢化 2.3.8.16
hydrogenation reactor 加氢反应器 2.2.40.10;2.2.41.5;2.2.69.12
hydrogen chloride 氯化氢 2.3.18.6
hydrogen cooler 氢冷却器 25.4.4.7
hydrogen,electric power 氢气,电力 2.2.3.17
hydrogen gas compressor 氢气压缩机 17.1.2.5
hydrogen generation 发生氢气 2.2.92.4
hydrogen making 制氢 2.1.1.2
hydrogen plant 氢气生产装置 2.2.6.21
hydrogen production 制氢 2.2.1.8
hydrogen recovery 氢回收 2.2.98.16

hydrogen recycle 氢循环 2.2.6.9
hydrogen-rich gas 含氢富气 2.2.101.1
hydrogen-rich recycle gas 富氢循环气 14.2.6.14
hydrogen sulfide 硫化氢 2.2.94.12
hydrogen sulphide scrubber 硫化氢洗涤塔 2.3.14.22
hydrolysis reactor 水解反应器 2.2.62.10
hydrophilic separatory membrane 亲水分离膜 6.1.8.1
hydrophobic separatory membrane 憎水分离膜 6.1.8.5
hydrostatic head 液柱静压头 1
hydrostatic testing 流体静压试验 21.1.2.15
hydrotreating 加氢处理 2.2.2.26
hydroxylamine 羟胺 2.3.8.21
hydroxylation 羟基作用 25.2.8
hypersorption 超吸附 1
hypersorption adsorber vessel 超吸附塔 6.3.6
hypothetical tangent circle 假想的相切圆 11.5.1.4

I

ice bath 冰浴 3.4.11.13
ICI methanol reactor ICI甲醇反应器 14.3.2.(a)
ICI valve ICI阀 14.1.7.(c)
ideal gas 理想气体 1
ID fan 引风机 25.5.3.18
idler 导轮 9.5.11.3
idler roll 惰辊 23.2.1.16
ignition chamber 起燃室 3.3.12.8
ignition source 点火装置 3.3.11.8
ignition tube 点火管 24.2.9.2
ignitor 点火器 3.5.21.5
IGT(Institute of Gas Technology) (美)煤气工艺研究所 2.2.6
IGT process (美)煤气工艺研究所工艺 2.2.6
IIM(insoluble inorganic matter) 不溶性无机物质 3.2.1.5
illumination 照明设备 3.1.16.9
imbibition 浸润;吸液 1
immobilization technology 固定化技术;固相化技术 1
immobilized enzyme reactor 固定化酶反应器 1

impact chamber 撞击室 11.5.2.4
impact crusher 冲击式破碎机 11.1.6
impact pin 冲击销 11.1.6.11
impact spray thrower 冲击式淋洒器 5.3.4.4
impact testing 冲击试验 21.1.2.15
impeller 搅拌器 14.4.2
impeller 桨叶 6.1.5.7
impeller and 叶轮端 19.1.8.3
impeller bushing 叶轮衬套 16.4.25.6
impeller diam 叶轮直径 16.4.3.19
impeller diameter 叶轮直径 16.4.6.9
impeller end 叶轮端 19.2.3.14
impeller hub 叶轮轮毂 16.4.5.4
impeller key 叶轮键 16.4.9.28
  叶轮(轴)键 16.4.12.17
impeller labyrinth 叶轮迷宫密封 20.2.1.5
impeller magnet 叶轮磁体 16.5.5.5
impeller nut 叶轮(紧固)螺母 16.4.29.2
impeller of pump 泵叶轮 16.4.22.2
impeller ring 叶轮口环 16.4.9.36
impeller shaft 旋转混合器轴 12.8.1.9
impeller shaft 叶轮轴 16.4.25.16
impeller shaft coupling 叶轮轴联轴器 16.4.26.1
impeller spacer 叶轮空距环 17.4.2.21
impeller vane 叶轮叶片 16.4.5.3
impeller width 叶轮宽度 16.4.3.20
impervious graphite 不透性石墨 22.6.2.②
impervious liner 不渗透的(混凝土)衬砌 3.3.12.3
impingement baffle 防冲板；缓冲挡板 4.1.1.29
impingement baffle plate stages 撞击挡板级 10.5.1.7
impingement plate 撞击板 10.5.1.11
impingement-plate scrubber 撞击板式涤气器 10.5.1
impingers 冲击器 3.5.11.14
  撞击器 3.5.14.9
improper filtration 不适当过滤 6.3.2.23
impure water 不纯净的水 2.4.1.1
impurity removal 杂质脱除 2.2.3.12
in 进(口) 16.5.6.1
inactivation 失活 1
in adjustment 调节位置 11.4.3.5
inboard 内侧 16.4.9.20
inc. or incr. (increase) 增加；增大；增多
incineration process 焚烧 13.4.1(b)
incineration with steam generation 焚化伴随产生蒸汽 3.3.1.12
incinerator 焚烧炉 2.3.6.28；3.4.5.5

incinerator facility 焚烧炉装置 3.3.8
incinerator-feed surge pond 焚烧炉-进料缓冲池 2.3.6.24
incinerator-feed surge tank 焚烧炉-进料缓冲槽 2.3.6.26
incipiently fluidized bed 初始流化床 14.1.3.7
incipient stage 流态化初期阶段 4.3.3.26
incline 斜面 23.3.5
inclined "tumbling" wall self-discharge 倾斜滚光壁自动卸料离心机 9.5.1.(k)
inclined bowl discharge 倾斜转鼓卸料 9.5.2.13
inclined hole tray 斜孔塔板 5.2.10.17
inclined vibrating wall self-discharge 振动斜壁自动出料离心机 9.5.1.(j)
inclined wall scroll discharge 倾斜壁卷轴卸料离心机 9.5.1.(l)
inclined wall self-discharge 斜壁自动卸料离心机 9.5.1.(i)
included angle 夹角 21.2.5
incorporated combustion chambers 结合的燃烧室 25.3.1.17
indermediate crosshead frame 中体十字头拒 17.2.15.16
indicator dial 指示盘 15.6.4.2
indirect evaporative cooler 间接蒸发冷却器 6.2.8.10
indirect-heated continuous plate dryer 间接加热连续操作圆盘干燥器 8.9.4
indirect heat-transfer equipment 间接传热设备 4.3.4
indirect liquefaction 间接液化 2.2.1.15；2.2.89.11
indirectly fired 间接燃烧 3.3.15.(b)
indirect-process heater 间接过程加热器 25.5.1.(c)
indirect rotary 间接加热回转 8.1.1.⑨
Indox V magnet assembly Indox V 型永久磁铁组件 12.6.10.5
induced-air flotation 导气浮选 3.1.13
induced draft fan 引风机 3.1.21.16；3.3.24.12
induced flow 导流系 10.2.3.18
induced-roll separator 磁感应分离器 12.6.11
industrial boiler 工业锅炉 3.3.1.13
industrial furnace 工业炉 24.1
industrial horizontal tubular filter 工业用水平列管式过滤机 9.4.4
industrial incinerator 工业焚烧炉 24.1.4
industrial processes 工业过程 25.1.1.14

553

industrial wiped-film stills　工业刮膜式分子蒸馏釜　5.5.2
inert-gas connection　惰性气体接口　4.3.4.2
inert-gas makeup　惰性气体补充口　8.9.2.14
inert-gas supply to burden chamber　惰性气体供入装料室　13.5.2.4
inert hydrocarbon liquid　惰性烃类液体　2.2.39.d
inert liquid　惰性液体　2.2.9.4
inert particle　惰性粒子　8.5.3.12
infeed　进料　3.3.29.2；3.3.29.16
infeed conveyer　进料输送机　3.3.30.7
inflation film manufacturing machine　吹膜机　15.7.5
inflow　流入　3.3.30.1
inflow traverse conveyor　流入水平往返移动输送机　3.3.29.8
influent pump　进液泵　3.1.39.1
influent　流入物［液］　15.5.4.1
　　流入液体　15.5.1.1
　　流入　15.6.1.33
　　进水　3.1.32.8；3.1.33.2；3.1.42.1；3.2.1.1
　　进料　6.2.5.3
　　进气　3.4.3.4
influent injection　进水注入　3.1.32.2
influent inlet　流入物进口　9.4.4.11
influent lines　进水管道　3.1.32.25
influent pump　进液泵　3.1.39.1
influent raw wastewater or primary effluent　原废水流入或一次水流出　3.1.46.4
information flow diagram　信息流图　1
infrared drier　红外线干燥器　2.3.11.32
infusion　浸渍；浸泡
in.gage．　表压　10.3.5
in hearth　炉膛内　3.3.26.4
inhibitors　抑制剂　2.3.19.1
initial condition　初始条件　1
initial dewatering　初始脱水　9.3.9.1
initial separation　初级分离　8.6.3.9
injected water　喷射用水　6.2.3.5
injection cylinder　注塑缸　15.7.3.4
injection moulding machine　注塑成型机　15.7.3
injection port　注入孔　19.1.7.15
injection pump　注射泵　2.2.105.6
injection pump assembly　喷射泵组件　16.2.7.8
injector tee　T形喷射器　23.1.8.1
inlet　进口，入口　10.1.4.2
inlet air　进空气　3.3.24.16
inlet air casing　空气进口壳体　25.3.1.2

inlet baffle　进口挡板　3.2.15.11
inlet check valve　进口止逆阀　16.2.7.1
inlet damring　进口挡料圈　13.3.1.2
inlet flushing system　冲洗装置　10.3.4.15
inlet guide vane　进口导流叶片　17.4.2.13
inlet head　进料头　8.8.1.1
inlet hole　进气孔　18.4.1.4
inlet passage　进气通道　17.2.7.11
inlet pipe　入口管　10.2.1.31
inlet pipe　进口［料，气］管　14.1.11.4
inlet port　进气孔　17.3.3.2
inlet portion　进口段　16.4.18.1
inlet pressure seal　进口加压密封件　6.3.8.4
inlet seal　进口密封　8.8.1.24
inlet sealing　进口密封件　12.2.3.13
inlet separator　进气分离器　6.3.3.3
inlet sluice　入口斜槽　12.5.2.8
inlet tube sheet　入口管板　10.1.2.35
inlet valve　吸气阀　17.2.12.11
　　进气阀　17.2.6.2
inlet valve unloader　（顶开）吸气阀卸荷器　17.2.12
inlet wastewater　废水进　3.1.25.1；3.1.27.1
inline gas scrubber　管道气液分离器　5.3.3.9
inline multiple-chamber incinerators　直列式多室焚烧炉　3.3.13
in line pump　管线泵　16.1.2.3.53
inline pump　管道泵　16.4.20
inner bore hole　内钻孔　3.1.24.7
inner bowl space　内转鼓空间　9.5.16.6
inner closing chamber　内密闭室　9.5.13.9
inner cover　内盖　22.2.1.4
inner dead point　内止点（死点）　17.2.2.7
inner drum　内转鼓　9.3.2.17
inner perforated bowl　多孔内转鼓　9.5.17.30
inner pipe　内管　4.2.2.4
inner pipe nozzle　管程接管口　4.2.2.9
inner shell　内壳层［体］　14.3.2.13
inner skirt　内裙罩　23.1.13.3
in-process　操作过程中　4.3.4.23
input　进口　3.1.24.1
　　进料　4.3.3.14
input material CO　原料CO　2.2.55.1
input material methanol　原料甲醇　2.2.55.2
input-output　投入产出　1
insert　衬垫　22.4.4.3
　　衬芯　24.2.7.4
insert pump　插入式泵　16.1.2.2.56
inside-corner radius　内转角半径　21.4.1.9
inside diameter　内径　21.1.1.64
inside drum cover　内筒盖　12.3.4.10

inside drum liner 内筒衬板 12.3.4.12
inside pinion bearing 小齿轮轴内侧轴承 12.3.5.25
in-situ retorting 就地干馏 2.2.85
inspection 检查 3.3.7.18
inspection and clean out door 检查与清扫门 8.8.10.10
inspection cover 观察罩 14.1.11.6
inspection door 检查门 10.5.4.12
inspection hole for exhaust gas pressure 排气压力检查孔 24.1.1.28
inspection port 观察孔 12.2.1.9
inspirator body 引射器本体 24.2.7.7
installed vertically 垂直安装 15.5.2
instrument connection 仪表接口 4.1.1.34
insulated cover 绝热顶盖 24.1.7.11
insulated handle 绝缘柄 25.4.1.54
insulated housing 绝热外壳 8.9.1.5
insulating block 绝热砌块 24.1.7.8
insulating concrete 保温混凝土 4.4.2.3
insulating ring 保温圈 5.2.1.5
insulating tape 绝缘(胶)带 25.4.1.34
insulation 保温层 3.1.16.12
insulation board 保温板 24.1.2.12
insulation cover 绝热罩 22.5.1.(h)
insulator 绝缘体 3.5.1.21
intact tray 整块式塔板 5.2.10.27
intake 进气 10.4.3.14
进(气)口 17.2.4.1
进料 9.4.4.20
intake cone liner 进口锥体衬板 12.3.4.5
intake end thermometer 进气端温度计 17.4.2.12
integral cover 整体盖板 4.1.2.(b)
integral fan 整体式风机 24.2.9.8
integral fittings 整体接头 15.5.1.2
integrally clad plate 整体复合板 21.1.1.12
integral splash baffle 整体式防喷溅挡板 18.2.4.1
integral type flange 整体法兰 22.4.2.(d)
integrated-sample setup 积分采样装置 3.4.9
integration 整合 2.2.93
interbraid packing 交叉编织填料 19.2.3.4
intercondenser 级间冷凝器 18.3.4.21
中间冷凝器 20.1.10.74
intercooler 级间冷却器,中间冷却器 17.6.2.12
interface 界面 6.1.12.6
intermediate-alloy steel 中合金钢 22.6.2.⑤
intermediate casing 中间壳体 25.3.1.16

intermediate catalytic gas oil 中间催化裂化粗柴油 2.1.5.8
intermediate conical casing with diffusor 带扩压段的中间锥形外壳 25.3.1.10
intermediate feed 塔中部液体入口 5.2.8.9
intermediate gasket 中间垫片 4.4.27.10
intermediate guide 中体导向装置 17.2.15.14
intermediate platform 中间平台 2.1.2.5
intermediate portion 中段 16.4.18.2
intermediate pressure steam 中压蒸汽 2.2.46.11;2.2.47.10;2.2.111.10
intermediate pressure steam turbine & generator 中压蒸汽涡轮机和发电机 2.2.111.11
intermediate-product 中间产物 1
intermediate rod 中体导杆 17.2.15.15
intermediate shaft sleeve 中轴套 16.4.27.17
intermediate-size material 中等尺寸物料 3.3.16.27
intermediate-temperature CO-shift 中温一氧化碳变换 2.3.2.3
intermediate-temperature liquid 中间温度的液体 20.1.3.13
intermediate temperature heat exchanger/steam filter 中温换热器/蒸汽过滤器 2.2.34.3
inter-meshing type 交叉型 16.4.8
intermittent feed 间歇投料 3.2.5.(a)
intermittent overhead spray nozzle 间歇式顶喷嘴 10.3.4.2
internal circulation 内部循环流动 6.1.9
internal circulator drive 内循环器的驱动装置 7.2.9.8
internal-combustion heat engine 外燃式热机 25.1.1.9
internal cyclone 内部旋风分离器 14.1.4.(c)
internal diffusion 内扩散 1
internal distributor 内部分布器 14.3.2.(a)
internal floating head 内浮头 4.1.4.9
internal-floating-head exchanger 内浮头换热器 4.1.1.(b)
internal heat exchanger 内部换热器 14.3.2.7
internal heat transfer 内部热交换器 14.3.4.(h)
internal mechanical seal 内装式机械密封 19.1.3
internal pipe 内部管 9.3.2.22
internal piping typical of all sections 内部管线各区段相同 9.3.10.9

internal recycle pump 内部循环泵 3.1.35.8
internals 内部结构 6.2.3
internal skirt baffle 内部裙形挡板 7.2.8.3
internal structures 内部结构 5.1.2
internal whizzer classifier 内离心分级器 11.2.12
interrupter 断续器 25.4.9.5
interstage 级间 18.3.4.(b)
interstage baffle 级间挡板 3.2.11.6
interstage labyrinth seal 级间迷宫密封 17.4.2.20
interstitial velocity 空隙速度 1
int.pressure conical head 受内压锥形封头 21.1.1.57
int pressure ellipsoidal head 受内压椭圆形封头 21.1.1.6
int.pressure hemispherical head 受内压半球形封头 21.1.1.39
int.pressure torispherical head 受内压碟形封头 21.1.1.63
int.pressure tube 内压管 21.1.1.50
intrinsic barrier 内阻层 23.4.1.11
inventory conveyor 库存料输送器 23.4.2.19
inventory hopper 贮料斗 10.4.1.1
inverted bucket 倒吊桶 22.2.3.14
inverted bucket steam trap 倒吊桶式蒸汽疏水阀 22.2.5
inverter 逆变器(直流变交流) 2.2.113.1
IOM(insoluble organic matter) 不溶性有机物质 3.2.1.3
ion exchange 离子交换 1
ion-exchange bed 离子交换床 3.5.22.9
ion-exchange equipment 离子交换设备 6.2
ion-exchanger regeneration 离子交换器再生 6.2.2
ion exchange tank 离子交换槽 6.3.10.11
ionic bombardment 离子辐射 12.7.2
ionic electrode 离子电极 12.7.2.7
ionizing unit 离子化单元 10.3.1.21
iron 铁 11.6.3.19
iron mine concentrator 铁矿石富集器 11.6.3
iron oxide filter 氧化铁过滤器 3.3.20.13
irreversible process 不可逆过程 1
irrigation pump 灌溉用泵 16.1.2.3.7
isenthalpic process 等焓过程 1
isentropic process 等熵过程 1
isobaric process 等压过程 1
isobutane 异丁烷 2.1.1.67
isobutene 异丁烯 2.1.1.68
isochoric process 等容过程
isolating valve 隔断阀 22.1.9.14

isothermal process 等温过程 1
ITM(ion transport membranes) 离子迁移膜 2.2.114.3

J

jacket 夹套 4.3.3.8
jacket coolant 夹套冷却液 4.3.3.11
jacket drain 夹套排液口 5.5.2.7
jacketed for coolant spraying 用冷却液喷射的夹套式 4.3.4.(b)
jacketed type exchanger 夹套式换热器 4.2.5.15
jacketed vessel 夹套容器 21.1.1.19
jamming 阻塞 23.2.2.4
jaw crusher 颚式破碎机 12.8.2.5
jet 喷嘴 11.5.5.4
喷射式 16.1.1.48
jet annular disk 喷嘴环轮 11.5.5.5
jet assembly 喷嘴组合件 18.2.4.5
jet axes 射流轴线 11.5.1.3
jet condenser 喷射冷凝器 2.3.10.16
jet direction of airstream 气流喷射方向 11.5.5.8
jet ejector 喷射泵 18.4.2.6
jet mill 气流粉碎机 11.5
jet mill with flat chamber 扁平室气流粉碎机 11.5.5
jet nozzle 喷射嘴 25.4.4.2
jet orifice 射流孔 11.5.1.11
jet pump 喷射泵 16.1.2.2.45
jet reactor 射流反应器 1
jet tray 舌形塔盘,喷射式塔板 5.2.10.15
jog switch 啮合开关 8.8.11.8
joint efficiency 焊缝系数 21.1.2.3
JO type polytetrafluoroethylene seal ring JO型聚四氟乙烯密封环 19.4.1.20
juice feed 糖汁进料 6.3.1.1
juice pan 榨汁盘 9.6.1.1
juice shield 榨汁挡板 9.6.1.2
junior hacksaw 轻型钢锯 25.4.1.52
jute serving 黄麻护层 25.4.8.7
J valve J阀 14.1.7.(b)

K

Kellogg orthoflow unit Kellogg正流式装置 14.2.6.(g)
kelly 方钻杆 2.1.2.14

kerosene 煤油 2.1.4.11
Kerr-McGee multistage mixer-settler Kerr-McGee 多级混合澄清器 6.1.12
kettle 釜体 14.4.1.24
kettle-type floating-head reboiler 釜式浮头再沸器 4.1.1.(f)
kettle type reboiler 釜式再沸器 4.1.2.(k)
key 键 13.1.2.12
key component 关键组分 1
key for driving sheave 传动皮带轮键 11.1.4.3
kicker 抖动器 8.3.4.22
kieselguhr filter 硅藻土过滤器 2.3.13.4
kiln 窑 2.2.87.1
 窑炉 3.3.7.22
kiln burner 窑炉燃烧器 3.3.7.21
kiln dust chamber 炉子除尘室 11.6.4.21
kiln gas 炉气 11.6.4.22
kilning floor 烘干层 2.3.12.16-18
kinematic viscosity 运动黏度 1
kinetic 动力(式) 16.1.1.29
kinetic head 动压头 1
knife 刮刀 8.7.1.2
knife edge 刀刃 15.4.2.4
knife hammer 刀头锤片 11.1.5.9
knob operates take-up 旋柄控制拉紧 23.3.7.12
knocker 锤击器 8.8.1.5
knock-out drum 气液分离罐,缓冲罐 5.3.3.13
knock pin 定位销 22.2.1.2
knotter 除节器 2.3.11.13
knowledge base 知识库 1
knuckle radius 过渡半径 21.1.1.62
Knudsen number (克)努森数 1
Koppers-Totzek Koppers-Totzek 气化炉 25.3.3

L

labeling 贴标签 3.3.7.1
labyrinth 迷宫式(密封) 16.4.8
labyrinth materials 迷宫材料 19.3.3.(a)
labyrinth nut 迷宫密封螺母 11.1.8.22
labyrinth seal 迷宫(式)密封 19.1.7.13
labyrinth type seal 迷宫式密封 8.8.5.10
ladder 梯子 25.4.1.1
 直梯 5.2.1.11
ladder and cage 梯子和罩 24.1.1.8
ladder to manhole 人孔梯子 24.1.1.12

ladle-type thermometer 勺式温度计 2.3.12.53
lager cellar (贮酒的)大地窑 2.3.13.12
lager of product 物料层 3.1.16.10
lager of resin 干层树脂 6.2.2.(b)
laminar flow 层流;滞流 1
laminated cloth packing 叠层织物填料 19.3.5
lamp 信号灯 3.5.6.1
lampholder 灯座 25.4.1.60
land disposal 埋入地下 3.3.1.16
landfill 掩埋 3.3.16.10
lantern gland 润滑环密封垫 16.2.4.16
lantern ring 液封环 16.4.12.15
 套环;灯笼环 4.1.1.26
lap joint flange 松套法兰 22.4.2.(g)
lap jt.stub end 搭接短管端 21.1.1.4
large ball 大球 11.2.3.6
large hole sieve tray 大孔筛板 5.2.10.7
large material 大物料 3.3.16.7
large particle 大颗粒 11.5.2.7
large ring 大环管 24.2.9.(d)
large pump sizes 大型泵 18.2.4.4
large scale 工业规模 11.1.1.25
large scale vessel 大型容器 5.3.3.14
large spiral 粗螺旋 4.3.3.(e)
larry car 装煤车 2.3.14.6
latch cylinder 排料启闭器汽缸 15.7.1.3
latent heat 潜热 20.2.4.5
launder 溜槽 15.6.3.17
lauter battery 过滤器组 2.3.12.51
lauter tun 过滤槽 2.3.12.50
Laval nozzle 拉瓦尔喷管[嘴] 18.2.3.8
4′ LAVA rock (expandable) to 6′ deep LAVA rock 4 英尺火山石(可膨胀)到 6 英尺深火山石 3.4.3.5
law of conservation of energy 能量守恒定律 1
layer-by-layer loading 分层进料 12.1.4.(d)
LC (level controller) 液位控制器 3.1.6.20
LC-Fining 液位控制澄清 2.2.91.3
LCGO(light coking gas oil) 轻焦化粗柴油 2.1.6.8
leached solid 浸提后的渣料 6.1.17.2
leaching 浸取 1
lead concentrate 浓铅 12.8.1.1
lead glass window 铅玻璃窗 25.2.5.3
lead-in tube 铅封管 25.4.5.7
lead pipe 前管 9.3.10.10
lead plug 铅塞 3.1.3.1

lead seal　铅封　22.1.8.17
lead sheath　铅包皮,铅护套　25.4.5.6
lead wire　引出线　16.4.28.11
leaf　过滤叶片　9.2.5.12
leaf driving motor　过滤叶片驱动电机　9.2.5.3
leaf driving unit　过滤叶片驱动装置　9.2.5.21
leaf fastener　叶片紧固器　9.2.7.7
leaf nozzle　过滤叶片接口　9.2.5.28
leaf support　过滤叶片支架　9.2.5.26
leakage fluid drain　排液口　16.4.20.24
leakage path　泄漏途径　19.3.3.3
leakage steam pipe　漏泄蒸汽通道　25.4.4.8
leak detection port　泄漏检测孔　16.2.7.14
leak-free connections　无泄漏联结　3.4.2.4
lean absorbent　贫吸收剂　2.2.106.36
lean gas　贫气　2.2.1.5
lean gas riser　贫气提升管　5.4.2.9
lean heavy solvent　贫的重溶剂　6.1.9.17
lean light solvent　贫的轻溶剂　6.1.9.3
lean phase　贫相　1
lean-phase fluidized bed　稀相流化床　14.1.3.23
lean solvent　贫溶剂　2.2.32.6
leg　支腿　1
Leine rock salt　莱茵岩盐　25.2.4.8
length of cylinder　筒体长度　21.2.5
Leonard process for the indirect synthesis of formic acid via methyl formate　伦纳德工艺;甲醇纯甲酸甲酯制甲酸　2.2.62
Letdown tank　排放槽　2.2.44.14
lethal substance　致死物质　21.1.2.4
level　水平坑道　25.2.5.1
level detectors　液面测定仪　2.1.10.2
leveler　平料器　8.9.1.7
level floor　水平底面　3.5.4.2
level indication pipe　液面指示管　13.1.2.18
level indicator　料面指示器　8.9.3.3
level of material to be filtered　过滤物料液面　9.3.5.8
level ring　水平环板　9.5.12.4
level sensor　物位传感器　23.1.2.21
lever for pan position　固定锅体手把　15.7.4.1
lever key　杠杆键　16.2.4.34
lift　扬程　1
lifter-roof　升降顶贮罐　21.2.1.4
lift gas　提升气体　6.3.11.7
lift gas return　提升气体返回口　6.3.11.1
lifting flights　提升抄板　8.8.1.12
lifting lug　吊耳　4.1.1.36

lifting magnet　提升磁体　12.6.1.(a)
lifting mechanism　提升机械　12.4.2.10
lifting screw　升降螺杆　22.1.3.4
lifting yoke　吊钩　13.1.2.41
lift line　提升管　14.2.3.7
lift pipe　提升管　2.2.4.4
lift platform　提升台　2.3.11.40
lift stop of a valve　阀升程限制器　17.2.11.5
light bulb with bayonet fitting　卡口灯泡　25.4.1.69
light catalytic gas oil　轻催化裂化粗柴油　2.1.5.9
light component　轻组分　15.5.4.10
light component removal column　脱烃组分塔　2.2.56.3
light distillate　轻馏出物　2.1.4.25
　轻馏分(油)　2.2.94.18
light-duty jacketed construction　轻负荷夹套式结构　4.3.4.(c)
light ends　轻馏分　2.2.51.7
lighter organic chlorides　轻质有机氯化物　3.1.7
light fraction　轻部分(成分)　15.4.2.3
light fraction collection launder　轻成分聚集洗矿槽　15.5.3.8
light gas　轻质气　2.2.75.8
light gasoline　轻质汽油　2.2.75.18;2.2.77.9
light liquid　轻液(体)　6.1.14.3
light liquid dispersion pipe　轻液分布管　6.1.18.12
light liquid outlet　轻液出口　6.1.18.5
light mineral collection tank　轻质矿石收集池　15.5.3.5
light mineral ripple　轻的矿石粒子　15.5.3.9
light naphtha　轻石脑油　2.1.1.16
lightning rod　避雷针　21.2.3.1
light oil　轻油　2.2.6.7
light-phase effluent　轻相流　9.5.14.2
lights　轻组分　2.3.18.21
lights and benzene　轻组分和苯　6.1.3.11
light shale oil　轻质页岩油　2.2.6.17
lignite　褐煤　2.2.107.6
lime slaker　化灰器　2.3.10.51
limestone　石灰石　3.3.12.5
liimestone,water　石灰石,水　2.2.115.21
limestone feed pipe　石灰石加料管　14.2.2.2
limit rod　限位杆　22.5.3.1
limit switch　极限开关　9.5.6.19
Linde cycle　林德循环　1
Linde sieve tray　林德筛板　5.2.10.12
linearizer　线性化电路　3.5.18.4

lined blind flange 衬里法兰盖 22.4.2.(i)
liner 衬板 11.1.5.1
liner plate 衬板 11.1.5.7
liner ring 阻漏环,衬环 16.4.29.6
line shaft 中间轴 16.4.23.12
lining 衬里 1
lining plate 内衬板 11.1.8.10
linings 衬里 21.1.1.11
Liprotherm rotating thin film evaporator Liprotherm 旋转床膜蒸发器 5.5.1.(d)
liq penetration exam 液体渗透探伤 21.1.2.15
liquefaction 液化 2.2.97.7;3.3.18.18
liquefaction product 液化产品 2.2.90.15
liquefied gas 液化气 2.1.1.25
liquefied gas pump 液化气泵 16.1.2.3.33
liquid 液体 8.1.1;3.4.4.8
liquid ammonia 液氨 14.3.1.25;2.2.37.11; 2.2.38.15
liquid collector 液体收集器 5.3.1.15
liquid connection 液体接口 4.3.1.6
liquid cylinder 泵缸 16.2.1.2
liquid-cylinder foot 泵缸支腿 16.2.3.23 液缸底盖 16.2.4.18
liquid-cylinder head 泵缸端盖 16.2.3.29
liquid-cylinder lining 泵缸套 16.2.3.30
liquid dam 液体堰 5.5.2.5
liquid discharge 滤液出口 9.5.6.27
liquid dispenser 液体分配器 2.2.39.6
liquid-distribution 液体分布 10.5.11.(a)
liquid drain 液体排放口 3.4.4.5
liquid drainage 液体排泄 10.4.2.13
liquid draw off 液体排出 9.5.5.14
liquid flow 液(体)流 6.1.21.5
liquid fluidized bed 液体流化床 14.1.3.12
liquid fuel 液体燃料 3.3.1.9
liquid head 泵缸盖 16.2.8.2
liquid heights in condensate 冷凝液柱高度范围 25.6.2
liquid holdup 持液量;持液率 1
liquid hourly space velocity(LHSV) 液态空速 1
liquid hydrocarbons 液态碳氢化合物 2.3.5.18
liquid in 液体进入 3.4.4.9
liquid injection incineration 液体喷射焚烧过程 3.1.22
liquid inlet 液体进[入]口 10.5.10.3;3.4.4.12 进液 17.2.14.1
liquid level 液面 12.4.2.5;液位 2.2.39.7
liquid level connection 液面计接口 4.1.1.39
liquid level gauge 液面计 21.2.4.5

liquid-liquid equilibrium(LLE) 液液平衡 1
liquid-liquid extraction of uranium 铀的液-液萃取 6.1.2
liquid metal 液态金属 19.4.1.8
liquid methanol 液体甲醇 2.2.28.8
liquid or condensate connection 液体或冷凝液接口 4.3.1.8
liquid or steam jacket 液体或蒸汽夹套 4.3.1.2
liquid out 流体出口 10.1.4.7 出液 17.2.14.7
liquid outlet 液体排出口 5.2.1.6 排[出]液口 9.5.4.2
liquid overflow pipe with screen 带筛网的溢流管 10.5.8.3
liquid paraffin 液蜡 2.1.1.23
liquid path 液体通道 15.3.1.8
liquid piston body 泵缸活塞 16.2.3.22
liquid piston bull rings 泵缸活塞耐磨环 16.2.3.21
liquid piston fibrous packing rings 液缸活塞纤维填料环 16.2.3.24
liquid piston follower 泵缸活塞随动板 16.2.3.25
liquid piston rod 液缸活塞杆 16.2.3.15
liquid piston-rod nut 泵缸活塞杆螺母 16.2.3.26
liquid piston-rod stuffing box 泵缸活塞杆填料函 16.2.3.18
liquid piston-rod stuffing box bushing 泵缸活塞杆填料函衬套 16.2.3.19
liquid piston-rod stuffing box gland 泵缸活塞杆填料函压盖 16.2.3.17
liquid piston-rod stuffing box gland lining 泵缸活塞杆填料函压盖衬套 16.2.3.16
liquid piston snap ring 泵缸活塞弹力环 16.2.3.20
liquid-piston type of rotary compressor 回转式液环压气机 17.3.4
liquid product 液态[体]产品 2.2.6.18
liquid redistributor 液体再分布器 5.3.4.10
liquid-ring 液环(式) 16.1.1.28
liquid-ring 液体环 18.4.1.3
liquid ring compressor 液环鼓风机 17.3.1.(b)
liquid ring pump 液环泵 16.1.2.2.44
liquids 液体 2.2.79.14
liquid seal and flushing water supply 封与冲洗水供入口 9.5.12.20
liquid seal pot 液封槽[筒] 5.2.1.21
liquid seal to lantern ring 密封液通入封环

16.4.4.20
liquid slag 液态炉渣 24.3.1.20
liquid-solid extraction 液固萃取 1
liquid-solid separator 固液分离器 2.2.98.26
liquid state 液态 13.5.2.1
liquid stream 5.2.1.15
liquid subcooling 液态过冷 20.2.4.17
liquid supply 供入液体 13.4.1.3;3.4.4.3
liquid tin-base alloy 液态锡基合金 19.4.1.16
liquid to be treated 待处理的料液 2.2.56.11
liquid-vapor interface 液-气相界面 2.2.7.6
liquid volume 液体的容积 21.2.5
liquid waste 废液 3.1.22.7
liquid waste feed 废液进料 3.3.11.5
liquid zone 液相区 20.2.4.16
liquor discharge 滤液出口 9.5.6.15
liquor inlet 液体加入 7.2.11.6
liquor preheater 液体顶热器 2.3.10.8
liquor pump 液泵 2.3.10.33
lithium bromide absorption cycle 溴化锂吸收循环 20.1.8
little amount of CO, CO_2, N_2, H_2 少量 CO, CO_2, N_2, H_2 2.2.38.9
LL(1)(light liquid) 轻液 6.1.15.1
LLR(light liquid recycle) 轻液循环 6.1.15.3
Lug support 耳式支座
load cell scale platform & roller conveyor section 负载传感秤,平台和辊式输送机段 23.4.1.4
load indicator 负荷指示 15.6.1.8
loading 负载 21.1.2.5
loading end 装料端 23.3.7.(c)
loading facilities 装货设备 23.4.2
loading point （承）载点 23.2.1.4
装料口 23.3.3.1
loading skirt 加料裙罩 23.2.1.3
loading spout 装料嘴 23.1.2.30
lobe 凸轮（式）；凸叶（式） 16.1.1.26
lobe pump 凸轮泵 16.1.2.2.10
locating ring 定位环 16.4.23.19
lock hopper 闭锁式料斗 2.2.106.20;2.2.20.2
煤锁斗 2.2.13.7
灰锁斗 2.2.29.12;2.2.30.9
locking bar 锁紧棒 11.1.7.11
locking nut 锁紧螺母 16.4.15.18
locking plate 锁紧板 9.6.1.7
lock ring 防松环 22.1.4.12
lock screw 闭锁螺旋 11.4.3.5
lock washer 锁紧垫圈 16.4.20.4
logarithmic mean temperature difference 对数平均温差 1
log pressure gradient 对数压降 14.1.1.1
log velocity 对数速度 14.1.1.2
long arm 长臂 15.6.1.7
long connecting rod 长连杆 16.2.3.39
long crank upper rock shaft 长曲柄上（面）摇臂轴 16.2.3.45
longitudinal baffle 纵向折流板 4.1.1.30
longitudinal joints 纵缝接头 21.1.1.16
longitudinal vortex 纵向漩涡 16.4.15.20
long pitch mixing ribbon 长节距混合螺条 4.3.3.19
90° long radius elbow 90°长半径弯头 22.3.2.2
long residuum 长渣油 6.3.2.13
long-tube evaporator 长管蒸发器 7.1.4
loop-feed elevator 环形加料提升机 23.3.3.(c)
loop-feed elevator used for dewatering 用于脱水的环形加料提升机 23.3.3.(c)
loop reactor 环流反应器 1
loose type flange 活套法兰 21.1.1.5
Lorain liner Lorain 型衬里 11.2.2.2
lost-motion adjusting nut 空转调节螺母 16.2.4.39
lost-motion locknut 空转锁定螺母 16.2.4.38
louvre damper 气门 11.6.4.14
low air velocity 低空气速度 4.3.3.25
low-alloy steel 低合金钢 22.6.2.⑥
low-Btu 低热值 2.2.2.8
low Btu fuel gas 低热值燃料气 2.2.104.1
low capital cost 低投资成本 8.1.1.35
low-carbon olefins 低碳烯烃 2.2.70.12
lower air hood 下部空气集气罩 4.2.4.11
lower ball cover 下轴承盖 13.3.5.12
lower bearing 下轴承 12.1.1.27
lower bearing housing 下轴承箱 12.3.5.11
lower bearing housing cap 下轴承箱盖 12.3.5.12
lower belt tensioning 下部滤带张紧器 9.6.2.13
lower belt wash 下部滤带冲洗 9.6.2.12
lower case 底座 11.1.5.24
lower casing 下壳体 16.4.12.4
lower distribution plate 下分配板 12.3.5.15
lower gasket 下垫片 16.4.27.33
lower grating 下栅板 5.2.9.6
lower head 下盖 16.2.7.6
lower hold 底货仓 23.3.9.8
lower inlet for medium 介质下部入口管

14.4.1.9
lower jacket 下部夹套 14.4.1.22
lower jet 下部喷嘴 18.2.1.5
lower Keuper 下[旱]考依波（泥灰岩及砂岩） 25.2.4.1
lower lift drum 下部提升筒 6.3.11.8
lower Muschelkalk 下[旱]壳灰岩层 25.2.4.3
lower pebble bed 下层卵石床 14.3.3.14
lower return nozzle for medium 介质下部回流管口 14.4.1.11
lower rock shaft 下（面）摇臂轴 16.2.3.44
lower rotor bearing 转子下轴承 5.5.1.24
lower screen 下层筛 12.2.3.10
lower shaft sleeve 下轴套 16.4.27.13
lower sleeve 下套筒 16.2.5.10
lower sleeve bearing 下套筒轴承 16.4.27.3
lower tracking roll 底部跟踪辊 9.6.2.11
lower weight 下部重块 11.3.1.9
low ester by products 低酯副产物 2.2.58.2
low ester recovery column 低酯回收塔 2.2.58.12
low ester recovery tower reflux tank 低酯回收塔回流罐 2.2.58.7
low-freezing oil 低凝点油 2.1.1.22
low-gravity compartment 低密度室 12.5.2.17
low-gravity medium 低密质介质 12.5.2.12
low N products 低含氮量产品 2.2.91.10
low pressure 低压 19.3.3.1
low pressure absorbing tower 低压吸收塔 2.2.55.19
low-pressure cylinder 低压汽缸 25.4.4.5
low-pressure gas 低压气体 19.3.2.18
low-pressure gas main 低压煤气总管 2.3.14.44
low-pressure pump 低压泵 16.1.2.1.2
low-pressure return line 低压回流线 2.1.7.3
low pressure separator 低压分离器 2.2.54.4
low-pressure side 低压侧 20.1.3.6
low-pressure stream 低压气流 17.2.2.12
low pressure steam superheater 低压蒸汽过热器 2.2.34.6;2.3.34.7
low pressure steam 低压蒸汽 2.2.43.4;2.2.111.15
low pressure steam turbine & generator 低压蒸汽涡轮机和发电机 2.2.111.16
low-pressure turbine 低压汽轮机 25.3.2.9
low-pressure vapor 低压蒸气 20.1.2.1
low pressure waste heat boiler 低压废热锅炉 2.2.34.9

low rate anaerobic process 低速厌氧工艺 3.2.7
low shaft 下段轴 14.4.3.21
low-speed agitators 低速搅拌器 14.4.4
low S products 低含硫量产品 2.2.91.10
low-stage compressor 低压级压缩机 20.1.3.15
low-stage evaporator 低压级蒸发器 20.1.3.20
low-strength wastewater 低浓度废水 3.1.31.6
low temperature 低温 8.7.2.7
low temperature separation 低温分离 2.2.107.10
low temperature shift 低温变换 2.3.2.12
low temperature shift 低温变换 2.3.3.10
low-temperature vessel 低温罐 25.6.4.6
low temperature zone 低温区 8.7.2.10
low vacuum 低真空 19.4.1.23
low-vapor pressure 低蒸气压 19.4.1.12
low-voltage safety current 低压安全电流 25.4.1.2
low-voltage winding 低压绕阻 25.4.3.6
LOX storage(stand-by) 液氧储罐（备用） 3.1.46.2
LPG(liquefied petroleum gas) 液化石油气 2.2.77.8
L.P.steam 低压蒸汽 3.1.7.6
LSR naphtha LSR 石脑油 2.1.4.16
lt.ends 轻质物 2.2.90.8
L type fixed ring L型固定环 19.3.2.2
L-type reciprocating compressor L型往复压缩机 17.2.1.5
lube pump 油泵 6.3.2.15
lubricated or dry friction 润滑式干式摩擦 8.8.5.1
lubricated plastic packing 润滑塑料填料 19.2.3.14
lubricating device 注油器 6.1.18.11
lubricating oil 润滑油 19.4.1.19
lubricating oil pump 润滑油泵 16.1.2.3.25
lubrication 润滑 22.5.3.2
lubrication oil 润滑油 17.2.9.24
lubricator 油杯 22.1.2.15
Lurgi Lurgi气化炉 24.3.1
Lurgi dry ash gasifier 鲁奇干煤灰气化炉 2.2.13
Lurgi company methanol synthesis reaction process 鲁奇公司甲醇合成反应工艺流程 2.2.42
Lurgi company methanol synthesis and rectifi-

cation process 鲁奇公司甲醇合成和精馏工艺流程 2.2.43
Lurgi gasifies 鲁奇煤气化炉 2.2.11
Lurgi's Mega DME process 鲁奇公司大型二甲醚制备过程 2.2.51
Lurgi process 鲁奇(煤气化)法 2.2.8
Lurgi-Ruhrgas process 鲁奇-鲁尔公司工艺 2.2.4
lute cap 封泥盖 3.3.26.1
L valve L型阀 14.1.6.(f)
LVGO (light vacuum gas oil) 轻减压粗柴油 2.1.4.17
lyophilization 冷冻干燥 1

M

macerator 浸解池 3.3.21.2
machine-base 机座 15.7.2.12
machine body 机身 11.1.8.13
machine chest 机械浆料池 2.3.10.86
machine frame 机身 17.2.9.25
machine hood 机(器)罩 2.3.11.27
machine supporter 机座 11.1.8.17
Mach number 马赫数 1
macrofluid 宏观流体 1
magnet 磁铁 12.6.1.6
magnet cover 磁铁罩 14.4.3.4
magnetic bearing 磁力轴承 9.5.21.10
magnetic cobber 磁选组 11.6.3.10
magnetic concentrate 磁力分离的富集物 12.5.5.1
magnetic drum 磁鼓 12.6.9.6
magnetic drum operating as a lifting magnet 磁性鼓用作提升磁体 12.6.1.(a)
magnetic drum separators 磁性鼓式分离器 12.6.3
magnetic induction part 磁感应元件 14.4.3.9
magnetic materials 磁性材料 12.6.1.5
magnetic material 磁性材料 12.6.9.2
magnetic meter 磁测仪表 3.5.18.(a)
magnetic pulley 磁性转轮 12.6.1
magnetic-return circuit 磁回路 16.5.4.4
magnetic roll 磁鼓 12.6.11.5
magnetics 磁性 12.6.4.8
magnetic separation 磁分 3.3.16.10
magnetic sepa rator 磁力[选]分离器 12.6.2;2.3.16.19
magnetizing block 磁铁 12.5.1.12
magnet receiver 磁铁座 14.4.3.7

mag.particle exam 磁粉探伤 21.1.2.15
main 总管 17.6.2.5
main air blower 主空气风机 14.2.7.21
main air inlet 主空气入口 14.3.4.22
main bearing 主轴承 11.2.4.4
main body 机体 11.4.1.10
main case 机壳 13.3.5.24
main compressor 主压气机 25.3.2.17
main condenser 主冷凝器 18.3.5.2
main feed belt 主进料皮带 11.6.4.3
main flow to cyclonic separator unit 主流向旋风分离机组 10.1.3.1
main flow to exhauster 主流向排气装置 10.1.3.3
main fractionator 主分馏塔 2.1.6.3
main-inlet shutoff valve 主入口截止阀 2.1.7.2
mainly organic fraction 主要的有机物料 3.3.14.19
main methanation reactor 主甲烷化反应器 2.2.10.8
main motor 主电动机 9.5.4.18
main operating valve 主操作阀 25.4.9.2
main parts 主要部件(图) 18.3.4.(a)
main pulley 主动V型皮带轮 11.1.5.22
main shaft 主轴 11.1.3.6
main shaft bearing 主轴轴承 9.2.1.2
main shaft key 主轴键 9.5.4.41
main steam supply 主供蒸汽 18.3.5.9
main stream 主物流 12.7.4.10
main turbine 主气轮机 25.3.2.18
main valve 主阀 3.4.11.16
Majac jet pulverizer Majac型射流粉碎机 11.5.3
major beam 主梁 5.2.2.14
major beam clamp 主梁固定板 5.2.7.5
major component 主要零部件 23.2.1
make-up earthworms 补充用的蚯蚓 3.3.28.5
make-up gas 补充气体 2.2.42.1
makeup H_2 补充氢气 2.2.79.1
makeup H_2 and feed oil 补充氢气和油进料 2.1.10.4
makeup hydrogen 补充氢气 2.2.6.2
make-up steam 补充蒸汽 7.1.14.10
make-up tank 补给罐 6.3.10.3
make-up water 补充水 20.1.10.29;2.2.30.2
make-up water inlet 补充水进水管 4.6.4.11
make-up water supply 补充供水口 9.5.12.15
making molecular sieve adsorbents 制备分子筛吸附剂 6.3.10

male 插入式配件 3.5.16.23
male-female seal contact face 凹凸密封面 22.4.3.(c)
male fitting 凸形管件 22.3.1.21
male flange 凸面法兰 22.4.3
male screw 阳螺杆 17.3.2.3
malleable cast-iron 可锻铸铁 22.6.2.②
malt elevator 麦芽提升机 2.3.12.35
malting 麦芽制造 2.3.12
malting tower 麦芽制造塔 2.3.12.1
malt silo 麦芽筒仓 2.3.12.37
man door 人孔 10.2.1.5
manganese-molybdenum-vanadium steel 锰钼钒钢 22.6.2.⑫
manganese-vanadium steel 锰钒钢 22.6.2.⑨
man hole 人孔 10.1.2.26 检修孔 3.3.20.14
manhole cover 人孔盖 11.2.1.5
manhole cover plate 人孔盖板 21.1.1.29
manhole flange 人孔法兰 8.9.5.1
manhole to the storage tank 贮酒罐的人孔 2.3.13.13
manhole with sight glass 带视镜人孔 24.1.1.11
manifold 集管箱 8.8.7.(b) 总管 9.1.3.13 歧管
manifold classifier 歧管粒料分选器 8.4.6.9
manifolds steam or hot water 蒸汽或热水集管 8.2.2.13
Mannheim-type mechanical hydrochloric acid furnace Mannheim型机械盐酸炉 24.1.11
manometer 测压计 3.4.14.33
mantle 套筒 11.1.3.4
manual 人工的 9.5.2.19
manual dry divider 手动干料分流器 3.3.24.5
manufacture of synthetic fibres 合成纤维的生产 2.3.7
manufacturing plant 制造厂 3.3.7.11
manway 人员专用道,走道 25.2.2.12
manway plate 通道板 5.2.7.7
maphtha 粗汽油 2.1.9.19
marbles packed bed 卵石填充床 10.5.8.4
Marcy grate-type continuous ball mill Marcy型栅板式连续球磨机 11.2.1
marine propeller 船舶螺旋桨;推进式 14.4.2.(c)
marine pump 船用泵 16.1.2.3.50
marume plate 造粒板 13.3.5.27
marumerizer 球形造粒机 13.3.6
mash 麦芽浆 2.3.10.10

mashhouse 酿酒捣碎车间 2.3.12.42-53
mashing kettle 麦芽浆蒸煮锅 2.3.12.44
mashing process 捣浆过程 2.3.12.42-53
mashing the malt 捣碎麦芽 2.3.12.43
mash tub 捣浆槽 2.3.12.43
mass-diffusion column 质量扩散塔 15.4.1
mass-diffusion screen 质量扩散筛 15.4.1.3
mass-flow bin 密相流料仓 23.1.2
mass-flow hopper 密相流料斗 23.5.1
mass flow rate 质量流率;质量流量 1
mass transfer zone(MTZ) 传质区 1
master switch 主[总]开关 25.4.2.2
matched upper and lower stones 相配的上下磨石 11.4.3.3
material 物料 23.1.13.6
material balance 物料衡算;物料平衡 1
material bed 物料床[层] 23.2.5.8
material destination 物料收集器 23.1.2.6
material discharge 下料 10.2.3.17 卸料 3.3.5.4
material extraction mechanism 物料抽提机构 3.3.29.12
material feed 进料 12.2.1.10
material feed conveyor 供料输送机 3.3.29.7
material-handling sequence 物料处理工序 3.3.7
material in 进料 23.1.1.8 物料进口 8.4.1.30
material in carrying run 载运的物料 23.3.7.9
material inlet 进料口 23.1.6.3
material input 物料加入 11.5.2.1
material-into-air 物料进入空气中 23.1.8.③
material line 物料管线 23.1.2.8
material loading 装料 23.3.5.2
material receiver 物料接收器 23.1.2.20
material removal 物料卸出 3.3.30.4
material removal system 物料去除系统 3.3.29.11
materials 材料 21.1.12.7
material source 物料仓 23.1.2.14
matrix 型块 8.1.1.16
matrix mimic board 电路模拟板 25.4.2.8
maturing 熟化 2.3.7.7
max.(allowable working pressure) 最大许用操作压力 21.1.2.8
max.allow.stress value 最大许用应力值 21.1.2.14
maximum and minimum liquid levels in vaporizer drum 蒸发器鼓内最高与最低液面 25.6.2
maximum mixedness 最大混合度 1

maximum O.A.damper 最大室外空气调节风门 20.1.14.8
maximum opening 最大开度 11.1.3.12
maximum water level 最高水平面 15.6.3.14
max liquid level 最高液位 5.4.1.13
max. LQ. level 最高液面 2.1.10.6
mazut 重油 2.1.1.35
MBR(membrane bioreactor)膜生物反应器 3.1.30.A
MBR(Membrane bioreactor) process system with membrane module situated outside the bioreactor 生物反应器外设膜组件的膜生物反应器系统 3.1.28
MCGO(intermediate coking gas oil) 中间焦化粗柴油 2.1.6.9
mean residence time 平均停留时间 1
MEA solution 一乙醇胺溶液 2.2.27.8
measured feed 计量后进料 23.3.3.8
measurement cell 测量元件 3.4.19.4
measuring and resin cleaning chamber 计量及树脂清洗室 6.2.6.4
measuring instrument 测量仪表 25.4.4.1
measuring vessel for the kieselguhr 板状硅藻土计量器 2.3.13.3
mechanical 机械 8.1.1.32
mechanical air classifier 机械式空气分级器 11.6.5.1
mechanical backing pump 前级机械泵 5.5.3.1
mechanical contact shaft seal 机械接触式轴封 19.1.7
mechanical dash pot 机械减振器 23.3.2.4
mechanical float 机械浮动式 3.4.18.(c)
mechanically actuated 机械作用的 16.2.8
mechanically fired industrial incinerator 机械添煤工业煅烧炉 24.1.5
mechanical packer 机械密垫 3.1.24.12
mechanical recirculation of heat carrier 热载体机械循环 3.3.15.(d)
mechanical refrigeration 机械压缩制冷 20.1.1.(a)
mechanical scrubber 机械式涤气器 10.5.9
mechanical seal 机械密封 19.1
mechanical seal components 机械密封部件 19.1.1
mechanical spray thrower 机械式喷嘴 5.3.4.6
mechanical vacuum pump 机械(真空)泵 19.4.1.22
mech.operated 机械驱动的 16.1.1.16
med-Btu fuel gas 中热值燃料气 2.2.89.15

media 介质 3.2.13.2;3.1.38.5
media and biomass to separation 介质和生物质去分离 3.2.9.3
media/dust deentrainment chamber 介质/粉尘减少夹带室 10.4.1.14
media filter bed 介质过滤床层 10.4.1.9
media flow air injector 介质流动空气注入器 10.4.1.6
media left air blower 介质提升气风机 10.4.1.4
media level 介质的液面 12.5.2.3
media retainer (研磨)介质挡板 11.3.1.11
media return 介质返回 10.4.1.20
medium 介质 12.5.2.18
medium-Btu 中热值 2.2.2.10
medium-BTU gas 中等热值可燃气体 3.3.23.10
medium concentrate 中间富集物 12.5.5.8
medium discharge 介质排出口 14.4.1.6
medium inlet 热载体入口 24.1.1.4
medium outlet 热载体出口 24.1.1.17
medium pattern return bend 普通型180°回头 22.3.2.16
medium-pressure cylinder 中压汽缸 25.4.4.4
medium pressure pump 中压泵 16.1.2.1.3
medium pump 介质泵 12.5.1.18
medium scale 中等规模 8.1.1.24
medium scmp 介质储槽 12.5.1.17
medium temperature 中温 8.7.2.6
medium thickener 介质沉降槽 12.5.1.14
medium washing 介质洗涤 9.3.7
mellapak 板波纹填料 5.3.4.48
melt 熔融物(料) 13.5.1.14
melted burden 熔融物料 13.5.2.1
melting bath 熔化池 2.3.9.14
melting pot 熔化罐 2.3.8.30
melting the polyamide 熔化聚酰胺 2.3.8.41
melt thermocouple 熔融物料热电偶 13.2.1.11
membrane 膜 15.2.2.5 滤膜 3.1.30.A.5;3.1.31.7
membrane bioreactor 膜生物反应器 1
membrane module 膜组件 1;3.1.28.7;3.1.29.8
membrane permeation module 膜渗透单元 15.2.2
membrane processes 膜分离过程 15.2
membrane separator 膜分离器 6.1.8
MeO$_x$—metal oxide 金属氧化物 2.2.80.3
mercury diffusion pump 水银扩散泵 16.1.2.

mercury pump 水银泵 16.1.2.3.32
mesh 目 11.6.3.17
mesh connection 网形连接 25.4.3.3
mesh separator 筛网分离器 7.1.14.8
mesh wash 筛网洗涤液 7.1.14.7
metabolic products and excess cell growth 代谢产物和多余的细胞产物 3.1.47.10
metal 金属 22.6.2.1;2.2.80.6
metal band 金属带 8.8.5.20
metal chamber 金属小室 19.3.1.3
metal collector 金属捕集器 3.3.2.2
metal core packing 金属芯填料 19.2.3.12
metal flange 金属法兰 22.4.4.6
metal foil crinkled and twisted packing 绉状金属箔卷制填料 19.2.3.11
metal foil spiral wrapped packing 金属箔螺旋卷制填料 19.2.3.10
metal for product discharge 排料接口 11.1.8.18
metal Intalox saddle 金属环矩鞍 5.3.2.(i)
metallic foundation 金属底座 25.3.1.1
metallic packing 金属填料 19.2.3.9
metallic packing of crosshead side 十字头侧金属填料 19.2.1.9
metallic packing of pressure side 压力侧金属填料 19.2.1.10
metallic paper 金属化纸 25.4.8.2
metallized paper 镀金箔纸 25.4.8.2
metallurgical coke 冶金焦炭 2.2.91.11
metal Pall ring 金属鲍尔环 5.3.2.(d)
metal radiating cone 金属辐射锥 25.1.3.6
metal retainer 金属定位块 19.3.1.10
metal screen 金属筛板 6.3.9.13
metal shroud 金属套筒,金属护罩 4.4.1.5
metal stack 金属烟囱 24.1.3.10
metal system 金属管路 22.4.4
metal-to-metal contact 金属与金属接触 18.2.4.1
metal tripak 金属球形填料 5.3.2.(l)
metal valve 金属阀 16.2.4.20
metastable region 亚稳区;介稳区 1
metering conveyor 计量输送器[机] 23.4.2.5
metering inlet pump 进料计量泵 5.5.1.2
metering section 计量段 13.2.1.14
methanation 甲烷化(作用,反应) 2.2.36.10;2.2.101.15
methanator 甲烷转化器 2.2.6.15 甲烷化反应器 2.3.2.14

methane 甲烷 2.2.68.19;3.3.18.16
methane formation 甲烷形成 3.3.18.22
methane SNG 甲烷合成天然气 2.2.104.13
methanol 甲醇 2.2.1.14;2.2.2.20;2.2.28.2;2.2.40.3;2.2.43.16;2.2.45.14;2.2.50.10;2.2.52.1;2.2.54.12;2.2.61.13;2.2.62.2;2.2.64.11;2.2.65.2;2.2.67.15;2.2.68.1;2.2.69.5;2.2.71.12;2.2.76.7;3.1.37.8;3.1.38.4;3.1.39.2;3.1.40.7
methanol carbonylation 甲醇羰基化 2.2.40.4
methanol + catalyst 甲醇+催化剂 2.2.62.6
methanol + CO_2 甲醇+二氧化碳 2.2.28.14
methanol column 甲醇塔 2.2.64.3;2.2.65.d
methanol, dimethyl ether hydrocarbons 甲醇,二甲醚.碳氢化合物 2.2.3.19
methanol extraction 甲醇萃取 2.2.65.c
methanol feed 甲醇进料 3.1.36.4
methanol feed pump 甲醇进料泵 3.1.36.5;3.1.38.2
methanol feedstock 原料甲醇 2.2.51.1
methanol + H_2S 甲醇+硫化氢 2.2.28.11
methanol-methyl formate separation 甲醇-甲酸甲酯分离 2.2.62.15
methanol preheater 甲醇预加热器 2.2.51.2
methanol recycle 甲醇循环 2.2.61.10
methanol storage tank 甲醇储罐 3.1.36.6;3.1.38.3
methanol synthesis 甲醇合成 2.2.2.17;2.2.67.13
methanol synthesis converter 甲醇合成塔 2.2.46.1;2.2.46.2;2.2.47.1;2.2.47.2
methanol synthesis flow sheet based on a quench reactor 基于急冷反应器的甲醇合成工艺流程 2.2.48
methanol synthesis using a double-tower series 串塔合成甲醇工艺流程示意 2.2.46
methanol to olefins 甲醇制烯烃 2.2.67.16
methanol to olefins(MTO) process 甲醇制烯烃(MTO)工艺流程示意图 2.2.69
methanol tower 甲醇塔 2.2.49.13;2.2.50.8
methanol vapor 甲醇蒸气 2.2.28.9
methanol vapouriser 甲醇汽化器 2.2.51.5
method of operation 操作方法 8.1.1(a)
methylbenzene 甲苯 2.1.1.52
methyl formate 甲酸甲酯 2.2.62.7
methyl formate production by dehydrogenation of methanol 甲醇脱氢制甲酸甲酯 2.2.61
methyl formate reaction 42 bar 42巴甲酸甲酯反应器 2.2.62.5

methyl mercaptane production process 甲硫醇生产工艺 2.2.63
microcapsule 微胶囊 1
microfiltration 微(孔过)滤 1;3.1.30.B.11.
micronizer fluid-energy (jet) mill Micronizer 流能磨(气体粉碎机) 11.5.1
microorganism 微生物 1
microwave drying 微波干燥 1
middle case 中间泵壳 16.4.28.31
middle casing 中间壳体 16.4.19.31
middle cone 中圆锥 10.1.2.15
middle cylinder 中圆筒 10.1.2.10
middle distillate 中(间)馏分(油) 2.2.94.19
middle frame 中部框架 12.2.1.14
middle-frame discharge 中部框式卸料口 12.2.1.8
middle Muschelkalk 中壳灰岩层 25.2.4.3
middling 中间物 12.5.2.15
　　中级品 12.6.4.24
middling 中间产物 2.1.9.10
midget bubbler 小型鼓泡器 3.4.14.8
Mikro-ACM pulverizer Mikro-ACM 型粉磨机的截面图 11.2.10
Mikro-atomizer Mikro 型粉碎机 11.2.9
Mikro-pulverizer Mikro 型粉磨机 11.2.6
Mikro-pulverizer hammer mill Mikro 型锤磨机 11.2.5
　　mill 磨(机) 11.6.4.6
　　磨碎机 11.6.5.6
　　碾磨机 2.2.29.4;2.2.30.4
mill base 粉碎机底座 11.3.2.8
mill cover 粉碎机盖 11.3.2.6
mill drive 磨碎机传动装置 11.2.12.1
mill feed bin 磨碎机料仓 11.6.5.4
mill housing cover 粉碎室上盖 11.1.11.12
milling chamber 粉碎室 11.5.5.2
milling medium 研磨介质 11.3.3.8
mined portion 坑道地段 2.2.85.3
minehead buildings 井口建筑物 25.2.4.14
mine iron-ore concentrator 铁矿石富集器 11.6.2
mineral residue 矿物废渣 2.2.92.24
miniature circuit breaker 小型断路器(熔断器) 25.4.1.19
minimum fluidization 临界流态化 14.1.3.8
minimum fluidizing velocity 最小流化速度 1
minimum head loss 最小压头损失 3.5.4
minimum O.A.damper(manual) 最小室外空气手动调节风门 20.2.1.10
minimum opening 最小开度 11.1.3.13
minimum reflux ratio 最小回流比 1

mining 采矿 2.2.85.1
　　混合 3.1.12.2
minispring 小弹簧 19.1.1.7
minor beam support clamp 支果固定板 5.2.7.2
minute bubbles 小气泡 3.1.13.3
mist eliminator 脱湿器 10.5.4.18
　　捕沫器 3.1.9.14
　　除雾器 8.9.2.11;除沫器 3.4.4.7
mix 混合物 6.1.7.9
mixed-bed ion exchange 混合床离子交换 6.2.9
mixed C_4 混合 C_4 2.2.67.20;2.2.69.21
mixed C_5 混合 C_5 2.2.67.21
mixed flow 混流(式) 16.1.1.32
mixed-flow impeller, single-suction, enclosed & open design 混流式叶轮,单吸,闭式与开式结构 16.4.5.(c)
mixed flow pump 混流泵 16.1.2.2.25
mixed liquor 混合液 3.1.29.4;3.1.34.9;3.1.46.13;3.3.18.7
mixed liquor flow pattern 混合液流型 3.2.13.9
mixed liquor inlet 混合液入口 3.1.29.1
mixed liquor to clarifier 混合液去澄清器 3.1.45.8
mixed material 混合物料 12.6.10.3
mixed resin 混合树脂 6.2.9.1
mixer 混合器 2.2.57.1;2.2.88.10;3.1.35.7
mixer 搅拌器 3.2.2.15;3.3.29.1
　　搅拌机 3.3.28.4
mixer 混合器 6.1.6;2.2.66.4;3.3.24.3
mixer breaks up material 混合器打碎物料 11.4.3.2
mixer for coal slurry discharge tank 煤浆出料槽搅拌器 2.2.35.9
mixer for coal slurry tank 煤浆槽搅拌器 2.2.35.11
mixer/granulator 混合/造粒机 3.1.16.3
mixer screw 混合器螺杆 8.9.5.9
mixer screw drive unit 混合器螺旋(自转)驱动装置 8.9.5.14
mixer support level 混合器支承面 8.4.1.10
mixing agitator 混合搅拌器 6.3.2.12
mixing and separation equipment for solids 固体混合与分离设备 12
mixing blade 搅拌叶片 12.1.1.14
mixing cellulose sheet 混合纤维素板 2.3.7.2
mixing chamber 混合室 15.6.4.15
mixing chest for stuff 浆料混合池 2.3.11.1

mixing coil 混合盘管 2.2.57.8
mixing drum 混合转鼓 9.6.2.4
mixing/elutriation tank 混合/淘析槽 3.2.10.2
mixing/flocculation 搅拌/絮凝 2.4.5.4
mixing & grinding machine 混合研磨机 15.7.4
mixing kettle wall 混合釜壁 4.3.1.7
mixing lenght 混合长 1
mixing machine 混合机械 15.7
mixing-ribbon spirals 混合螺带螺旋 4.3.1.4
mixing screw 混合螺杆 12.1.1.26
mixing section 混合段 18.3.3.4
mixing tank 混合槽 2.3.10.35
mixing tank 混合槽 6.1.1.3
mixing zone （气化炉）混合区 2.2.24.8 混合区 3.1.32.12
mix muller 混合研磨机 6.3.10.15
mixture outlet 混合器出口 25.2.7.8
ML circulating pipe 母液循环管 7.2.8.15
MLR (mixed liquor recirculation) 混合液循环 2.2.2.12
ML recycle pump 母液循环泵 7.2.8.18
ML return 母液回流 7.2.8.16
mockup experiment 冷模试验 1
model Ⅰ catalytic-cracking unit Ⅰ型催化裂化装置 14.2.7
model M colloid mill M型胶体磨 11.4.3
mode of operation 操作方式 6.2.4
moderate clearance 适度间隙 19.3.1.4
moderate pressure 中压 2.2.100.2
modern chemical processing of coal 现代煤化工 2.2.1.23
modern FCC unit 现代催化裂化装置 14.3.4.(b)
modern materials handling 现代物料输送方法 23.3.2
moist material （半）湿物料 8.8.9.4
moisture 含湿量 2.3.1.34
moisture-carrying exhaust air 携带水分的废气 8.9.6.12
moisture content 含湿量 13.1.1
moisture control 水分控制 3.3.16.31
moisture-sample train 湿含量测定系统 3.5.7
moisture separator 去湿器，水分分离器 17.6.1.9
moisture trap 水分分离器 9.3.1.7
molecular distillation 分子蒸馏 5.5.1
molecular distillation still 分子蒸馏釜 5.5
molecular pump 分子泵 9.5.21.8

molecular rearrangement 分子重新排列 2.3.8.23
molecular sieve 分子筛 1
molecular still 分子蒸馏器[釜] 18.5.1.12
molecular thermodynamics 分子热力学 1
molecular-weight determination 测定分子量 3.5.8
molten chemical 熔融化学品 13.5.3.1
molten glass 熔化玻璃 2.3.9.50
molten plastics 熔融塑料 13.2.1.(a)
molten salt 熔盐 3.3.11.11
molten-salt chamber 熔盐室 3.3.11.3
molten-salt demister 熔盐除沫器 3.3.11.8
molten-salt incinerator design 熔盐焚烧炉结构 3.3.11
molten-salt level control 熔盐位面控制器 3.3.11.2
molten tin 熔化的锡 2.3.9.17
momentum transfer 动量传递 1
monitoring control for the switching systems 开并系统监控盘 25.4.2.6
monitoring well 监挖井 3.3.12.1
monochlorobenzene 单氯苯 2.3.8.9
mono ethanol amina (MEA) based process for CO_2 removal from flue gas 乙醇胺法脱除烟气中的 CO_2 2.2.27
monorail beam 单轨梁 24.1.2.1
Monsanto/BP process for acetic acid production Monsanto/BP 制乙酸工艺流程 2.2.56
MOR—metal oxidation reactor 金属氧化反应炉 2.2.80.2
mother liquid 母液 9.5.19.6
mother liquor outlet 母液出口 7.2.1.6
motion （往复）运动 16.2.1.7
motion caused by rake 耙子使物料（向上）移动 12.4.2.15
motions of screen 筛子的振动方式 12.2.1
motive gas 动力气体 18.3.3.1
motor 电机 10.2.1.20
motor and gear 电机与减速机 3.1.14.20
motor base 电机座 9.5.4.16
motor casing 电机壳体 16.4.20.7
motor end 电机端 19.1.8.4
motor end 电机端 19.2.3.15
motor for conveyer 输送机用电机 8.3.4.19
motor frame 电机机架 16.4.28.20 电机座 20.2.1.20
motor/gear reducer assembly 电机/齿轮减速机组件 3.1.46.8
motor half coupling lock nut 电机侧联轴器锁

567

紧螺母 16.4.27.22
motorized lifting device 动力升举机构 15.6.1.9
motor reducer drive 电机减速传动装置 8.8.10.1
motor reducer variator 电机减速机 13.1.2.1
motor rotor 电动机转子 20.2.3.13
motor stand 电机机架 16.4.21.2
motor stator 电动机定子 20.2.3.14
motor support 电机支架 9.5.10.15
motor support column 电动机支架 16.4.27.21
motor with reducer 电机减速机 24.1.13.4
motyer liquor 母液 1
mould 霉菌 1
模(盘) 2.3.11.48
moulded corrosion-resistant fibreglass envelope 模制耐腐蚀玻璃纤维容器 15.5.1.3
mounting hinge 支架 24.2.4.11
mounting position 安装位置 18.2.4.9
movable die plate 可拆模板 15.7.3.11
movable head 活动端板 9.1.2.7
movable partition 可动隔板 3.8.1.8
moveable end cover 可移动端盖 4.2.1.3
moving bed 移动床 14.1.3.25
moving bed catalytic cracker 移动床催化裂化装置 14.3.3.(c)
moving bed sorber 移动床吸附器 6.3.11
moving-packed-bed reactor 移动填充床反应器 3.3.15.(a)
moving-stirred-bed staged reactor 移动搅拌分段式反应器 3.3.15.(c)
MP steam 中压蒸汽 2.2.31.16
MTBE 甲基叔丁基醚 2.2.65.5
MTO(methanol to olefins) reactor 甲醇制烯烃反应器 2.2.68.4
M-type balanced opposed reciprocating compressor M型对称平衡式压缩机 17.2.1.10
mud drum 泥浆包 14.2.2.10
mud pump 泥浆泵 2.1.2.16
muffle furnace 隔焰炉 24.1.10
muller 滚轮 12.1.3.10
muller turret 滚轮 12.1.3.9
multiclone 多管式旋风分离器 10.1.4.(a)
multiclone collector 多管式旋风分离器 10.1.2.(e)
multi clones 多管式旋风分离器 10.1.1.(b)
multi-compartment vibro bag filer 多室振打袋滤器 10.2.2
multicomponent ion-exchange process 多元离子交换过程 6.2.7

multicomponent mixture 多元混合物;多组分混合物 1
multicyclones 多管式旋风分离器 14.2.7.17
multiflow pump 多流式泵 16.1.2.2.60
multifunction retort 多功能干馏釜 2.2.84
multi-holed spinneret 多孔纺纱头 2.3.7.14
multiphase flow 多相流 1
multiple bed dryer 多层流化床干燥器 8.5.4.(b)
multiple conveyors 多路输送器 23.2.3.(e)
multiple cyclone 并联旋风分离器 8.4.1.16
multiple discharge 多头卸料 23.1.1.(b)
multiple downcomer sieve tray 多降液管筛板 5.2.10.13
multiple drying machine 多层烘干机 2.3.7.31
multiple-effect evaporation 多效蒸发 1
multiple-effect evaporator 多效蒸发器 2.3.6.25
multiple-hearth furnace 多膛炉 3.3.15.(c)
多膛燃烧炉 3.3.3
multiple-hearth furnace for carbon reactivation 活性炭再生用多层膛式炉 6.3.5
multiple-hearth incinerator 多室焚烧炉 3.3.9
multiple induced-roll magnetic separator 多级感应磁辊分离器 12.6.4
multiple intertube 复式内管 24.2.2.5
multiple-loop recirculation blending system 复合回路循环混合系统 12.1.5
multiple meter 多用电表 25.4.1.41
multiple opening 多孔 21.1.1.37
multiple rotor 多转子 16.1.1.24
multiplex 多列,多段 16.1.1.11
multipurpose incinerator 多功能焚烧炉 3.3.5
multistage 多级 16.1.1.40
multistage centrifugal pump 多级离心泵 16.1.2.2.15
multi-stage centrifugal pump 多级离心泵 16.4.18
multistage centrifugal system 多级离心压缩制冷系统 20.1.6
multistage compressor 多级压缩机 1
multistage fluized-bed dryer 多层流化床干燥器 8.5.1
multitubular column 多层塔 5.1.2.6
muslin sack 棉布包 23.1.9.16

N

N₂ 氮气 2.2.25.1;2.2.33.18;2.2.115.2;2.2.117.11;2.2.118.3;2.2.119.3

Na$_2$CO$_3$ 碳酸钠 3.4.6.5
NaOH 氢氧化钠 2.2.41.12;3.4.6.4
N$_2$+CO$_2$ 氮气+二氧化碳 2.2.33.19;2.2.77.3
N$_2$+H$_2$ 氮气+氢气 2.2.33.12;2.2.38.11
N$_2$,O$_2$ 氮气,氧气 2.2.77.1
name plate 铭牌 22.2.1.3
name plate set 铭牌位置 22.2.1.18
nanofiltration 纳米过滤 1
naphtha 石脑油 2.2.79.10
naphtha feed 石脑油进料 14.2.6.15
naphtha fraction 石脑油馏分 2.2.93.3
naphtha hydrotreater 粗汽油加氢处理装置 2.1.9.36
naphthalene 萘 2.2.91.9
naphtha recovery tank 粗汽油回收罐 2.1.9.24
Nash pump 纳氏泵 16.1.2.2.53
natural circulation fire-tube boilers 自然循环火管锅炉 4.4.2
natural-draft stack 自然抽风筒 8.8.6.11
natural frequency conveyor 固有频率输送器 23.1.7.2
natural gas 天然气 2.2.3.8;2.2.15.1;2.2.45.1;2.2.78.7;2.2.102.6;2.3.17.5
natural-gas pump 天然气泵 10.3.5.1
natural gas to pipeline network 天然气去管网 2.2.36.12
(4)(normally closed) 常闭 20.2.2.8
near end 近端 4.3.4.27
needle 阀针 22.2.3.3
needle and bottom damping assembly 针与底部阻尼系统 9.5.21.5
needle-point HV electrode 针状高压电极 15.3.1.6
needle valve 针形阀 3.4.11.25
neg.press. 负压 3.2.6.12
neoprene gasket 氯丁橡胶垫圈 8.2.2.9
neoprene vane 橡胶叶轮 16.3.4.1
net positive suction head(NPSH) 汽蚀余量;净正吸压头
net power generation 净发电量 2.2.107.13
neutral conductor 中性导体 25.4.1.13
neutralization pond 中和池 3.1.2.4
neutralize 中和 2.3.8.24
neutralizer 中和塔 2.3.15.2
neutral point 中(性)点 25.4.20.4
neutral sulfite semi-chemical waste liquor 中性亚硫酸盐半化学废液 3.3.4
Newtonian fluid 牛顿流体 1
NG(natural gas) 天然气 2.2.113.13;2.

114.2;3.4.6.13
NH$_3$ 氨 2.2.38.13;2.2.111.12
NH$_3$ product 氨成品 2.3.3.18
nickel-base alloy 镍基合金 22.6.2.⑱
nickel-copper alloy 镍铜合金 22.6.2.⑲
nipping roll(lower) 下部压紧胶辊 15.7.5.10
nipping roll(upper) 上部压紧胶辊 15.7.5.8
nitric acid 硝酸 14.3.1.9
nitrification 硝化 3.1.37.7
nitrified influent 硝化污水进 3.1.38.3
nitrified effluent 硝化污水 3.1.36.1
nitrified recycle(400%Q) 硝化循环(400%Q) 3.1.33.3
nitrogen 氮(气) 2.2.8.2;2.2.31.4;2.2.76.6
nitrogen+air 氮气+空气 2.2.76.4
nitrogen/air 氮气/空气 2.2.75.1
nitrogen compressor 氮气压缩机 17.1.2.4
nitrogen fertilizer 氮肥 2.2.1.16
nitrogen fixation 固氮[作用] 1
nitrogen oxide absorption 氧化氮吸收 14.3.1.(e)
nitrogen oxide absorption in packed columns 填料塔内进行氧化氮吸收 14.3.1.(b)
nitrogen oxide sample train 氮氧化物采样系统 3.5.15
nitrogen wash 氮气洗涤 2.3.1.13 氮洗 2.2.33.10
nitrogen wash tailgas 氮洗尾气 2.2.33.11
NKK one-step liquid phase dimethylether synthesis NKK液相一步法二甲醚工艺流程 2.2.50
N.O.(normally open) 常开 20.2.2.7
no code 无规范 21.1.1.49
no dust carryover 无粉尘带出 4.3.3.25
No.2 fuel oil 2号燃料油 3.3.2.9
noise level 噪声水平 1
nomenclature 命名规定 10.1.3.11
non-catalytic fluidized bed system 非催化流化床系统 14.1.12
non-condensable gas 不凝气出口 7.2.5.7
non-destructive testing 无损检验 21.1.2.15
non-directional 非定向定式 23.3.2.14
non-dispersive infrared analyzer(NDIR) 非色散型红外线分析仪 3.5.11.27
nonenergy resource recovery 非能量资源回收 3.5.11.9
non-equilibrium stage model 非平衡级模型 1
non-equilibrium system 非平衡系统 1
non-flowing region 死区 23.2.2.1
non-lubricated cylinder 无油润滑气缸 17.

2.10
non-magnetic coarse tailings 非磁性粗尾砂 12.6.4.11
nonmagnetic concentrate 非磁性富集物 12.6.1.14
nonmagnetic fine failings 非磁性细粒尾砂 12.6.4.10
nonmagnetic materials 非磁性材料 12.6.10.6
nonmagnetics 非磁性 12.6.4.9
nonmagnetic tailings 非磁性尾砂 12.6.4.6
nonmagnetic tailings discharge 非磁性渣料卸出 12.6.9.9
nonmetallic 非金属 19.3.3.(c)
nonmetallic pipe and lined pipe systems 非金属管系和非金属衬里管系 22.6.2.3
nonmetals 非金属 22.6.2.2
non-Newtonian fluid 非牛顿流体 1
non-positive drive 非正位移驱动装置 23.3.2.3
non pressure parts 非受压件 21.1.1.38
nonpumpable 不可泵送的 3.3.7.16
nonpumpable waste 不可泵送的废料 3.3.7.6
non-returnable bottle 不回收的瓶子 2.3.13.29
non-return valve 止逆阀 22.1.10
non-rigid 非刚性 22.6.1.(c)
non-rotary ball 不旋转的球磨机 11.3
non-self-priming 非自吸式 16.1.1.38
non-wash plate 非洗涤板 9.1.1.1
no radiographic exam 不需要射线检验 21.1.2.11
No 1 riding ring 1号滚圈 8.8.1.15
normal butane 正丁烷 2.1.1.66
normal level 正常液面 14.3.4.23
normal staple lengths 标准纤维长度 2.3.8.60
north temperate zone plate 北温带板 21.2.3.5
No 1 runner 1号碾碎头 11.1.6.12
No12 sweetland filter No12过滤器 6.3.2.28
note-remix 再混合 23.3.4.6
nozzle 喷嘴 3.1.22.1
nozzle centrifuge 喷嘴卸料离心机 9.5.16
nozzle discharge 喷嘴卸料 9.5.14.(b)
nozzle for compound pressure and vacuum gauge 压力真空表接管口 14.4.1.8
nozzle for reflux condenser 回流冷凝器接管口 14.4.1.12
nozzle for thermometer 温度计接管口 14.4.1.15
nozzle for thermometer for gas in furnace 测炉内气温用温度计接管 24.1.1.9
nozzle for thermometer of exhaust gas 排气温度计接管 24.1.1.1
nozzle head 出口接头 16.4.27.32
nozzle head bushing 出口接头衬套 16.4.27.30
nozzle plate 喷嘴固定板 18.3.1.4
nuclear energy 核能 25.1.1.2
nuclear power plant 核动力站 25.2.2.2
nuclear reactor 核反应堆 25.2.2
nucleate boiling 泡核沸腾 1
nucleic acid 核酸 1
number of (mass) transfer units (NTU) 传质单元数 1
Nusselt number 努塞特数 1
nut 螺母 5.2.4.8
nutrient 营养素 3.2.11.10
nylon 尼龙 22.6.2.⑤
nylon insert 尼龙衬垫 19.1.5.12

O

O (overflow product) 溢流产品 12.4.1
O_2 氧气 2.2.25.5; 2.2.29.9; 2.2.112.1; 2.2.114.4; 2.2.115.4; 2.2.117.4; 2.2.118.4; 2.2.119.5; 2.3.19.4; 3.1.47.8
O_3 臭氧 2.4.4.4
O.A (outdoor air) 室外空气 20.2.2.24
objective function 目标函数 1
observation port 观察孔 25.1.3.4
observation window 观察窗 24.1.13.8
O_2 depleted air 氧气耗尽的空气 2.2.80.9; 2.2.114.7
O_2 diffusion 氧气扩散 3.1.46.12
odorous exhaust 有气味的废气 10.5.13.6
odorous exhaust air 有臭味的废气 3.1.21.3
offgas 废气 2.2.52.6; 2.3.5.15; 3.3.29.10
排出气体 25.5.1.7
尾气 14.3.1.11
off gas absorber 尾气吸收塔 2.2.52.5
offline MTG reactor burning coke off catalyst 独立的MTG反应器,烧去催化剂表面附着的焦质 2.2.77.2
O_2 from ASU 从空气装置来的氧气 2.2.109.1
offset 偏置距离 22.5.2.5
off-tank support 槽外支撑物 15.6.1.16
off with light constituents tower 脱轻塔 2.2.55.10
oil 油 2.2.15.3; 2.3.17.2; 15.5.4.7
oil and dusttight chain casing 链条防油防尘

罩 8.8.11.4	
oil attracting tubes 吸油管 15.5.1.4	
oil booster pump 油增压泵 18.2.3	
oil burner 油燃烧器 24.1.12.7	
oil cap 油杯盖 16.4.29.13	
oil conservator 储油柜,油箱 25.4.3.9	
oil cup 油杯 15.7.4.1	
oil cup(L type) L型油杯 15.7.4.12	
oil diffusion pump 油扩散泵 16.1.2.2.48	
oil diffusion pump 油扩散泵 18.2.2	
oil drain 排油口 18.2.4.15	
oil draw out 排油口 15.7.4.4	
oil drip box 盛油箱 9.6.1.18	
oil exit pipe 排油管 20.2.1.27	
oil feed 油供入 14.3.4.13	
oil feed bearing 给油轴承 9.5.10.12	
oil filler pipe 注油管 19.2.1.4	
oil filler plug 滤油器丝堵 18.1.4.3	
oil filler point 注油孔 19.2.1.5	
oil filling port 注油口 15.7.4.15	
oil from elevated tank 高位槽来油 19.3.2.16	
oil-gas seperator 油气分离器 5.3.3.3	
oil gauge 油标 13.3.5.15	
油位计 25.4.3.11	
oil globules 油珠 3.1.4.8	
oil-hole 油孔 19.2.3.10	
oil hole cover 给油口盖 12.3.5.28	
oil inlet 注油口 15.7.4.2	
oil intake port 进油孔 19.3.2.13	
oil layer 油层 3.1.4.9	
oil level sight glass 油位视镜 18.1.4.11	
oil or gas burner 燃油或燃气器 8.4.1.11	
oil out 油出口 6.1.9.14	
oil outlet 出油口 15.7.4.11	
oil outlet for reducer 减速机出油口 15.7.4.11	
oil pan 油槽 9.5.10.15	
oil pipe 油管 12.3.5.9	
oil preheater 油预热器 2.3.16.6;2.3.17.1	
oil pressure gauge 油压表 15.7.3.15	
oil product 油产品 2.2.6.17	
oil production 采油 2.1.2	
oil pump 油泵 2.3.16.4;16.1.2.3.24	
oil recovery and filtration 油回收与过滤 2.2.30.5	
oil reservoir 油腔 16.4.29.10	
oil retainer 挡油环 17.3.2.13	
oil-retention baffle 截油挡板 3.1.3.14	
oil return cavity of low pressure side 低压侧回油空腔 19.3.2.14	
oil return pipe 回油管 18.2.3.9	
oil return tube 回油管 18.2.1.14	
oil sand 油沙 2.1.9.2	
oil seal 油封 11.1.5.3	
oil sealed ring 油封环[圈] 13.1.2.25	
oil-sealed stuffing box 油封填料函 5.5.2.4	
oil seal ring 油封圈 16.4.22.8,9	
轴封环 17.3.2.14	
oil separator 油分离器 18.2.3.6	
oil shale 油页岩 2.2.84.13;2.2.87.7	
oil-shale mine 油页岩矿 2.2.82.1	
oil-shale process 油页岩加工 2.2.84	
oil-shale semicoking chamber 油-页岩半焦化室 2.2.84.11	
oil skimmer 撇油器 3.1.4.7	
oil skimmer outlet 撇油器出口接头 15.5.1.10	
oil-skimming pipe 撇油管 3.1.3.13	
oil slinger 甩油环 19.3.2.7	
oil spray arrester 除油雾装置 18.1.4.13	
oil-storage tank 储油罐 2.1.7.12	
oil thrower 挡油圈 16.4.1.8,11	
oil vap 油气 2.2.84.14	
oil-vapor collection 油气收集 2.2.84.9	
oil vapors to condensation section 油气去冷凝工段 2.2.88.17	
oil/water separator 油水分离器 3.1.3	
oil wiper rings 刮油环 17.2.6.4	
oily waste 含油废水 3.1.13	
oily water 含油的水 2.3.6.12	
oily-water holding pond 含油水储存池 2.3.6.10	
oily-water influent 含油废水进入 3.1.14.19	
oily water sewer 含油污水 3.1.1.2	
O jet O射流 11.5.2.11	
olefin 烯烃 6.1.5.19	
olefin hydrocarbon 烯烃 2.2.1.17	
olefin feed 烯烃原料 6.1.5.10	
olefin separation 烯烃分离 2.2.67.19	
oligomerization 低聚 2.2.78.10	
Oliver continuous vacuum drum filter 转筒真空过滤机 9.3.4	
one-dimensional model 一维模型 1	
one pass shell 单程壳体 4.1.2.(f)	
one rotor 单转子 17.1.1.6	
one stage 单级 17.1.1.14	
onical pusher with de-watering cone 锥形推料脱水离心机 9.5.1.(o)	
online MTG reactor 流程中的MTG反应器 2.2.77.7	
on-load tap changer 有载换接器 25.4.3.13	
on-off control 通断控制 1	
open 开式 16.4.5.7	
open bucket steam trap 浮桶式蒸汽疏水阀

22.2.3
open circuit grinding system 开路磨碎系统 11.6.1
open design 开式结构 16.4.5.(d)
open-drop discharge 敞口卸料 23.3.4.1
open end 开口端 4.3.2.7
open-end downcomer 下端开口的下降管 24.1.6.5
open-end downcomer from burner 上接燃烧器下端开口的下降管 24.1.6.5
open ends 开放末端 15.2.1.9
open impeller 开式叶轮 16.1.1.41
opening 开孔 21.1.1.36
opening chamber 敞开室 9.5.13.10
opening water 常开水管 9.5.13.13
open pattern return bend 敞开式180°回弯头 22.3.3.14
open seperator 敞开式气液分离器 5.3.3.4
open shelf liquid distributor 开口搁架式液体分布器 10.5.13.1
operating 操作中 6.3.2.17
operating chamber 工作室 18.4.1.5
operating characteristic 操作特性 16.1.2.1
operating floor 操作地面 4.3.4.24
operating fluid 工作液体 19.1.1.11
operating gas(air) 工作气体(空气) 18.4.2.1
operating handle 操作手柄 23.3.8.8
operating handle locking bolt 操作手柄固定销 23.3.8.5
operating liquid 工作液体 19.1.3.2
operating mechanism housing 操作机械盒 25.4.9.10
operating nozzle 工作喷嘴 18.4.2.2
operating platform 操作平台 15.6.3.5
operating principle 操作原理 11.2.9
operating slide 操作滑板 9.5.12.11
operating steam 工作蒸汽 18.3.1.1
operating steam inlet 工作蒸汽入口 18.3.1.1
operating water paring disc 操作弧形水盘 9.5.12.13
operating water supply 开式供水口 9.5.12.14
operating zone 操作区 9.3.10
operational variable 操作变量 1
operation of pastillization 制锭操作 13.5.3
operation of the falling film still 降膜式分子蒸馏设备 5.5.1.(a)
operation process chart 工作流程图 18.4.1
optical separator 光子分离器 12.7
option 挑选 3.3.16.4

optional 任选的 11.2.7.6
optional processing steps 供选择的加工步骤 3.3.16.34
optional type flange 任意型式法兰 21.1.1.27
option 附加的系统 3.1.2.29
option for upgrading 提质方案 2.2.104
or 或者 3.4.18.15
orbiting arm 旋转臂 12.1.3.5
orbiting type 环行式 12.1.3.(f)
ore 矿石 11.6.2.1
organic 有机相 6.1.12.2
organic acids 有机酸 3.3.18.14
organic effluent 有机相流出物 6.1.12.10
organic phase 有机相 6.1.2.1
organic pollutant 有机污染物 3.1.47.7
organic wastes 有机废液 3.1.23
orientation 定向 12.1.5.11
orifice 锐孔 10.2.3.6
孔板 3.4.11.19
orifice manometer 孔板测压计 3.5.16.20
orifice-type distributor 锐孔型分布器 5.3.4.12
original fill line 初始装填线 23.3.4.4
O-ring O形(密封)圈 13.1.2.24,30,39
O形环 16.4.20.14
O形环 22.1.1.14
O-ring seal O形密封圈[环] 15.2.1.2
O-ring sealed joint O形环密封接头 18.2.4.1
oscillating spider 摆动轮 9.3.3.12
Oslo evaporative crystallizer Oslo蒸发结晶器 7.2.3
Oslo surface-cooled crystallizer Oslo表面冷却结晶器 7.2.4
Oslo-type crystallizer 奥斯陆型结晶器 7.1.10
osmotic coefficient 渗透系数 1
other fuels and chemicals 其他燃油风化学品 2.2.73.12
other gas 其他气体 2.2.3.9
O tube O管 11.5.2.12
out 出(口) 16.5.6.8
outboard 外侧 16.4.9.9
outboard bearing 外置轴承 8.8.10.3
outdoor air W.B.limit thermostat 室外空气湿球恒温限值 20.2.2.11
outer annular surface 外侧环形表面 15.5.2
outer bore hole 外钻孔 3.1.24.4
outer closing chamber 外密闭室 9.5.13.8
outer cone 外锥体 12.3.3.5
outer cover 外盖 22.2.1.1

outer diameter 外径 21.1.1.64
outer drum 外转鼓 9.3.2.19
outer pipe 外管 4.2.2.6
outer safety ring 安全外挡圈 13.1.2.22,31
outer shell 外壳(体) 21.2.2.5
outer shell steam-jacket 带有蒸汽夹套的外壳 8.9.6.10
outer tank shell 容器外壳 11.3.1.13
outfall 流出口 3.1.2.28
outfall flow measurement 排水流量测量 3.5.18
outfeed 出料 3.3.29.6
outfeed conveyer 出料输送机 3.3.30.6
out flow condenser 外流凝结器 2.3.12.4
out going conveyor 外运输送器 23.2.5.5
outhearth 炉膛外 3.3.26.2
outlet 出口 10.1.4.1
　出口,排出口 12.1.1.17
　排泄口 3.3.17.7
outlet arm handle 出口臂手柄 13.3.5.21
outlet case 出口支架 13.3.5.22
outlet casing 出口外壳 25.3.1.27
outlet diffusor 出口扩压器 25.3.1.28
outlet duct for fine material 细粉排出管 12.3.2.2
outlet duct for tailing 粗粉排出管 12.3.2.4
outlet from the kiln 烘窑出气口 2.3.12.19
outlet head 出口端帽 10.1.2.18
outlet launder 出液溜槽 15.6.3.17
outlet passage 排出通道 16.5.6.3
outlet pipe 出口管 10.2.1.29
outlet pocket 卸出槽 15.6.3.7
outlet portion 排出段 16.4.18.12
　出口段 16.4.18.13
outlet ports 出料口 3.3.17.8
outlet radial 径向出口 15.6.3.6
outlet sluice 出口斜槽 12.5.2.7
outlet valve 出口阀 23.1.2.34
output device 输出设备 3.4.20.7
outside air 外界空气 20.2.2.9
outside casing liner 外壳内衬 12.3.4.11
outside casing plate 外壳板 9.5.2.20
outside packed floating head 外填料函式浮头 4.1.4.(p)
outside-packed floating-head exchanger 外填料函浮头换热器 4.1.1.(d)
outside storage 外部贮存 6.3.2.33
outside storage tank 外部贮罐 6.3.2.34
oven 恒温器 3.5.20.4
overall heat transfer coefficient 总传热系数 1
overburden 表土层 2.2.85.6

overfire air port 过烧空气通道 3.3.12.9
overflash 超闪蒸 2.1.4.27
overflash set 超闪蒸 2.1.4.10
overflow and fines 细粒随溢流而出 12.4.2.13
overflow discharge 溢流卸出 12.6.9.1
overflow launder 溢流溜槽 15.6.1.10
overflow launder outlet 溢流溜槽出口 15.6.1.4
overflow level control 溢流口液面控制 7.2.9.3
overflow plate 溢流板[堰] 8.5.1.15
overflow tube 溢流管 20.1.8.4
overflow vessel 溢流槽 10.4.1.12
overflow weir 溢流堰 14.2.2.8
overflow well 溢流管 14.3.4.5
overhead 塔顶馏出物 2.1.4.23
overhead spray washing plant 喷淋水洗设备 2.3.7.29
overhead tanks 压力罐 2.1.7.6
overload release 过载断路器 9.5.5.12
oversize 大粒度 11.6.4.8
　筛上物料 12.7.2.14
oversize material 尺寸过大物料 3.3.16.21
oversize outlet 筛上物料排出口 12.2.6.6
oversize particles 过大颗粒 12.3.3.6
over-temperature cutout 超温切断 25.6.7.2
oxidation 氧化 8.1.1.33
oxidation(dehydrogenation) 氧化(脱氢作用) 2.3.8.19
oxidation/nitrification tank 氧化/硝化池 3.1.25.6
oxidation reactor 氧化反应器 2.2.64.1
oxidation zone 氧化区 3.1.32.11;3.4.5.3
oxidized asphalt 氧化沥青 2.1.1.76
oxychlorination 氧氯化 2.3.18.3
Oxy-combustion power plant based on design by Haslbeek et al. 哈斯贝克设计的氧燃烧动力装置 2.2.115
oxygen 氧(气) 2.1.8.1;2.2.30.1;2.2.33.4;2.2.67.3;2.3.18.3
oxygen/air 氧气/空气 2.2.19.2
oxygen/air/steam 氧气/空气/蒸汽 2.2.110.4
oxygen bottle 氧气瓶 3.4.7.3
oxygen compressor 氧气压缩机 17.1.2.6
oxygen enriched air 富氧空气 2.2.53.8
oxygen feed gas 氧气供气 3.1.45.1
oxygen generator 氧气发生器 3.1.46.1
oxygen lance 氧气吹管 10.5.18.1
oxygen/steam 氧气/蒸汽 2.2.31.8

oxygen supply 供氧 3.1.46.5
ozone treatment 臭氧处理 2.4.4.5

P

PAC（powdered activated carbon）粉末活性炭 2.4.4.6
pacific multiple-hearth furnace pacific 多层床炉 24.1.13
packaging 包装 23.4
装箱 3.3.7.4
packed-bed scrubber 填充床涤气器 10.5.8
packed column 填料塔 5.3.1;3.4.1.1;
packed column filled with liquid 充液填料塔，乳化塔 5.1.2.9
packed column with air pulser 空气脉冲填料塔 6.1.16.(b)
packed tower 填料塔 5.3.3.6
packer conveyor 包装机输送器 23.4.2.2
packing 填料 16.2.6.8
填料函 17.2.2.9
packing box 填料函[盒] 4.1.4.8
packing box flange 填料函法兰 4.1.1.23
packing case 填料盒 19.2.1.3
packing draw-off hold 填料卸出孔 5.3.4.26
packing flange 填料压盖 22.1.2.10
packing flange stud 填料压盖双头螺柱 22.1.2.9
packing follower 填料压盖[环] 16.4.25.30
packing follower ring 填料压盖 4.1.1.25
packing for cover 顶盖垫片 9.2.1.20
packing for discharge cover 卸渣管法兰垫片 9.2.1.26
packing gland 填料压盖 14.4.3.16
packing lubricator assembly 填料润滑装置部件 22.1.9.13
packing media 填充介质 3.4.4.6
packing press ring 填料压环 5.3.4.25
packing ring 填料隔圈 16.4.12.13
填料环 16.4.19.7
packing seal 填料密封 7.1.3.13
packing support 填料支承 5.3.4.21
packing support plate 填料支承板 5.3.4.20
packing take-up device 填料压紧装置 16.2.6.7
pad plate 垫板 21.3.3.1
pad recirculation line 团块再循环轨迹 3.2.13.13
paint residues 油漆废料 3.3.5
pan 槽 12.1.3.8

盆 13.3.1.4
锅 15.7.4.6
pan granulation 盘式固化 13.3.5
pan granulator 盆[盘]式造粒机 13.3.4
paper machine production line 造纸机生产线 2.3.11.13-28
papermaking 造纸 2.3.11
papermaking by hand 手工造纸 2.3.11.46-51
paper mill waste 纸浆废液 13.4.1(b)
paraffin 烷烃 6.1.5.18
paraffin feed 烷烃原料 6.1.5.9
paraffins removal column 脱烷塔 2.2.56.5
paraffin 链烷烃 6.1.6.(a)
paraffin wax 石蜡 2.1.1.79
parallel current tunnel dryer 并流隧道式干燥器 8.3.1.(b)
parallel feed 平行进料 1
parallel flow 并流 17.4.3.2
parallel row 并联排列 10.5.12.5
paring disc 弧形油盘 9.5.12.3
parison 长颈瓶 2.3.9.47
parison mould 玻璃瓶模 2.3.9.26
parshall flume 帕里斯霍尔水槽 3.4.4
parson process parson 工艺 2.2.10
partial 部分的 2.2.104.8
partial cross section 局部剖面图 8.8.10.16
partially filled horizontal tank 部分充填卧式贮罐 21.2.5
partial molar quantity 偏摩尔量 1
partial oxidization 部分氧化 2.2.33.3;2.2.78.8
partial recycle 部分循环 8.4.3
particle collection 粒子(的)收集 15.3.1.7
particle concentrate 粒子富集 10.4.3.9
particle counter monitor 粒子计数器 3.5.6
particle fountain 颗粒喷泉 13.4.1.22
particle size 粒度 1
particle size distribution 粒度分布 1
particle trajectory 颗粒轨迹 23.3.1.4
particulate control device 微粒控制设备 2.2.24.13
particulate fluidization 散式流态化 1
particulately fluidized bed 散式流化床 14.1.3.9
particulate removal 除尘 2.2.89.6
除微粒 25.5.2.4
颗粒去除 2.2.109.6
particulate-sample apparatus 颗粒物采样仪器 3.5.11
partition 隔板 14.1.9.5

partition wall 隔板 20.2.3.4
parts of a double-suction impeller 双吸叶轮部件 16.4.6
passage 通道 18.1.2.8
pass partition 分程隔板 4.1.1.31
paste 膏糊状 8.1.1
pasteurized water T_2 巴氏消毒过的水(温度 T_2) 2.4.1.3
pastillization 制锭 13.5.3
path 通道 9.1.2.3
path of wash water 洗水路径 9.1.1.9
pattern of coarser dust main stream 粗粉尘主流线 10.1.2.8
pattern of dust stream 粉尘流型 10.1.2.20
PBR (packed bed reactor) system with coarse media denitrification column 带粗介质反硝化塔的填料床反应器 3.1.39
PE 聚乙烯产品 2.2.67.26
pearlite 珍珠岩 21.2.2.7
pebble elevator 卵石提升机 14.3.3.4
pebble feed 卵石进料 11.6.3.13
pebble feeder 卵石加料器 14.3.3.11
pebble hopper 卵石料斗 14.3.3.6
pebble reactor 卵石反应器 14.3.3.(a)
peeling knives intermittently operated 断续操作的剥离刮刀 4.3.1.13
peep hole 视孔 7.1.2.14
peep sight glass 窥视视镜 5.4.2.6
pellet extruder 挤压造粒机 6.3.10.16
pelletized 成粒 3.3.16.32
pelletized(densified)RDF 成粒(压实)垃圾废料 3.3.16.33
pelletizer 造粒机 11.6.3.18;2.3.16.12;2.3.17.11
pelletizing 造粒 1
pelletizing by solidification method 固化法造粒 13.3.5
pelletizing plant 造粒设备 13.2.4
pen arm 笔尖杆 3.5.5.3
pendulum feed pipe 摆式进料管 8.7.1.11
percolation 渗滤 1
perfect mixing 全混 1
perforated baffle 多孔挡板 15.6.5.6
perforated basket 多孔吊篮(转鼓) 9.5.5.21
带孔转鼓 9.5.6.18
perforated cylindrical screen discharge 多孔圆筒形排料筛 11.2.5.4
perforated metal cloth-support 多孔金属滤布及支架 9.4.2.9
perforated pipe distributor 多孔管分布器 5.3.4.11

perforated piping 多孔管 3.3.20.10
perforated-plate column 筛板塔 6.1.16.(a)
perforated plate for cooling 冷却段多孔板 8.5.1.7
perforated plate for drying 干燥段多孔板 8.5.1.5
perforated plate tower 孔板塔,穿流筛板塔 5.1.2.13
perforated tray 多孔筛板 5.2.10.6
perforated tube with liner 带衬的多孔管 9.4.4.9
perform tray 网孔塔盘 5.2.10.20
perform well 运行良好 23.3.5
peripheral 圆周(式) 16.1.1.44
peripheral pump 漩涡泵 16.1.2.2.19
peripheral ring clamps 周边圆环固定板 5.2.7.10
peripheral speed 圆周速度 1
peristaltic 蠕动 16.1.1.47
peristaltic pump 蠕动泵 16.1.2.2.65
Permasep permeator Permasep 反渗透器 15.2.1
permeable stainless steel slope sheets 可穿过的不锈钢斜板 23.3.8.11
permeate 渗透液 1
permeate out 渗透液流出 15.2.2.2
PERT (project evaluation and review technique) 项目评审技术 1
per unit of output 单位产量 8.1.1.35
pesticide disposal 农药废弃物 3.3.11
pestle 捣棒 15.7.4.5
petcoke 石油焦 2.2.3.6
peterson roto-disc clarifier peterson 转盘沉降器 9.4.3
petreco cylectric coalescer petreco cylectric 电聚结器 6.1.9
petri dish 有盖玻璃皿 2.3.11.7
petrochemical building block 石油化工基本原料 2.2.91
petrochemical processing 石油加工 2.1
petrogas 石油气 2.1.1.29
petrol coke 石油焦 2.1.1.46
petroleum refinery 炼厂加工 2.1.1
petrosix process petrosix 工艺 2.2.12
PFM 加压过滤膜 3.1.42.8
pH 酸碱值 3.1.2.3
pH adjust 调节 pH 值 3.1.31.2
phase diagram 相图 1
phase equilibrium 相平衡 1
phenol (苯)酚 2.3.8.14

575

phenolics 酚醛塑料 2.3.6
phenol still 苯酚蒸馏塔 3.1.5.9
phostrip 侧流除磷 3.1.37.11
photocell dropout alarm 光电物料下落板凳设备 23.1.7.13
photodiode 光电二极管 3.5.6.7
photomultiplier tube 光电倍增管 12.7.5.12
pH sensor pH值传感元件 3.1.12.3
pH 传感器 3.5.10.2
physical unit operation 物理单元操作 3.2.1.10
$p\text{-}h$ diagram $p\text{-}h$ 图 20.2.3
PI(pressure indicator) 压力表 3.1.6.19
pickup hoods 抽风罩 10.4.4.9
pickup the solid 吸取物料 23.1.1.(d)
pickup device 吸取装置 23.1.1.1
pier 码头 23.3.9.6
pilot burner 导燃烧嘴 3.3.10.10
pilot plant 中间试验装置 1
pilot-plant fluid-bed reactor system 流化床反应器中试装置 2.3.5
pilot-scale 中试装置 10.3.5
pilot system 中试装置 3.1.19
pin 销钉 19.3.2.3
销(子) 22.1.3.5
pinch-bottom field closer 压紧底部现场封口机 23.4.3.11
pinch-bottom system 黏压底部系统 23.4.3
pinion 小齿轮 12.3.5.26
pinion shaft 小齿轮轴 12.3.5.24
pinion shaft outside bearing 小齿轮轴外侧轴承 12.3.5.23
pin plate 销轴垫板 11.1.8.28
pin rotor 钢枪转子 11.2.10.8
pipe 管 9.5.4.44
pipe bend 管路拐弯处 23.1.7
pipeline 管线 2.1.9.41
pipeline fuel to power plant 燃料管道通发电站 2.2.102.12
pipeline pump 管道泵 16.1.2.3.52
pipe plate 管板 9.3.3.13
pipe precooler 管式预冷器 6.3.2.26
pipe segments 弓形管段 22.5.1.(c)
pipe still 管式炉 6.3.2.18
pipe-systemsmaterials 管系材料 22.6.3
piping pump 管式泵 16.1.2.3.54
piping system 管系 22.6.1
piston 活塞 16.2.1.14
气缸 17.2.2.6
piston equipped with carbon piston and wearing rings 配有石墨活塞环和承磨导向环

的活塞 17.2.10
piston foot 活塞底座 16.5.6.4
piston packing 活塞密封环 16.2.1.18
piston pump 活塞泵 16.1.2.2.4
piston ring 活塞环 17.2.7.21
piston rod 活塞杆 16.2.1.15
piston-rod packing 活塞杆填料 16.2.1.16
piston-rod spool bolt 活塞杆接头螺栓 16.2.3.13
piston-rod stuffing box bushing 活塞杆填料函衬套 16.2.4.9
piston valve 活塞阀 16.2.4.47
pitched bladed turbine 斜叶开启涡轮式 14.4.2.(b)
pitch feed 沥青进料 14.2.6.19
pitch with flow 流动坡度 17.6.2.7
pit furnace 槽式炉 25.1.7
pitman 连杆 11.1.1.8
pitot manometer 皮托测压[压力]管[计] 3.5.11.5
pitot tube 皮托管 1
pittsburgh moving-bed system pittsburgh 移动床系统 6.3.1
pivot 摆动中心(枢轴) 11.1.1.15
pivoted flight 枢轴刮片 15.6.1.37
P jet P 射流 11.5.2.2
placement of fixed media in combined suspended and attached growth systems 固定介质悬浮与附着生长相结合的模式 3.2.13
plain 光滑式 4.3.2.(a)
plain nickel steel 普通镍钢 22.6.2.⑦
plaited packing 折叠编织填料 19.2.3.2
plant capacity 生产能力 3.1.6
plant effluent 厂出水 3.1.34.14
plant influent 厂进水 3.1.34.1; 3.1.41.1
进厂污水 3.1.37.1
plan view 俯视图 17.2.3
平面图 3.2.13.6
plastic Intalox saddle 塑料矩鞍 5.3.2.(h)
plastic lined steel pipe 衬塑料钢管 22.6.2.⑫
plastic packing 塑料填料 4.6.1.4
plastic Pall ring 塑料鲍尔环 5.3.2.(e)
plastic pipe 塑料管 22.6.2.⑭
plastic pump 塑料泵 16.1.2.5.1
plastic tripak 塑料球形填料 5.3.2.(k)
plate 法兰板 16.4.4.1
料盘 8.9.4.5
滤板 9.1.1.4
plate above 上一层塔板 5.2.6.1
plate and frame filter press 板框压滤机 9.

plate-and-frame heat exchanger 板框式换热器 4.2.1
plate-and-frame reverse osmosis system 板框型反渗透器 15.2.3
plate-and-frame type filter press 板框压滤机 9.1
plate below 下一层塔板 5.2.6.10
plate column 板式塔 5.2.1
plate efficiency ［塔］板效率 1
plate fin-and-tube type 板翅管式 4.2.5.5
plate-fin cooler 散热片式冷却器 4.2.5.4
plate-fin heat exchanger 翅片式换热器 4.5.2.8
plate joint 滤板接缝 9.1.3.2
plate No.1 第一块板 15.2.3.6
plate support ring 钢板支承圈 5.2.7.9
plate-type exchanger 板式换热器 4.2.5.1
plate-type flat welding flange 板式平焊法兰 22.4.2.(a)
platform 平台 21.2.3.3
plating rinse water 喷镀漂洗水 6.2.7
plenum 风斗 4.5.1.14
 通风管道 3.3.30.10
 充气室 8.5.6.4
plenum air 加压空气 3.3.8.15
plenum chamber 强制通风室 8.2.1.4
 气室 14.1.12.3
plicate pump 复式泵 2.1.7.17
ploughs 搅拌 24.1.11.10
plug 丝堵 13.1.2.11
plug 螺塞 13.3.5.13
 堵头 24.2.7.6
plug cock 旋塞 22.1.1.(d)
plug flow 平推流;活塞流 1
Plug-flow in-vessel composting bioreactors；(a)cylindrical；(b)rectangular；(c)tunnel 容器内活塞流复合肥生物反应器；(a)圆筒形；(b)矩形；(c)隧道式.3.3.29
plug for air vent 放气螺塞 16.4.19.16
plug for rotor cleaning 转鼓清洗螺塞 6.1.18.6
plug in type lubro-control unit 组合式润滑油控制装置 17.6.2.8
plug in type vitalized unit 插入式重要部件 17.6.2.4
plug valve 旋塞阀 22.1.6
plug welds 塞焊 21.1.1.20
plunger pump 柱塞泵 16.1.2.2.3
plunger 柱塞(式) 16.1.1.5
plunger gland flange 柱塞填料压盖法兰 16.2.4.14
plunger gland lining 柱塞填料压盖衬套 16.2.4.15
plunger lining 衬套 16.2.4.17
plunger nut 柱塞螺母 16.2.4.13
plunger pump 柱塞泵 16.2.6
plunger type metering pump 柱塞式计量泵 16.2.5
pneumatic 气流 8.1.1.⑬
pneumatic classifier 气力分级器 12.3.2
pneumatic conveying dryer 气流输送干燥器 8.4.5
pneumatic conveying drying pipe 气流输送干燥管 8.5.1.14
pneumatic conveying system 气流输送系统 23.1
pneumatic dryer 气流式干燥器 8.4
pneumatic plant 气动设备 2.3.12.12
pneumatic test 气压试验 21.1.2.15
pocket torch （袖珍）手电筒 25.4.1.26
point efficiency 点效率 1
pointer of balance disk 平衡盘指针 16.4.18.18
point of clearance 间隙点 8.8.5.14
point of entrance 入口点 16.4.3.4
polar crane 极性起重机 25.2.2.18
polarizer electrode 极化器电极 3.4.21.4
pole face 极面 16.5.4.2
polish(ed) rod 抛光杆 2.1.2.27
polishing 精制 3.1.32.5
pollutant analyzer 污染物分析器 3.4.19.9
pollutant monitoring system 污染物监控系统 3.4.19
polluted water 污水 2.2.74.13
polyamide cone 聚酰胺锥形筒 2.3.8.51
polyamide fibre 聚酰胺纤维 2.3.8
polyamide staple 聚酰胺(短)纤维 2.3.8.61
polyamide thread 聚酰胺丝线 2.3.8.46
polybenzine 叠合汽油 2.1.1.71
polyethylene 聚乙烯 22.6.2.③;2.2.67.24
poly-fluid theory 多流体理论 1
polyforming 叠合重整 2.1.1.70
polymer 多聚体 3.1.12.8
polymer and sludge mixing tank 聚合物和污泥搅拌罐 3.3.19.5
polymer dissolving tank 聚合物溶解槽 3.3.19.7
polymer feed pump 聚合物进料泵 3.3.19.4
polymerization in the autoclave 在高压釜内聚合 2.3.8.33

polymer powder 聚合物粉末 3.3.19.2
polymerization under vacuum 真空下聚合 2.3.8.54
polypropylene 聚丙烯 2.26.7.25;22.6.2.④
polyreaction 叠合 2.2.78.10
polytetrafluoroethylene seal ring 聚四氟乙烯密封环 19.4.1.20
poor-flowing 流动不良 23.2.2.1
poorly designed 拙劣设计 23.2.2.(a)
popular method 普遍(使用)方法 12.8.1
porcelain 陶瓷筒 14.4.3.20
"porcupine"medium shaft "porcupine"中间轴式 4.3.3.(c)
porous backup disk 多孔状支撑片 15.2.1.11
porous-feed distributor tube 多孔状进料分配管 15.2.1.13
porous matrix 多孔基质 15.3.1.7
porous medium 多孔介质 23.1.2.28
porous piston 多孔活塞 7.2.14.3
porous sheet 多孔片 15.2.2.6
porous sintered stainless plate 烧结的多孔不锈钢板 4.3.3.27
porous support 多孔支承板 6.3.3.9
porous wall 多孔壁 15.4.1.4
port for outgoing pipe 引出管孔 21.3.1.9
portion of a RECTISOL plant designed to process syngas prior to ammonia synthesis and subsequent conversion of ammonia and CO_2 to urea 低温甲醇洗净化技术(RECTISOL)中,还有氨合成以及氨与二氧化碳制成尿素的工艺(未展示) 2.2.28
position retaining handle 固位手柄 23.3.8.9
positive displacement 容积式(正位移) 17.1.1.2
positive displacement compressor 容积式压缩机 17.2.11
positive displacement oil-sealed liquor pump 油封式正位移式打液泵 5.5.3.2
positive displacement pump 容积泵,正位移泵 16.1.2.2.1
positive drive 正位移驱动装置 23.3.2.1
positive-pressure conveying 正压输送 23.1.5
positive pressure system 正压输送系统 23.1.8
post—aeration 后曝气 3.1.37.13
post-heating 后加热段 24.1.12.3
post ready for pressing 准备压制的位 2.3.11.50
potable water production from a river 河水制备饮用水流程 2.4.4

potable water to disinfection and distribution 饮用水去消毒及配送 2.4.4.7
potash salt bed 钾盐床 25.2.4.9
potash salt seam 钾盐层 25.2.4.9
potential transformer 变压器 25.4.9.9
pot with hearth 炉膛底 24.1.11.8
powder 粉料 12.1.5.13
powdered activated carbon activated sludge process (PACT) 粉末活性炭处理活性污泥工艺 3.1.43
powder RDF 垃圾废料粉末 3.3.16.28
powder stock 粉末原料 12.1.5.7
powder technology 粉末技术;粉体工程 1
power 电力,动力 16.1.1.13
power block 动力单元 2.2.109.12
power connection 电源接头 10.3.5.16
power-conversion-system 电力转换系统 25.1.1.18
power generation 发电 2.3.1.24
power line 电源线 25.4.1.23
power plant 动力装置 27 发电站 2.2.108.3
power plant and electric parts 发电厂与电器件 25.4
power-plant fuel 动力装置燃料 2.2.93.6
power supply 电源 3.4.7.1
power turbine adjustable rotor blade 动力涡轮可调转子叶片 25.3.1.23
power turbine radial-axial bearing 动力涡轮机径向止堆轴承 25.3.1.29
power turbine radial bearing 动力涡轮机径向轴承 25.3.1.21
power turbine rotor 动力涡轮机转子 25.3.1.22
power turbine rotor blade 动力涡轮机转子叶片 25.3.1.26
power turbine stator blade 动力涡轮机定子叶片 25.3.1.25
power turbine stator blade carrier 动力涡轮机定子叶片托架 25.3.1.24
power unit 动力装置 23.1.2.13
PP 聚丙烯产品 2.2.67.27
Prandtl number 普朗特数 1
prechlorination 预加氯消毒 2.4.5.3
precious metal catalyst 贵金属催化剂 14.3.4.(c)
precipitator plate cover 除尘器平板盖 10.3.1.4
pre—cleaner cyclone 预净化旋风分离器 2.2.23.6
precoat 预覆盖层 9.1.3.10
precoated 预覆盖 9.1.4

precoat filter medium 预敷层过滤介质 9.3.8.6
precoat filtration system 预涂过滤系统 3.1.9
precoat liquid 预涂助滤液 3.1.9.5
precoat material 预涂助滤剂 3.1.9.3
precoat mix tank 预涂助滤剂混合槽 3.1.9.4
precoat pump 预涂助滤剂泵 3.1.9.1
preconditioning screen 预调节筛 12.5.5.6
precooler 预冷(却)器 13.3.5.3
preform 预成型物 8.1.1
preformed paste 预成型的膏状物 8.1.1.15
pregnant strip solution 富集萃取液 6.1.2.4
preheat 预热 2.2.79.8;2.2.90.4
preheat burner 预热燃烧器 3.3.2.7
preheated gas 已预热的气体 3.4.5.1
preheated shale 已预热油页岩 2.2.5.5
preheated vapor 预热后的蒸气 3.1.21.21
preheater 预热器 2.2.54.3;2.2.92.23;2.3.17.6
preheating 预热段 25.1.12.2
preheating compartment 预热室 14.2.5.8
preheating compartment No1 第1预热室 14.2.5.8
preheating tube 预热管 6.3.7.12
preliminary physical unit operations 初步物理单元操作 3.2.1.6
preliminary stretching 预拉伸 2.3.8.45
preliminary treatment 初步处理 3.1.37.2
premasher 初捣 2.3.12.42
premix burner 预混燃烧器 24.2.7
preparation 制备 2.2.2.2
preparation of malt 制麦芽 2.3.12.1-41
preprocessing 预处理 3.3.1.2
per-reforming 预重整 2.2.45.4
presorter 预选料器 2.3.10.21
press 压制品 8.1.1.13
press-and-blow process 压-吹法 2.3.9.30
press filter 压滤机 3.1.2.18
press fit 压入配合 8.8.5.26
pressing 压制 2.3.9.33
pressing device 压紧装置 17.2.11.9
press mould 压制模 2.3.9.34
press ring 压圈 5.2.4.10
press rolls 压榨滚轮 2.3.11.37
pressure 压力 20.2.3.2
pressure-adjusting valve 压力调节阀 15.7.2.5
pressure control pump 压力控制泵 3.3.22.9
pressure drain plug 加压排液螺塞 18.1.4.12
pressure drain tube 加压排液管 18.1.4.7
pressure drop across air lift line 空气提升压力损失 14.3.3.(c)
pressure-enthalpy diagram 压焓图 1
pressure equalizing springs 压力平衡弹簧 11.1.2.4
pressure filter 加压过滤器 3.1.2.16
pressure filter 压力式过滤机 2.2.92.16
pressure gas 加压煤气 2.3.14.28
pressure ga(u)ge 压力表 16.5.1.2
pressure ga(u)ge connection 压力表接管头 6.1.8.2
pressure gradient 压力降 14.1.2
pressure hemispherical head 受压半球形封头 21.1.1.39
pressure leaf filter 加压叶滤机 9.2.1
pressure-let-down 减压 2.2.90.9
pressure of liquid 液体压力 20.1.2.23
pressure regulating valve 压力调节阀 6.2.1.2
pressure-relief valve 泄压阀 2.1.7.17 减压阀 8.2.2.12
pressure shell 受压外壳 2.2.14.4
pressure swing adsorption 变压吸附 2.2.43.7
pressure system 压送系统 23.1.1.(b)
pressure tap 压力计接口 3.3.27.11
pressure-tap nozzle 测压管口 14.1.9.6
pressure testing 压力试验 1.1.2.15
pressure test pump 试压泵 16.1.2.3.49
pressure-tight cover 压力密封盖 4.3.4.3
pressure transmitter 压力变送器 3.4.18.6
pressure type waterproof lubricating device 压力式防水注油器 15.6.4.12
pressure-vacuum 压力-真空系统 23.1.2.(c)
pressure-vacuum unloading and transfer 压力-真空卸料与输送系统 23.1.2.(d)
pressure vessel 压力容器 24.2.10.3 受压容器 21.1.2.10
pressure vessel code 压力容器规范 21.1.2.21
pressure water inlet 压力水进口 2.2.26.6
pressure water outlet 压力水出口 2.2.26.4
pressurization 增压 3.1.14
pressurization system 加压系统 3.1.15.3
pressurized 加压下 8.9.6.7
pressurized adsorber vessel 加压吸附器 6.3.9
pressurized combustor 加压燃烧室 25.5.2.3
pressurized feed system for the high tempera-

ture winkler gasifier 用于高度温克勒气化炉的加压进料系统 2.2.20
pressurized fluid bed combustion 增压流化床燃烧 2.2.116.10
pressurized-fluidized-bed-combustion 加压流化床燃烧 25.5.2
pressurized fluidized-bed combustor 加压流化床燃烧室 24.2.10
pressurized-water reactor 压水反应堆 25.2.2
pressurized-water system 压水系统 25.2.2.20
pressurizing air delivery 压缩空气出口 17.2.9.21
press zone 压榨区 9.6.2.3
pretreater 预热器 2.2.106.21
pre-treatment (aeration, etc.) 预处理（曝气等） 3.3.28.6
pretreatment facilities 预处理设施 3.3.20.3
pretreatment to prevent caking 防止结块处理 2.2.101.6
preweigh scale 预称量秤 23.4.3.12
pre-wet screen 预湿筛 12.5.1.2
prill bucket 造粒筒 13.1.2.15
prilling tower 造粒塔 13.1.1.4
prilling tower for ammonium nitrate 硝酸铵造粒塔 13.1.1
prill spray assembly 造粒喷头组件 13.1.2
prill tower 造粒塔 2.2.92.20
primary air 一次空气 25.2.3.2
primary air and coal 一次空气与煤 24.2.2.2
primary air inlet 主空气入口 14.1.9.8
primary and secondary sludge 一次和二次污泥 3.1.26(a).1
primary augmentor 一次扩压器 18.3.5.10
primary circuit[loop] 一次电路[回路] 25.2.2.22
primary circulating pump 主循环泵 16.1.2.3.40
primary clarification 一次澄清 3.1.37.3
primary clarifier 一次沉降池 3.1.2.6
一次澄清池 3.2.10.3
primary column 主塔 2.2.28.13
primary combustion chamber 主澄清器 3.1.41.3
主要燃烧室 24.1.4.3
primary condenser 主冷凝器 20.1.10.5
primary coolant pump 一次冷却剂泵 25.2.2.5
primary cyclone 第一级旋风分离器 23.1.13.8

primary discharge 主要排料口 11.2.7.7
primary discharge baffle 一级排液挡板 10.1.4.11
primary discharge lock 第一级卸料阀 23.1.13.7
primary dust hopper 一级集尘箱[斗] 10.1.3.7
primary effluent 一次排放液 3.2.10.11
一次污水 3.1.34.7
primary effluent feed 一次污水进 3.1.44.1
primary effluent with recycles 多次循环的一次污水 3.1.35.2
primary ejector 主喷射器 20.1.10.3
primary entrainment separator 预雾沫分离器 10.5.4.22
primary extraction-solvent 主级萃取溶剂 6.1.2.6
primary froth 一次泡沫 2.1.9.14
primary furnace slag-tap hole 主炉膛出渣孔 24.2.3.6
primary gas cooler 主煤气冷却器 2.2.24.11
primary magnet pole 主要磁极 12.6.11.3
primary reaction zone 一次反应区 3.2.6.16
primary reformer 一段转化炉 2.3.2.8
primary reforming 一段转化 3.3.3.7
primary separation 一次分离 2.1.9.1
primary separation tank 一次分离罐 2.1.9.9
primary settler 主要沉降池 3.1.31.3
一次沉降池 3.1.34.5
primary settling thank 一次沉降池 3.1.27.6
一次沉降池 3.1.25.4
primary shaper 一次整形机 13.2.4.23
primary sludge 一次污泥 3.1.25.11;3.1.26.(b).1;3.1.27.8;3.1.34.6;3.2.10.9
primary sludge recycle 一次污泥循环 3.2.10.15
primary sodium pump 一次钠泵 25.2.1.5
primary thickened sludge 一次浓缩污泥 3.1.26.(b).4
primary whirling vane 一次旋流叶片 24.2.4.6
primary winding 初级绕阻（高压绕阻） 25.4.3.5
prime steam line 主蒸汽管道 25.2.2.11
priming cup 灌液漏斗 16.4.19.21
priming pump 灌液泵 16.1.2.7
principle of entropy increase 熵增原理 1
printer recessed in panel 遥控板内打印机 ·23.4.1.15

probe 探头 3.5.11.3
probe-pitot tube 皮托管探测器 3.4.7.18
process 工艺过程 19.3.1.19
　加工 2.2.105.7
　过程 20.1.9.5
process air 工艺空气 3.1.41.10;3.2.13.8
　过程空气 6.2.8.7
process bins 工艺过程料仓 23.1.6.11
　生产过程料仓 23.1.8.23
process cooling towers 工艺冷却塔 2.3.6.17
process equipment engineering 化工机械工程 1
process for a demonstration project of polyolefin product from coal 煤制聚烯烃示范项目工艺流程 2.2.67
process flow for Davy company improved low pressure methanol synthesis Davy 公司改进低压甲醇合成工艺流程 2.2.45
process flow for ethanol synthesis from syngas 合成气制乙醇工艺流程 2.2.41
process fluid inlet and outlet 工艺流体进出口 24.1.3.2
process-fluid inlet nozzle 过程流体进口接管 4.5.1.12
process-fluid outlet nozzle 过程流体出口接管 4.5.1.13
process gas （工艺）过程气体 19.3.1.12
　工业废气 2.1.9.42
process gas(countercurrent) flow 加工气（逆流）流 15.4.1.2
process gas flow 加工气流 15.4.1.2
process gas inlet 过程气进口 4.4.2.1
process gas outlet 过程气出口 4.4.2.7
process gas pump 加工气体泵 15.4.1.6
process heat 过程加热 24.1.1.2
process of continuous production of ethyl acetate 连续法生产乙酸乙醇工艺流程 2.2.57
process of ethanol production from hydrogenation via ethyl acetate 经乙酸乙酯加氢制备乙醇的工艺流程 2.2.59
process of intermittent production of butyl acetate 间歇法生产乙酸丁酯工艺流程 2.2.60
process of methane production from coal via syngas 煤经合成制甲烷工艺流程 2.2.36
process machine 工艺过程机械 23.1.1.16
process of syngas conversion unit 合成气变换单元工艺流程 2.2.34
process optimization 过程优化 1
process pump 化工工艺用泵;流程泵 16.1.2.3.36
process schematic for Coxsackie sewage treatment plant, New York 纽约 Coxsackie 下水道污水处理厂工艺简图 3.1.34
process simulation 过程模拟 1
process solvent 过程用溶剂 2.2.95.16
process steam 工艺（过程）蒸汽 25.3.1.11
process stream in 过程流体加入 4.5.2.5
process stream out 过程流体出口 4.5.2.6
process system engineering 过程系统工程,化工系统工程 1
process vessel 工艺容器 21.2
process water 工业用水 2.2.107.19
　工艺用水 2.2.51.11
produce outlet 产品出口 7.2.1.24
product 成品 12.1.1.5
　产品 2.1.9.29;2.2.71.14;2.3.17.15
product acetic acid 成品乙酸 2.2.55.14
product coke 焦炭产品 2.1.9.29
product collection 产品收集管 6.3.8.7
product collector 成品收集器 8.6.1.5
product cooler 产品冷却器 13.3.5.6
product crystal 产品晶体 7.2.1.7
product discharge 产品排出 7.2.5.12
product discharge valve 产品出口阀 8.9.5.12
product distillation 产品精馏 2.2.62.13
product feed 物料进口 8.5.6.17
product flaking belt 片状产品输送带 2.2.92.22
product flowmeter 产品流量计 5.5.3.12
product fractionation 产物分馏 2.2.95.7
product gas 产[成]品气体 2.2.99.8
　产品煤气 2.2.85.9
　成品气 2.2.21.6
product gasoline and wax 产品汽油和蜡 2.2.78.11
product inlet flange 物料进口法兰 8.9.5.4
product water 成品水 2.4.2.5　2.4.3.10
production of ammonium sulphate 硫铵产出 2.3.14.35
production of clean fuels 洁净燃料制造 2.2.9
production of clean fuels from coal 从煤制造洁净燃料 2.2.9
production of crude benzol 粗苯产出 2.3.10.41
production of glass fibre 玻璃纤维的生产 2.3.9.48-55
production of sulphuric acid 硫酸产出 2.3.14.34
production well 产出井 2.2.85.21

product oil 产品油 2.2.82.14
product outlet 产[成]品出口 11.1.3.8
product outlet to exhaust fan 产品出口接排风机 11.2.13.15
product receiver 产品接收器 3.1.6.16
product refinement 产品精制 2.2.89.19
product refinery 产品精制 2.2.72.5
product separation 产品分离 2.2.72.4
product separator 产品分离器 2.3.4.21;2.2.75.6
product size control 产品大小控制器 11.1.3.11
product slurry 煤浆产品 2.2.102.15
product slurry discharge 浆状产品排出 7.2.8.17
products 产品 2.2.3.4
product separation zone 产品分离区 2.2.71.B
products outlet 产品出口 13.5.1.5
products to olefin separation 产品去烯烃分离 2.2.72.12
product storage tank 产品储槽 2.2.57.11
product tank 产品罐 2.3.16.16
product urea 产品尿素 2.2.33.15
product vapor 产品蒸气 5.5.1.15
product water 生成水 3.1.18.8
project evaluation and review technique (PERT) 项目评审技术 1
promoted iron catalyst 激励铁催化剂 14.3.2.(c)
propagas 丙烷 2.1.1.65
propane 丙烷 2.2.71.20;2.2.75.11
propeller 搅拌浆 14.4.1.18
叶轮,螺旋浆 16.4.24.2
propeller 螺旋浆式搅拌器 7.2.8.20
propeller agitator 螺旋浆式搅拌器 7.1.12.7
propeller calandria 螺旋浆排管式 7.1.12.(a)
propeller drive 浆叶驱动器 7.2.5.22
螺旋浆驱动装置 7.2.8.19
propeller key 叶轮键,螺旋浆键 16.4.24.3
propeller or axial-flow impeller, single-suction, open design 螺旋浆或轴流式叶轮,单吸,开式结构 16.4.5.(d)
propeller pump 螺旋浆泵 16.1.2.2.23
proportional mixer 比例混合器 24.2.7
proportioning valve 比例调节阀 23.1.8.16
propylene 丙烯 2.1.1.64;2.2.67.23;2.2.68.14;2.2.71.19;2.2.75.9
propylene fractionator 丙烯精馏塔 2.2.71.3
protecting collar 防护环 16.4.23.4
protection casing 保护层 3.1.24.14
protective inert-gas atmosphere 惰性气体保护 2.3.9.16
protective screen 保护壁 25.2.5.2
protector for water swelling 防水胀保护器 16.4.28.26
protein engineering 蛋白质工程 1
prototype experiment 原型试验 1
PSA(pressure swing adsorption) 变压吸附 2.2.67.14
PTFE coated metal mesh 涂覆聚四氟乙烯的金属筛网 15.5.5.13
PTGL processes 热分解、热气体生成和熔解过程 3.3.1.11
P tube P管 11.5.2.3
P-type ring dryer P型环形干燥器 8.4.6
pulley 转轮 12.6.1.(b)
pulley nut 皮带轮螺母 11.1.5.21
pull-push system 拉-推系统 23.1.1.(d)
pull-switch 拉线开关 25.4.1.16
pull through floating head 牵拉浮头式封头 4.1.4.(r)
pulp-drying machine 纸浆干燥机 2.3.10.61
pulper 碎浆机 2.3.1.80
pulping 制浆 2.3.10
pulp level 浆状物料面 12.6.4.7
pulp pump 纸浆泵 16.1.2.3.29
pulp water pump 纸浆水泵 2.3.10.55
pulsed column 脉冲塔 5.1.2.7
pulsing piston 脉冲活塞 16.2.7.12
pulverized clay 粉碎黏土 6.3.2.10
pulverized clay blower 粉碎黏土吹送机 6.3.2.9
pulverized coal combustion 碎煤燃烧 2.2.115.6
pulverized-coal firing 粉煤燃烧 25.2.2
pulverized material 磨碎物料 11.2.6.8
pulverizer 粉磨机 2.2.31.2
粉碎机 2.2.102.3;2.3.16.9
pulverizer 磨煤机 2.2.35.3
pulverizer discharge tank pump 磨煤机出料槽泵 2.2.35.11
pulverizer rotor 粉碎机转子 11.2.6.7
pulverizing zone 粉碎区 11.5.3.2
pump 泵 16,2.2.57.2;2.2.76.18;2.2.81.9;2.2.115.12;3.4.1.7;3.4.2.4
pumpable 可泵送的 3.3.7.15
pumpable liquid 可泵送液体 3.3.8.1
pumpable waste 可泵送废料 3.3.7.5
pump around 泵循环回流 2.1.4.20
pump bearing frame 泵轴承架 16.4.21.6

pump body 泵体 16.3.6.2
泵壳 16.3.7.2
pump body wearing ring 泵体耐磨环 16.4.2.8
pump bonnet 泵盖 16.4.2.3
pump casing 泵壳 16.3.1.5
泵体 16.3.2.6
pump casing cover 泵盖 16.4.15.2
pump cover 泵盖 16.4.11.2
pump cylinder 泵缸 16.2.5.1
pump discharge 泵出口 16.4.1.20
pumped from drum 从桶泵送出 3.3.7.17
pumped system 泵送系统 3.4.10.(a)
pump flange 泵法兰 18.2.1.2
pump half 泵侧 16.4.19.21
pump half coupling lock nut 泵侧联轴器锁母 16.4.27.23
pumping rod 泵杆 2.1.2.25
pumping unit 抽油装置 2.1.2.22
pump line bearing 泵径向轴承 26.3.5.12
pump-mix mixer-settler 泵混(合)式混合-澄清槽 6.1.11
pump out 泵出 6.1.7.13
pump power 泵动力 25.1.2.8
pump pulse generator 泵送脉冲发生器 6.1.16.1
pump room (水)泵房 2.3.12.11
pump shaft 泵轴 16.3.1.3
pump sink removal 泵排出下沉产品 12.5.4.(a)
pump station 泵站 3.1.41.4
pump suction nozzle 泵吸入口 16.4.1.19
pump up 泵送 25.3.2.1
pump valve 泵阀 3.5.14.24
Purasiv HR adsorber vessel Purasiv HR 吸附塔 6.3.7
pure culture plant for yeast 酵母纯培养设备 2.3.13.6
pure cyclohexanol 纯环己醇 2.3.8.18
pure methanol 纯甲醇 2.3.5.4
purge 吹扫 3.4.14.21
放出 7.1.14.6
清洗气 8.9.2.12
purge fittings & devices 吹扫配件和组件 23.4.1.13
purge gas 排出气 2.2.95.12
弛放气 2.2.44.15;2.2.46.15;2.2.47.13;2.2.48.9;2.2.49.9;2.2.77.6
purge gas and solid 排出气体与固体 10.1.6
purge steam 冲洗蒸汽 14.3.3.25

purge vapor 冲洗蒸气 14.3.3.31
purifax unit 纯化装置 3.3.21.8
purification 净化 2.2.2.7;2.2.78.5
purification treatment of water 水净化处理 2.4
purification unit 提纯装置 2.2.6.11
purification zone 提纯区 7.2.14.7
purified syngas 净化合成气 2.2.28.16;2.2.32.5
purified water to heat exchanger 净化水至换热器 2.2.72.15
pusher centrifuge 活塞推料离心机 9.5.9
pusher control with pusher piston 带推料活塞的推送控制器 9.5.10.8
pusher piston 推料活塞 9.5.10.8
pusher plate 推料板 9.5.10.6
pusher ram 推进溜板 2.3.14.7
pusher shaft 推料轴 9.5.10.7
push out plate 推料板 23.3.7.3
push-pull jacking screws 推挽式起重螺旋 9.6.1.16
pyramid support 锥型支持板 5.3.4.19
pyrex wool 耐热玻璃棉 3.5.14.4
pyritic sulfur 黄铁矿的硫 2.2.2.33
pyrolysis 热解 2.2.2.22
热分解 3.3.15.3
pyrolysis drum 裂解转筒 2.2.14.19
pyrolysis gas 热解气 2.2.99.4
pyrolysis vessel 热解容器 2.2.6.5

Q

Q 进水 3.1.40.1
出水 3.1.40.10
Q_L (latent heat flow) 潜热流 2.4.1.10
Q_S (sensible heat flow) 显热流 2.4.1.7
quantiment 图像分析仪 1
quantity meter [累计]总量表 1
quantity ton/wk 处理量(吨/星期) 3.3.5.2
quantum effect 量子效应 1
quartz 石英 3.4.14.4
quartz jet 石英喷嘴 3.4.21.3
quasilinearization 拟线性化 1
quasi-static process 准静态过程 1
quench 冷激,骤冷 1
quench chamber 急冷室 3.3.8.7;2.2.29.11
quench cooler 急冷室 2.2.13.4
quench elutriator 急冷淘析器 14.3.4.37
quencher 急冷塔 2.3.18.7
quench gas 淬冷气体 15.1.1.11

急冷气　2.2.25.9;2.2.31.6;2.2.37.8;2.2.48.7
quenching car　淬火车　2.3.14.10
quenching tower　淬火塔　2.3.14.11
quenching tower　急冷塔　2.2.69.7;2.2.71.7;2.2.72.11
quenching water　急冷水　2.2.15.5
quench liquid　急冷液　2.2.13.3
quench tank　急冷槽　14.1.11.10
quench tower　骤冷塔　2.2.117.3
quench water　急冷水　2.2.17.2;2.2.22.1;2.2.26.7;14.2.6.20
quench water in　急冷水进入　2.2.16.3
quench-water inlet　急冷水进口　14.1.11.1
quick connect female　快装管接头　3.5.16.31
quick disconnect　快速切断　3.4.7.23
quick release joint　快速松脱接头　17.6.2.13
quiet zone　静止区　6.1.9.13

R

R-12(chlorodifluoromethane)　一氯二氟甲烷　20.2.4
R-13(chlorotrifluoromethane)　一氯三氟甲烷　20.1.7.注1.
R-22(dichlorodifluoromethane)　二氯二氟甲烷　20.1.7.注2.
rabble arm　搅拌耙臂　3.3.9.6
耙臂　3.3.26.5
rabble-arm drive　搅拌耙臂驱动装置　3.3.9.12
rabble arm teeth　耙臂齿　3.3.26.10
rabble teeth　搅拌耙齿　6.3.5.5
R.A damper　回风调节风门　20.5.2.2.7
radial arms　径向臂　11.3.4.6
radial ball bearing　径向球轴承　11.1.5.19
radial bearing　径向轴承　16.4.4.11
radial flights　直立式(径向)抄板　8.8.1.17
radial flow　径流(式)　16.1.1.31
radial flow reactor　径向反应器　1
radial inlet damper　径向进口风门　8.4.6.24
radial launder　径向溜槽　15.6.3.11
radial piston pump　径向活塞泵　16.5.6
radial pump　径流泵　16.1.2.2.13
radial vane　径向叶片　16.4.5.3.2
radial vent　径向通气　19.3.1.6
radiation intensity　辐射强度　1
radiation loss　热辐射损失　18.2.4.6
radiation shield　辐射防护屏　18.2.1.7
radioactive waste　放射性废料的废盐矿的地质构造　25.2.4.1～12
radioactive waste storage in salt mine　放射性废料存放在盐矿中　25.2.4
radiographic exam　射线照相检验　21.1.2.11
radius　弧度　22.5.2.4
radius of cylinder　筒体半径　21.2.5
raffinate　萃余[残]液;抽余液　1;2.2.65.3
raffinate oil　抽余油　2.1.1.50
raffinate phase　萃余相　6.1.1.8
rail car　铁路货车　23.1.2.7
机动有轨车　23.1.7.3
railroad boxcar　火车闷罐车　23.4.2.16
raised face　突面　22.4.3.(b)
rake　耙(子)　9.2.2.10
rake arm　耙臂　15.6.3.23
rake classifier　耙式分级器　12.4.3
rake drive motor　耙子驱动马达　15.6.3.3
random packing
散装填料　5.3.2
random process　随机过程　1
Raoult's law　拉乌尔定律　1
rapper bar　振打棒　10.3.1.14
rapping mechanism　振打机构　10.3.5.15
rapping mechanism(tubes)　振打机构(管)　10.3.5.11
RAS　循环活性污泥　3.1.30.B.5
RAS(return activated sludge)　回流活性污泥　3.2.13.5;3.1.35.1;3.1.37.10;3.1.40.4
Raschig ring　拉西环　1
RAS(100%Q)　回流活性污泥(100%Q)　3.1.33.4
rate meter　流量计　3.4.11.26
rate of return on investment(ROI)　投资收益率　1
rat holes　鼠洞　23.1.3
raw coal　原煤　2.2.92.8
raw coal storage hopper　原煤贮仓　2.2.106.17
raw cyclohexanol　粗环己醇　2.3.8.16
raw gas　粗制煤气　2.2.19.5
未加工[处理]煤气[气体]　2.2.35.5
原料气体　6.3.7.5
raw gasoline　粗汽油　2.1.1.34
粗汽油　2.3.4.22
raw influent　原水　3.2.10.1
raw lignite　未加工褐煤　2.2.107.2
raw lubricating oil　润滑油料　2.1.1.37
raw material　原料　1
raw material inlet　原料入口　14.4.1.7
raw materials for catalytic cracking　催化裂化原料　2.1.1.38

raw oil charge 原料油加入 14.3.4.1
raw paper 粗纸 2.3.11.29
raw septage 原腐化物 3.3.20.1;3.3.22.6
raw shale 未加工页岩 2.2.88.6
raw shale feed 未加工页岩加料 2.2.82.10
raw stuff coal 原料煤 2.2.67.5
raw synthesis gas 粗制合成气 2.2.11.6
raw waste 原废液 3.1.12.4
raw wastewater 原污水 3.1.23.2
 原废水 3.1.28.1;3.1.30.B.1;3.1.45.2
 未经处理的废水(原废水) 3.1.30.A.1
raw water 原料水 6.2.1.1
Raymond flash dryer 雷蒙闪急干燥器 8.4.5.(a)
Raymond flash dryer.(Courtesy of Raymond Division, Combustion Engineering Inc.) 雷蒙德闪蒸干燥机(源自雷蒙德分部,燃烧工程公司) 3.3.24
Raymond high-side mill 高边雷蒙磨碎机 11.2.12
Raymond lmp mill 冲击型雷蒙磨碎机 11.2.8
Raymond mill 雷蒙磨 6.3.2.8
Raymond vertical mill 立式雷蒙磨 11.2.11
RBC(rotating biological contactor) 回转生物接触器 3.2.11.4
PDC(1)(rotating disc contactor) 转盘塔 1
reaction fluid 反应液 2.2.55.5
reaction gas 反应气体 3.4.22.1;2.2.67.17
reaction kettle 反应釜 1
reaction kinetics 反应动力学 1
reaction mechanism 反应机理 1
reaction order 反应级数 1
reaction section 反应段 14.3.3.29
reaction well 反应井 15.6.3.18
reaction zone 反应区 2.2.71.A
reactivation air heater coil 再活化空气加热器盘管 6.2.8.4
reactivation and carbon-handling system 再活化和活性炭处理系统 3.1.17
reactivation carbon storage 再生活性炭储仓 3.1.17.3
reactivation furnace 再生加热炉 3.1.17.6
reactor 反应器 2.1.5.3;2.2.29.4;2.2.52.4; 2.2.54.1;2.2.56.1;2.2.61.2;2.2.63.8;2.2. 65.a;2.2.66.5;2.2.72.9;2.2.74.2;2.2.76. 11;2.3.16.21;2.3.17.4;2.3.18.4;2.3.19.2; 2.4.7.6
 反应堆 25.2.1.1
reactor 反应釜 2.2.55.6

reactor building 反应堆建筑物 25.2.2.29
reactor casing 反应器外壳 3.1.32.23
reactor-clarifier 反应澄清器 15.6.3
reactor-clarifier of the high-rate,solids-contact type 高速率固体接触式反应澄清器 15.6.3
reactor effluent 反应器流出物 2.3.4.10
reactor housing 反应器壳体 2.2.39.3
reactor pressure vessel 反应堆压力容器 25.2.2.2
reactor product splitter 反应产物分离器 2.2.51.10
reactor pump 反应堆泵 16.1.2.3.38
reactors for specific liquid-gas processes 特定气-液过程反应器 14.3.1
reactors with moving bed 移动床反应器 14.3.3
real gas 真实气体 1
rear bearing 后轴承 9.3.3.3
rear casing 后壳体 16.5.5.6
rear cover 后盖 16.4.18.14
rear cross head of plunger 柱塞后十字头 16.2.4.23
rear end head types 后端封头型式 4.1.4
rear head end shell flange 浮动端壳体法兰 4.1.1.11
rear labyrioth seal 后迷宫密封 17.4.2.26
reboiler 再沸器,重沸器 1;2.2.27.10;2.3.19.12
reboiler return 再沸器返回口 5.2.1.8
receiver 集尘器 10.1.4.4
 贮气罐 17.6.2.1
 接收装置 23.1.8
 接收器 25.6.3.7
receiving area 接受区 3.3.16.2
receiving pan 受液盘 5.2.1.17
receptacle 接受器 10.1.2.12
reciprocating 往复式 17.1.1.4
reciprocating compressor 往复式压缩机 17.2
reciprocating-conveyor continuous centrifuge 活塞推料离心机 9.5.7
reciprocating feed tray 往复加料盘 6.3.6.5
reciprocating floor scraper 往复式平台刮刀 13.3.4.3
reciprocating grate 往复炉算 24.1.5
reciprocating piston rod 往复式活塞杆 9.5.6.2
reciprocating pump 往复泵 16.2
reciprocating push discharge 往复推料 9.5.2.8

reciprocating scraper 往复式刮刀 13.3.3.7
reciprocating shaft 往复轴 19.2.1.1
recirculated hot air 再循环热空气 4.5.2.3
recirculated material 循环物料 8.4.5.20
recirculated methanol 甲醇循环 2.2.62.1
recirculated methly formate 甲酸甲酯循环 2.2.62.11
recirculating baffle collector 循环式挡板集尘器 10.4.4
recirculating duct 再循环管道 4.3.4.25
recirculating fan 再循环风机 8.9.2.5
recirculating heat carrier 循环的热载体 3.3.15.17
recirculating pipe 循环管 11.4.1.9
recirculating pump 再循环泵 2.2.105.5
recirculating water 循环水 3.4.1.8
recirculation 循环 3.2.11.12
recirculation drum 循环筒 15.6.3.19
recirculation duct 循环管 8.4.5.13
recirculation fan 循环风机 3.1.21.13
recirculation pipe 再循环管 7.2.1.5
recirculation pump 再循环泵 25.2.2.23
循环泵 3.3.22.1
recirculation water flow paths 再循环水流径 25.2.3.1
reclaim conveyor 再堆放运送器 23.2.5.2
reclaiming 再装料操作 23.2.4
再堆放 23.2.5.4
reclaim valve 回收阀 20.1.8.8
recleaner flotation 再净化器浮选 12.8.2.16
recombinant DNA 重组 DNA 1
reconverted lime 再生石灰 2.3.10.52
record 自动记录 3.4.18.10
recorder 记录器 3.4.22.14
recovered chromic acid 回收的铬酸 6.2.7.1
recovered fines 回收的粉末 2.2.19.8
recovered organic chloride 回收的有机氯化物 3.1.7.5
recovered solute 被回收的溶质 5.4.4.11
recovered solvent 回收的溶剂 6.3.7.10
recovering 回收 2.3.14.37
recovery 回收[率] 1
recovery plant 回收装置 2.3.14.37
recovery quench for oil 油急冷回收 3.3.15.15
rectangular clarifier 矩形澄清器 15.6.2
rectangular concrete structure 长方形混凝土结构 3.2.7
rectangular plate 矩形板 5.2.2.2
rectangular sedimentation tank 矩形沉降槽 3.2.16
rectangular valve tray 条形浮阀塔盘 5.2.10.4
rectification 精馏 1
rectification column 精馏塔 2.2.54.10
rectifying section 提纯段 6.3.11.14
rectisol acid-gas removal 甲醇法脱除酸性气体 2.2.104.3
rectisol process 甲醇法 2.3.1.12
Rectisol wash 低温甲醇洗 2.2.33.7
recycle 再循环 11.6.3.9
循环(物)料 13.3.4.2
recycle 循环 2.2.52.2;2.2.75.4;2.2.106.40;2.2.114.1;3.1.42.2
recycle acid 循环酸 6.1.5.12
recycle brine 循环盐水 2.4.2.8
recycle clarifier 再循环澄清器 9.5.14.(b)
recycle CO_2 循环 CO_2 2.2.106.14;2.2.112.4
recycle compressor 循环压缩机 2.2.25.10;2.2.37.3;2.2.41.2;2.2.42.2;2.2.48.3;2.2.74.1;2.2.79.6
recycle control baffle 控制循环导流板 8.4.1.28
recycle cooling 循环冷却 14.3.1.16
recycle cyclone 循环旋风分离器 2.2.23.3
recycled earthworms, uneaten sludge particles, and (ifused) bulking agent 循环的蚯蚓,吃剩的污泥颗粒和填充剂(如果使用的话) 3.3.28.10
recycled rinse water 再循环淋洗水 12.5.5.24
recycled sludge 循环污泥 3.1.12.7
recycled H_2 循环氢气 2.2.79.5
recycled solids 循环固体 2.2.29.3;2.2.30.5
recycled solvent 循环溶剂 2.2.79.3
recycle EDC 循环 EDC 2.3.18.14
recycle extraction-solvent 循环萃取溶剂 6.1.3.6
recycle fine 循环回的细粉末 13.3.3.8
recycle flow 循环液流 3.1.36.2
recycle-flow 循环流动 3.1.14
recycle gas 循环气 2.2.37.4;2.2.45.2;2.2.48.4;2.2.73.1;2.2.77.5;2.2.84.3
recycle gas inlet 循环气入口 2.2.84.8
recycle gas compressor 循环气压缩机 2.3.18.9
recycle heavy distillate 循环重馏分 2.2.29.30
recycle hydrogen 循环氢气 2.2.94.7
recycle-medium concentrate 再循环的中间富集物 12.5.5.8
recycle pump 循环泵 2.3.4.15

recycle-rinsewater pump 再循环的淋洗水泵 12.5.5.11
recycle slurry 循环煤浆 2.2.95.26
recycle solvent 循环溶剂 2.2.95.15 溶剂循环 3.1.5.10
recycle syngas 循环合成气 2.2.24.2
recycle syngas compressor 循环气压缩机 2.2.44.9
recycle tank 循环槽 3.3.4.11
recycle tube 回流管 2.2.98.7
recycle water 循环水 3.1.14.13
recycling coke quench water 炼焦用循环冷却水 16.4.4
recycling oil 循环油 2.1.10.10
redistributor 再分布器 5.3.1.12
reduce backstreaming 减小回流 18.2.4.4
reduced temperature 对比温度 1
reducer
reducer for conveyer 输送机用减速机 8.3.4.18
reducer for doffer 滚筒减速机 8.3.4.21
reducer with motor 电机减速机 8.8.12.1
reducing coupling 缩径管接头 22.3.2.7
reducing cross on both outlet 出口两端均为异径的四通 22.3.1.15
reducing cross on one outlet 出口一端为异径的四通 22.3.1.14
90° reducing elbow 90°异径弯头 22.3.1.3
reducing gear 减速机 9.5.4.11
reducing lateral on branch 支管为异径的斜三通 22.3.1.10
reducing lateral on one run 主管一端为异径的斜三通 22.3.1.11
reducing street elbow 异径螺纹弯头 22.3.3.9
reducing tee on one run 主管一端为异径的三通 22.3.1.9
reducing tee on outlet 出口为异径的三通 22.3.1.8
reduction drive-gear for filter 过滤机减速驱动装置 9.4.2.1
reduction gear 减速齿轮 2.3.10.69
reed switch 簧片开关 20.1.8.15
reel of coated paper 涂布的卷筒纸 2.3.11.35
re-entrainment 二次夹带 1
reentry opening 重返通道 10.1.2.37
reference cell 参比池 2.3.2.8
refined cellulose 精制纤维素 2.3.10.78
refined methanol 精甲醇 2.2.49.11
refiner 精磨机 2.3.10.84
refinery pump 炼厂泵 16.1.2.3.37

refinery wastewater 炼厂废水 3.1.3
refining bath 精炼池 2.3.9.4
refining tower 精制塔 2.2.58.13
reflotation 再浮选 11.6.2.21
reflux 回流 14.3.4.29
reflux dephlegmator 回流分凝器 2.2.57.5
reflux from condenser 回流液来自冷凝器 5.3.1.20
reflux inlet 回流口 5.2.1.2
reflux ratio 回流比 1
reflux return nozzle 回流管口 14.4.1.1
reflux tank 回流罐 2.2.76.16
reflux tank of concentration tower 提浓塔回流罐 2.2.58.6
reflux tank of esterification tower 酯化塔回流罐 2.2.58.5;2.2.59.9
reflux tank of ethylacetate tower 乙酸乙酯塔回流罐 2.2.59.10
reflux tank of ethanol tank 乙醇塔回流罐 2.2.59.11
reflux tank of refining tower 精制塔回流罐 2.2.58.8
reflux tank of regeneration tower 再生塔回流罐 2.2.59.12
reflux tank of wastewater recovery tower 废水回流塔回流罐 2.2.58.9;2.2.59.13
reformate 重整油 2.1.1.48
reformed gas 重整气 14.2.6.13
reformed gas waste-heat boiler 重整气废热锅炉 4.4.1
reformed natural gas 重整天然气 2.2.102.1
reforming 重燃 2.2.78.12
reforming process 重整工序 2.2.45.5
reformer 重整器 2.2.119.4
refractory 耐火材料 2.2.16.5;24.1.3.7
refractory arch 耐火拱顶 3.3.27.3
refractory brick 耐火砖 24.1.2.13
refractory concrete 耐火混凝土 4.4.2.2
refractory insulating brick 耐火绝热砖 24.1.7.7
refractory lining 耐火材料衬里 2.2.14.10 耐火层 2.2.84.4
refrigerant 制冷剂 2.2.28.12;2.2.37.10; 20.1.8.18
refrigerant inlet 冷冻剂入口 7.2.1.26
refrigerant reducing orifice 制冷剂节流孔板 20.2.3.9
refrigerant return pipe 制冷剂回流管 20.2.3.8
refrigerated dryer 冷冻式干燥机 2.2.53.4
refrigerated storage tank 低温贮罐 21.2.2

refrigerated trap 冷阱 18.5.1.3
refrigerating 冷冻 2.3.3.16
制冷 20
refrigeration cycle 制冷循环 1
refrigeration machine 制冷机 20.2.2.23
refrigeration system 制冷系统 2.3.12.11
制冷系统 20.1.2
refuse 废料 2.2.101.4
垃圾 3.3.1.1
refuse-derived fuels(RDF) 垃圾燃料 3.3.16
refuse discharge 残渣卸出口 12.5.2.6
regenerant 再生剂 6.2.4.3
regenerant inlet 再生剂进口 6.2.3.1
regenerant solution 再生液 6.2.2.(b)
regenerated carbon 再生后的活性炭 3.1.18.2
regenerated catalyst 再生催化剂 14.1.12.18
循环催化剂 2.2.98.11
regenerated catalyst hopper 再生催化剂料斗 14.2.7.3
regenerated gas 再生气体 2.2.76.1
regenerated glycol 再生乙二醇 2.2.59.7
regeneratine turbine pump 再生泵操作 16.4.16
regenerating 再生 6.3.3.5
regenerating catalyst 再生剂 2.2.72.3
regenerating furnace 再生炉 6.3.1.11
regenerating gas 再生气 14.1.12.15 1;6.2.9.15
regeneration air 再生空气 6.2.8.1
regeneration cycle 再生循环 6.2.3.3
regeneration equipment 再生设备 6.3.4
regeneration gas 再生气体 6.3.3.2
regeneration or disposal 再生或处置 3.1.43.9
regeneration process 再生过程 6.2.3.11
regeneration tank 再生罐 6.2.5.8
regeneration tower 再生塔 2.2.59.17
regenerative 再生(泵);漩涡(泵) 16.1.1.45
regenerative-cycle gas turbine 再生回热循环的燃气轮机 25.3.3
regenerative heat exchanger 蓄热式换热器 2.2.113.21
regenerative pump 再生泵 16.1.2.2.20
regenerator 再生器 2.1.5.2;2.2.68.3;2.2.69.2;2.2.71.4;2.2.72.5;2.2.75.2;2.2.76.3
回热器 25.3.2.4
regen gas compressor 再生气体压缩机 6.3.3.6
regen gas cooler 再生气体冷却器 6.3.3.7
regen gas heater 再生气体加热器 6.3.3.10
regen.recycle compressor 再生循环压缩机 2.3.4.11
regen./recycle exchanger 再生/循环换热器 2.3.4.9
regime 工况 14.1.1
regression analysis 回归分析 1
regrind ball mill 再磨碎球磨机 11.6.2.20
regular packing 规则填料 5.3.4.45
regular-solids bed 规则的固体床层 4.3.3.1
reheat 再加热 2.2.115.18
reheat coil 再加热盘管 20.2.2.22
reheater 再加热器 20.2.2.16
reheating 再加热 2.3.9.27
reheating combustion chamber 再热燃烧室 25.3.2.6
reinforced thermosetting resins 增强热固性树脂 22.6.2.⑥
reinforcement pad 加强圈 21.1.1.3
reinforcing ring 补强环 9.5.4.19
增强环 22.5.1.(h)
reject 筛余粗料 11.6.4.13
reject chest 废料池 2.3.10.59
reject converyor 不合格品输送器[机] 23.4.2.8
不合格品输送机 23.4.3.6
reject gas N₂ 废气氮气 2.2.104.14
relative volatility 相对挥发度 1
relay lens 中继透镜 3.5.6.4
released gas 放出气 2.2.43.6
reliability 可靠性 1
relief chamber 节流降压室 16.4.6.14
relief devices 泄压装置 21.1.2.12
relief port 放气口 16.4.26.23
relief spring 保险弹簧 11.1.3.10
relief valve 安全阀,减压阀 17.6.1.5
relief vent 安全放空口 3.3.24.8
remaining acid gas 剩余酸性气体 2.2.89.7
remote-control panel 遥控板 23.4.1.14
remote manual dampers 遥控手动风门 3.3.24.17
removable 可拆的 15.5.1.4
removable baffle 可拆挡板 10.2.1.4
removable cover 可拆盖板 4.1.2.(d)
removable downcomer 可拆降液管 5.2.2.10
removable jet assembly 可拆喷嘴组合件 18.2.4.5
removable tube plugs 可拆管丝堵 4.5.1.9
removable valve plate 提升式(卸料)阀板

9.5.5.22
remove gases 脱气 2.2.90.7
removing heat 移出热量 20.1.2.6
removing residual water 除去残留水分 4.3.3.24
renewable resources 可再生资源 1
renewable seal ring 可更换密封圈 22.1.2.3
repair 维修 21.1.2.13
replaceable liners 可更换衬里 11.2.8.6
repulping box 二次浆化箱 12.6.9.3
reradiating cone 辐射锥 24.1.2.5
Resbon tube 不透性石墨管 5.4.2.8
research and development(R&D) 研究与开发 1
reserve liquid tank 储流罐 2.2.49.6
resetting button 复位[重接]按钮 25.4.1.20
RESID(residuum) 残渣油 2.1.4.19
resid 残液 2.2.5.18
resid hydrocracker reactor 渣油加氢裂化反应器 2.1.10
residual bitumen 剩余沥青 2.1.9.5
residual cake valve 排渣阀 9.2.2.13
residual char 剩余炭 2.2.107.1
residual chlorine 残余氯 3.1.2.22
residual enthalpy 残余焓,剩余焓 1
residual error 残差 1
residual fuel 残渣燃料 2.2.98.29
residual oil 渣油 2.1.1.32
residue 残液;釜液 1
残渣 15.1.1.5
残余物 3.3.16.14
residue discharge 残液排出口 5.5.4.22
residue dryer 滤渣干燥器 2.2.92.15
residue heat exchanges 残液换热器 18.5.1.6
residue line 釜残液管路 5.5.4.16
residue of leached Zechstein 淋溶镁灰残留物,沥滤的苯希斯坦残渣 25.4.6
residue outlet 釜残液出口 5.5.2.15
residue recycling tanks 残液循环槽 18.5.1.8
residues 残渣 3.3.1.7
residue tank 釜残液罐 5.5.4.17
residuum feed 渣油进料 14.3.4.27
resin 树脂 13.2.1.2
resin bed 树脂床 6.2.2.(b)
resin hopper 树脂料斗 6.2.5.6
resin mixing 树脂混合 6.2.9.(e)
resin movement period 树脂移动期 6.2.4.9
resin particles 树脂粒子 6.3.3.3
resin pump 树脂泵 6.2.4.4
resin pump stopped 停树脂泵 6.2.4.7

resin transfer lines 树脂转移路线 6.2.6.7
resin valve 树脂阀 6.2.4.5
resistor 电阻器 25.4.9.6
resolution 分辨率 1
restricting orifice 节流降压装置 16.4.6.22
retainer lug 挡板凸缘 10.4.2.4
retainer plate 挡板 10.4.2.6
retainer plate screw 挡板螺钉 16.4.26.21
retaining grid 固定栅板 10.5.4.15
retaining ring 转轴定位 16.4.19.29
retentate 渗余物 1
截留物 3.1.29.6
retentate out 渗余液流出 15.2.2.3
retention 截留;保留 1
retort 干馏 2.2.82.4
干馏器 2.2.87
retorted shale slurry 干馏过的页岩淤浆 2.2.6.21
retorting 干馏 2.2.85.17
retorting zone 干馏段 2.2.85.14
retort multiple-chamber incinerators 甑式多室焚烧炉 3.3.12
retractable filling lance valve 伸缩式装料长杆阀 23.4.1.19
return 回流 25.6.7.7
返回口 8.8.11.6
return activated sludge 回流活性污泥 3.1.25.9
return air 返回空气(回风) 20.2.2.6
回流空气 3.3.3.7
return biosolids 回流生物固体 3.1.31.5
return bend 回流弯道 17.4.2.16
180°弯头 17.6.2.2
回弯头 4.2.2.2
return channel 回流室 20.2.3.11
returned oil 回油 25.2.4.8
return guide vane 回流导叶 20.2.1.7
return hydrogen 返氢气 2.2.46.16;2.2.47.14
return idlers (回程)下托辊 23.2.1.11
returning pump fluid 泵内回流液 18.2.4.11
return line of cooling water 冷水返回管线 4.6.2.10
return opening of high pressure side 高压侧回流孔 19.3.2.11
return passage 回流器 17.4.2.17
return passage inlet 回流器进口 17.4.1.5
return passage outlet 回流器出口 17.4.1.6
return run 回程 23.3.7.(b)
return sludge 污泥回流 3.1.20.4
回流的污泥 3.1.32.17;3.1.34.10;3.1.45.3;3.1.44.5;3.1.46.6

589

return spring 复位弹簧 16.2.8.5
return water 回水 20.1.10.15
reverse air cleaning 反向空气净化(反吹) 10.4.4.8
reverse extraction 反萃取 1
reverse flow tray U形流塔盘 5.2.10.26
reverse osmosis 反渗透 1
reverse-pulse fabric filter 逆向(反吹)脉冲纤维过滤器 10.2.3
reverse-type Pitot tube 反向皮托管 3.5.11.4
reversible conveyor 可逆输送器 23.4.2.13
reversible reaction 可逆反应 1
reversible shuttle conveyor 振荡输送机 23.2.3.(b)
reversing rotational drive 回转(公转)驱动装置 8.9.5.15
revertive signal panel 反回信号盘 25.4.2.7
revolving cylinder 旋转圆筒体 12.6.10.2
revolving drum 转鼓,滚筒 8.7.1.1
revolving-drum-type dense-media separatory vessel 转筒式稠密介质分离器 12.5.3
revolving scum skimmer 转动除沫器 15.6.1.23
revolving shell 旋转壳体 12.6.1.10
revolving vaporizer body 旋转蒸发器机体 5.5.1.40
revolving whizzer 旋转离心机 11.2.12.6
rewinding 复绕 2.3.8.50
rewind station 回卷台 2.3.11.41
Reynolds number 雷诺数 1
rheological property 流变性质 1
ribbon mixer 螺条混合器 12.1.3.(e)
ribonucleic acid(RNA) 核糖核酸 1
ribose 核糖 1
rich absorbent 富吸收剂 2.2.106.35
rich gas 富(煤)气 2.2.81.4
rich phase 富相 1
rid 肋板 21.3.1.7
riding ring 滚圈 13.3.1.10
Rietz disintegrator Rietz型破碎机 11.2.7
rigid container 刚性容器 3.5.7.25
rigid flights 刚性刮片 15.6.1.24
rigid leak proof container 刚性防漏容器 3.4.16.28
rigid truck loader 刚性滚轴搬运器 23.4.2.14
ring (圆)环 16.4.4.2
ring dam 环形堰板 9.5.11.5
ring duct 环形风管 8.4.5.25
环形管道 8.4.6.10
ringed channel 环形沟 5.3.1.14
ring extruders 圆环挤压机 13.2.1.(b)

ring-section pump 环段泵 16.1.2.2.61
ring-shaped path 环形风管 8.4.5.(c)
ring type joint face 环槽密封面 22.4.3.(e)
ring welded to rotating shell 焊在四转壳体上的环 8.8.5.3
rinse 洗涤 12.5.1.6
冲洗 6.2.6.10
rinse screen 淋洗筛 12.5.5.19
rinse water 冲洗水 6.2.6.8
rinsewater pump 淋洗水泵 12.5.5.11
ripple tray 波纹(穿流)塔盘 5.2.10.23
riser 提升管 14.3.2.20;2.2.24.9
riser pipes 上升管 4.4.2.9
riser reactor 提升管反应器 14.3.4.16
rising knife 升降刮刀 9.5.2.20
RNA(ribonucleic acid) 核糖核酸 1.
rocker arm 摇臂 10.2.1.10
rocker shaft 摇臂轴 10.2.1.11
rock pebble mill 卵石磨机 11.6.3.15
roding hole 磨棒装入口 11.2.4.2
rod mill 棒磨机 11.6.2.14
ROI(rate of return on investment) 投资收益率 1.
rolex-screen 转动筛分机 12.2.5
roll 滚轮 9.3.5.7
roll cutter 滚轮切纸机 2.3.11.42
rolled cloth packing 卷压织物填料 19.2.3.7
roller 滚子 4.2.4.8
辗辊 4.3.4.23
辊子 24.1.9.9
roller-type pump 滚柱式泵 16.1.2.2.55
roll feeder 滚筒加料器 11.2.6.9
rolling drum granulator 滚筒式造粒机 13.3.1
Rol－shift converter Rol-变换炉 2.2.34.19
roof 顶盖 10.2.1.8
room 室内 20.2.2.3
root-mean-square error 均方根误差 1
Roots 罗茨式 17.1.1.10
Roots blower 罗茨鼓风机 17.3.1.(c)
Roots pump 罗茨泵 16.3.7
rotary 回转式 17.1.1.5
rotary airlock 旋转气闸 10.4.1.17
旋转气塞阀 23.1.2.4
气密旋转出料器 23.1.7.8
旋转式空气闭锁器 8.4.1.27
rotary blower 旋转式鼓风机 17.3
rotary compressor 回转压缩机 17.3.1
rotary degasser 回转式脱气器 5.5.3.6
rotary-drum vacuum filter with scraper discharge 带刮刀的转鼓真空过滤机 9.3.10
rotary dryer 回转干燥器 6.3.10.17;8.8

回转烘干机 3.3.25.5
rotary feeder 旋转加料器 11.2.8.2
rotary gas seals 旋转气体密封； 8.8.4.(a)
rotary hearth 旋转炉床 24.1.9.8
rotary-hearth furnace 旋转膛式炉 24.1.9
rotary hose 旋转泥浆管 2.1.3.13
rotary joint 旋转接头 8.8.12.3
rotary joint for steam inlet 蒸汽进口旋转接头 8.8.10.2
rotary kiln 回转窑 3.3.8.4
rotary knife 旋转刮刀 9.5.2.21
rotary lobe pump 旋转凸轮泵 16.3.6
rotary mixer 回转圆筒混合器 12.1.1
rotary moving blade vacuum pump 旋转刮板真空泵 18.1.4
rotary oil sealed mechanical pump 油封式旋转机械真空泵 18.1
rotary pipe skimmer 旋转管式撇油器 15.5.1.6
rotary screen 旋转筛 2.3.10.2
rotary shaft 回转轴 19.1.1.5
rotary shaft set screw 转轴固定螺钉 15.7.6.8
rotary sorter 旋转筛浆机 2.3.10.23
rotary steam joint 旋转蒸汽接头 8.8.7.1
rotary table 转盘 2.1.2.15
rotary tube dryer 旋转管束干燥器 2.2.107.3
rotary type sampling tube 旋转式采样管 15.6.4.1
rotary vacuum drum filter 转筒真空过滤机 9.3
rotary vacuum filter 旋转真空过滤机 3.1.11.6
rotary valve 旋转阀 23.1.1.7
rotary vane feeder 回转叶片加料器 12.1.5.16
rotary vane type pump 旋转式叶片泵 16.1.2.2.11
rotated equilateral triangle pitch 转角正三角形排列 4.1.5.4
rotated square pitch 转角正方形排列 4.1.5.2
rotates clockwise 顺时针转动 12.1.3.8
rotates counterclockwise 反时针方向转动 12.1.3.9
rotating annular chromatograph 回转式环形色谱仪 6.3.8.11
rotating arm 旋转臂 24.1.11.10
rotating arm with ploughs 装有搅拌耙的旋转臂 25.1.11.10

rotating bladed cone 旋转刀片的锥部 2.3.10.75
rotating bowl 转鼓 9.5.11.8
rotating brush 回转刷子 7.1.3.11
rotating carbon ring 碳石墨动环 19.1.7.7
rotating casing 旋转壳体(泵) 16.1.1.53
rotating casing pump 旋转壳体泵 16.5.3
rotating cell 回转隔池 6.1.17.6
rotating cup 旋转杯 3.5.1.4
rotating disc column 转盘塔 5.1.2.10
rotating disc contactor(RDC) 转盘塔 1
rotating dish granulator 转盘造粒机 13.3.3
rotating dish granulators 转盘造粒机 13.3
rotating disk 旋转叶轮 9.4.9.10
rotating-disk(RDC) extractor 转盘萃取塔(RDC) 6.1.14
rotating(double-cone)vacuum dryer 双锥型回转式真空干燥器 8.8.11
rotating drum 转鼓 12.6.4.3
旋转磁鼓 12.6.9.5
转筒 15.6.3.21
rotating drum dryer 滚筒干燥器 8.7.1
rotating element 旋转部件 9.3.5.12
回转件 8.8.5.12
旋转元件 9.4.9
rotating extractor 转盘萃取器 1
rotating filter leaves 旋转滤叶 9.4.9.2
rotating grate 旋转凸形栅 2.2.11.10
旋转凸栅 2.2.11
旋转炉栅 2.2.18.11
rotating labyrinth 旋转的迷宫 19.3.3.(d)
rotating labyrinth type 迷宫旋转型 19.3.3.(c)
rotating metal shell 回转的金属壳体 4.3.2.3
rotating plate 旋转板 9.3.2.24
rotating pump 回转泵 16.3
rotating rake 转动耙 15.5.3.11
rotating scraper arm 旋转刮板臂 3.1.14.2
rotating seal ring 密封动环 19.1.1.3
动环 19.1.3.3
rotating segment feeder 旋转星形加料器 2.2.87.6
rotating shaft 转轴 16.4.23.1
旋转轴 19.4.1.5
rotating shaft seal for liquid metal 液态金属转动轴密封 19.4.1
rotating shaft seal for very high vacuum 超高真空转轴密封 19.4.3
rotating shell 回转壳体 8.8.5.16
旋转壳体 8.8.5.20
rotating shell as indirect heat-transfer equip-

ment 回转壳体用作间接传热设备 4.3.2
rotating skimmer blade 旋转撇油器叶片 3.1.14.9
rotating stream tray 旋流塔盘 5.2.10.21
rotating surface 转动表面 19.3.3.7
rotating tube bundle dryer 旋转管束干燥机 8.8.12
rotation 回转 13.3.3.9
 转向 9.3.10.17
 旋转方向 9.3.9.4
Rotocel extractor 洛特塞萃取机 6.1.17
rotor 转盘 11.4.1.12
rotor assembly 转子组件 17.4.2.22
rotor bearing 转子轴承 5.5.1.16
rotor blade 转子叶片 12.3.2.10
rotor core 转子芯 16.4.28.22
rotor cover 转鼓外罩 6.1.18.14
rotor disk 转盘 6.1.14.5
rotor drive 转子驱动装置 8.9.6.13
rotor guide cone 转子导流锥体 12.3.2.11
rotor housing 转子外壳 16.5.3.7
rotor shaft (旋)转轴 11.2.9.7
rougher flotation 粗分器浮选 12.8.2.15
roughness 粗糙度 1
round-bottom boiling flask 圆底长颈烧瓶 3.5.14.29
round-nose pliers 圆嘴钳 25.4.1.50
round nut 圆螺母 16.4.1.15
RPM(r/min) 16.4.3.16
R tracks R 轨道 6.3.2.1
rubber-lined pump 橡胶衬里泵 16.1.2.5.3
rubber lined steel pipe 衬橡胶钢管 22.6.2.⑬
rubber lined steel shell 衬橡胶钢制壳体 5.4.2.7
rubber pipe 橡胶管 22.6.2.⑮
rubber ring 橡胶圈 17.2.12.10
rubber seal ring 橡胶密封圈 19.4.1.6
rubber support 橡胶垫 12.2.3.18
rubbing contact 接触摩擦 8.8.5.10
rubblizing 粗碎 2.2.85.2
rundown tank 降温罐 6.3.2.32
runner 碾轮 11.4.1.6
runoff 逸出 5.5.1.43
rupture disk 防爆膜 3.1.6.5

S

S(1)(sand product) 砂浆产品 12.4.1
S_8(liquid) 单斜晶硫(液相) 2.1.9.48

saccharification 糖化作用 1
sacker 装袋器 2.3.12.39
saddle 鞍形填料 5.3.1.10
saddle clip 撑棍 9.3.3.37
saddle-point azeotropic mixture 鞍点共沸物 1
saddle support 鞍式支座 21.3.2
safety factor 安全系数 1
safety guard 安全护罩 15.7.3.8
safety seal 安全封 14.1.11.2
safety valve 安全阀 9.2.7.9
salt 盐 25.1.11.3
salt cake 盐饼 24.1.11.7
salt cake storage tank 盐饼贮槽 2.3.10.36
salt level 盐液面 25.6.5.2
salt mine 盐矿 25.2.5
salt outlet 盐出口 7.2.12.13
salt quenching chamber 盐急冷室 3.3.11.13
salt recovery 回收盐 3.3.11.1
salt suspension 盐悬浮液 7.2.12.14
sample 取样 4.11.20
sample bottle 取样瓶 3.5.3.6
sample container 样品存储器 3.5.10.7
sampled data control system 采样控制系统 1
sample in 样品进入 3.1.19.4
sample manifold 取样歧管 3.5.19.1
sample out 取样口 3.1.19.8
sample pump 采样泵 3.5.10.1
samples extraction 样品提取 3.4.7.7
sample valve 取样阀 3.1.19.9
sampling apparatus for CO CO 采样仪 3.5.13
sampling apparatus for fluoride 氟化物采样装置 3.5.16
sampling line 取样管线 3.5.3.2
sampling system for organics and hydrocarbons 有机物与烃类的采样系统 3.5.17
sampling tube 采样管 15.6.4.5
sand 砂 2.1.9.12
 砂子 3.3.23.7
sand-bed filter 砂滤器 1
sand feed 砂子进口 3.3.27.6
sand & char 砂子和木炭 3.3.23.8
sand filter 砂粒过滤器 3.1.1.7
 砂滤器 3.1.30.B.9
sand pump 砂泵 16.1.2.3.56
sands discharge 排砂 12.4.2.7
sand seal 砂封 3.3.26.11
sand table 砂滤器 2.3.11.13
sanitary shield 防尘罩 23.3.8.6
SASOL-l Fischer-Tropsch process SASOL-I 费-托合成过程 2.2.100
saturated liquid curve 饱和液体曲线 20.2.

4.15
saturated liquid line 饱和液体线 20.2.4.22
saturated steam 饱和蒸汽 25.5.3.5
saturated-steam boiler 饱和蒸汽锅炉 25.5.1.(a)
saturated-steam outlet 饱和蒸汽出口 25.5.1.8
saturated vapor curve 饱和蒸汽曲线 20.2.4.12
saturated vapor line 饱和蒸汽线 20.2.4.18
saturation zone 饱和区 3.1.32.13
Sauter mean diameter 索特平均直径；当量比表面直径 1
scale 污垢；结垢 1
scale factor 标度因子 1
scale of production 生产规模 8.1.1（c）
scale up 放大 1
scavenger 清洗物 9.4.4.17
scavenger plate 清洗板 9.4.4.16
scheduling of production 生产排序 1
Schematic diagram of a conventional bioscuber 传统的生物涤气示意图 3.4.2
Schematic diagram of anaerobic digestion system 厌氧消化系统示意图 3.3.17
Schematic of a septage chlorination system (US EPA) 腐化物氯化系统示意图（美国EPA）3.3.22
Schematic diagram of a triple combined cycle including SOFC (solid oxide fuel cell), gas, and steam turbinse. 固体氧化物燃料电池、燃气轮机、蒸汽轮机三结合联合循环示意图 2.2.113.
Schematic diagram of conventional trichkle bed biofilter 传统的滴流床生物过滤器示意图 3.4.1
Schematic diagram of pressure swing adsoption system 变压吸附系统示意图 2.2.53
Schematic diagram of two-stage anaerobic digestion of sludge 污泥的两级厌氧消化示意图 3.3.18
Schematic diagem of the oxyfuel combustion process for CO_2 capture 富氧燃烧工艺中二氧化碳捕集示意 2.2.118
Schematic diagram of trickling filter process 滴流式过滤工艺示意图 3.1.47
Schematic drawing of oil product synthesis 合成油产品路线 2.2.73
Schematic flow diagram of the fluid-bed methanol-to-gasoline process 甲醇制汽油的流化床工艺流程示意图 2.2.75
Schematic of acetic acid synthesis flom low pressure methanol carbony lation 低压甲醇羰基化制乙酸流程简图 2.2.55
Schematic of an ECUST gasifier ECUST 气化炉示意 2.2.17
Schematic of a multiple-effect distillation (MED) plant 多效蒸馏法（海水淡化）厂示意 2.4.3
Schematic of a multistage flash distillation process （海水）多级闪速蒸馏工艺示意 2.4.2
Schematic of a typical GE coal gasification system with quench chamber 典型的带急冷室的 GE 煤气化系统示意 2.2.29
Schematic of the screw press dewatering system(courtesy of Hollin iron works) 螺旋压榨脱水系统示意图 （Hollin 钢铁厂）3.3.19
Schematic of typical GE gasification system with radiant system cooler 典型的带辐射合成气冷却器的 GE 煤气化系统示意 2.2.30
Schematic of a typical GSP gasifier 典型的 GSP 气化炉示意 2.2.15
Schematic of a typical GSP plus gasifier 典型的 GSP+ 气化炉示意 2.2.16
Schematic of a typical Winkler gasifier 典型的温克勒气化炉示意 2.2.19
Schematic of a waste sludge chlorination process(BIF) 废污泥氯化工艺系统简图（BIF 公司）3.3.21
Schematic of coal slurry preparation unit 煤浆制备单元示意 2.2.35
Schematic of methanol to olefins(MTO) process 甲醇制烯烃（MTO）工艺装置简要流程 2.2.72
Schematic of typical Shell gasification process MP medium Pressure 典型的壳牌气化工艺示意（MP,中压）2.2.31
Schmidt number 施密特数 1
Scott Smith molecular still Scott Smith 分子蒸馏釜 5.5.3.14
scraped-surface exchanger 刮面式换热器 4.2.5.14
scraper 刮板,刮刀,刮棒 13.3.5.26 刮料装置 9.5.4.29
scraper adjuster 刮刀调节器 9.3.3.31
scraper bar 刮刀杆 13.3.1.3
scraper bearing 刮刀轴承 9.3.3.33
scraper blade 刮板 8.8.12.15 刮刀（片）9.3.3.30
scraper boss 刮板轮毂 13.3.5.29
scraper plate 刮刀压板 9.3.3.32

scraper tip 刮刀尖 9.3.3.29
screen 挡板 11.1.8.31
筛网 12.2.5.9;2.4.5.1;3.1.28.2;3.1.30.A.
 3;3.1.30.B.3
筛浆机 2.3.10.57
滤网 22.2.1.9
筛子 2.2.30.11
screen analysis 筛析;筛分 1
screen beater 筛网击打器 10.2.1.13
screen channel 滤网通道 3.1.34.2
screen cradle 筛摇架 12.2.3.14
screen deck 筛板 12.2.1.5
screened wellhead 多孔套管 3.1.24.11
screen frame 筛架 12.2.5.7
screen guide 护板定位框 11.1.5.11
screen guide 挡板导架 11.1.8.30
screen holder 滤网架 22.2.1.10
screening 筛分 3.3.16.9
screening machines 筛分机械 12.2
screening of lump coal and culm 块煤和屑煤
 筛分台 2.3.14.14
screening unit 筛选机组 13.2.4.17
screening unit for finished product 成品筛
 13.2.4.24
screen pack 筛组 13.2.1.8
screen plate 筛板 12.2.3.16
screen retaining ring 筛板固定环 9.5.10.5
screen with cloth removed 包有可拆滤布的
 筛网 10.2.1.14
screw 螺杆,螺钉 11.1.5.16
screw base 螺旋灯头 25.4.1.59
screw case 螺杆箱体 11.1.5.6
screw centrifugal 螺杆离心(式) 16.1.1.52
screw conveyer 螺旋输送机 13.5.1.4
screw conveyer[conveyor] 螺旋输送机[器]
 8.3.4.24
screw feeder 螺旋加[送]料器 13.2.4.16
 螺旋给料机 2.2.18.2
screw pump 螺杆泵 16.1.2.2.6
screw though pump 槽式螺旋泵 16.1.2.2.7
screw-type extruder 螺杆式挤压机 13.2.
 1(a)
scroll conveyor 涡旋式输料器 9.5.5.4
scroll discharge 卷轴(螺旋)卸料 9.5.2.6
scroll-type centrifuge 涡旋式离心机 2.1.9.20
scrubbed exhaust 洗涤后排出气 10.5.13.9
scrubbed gases 洗涤后的气体 3.1.22.13
scrubber 涤气塔 3.3.24.24;10.5
 涤气塔 2.2.51.8
 洗涤器 3.3.25.17
 洗气塔 2.2.49.5

scrubber copartment 涤气塔空间 3.4.2.1
scrubber equipped with vertical rotor 装有立
 式转子的涤气器 10.5.2
scrubber equipper 涤气器 10.5.2
scrubber water 涤气器用水 3.1.21.18
scrubbing 洗涤;水洗 1
scrubbing agents 洗涤液 2.3.14.37
scrubbing cooler 涤气冷却器 2.2.12.5
scrubbing liquid 洗涤液 2.2.11.5
scrubbing liquid inlet 洗涤液入口 10.5.15.2
scrubbing oil tank 洗涤油罐 2.3.14.43
scrubbing vanes 洗气叶片 10.5.15.6
scrubbing water inlet 涤气水入口 10.5.1.4
scum baffle 浮渣挡板 15.6.7.13
scum baffle 浮渣挡板 3.2.15.6
scum box and baffle 浮渣室和挡板 15.6.5.5
scum collector 浮渣收集装置 15.6.5.3
scum layer 浮渣层 3.2.6.15;3.3.17.3
scum pipe 浮渣导管 15.6.5.4
scum port 浮渣出口 3.2.6.6
SDO (solvent deoiling) 溶剂脱油 2.1.1.82
SDTO process principle for methanol to low-
 carbon olefins 甲醇制低碳烯烃 SDTO 工
 艺原则流程 2.2.70
seal 密封(件,装置) 14.2.4.8
seal assembly 密封组件 16.3.2.12
seal by special rubber ring 特殊橡胶密封圈
 16.3.6.3
seal cage 密封隔环,液封环,隔离环 16.4.
 27.18
seal cover 密封压盖 9.2.1.4
sealed top jet cap 上喷嘴密封帽 18.2.4.3
seal gas 密封气 2.2.6.2
seal housing 密封盒 19.3.1.14
sealing box 密封盒 19.2.1.2
sealing cavity 密封腔 19.1.3.1
sealing gasket 密封垫片 9.1.3.1
sealing of rotating shafts 旋转轴密封 19
sealing ring 密封环 16.4.1.16
sealing valve 密封阀 23.1.8.14
sealing water pipe 水封管 16.4.19.19
sealing water pipe connection 水封管接头
 16.4.19.20
seal labyrinth 密封梳齿 19.3.2.9
seal leg 密封腿 10.4.1.2
seal oil 密封用油 19.1.7.10
seal oil drain line 密封油排放管 19.1.7.18
seal packing 密封垫 4.2.1.13
seal pipe 密封管 14.1.11.5
seal piping[tubing] 密封液管 16.4.19.16
seal pot 液封槽 18.3.5.14

受液盘 5.2.2.6
seal ring 密封环[圈] 17.2.12.3
seal ring set 密封环组 17.2.9.36
seal ring spacer 密封环隔圈 16.4.25.14
seal spacer 密封环隔圈 16.4.26.5
seal system 密闭系统 2.2.6.19
seal trough 密封槽 25.1.9.4
seal wiper ring 封闭刮油环 19.1.7.2
seat for heating tube support 炉管支承架底座 24.1.1.18
seat of bolt 螺栓座 21.3.1.3
seat ring 阀座圈[环] 22.1.4.15
seawater 海水 2.4.3.7
seawater in 海水进入 2.4.2.7
sea water pump 海水泵 16.1.2.3.13
secondary activated sludge 二次活性污泥 3.1.25.10
secondary air 二次空气 3.3.8.3
secondary air inlet 二次空气入口 14.1.9.7
secondary air inlet pipe 二次空气入口管 12.3.2.8
secondary air port 二次空气通口 3.3.12.5
secondary anoxic zone 第二缺氧区 3.1.35.5
secondary augmentor 二次扩压器 18.3.5.7
secondary burner 二次燃烧器[炉] 3.3.12.15
secondary circuit[loop] 二次电路(回路) 25.2.1.7
secondary clarifieation 二次澄清 3.1.37.5
secondary clarifier 二次沉降池 3.1.2.13
  二次澄清器 3.1.8.8
secondary collector 二级集尘器 8.4.1.5
secondary combustion chamber 二次燃[焚]烧室 3.3.8.6
secondary condenser 二次冷凝器 3.1.6.10
secondary cone crusher 二极圆锥轧碎机 12.8.2.8
secondary cyclone 第二级旋风分离器 23.1.13.2
secondary discharge 二次排料 11.2.7.8
secondary discharge lock 第二级卸料阀 23.1.13.4
secondary discharge valve 后级排气阀 18.1.2.12
secondary drain 二级排液口 10.1.4.12
secondary effluent 二次废水 3.1.8.9
  二次出水 3.1.44.9
secondary ejector 二次喷射器 20.1.10.10
secondary-extract 辅助萃取液 6.1.2.4
secondary froth 二次泡沫 2.1.9.15

secondary heat 第二热源 8.7.2
secondary pump 后级泵 18.1.2.10
secondary reaction 二次反应 1
secondary reaction zone 二次反应区[段] 3.2.6.17
secondary reformer 二段转化炉 2.3.2.9
secondary reforming 二段转化 2.3.3.8
secondary screen 二道筛 2.3.10.58
secondary separation 二次分离 2.1.9.13
secondary separation tank 二次分离罐 2.1.9.16
socondary settling tank 第二沉淀池 3.1.25.7
secondary shaper 二次整形机 13.2.4.25
secondary shredder 二次撕碎机 3.3.16.22
secondary sludge 二次污泥 3.1.26.(b).13
secondary sludge clarifier 二次污泥澄清器 3.1.20.5
secondary sodium pump 二次钠泵 25.2.1.8
secondary-solvent 辅助溶剂 6.1.2.5
secondary steam 二次蒸汽 7.1.3.9
secondary thickened sludge 二次浓缩污泥 3.1.26.(b).14
secondary treatment 二次处理 3.1.27.7
secondary trommel 二次转筒筛 3.3.16.23
secondary vortex breaker 二次涡流破碎器 10.4.3.18
secondary winding 次级绕阻 25.4.3.6
second ejector 二级喷射泵 18.3.5.5
second law of thermodynamics 热力学第二定律 1
second-order phase transition 二级相变 1
second press 第二道压榨 2.3.11.20
second reactor 第二反应器 3.3.22.8
second regeneration zone 第二再生层段 14.3.3.37
second stage 第二级 17.2.4.5
  二段 2.2.21.5
second-stage aerobic(nitrification) 二段有氧处理(硝化) 3.1.40.6
second stage combining throat 第二级混合喉管 18.3.4.12
second stage cylinder (第)二级气缸 17.2.9.14
second stage discharge valve 第二级排气阀 17.2.9.18
second stage impeller 第二级叶轮 16.4.27.7
second stage piston 第二级活塞 17.2.7.15
second stage piston rod 第二级活塞杆 17.2.9.16
second stage snap ring 第二级开口环 16.

4.27.35
second stage steam chest 第二级蒸汽室 18.3.4.8
second stage steam nozzle 第二级蒸汽喷嘴 18.3.4.11
second stage suction head 第二级吸气接头 18.3.4.10
second stage suction valve 第二级吸入阀 17.2.9.15
second synthesis column 第二合成塔 2.2.44.12
second wet felt 第二道湿毛布 2.3.11.18
section 剖面 11.2.10
sectional end elevation 端部剖视 10.3.1.19
sectional tower shell 塔节 5.3.4.33
sectional view 剖视[面]图 10.2.1.(a)
section through 通流截面 17.4.1
section through flight 通过刮板的剖面 4.3.3.5
section through stage inlet 通过级进口的截面 17.4.1.7
section through steam manifold 蒸汽集管箱部面 8.8.6.7
secure burial pit 可靠的深埋坑 3.3.14
security filter 安全过滤器 3.1.28.6
sediment 沉淀物 3.1.4.12
sedimentation 沉降 2.4.4.2；3.2.1.7
sedimentation operations 沉降操作 15.6
sediment ejection ports 沉积物排出口 9.5.13.6
sediment holding space 沉积物聚集空间 9.5.13.5
sediment trap 沉淀物捕集器 3.1.4.13
seed crystal 晶种 1
seed particles 晶种粒子 13.3.4
segmental plate 弓形板 5.2.2.3
segmented 扇形 19.3.1.10
segmented carbon seal 扇形的石墨密封块 19.3.1.7
segregate 离析 23.1.3
segregation 离析 1
分开 3.3.7.7
segregative tray 分块式塔板 5.2.10.28
segregative tray construction 分块式塔盘结构 5.2.7
selective control 选择性控制 1
selectivity coefficient 选择性系数 1
selector 选择器 3.2.2.7
selector activated sludge(SAS)process 选择器活性污泥工艺 3.2.3
SELEXOL process for the selective removal of H_2S and CO_2 选择性脱除硫化氢和二氧化碳的 SELEXOL 工艺流程 2.2.32
self-aligning bearing 自位轴承 15.6.1.29
self-aligning roller bearing 自定心滚柱轴承 11.1.1.6
self-beam tray 自身梁式塔板 5.2.10.29
self-centering flange 自动对中法兰 18.3.1.7
self-cleaning bowl 自动卸料转鼓 9.5.13
self-feeding 自动加料 23.3.7.10
self-induced spray scrubber 自导式喷雾涤气器 10.5.6
self-priming 自吸式 16.1.1.37
self-priming pump 自吸泵 16.1.2.2.58
self-wiping vacuum tight discharge valve 自清洗式真空密封卸料阀 8.8.11.14
semiautomatic palletizing 半自动码垛 23.4.3
semiclosed, internally fired gas-turbine cycle 半闭式内部燃烧型燃气轮机循环 25.3.5
semi-continuous process 半连续过程 1
semi-empirical model 半经验模型 1
semi-ideal solution 半理想溶液 1
semi-open impeller 半开式叶轮 16.1.1.42
semipermeable membrane 半透膜 1
semipositive drive 半正位移驱动装置 23.3.2.2
semiradial reciprocating compressor 扇型往复压缩机 17.2.1.8
semirigid 半刚性 22.6.1.(b)
sensing liquid clarifying discs 感液澄清盘 9.5.13.17
sensing liquid pump 传感液泵 9.5.13.18
sensing pressure 传感压力 19.3.1.18
sensing volume （信号）传感体积 3.4.6
sensing zone disc 传感层圆盘 9.5.13.16
sensitive product 敏感性物料 8.1.1.30
separation 分离 2.2.36.11
separated constituents 分开的组元 6.3.8.6
separated tailing trough 分出的粗粉料槽 12.2.3.19
separate liquefaction product 分离液化产品 2.2.21.14
separate power turbines 功率燃气轮机 25.3.3
separating column 分离塔 2.2.76.17
separating solid 分离固体 12.8.1
separating unit 分离装置 2.2.61.6
separating vessel 分离槽 12.5.5.13
separation 分离 2.2.85.8

separation and reforming of aromatics 芳烃分离与转化 2.1.1.49
separation bin 分离仓 2.2.4.9
separation by incinerator operator 由焚烧炉操作者分离 3.3.7.14
separation-element tube 分离元件管 15.4.3
separation factor 分离因子 1
separation nozzle 分离喷嘴 15.4.2
separation of earthworms from bulking agent(if required) 从填充剂中分离出蚯蚓（如需要） 3.3.28.13
separation of water and 85% for micacid 分离的水和85%的甲酸 2.2.62.14
separation tank 分离罐 2.2.49.7;2.2.50.5
separation vessel 分离器 2.2.64.4
separation zone 分离段 7.1.2.6
separator 分离器[室,机] 2.2.72.6;10.5.13.10.4
separator body chamber 分级器筒体 12.3.2.5
separator cell 分离器元件 3.4.22.12
separator channel 分离器流道 3.1.3.8
separator column 分离塔 2.2.54.7
septage odor control 腐化物除臭 3.3.20
sequencing valve 顺序阀 23.1.2.11
sequential modular approach 序贯模块法 1
sequestration 储存 2.2.3.16
serial correlation 序列关联 1
series grinding inlet 连续研磨入口 11.3.1.3
series operation flowchart 串联操作流程图 18.4.2
serubbing liquor 涤气液 10.5.4.17
service 供应 6.2.5.5
service bin 供油器 2.1.8.3
service building 服务大楼 25.2.2.16
service bunker 煤仓 2.3.14.3
service condition of pumps 泵的使用情况 16.1.2.3
service period 工作期 6.2.9.(a)
servomotor 伺服电动机 9.5.6.1
set screw 止动螺钉 16.4.6.30
  定位螺钉 19.1.1.12
  紧定螺钉 19.1.5.9
setting bolt 固定螺栓 14.4.3.12
setting zone 澄清区 7.2.5.19
settleable solid 沉淀的固体 15.5.4.3
settled catalyst level 沉积催化剂液面 2.1.10.8
settled solid 沉积固体 12.4.2.6
settler 澄清器 6.1.12.5
  沉降池 7.2.9.4
settling 沉降 1;2.4.5.6
settling and demixing 沉降分层 6.1.1.4
settling chamber 沉降室 24.1.4.7
settling the draff 沉淀残渣 2.3.12.50
settling zone 沉降区 7.2.10.7
  澄清区 7.2.1.12
sewage 污水 19.1.5
sewage pump 污水泵 16.1.2.3.18
sewage treatment plant 污水处理厂 3.1.21.4
shaft 矿井,立井 25.2.4.13
shaft cap 轴盖 16.4.28.3
shaft collar 轴迷宫圈 16.4.9.34
shaft column 轴承柱 9.5.17.21
shaft cooling air fan 主轴冷却风机 3.3.26.15
shaft coupling 轴连接件 16.4.23.11
  轴联结器 9.3.3.22
shaft coupling link 轴连接链 9.3.3.40
shaft-cup 轴盖 17.5.1.5
shaft enclosed tube 轴的密封套管 16.4.23.10
shaft enclosing tube 泵轴套管 16.4.24.6
shaft end nut 轴端螺母 17.2.9.31
shaft for hinged lever 铰链轴 9.2.1.23
shaft furnace 竖式炉 3.3.15.(a)
shaft furnace for lime production 竖式石灰窑 24.1.12
shaft guide 轴导套 14.4.3.6
shaft labyrinth 主轴迷宫密封 20.2.1.11
shaft nut 轴(套定位)螺母 16.4.19.6
shaft protecting sleeve 轴套 16.4.20.12
shaft seal 轴封 16.4.1.6
  转轴密封装置 7.1.11.8
shaft sealing 轴封 7.1.2.4
shaft sheave 轴套 19.3.2.5
shaft-sleeve nut 轴套螺母 16.4.11
shaft to be "dimpled" 被打有微坑的轴 19.1.5.6
shaft tube 轴套管 16.4.25.34
shaft tube stabilizer 轴管固定件 16.4.25.18
shaft tube tension bearing 轴套管压力支承块 16.4.25.27
shaft with key 带键的轴 16.4.27.34
shake-flask culture 摇瓶培养 1
shaker 振动器 8.9.6.8
shaking feeder 振动送料机 8.9.3.6
shaking mechanism 抽打机构 10.2.1.25
shale 页岩 3.1.24.6
shale distributor 油页岩分布器 2.2.6.4
shale feed 油页岩加料 2.2.4.5;2.2.5.4
shale oil 页岩油 2.2.87.9

597

shale pillar 油页岩矿柱 2.2.85.4
shale preheating zone 页岩预热段 2.2.82.5
shale retorting zone 页岩干馏段 2.2.85.6
shallow bed 浅床 1
shallow dished head 浅碟形封头 21.4.1.2
shape factor 形状系数 1
share-off-dust channel 粉尘刮沟 10.1.2.21
share-off-reentry opening 所刮下的重返通道 10.1.2.23
sheared media & return biomass 切变的介质和回流生物质 3.1.36.12
shearing pin 剪切销 11.1.7.25
shearing pin busher 剪切销套 11.1.7.24
shear shredder 撕剪机 3.3.16.3
sheet 片状 8.1.1.21
sheet glass production 平板玻璃制造 2.3.9.1～20
sheeting machine 压片机 13.2.4.21
shelf dryer 厢式干燥器 1
shell 外壳 10.1.2.29
    壳体 11.2.1.4
    筒体 11.2.4.8
    机壳 6.1.7.7
shell-and-tube heat enchanger 管壳换热器，列管换热器 4.1
shell－and－tube reactor 列管式反应器 2.2.70.3
shell cover 封头 4.1.1.9
shell cover flange 封头法兰 4.1.1.13
shell liner 筒体衬板 11.2.4.9
shell nozzle 壳程接管(口) 4.2.2.8
shell of convection chamber 对流段外壳 25.1.1.2
shell of radiation chamber 辐射段外壳 25.1.1.20
shell(side)pass 壳程 4.1.1.42
shell thickness 壳壁厚度 21.1.1.41
shell unit 壳牌装置 14.2.6.(e)
shell with diaphragm 内装隔板的筒体 8.8.1
shielded container 屏蔽好的容器 25.2.5.10
shift 变换 2.2.89.8
shift conversion 变换(反应) 2.3.1.9
shift converter 变换炉[器] 2.2.101.10
E13－shift gas water cooling E13－变换气水冷器 2.2.34.12
shift reaction 变换反应 2.2.104.7
shiplap liners 搭叠衬里 11.2.2.3
ship-loading system 轮船装料系统 23.3.9
shock-resisting safety helmet 防震安全帽 25.4.1.48

short arm 短臂 15.6.1.5
short connecting rod 短连杆 16.2.3.38
shortcut method 简捷法 1
short-tube evaporator 短管蒸发器 7.1.3
shower 喷淋管 9.2.5.24
shower header 喷管 9.2.5.17
shower nozzle 喷头 9.2.5.16
shower nozzle thrower 莲蓬头式喷淋器 5.3.4.3
shower pipe 喷淋管 12.2.3.17
shredder 撕碎机 3.3.16.6
    切碎机 3.3.2.4
shredding the cellulose sheets 粉碎纤维素板 2.3.7.6
shrinking core model 缩核模型 1
shroud 箍环 11.2.10.3
    轮盖 16.4.5.2
shrouded turbine impeller 蔽式涡轮搅拌器 14.4.4.4
shroud ring 遮环 6.1.12.11
shut(-)off valve 切断阀 17.6.2.3
    截止阀 2.1.7.5
shutter 空气调节器 24.2.6.7
shutter adjustment 风门调节器 12.3.1.8
shuttle block pump 滑块泵 16.1.2.2.66
side box pullback bolt 侧箱拉回螺栓 9.6.1.5
side-by-side loading 两侧进料 12.1.4.(c)
side cap 侧盖 9.6.1.8
side cap key 侧盖开关 9.6.1.9
side channel 边槽 23.5.4.10
side cooler 中间冷却器 1
side discharge 侧卸料 23.1.2.31
side elevation 侧视图 10.3.1.20
side liners for casing 壳体侧衬板 11.1.7.8
side lug 支耳 9.1.1.5
side pass 管程 1
side plate 侧板 11.1.8.9
side-pull horizontal recirculating conveyor 侧边拖曳水平再循环输送机 23.3.3.(d)
side rail 侧轨，横杆 9.1.2.10
side reaction 副反应 1
side rod 侧杆 16.2.4.24
side-rod guide 侧杆导轨 16.2.4.25
side-rod nut 边杆螺母 16.2.4.22
side roll box 侧辊箱 9.6.1.19
side roll juice collar 侧辊糖汁环 9.6.1.10
side stream with drawl 侧流抽出口 5.2.8.10
sieve 筛板 5.2.6
Siemens gasifier 西门子气化炉 2.2.26
sieve basket 筛孔篮筐 9.5.18.6

sieve diameter 筛孔直径 1
sieve plate 筛孔板 10.5.1.12
sieve tower 筛板塔 3.3.8.10
sieve tray 筛板 5.2.10.5
sieve tray column 筛板塔 5.1.2.3
sight feed valve 可视给油阀 16.4.25.29
sight glass 视镜 3.3.27.14;15.5.2.5
sight hole 视孔 24.1.1.6
sight indicator 观察罩 16.5.2.3
signal light 信号灯 25.4.2.3
signal processing 信号处理 3.4.19.7
signal processor 信号处理系统 3.5.20.6
silencer 消音器 18.3.5.6
silica gel drying tube 硅胶干燥管 3.5.14.11
silica-gel tube 硅胶管 3.5.7.6
silicon iron pump 硅铁泵 16.1.2.5.6
silicon-killed carbon steel 硅镇静碳钢 22.6.3.⑭
silo 料库 23.1.2.3
贮仓 23.1.6.14
筒仓 23.1.8.10;2.2.31.3
SilvaGas© refuse gasification system flow sheet SilvaGas 垃圾气化系统工艺流程 3.3.23
SIM(soluble inorganic matter) 可溶性无机物质 3.2.1.4
similar smaller containers 类似较小容器 23.4.1
simple balancing disk 普通平衡盘 16.4.7
simple distillation 简单蒸馏 1
simple refrigeration cycle 简单制冷循环 20.2.4.13
simple sublimation 简单升华 15.1.1
simple washing 简单洗涤 9.1.1.12
simplex 单缸 16.1.1.7
simplex horizontal reciprocating compressor 单缸卧式往复压缩机 17.2.1.1
simplex plunger pump 单缸柱塞泵 16.2.4
simplified block diagram of a modern Integrated Gasification Combined Cycle(IGCC)plant 现代整体煤气化联合循环(IGCC)工厂的简化框图 2.2.109
simplified diagram of the advanced zero-emission power plant(AZEP) 创新的零排放电厂概念示意 2.2.114
simplified flow sheet 简化流程 11.6.2
simplified IGCC (integrated gasification combined cycle) with pre-combustion CO_2 capture process 带燃烧前二氧化碳捕集的整体煤气化联合循环简化流程 2.2.117
simplified process diagram of an atmospheric CO_2 capture system (packed bed tower option) 大气中二氧化碳捕集系统的简化工艺流程图(填料塔方案) 3.4.6
simplified schematic diagram of chemical looping combustion system 化学循环燃烧系统简图 2.2.80
simply supported belt conveyor 简单支承的皮带输送机 23.2.1.12
simulated annealing 模拟重结晶法 1
simulation 模拟;仿真 1
single acting 单作用 16.1.1.7
single acting cylinder 单作用气缸 17.2.2
single-compartment 单室 14.2.4
single crystal 单晶 1
single diameter 单直径式 8.8.5.(a)
single drum dryer with dip-feed 浸没布料式单滚筒干燥器 8.7.1.(a)
single-gravity two-product system 单纯重力式两产品系统 12.5.4.(a)
single labyrinth 单一迷宫式密封 16.4.8
single pass tray 单流塔盘 5.2.2
single-phase flow 单相流 1
single-reversing knife rising knife 单一刮刀可升降换向离心机 9.5.1.(f)
single rotor 单转子 16.1.1.19
single-screw extruder 单螺杆挤压机 13.2.3
single-sludge biological nutrient removal process 单一污泥生物营养素脱除工艺 3.2.4
single-speed automatic rotary knife 旋转刮刀单速自动离心机 9.5.1.(g)
single-speed automatic traversing knife 单速横移刮刀自动离心机 9.5.1.(h)
single stage 单级 16.1.1.39
single stage centrifugal pump 单级离心泵 16.4.1
single stage centrifugal vapor compressor 单级离心式蒸气压缩机 7.1.14.15
single-stage, double-acting water-cooled compressor 水冷式单级双作用压缩机 17.2.5
single-stage ejector 单级喷射泵 18.3.3
single-stage end-suction volute pump 单级侧面进口蜗壳泵 16.4.3
single-stage external cyclone 单级外旋风分离器 14.1.5.(c)
single-stage fermenter/thickener 一级发酵罐/增稠器 3.2.10.(c)
single-stage internal cyclone 单级内旋风分离器 14.1.5.(a)
single-stage pneumatic-conveyor dryer 单级气流输送干燥器 8.4.1

single stage rotating blade vacuum pump 单级旋片式真空泵 18.1.2
single-stage with conical pusher screen 单级带网板锥形推料机 9.5.6.(a)
single-step-type cylinder 单作用活塞 17.2.4
single suction 单吸 16.1.1.34
sink 废弃 12.5.1.8
下沉物 12.5.2.14
sink-material rinse screen 下沉物料淋洗筛 12.5.5.18
sink product 下沉的产品 12.5.5.25
sintering 烧结 1
siphon 虹吸管 16.5.2
siphon pipe 虹吸管 16.5.2.4
siphon tube 虹吸管 6.2.3.9
Sirocco type D collector Sirocco D 型分离器 10.1.2.(b)
size classification 粒度分级 1
size press 压榨机 2.3.11.24
size reduction 粉碎,磨细 1,粉碎 11
sketch 示意图 13.3.1
skimmed-oil 撇出的油 3.1.14.4
skimmer 脱尘板 10.1.4.8
撇沫板 15.6.5.14
skimmer edge 脱液极边缘 10.1.4.10
skimmer tube 除沫管 5.5.1.52
skimmings hopper 撇油槽 3.1.14.8
skip 翻斗 23.3.6.2
skip bucket 翻斗式抓斗 23.3.6.5
skip-hoist paths 倒卸式起重机运行路径 23.3.6.(d)
skip hoists 倒卸式起重机 23.3.6
skirt 套筒 16.5.6.6
裙罩 23.2.2.7
skirt baffle 裙式挡板 7.2.1.4
skirt support 裙(式支)座 21.3.1
slack wax 蜡膏 2.1.1.81
slag 炉渣 2.2.15.8;2.2.25.7;2.2.29.6;2.2.30.10;2.2.31.11;2.2.109.3;3.3.15.5
slag crusher 碎渣机 2.2.31.9
slagging gasifier 熔渣气化炉 2.2.14
slag hopper 渣斗 2.2.110.9
slag lock-hopper 密封渣斗 2.2.14.15
slag quench-chamber 熔渣急冷室 2.2.14.14
slag quench tank 炉渣猝灭槽 24.3.1.21
slag quench water 炉渣急冷水 2.2.21.3
slag removal 除渣器 2.2.31.10
slag screen 炉渣筛 2.2.29.11
slag tap 渣排口 2.2.14.3
slag/water slurry 水炉渣浆 2.2.21.7
水炉渣混合浆 2.2.22.8

slag water treatment 渣水处理 2.2.31.12
slaker 消化器 3.4.6.10
SLE(solid-liquid equilibrium) 固液平衡 1
sledging crushing 锤碎 11.1.2.3
sleeve 轴套 16.4.28.12
sleeve bearing 轴瓦 16.4.28.17
滑动轴承 17.4.2.10
sleeve construction 轴套结构 16.4.11
sleeved glass yarn 玻璃丝卷 2.3.9.57
sleeve material 衬套材料 19.3.3.(c)
sleeve nut 轴套螺母 16.4.12.22
牵紧螺母 16.4.18.16
sleeves for thermocouples 热电偶套 2.2.39.1
slewing boom 旋转尾端运送器 23.2.5.6
slide gate 滑板闸门 23.3.3.2
slide-key 滑键 16.2.5.9
slide valve 滑阀 16.2.3.54
slide valve ring 滑阀环 18.1.1.4
slide valve spindle 滑阀杆 18.1.1.5
slide valve vacuum pump 滑阀式真空泵 18.1.1
sliding bowl bottom 滑板型转鼓底 9.5.12.9
sliding door 拉门 2.3.12.47
sliding saddle 滑动鞍座 21.3.1.3
sliding vane 滑片 16.3.5.1
sliding vane compressor 滑片式压缩机 17.3.1.(a)
sliding-vane pump 滑片泵 16.3.5
sliding-vane type of rotary compressor 回转式滑片压气机 17.3.3
sliding-vane vacuum pump 滑板式真空泵 16.1.2.42
sliding way 导轨 18.1.1.7
slimes overflow 泥浆溢流 12.4.2.4
slinger 轴承罩 9.5.17.32
slip blade 滑动托板 9.5.4.13
slip blade pin 滑动托板销 9.5.4.14
slip boss 滑动轮毂 9.5.4.17
slip lining 滑动衬板 9.5.4.12
slip-on backing flange 活套靠背法兰 4.1.1.20
slip-on weld hubbed flange 带颈平焊法兰 22.4.2.(b)
slipper 滑块 17.2.15.5
slope 倾斜角 23.2.2.8
sloping bottom tank 斜底槽 12.4.2.12
slop oil 不合格石油产品,废油 2.3.6.1
slop-oil tank 不合格石油产品贮槽 2.3.6.1
slop oil to boilers 废油去锅炉 2.3.6.2
slot 槽沟 2.3.9.9
slot for channel gate 流道口缝隙 3.1.3.4

slot screen 长眼筛板 9.5.10.3	sluice separation 淘析 1
slotted hole 长圆孔 21.3.1.8	slurried cake discharge 稀渣排放 9.2.3.6
slotted sieve tray 导向筛板 5.2.10.9	slurry 浆状[料] 8.1.1.1
slow rinse 慢洗 6.2.1.5	slurry bed reactor 浆态床反应器 2.2.50.4
slow rinse supply 慢洗供应 6.2.1.3	slurry blend 煤浆调和 2.2.92.9
sludge 污泥;淤泥[渣] 1;2.4.5.5;3.1.4.2; 3.3.21.11;3.1.26(a).3	slurry blend tank 煤浆调合罐 2.2.105.4
	slurry coal 煤浆 2.2.90.3
sludge cake 泥饼 3.3.19.11	slurry discharge pump 浆液排出泵 7.2.13.13
sludge cake to disposal 淤泥滤饼去处理 2.3.6.5	slurry feed 加料口 9.2.3.5
	浆态进料 14.2.4
sludge collecting board 集泥板 15.6.5.10	slurry feed gun 淤浆进料枪 14.2.4.4
sludge-collecting hopper 集泥斗卸出口 3.1.3.1	slurry feed pipe 浆液进料管 15.5.3.1
	slurry feed system 煤浆进料系统 2.2.105
sludge collector 污泥沉积室 15.6.4.13	slurry inlet 矿浆入口 12.6.13.1
集泥机 15.6.5	滤浆进口 9.1.2.1
污泥收集器 3.2.15.3	料浆进口 9.2.6.30
sludge conditioning tank 污泥老化罐 3.1.11.5	slurry level 浆面 2.2.98.10
	料浆液面 9.3.10.6
sludge discharge 滤渣卸料 3.1.9.8	浆料液面 9.3.9.9
sludge drawoff 污泥排出管 3.3.18.11	slurry makeup tank 煤浆制备罐 2.2.94.3
sludge feed 污泥进料 3.3.6.6	slurry mixing tank 煤浆混合罐 2.2.102.4
sludge feed pump 污泥进料泵 3.3.19.9;3.3.21.3	slurry-mix tank 煤浆混合槽 2.2.95.2
	slurry out 淤浆出口 10.5.11.6
sludge for agricaltural utillisation 农业可利用的污泥 3.1.26(b).15	slurry outlet 泥浆出口 10.5.15.7
	slurry preheater 煤浆预热器 2.2.94.5
sludge from clarifier 澄清器污泥 3.1.11.3	slurry preheater dissolver 煤浆预热溶解器 2.2.92.10
sludge hearth 泥状沉淀炉膛 3.3.5.15	
sludge hopper 泥浆料[贮]斗 15.6.1.36	slurry preparation 煤浆制备 2.2.2.29
sludge hopper 污泥接受器 3.2.15.4	slurry pump 淤浆泵,泥浆泵 16.1.2.3.55
sludge inlet 污泥进入 3.3.9.10	煤浆泵 2.2.29.6
污泥入口 3.3.17.1;3.3.18.2;3.3.27.4	slurry reactor 浆料反应器 1
sludge outlet 滤渣出口 9.2.7.17	slurry recycle 浆料循环 14.2.6.18
sludge pipe 泥浆管 15.6.1.35	slurry tank 煤浆槽 2.2.29.5;2.2.30.6;2.2.79.4
污泥管 15.6.5.2	
sludge pit 污泥槽池 3.1.4.1	slurry trough 滤浆槽 9.3.2.16
sludge port 污泥出口 9.5.12.19	small ball 小球 11.2.3.5
sludge preparation 污泥制备 3.1.16.6	small end bearing 小头轴承盖 17.2.15.25
sludge pump 污泥泵 16.1.2.3.31	smaller units 小设备 15.5.2
sludge pump suction pipe 污泥泵吸入管 3.1.3.2	small industrial [commercial] incinerator 小型工业焚烧炉 24.1.4
sludge return 污泥回流 3.3.18.3	small muffle furnace 小型隔焰炉 24.1.10
sludge recycle 循环污泥 3.2.10.5	small radial clearanece 小的径向间隙 19.3.1.8
sludge settling tank 污泥沉淀池 3.4.1.5	
sludge storage 污泥储存池 3.3.21.1	small ring 小环管 24.2.9.(b)
sludge storage tank 污泥储存罐 3.3.19.6	small scale 小批量 8.1.1.23
sludge sump 污泥槽坑 15.6.3.29	small spiral large shaft 小螺旋粗轴式 4.3.(b)
sludge sump scraper 泥浆坑刮片 15.6.3.27	
sludge thickening tank 污泥增稠罐 3.1.11.1	small welded fittings 小型焊接管件 21.1.1.58
sludge wastage 废污泥 3.2.6.29	smoothly fluidized bed 平稳流化床 14.1.3.11
slugging 节涌;腾涌 1;14.1.3.21	snap lid 铰链盖 25.4.1.24
sluice opening 斜槽口 12.5.2.2	snap ring 弹性挡环,止动环,开口环 15.2.

socket-welding flange 承插焊接法兰 22.4.2.(e)
SOD hydroformer SOD 临氢重整装置 14.2.6.(f)
sodium aluminate 铝酸钠 6.3.10.1
sodium chloride 氯化钠 2.3.8.13
sodium chromate 铬酸钠 6.2.7.11
sodium silicate 硅酸钠 6.3.10.2
sodium sulfate pellet product 硫酸钠颗粒成[产]品 3.3.4.16
SOD model Ⅱ catalytic cracking unit SOD Ⅱ型催化裂化装置 14.2.6.(a)
SOD model Ⅲ catalytic cracking unit SOD Ⅲ型催化裂化装置 14.2.6.(b)
SOD model Ⅳ catalytic cracking unit SOD Ⅳ型催化裂化装置 14.2.6.(c)
SOFC 固体氧化物燃料电池 2.2.113.3
softened makeup water 补给软化水 2.3.6.16
soft materials 软材料 19.3.3.(c)
soft-packed 软填料 19.2.3
soft-packed stuffing box 软填料填函箱 16.4.20.18
soft packing 填料 19.2.3.5
soft paste 软的膏状物 8.1.1.12
soft wax 软蜡 2.2.74.14
soil 土壤 3.3.12.8
soil conditioner 土壤调理剂 3.3.1.6
soil filter 土壤过滤器 3.3.20.7
solar energy 太阳能 25.1.1.1
solar power 太阳能电力 2.4.1.6
solenoid oiler 圆筒形油杯 16.4.25.28
solenoid valve 电磁阀 10.2.3.7
solenoid vent valve 电磁排气阀 3.4.22.4
solid 10.1.1.3
solid and/or liquid feed 固体和(或)液体进料 11.2.7.1
solid boiler fuel 固体锅炉燃料 2.2.91.11
solid collecting chamber 固体存积腔 9.5.17.27
solid column of material 物料竖筒 23.3.7.5
solid feed 固体加料口 6.1.17.4
solid feed chute 固体进料斜槽 13.3.1.1
solid flow control valve 固体流量控制阀 6.3.11.6
solidification 固化成型 2.2.2.35
solidification of the polyamide 聚酰胺的固化 2.3.8.35
solidification of the polyamide filaments in the cooling tower 聚酰胺丝于冷却塔中凝固 2.3.8.43
solidifys 造粒 13.3.5
solidifying burden 凝固物[粒]料 13.5.2.3
solid level 固体料面计 6.3.11.2
solid level controller 固体料面调节器 6.3.11.5
solid liquid 固体液体 2.2.98.13
solid-liquid equilibrium(SLE) 固液平衡
solid-liquid separation 固液分离 1
solid mixing machines 固体物料混合机械 12.1.3
solid organic rubbish 固体有机垃圾 3.3.5.6
solid outlet 滤渣出口 9.5.4.9
solid pastille 固体颗粒 13.5.3.3
solid-plate conveying surface 固体平板输送表面 4.3.4.28
solid product 固体[态]产品 2.2.102.24
solid 固体 2.2.90.11
solid cake 固体滤渣 9.5.5.20
solid charge opening 固体颗粒加料孔 4.3.1.1
solid collecting housing 固体聚集箱 9.5.10.17
solid collecting housing with volute race discharge 带蜗壳卸料口的固体聚集箱 9.5.10.17
solid-contact type 固体接触式 15.6.3
solid discharge 固体卸料[排出] 14.1.11
solid discharge housing 固体排出壳体 9.5.18.10
solid-discharge port 固体排出口 9.5.5.2
solid discharge provision 固体排料装置 4.3.2.4
solid feed 固体进[加]料 14.1.12.1
solid feeder 固体加料器 13.4.1.14
solid fermentation process 固体发酵工艺 3.2.10
solid-flow channel 固体流通道 23.2.2.3
solid-flow-control devices 固体流量控制装置 14.1.6
solid flow control valve 固体流量调节[控制]阀 6.3.6.4
solid shaft 实心轴 9.4.9.9
solid handling system 固体处理系统 3.1.11
solid holding space 固体持留空间 9.5.14.4
solid in 固体进口 4.3.2.13
solid liquid separation 固液分离 2.2.90.10
solid mixing machine 固体物料混合机械 12.1
solid solvent refined coal 固体溶剂精制煤 2.2.91.2
solid outlet 固体出口 4.3.2.15

solid processing flow diagram considered in the study of waste water treatment plant in Avellino, Italy 意大利 Avellino 废水处理厂研究中的固体处理工艺流程图 3.1.26
solid sampling 固体取样 11.7
solid separation 固相分离 2.2.95.18
solid separation unit 固体分离器 2.2.24.10
solid SRC 固相溶剂精制煤 2.2.91
solid stand pipe 固体煤粒立管 2.2.23.4
solid surge bin 固体颗粒缓冲仓 2.2.4.2
solid-state weight controller 固态电路质量控制器 23.4.1.12
solid valve 颗粒阀 14.1.10.3
solid wall 整体器壁 9.5.14.(a)
solid waste 固体废(弃)物 3.3.1.1
solid waste feed 固体废物进料 3.3.11.4
solid waste treatment 固体废物处理 3.3
solubility 溶解度 1
soluble waste 可溶性废弃物 3.1.18.4
solute 溶质 1
solution 溶液 1
solution flow period 溶液流动期 6.2.4.7
solution polymerization 溶液聚合 1
solution return 溶液回流 7.2.10.11
solution spray 溶液雾滴 13.3.3.6
solution to be treated 待处理溶液 6.2.4.1
solvent 溶剂 2.2.90.16
solvent container 溶剂容器 2.2.64.6
solvent containing chlorine 含氯的溶剂 3.1.21.2
solvent deasphalting 溶剂脱沥青 2.1.1.36
solvent extraction 溶剂萃取 1
solvent hydrogenation 溶剂加氢 2.2.90.17
solvent pumping tank 溶剂泵前贮罐 3.1.5.5
solvent recovery 回收溶剂,溶剂回收 8.9.2
solvent recovery column 溶剂回收塔 3.1.5.12
solvent-recovery distillation 溶剂回收蒸馏 2.2.92.19
solvent recovery still 溶剂回收蒸馏釜 2.2.33.16
solvent refined coal 溶剂精制煤 2.2.2.36
solvent-refined-coal product storage 溶剂精制煤产品仓 2.2.92.21
solvent removal 脱溶剂 2.2.2.32
solvent removing 脱除溶剂 6.1.1.5
solvent reservoir 溶剂储存 3.4.22.11
solvent stripping column 溶剂汽提塔 3.1.5.6
SOM(soluble organic matter) 可溶性有机物质 3.2.1.2
sonic 声响式 3.5.18.(f)
soot 烟灰 2.3.1.18

soot blower 吹灰装置 4.2.4.17
soot/water 煤灰水 2.2.15.7
sorbent 吸收剂 25.5.2.2
sorption section 吸附段 6.3.11.15
Sortex 711M optical separator Sortex 771M 光学分离器 12.7.4
source 来源 3.3.5.1
south temperate zone plate 南温带板 21.2.3.8
spaced idlers 托辊 23.2.1.4
space heat 空间加热 25.1.1.11
space of plates 板间距 5.2.4.14
spacer 定距片[管] 12.3.5.6
spacer column 隔离竖筒 16.4.27.8
spacer ring 定距环 9.2.2.9
spacer sleeve 挡套 16.4.1.12
spacer tube 定距管 5.2.4.4
space time yield(STY) 空时收率 1
space velocity(SV) 空间速率;空速 1
spacing plates 定距板 14.3.1.29
spacing tube 隔离套筒 16.4.1.10
span 刻度间隔 3.4.11.22
spare nozzle 备用管口 14.4.1.2
  备用接管 24.1.1.7
spare nozzle for furnace inspection 炉内检查孔备用 24.1.1.10
spare nozzle for thermometer for gas in furnace 测炉内气温用温度计备用接管 24.1.1.7
spark gap 火花(放电)间隙 25.4.9.12
Sparkler horizontal plate filter Sparkler 水平板式过滤机 9.4.6
special 特种 16.1.1.46
special burner on boiler plant 锅炉厂专用燃烧炉 3.3.5.12
special conveyor dryer 特殊输送式干燥器 8.3.2
special features 附加特征 8.1.1.(d)
special form 特殊形状 8.1.1.34
special high pressure closure 特殊高压密封端盖 4.1.2.(e)
special pump 特种泵 16.5
specific death rate 比死亡速率 1
specific liquid rate 喷淋密度 1
specific speed 比转速 1
specific surface area 比表面积 1
speed change gear 变速箱 9.5.19.3
speed reducer 减速装置 12.1.1.23
  减速机[器] 14.4.1.16
spent acid 废酸 6.1.5.13
spent bed 废料床 25.5.3.11
spent carbon 用过的活性炭 6.3.1.8

603

spent carbon storage　失效活性炭储仓　3.1.17.4
spent carbon thickener　失效活性炭增稠器　3.1.18.9
spent catalyst　用过的催化剂　14.1.12.19
　　待生剂　2.2.72.2
　　废催化剂　2.2.98.5
spent-catalyst hopper　用过的催化剂料斗　14.2.7.13
spent clay to disposal　废黏土排放　6.3.2.27
spent fluid　废流体　11.5.4.6
spent liquor　废液　2.2.9.16
spent scrubber water　涤气器废水　10.5.13.11
spent shale　用过的(油)页岩　2.2.6.1
　　废(油)页岩　2.2.88.2
spent-shale cooling zone　废页岩冷却段　2.2.82.8
spent-shale discharge device　废页岩卸料装置　2.2.84.2
spent-shale solid　废页岩固体颗粒　2.2.82.9
spherical cover　球面封头　21.1.1.28
spherical double-wall tank　双壳球罐　21.2.2.4
spherically dished cover　球面碟形盖　21.4.1.6
spherical seperator　球形分离器　5.3.3.12
spherical storage tank　球形贮罐　21.2.3
sphericity　球形度　1
spider cap　多幅架帽　11.1.4.15
spider suspension　机架悬挂　11.1.4
spider vane　辐射形叶片　25.2.9.7
spindle　主轴　16.4.22.6
　　阀杆　22.2.1.17
spindle sleeve　轴套　16.4.22.5
spin flash dryer　旋转闪蒸干燥器　1
spinneret holes　喷丝孔　2.3.8.42
spinner thrower　漩涡式喷淋器　5.3.4.7
spiral　螺旋式　10.1.1.(e)
spiral baffle　螺旋挡板　15.2.3.3
spiral conveyor adaptations　螺旋输送设备　4.3.3
spiral dryer　螺旋干燥器　8.9.7
spiral flight　螺旋抄板　8.8.1.16
spiral-plate exchanger　螺旋板换热器　4.2.5.2
spiral-shaped rotor　蜗壳形转子　15.7.1.2
spiral-tube exchanger　蛇管式换热器,螺旋管换热器　4.2.5.10
splashing device　溅料轮　8.7.19
split end seal　对开式端面密封　19.1.5.10
split flow　分隔流动　4.1.2.(h)
split pan　分离盘式　23.3.2.8
split pin　开口销　22.1.10.8
split shear ring　对开挡圈;对开环　4.1.1.19
splitter　分流器　1;12.6.1.4
　　分离器　2.2.41.4;2.2.44.11;2.2.46.5;2.2.47.5;2.2.57.9;2.2.60.4;2.2.75.16
splitter valve　多通阀,分流阀　8.4.1.17
spontaneous process　自发过程　1
spool　卷线轴　2.3.9.55
spot exam.of welded joint　焊缝接头局部检验　21.1.2.11
spout　喷发　13.4.1.21
spouted bed　喷动床　1
spouted bed granulators　喷动床造粒机　13.4
spout feeder　进料管　11.2.4.1
spouting gas　喷动用气　13.4.1.19
spray　雾沫　10.5.8.5
spray　雾沫　5.2.6.4
　　喷雾　8.1.1.⑩;10.5.15.5
　　喷散物　8.6.3.7
spray apparatus　喷淋装置　5.3.4.1
spray chamber　喷洒室　4.3.4.10
spray-column　喷洒塔　10.5.11.(a)
spray density　喷淋密度　1
spray dryer　喷雾干燥器　8.6.2
spray drying tower　喷雾干燥塔　8.6.1
sprayed in　喷入　2.1.9.26
sprayer　喷淋[洒]器　10.5.12
spray head　喷头　8.9.2.6
spray header　喷淋水管　4.2.3.2
　　喷淋头　5.4.1.4
spray head　喷头　10.5.12.5
spraying nozzle　雾化喷嘴　8.6.1.2
spray jet　喷头　10.5.1.26
spray manifold　多头喷嘴集管　10.5.4.8
spray nozzle　喷嘴　12.1.3.1
spray or jet　喷雾式喷射　3.4.4.14
spray pan　雾水盆　20.2.2.15
spray pipe　喷水管　15.7.1.7
spray pipe　喷淋管　2.3.10.72
spray pump　喷水用泵　20.2.2.19
sprays　雾滴　13.3.1.4
　　喷头　3.4.1.14
spray tower　喷淋塔　10.5.3
spray-type coil heat exchanger　喷淋式盘管换热器　4.2.3
spray zone　溅雾区　10.5.1.20
spreader　布料器,膜厚控制器　8.7.1.3
spreader coal feeders　布料器煤加料器　14.2.2.1
spreader stokers　布煤机炉排　25.2.4
spring adjustment bolt　弹簧调整螺栓　19.1.

spring adjustment nut 弹簧调整螺母 19.1.5.4
spring adjustor 弹簧调整装置 22.1.9.9
spring assembly 弹簧组件 12.2.1.3
spring-loaded take-up assembly 弹簧负载拉紧装置 23.5.4.5
spring retainer 弹簧(导)座 19.1.7.4
spring safety-relief valve 弹簧安全泄压阀 22.1.8
spring seat 弹簧座 17.2.12.4
spring to maintain tension on rubbing surfaces 保持摩擦力的螺钉 8.8.5.8
spring washer 弹簧垫圈 22.1.10.9
sprinkler pump 喷淋泵 16.1.2.3.20
spud 锥体 24.2.6.4
spud holder 锥固定器 24.2.6.8
square braided packing 方形编织填料 19.2.3.1
square head plug 方头丝堵 22.3.2.19
square pitch 正方形排列 4.1.5.1
squeeze bulb 挤球 3.5.14.23
squeeze pump 挤压泵,胶管泵 16.1.2.2.64
SRC-Ⅰ process SRC-Ⅰ过程 2.2.95
SRC-Ⅱ process SRC-Ⅱ过程 2.2.96
SRC demonstration plant 溶剂精制煤实验厂 2.2.93.4
SRC process 溶剂精制煤工艺 2.2.94
SRC product 溶剂精制煤产品 2.2.93.5
S-shape tray S形塔盘 5.2.10.14
stabiliser 稳定器 2.2.63.13
stability analysis 稳定性分析 1
stabilized primary sludge 稳定后的一次污泥 3.1.26.(b).7
stabilized sludge 稳定后的污泥 3.1.26.(a)5
stable residue and biomass 稳定的残渣与生物质 3.2.1.19
stable state 稳态,稳定状态 1
stack 烟囱 2.2.81.12;2.2.110.22;2.2.116.6;2.2.111.21;3.3.24.10
stack connection 烟道连接 8.8.8.6
stacker 码垛机 23.4.2.17
stack gas 烟道气 25.1.2.3
烟囱气体 25.5.3.17
stacking machines 堆置机 2.3.9.20
stack of beer crates 啤酒箱堆 2.3.13.14
stack wall 烟囱壁 3.5.11.6
stage baffle 级间隔板 3.1.45.9
stage-by-stage method 逐级计算法 1
1st stage coal/water slurry, O_2 一段水煤浆,氧气 2.2.22.2

2nd stage coal/water slurry 二段水煤浆 2.2.22.3
2-stage complete mix/thickener fermenter 二级全混/增稠器发酵罐 3.2.10.(d)
stage for operation 操作台 24.1.1.23
4-stage(nitrogen removal)process 4级(除氮)工艺 3.1.33.11
5-stage (phosphorus and nitrogen removal) process 5级(脱磷和脱氮)过程 3.1.33.8
2-stage Rectisol wash for ammonia/urea plant 用于合成氨/尿素厂的两级低温甲醇洗工艺流程 2.2.33
stainless-clad steel 钢表面覆盖不锈钢 19.4.1.10
stainless steel ducting and PVC laterals 不锈钢主管道和聚氯乙烯支管 3.4.3.6
stainless steel grating 不锈钢格栅 3.4.3.11
stainless steel slope sheets 不锈钢斜板 23.3.8.11
stairway 盘梯 21.2.3.12
stamping 标志 21.1.2.16
standard chemical pump 标准化工泵 16.1.2.3.34
standard dishes head 折边标准碟形封头 21.4.1.3
standard error 标准误差 1
standard heavy media circuit 标准重介质分离系统 12.5.1.19
standard jacketed solid flight 标准夹套固定刮板式 4.3.3.(a)
standard pump 标准泵 16.1.2.1.1
standard state 标准态 1
standard steam-tube dryer 旋转蒸汽接头 8.8.7
standby 闲置中 6.3.2.20
standing wave 驻波 3.4.4.3
standing wave principle 驻波原理 15.5.3
stand leg 机架 9.1.2.12
stand pipe 立管 14.2.6.10;2.2.24.7
竖管 2.1.2.13
star connection[network] Y连接 25.4.3.2
start 开工,开始 14.3.4.40
startup burner 开工燃料烧嘴 14.2.4.7
开工烧嘴 2.2.24.3
startup heater 开工预热器 14.2.7.22
startup preheat burner for hot windbox 开工预热风箱的燃烧器 3.3.27.9
startup/regen.furnace 开工/再生炉 2.3.4.8
start-up stack 开工烟囱 8.5.6.33
star valve 星形阀 14.1.6.(b)
Stassfurt salt 斯塔斯弗盐 25.2.4.10

Stassfurt seam 斯塔斯弗矿层(钾盐层,钾盐床) 25.2.4.9
state point 状态点 6.2.8.12
state variable 状态变量 1
static electrode 静电极 12.7.2.4
static head 静压头 1
static mixer 静态混合器 1
static seal 静密封圈 19.1.1.1
stationary 静止的 19.3.3.4
stationary bladed shell 固定刀片壳 2.3.10.76
stationary comb 固定式梳状物组件 9.4.7.4
stationary die plate 固定模板 15.7.3.7
stationary element 固定件 8.8.5.13
stationary eluent waste collection 固定式洗脱废液收集槽 6.3.8.10
stationary filter plate 静止滤板 9.4.9.3
stationary head 固定封头 4.1.4.(n)
stationary head bonnet 固定端端盖 4.1.1.2
stationary head channel 固定端管箱 4.1.1.1
stationary head end shell flange 固定端壳体法兰 4.1.1.10
stationary header gear 固定的集管装置 4.3.2.11
stationary head flange 固定端法兰 4.1.1.3
stationary head nozzle 管箱接管 4.1.1.5
stationary hood (固定的)机罩 4.3.4.14
stationary Indox V magnet assembly 固定的 Indox V 型永久磁铁组件 12.6.10.5
stationary inlet distributor 固定式进口液分配器 6.3.8.3
stationary intermediate plate 中间固定板 9.3.2.25
stationary magnet 固定磁铁 12.6.1.9
stationary magnet assembly 固定磁性装置 12.6.9.8
stationary-magnets position adjustable by mounting shaft 通过安装的轴使固定的磁性状态可调 12.6.4.2
stationary permanent-magnet assembly 固定永磁体组合体 12.6.4.21
stationary plate 固定板 9.3.2.28
stationary sample systems 静态采样系统 3.5.10
stationary seal ring 密封静环 19.1.1.2
stationary shell 固定壳 12.1.3.(h)
stationary sidewall scraper 固定的侧壁刮刀 13.3.4.4
stationary sleeve 固定轴[衬]套 19.1.7.8
stationary tubesheet 固定端管板 4.1.4.21
statistical model 统计模型 1

stator 定子 4.2.4.15
stator coil 定子线圈 16.4.28.18
stator core 定子芯 16.4.28.21
stator ring 静止环 6.1.14.6
stay 支撑 21.1.1.21
staybolt 拉螺栓 16.4.18.26
支撑螺栓 21.1.1.22
stayed surface 支撑表面 21.1.1.21
steady state 定态;稳态;定常态 1
steam 蒸汽 2.2.5.13;2.2.18.7;2.2.25.8;2.2.28.6;2.2.30.12;2.2.37.5;2.2.44.8;2.2.46.13;2.2.59.15;2.2.62.9;2.2.74.5;2.2.78.2;2.2.109.7;2.2.110.17;2.2.116.7;2.2.118.8;2.3.18.5;2.4.3.1;3.3.23.5;14.2.6.5
水蒸气 2.1.5.5
steam air heater 蒸汽式空气加热器 8.4.1.32
steam + air or O₂ 蒸汽+空气或氧气 2.2.18.4
steam boiler 蒸汽锅炉 2.3.1.16
steam bubble 蒸汽泡 22.2.3.16
steam chest 进汽室 16.2.3.55
蒸汽室 16.2.4.48
steam-chest cover 蒸汽室顶盖 16.2.3.53
steam-chest head 蒸汽室端盖 16.2.4.50
steam coil 蒸汽盘管 10.3.4.3
steam collector 蒸汽包 14.3.2.16
steam-cooled tubes in bed 床内蒸汽冷却管 25.5.2.(a)
steam cylinder 蒸汽缸 16.2.1.1
steam cylinder drain valve 汽缸放液阀 16.2.4.2
steam-cylinder foot 汽缸支腿 16.2.3.6
steam cylinder head 蒸汽缸端盖 16.2.4.1
steam-cylinder with cradle 带托架的汽缸 16.2.3.2
steam distillation 水蒸气蒸馏 1
steam dome 蒸汽室 18.3.1.3
steam drum 汽包 2.2.44.7;2.2.46.7;2.2.47.7;2.2.74.4;14.2.2.11
steam drum blow-off 汽包排污 2.2.44.6
steam dryer 蒸汽干燥器 18.3.5.12
steam ejector 蒸汽喷射泵 18.3.5
steam exhaust 乏汽出口 16.2.1.4
steam for desorption 解吸用蒸汽 6.3.7.7
steam for heating 加热用蒸汽 6.3.7.6
steam for preheating 预热用蒸汽 3.1.23.6
steam generation 发生蒸汽 2.2.45.10
steam-generation system 蒸汽发生系统 25.5
steam generator 蒸汽发生器 3.1.23.8;2.2.

22.4
steam-heated drum 蒸汽加热鼓 8.7.2.4
steam-heated rotor 被蒸汽加热的转子 8.9.6.15
steam-heated tube 蒸汽加热管 8.8.7.5,6
steam heater 蒸汽加热器 2.3.10.38
steam-heating unit 蒸汽加热装置 8.3.2.5
steam in 蒸汽进入 2.4.2.1
steam inlet 蒸汽进[入]口 18.3.4.14
steam inlet for snuffing 熄火蒸汽吹入管 24.1.1.13
steam inlets to jacket 夹套蒸汽进口 8.8.10.15
steam jet 蒸汽喷射 20.1.2.20
steam-jet ejector 蒸汽喷射泵 18.3.1
steam-jet refrigeration cycle 蒸汽喷射制冷循环 20.1.1
steam-jet system 蒸汽喷射系统 18.3.6
steam jet water cooling unit 蒸汽喷射冷却单元 20.1.10.19
steam joint 蒸汽接头 8.8.7.(b)
steam line 蒸汽管线,蒸汽管路 25.2.1.11
steamline 蒸汽管线 25.6.6.1
steam manifold 蒸汽总管;蒸汽分配盘 8.8.7.4
steam neck 蒸汽缩颈管 8.8.6.1
steam nozzle 蒸汽喷嘴 18.3.1.2
steam out 蒸汽出口 10.1.4.14
steam outlet 蒸汽出口 15.5.5.15
steam/oxygen 蒸汽/氧气 2.2.13.9;2.2.19.4
steampipe 蒸汽管 2.3.10.64
steam piston 蒸汽缸活塞 16.2.3.4
steam piston nut 蒸汽活塞螺母 16.2.3.3
steam piston ring 蒸汽活塞环 16.2.3.5
steam piston rod 蒸汽活塞杆 16.2.4.8
steam piston-rod jam nut 蒸汽活塞杆锁紧螺母 16.2.4.11
steam piston-rod nut 蒸汽活塞杆螺母 16.2.4.3
steam piston-rod stuffing box bushing 蒸汽活塞杆填料函衬套 16.2.3.9
steam piston-rod stuffing box gland 蒸汽活塞杆填料函压盖 16.2.3.11
steam piston-rod stuffing box gland lining 蒸汽活塞杆填料函压盖衬套 16.2.3.10
steam piston spool 蒸汽活塞杆接头 16.2.3.12
steam pressure 蒸汽压力 16.2.1.5
steam reformer 蒸汽重整炉 2.2.107.17
steam reforming 蒸汽重整 2.2.3.11;2.2.97.5
steam regulator 蒸汽调节器 7.1.2.18

steam section 蒸汽解吸段 6.3.11.12
steam strainer 滤汽器[网] 18.3.4.2
steam stripping 汽提 1
steam supply 供入蒸汽 14.3.3.2
steam trap 疏水阀 7.1.2.15
steam-tube rotary dryer 蒸汽管式回转干燥器 8.8.6
steam turbine 汽轮机,蒸汽涡轮机 25.2.1.14
steam turbine 汽轮机 2.2.113.9;2.2.114.10;25.4.4
蒸汽轮机 2.2.110.18;2.2.111.21;2.2.118.11
蒸汽透平 25.5.2.8
steam turbines/generator/condenser 蒸汽涡轮机/发电机/冷凝器 2.2.112.7
steam valve 蒸汽阀 20.2.2.24
steam/water out 蒸汽/水出口 4.4.1.4
steel 钢 19.3.3.(a)
steel belt 钢带 13.5.3.2
steel casing 钢质框架 23.3.7.6
steel containment (steel shell) with air extraction vent 带抽气孔的钢(安全)壳 25.2.1
steel door 钢门 8.2.2.6
steel draw-in wire 钢拉线 25.4.1.30
steel drum 钢圆桶 23.4.1
steel grate 钢栅板 11.1.8
steel liner 钢衬里 11.2.1.3
steel pipe 钢管 22.4.4.8
steel shell 钢制炉体 25.1.2.11
钢外壳 3.3.26.3
steel tank 钢槽 15.6.1.20
steel tape or steel wire armour 钢带或钢丝护套 25.4.8.8
steel track 钢轨 23.3.7.8
steeping floor 浸渍层 2.3.12.28
steep liquor 浸液 2.3.12.6
stem 阀杆 22.1.4.2
stem connector 阀杆接头 22.1.9.12
stem gasket 阀杆垫片 22.1.7.5
stem packing 阀杆填料 22.1.3.15
stem sleeve 阀杆衬套 22.1.2.13
stepped labyrinth 阶梯形迷宫 19.3.3.12
step plate 阶梯状平板 13.3.5.31
step-type cylinder 阶梯式气缸 17.2.4.7
sterile operation 无菌操作 1
stiff 刚性 22.6.1.(a)
stiffener plate 加强垫板 21.1.1.14
stiffener ring 加强圈 21.2.2.10
stiffening pad 加强垫板 21.3.1.4
stiffening ring 加强圈 21.1.1.42

stiff equation 刚性方程 1
still 蒸馏柱 2.2.28.21
still bottom 釜底残留物 3.3.5.4
stilling well 沉降槽 3.5.4.7
stirred tank reactor 搅拌釜式反应器 14.3.1.(d)
stirred type crystallizer 搅拌结晶器 1
stirred vessel 搅拌容器 13.2.4.20
stirrer 搅拌器 3.4.2.5;3.4.7.5
stirring 搅拌 1
STM （steam）水蒸气 2.1.4.9
stochastic control 随机控制 1
Stokes diameter 斯托克斯直径 1
stone-dressing device 刻石装置 2.3.10.70
stone media 石子介质 3.1.47.4
stop plate 挡板 11.1.7.28
stop plate 升程限制器 17.2.12.9
storage 贮罐 21.2
　　　　贮存 2.2.106.6
storage bin 料仓 23.3.4
storage building 储料仓 23.2.4.6
storage chamber 贮存室 25.2.4.15
storage compartment 贮料层段 3.1.10.7
storage hopper 贮斗 2.2.92.7
storage of medium-active waste 中放射性废料 25.2.5
storage sump 贮槽 16.5.2.6
storage tank 贮槽 2.3.6.15
　　　　贮罐 3.3.4.1
storage tank for condensate 凝结液贮存槽 2.3.10.17
storage tank for the cooking liquor 蒸煮液贮槽 2.3.10.46
storage tank for the uncleared green liquor 不洁绿液贮槽 2.3.10.41
storage tank for the weak liquor 稀释液贮槽 2.3.10.45
storm-water retention 雨水积池 3.1.1.1
straight-blade 直叶片 17.5.1.8.1
straight chain molecules 直链分子 14.3.4.(c)
straight labyrinth 平直迷宫 19.3.3.(f) 19.3.3.11
straight-line reciprocating compressor 直线型往复压缩机 17.2.1.2
straight lining 直线段衬里 11.1.8.6
straight-lobe 直叶瓣，直凸轮 17.3.3.11
straight-run diesel oil 直馏柴油 2.1.1.9
straight-run heavy diesel fuel 直馏重柴油 2.1.1.10
straight side wiper 平直型塔侧收集器 5.3.4.17
strainer 过滤器 16.4.25.1
　　　　滤网 16.4.28.7
　　　　滤筛 2.3.10.54
　　　　滤器 3.1.19.13
strand 玻璃纤维线 2.3.9.54
Stratco contactor Stratco 混合器 6.1.7
stratification 分层现象 13.3.3(b)
stratification of particle sizes 粗细粒子分层现象 13.3.3(b)
stream 物流 1
stream flow 水流 3.5.5
streamline flow 层流；滞流 1
45° street elbow 45°异径弯头 22.3.2.10
street elbow 带内外螺纹的弯管接头 22.3.2.8
street tee 异径三通 22.3.2.11
stretching 拉伸 2.3.8.55
Stretford desulfurization unit 蒽醌二磺酸钠法脱硫装置 2.2.92.14
Stretford desulfurizer 蒽醌二碘酸钠法脱硫器 2.2.106.34
strike level 抛起界面 7.2.11.8
string-discharge filter 绳索卸料过滤机 9.3.5
string returning to drum 绳索绕面转鼓 9.3.5.6
strip of sheet metal 薄金属片 25.4.1.17
strip of thermoplastic connectors 热塑连接器的插座条 25.4.1.29
stripped slurry 汽提后煤浆 2.2.95.27
stripped water 溶脱过的水 15.5.2.3
　　　　汽提工艺水 3.1.7.7
stripped-water cooler 汽提工艺水冷却器 3.1.7.8
stripper 汽提段 14.3.4.19
　　　　限流环 16.4.15.12
　　　　刮料板 23.3.7.14
stripper 汽提塔[器] 2.2.63.12;2.2.72.14;2.3.15.4;3.1.7.9
　　　　解吸塔 2.2.27.11
stripper elements 溶脱器元件 15.5.2.8
stripping 提馏；解吸；反萃取 1
stripping column 汽[气]提塔 2.3.19.14;5.1.1.8
stripping factor 解吸因子 1
stripping process 汽提过程 3.1.7
stripping section 提馏段 1
stripping steam 汽提蒸汽 14.2.6.12
　　　　解吸用蒸汽 6.3.11.4
stripping tower 提馏塔，汽提塔 5.1.1.4
stroke length 行程长度 17.2.2.11

strong aqua 浓液 20.1.9.9
Strong-Scott flash dryer Strong-Scott Strong-Scott 闪急干燥器 8.4.2
strong solution 浓溶液 20.1.8.17
strong wash liquor 强制洗涤液 9.4.1.12
structural foundation 结构基础 8.8.11.9
structural representation 结构示意图 18.2.2
structural shapes 棒 21.1.1.21
structural support 设备支座 10.2.1.21
structured packing 整装填料;规整填料 1
stud 双头螺栓 19.2.3.6
 螺柱 21.1.1.10
studded connections 螺柱连接 21.1.1.26
stud end 双头螺柱 13.1.2.2
stuff chest 纸料池 2.3.10.85
 贮浆池 2.3.12.1
stuffing box 填料箱[函,盒] 19.2
stuffing box area 密封盒外表面 19.1.5.15
stuffing box bushing 填料函衬套 16.4.25.15
stuffing box cover 填料函盖 16.4.9.32
stuffing box cover ring 填料函盖圈 16.4.9.37
stuffing box gland 填料压盖 16.4.20.25
stuff preparation plant 备料设备 2.3.10.79~86
STY(space time yield) 空时收率 1
styrene 苯乙烯 2.1.1.58
sub-assembly 分部装配 6.2.8.14
subcooled liquid 过冷液体 20.2.4.24
sublimation 升华 15.1
sublimer 升华器 15.1.1.2
submerged combustor 浸没式燃烧炉 3.3.10
submerged-exhaust burner 沉浸式排气燃烧炉 25.1.6
submerged length 沉浸深度 16.5.1.7
submerged-pipe coil exchanger 沉浸式盘管换热器 4.2.5.16
submerged propeller(optional) 液下搅拌叶片(任选) 3.1.45.10
submerged pump 液下泵 16.1.2.2.35
submerged-tube F.C.(forced-circulation)evaporator 浸没管束强制循环蒸发器 7.1.8
submersible motor pump 潜水泵 16.4.28
substitute natural gas, Fischer-Tropsch hydrocarbons 替代天然气,费-托法合成烃类化合物 2.2.3.20
substrate 底物;基质 1
 底生 15.2.3.1
substructure 井架底座 2.1.3.2
substructure columns 构架 4.5.1.8
subsupport angle ring 辅助支承角钢圈 5.2.7.1
subsupport plate ring 辅助支承环板 5.2.7.3
subsystem 子系统 1
sucker rods 抽油杆 2.1.2.25
suction 吸入(口) 16.2.6.1
suctional bell 喇叭形吸入口 16.4.27.2
suction bell 吸水喇叭口 16.4.23.14
suction bell bearing 吸入喇叭口轴承 16.4.23.15
suction bell ring 喇叭形吸入罩 16.4.27.40
suction bowl 吸入筒 16.4.24.1
suction branch 吸入支管 18.3.5.11
suction casing 进口壳体 16.4.19.26
 吸入口机壳 20.2.1.6
suction chamber 吸入腔[室] 16.3.2.2
suction cover 吸入端盖 16.4.1.2
suction diam 吸入口直径 16.4.3.17
suction duct 吸入管 20.2.3.17
suction eye 吸入口 16.4.6.4
suction flange 吸入口法兰 16.4.3.7
suction gas 吸入气体 20.1.3.14
suction gas chamber 气体吸入腔 19.1.7.11
suction head 吸入(口)接头 16.4.25.3
suction head bearing 吸入接头轴承 16.4.25.4
suction head cap 吸入接头盖帽 16.4.26.14
suction head plug 吸入接头堵塞 16.4.26.17
suction head ring 吸入接头套圈 16.4.9.38
suction in center 中央进气 17.4.3.3
suction in end 两端进气 17.4.3.2
suction inlet 吸入口 16.3.2.1
 进气口 18.2.2.1
suction line 吸入管 2.1.7.15
suction manifold 吸入腔 16.2.1.8
suction nozzle 吸气口 18.1.1.8
suction pipe 吸入管 16.2.1.12
suction port 吸入口 16.3.7.1
 进气口 17.2.15.35
suction pressure 吸入压力 16.2.1.9
suction roll 吸水滚轮 2.3.11.16
suction side 吸入端 16.3.6.10
suction side bearing cover 吸入侧轴承压盖 16.4.19.28
suction side sleeve 进口侧轴套 16.4.19.23
suction valve 吸气阀 17.2.2.13
suction-valve plug 吸入阀螺塞 16.2.4.27
suction vane 入口导流叶片 20.13.3
suction vane edge or tip 吸入叶片之前缘 16.4.6.3
suction vane gear 入口叶片齿轮 20.2.1.2
suction vane tip 吸入叶尖 16.4.3.11

sudden contraction 骤缩;突然缩小 1
sudden enlargement 骤扩;突然扩大 1
suitable for recycle/reuse within the plant 适合在厂内循环/再利用 3.1.30.A.9
suitability 适应性 8.1.1(d)
sulfate pulp mill 硫酸盐纸浆厂 2.3.10.1~52
sulfur 硫黄 2.2.5.13;2.2.67.11;2.2.109.11
sulfur absorber 硫吸收塔 2.2.32.2
sulfur absorption 吸收硫 2.2.103.5
sulfur dioxide sample train 二氧化硫采样系统 3.4.14
sulfur guard reactor 硫保护反应器 2.2.106.39
sulfuric acid 硫酸 24.1.11.4;2.2.57.13
sulfuric acid alkylation 硫酸烷烃化 6.1.6.(a)
sulfur, mercury and trace elements removal 硫,汞和微量元素脱除 2.2.110.11
sulfur or sulfuric acid 硫或硫酸 2.2.36.9
sulfur plant 硫加工设备 2.2.102.22
sulfur recovery 硫回收 2.2.36.8;2.2.67.10;2.2.95.22
sulfur recovery unit 硫回收装置 2.2.95.14
sulfur removal 脱硫 2.2.103.4
sulphuric acid supply 硫酸供应 2.3.14.33
summary of CO_2 capture strategies,(a) pre-combustion capture,(b) post-combustion capture,and (c) oxyfuel combustion 二氧化碳捕集方法简介:(a)燃烧前捕集;(b)燃烧后捕集;(c)富氧燃烧 2.2.119
sump 槽 11.6.3.6
sump pump 底料泵 16.1.2.3.30
sunken floor socket 凹入地面的插座 25.4.1.24
supercritical fluid extraction 超临界[流体]萃取 1
superheated low pressure steam 过热低压蒸汽 2.2.2.73
superheated steam 过热蒸汽 2.2.31.14;25.5.3.6
superheated vapor 过热蒸气 20.2.4.19
superheater 过热器 2.2.81.10
superheater-steam generator 过热蒸汽发生器 25.5.1.(b)
superheating 过热 25.5.1.10
superheat zone 过热区 20.2.4.10
supernatant 上清液 3.2.2.16;3.3.17.4;3.3.18.8
supernatant removal 上清液去除 3.3.18.10
supernatant return to plant 上清液回厂 3.3.21.10
supernatant withdrawal 上清液出料 3.2.6.8
superposed tray 重叠式塔板 5.2.5
superstructure 上层构架 15.6.1.1
suport 支柱 21.2.2.3
supplementary gas 补充气 2.2.46.14;2.2.47.12;
supply air 供应空气 6.2.8.8
supply and return lines 供油与回油线 2.1.7.1
supply pump 供料泵 3.3.22.7
supply water piping 供水管 10.5.1.27
support 支撑体 3.1.29.3
支柱 4.2.2.7
支座 5.2.2.15
support bar 支承扁钢 11.1.7.13
support belt 支承带 9.4.8.7
support bracket 悬挂式支座 4.1.1.37
support column 支承管 16.4.22.7
支柱 21.2.3.7
supporter 载体 1
支柱 9.5.4.37
supporter cap 支柱帽 9.5.4.33
support flange 支座法兰 8.9.5.7
support for drip-piping wash 洗涤水水管支架 9.4.2.3
support fuel 辅助燃料 3.1.22.4
support gas 辅助气 3.1.22.3
support gravel 砾石支撑层 3.1.38.6
support grid 支承格栅 5.3.1.16
supporting block 支座 21.2.4.6
supporting construction 支承结构 25.1.13.2
supporting plate 支承板 5.1.5.12
supporting ring 支持圈 5.2.2.4
supporting roll 支承滚轮 12.1.1.20
supporting saddle 鞍式支座 8.8.10.11
support insulator 支架绝缘子 25.4.9.4
support leg 支柱 5.2.4.18
support level 支承平面 8.4.1.7
support lug 耳式支座 21.1.1.15
supportor 支架 13.5.1.3
support piece 支承块 4.2.2.5
support plate 支承板 10.4.2.8
支持板 15.2.3.5
support prill bucket 造粒筒支承套 13.1.2.19
support ring 支承圈 5.2.4.21
鞍式支座 4.1.1.35
support skirt 裙座,裙式支座 21.1.1.24
surface aerator 表面曝气器 1
表面曝气装置 3.1.45.5
surface casing 外壳 3.1.24.18

surface-coal gasification 地上煤气化 2.2.89
surface condenser 表面冷凝器 7.2.8.8
surface diffusion 表面扩散 1
surface-direct retorting 地面直接干馏 2.2.82
surface-indirect retorting 地面间接干馏 2.2.83
surface-mounted socket outlet 表面安装式插座引出线盒 25.4.1.25
surface-mounted three-pole earthed socket 表面安装式三眼接地插座 25.4.1.13
surface-mounted three-pole earthed socket with neutral conductor 带有中性导体的表面安装式三眼接地插座 25.4.1.13
surface reaction control 表面反应控制 1
surface renewal theory 表面更新理论 1
surface retort 地面干馏 2.2.85.5
surface type 表面型 20.1.10.23
surface-type condenser 表面型冷凝器 20.1.11
surface wash 表面洗涤 6.3.9.10
surface water-bearing strata 地表水层 3.1.24.3
surface work 表面功 1
surfacing hard alloy 堆焊硬质合金 22.1.8.18
surge 喘振
surge hopper 缓冲装料斗 2.2.5.3
surge tank 缓冲罐 2.2.106.27;3.1.34.4
均压箱 3.5.14.15
surplus earthworms for sale 过剩的蚯蚓可出售 3.3.28.15
surplus earthworms(if any) 过剩蚯蚓(如果有的话) 3.3.28.11
surplus steam 剩余蒸汽 2.1.8.7
surrounding [热力学]环境 1
suspended combustor 悬浮式燃烧器 2.2.16.2
suspended growth reactor 悬浮生长反应器 3.2.13.4
suspended solids 悬浮固体 3.1.13;3.3.18.12
suspension 悬浮液;悬浮 1
悬挂装置 11.1.4.14
suspension bushing 悬挂轴衬 11.1.4.13
suspension chamber 悬浮室 7.2.1.23
suspension polymerization 悬浮聚合 1
suspension weight 悬挂重物 10.3.5.22
SV(1)(space velocity) 空间速率;空速 1
swash plate pump 斜盘式泵,甩板泵 16.1.2.2.54
sway bar 稳定杆 21.2.2.11
sweeping gas 吹扫气 2.2.42.3
sweep vapor 扫掠蒸气 15.4.1.5

sweep vapor boiler 扫掠蒸气锅炉 15.4.1.1
sweeten-off tank 除甜罐 6.3.1.6
sweet water 甜水 6.3.1.7
Swenson atmospheric reaction-type DTB crystallizer 常压反应型DTB结晶器 7.2.9
Swenson reaction type DTB crystallizer 反应型DTB结晶器 7.2.8
Swenson single-stage recompression evaporator Swenson单级再压缩蒸发器 7.1.14
swing arm 摆动臂 12.1.1.25
swing bolt 活节螺栓 9.5.17.1
swing check valve 旋启式止逆阀 22.1.1.(1)
swing hammer 摆动锤 11.2.8.5
swinging inlet damper 旋转式入口气流调节器 10.5.4.4
swinging vane pump 转叶泵 16.1.2.2.9
swing-jaw plates 摆动颚板 11.1.1.3
swing pipe 摆动管 2.3.10.10
swirl breaker 破旋器 7.2.5.10
swirl vanes 旋转叶片 10.1.1.4
switched socket 带开关的插座 25.4.1.7
switch station 切换装置 23.1.1.9
swivel 旋转龙头 2.1.2.10
活节 22.5.4.1
symmetric membrane 对称膜 1
Symons standard cone crusher Symons标准圆锥破碎机 11.1.3
synchro 同步器 25.4.6
synchromesh gear 同步齿轮 17.3.2.7
synchronizing gear 同步传动齿轮 16.3.6.6
syn-crude 合成原油 2.2.2.34
synfuel 合成燃料 25.1.1.5
syngas liquids 氨汽提塔 2.3.6.13
synthesis 合成 2.3.3.15
合成反应 2.2.73.3
syngas 合成气 2.2.1.12;2.2.22.5;2.2.23.8;2.2.24.12;2.2.25.14;2.2.29.14;2.2.30.13;2.2.31.20;2.2.32.1;2.2.40.2;2.2.41.1;2.2.43.5;2.2.49.1;2.2.50.3;2.2.70.1;2.2.74.3;2.2.109.4;2.2.119.7
syngas compressor 合成气压缩机 2.2.44.2
syngas cooler 合成气冷却器 2.2.23.5;2.2.25.12;2.2.31.15
syngas feed 合成气进料 2.2.28.1
synthesis converter 合成塔 2.2.49.2
syngas out 合成气出口 2.2.16.6
syngas processing 合成气加工 2.2.3.3
syngas purification(Rectisol process) 合成气净化(低温甲醇洗) 2.2.67.7
synthesis gas 合成气 2.1.8.12
synthesis gas compressor 合成气压缩机 2.3.3.14

611

synthesis process 合成工序 2.2.45.8
synthesis gas/recycle gas compressor 合成气/循环气压缩机 2.2.43.2
synthesis reactor 合成反应器 2.2.37.7;2.2.40.8;2.2.41.3;2.2.48.6
synthetic ammonia 合成氨 2.2.1.13
synthetic crude 合成的粗产物 2.1.9.41
synthetic crude oil 合成原油 2.2.99.11
synthetic fuel 合成燃料 14.3.2
synthetic light oil 合成轻油 2.2.74.12
Synthol fluidized bed continuous reactor system Synthol 流化床合成燃料连续反应器系统 14.3.2.(d)
system controls 系统控制装置 23.1.2.25
system fan 系统风机 8.4.1.15

T

table printer 说明打印机 23.4.1.9
table feeder 进料台 11.1.6.1
盘式加料器 14.1.6.(c)
tail gas 废气,尾气 2.3.1.25;2.2.54.13;2.2.55.15;2.2.74.7;2.2.78.13;2.3.15.14
tail gas scrubber 尾气洗涤塔 2.2.54.5
tailing chute 粗粉溜槽 12.3.4.15
tailing chute flange 粗粉溜槽法兰 12.3.4.17
tailing cone elbow 弯头 12.3.4.16
tailing elbow clamp ring 粗粉溜槽夹紧环 12.3.4.18
tailing outlet 尾砂出口 12.8.1.7
tailings 选余物 11.6.3.11
tailings 尾渣[砂] 11.6.5.2
tailings cone 粗粉锥筒 12.3.1.11
tailings discharge 粗粉排出口 12.3.1.12
tail gas to reformer or fuel gas 尾气去重整或作燃料气 2.2.73.6
tail oil 尾油 2.1.1.20
tail piece 尾管 10.1.2.11
tail pipe 尾管 18.3.6.4
tail pulley 尾轮,尾部皮带轮 23.2.1.1
tail vane 尾部叶片 3.5.1.2
tail wheel 尾轮 23.3.7.13
take-up 松紧装置 15.6.1.25
take-up tray 接料盘 8.3.4.3
take up unit 松紧装置 15.7.5.9
tangent circle 相切圆 11.5.1.4
tangential gas inlet 气体切向入口 10.5.4.5
tangential secondary air 切向二次空气 25.2.3.4

tank coil heater 盘管式油罐加热器 4.2.5.18
tank cover 槽盖 14.1.11.8
tank farm pump 罐区泵 16.1.2.3.43
tank farm submersible pump 罐区潜水泵 16.1.2.3.44
tank-self-supported roof 自撑顶 21.2.1.2
tanks-in series model 多釜串联模型 1
tank suction heater 油罐吸入加热器 2.1.7.14
tank-supported roof 支撑顶 21.2.1.1
tank-Wiggins dry-seal type 干封型 21.2.1.4
tank-Wiggins-Hidek type W-H 型 21.2.1.3
tappet 挺杆 16.2.4.36
tap(ping) 分接头 25.4.3.8
tapping hole 螺纹孔 5.4.2.18
tapping mechanism 敲击机构 10.2.1.18
tar 焦油 2.2.108.15
tar removal 除焦油 2.2.89.6
tar remover 脱焦油 2.2.108.14
tar/slurry 焦油浆 2.2.15.2
tar to gasifier 焦油去气化器 2.2.106.24
TDH(transport disengaging height) [输送]分离高度 1
techniques for chemical processing of coal 煤化工技术路线 2.2.1
techniques for ethanol synthesis from coal via syngas 煤制乙醇的主要技术路线 2.2.40
tee rail T 型轨 15.6.1.27
teflon heat exchanger 聚四氟乙烯换热器 4.2.5.13
teflon sample line 聚四氟乙烯采样管 3.5.16.22
telescoping chute 套叠式斜槽 23.3.9
television camera 电视摄像机 25.2.5.6
Teller Bosette packing 特勒花环填料 5.3.2.(j)
tell tale holes 指示孔 21.1.1.17
temp.(temperature) 温度 20.1.7.4
temperature 温度 3.1.2.9
temperature-control electronics 温度控制电子元件 3.4.20.8
temperature gradient 温度梯度 1
temperature indicator 温度显示器 8.5.3.7
temperature profile 温度(分布)剖面(图) 1
temperature sensor 温度传感器 3.4.2.1
temperature swing adsorption(TSA) 变温吸附 1
temper handle 缓启闭手轮 9.2.5.23
tension rod 拉杆 11.1.1.14
tension rod spring 拉杆弹簧 11.1.1.10
terminal block 接线盒 18.2.1.13
terminal box 接线盒 16.4.20.16

terminal velocity 终端速度 1
tertiary air 三次空气 24.2.2.3
tertiary augmentor 三次扩压器 18.3.5.1
testing door 检修门 8.3.4.13
test tube rack 试管架 2.3.11.8
Texaco coal-gasification process 德士古煤造气工艺 2.3.2
thawing 融化 1
The Foster-wheeler partial gasifier Foster-wheeler 部分气化炉 2.2.23
theoretical plate 理论[塔]板 1
thermoplastic resin 热塑性树脂 15.7.3.(a)
thermal 热量 2.2.107.18
thermal cracking 热裂化 2.1.1.28
thermal cut-out 热熔断路器 18.2.4.9
thermal diesel oil 热裂化柴油 2.1.1.31
thermal diffusivity 热扩散系数;导温系数 1
thermal efficiency 热效率 1
thermal gasoline 热裂化汽油 2.1.1.30
thermal insulation 隔热;保温 1
thermal medium evaporator 热载体蒸发器 7.1.12
thermal medium heater 热载体加热炉 24.1.1
thermal pyrolysis 热分解过程 3.3.1
thermal recovery unit 热回收装置 2.3.3.19
thermal(SRC & EDS) 热过程(SRC & EDS) 2.2.90.6
thermal stability 热稳定性 1
thermistor probes 热敏电阻探针 7.2.14.8
Thermix ceramic tube Thermix 陶瓷管分离器 10.1.4.(b)
thermocouple 热电偶 14.3.2.4;3.3.27.5
thermodynamic analysis of process 过程热力学分析 1
thermodynamic characteristic function 热力学特性函数 1
thermodynamic consistency test 热力学一致性检验 1
thermodynamic equilibrium 热力学平衡 1
thermodynamic trap 热动力式疏水阀 22.2.1
thermometer 温度计 3.4.11.8
thermoplastic moisture-proof cable 热塑防潮电线 25.4.1.42
thermoplastics 热塑性塑料 22.6.2.①
thermoset resin 热固性树脂 15.7.3.(b)
thermostat 恒温器 8.3.4.15
thermostat-controlled valve 定温控制阀 25.6.4.10
thermostatically controlled steam valve 蒸汽恒温控制阀 15.5.5.11

thermostatic trap 恒温式疏水阀 22.2.6
thermowell 热电偶套管 5.4.2.11
The sargas process combines a pressurized fluidized bed combustion process with a post-combustion CO_2 removal process, shown as the Benfield process 本菲尔德工艺—增压流化床燃烧与燃烧后脱碳联合的 Sargar 工艺流程 2.2.116
The shell coal gasifier 壳牌煤气化炉 2.2.25
The transport gasifier 输送式气化炉 2.2.24
The VERTREAT™ process (Courtest NORRAM engineering and construction Ltd.) VERTREAT(courtesy NORRAM 工程建筑有限公司)废水处理工艺 3.1.32
The Winkler gasifier 温克勒气化炉 2.2.18
thickened slurry 已增浓的料浆 9.4.9.5
thickener 增稠器;浓密机;浓缩器 1
thickening & dewatering 增稠与脱水 3.2.1.20
thickening zone 增浓区 7.2.10.1
thickness 稠度 1
thimble 套管 23.1.13.10
thin cake on filter cloth 滤布上的薄层滤饼 9.4.9.8
thin film evaporator 薄膜蒸发器 1
thin film of solids 固体薄片 9.3.8.5
thin film reactor 薄膜反应器 14.3.1.(f)
thin-layer evaporator with rigid wiper blades 刚性刮板薄层蒸发器 5.5.1.(b)
third stage cylinder 三级缸 17.2.3.3
third tower 3 塔 2.2.41.9
thixotropy 触变性 1
Thormann tray 索曼塔盘,槽形泡罩塔盘 5.2.10.2
thorough washing 穿过式洗涤 9.1.1.11
threaded cross 带螺纹四通 22.3.2.3
45° threaded elbow 45°螺纹弯头 22.3.2.4
90° threaded elbow 90°螺纹弯头 22.3.2.1
threaded fitting 螺纹管件 22.3.2
threaded flange 螺纹法兰 22.4.2.(f)
threaded opening 套扣开孔(螺纹孔) 21.1.1.59
threaded reducing tee 带螺纹异径三通接头 22.3.2.5
threaded tee 带螺纹三通 22.3.2.2
threading key 插入键 25.4.1.30
three-column basket centrifuges 三足式转鼓离心机 9.5.3
three-core cable 三芯电线 25.4.1.44
three-phase fluidization 三相流态化 1
three-phase generator 三相发电机 25.4.4.6

three-phase plug 三相插头 25.4.1.14
three-pin plug 三脚插头 25.4.1.67
three-pin socket 三眼插座 25.4.1.66
three-roll sugar mill 三辊榨糖机 9.6.1
three-screw pump 三螺杆泵 16.3.3
three-stage condenser 三级冷凝器 2.3.5.9
three-stage fluidized bed granulator 三级流化床造粒机 13.4.1(c)
three-way chute 三通滑槽 2.3.12.34
throat 喉部 18.3.1.7;3.4.4.13
throat ring 喉部垫环,填料衬环 16.4.20.22
throat section 喉部截面 3.5.4.6
throttle bushing 节流衬套 19.1.5.20
throttle ring 节流圈 16.4.22.4
throttle valve 节流阀 25.6.2.5
throttling process 节流过程 1
throttling valve 节流阀 20.1.2.15
through-circulation 穿流循环 8.1.1.④
through-flow rotary dryer 穿流回转干燥器 8.8.9
throughput 通过量;产量 1
through to 相通 16.3.2.4
throw angle 抛射角 23.3.1.7
thrower 甩油环 16.4.20.26
thrust ball bearing 止推滚珠轴承 13.3.5.5
thrust bearing 止推[推力]轴承 13.2.3.4
thrust collar 止推环 16.4.24.11
thrust metal 推力轴承衬瓦 16.4.28.15
thrust pad 止推轴承调节垫 20.2.1.16
thrust ring 推环 19.1.1.8
thrust roll 止推辊 8.8.5.31
thrust roll bearing 止推滚动轴承 8.8.5.32
thrust roll bushing 止推辊轴承 8.8.5.30
thrust roll shaft 止推辊轴 8.8.5.29
thrust runner 推力轴承滑道 16.4.28.16
thrust washer 止推垫圈 16.4.1.14,18
tie bolts 拉杆螺栓 11.1.7.3
拉杆 21.2.2.2
tie(-)rod 长螺栓 19.2.1.7
tighter seal 密封效果 19.3.3.(b)
tilting-disk check valve 斜盘式止逆阀 22.1.1.(k)
timer 定时器 10.2.3.8
timing unit 守时装置 9.5.13.7
tins 白铁罐 3.3.5.19
tip speed 浆尖速度 1
tire 滚圈 12.1.1.11
titanium and titanium alloys 钛和钛合金 22.6.2.㉔
to ambient 排空 6.2.8.6
to atmosphere 通大气 11.6.4.20

to bioreactor 去生物反应器 3.1.30.A.6
to clear well 去清水池 3.1.38.11
to decarbonization tower 去脱碳塔 2.2.49.4
to desaturation tower 去饱和塔 2.2.43.15
to gasifier 去气化炉 2.2.35.18
toggle plates 肘板 11.1.1.13
to grit removal 移走砂粒 3.1.8.4
to heat exchanger 去热交换器 2.2.110.20
Tol-ammonia washing tower Tol-洗氨塔 2.2.34.20
tolerance 允差;公差 21.1.2.19
tongs 夹钳 2.3.9.44
tongued flange 榫面法兰 22.4.3
tongue-groove seal contact face 榫槽密封面 22.4.3.(d)
tool case 工具箱 25.4.1.49
tooth 齿轮 13.1.2.10,28
top angular down 上倾角出风 17.5.2.8
top angular up 上仰角出风 17.5.2.6
top annular plate 顶部环板 9.5.4.22
top bearing 上轴承 4.2.4.2
top bearing adjusting nut 上轴承调节螺母 12.3.5.3
top bearing cover 上轴承盖 12.3.5.30
top bearing housing 上轴承箱 12.3.5.4
top bearing nut lock 上轴承锁紧螺母 12.3.5.31
top bearing shaft nut 上轴承螺母 12.3.5.32
top bowl 上(部圆)筒 16.4.25.11
top bowl bearing 上部圆筒支架;上筒轴瓦 16.4.25.13
top bowl connector bearing 上部圆筒连接器支座 16.4.25.15
top cap 上盖板 9.6.1.13
top cap key 上盖开关 9.6.1.14
top column pipe 上部圆筒管 16.4.25.22
top crown 上极板 21.2.3.4
top disc 顶层盘 9.5.12.6
top discharge 顶部出料 9.5.2.14
top drive 顶部驱动 9.5.2.11
top drive bottom discharge batch basket 顶部驱动底部出料间歇式离心机 9.5.1.(b)
top-fill valve & nozzle 顶部装料阀和接管 23.4.1.20
top frame 顶部框架 12.2.1.13
top-frame discharge 顶部框架卸料口 12.2.1.12
top grating 上栅板 5.2.9.3
Topham centrifugal pot[box] 托范式离心罐 2.3.7.17
top heating shelf 顶部加热架 8.2.2.5

614

top horizontal 上水平出风 17.5.2.7
top of spinneret 喷丝头 2.3.8.41
top-outlet tubular filter 顶排液的列管过滤机 9.4.5
topping distillation column 拔顶蒸馏塔 2.2.43.12
top product 塔顶产品 6.3.6.8
to process 去下道工序 23.1.7.11
top roll flange 上辊法兰 9.6.1.15
Topsoe DME synthesis process(two pot synthesis) 托普索公司二甲醚合成工艺（两步合成法） 2.2.52
top shaft 上部轴 16.4.25.25
top shaft tube 上部轴套管 16.4.25.32
topsoil 表土 3.3.20.8
top-suspended basket centrifugal 上悬式转鼓离心机 9.5.6
to Rectisol process 去低温甲醇洗 2.2.34.17
toriconical head 折边锥形封头 21.1.1.25
torispherical head 碟形封头 21.1.1.63
准球形封头 21.4.1.4
toroidal 环形接头 22.5.1.(f)
torque wrench 扭力扳手 23.4.1.8
tortuosity 曲折因子 1
TOSCO (the Oil Shale Corporation of America) 美国油页岩公司 2.2.5
TOSCO Ⅲ process 美国油页岩公司Ⅲ工艺 2.2.5
to secondary clarifiers 去二次澄清器 3.1.35.3
to shipping 发运 23.4.1.7
to steam turbine 去蒸汽轮机 2.2.110.6
to storage 去贮存 6.2.1.6
to suction 接吸入口 16.4.6.21
to tailings pond 去尾渣池 2.1.9.7
total capacity 总产量 2.2.88.1
total head 总压头 16.4.3.15
totalize 累计器 3.4.18.11
totalizing meter 总计表 6.2.1.11
total reflux 全回流 1
total solids 含固形物 13.4.1.10
固含量 3.3.4.3
total waste-treatment system 总废物处理系统 3.1.21
to torch 去火炬排放 2.2.55.20
去往火炬 2.2.56.9
to treatment process 去处理流程 3.3.20.6
toughed-belt conveyor 槽式皮带输送机 23.2.1.17
to vent 去放空 2.2.110.21
tower body 塔体 4.6.1.6

tower cone 下圆锥 10.1.2.14
tower inside diameter 塔内径 5.4.1.2
tower internal diameter 塔内径 5.2.4.13
tower internals 塔内件 5.3.4
tower shell 塔体 5.2.1.9
toxic 毒品 8.1.1.28
T-pipe outlet for effluent 废液T形出口管 15.5.1.8
TR (temperature recorder) 温度记录仪 3.1.6.8
trace organics removal 脱除微量有机物 2.2.101.14
tracer 描绘器 25.4.8.3
tracer element 描绘器元件 25.4.8.3
trace sulfur removed 脱除微量硫 2.2.101.13
traditional chemical processing of coal 传统煤化工 2.2.1.22
traditional multi-stage pusher 传统的多级推料离心机 9.5.1.(n)
traditional single-stage pusher 传统的单级推料离心机 9.5.1.(m)
trail pipe 后管 9.3.10.11
trains 辅料 13.3.5.5
tramp discharge 杂质排出口 8.4.1.21
transfer function 传递函数 1
transfer line reactor 传输线反应器 14.2.7.25
transfer pipe 输液管 3.1.10.6
transformer 变压器 25.4.3
transformer connection 变压器连接方式 25.4.3.1
transformer rectifier set 整流高压器机组 10.3.1.10
transformer station 变电所 2.3.12.21
transition diffuser 过渡段扩压器 16.4.23.8
transmission line 输电线路 25.2.1.10
transmitter 变送器 3.4.18.1
transparent disk 透明圆盘 11.7.3.7
transport 输送床 14.1.4.(g)
transportation 输送 3.3.7.12
transport disengaging height(TDH) （输送）分离高度 1
transport reactor 输送反应器 3.3.15.(d)
transverse baffle 横向折流板 4.1.1.28
trap 疏水器；汽水分离器；集水器；阱；弯管 捕集器 1
trap to drain 疏水阀排水 3.1.6.1
travel 移动；传送 4.3.4.30
travel away 向后传送 4.3.4.19
travel forward 向前传送 4.3.4.18
travel indicator 行程指针 22.1.9.10

travel indicator scale 行程标尺 22.1.9.11
traveling gantry 移动式吊架 23.3.9.2
traveling tripper 移动式卸料器 23.2.3.(c)
travelling block 游动滑车 2.1.2.8
traversing knife 横移刮刀 9.5.2.22
tray 塔板,塔盘 1
tray and compartment dryer 厢式干燥器 8.2
tray column 板式塔 1
tray dryer 厢式干燥器 8.2.1
tray ring 塔板[盘]圈 5.2.4.19
tray sheet 塔板 5.2.4.22
tray spacing 板间距 5.2.6.11
tray stiffer 塔盘支承 5.2.8.5
tray support ring 塔盘支持圈 5.2.8.4
trays with spacer tubes 带定距管塔盘 5.2.4
TRC(temperature recording controller) 温度记录控制器 3.1.6.17
treated 处理过的 6.2.9.10
treated air 处理后的空气 3.4.3.3
treated air(effluent) 处理后的空气(废气) 3.4.3.1
treated effluent 处理后废水 3.1.12.10
treated effluent 处理后出水 3.1.28.8
处理后的废液 3.1.29.5
处理后的中水 3.1.30.A.7;3.1.30.B.13
treaded heavy gasoline 处理后的重质汽油 2.2.75.19
treated juice return 处理过的糖汁返回 6.3.1.3
treated solution 经过处理的溶液 6.2.4.2
treated wastewater 处理后的废水 3.1.23.3
treated water 净化水 15.5.4.11
清水 15.6.4.17
处理后的水 3.1.22.18
treated water conduit 清水导管 15.6.4.18
treated water outlet(permeate) 处理后的水出口(渗透) 3.1.29.7
treating of cake to give filaments softness 丝饼柔软处理 2.3.7.22
treatment for recovery 回收处理 12.5.5.10
Treybal extractor Treybal 萃取器 6.1.15
trickle bed 滴流床;涓流床 1
trickle bed hydrotreater 滴流床加氢装置 2.2.79.7
trickle reactor 滴流床反应器 14.3.1.(a)
trimethylbenzene 三甲苯 2.1.1.61
trimmer 堆料器 23.3.9
trimming tool 修整工具 2.3.9.43
triple point 三相点 1
triplex 三列,三缸 16.1.1.10
triplex vertical compressor 三缸立式压缩机 17.2.1.4
tripod 三脚架 2.3.11.6
tripper stacker 自动转换堆放器 23.2.5.7
trommel 滚筒回转筛 2.2.5.16
转筒筛 3.3.16.5
trommel screen 转筒筛 14.3.3.7
Trost jet mill Trost 型气流粉碎机 11.5.2
trough 输送槽 23.3.3.1
troughed conveyor belt 槽式输送机皮带 23.2.1.6
troughing angle 槽形角度 23.2.1.13
troughing carrying-idlers 槽形承载托辊 23.2.1.5
trough-type distributor 横型分布器 5.3.4.13
truck 货车 23.4.2
小车 8.3.1.6
true in-situ retorting 真实就地干馏 2.2.6
trunnion 耳轴 11.2.1.2
trunnion and thrust roll assembly 枢轴与止推辊装配体 8.8.1.14
trunnion block adjusting screw 轴颈座调节螺钉 9.6.1.22
trunnion roll assembly 枢轴装配体 8.8.1.10
trunnion roll bearings 滚动轴承 8.8.4.(b)
TS(1)(total solids) 总various固量 3.1.21
TSA(temperature swing adsorption) 变温吸附 1
T-shaped hammers T形锤子 11.2.5.3
Tsinghua university one-step liquid phase dimethyether(DME)synthesis 清华大学液相一步法二甲醚工艺流程 2.2.49
tube 管(子) 9.4.4.5
tube arrangement 管子排列形式 4.1.5
tube bundle 管束 1
tube bundle column 多管塔 5.1.2.6
tube coils 盘管 24.1.3.5
tubed 管式 4.3.2.(c)
4-tube force flow bag packer 4 管强制流袋包装机 23.4.2.1
tube pass 管程 1
tube pitch 管心距 4.1.5.5
tube sheet 管板 10.1.2.28
tube sheet support 管板支承板 24.1.1.24
tube(side)pass 管程 4.1.1.41
tubes rotate with shell 管子随壳体回转 4.3.2.10
tubing 管道 18.4.1.11
井管,油管 2.1.2.24
tubular-bowl centrifuge 管式高速离心机 9.5.11
tubular casing pump 管壳泵 16.1.2.2.28

tubular furnace 管式炉 24.1.2
tubular membrane ultrafiltration 管式膜超滤 15.2.4
tubular precipitator 管式电除尘器 10.3.5
tubular shaft 管式轴,空心轴 16.4.21.4
tubular sprayer 管式喷淋器 5.3.4.2
tumbler 转鼓 12.1.3.2
tumbler centrifuge 转鼓离心机 9.5.18
tumbler mixer 转筒混合机 2.1.9.3
tumbling agglomerators 滚动团聚器 13.3.1.13.4.2
tungsten carbide plunger 碳化钨柱塞 17.2.15.9
tunnel 通道 25.2.6.11
tunnel dryer 隧道干燥器 1
turbine 涡轮(机) 2.2.81.11;2.2.111.5;2.2.116.8
汽轮机 25.3.2.2;2.2.80.7
turbine agitator 涡轮搅拌机 14.4.1.19
turbine drive motor 涡轮驱动马达 15.6.3.4
turbine driven pump 透平驱动泵 16.1.2.4.2
turbine fuel 透平机燃料 2.2.91.8
turbine generator 涡轮发电机 2.2.119.11
turbine impeller 涡轮式搅拌器 14.4.2.(a)
turbine monitoring panel 汽轮机监控盘 25.4.4.1
turbine pump 涡轮泵 16.1.2.2.17
turbo dryer 涡轮干燥器 8.9.2.1
turbo expander 透平膨胀机 3.1.23.15
turbo-fan 涡轮风机 8.9.1.2
turbo generator 汽轮发电机 25.2.2.14
turbo generator set 汽轮发电机组 25.2.2.13
turbo generator unit 涡轮发电机组 25.4.4
turbo grid tray 穿流栅板 1
turbo-mixer agitators 涡轮混合搅拌器 6.1.9.18
turbo refrigeration machine(closed type) 封闭式透平制冷机 20.2.3
turbo refrigerator 透平制冷压缩机 20.2.1
turbo-tray dryer 涡轮转盘式干燥器 8.9.1
turbo-tray dryer in closed circuit for continuous drying 连续密闭循环涡轮转盘式干燥器 8.9.2
turbulent 湍流 14.1.1
turbulent ball column 湍球塔 5.1.2.8
turbulent bed 湍流床 14.1.2.4
turbulent flow 湍流,紊流 1
turbulent fluidized bed 湍动流化床 1
turbulizer-backmixer 涡流返混合器 8.4.1.19
turnbeam adjusting level and screws 导向板

调节杆与螺钉 9.6.1.21
turner 翻拌器 2.3.12.24
turning vanes 导向叶片 8.2.1.3
turn tube 弯管 13.1.2.37
turpentine separator 松节油分离器 2.3.10.14
tuyeres 吹风管嘴 2.2.14.11
风口 3.3.27.12
tween 两甲板间的空间 23.3.9.4
twin primary cyclone 并联两个一级旋风分离器 8.4.5.23
twin rotor 双转子混合器 12.1.3.(i)
twin-screw feeder 双螺杆加料器 12.1.5.15
twin shell(vee)mixer 双筒混合器(V型) 12.1.3.(b)
twisted packing 麻花填料 19.2.3.3
two autogenous wet-grinding stages 两段湿式自磨机 11.6.3
two-bed deionizing system 双床脱离子系统 6.2.1
two-bed TSA system 双床TSA系统 6.3.4
two-deck screen separator 双层筛分离器 12.2.2
two diameter 双直径式 8.8.5.(b)
two-film theory 双膜理论 1
two-impeller type of rotary positive-displacement blower 双叶型旋转正位移式(容积式)鼓风机,罗茨鼓风机 17.3.5
two pass shell with longitudinal baffle 双程壳体带纵向隔板 4.1.2.(g)
two pass tray 双流塔盘 5.2.10.25
two-phase flow 两相流 17.1.1.7
two rotor 双转子 17.1.1.7
two-screw pump 双螺杆泵 16.3.2
two-shaft arrangement 双轴布置 25.3.3
two-stage air stream and cage mill pneumatic-conveyor dryer 两级气流和笼式磨机气流输送干燥器 8.4.4
two-stage cascade system 两级逐级系统 20.1.5
two-stage cyclone receiver 两级旋风分离接收器 23.1.13
two-stage double-acting air compressor 两级双作用空气压缩机 17.6.1.8
two-stage double-acting compressor cylinders 两级双作用压气机汽缸 17.2.7
two-stage ejector 两级喷射泵 18.3.4 (a)
two-stage electrical-precipitation 两级电除尘器 10.3.3
two-stage external cyclone 双级外旋风分离器 14.1.5.(d)

two-stage internal cyclone 双级内旋风分离器 14.1.5.(b)
two-stage multiple cyclonic separator 两级组合式旋风分离器 10.1.3
two stage regenerator 两级再生器 14.3.4.18
two-stage single-acting opposed piston 两级对置的单作用活塞 17.2.4
two-stage single-acting opposed piston in a single-step-type cylinder 级差式气缸中两级对置的单作用活塞 17.2.4
two-way diverter 双向分流器 23.1.6.6
Tyler standard sieve 泰勒标准筛
type A agitator drive A 型搅拌器驱动装置 9.3.3.21
type A drum drive A 型转鼓驱动装置 9.3.3.39
type of flange facing 法兰密封面型式 22.4.3
type of packing 填料类型 19.2.3
type of towers 塔器类型 5.1
type of trays 塔盘类型 5.2.10
type RA line separator RA 型在线分离器 10.4.3.(e)
types of heads 封头类型 21.4.1
types of impellers 叶轮类型 16.4.5
type-S Pitot tube S 型皮托管 3.5.2.2
typical continuous-treatment system 典型连续处理系统 3.1.12
typical open basin 典型的敞口池 3.1.46.10
typically sedimentation 一般沉降 3.2.1.10

U

U bends U 形弯管 14.3.4.26
U-bolt U 形螺栓 4.2.2.1
Udex process Udex 工艺过程 6.1.4
UF capillary modules 超滤毛细管膜组件 2.4.4.8
U-gauge pressure U 形管压差计 8.5.3.11
ultimate disposal 最终处置 3.2.1.14
ultimate disposal site 最终处置场地 3.1.11.7
ultracentrifuge 超速离心机 1
ultrafiltration 超滤 1
ultrasonic exam 超声波探伤 21.1.2.15
unbaffled jacket 无挡板夹套 4.3.3.6
unbalanced 非平衡式 23.3.2.6
uncleared green liquor 不洁绿液 2.3.10.41
underdrain 聚水系统 3.1.38.7

undeflected position 未变形位置 22.5.2.1
underfloor sockets 地板下的插座 25.4.1.22
underflow 底流；下漏 1
underflow primary sludge 一次底流污泥 3.2.1.15
underflow secondary sludge 二次底流污泥 3.2.1.17
underflow to plant influent 浓缩至厂流入液中 3.1.26.(a).9；3.1.26.(b).3；3.1.26.(b).11
under frame 机座 12.2.5.6
底架 17.5.1.12
undergrate air port 炉条下面空气通道 3.3.12.13
underground tank for additives 添加剂地下槽 2.2.35.6
undersize 筛下物料 12.7.2.13
细粒 13.3.3.10
underwater motor pump 水下电机泵 16.1.2.2.37
underwater pump 水下泵 16.1.2.2.36
undiluted mother liquor 过滤母液 9.4.1.13
unequal thick ness 不等厚度 21.1.1.40
unfiltered liquid 未过滤液体 9.1.3.7
unfired steam boilers 非火蒸汽锅炉 21.1.2.17
unheated downcomers 不加热下降管 25.6.1.7
uniflow cyclone 单向流动旋风分离器 10.1.1.(a)
uniform conversion model 均匀转化模型 1
uniform waterfilm 均匀水膜 4.2.3.3
union nipple 接管接头 22.2.3.24
unit heater 单独的加热器 4.3.4.26
unit operation 单元操作 1
units in parallel 装置并联 14.3.3.(a)
unit thickener 单室增稠器 15.6.1
United States design 美国型 9.4.9.(b)
universal joint 万向接头 16.4.21.5
universal-joint pump 万向接泵 16.1.2.2.67
universal test meter 多用途电表，万用表 25.4.1.41
unloaded valve 减荷阀 17.2.9.2
unloader control cylinder 卸料装置控制气缸 9.5.6.20
unloader hydraulic pump 卸料装置用的液压泵 9.5.6.16
unloader knife 卸料刀具 9.5.6.17
unloading connection 填料卸出口 5.4.1.18
unreacted core model 非反应核模型 1

618

unreacted gases, and methanol and water vapors 未反应的气体,以及甲醇和水蒸气 2.2.39.a
unreacted syngas 未反应之合成气 2.2.70.4
unrefined cellulose 未精制纤维素 2.3.10.77
unstabilized gasoline 不稳定汽油 2.1.5.10
unsteady state 非定态;非稳态 1
underweighted 无平衡块 23.3.6.(a)
unwind station 退纸装置 2.3.11.39
UOP/HYDRO MTO (methanol to olefins) process UOP/HYDRO MTO(甲醇制烯烃)工艺
UOP/Hydro MTO process (production of polymergrade olefins) UOP/Hydro MTO 工艺流程(生产聚合级烯烃) 2.2.71
UOP stacked unit UOP 烟囱式装置 14.2.6.(d)
up blast 自下往上送风 17.5.2.5
upflow anaerobic sludge blanket bioreactor 厌氧污泥上流过滤层生物反应池 3.2.8
upflow fluidized-bed denitrification 上流式流化床去硝化 3.1.37.9
upflow regenerated unit 往上流再生设备 6.2.3
upgrading 改质 2.1.9.34
upper air hood 上部空气分布罩 4.2.4.7
upper ball cover 上轴承盖 13.3.5.10
upper bearing 上轴承 9.2.2.4
upper belt tensioning 上部滤带张紧器 9.6.2.8
upper belt wash 上部滤带冲洗 9.6.2.6
upper case 上盖 11.1.5.10
upper casing 上壳体 16.4.12.20
upper cone 上圆锥 10.1.2.16
upper crosshead frame 十字头上体 17.2.15.8
upper cylinder 上圆筒 10.1.2.9
upper end cover 上端盖 14.4.28.9
upper gasket 上垫片 16.4.27.16
upper half 上半部 16.4.9.13
upper head 上盖 16.2.7.3
upper inlet for medium 介质上部入口 14.4.1.14
upper jet 上部喷嘴 18.2.1.4
upper lift drum 上部提升筒 6.3.11.19
upper metal 上(滑动)轴承 9.5.4.31
upper motor weight 电机上端重块 11.3.2.3
upper Muschelkalk 上[晚]壳灰岩层 25.2.4.2

upper pebble bed 上层卵石床 14.3.3.18
upper plenum 上部气室 10.2.3.14
upper return nozzle for medium 介质上部回流管 14.4.1.13
upper screen 上层筛 12.2.3.6
upper shaft 上段轴 14.4.3.8
upper shaft sleeve 上轴套 16.4.27.24
upper sleeve 上套筒 16.2.5.19
upper sleeve ring 上套筒环 16.4.27.37
upper tracking roll 上部跟踪辊 9.6.2.7
upper weight 上部重块 11.3.1.12
upright column 立柱 4.2.1.1
upright cylindrical pressure tank 立式圆筒形压力罐 9.2.5.31
upstack 上升管 11.5.2.5
upward-flow evaporator 升膜蒸发器 7.1.9
upward ventilator 向上通风装置 25.2.5.9
urea dewaxing 尿素脱蜡 2.1.1.21
urea synthesis 尿素合成 23.3.14
used with simultaneous 配合使用 23.5.1
user 用热设备 25.6.7.4
utility 公用工程 3.3.1.13
U-tube bundle U 形管管束 4.1.4.(s)
U-tube heat exchanger U 形管换热器 1

V

vacuum 真空 19.4.1.1
vacuum applied 施真空 9.3.10.8
vacuum balancing line 真空平衡管线 18.5.1.4
vacuum band 真空带式 8.1.1.⑪
vacuum box 真空箱 2.3.11.15
vacuum break valve 消除真空用阀 8.2.2.3
vacuum connection 真空接口[头] 8.8.11.1
vacuum diesel oil 减压柴油 2.1.1.11
vacuum distillation 真空蒸馏;减压蒸馏 1;2.2.66.10;2.2.79.13
vacuum distillation column 真空精馏塔 2.2.94.17
vacuum distributor 真空分布器 9.4.1.14
vacuum double drum dryer 双滚筒真空干燥器 8.7.1.(e)
vacuum equipments 真空设备 18
vacuum filter 真空过滤机 9.4
vacuum flash 减压闪蒸 2.2.92.18
vacuum flash tank 真空闪蒸罐 2.2.102.10
vacuum gauge 真空表[计] 3.5.16.12
vacuum ga(u)ge connection 真空表接口

18.3.1.8
vacuum hopper 真空罐 23.1.2.10
vacuum line 真空管线 3.5.11.12
vacuum pan for crystallization of sugar 糖真空结晶器 7.2.11
vacuum precoat filter 预敷层转鼓真空过滤机 9.3.3.8
vacuum-pressure & material receiver 真空-压力和物料接收器 23.1.2.20
vacuum pump 真空泵 1;16.5.2.1
vacuum receivers 真空受液罐 9.3.1.11
vacuum residue 减压残渣 2.2.95.28
vacuum residuum 减压渣油 2.1.1.12
vacuum ripening tanks 真空成熟槽 2.3.7.11
vacuum rotary dryer 真空回转干燥器 8.8.10
vacuum seal 真空密封(装置) 19.4
vacuum sequencing system 真空顺序系统 23.1.2.12
vacuum-shelf dryer 真空盘架式干燥器 8.2.2
vacuum sourcer 真空源 15.1.1.10
vacuum still 减压蒸馏 2.2.98.28
vacuum system 真空系统 18.5
vacuum tower 减压塔 2.1.4.2
vacuum tray 真空盘式 8.1.1.⑫
vacuum unit 真空装置 25.6.3.3
valve 阀,阀门 22.1
valve-adjusting pivot 阀调节铰链 9.3.3.41
valve-adjusting rod 阀调节杆 9.3.3.46
valve body 阀体 22.1.10.2
valve catcher case 进气阀压罩 17.2.7.2
valve clack 阀瓣 22.1.8.2
valve closed 阀关闭 22.1.10.11
valve disc 阀盘 22.1.10.7
valve fully open 阀全开 22.1.10.13
valve guard 阀挡 16.2.3.34
valve handle 活门把手 11.3.1.10
阀(门)手轮 22.1.3.3
阀柄 22.1.13
valve head 阀盖 17.2.12.2
valve housing 阀体 22.1.4.6
valve open 阀开 6.3.3.13
valve plate 阀片 16.2.3.31
阀极 17.2.12.8
valve plug 阀芯 22.1.4.16
valve plug stem 阀杆 22.1.2.7
valve ring 阀环 22.1.3.11
valve rod 阀杆 12.3.4.8
valve-rod head 阀杆接头 16.2.3.49
valve-rod head pin 阀杆接头销 16.2.3.48
valve-rod head-pin nut 阀杆接头销螺母 16.2.3.47
valve-rod link 阀杆连(接)杆 16.2.4.40
valve-rod link head 阀杆连杆接头 16.2.4.41
valve-rod nut 阀杆螺母 16.2.3.52
valve-rod stuffing box gland 阀杆填料函压盖 16.2.3.50
valve seat 阀座 16.2.3.37
valve seat ring 阀座圈 22.1.3.12
valve spindle 阀杆 22.1.8.9
valve spring 阀弹簧 16.2.3.35
valves reversed 阀反向 6.2.4.8
valve system 阀门组件 17.2.11
vanadium extraction 钒萃取 6.1.12.(c)
vane 叶片 16.4.15.15
vane control motor 叶片控制电机 20.2.3.16
vanes 导流叶片 12.3.3.4
vane wheel 叶轮 18.4.2.10
van Tongeren cyclone van Tongeren(旋风)分离器 10.1.4.(c)
vapor 蒸气 2.2.98.12;2.4.3.8
vapor body 蒸发器壳体 7.1.14.5
vapor body 蒸发室主体 8.8.4
vapor compression 蒸气压缩 20.1.2.19
vapor-compression cycle 蒸气压缩循环 20.2.5
vapor condensation zone 蒸气冷凝段 2.2.85.15
vapor connection 蒸气接口 4.3.1.3
vapor desuperheater 蒸气降温器 7.1.14.12
vapor fan 蒸气风机 3.3.24.13
vapor feed 蒸气进料 2.3.5.1
vapor flange 蒸气出口法兰 8.9.5.2
vapor flow 蒸气流 2.4.3.9
vapor hood 蒸气罩 8.7.1.10
vapor in 蒸气进入 10.4.3.4
vapor inlet 蒸气进[入]口 18.3.4.15
vaporized oils 气化的油 3.3.15.9
vaporizer 蒸发器 7.2.1.20
气化器 3.1.46.7
vaporizer(if required) 气化器(如需要) 3.3.21.6
vapor-liquid equilibrium(VLE) 汽液平衡 1
vapor-liquid equilibrium ratio 汽液平衡比 1
vapor out 蒸气出口 10.4.3.5
vapor outlet 排气口 13.5.1.13
蒸气出口 7.1.2.7
vapor path 蒸气路线 25.6.2.1
vapor product 蒸气产品 2.1.9.32
vapor riser 升气管 5.2.8.6
vapors 气化物 2.2.82.11

vapor separator 蒸气分离器 6.3.2.21
vapor stream 蒸气流 18.2.3.12
vapo(u)r feed 蒸气进口 5.3.1.7
vapo(u)r out 气体出口 5.4.1.16
vapo(u)r outlet 蒸气出口 5.2.1.1
vapour outlet 蒸气出口 7.2.10.15
vapo(u)r sample 气体取样口 5.4.1.19
vapo(u)r space 蒸气域 5.5.1.41
variable angle tumbler 可变角度转鼓 9.5.2.17
variable level 可变液位[面] 3.2.6.14
variable size 可改变尺寸 11.5.2.3
variable-speed and brake motor 变速电机 8.8.11.7
variable-speed drive 变速驱动装置 7.2.13.16
variable speed drive recycle pump 变速驱动循环泵 7.2.9.9
variable-speed driver 变速驱动器 8.9.1.1
variable speed reduction motor 电机减速机 13.5.1.2
variable-vapor-space 可变蒸气空间贮罐 21.2.1.5
vari-speed drive 变速驱动装置 24.1.9.1
varnished-cambric tape 漆布带,黄蜡布带 25.4.8.4
varying throat opening 可改变喉部开度 10.5.10.4
vat 受料槽 12.2.3.20
(纸)浆槽 2.3.11.47
vatman 捞工 2.3.11.46
V-belt drive V型带传动 4.5.1.2
V-belt driven product pump V形带传动的产品泵 7.2.9.7
VCM column 氯乙烯单体塔 2.3.18.18
VCM purification VCM净化 2.3.18.F
velocity adjustor 调速器 6.1.9.8
velocity-measurement system 流速测量系统 3.5.2
velocity profile 速度[分布]剖面[图] 1
vent 排[放]气口 10.3.6.1
放空 2.1.7.9;2.2.55.18;3.1.42.3;3.3.20.15
vent A 放空 A 2.3.19.7
vent air 放空 3.1.32.14
vent B 放空 B 2.3.19.13
vent chamber 放气室 23.1.9.2
vent condenser 放空冷凝器 25.6.2.6
vent connection 放气接口 4.1.1.32
vent fan 排(气)风机 23.1.6.5
vent gas 排出气 14.3.1.28;2.2.115.10

vent hole 通气孔,排气孔 21.3.1.7
ventilation shaft 通风管道 2.3.12.14
venting air heaters 排出空气加热器 4.3.4.22
vent line 放空(管)线 25.6.7.12
vent nozzle 排气管 21.3.1.8
vent pipe 通[排]气管 5.4.1.23
vent timer 排气定时器 3.4.22.3
vent to atmosphere 放空至大气 6.2.3.2
Venturi 文丘里(管) 10.5.13.7
Venturi feed 文丘里进料 23.1.1.(c)
Venturi scrubber 文丘里涤气器 2.2.106.30
venturi scrubber evaporator 文丘里涤气蒸发器 3.3.4.10
Venturi scrubber system 文丘里洗涤器系统 10.5.14
Venturi throat 文丘里管喉部,文丘里喉管 3.3.8.12
Venturi tube 文丘里管 1
vent valve 排气阀 23.1.8.9
vertical axial flow pump 立式轴流泵 16.4.30
vertical axis 垂直轴 3.4.1.3
vertical axis current meter 垂直轴流速计 3.4.1.(a)
vertical can-type pump 立式屏蔽泵 16.1.2.33
vertical-flow heavy-duty plate precipitator 垂直流动重负载板式除尘器 10.3.2
vertical-flow reactors 垂直流反应器 3.3.15
vertical lift 升降机,电梯 2.3.8.36
vertical lift check valve 直立升降式止逆阀 22.1.1.(a)
vertical multistage centrifugal pump 立式多级离心泵 16.4.27
vertical pressure leaf filter 立式叶滤机 9.2.2
vertical pump 立式泵 16.4.21
vertical rotor 立式转子 10.5.1.21
vertical screw 立式螺旋混合器 12.1.3.(f)
vertical screw mixer 立式螺杆混合器 12.1.2
vertical seperator 立式气液分离器 5.3.3.8
vertical-shaft end-suction pump 立式端吸离心泵 16.4.14
vertical sieve tray(VST) 垂直筛板 1
vertical-slot baffle 直立挡板 3.1.3.7
vertical spindle well pump 立轴式井泵 16.1.2.241
vertical split centrifugal pump 垂直剖分式离心泵 16.4.17
vertical tube coalescing 垂直管聚并 15.5.4
vertical tube coalescing separator 垂直管聚

621

并分离器 15.5.1
vertical tube condenser 立管冷凝器 5.5.2.9
vertical tube (perforated) coalescing separator 垂直管(多孔管)聚并分离器 15.5.4.5
vertical turbine pump 立式涡轮泵 16.4.25
vertical type dryer 立式干燥器 8.9
vertical type pressureleaf filter 立式加压叶片过滤机 9.2.6
vertical U-tube water-tube boiler 立式 U 形管水管锅炉 4.4.1
vertical wet-pit diffuser pump bowl 立式扩压型排水泵 16.4.24
very high pressure compressor for polyethylene process 聚乙烯超高压压缩机 17.2.15
very high pressure pump 超高压泵 16.1.2.1.6
very high vacuum 超高真空 19.4.1.14
vessel 容器 9.2.1.15
vessel diameter 容器直径 6.3.9.3
vessel flange 容器法兰,设备法兰 5.2.4.12
vessel heating 容器加热管 8.9.5.8
vessel hight 容器高 6.3.9.2
vessel lower section 容器下段 8.9.5.11
vessel upper section 容器上段 8.9.5.10
VFA(volatile fatty acid) 富挥发性脂肪酸 3.2.10.4
VFA rich fermenter supernatant 富挥发性脂肪酸发酵罐上清液 3.2.10.13
VFA rich primary effluent 富挥发性脂肪酸一次排放物 3.2.10.4
vibrated fluidized bed 振动流化床 1
vibrating conveyor 振动输送机 23.3
vibrating conveyor type 振动输送式 13.5.2
vibrating feeder 振动加料器 12.7.4.2
vibrating screen 振动筛 1
vibrating tray dryer 振动盘式干燥器 8.9.6
vibration-stopper 消振器 14.4.1.20
减振装置 25.1.1.22
vibrator 振动器 12.2.3.3
vibratory ball mill 振动球磨机 11.3.2
vibratory-conveyor adaptations 振动输送机 4.3.4
vibratory discharge 振动卸料 9.5.2.7
vibratory power unit 振动的动力装置 4.3.4.13
vibro-energy mill 振动研磨机 11.3.1
vibro frame 振打机架 10.2.2.4
vibro motor 振打电机 10.2.2.5
view A-A A-A 剖视 10.1.3.10
viewing and lighting windows 窥视和采光孔 13.5.2.7
viewing port 窥视孔 4.3.4.1
vinyl chloride monomer 氯乙烯单体 2.3.18.20
virgin-carbon makeup 新鲜活性炭补充 3.1.20.1
virgin kerosene 直馏煤油 2.1.1.8
visbreaking 减黏裂化 2.1.1.33
viscoelastic fluid 黏弹性流体 1
viscoplastic fluid 黏塑性流体 1
viscose rayon 人造丝 2.3.7.34
viscose rayon cake 黏胶丝饼 2.3.7.18
viscose rayon staple fibre 黏胶短纤维(人造丝) 2.3.7.28～34
viscose spinning solution 黏胶纺丝液 2.3.7.10
viscosity 黏度 1
visual exam 外观检验 21.1.2.11
vitrified-clay sewer pipe 釉面陶土排污管 22.6.2.⑤
Vitro uranium mixer-settler Vitro 混合-澄清器(用于铀萃取) 6.1.13
VLE(vapor-liquid equilibrium) 汽液平衡 1.
VOC destruction by catalytic oxidation with recuperative heat exchanger 带同流换热器的催化氧化法分解挥发性有机化合物 3.4.5
voidage 空隙率 1
volatile tops 挥发性馏分 5.5.3.7
voltage 电压 25.4.1.41
voltage transformer 变压器 25.4.9.9
volts(V) 伏特 10.3.1.29
volume 容积 21.2.5
volumetric flask 容量瓶 2.3.11.3
volumetric flow rate 体积流率;体积流量;体积流速 1
volumetric oxygen transfer coefficient 容积传氧系数 1
volute 蜗壳 1
volute casing 蜗(形机)壳 20.2.1.8
volute casing pump 蜗壳泵 16.1.2.2.27
volute race discharge 蜗壳卸料口 9.5.10.17
volute throat 蜗壳喉部 16.4.3.13
vortex 涡流 10.1.2.4
vortex breaker 防涡器 5.3.4.41
防涡流板 5.4.1.14
vortex pump 漩涡泵 16.1.2.2.18
vortex shield 涡流屏 10.1.2.5
votator apparatus 套管冷却结晶器 1
V-pulley V 型皮带轮 11.1.5.15
VST(vertical sieve tray) 垂直筛板 1.

V-type and double-cone mixers drum 转鼓型
 V 型和双锥型混合器 12.1.4
V-type mixer V 型混合器 12.1.4.(e)
V-type reciprocating compressor V 型往复
 压缩机 17.2.1.6

W

wafer type body 薄片型阀体 22.1.4.8
walkway 人行道 15.6.3.2
 走道 15.6.5.15
wall 壁 4.3.1.11
wall casting 墙固定板 24.2.6.3
wall effect 壁效应 1
warm air 暖空气 4.5.2.2
warm air to furnace 入炉热空气 25.5.3.13
warm balls 温热小球 2.2.5.21
warm product gas 热成品气 2.2.28.4
warm water inlet 热水进入 4.6.2.9
warm water outlet 温水出口 4.3.2.14
WAS(waste activated sludge) 废活性污泥
 3.2.2.20;3.1.30.B.6;3.1.33.7;3.1.40.5
waste water 废水 2.2.6.18
wash 洗液 9.3.1.13
 洗涤 9.4.8.2
wash basket 洗涤篮筐 9.5.10.2a
wash discharge 洗液出口 9.5.6.14
wash distributors 洗液分布器 9.3.9.3
washed coal 洗过的煤 2.2.93.1
washer 垫圈 13.1.2.3
 垫片 17.2.9.19
washer/cooler 洗涤塔/冷却塔 2.1.8.8
wash header 洗涤水总管 9.3.7.3
washing 水洗 2.3.7.19
 洗涤 9.1.1
washing floor 冲洗层 2.3.12.27
 洗涤层 2.3.12.3
washing nozzle 洗涤液接管 9.2.1.28
washing of yarn packages 丝包的洗涤 2.3.8.48
washing pipe 清液管 15.6.4.3
washings 洗涤液 1
washing tower 洗涤塔 5.1.1.9
washing water inlet 洗涤水入口 9.2.5.15
wash inlet 洗水入口 9.1.1.6
 洗涤入口 9.5.5.19
wash liquid 洗(涤)液 3.1.9.2
wash oil tank 洗油罐 2.2.106.28

wash pipe 洗液管 9.5.6.10
wash plate 洗涤板 9.1.1.3
washrinse tank 冲洗罐 6.2.5.9
wash solvent 洗涤溶剂,洗涤液 2.2.102.7
wash sprays 洗涤水喷头 9.3.1.3
wash vacuum connection 洗液通真空系统的
 接口 9.3.2.29
wash water 洗涤水 12.4.1.1
wash-water curtain 洗涤水水幕(水帘) 15.5.3.3
washwater inlet 洗水入口 6.2.6.6
wash water outflow 洗涤水流出 15.5.3.5
washwater outlet 洗涤水出口 6.2.6.5
washwater pump 洗涤水泵 9.3.1.1
wast 废料 6.2.1.10
waste 污水池 3.1.12.12
waste 废物 3.2.6.23;2.2.66.12.2.3.15.13
 废料 6.2.5.7
waste acid stripper 废酸汽提塔 2.2.56.6
waste bin 废料仓 12.7.5.9
waste biomass to solids handling 废弃生物质
 送去固体处理 3.1.36.10
waste carbon/biomass 废炭/生物质 3.1.43.8
waste conditioner 废弃物调节器 3.1.22.5
waste catalyst 废催化剂 2.2.72.7
waste disposal 排污 2.2.89.17
waste entrance 废物入口 3.3.11.12
waste feed 废弃物进口 2.2.12.10
waste feed 废弃物进料 3.3.3.6
waste gas 废气 3.3.10.12;2.3.17.9
 工业废气 3.4.1.2;3.4.2.2
waste gas combustor 废气燃烧炉 2.3.17.7
waste-gas cooler 废气冷却器 14.2.7.10
waste gas inlet 废热气入口 4.2.4.13
waste gas outlet 废气出口 4.2.4.18
waste gas treatment 废气处理 3.4
waste gas to be oxidized(inlet) 将被氧化的
 工业废气(入口) 3.4.5.2
waste heat 废热 20.1.9.11
waste-heat boiler 废热锅炉 2.1.8.9
waste heat boiler of shift converter 变换炉废
 热锅炉 2.2.34.4,5
waste heat recovery 废热回收 2.2.46.8
waste liquid 废液 3.3.10.8;3.1.47.5
waste gas combustor 废气燃烧炉 2.3.17.7
waste liquors 废液 2.2.106.26
waste regenerant 废再生液 6.2.7.7
waste sludge 废污泥 3.1.32.22;3.1.34.12;
 3.1.44.8

623

waste-sludge dewatering and carbon regeneration 污水污泥脱水和活性炭再生 3.1.20.9
waste-sludge disposal choices 污水污泥处理方法挑选 3.1.20.8
waste-sludge disposal systems 污水污泥处理系统 3.1.20.7
waste sludge to sludge handling 废污泥去污泥处理 3.2.10.6
waste solids 废固体颗粒 2.2.4.7
waste solvents 废溶剂 3.1.6
waste source 废物源 3.3.7.10
waste sump 废料坑 6.2.1.9
waste treatment 废物处理 3
waste water 废水 15.5.4.8;2.2.52.9;2.2.58.4;2.2.59.1;3.1.43.1
waste water discharge 废水排出 2.3.14.38
waste water feed 废水投入 15.2.4.1
waste water feed tank 废水料槽 3.1.7.2
waste water pump 废水泵 2.3.10.63
waste water recovery column 废水回收塔 2.2.58.14;2.2.59.18
waste water treatment 废水处理 3.1
wastewater-treatment facility 污水处理装置 3.1.1
wastewater treatment plant showing instrumentation 污水处理厂所用装置 3.1.2
wastewater treatment system the role of the biochemical operations 采用生化操作的废水处理系统 3.2.1
water 水 2.1.9.8;2.2.18.6;2.2.27.2;2.2.28.7;2.2.29.1;2.2.30.14;2.2.32.11;2.2.57.16;2.2.64.15;2.2.68.7;2.2.74.6;2.2.75.7;2.2.76.20;2.2.77.12;2.2.80.13;2.2.81.8;2.2.115.15;2.3.17.12;2.3.18.10;3.3.19.1
water addition for fines destruction 破损细粒用水 7.2.8.14
water and naphtha, etc 水,石脑油等 2.2.36.13
water ash 水灰 3.3.8.11
water body 水体 25.1.1.13
waterbox 水箱 18.4.1.10
water collecting orifice 集水孔 15.6.4.16
water collecting pond 集水槽 4.2.3.5
water-collecting tank 集水箱 2.3.12.5
water column 水塔 2.2.52.8
water condenser 水冷凝器 2.2.80.11
water-cooled 水冷却 13.5.3.2
water cooled gland 水冷式填料压盖 16.4.11.12

water-cooled reactor 水冷反应器 2.2.42.7
water-cooled synthesis reactor 水冷式合成反应器 2.2.43.10
water cooler 水冷器 2.2.44.3
water cooling stabilizer panel 水冷稳定板 15.7.5.7
water cooling tower 凉水塔 10.5.18.17 水冷塔 2.2.46.2;2.2.47.4
water depth variable 可变水深 15.6.1.26
water detection probes 水检测探头 15.5.2.6
water direction 水流方向 15.6.5.16
water discharge 水出口 18.3.4.20
water distributing system 配水系统 4.6.2.5
water distribution 淋水装置 4.6.1.5
water distributor 水分布器 5.4.2.15
water downcomer pipes 降水管 4.4.2.5
water drain 排水口 4.3.4.12
water drained by gravity 借助重力排水 8.9.6.3
water droplet 水滴 10.5.1.10
water film 水膜 4.3.2.1
water gas shift 水气变换 2.2.3.14
water gas waste heat boiler 水煤气废热锅炉 2.2.34.1,8
water hammer 水锤 1
water header 水总管 7.1.2.23
water jacket 水夹套 2.2.18.12
water jacket for cooling 冷却水夹套 17.2.5.12
water jet pump 水喷射泵 16.1.2.4.3
water knockout 气水分离器 6.3.3.8
water leg 水竖管 17.6.2.9
water level 液面 15.6.1.31 水平面 15.6.4.19 水位 24.1.5.7
water level detector 水位探测器 15.5.5.6
water-level recorders 水位记录仪 3.4.5
water line 水管线 25.6.6.3
water or nutrient feed 水或营养物进料 3.4.3.9
water out 水出口,出水 10.4.3.3
water outlet 水出口 10.5.4.2
water overflow 水溢流 11.6.6.7;2.2.26.10
water paring disc 弧形水盘 9.5.12.1
water-phase 水相 6.1.8.7
water pipe 水管 2.3.10.65
water pond 水盘 10.3.4.6
Water processing flow diagram considered in the study of wastewater treatment plant in Avellino, Italy （意大利 Avellino)废水处理

厂研究中的水处理工艺 3.1.25
water recycle 水循环 2.2.27.4
water-ring pump 水环泵 18.4.1
water removal 脱水 2.2.68.6
water-ring vacuum pump 水环真空泵 18.4.2.9
water scrubber 水洗塔 2.2.69.8;2.2.72.10
water seal 水封 6.2.3.10
water shower or air spray 水喷淋或空气喷雾 3.3.19.8
water splitter 水分离塔 2.2.28.5
water spray 喷水器 12.7.4.1
水喷头 14.2.4.5
water-spray cooler 洒水式冷却器 8.8.8.3
water-spray extended cooler 喷水式冷却器 8.8.8
water supply 供水 4.3.4.11
water supply pipe 供水管 2.3.12.48
water supply pump 供水泵 16.1.2.3.2
water surface 水表面 3.5.4.4
water table 地下水 3.3.12.9
water tank 水槽 4.3.2.5
water tank with grinder 研磨水槽 2.2.35.4
water to recycle 水去循环 2.2.106.5
water treatment 水处理 3.1.22.16
water tubes 水管 1.1.10.6
water vapor 水蒸气 7.1.14.9
water vapor outlet 水蒸气出口 6.3.2.22
waterwalls 水汽壁 25.5.3.7
waterworks pump 水厂泵 16.1.2.3.6
wax outlet 蜡出口 14.3.2.14
3-way stop cock 三通旋塞 3.5.14.32
3-way valve 三通阀 23.1.1.4
weak aqua 稀液 20.1.9.8
weak solution 稀溶液 20.1.8.16
weak wash liquor 循环洗液 9.4.1.11
wearing ring 耐磨环 16.4.22.21
wear plate 耐磨板 9.3.3.14
web 纸幅 2.3.11.30
卷筒纸 2.3.11.45
Weber number 韦伯数 1
web plate 筋板 5.2.2.5
wed 筋板 21.3.3.2
wedge bar liners 楔形挡板衬里 11.2.2.1
wedge compact set 楔形紧固件 5.3.4.28
wedge gate 楔形闸板 22.1.1.7
wedge zone 挤压区 9.6.2.2
wed plate 腹板 21.3.1.5
weep hole 泪孔 5.2.1.20
weigh beam 称量杆 23.1.2.22
weighing apparatus 称量设备[装置] 2.3.12.32
weighing equipment 称量设备 23
weighing of bulk solids 粉粒体的称量 23.5
weighing system 称量系统 23.4.1
weigh scale 秤 23.5.1.2
weight 负荷 3.5.1.1
weight cylinder 重锤导向筒 15.7.1.10
weighted mean 加权平均 1
weir 堰 15.5.1.7
weir height 堰高 1
weir plate 溢流堰板 8.5.6.21
weir-riser distributor 堰-上升管分布器 5.3.4.14
weir type overflow 堰式溢流 11.3.2.4
weld const 焊接结构 21.1.2.7
welded connection 焊接连接 21.1.1.2
welded ring support 焊接环支承 5.3.4.22
welding end construction 焊接端结构 22.5.3.3
well 钻井 2.1.2.17
well-designed 良好设计 23.2.2.(b)
wet-air oxidation 湿空气氧化作用 3.1.23
wet ash 湿灰 2.3.1.34
wet bulb 湿球温度 20.2.2.5
wet bulb temperature 湿球温度 1
wet classification machine 湿式分级设备 12.4.1
wet classifier 湿式分级器 12.4
wet-coal bin 湿煤仓 8.5.6.26
wet cottrell 湿式静电除尘器 10.3.4
wet drum separator 湿磁鼓分离器 12.6.9
wet feed 湿物料进入 8.4.6.19
湿进料 8.5.6.7
wet feeder 湿加料器 8.4.1.34
wet feed gas 湿气体 6.3.3.1
wet feed mixer 湿进料混合器 8.4.1.2
wet flammable sludge 含石油气易燃残渣 3.3.5.5
wet gas 湿气体 2.1.5.11
wet machine 湿式机器 12.7.4.1
wet material 湿物料 8.3.1.1
wet material feed 湿物料入口 8.8.6.4
wet oxidation lab equipment 湿氧化实验室设备 3.4.7
wet-pit pump 排水泵 16.1.2.3.14
wet product inlet 湿物料进口 8.9.6.9
wet scrubber 湿式涤气塔 2.2.118.13
wet scrubbing 湿法涤气器 2.2.31.21
wet sludge conveyor 湿污泥输送机 3.3.24.4
wet-stage cyclone 湿级旋风分离器 8.4.1.35

625

wet steam 湿蒸汽 10.1.4.13
wet steam in 湿水蒸气进入 10.4.3.1
wetted fan wheel 浸湿风扇叶轮 10.5.15.4
wetted perimeter 润湿周边 1
wetted wall column 湿壁塔 1
wet vapor 湿蒸汽 20.2.4.26
wet well 湿井 3.1.8.5
wet zone 湿区 20.2.4.8
WGS 水煤气变换炉 2.2.117.9
wheel nut 手轮螺母 22.1.3.1
when air 空气时 2.2.2.5
whirlpool separator 涡流分离器 2.3.13.2
white particles 白粒子 12.1.4.2
whizzed air classification 离心空气分级 11.2.8
whizzer blades 离心机叶片 11.2.8.7
whizzer drive 离心机传动装置 11.2.12.5
wide pattern 大半径 17.6.2.2
wier-web packing 网波纹填料 5.3.4.47
Wilson seal 威尔逊密封 19.4.1.13
windbox 风箱 24.2.4.13;3.3.27.2
winding machine 卷扬机 23.3.6.1
wind-up unit 卷扬机 15.7.5.11
Winkler Winkler气化炉 24.3.2
wiped-film still 刮膜式分子蒸馏釜 5.5.3
wiper 刮板 8.9.1.6
wiper blade 宽刮刀 9.4.9.6
wiper redistributor 收集型再分布器 5.3.4.15
wiper ring 刮油环 17.2.7.6
wire filter leaf 丝网滤叶截面图 9.1.4
wire mesh support 丝网支承 5.3.4.23
wire mesh walkway 金属丝网通道 10.2.1.6
wire net 多孔板或金属丝网 8.3.4.2
wire retainer 金属丝支承圈 10.2.3.2
wire 导线 10.3.5.15
wire strippers 剥线钳 25.4.1.64
wire winding 金属线材绕组 9.3.3.4
wiring center dolly box 绕线的中心滑动盒 9.3.3.20
wiring sprocket 绕线链轮 9.3.3.2
with bottom discharge by scraper 刮刀下部卸料式 9.5.3.(a)
withdrawal 提取 3.1.32.6
with manual top discharge 手工顶部卸料式 9.5.3.(c)
with pneumatic top discharge 气动顶部卸料式 9.5.4.(b)
wood 木质和衬 22.6.2.
wood flights 木制刮板 3.1.3.10
wood grid 木格栅 5.3.2.(m)
wood-lined steel pipe 衬木钢管 22.6.2.

wood screw 木螺钉 26.3.12.45
wood stave 木板条 9.3.3.27
work 作功 20.1.2.2
working baths 工作池 2.3.9.5
working face 工作表面 12.6.10.4
working fluid 工作液 18.2.3.4
working fluid inlet 工作液入口 25.5.1.11
working platform 工作平台 2.1.2.5
working principle of crossflow membrane filtration 错流膜过滤工作原理 3.1.29
working space 工作室 16.3.7.4
worm 蜗杆 11.1.8.15
worm beds 蚯蚓床 3.3.28.8
worm case 蜗轮箱 11.1.8.16
worm-drive gear 蜗杆传动蜗轮 9.3.3.44
worm gear 蜗轮 16.2.5.11
worm-gear-operated 蜗轮操作 8.8.11.14
worm shaft 蜗杆轴 9.3.3.47
worm speed reducer 蜗轮减速器 9.2.7.3
worm wheel 蜗轮 11.1.8.14
wort 麦芽汁 2.3.12.51
wort break removal 麦芽汁残渣的去除 2.3.13.1～5
wort cooler 麦芽汁冷却器 2.3.13.5
W-type compressor W型压缩机 17.2.1.7
wye connector 三通接头 17.6.2.11

X

X-quality X-质量 20.2.4.27
X-ray source X射线源 12.7.5.3
p-xylene 对二甲苯 2.2.64.12
xylenols 混合二甲酚 2.2.91.9

Y

45°Y-branches (straight size) 45°Y型三通（等径） 22.3.2.6
Y-connection Y连接 25.4.3.2
yeast 酵母 1
yield 收率 1
yoke 支架 21.1.1.31
yoke 轭架 22.1.4.3
       轭轭 25.4.3.14
yoke bolting 轭架螺栓 22.1.4.7
yoke bolt 支架螺栓 22.1.3.6
yoke plate 托架板 19.1.5.1

Z

Zechstein shale 蔡希斯坦页岩,镁灰油页岩 25.2.4.12
zeolite catalyst 沸石催化剂 1
zeolite crystals 沸石晶体 6.3.10.9
zero 零点 3.5.11.21
zero leakage 零泄漏 1
zig-zag entrainment separator 曲径式雾沫分离器 10.5.8.2
zinc flotation 锌的浮选 12.8.1.14
zirconium and zirconium alloys 锆和锆合金 22.6.3.㉕
ZnO reactor 氧化锌反应器 2.2.8.5
zone of mixing 混合区 3.3.18.5
Z-type conveyor-elevator Z型输送式提升机 23.3.3.(b)

汉英对照词汇部分

A

阿勒岩盐　Aller rock salt　25.2.4.7
安定区　calming area[zone]　5.2.7.8
安全防爆孔　explosion port　25.6.1.4
安全放空口　relief vent　3.3.24.8
安全封　safety seal　14.1.11.2
安全过滤器　security filter　3.1.28.6
安全护罩　safety guard　15.7.3.8
安全外挡圈　outer safety ring　13.1.2.22.31
安位销　dowel pin　17.2.11.8
安装位置　mounting position　18.2.4.9
氨　ammonia　2.3.15.10
氨　NH_3　2.2.111.12；2.2.38.13
氨成品　NH_3 product　2.3.3.18
氨,氮肥　ammonia,nitrogen fertilizers　2.2.3.18
氨或电或热　ammonia or electricity or heat　2.2.78.14
氨合成　ammonia synthesis　2.3.1.14；2.2.38.12
氨冷冻　ammonia refrigeratiom　2.2.38.14
氨合成回路,显示了急冷反应器　ammonia synthesis loop,showing a quench reactor　2.2.37
氨气洗涤塔　ammonia scrubber　2.3.14.23
氨-水吸收循环　ammonia-water absorption cycle　20.1.9
氨压缩机　ammonia compressor　17.1.2.1
氨压缩冷冻机　ammonia compression refrigerating machine　17.1.2.2
鞍点共沸物　saddle-point azeotropic mixture　1
鞍式支座　saddle support　21.3.2
支撑(ing)　saddle　8.8.10.11
凹面法兰　female flange　22.4.3.
凹凸密封面　male-female seal contact face　22.4.3.(c)
凹形拱墙　drop arch　25.1.4.6

奥氏体不锈钢　austenitic stainless steels　22.6.2.⑯
奥斯陆型结晶器　Oslo-type crystallizer　7.1.10

B

42巴甲酸甲酯反应器　methyl formate reactor 42 bars　2.2.62.5
巴氏合金　babbitt　19.3.3.(a)
巴氏消毒过的水（温度 T_2）　pasteurized water T_2　2.4.1.3
拔顶蒸馏塔　topping distillation column　2.2.43.12
白粒子　white particles　12.1.4.2
百叶窗循环挡板　circulation louvers　7.2.14.4
摆锤　swing hammer　11.1.8.8
摆锤销轴　hammer pin　11.1.8.20
摆动臂　swing arm　12.1.1.25
摆动锤　swing hammers　11.2.8.5
摆动颚板　swing-jaw plates　11.1.1.3
摆动管　swing pipe　2.3.10.10
摆动轮　oscillating spider　9.3.3.12
摆动中心(枢轴)　pivot　11.1.1.15
摆式进料管　pendulum feed pipe　8.7.1.11
班砂岩降侧　Bunter downthrow　25.2.4.5
班锋下落断层块　Bunter downthrow　25.2.4.5
Dollar板　Dollarplate　14.1.7.(e)
板波纹填料　mellapak　5.3.4.48
板翅管式　plate fin-and-tube type　4.2.5.5
板簧　spring　23.3.1.6
板间距　space of plates　5.2.14
　tray spacing　5.2.6.11
板框式换热器　plate-and-frame heat exchanger　4.2.1
板框型反渗器　plate-and-frame reverse osmosis system　15.2.3
板框压滤机　plate and frame（type）filter press　9.1.2

板式换热器 plate-type exchanger 4.2.5.1
板式加料器 apron feeder 12.8.2.3
板式平焊法兰 plate-type flat welding flange 22.4.2.(a)
板式塔 plate[tray] column 5.1.2.1
板状硅藻土计量器 measuring vessel for the kieselguhr 2.3.13.3
半闭式内部燃烧型燃气轮机循环 semiclosed, internally fired gas-turbine cycle 25.3.5
半顶角 1/2 apex angle 21.1.1.23
半刚性 semirigid 22.6.1.(b)
半接头 half coupling 10.4.2.11
半经验模型 semi-empirical model 1
半开式叶轮 semi-open impeller 16.1.1.42
半理想溶液 semi-ideal solution 1
半连续过程 semi-continuous process 1
半球形封头 hemispherical head 21.4.1.7
（半）湿物料 moist material 8.8.9.4
半透膜 semipermeable membrane 1
半正位移驱动装置 semipositive drive 23.3.2.2
半自动码垛 semiautomatic palletizing 23.4.3
棒锤 bar hammer 11.2.6.7
棒磨机 rod mill 11.2.4
包圈 binding 9.1.3.8
包套型 envelope type 10.2.1.(b)
包装 packaging 23.4
包装机输送器 packer conveyor 23.4.2.2
薄壁中空塑料管 hollow, thin-walled plastic tube 15.2.2.4
薄金属片 strip of sheet metal 25.4.1.17
薄膜反应器 thin film reactor 14.3.1.(f)
薄膜蒸发器 thin-film evaporator 1
薄片型阀体 wafer type body 22.1.4.8
饱和产物 heavy product 14.2.3.6
饱和区 saturation zone 3.1.32.13
饱和液体曲线 saturated liquid curve 20.2.4.15
饱和液体线 saturated liquid line 20.2.4.22
饱和蒸汽 saturated steam 25.5.3.5
饱和蒸汽出口 saturated-steam outlet 25.5.1.8
饱和蒸汽锅炉 saturated-steam boiler 25.5.1.(a)
饱和蒸汽曲线 saturated vapor curve 20.2.4.12
饱和蒸汽线 saturated vapor line 20.2.4.18
宝塔式淋洒器 cascade impact thrower 5.3.4.5
保持摩擦力的螺钉 spring to maintain tension on rubbing surfaces 8.8.5.8
保护壁 protective screen 25.2.5.2
保护层 protection casing 3.1.24.14
保护圈 guard ring 18.2.4.4
保留 retention 1
保温 thermal insulation 4.7
保温板 insulation board 24.1.2.12
保温材料 heat insulator 14.4.3.23
保温层 insulation 3.1.16.12
保温混凝土 insulating concrete 4.4.2.3
保温圈 insulating ring 5.2.1.5
保险弹簧 relief spring 11.1.3.10
保险线盒 fusible element 25.4.1.36
北温带板 north temperate zone plate 21.2.3.5
备料设备 stuff preparation plant 2.3.10.79～86
备用气口[接管] spare nozzle 14.4.1.2
背衬法兰 backing flange 22.4.4.2
背靠背叶轮泵 back-to-back impeller pump 16.1.2.2.26
背压力 back pressure 16.2.1.6
被抽气体 extracted gas 18.2.3.13
本菲尔德工艺——增压流化床燃烧与燃烧后脱碳联合的 Sargar 工艺流程 The Sargas process combines a pressurized fluidized bed combustion process with a post-combustion CO_2 removal process, here shown as the Benfield process 2.2.116
本菲尔德脱碳工艺 Benfield process 2.2.116.13
苯 benzene 2.1.1.51
（苯）酚 phenol 2.3.8.14
苯酚蒸馏塔 phenol still 3.1.5.9
苯氯化 benzene chlorination 2.3.8.8
苯洗涤塔 benzene scrubber 2.3.14.24
苯乙烯 styrene 2.1.1.58
泵 pump 16; 2.2.52.2; 2.2.76.18; 2.2.115.3; 3.4.1.7; 3.4.2.4
泵侧 pump half 16.4.12.21
泵侧联轴器锁母 pump half coupling lock nut 16.4.27.23
泵出 pump out 6.1.7.13
泵出口 pump discharge 16.4.1.20
泵动力 pump power 25.1.2.8
泵阀 pump valve 3.5.14.24
泵法兰 pump flange 18.2.1.2
泵盖 pump bonnet[cover] 16.4.2.3
　　　pump casing cover 16.4.15.2
泵杆 pumping rod 2.1.3.25
泵缸 pump[liquid] cylinder 16.2.5.1

泵缸衬套　liquid cylinder lining　16.2.3.30
泵缸端盖　liquid-cylinder head　16.2.3.29
泵缸盖　liquid head　16.2.8.2
泵缸活塞　liquid piston body　16.2.3.22
泵缸活塞螺母　liquid piston-rod nut　16.2.3.26
泵缸活塞杆填料函　liquid piston-rod stuffing box　16.2.3.18
泵缸活塞杆填料函衬套　liquid piston-rod stuffing box bushing　16.2.3.19
泵缸活塞杆填料函压盖　liquid piston-rod stuffing box gland　16.2.3.17
泵缸活塞杆填料函压盖衬套　liquid piston-rod stuffing box gland lining　16.2.3.16
泵缸活塞耐磨环　liquid piston bull rings　16.2.3.21
泵缸活塞随动板　liquid piston follower　16.2.3.25
泵缸活塞弹力环　liquid piston snap rings　16.2.3.20
泵缸支腿　liquid-cylinder foot　16.2.3.23
泵混(合)式混合-澄清槽　Pump-Mix mixer-settler　6.1.11
(泵)机架　frame　16.4.9.35
泵壳　casing　16.4.9.1
泵壳　pump body[casing]　16.3.7.2
泵壳口环　casing ring　16.4.9.12
泵壳耐磨环　casing wear ring　16.4.2.19
泵壳体　casing　16.4.29.7
泵内回流液　returning pump fluid　18.2.4.11
泵排出下沉产品　pump sink removal　12.5.4.(a)
泵驱动型式　drive type of pump　16.1.2.4
泵送的流体　fluid pumped　16.5.5.2
泵送或给料设备　pump up or charging set　25.3.2.21
泵送脉冲发生器　pump pulse generator　6.1.16.1
泵送系统　pumped system　3.5.10.(a)
泵体　pump body　16.3.6.2
　　bowl　16.4.26
　　cylinder　16.2.6.9
泵体耐磨环　pump body wearing ring　16.4.2.8
泵吸入口　pump suction nozzle　16.4.1.19
泵循环回流　pump around　2.1.4.20
泵叶轮　impeller of pump　16.4.22.2
泵站　pump station　3.1.41.4
泵轴　pump shaft　16.3.1.3
泵轴承架　pump bearing frame　16.4.21.6
泵轴套管　shaft enclosing tube　16.4.24.8

比表面积　specific surface area　1
比例抽样控制器　proportional sampling control　3.5.3.4
比例混合器　proportional mixer　24.2.7
比例调节阀　proportioning valve　23.1.8.16
比死亡速率　specific death rate　1
比转速　specific speed　1
笔尖杆　pen arm　3.5.5.3
毕托管　Pitot tube　16.5.3.5
闭路循环　closed circuit　11.6.5
闭式　enclosed　16.4.5.6
闭式结构　enclosed design　16.4.5.(b)
闭式循环燃气轮机　closed-cycle gas turbine　25.3.4
闭式叶轮　closed impeller　16.1.1.43
闭锁螺旋　lock-screw　11.4.3.5
闭锁式灰斗　ash lock　2.2.12.11
闭锁式料斗　lock hopper　2.2.106.20；2.2.20.2
闭锁式煤斗　coal lock　2.2.12.2
　　coal lock hopper　2.2.11.2
闭锁式煤渣斗　ash lock hopper　2.2.11.11
滗析槽　decantation tank　3.1.8.13
滗析器　decanter　2.2.106.25
蔽式涡轮搅拌器　shrouded turbine impeller　14.4.4.4
壁　wall　19.4.1.7
壁效应　wall effect　1
避雷针　Lightning rod　21.2.3.1
臂　arm　15.6.1.13
边杆螺母　side-rod nut　16.2.4.22
边界条件　boundary condition　1
扁平室空气流粉碎机　jet mill with flat chamber　11.5.5
变电所　transformer station　2.3.12.21
变换　conversion　2.2.33.8；2.2.36.4；2.2.38.5；2.2.78.6
CO变换　CO conversion　2.2.67.6
变换(反应)　shift conversion[reaction]　2.3.1.9
变换炉[器]　shift converter　2.2.101.10
R01-变换炉　R01-shift converter　2.2.34.19
E04-变换炉废热锅炉Ⅰ　E04-waste heat boiler of shift converterⅠ　2.2.34.4
E05-变换炉废热锅炉Ⅱ　E05-waste heat boiler of shift converterⅡ　2.2.34.5
E13-变换气水冷器　E13-shift gas water cooling　2.2.34.12
变速传动装置　variable speed drive　11.2.6.5
变速的驱动装置　variable speed drive　6.1.14.1

变速电机　variable-speed and brake motor　8.8.11.7
变速驱动器　variable-speed drives　8.9.1.1
变速驱动循环泵　variable speed drive recycle pump　7.2.9.9
变速驱动装置　variable-speed drive　7.2.10.16
变温吸附　temperature swing adsorption（TSA）　1
变压器　potential [voltage]transformer　25.4.9.9
变压器连接方式　transformer connection　25.4.3.1
变压吸附　PSA pressure swing adsorption　2.2.43.7；2.2.67.14
变压吸附系统示意图　Schematic diagram of pressure swing adsorption system　2.2.53
标定气体　calibration gases　3.5.11.23
标度因子　scale factor　1
标志　stamping　21.1.2.16
标准泵　standard pump　16.1.2.1.1
标准化工泵　standard chemical pump　16.1.2.3.34
标准夹套固定刮板式　standard jacketed solid flight　4.3.3.(a)
标准态　standard state　1
标准纤维长度　normal staple length　2.3.8.60
标准圆锥破碎机　Symons standard cone crusher symons　11.1.3
标准重介质分离系统　standard heavy media circuit　12.5.1.19
表观密度　apparent density　1
表孔　gauge well　2.1.7.7
表面安装式插座引出线盒　surface-mounted socket outlet　25.4.1.25
表面安装式三眼接地插座　surface-mounted three-pole earthed socket　25.4.1.13
表面反应控制　surface reaction control　1
表面更新理论　surface renewal theory　1
表面功　surface work　1
表面扩散　surface diffusion　1
表面冷凝器　surface condenser　7.2.8.8
Oslo 表面冷却结晶器　Oslo surface-cooled crystallizer　7.2.4
表面曝气器　surface aerator　1
表面曝气装置　surface aerator　3.1.45.5
表面洗涤　surface wash　6.3.9.10
表面型　surface type　20.1.10.23
表面型冷凝器　surface-type condenser　20.1.11
表土　topsoil　3.3.20.8
表土层　overburden　2.2.85.6
表压　in.gage.　10.3.5

冰浴　ice bath　3.5.14.17
冰浴冷凝器系统　condenser-ice-bath system　3.5.7.6
丙烷　propane　2.2.71.20；2.2.75.11
丙烷　C_3^0　2.2.69.23
丙烯　propylene　2.1.1.64；2.2.67.23；2.2.68.14；2.2.71.19；2.3.15.9
丙烯　$C_3^=$　2.2.69.20
丙烯腈-丁二烯-苯乙烯　ABS；acrylonitrile-butadiene-styrene　22.6.2.②
丙烯腈净化塔　acrylonitrile purification columns　2.3.15.8
丙烯精馏塔　propylene fractionator　2.2.71.3
并联排列　parallel rows　10.5.12.5
并联旋风分离器　multiple cyclones　8.4.1.16
并流　cocurrent flow　3.3.15.(b)
并流　parallel flow　17.4.3.2
并流喷嘴雾化式　concurrent-nozzle atomization　8.6.3.(d)
并流隧道式干燥器　parallel current tunnel dryer　8.3.1.(b)
并流型湿式鼓式磁力分离器　concurrent wet-drum separator　12.6.5
并流转盘雾化式　concurrent-disk atomization　8.6.3.(c)
波纹板油水分离器　corrugate-plate interceptor　3.1.4
波纹（穿流）塔盘　ripple tray　5.2.10.23
波纹套　bellows type　22.5.1.(g)
波纹管式差压流量计　bellows meter　24.1.31.27
波纹平行板　corrugated parallel plates　3.1.4.14
波形衬里　corrugated liners　11.2.2.4
玻璃板　glass ribbon　2.3.9.10
玻璃泵　glass pump　16.1.2.5.2
玻璃灯泡　glass bulb　25.4.1.57
玻璃钢罩　hood made of glass fibre plastics　12.2.3.15
玻璃工人坐椅　glassmaker's chair　2.3.9.45
玻璃管　glass pipe and tubing　22.6.2.⑦
玻璃管路　glass pipeline　22.4.4
玻璃管路与金属管路连接　connecting glass pipeline to metal systems　22.4.4
玻璃棉　glass wool　2.3.9.58
玻璃瓶模　parison mould　2.3.9.26
玻璃熔炉　glass furnace　2.3.9.49
玻璃丝　glass yarn　2.3.9.56
玻璃丝卷　sleeved glass yarn　2.3.9.57
玻璃纤维的生产　production of glass fibre　2.3.9.48-55

玻璃纤维丝　glass filaments　2.3.9.52
玻璃纤维线　strand　2.3.9.54
玻璃纤维制品　glass fibre products　2.3.9.56-58
玻璃油杯　glass oiler　12.3.5.8
玻璃制造　glass production　2.3.9
玻璃制造　glassmaking　2.3.9.38-47
剥线钳　wire stripper　25.4.1.64
伯努利方程　Bernoulli equation　1
补充供水口　makeup water supply　9.5.12.15
补充气　Supplementary gas　2.2.46.14;2.2.47.12
补充气体　make-up gas　2.2.42.1
补充氢气　makeup hydrogen　2.2.6.2;2.2.79.1
补充氢气和油进料　makeup H_2 and feed oil　2.1.10.5
补充水　makeup water　3.1.22.17;2.2.30.2
补充水进水管　make-up water inlet　4.6.2.11
补充用的蚯蚓　make-up earthworms　3.3.28.5
补充蒸汽　make-up steam　7.1.14.10
补给罐　makeup tank　6.3.10.3
补给软化水　softened makeup water　2.3.6.16
补强环　rein forcing ring　9.5.4.19
捕集器　collector　10.4.3.13
捕集筛网　collecting screen　7.2.11.2
捕集效率　collection efficiency　1
捕沫器　mist eliminator　3.1.9.14
不纯净的水(温度T_0)　impure water T_0　2.4.1.1
不带箍管接头　coupling without band　22.3.3.18
不带退刀槽的管帽　cap without recess　22.3.3.13
不等厚度　unequal thick ness　21.1.1.40
不合格品输送机　reject conveyor　23.4.3.6
不合格品输送器　reject conveyor　24.2.4.8
不合格石油产品贮槽　slop-oil tank　2.3.6.1
不回收的瓶子　non-returnable bottle　2.3.13.29
不加热下降管　unheated downcomers　25.6.1.7
不洁绿液　uncleared green liquor　2.3.10.41
不洁绿液贮槽　storage tank for the uncleared green liquor　2.3.10.41
不可泵送的　nonpumpable　3.3.7.16
不可泵送的废料　nonpumpable waste　3.3.7.6
不可逆过程　irreversible process　1
不凝气出口　non-condensable gas　7.2.5.7
不溶性无机物质　IIM;insoluble inorganic matter　3.2.1.5
不溶性有机物质　IOM;insoluble organic matter　3.2.1.3
不渗透的(混凝土)衬砌　impervious liner　3.3.12.3
不适当过滤　improper filtration　6.3.2.23
不透性石墨　impervious graphite　22.6.2.②
不透性石墨管　Resbon tube　5.4.2.8
不稳定汽油　unstabilized gasoline　2.1.5.10
不锈钢格栅　stainless steel grating　3.4.3.11
不锈钢斜板　stainless steel slope sheets　23.3.8.11
不锈钢主管道及聚氯乙烯支管　stainless steel ducting and PVC laterals　3.4.3.6
不需要射线检验　no radiographic exam　21.1.2.11
不旋转的球磨机　non-rotary ball　11.3
布袋捕集器　bag collector　8.5.3.3
布袋支撑　bag support　10.2.1.25
布料器　distributor　2.2.12.7
spreader　8.7.1.3
布料器煤加料器　spreader coal feeder　14.2.2.1
布煤机炉排　spreader stokers　24.2.4
布煤机炉排使用的炉栅　grates used with spreader stokers　24.2.4
布置　arrangement　12.6.3
部分充填卧式贮罐　partially filled horizontal tank　21.2.5
部分的　partial　2.2.35.8
Foster-Wheeler部分气化炉　The Foster-Wheeler partial gasifier　2.2.23
部分循环　partial recycle　8.4.3
部分氧化　partial oxidization　2.2.33.3;2.2.78.8
部件　component　6.2.8.13
部件配置图　component arrangements　8.8.1

C

采矿　mining　2.2.85.1
采样　sample　3.5.14.22
采样泵　sample pump　3.5.10.1
采样管　sampling tube　15.6.4.5
采样控制系统　sampled data control system　1
CO采样仪　sampling apparatus for CO　3.5.13
采用X射线在矿石中分离出金刚石工艺　diamond separation　12.7.5
采油　oil production　2.1.2
蔡希斯坦页岩　Zechstein shale　25.2.4.12
参比池　reference cell　3.5.22.8
残差　residual error　1
残液　resid　2.2.5.18
残液　residue　3.1.22.19
残液过滤叶片　filter leaf for residual　9.2.7.5
残液换热器　residue heat exchanges　18.5.1.6
残液循环槽　residue recycling tanks　18.5.1.8
残余焓　residual enthalpy　1
残余氯　residual chlorine　3.1.2.22
残余物　residue　3.3.16.14
残渣　residue　15.1.1.5
残渣燃料　residual fuel　2.2.98.29
残渣卸出口　refuse discharge　12.5.2.6
残渣油　RESID；residuum　2.1.4.19
操纵盘　control panel　2.3.13.20
操纵台　control desk　2.3.13.1
操作变量　operational variable　1
操作地面　operating floor　4.3.4.24
操作方法　method of operation　8.1.1 (a)
操作方式　mode of operation　6.2.4
操作过程中　in-process　4.3.4.23
操作弧形水盘　operating water paring disc　9.5.12.13
操作滑板　operating slide　9.5.12.11
操作机械盒　operating mechanism housing　25.4.9.10
操作盘　control panel　15.7.3.16
操作平台　operating platform　15.6.3.5
操作区　operating zones　9.3.10
操作手柄　operating handle　23.3.8.8
操作手柄固定销　operating handle locking bolt　23.3.8.5
操作台　stage for operation　24.1.1.23
(操作)弹性　flexibility　1
操作特性　operating characteristic　16.1.2.1
操作原理　operating principle　11.2.9
操作中　operating　6.3.2.17
槽　pan　12.1.3.8
　sump　11.6.3.6
　tank　24.1.6.6
槽盖　tank cover　14.1.11.8
槽钢框架　channel frame　4.5.1.7
槽沟　slot　2.3.9.9
槽面法兰　grooved flange　22.4.3.
槽式炉　pit furnace　24.1.7
槽式螺旋泵　screw though pump　16.1.2.2.7
槽式皮带输送机　toughed-belt conveyor　23.2.1.17
槽式输送机皮带　troughed conveyor belt　23.2.1.6
槽外支撑物　off-tank support　15.6.1.16
槽形承载托辊　troughing carrying-idlers　23.2.1.5
槽形角度　troughing angle　23.2.1.13
槽形泡罩塔盘　Thormann tray　5.2.10.2
侧板　end plate　19.3.1.2
侧板　side plate　11.1.8.9
侧边拖曳水平再循环输送机　Side-pull horizontal recirculating conveyor　23.3.3.(d)
侧衬板　end liner　11.2.4.6
　side plate　11.1.5.8
侧盖　side cap　9.6.1.8
侧盖开关　side cap key　9.6.1.9
侧杆　side rod　16.2.4.24
侧杆导轨　side-rod guide　16.2.4.25
侧轨　side rail　9.1.2.10
侧辊糖汁环　side roll juice collar　9.6.1.10
侧辊箱　side roll box　9.6.1.19
侧流抽出口　side stream with drawl　5.2.8.10
侧流除磷　phostrip　3.1.37.11
侧视图　edge；view　13.3.3(a)
　side elevation　10.3.1.20
侧箱拉回螺栓　side box pullback bolt　9.6.1.5
侧卸料　side discharge　23.1.2.31
测定分子量　molecular-weight determination　3.5.8
测量计　rate meter　3.5.7.21
测量仪表　measuring instrument　25.4.4.1
测量元件　measurement cell　3.5.19.4
测炉内气温用温度计接管　nozzle for thermometer for gas in furnace　24.1.1.9

633

测微计[尺] micrometer 2.3.11.10
测压管口 pressure-tap nozzle 14.1.9.6
测压计 manometer 3.5.14.33
测重秤 balance for measuring basis weight 2.3.11.9
层流 laminar flow; streamline flow 1
叉车 forklift truck 2.3.13.23
叉式连杆 fork connecting rod 17.2.15.23
差分电导测定电桥 differential conductivity bridge 3.5.22.13
插入键 threading key 25.4.1.30
插入式泵 insert pump 16.1.2.2.56
插入式配件 male 3.5.16.23
插入式重要部件 plug in type vitalized unit 17.6.2.4
拆卸用螺孔 bolthole for dismounting 16.4.15.9
柴油 diesel 2.1.4.12
$C_{13} \sim C_{19}$（柴油） $C_{13} \sim C_{19}$(diesel fuel) 2.2.73.9
掺和 blending 3.2.1.18
产出井 production well 2.2.85.21
产量 throughput 1
产品 P; product 2.2.3.4; 2.3.17.15; 2.2.71.14; 13.4.1.18
产品出口 produce outlet 7.2.1.24
产品出口阀 product discharge valve 8.9.5.12
产品储槽 product storage tank 2.2.57.11
产品大小控制器 product size control 11.1.3.11
产品分离 product separation 2.2.73.4
产品分离器 product separator 2.2.75.6; 2.3.4.21
产品分离区 product separation zone 2.2.71.B
产品罐 product tank 2.3.16.16
产品接收器 product receiver 3.1.6.16
产品晶体 product crystals 7.2.1.7
产品精馏 product distillation 2.2.62.13
产品精制 product refinement 2.2.89.19 product refinery 2.2.73.5
产品冷却器 product cooler 13.3.5.6
产品流量计 product flowmeter 5.5.3.12
产品煤气 product gas 2.2.82.13
产品尿素 product urea 2.2.33.15
产品排出 product discharge 7.2.1.15
产品气(体)(ES) product gas(ES) 2.2.99.8
产品收集管 product collection 6.3.8.7
产品炭 char product 2.2.99.12

产品油 product oil 2.2.82.14
产品油和蜡 product gasoline and wax 2.2.78.11
产品蒸气 product vapor 5.5.1.15
产物 product 14.2.6.7
产物分馏 product fractionation 2.2.95.7
90°长半径弯头 90° long radius elbow 22.3.2.2
长臂 long arms 15.6.1.7
长方形混凝土结构 rectangular concrete structure 3.2.7
长管蒸发器 long-tube evaporator 7.1.4
长节距混合螺条 long pitch mixing ribbon 4.3.3.19
长颈瓶 parison 2.3.9.47
长连杆 long connecting rod 16.2.3.39
长螺栓 tie-rod 19.2.1.7
长曲柄上(面)摇臂轴 long crank upper rock shaft 16.2.3.45
长眼筛板 slot screen 9.5.10.3
长圆孔 slotted hole 21.3.1.8
长渣 long residuum 6.3.2.13
常闭(的) N.C.; normally closed 14.2
常关水管 closing water 9.5.13.14
常规的活性污泥 conventional activated sludge 3.1.30.B
常减压蒸馏 atmospheric and vacuum distillation 2.1.1.4
常开(的) N.O.; normally open 20.2.2.1
常开水管 opening water 9.5.13.13
常压 atmosphere(atmos) 8.1.1.7
常压氮气 atmospheric nitrogen 14.3.3.(a)
常压反应型DTB结晶器 Swenson atmospheric reaction-type DTB crystallizer 7.2.9
常压工业气体燃烧器 atmospheric industrial gas-burner 24.2.6
常压甲醇蒸馏塔 atmospheric methanol distillation column 2.2.43.14
常压流化床燃烧锅炉 atmospheric-fluidized-bed-combustor 25.5.1
常压塔 atmospheric tower 2.1.4.1
常压贮罐 atmospheric storage tank 21.2.1
常用的 conventional 6.2.2.(a)
厂进水 plant influent 3.1.34.1; 3.1.41.1
厂出水 plant effluent 3.1.34.14
敞开式180°回弯头 open pattern return bend 22.3.2.14
敞开式气液分离器 open seperator 5.3.3.4
敞开室 opening chamber 9.5.13.10
敞口的纯氧活性污泥工艺流程图 Flow diagram for uncovered pure oxygen active

sludge process 3.1.46
敞口卸料 open-drop discharge 23.3.4.1
抄板式 flighted 4.3.2.(b)
超额焓 excess enthalpy 1
超高压泵 very high pressure pump 16.1.2.1.6
超高真空 very high vacuum 19.4.1.14
超高真空液体金属密封高速转轴 high-speed rotating shaft seal for liquid metal with very high vacuum 19.4.2
超高真空转轴密封 rotating shaft seal for very high vacuum 19.4.3
超临界(流体)萃取 supercritical fluid extraction 1
超滤 ultrafiltration 1
超滤毛细管膜组件 UF capillary modules 2.4.4.8
超闪蒸 overflash 2.1.4.27
超声波控伤 ultrasonic exam 21.1.2.15
超速离心机 ultracentrifuge 1
超温切断 over-temperature cutout 25.6.7.2
超吸附 hypersorption 1
超吸附塔 hypersorption adsorber vessel 6.3.6
朝前移动 front movement 2.2.85.22
车 truck 8.2.1.6
车底炉 car-bottom furnace 24.1.8
沉淀 sedimentation 2.4.4.2
沉淀残渣 settling the drafl 2.3.12.50
沉积催化剂液面 settled catalyst level 2.1.10.8
沉淀固体 settleable solid 15.5.4.3
沉淀物 sediment 3.1.4.12
sludge 8.1.1.12
沉淀物捕集器 sediment trap 3.1.4.13
沉积固体 settled solids 12.4.2.6
沉积物 sediment 1
沉积物聚集空间 sediment holding space 9.5.13.5
沉积物排出口 sediment ejection ports 9.5.13.6
沉降 sedimentation; settling 1;3.2.1.7;2.4.5.6
沉降操作 sedimentation operations 15.6
沉降槽 stilling wells 3.5.4.7
沉降池 settler 7.2.9.4
沉降分层 settling and demixing 6.1.1.4
沉降区 settling zone 7.2.10.7
沉降室 settling chamber 25.1.4.7
沉浸管 dip tube 3.5.18.8
沉浸深度 submerged length 16.5.1.7

沉浸式排气燃烧炉 Submerged-exhaust burner 25.1.6
沉浸式盘管换热器 submerged-pipe coil exchanger 4.2.5.16
陈化污泥 digested sludge 3.2.2.17
digesting sludge 3.2.6.10
衬板 liner(plate) 11.1.5.7
衬玻璃钢管 glass-lined steel pipe 22.6.2.⑧
衬垫 insert 22.4.4.3
衬环 liner ring 16.4.19.33
衬里 lining 1;21.1.1.11
衬里法兰盖 lined blind flange 22.4.2.(i)
衬圈 liner 23.1.10.13
衬套 plunger lining 16.2.4.17
衬套材料 sleeve materials 19.3.3.(c)
衬套止动螺钉 bush set screw 19.2.2.13
衬橡胶钢管 rubber lined steel pipe 22.6.2.⑬
衬橡胶钢制壳体 rubber lined steel shell 5.4.2.7
衬芯 insert 24.2.7.4
称量斗 weigh hopper 2.2.106.15;6.3.10.14
称量杆 weigh beam 23.1.2.22
称量设备 weighing apparatus 2.3.12.32
称量设备 weighing equipment 23
称量系统 weighing system 23.4.1
称重装置 weighing apparatus 2.3.8.29
撑棍 saddle clips 9.3.3.37
成粒 pelletized 3.3.16.32
成粒(压实)垃圾废料 pelletized; densified RDF 3.3.16.33
成粒液 granulating liquid 13.4.1.24
成品 (finished)product 8.4.1.3
成品出口 product outlet 7.1.2.13
成品罐 finished product tank 13.2.4.26
成品气体 product gas 2.2.8.11
成品容器 completed vessel 2.3.9.37
成品筛 screening unit for finished product 13.3.4.24
成品气 product gas 2.2.21.6
成品收集器 product collector 8.6.1.5
成品水 product water 15.2.3.8;2.4.2.5;2.4.3.10
成品塔 finished tower 2.2.55.13
成品乙酸 product acetic acid 2.2.55.14
成球转鼓 balling drum 13.3.1.5
成型接头 formed heads 22.5.1.(d)
承插焊法兰 socket-welding flange 22.4.2.(e)
承磨环 wearing ring 16.4.18.6

承压环[圈]　compression ring　21.1.1.55
承载点　loading point　23.2.1.4
承载梁　carrying bar　4.2.1.2
澄清的反洗水　clarified backwash water　3.1.39.7
澄清器[池]　clarifier;settler　2.3.6.8;2.2.29.16;2.2.30.15;3.1.43.5;3.1.44.7
澄清器污泥　sludge from clarifier　3.1.11.3
澄清区　clarification zone　3.2.6.18
澄清区　settling zone　7.2.1.12
澄清式离心机　centrifugal clarifier　9.5.13
澄清水排出　clarified effluent　3.1.32.21
澄清系统　clarifying system　12.5.1.11
澄清液流　clarified effluent　9.5.14.5
澄清液排出　clarified liquid discharge　9.5.16.8
秤　weigh scale　23.5.1.2
驰放　purge　2.2.77.6
驰放气　purge gas　2.2.44.15;2.2.46.15;2.2.47.13;2.2.48.9;2.2.49.9
持液量　liquid holdup　1
持液率　liquid holdup　1
尺寸过大物料　oversize material　3.3.16.21
齿轮　gear wheels　26.3.12.82-96
　tooth　13.1.2.10,28
齿轮泵　gear pump　16.3.1
齿轮传动泵　geared pump　16.1.2.4.1
齿轮架　gear mounting　8.8.1.26
齿轮减速器[机]　gear reducer　13.2.3.3
齿轮联轴器　gear coupling　16.4.11.6
齿轮(式)　gear　16.1.1.25
齿轮箱　gear box　13.3.5.11
　gear case　5.5.1.17
齿轮油泵　gear oil pump　17.2.9.40
齿圈　girt gear　13.4.1.9
赤道带板　equator zone plate　21.2.3.6
翅片管束　finned tube bundle　4.5.1.6
翅片加热器　fin heaters　8.2.1.7
翅片式换热器　plate-fin heat exchanger　4.5.2.8
翅片转鼓　finned drum　8.3.4.6
翅片转鼓给料机　finned drum feeder　8.3.4.4
充料　filling　9.1.1
充气缓冲器　aeration pad　23.1.8.15
充气室　plenum　8.5.6.4
充液填料塔　packed column filled with liquid　5.1.2.9
冲击板　impact plate　11.1.13.3
冲击器　impingers　3.5.11.14
冲击式淋洒器　impact spray thrower　5.3.4.4
冲击式破碎机　impact crusher　11.1.6
冲击试验　impact testing　21.1.2.15
冲击销　impact pin　11.1.6.11
冲击型雷蒙磨碎机　Raymond Imp mill　11.2.8
冲洗　rinse　6.2.6.10
冲洗层　washing floor　2.3.12.27
冲洗罐　washrinse tank　6.2.5.9
冲洗水　flush water　14.1.11.13
　rinse water　6.2.6.8
冲洗液　clear cut liquid　10.5.10.2
冲洗液出口　flushing fluid outlet　19.1.8.2
冲洗液进口　flushing inlet　19.1.5.21
冲洗液连通孔　flush connection port　19.1.5.5
冲洗液入口　flushing fluid in　19.1.8.1
冲洗蒸气　purge vapors　14.3.3.31
冲洗蒸汽　purge steam　14.3.3.25
冲洗装置　inlet flushing system　10.3.4.15
重叠式塔板　superposed tray　5.2.5
重返通道　reentry opening　10.1.2.23
重沸器　reboiler　1
重接按钮　resetting button　25.4.1.20
重整　reforming　2.2.78.12
重整工序　reforming process　2.2.45.5
重整气　reformed gas　14.2.6.13
重整气废热锅炉　reformed gas waste-heat boiler　4.4.1
重整器　reformer　2.2.119.4
重整天然气　reformed natural gas　2.2.102.1
重整油　reformate　2.1.1.48
重组 DNA　recombinant DNA　1
抽出油　extractive oil　2.1.1.80
抽风罩　pickup hoods　10.4.4.9
抽气　evacuate　3.5.15.20
抽汽-凝汽-蒸汽透平联合生产循环　extraction-condensing-steam-turbine cogeneration cycle　25.1.2
抽提输送机　extraction conveyor　3.3.30.8
抽提塔　extraction tower　5.1.1.7
抽油杆　sucker rods　2.1.2.25
抽油装置　pumping unit　2.1.2.22
抽余液　raffinate　1
抽余油　raffinate oil　2.1.1.50
稠度　thickness　1
稠密介质分离　dense-media separation　12.5
稠密介质分离流程　dense-media flow sheet　12.5.5
臭氧　O_3　2.4.4.4
臭氧处理　ozone treatment　2.4.4.5

出风口　delivery port　17.5.1.11
出口　out　16.5.6.8
　outlet　10.1.4.1
出口泵壳　discharge case　16.4.28.33
出口臂手柄　outlet arm handle　13.3.5.21
出口侧轴承压盖　discharge side bearing cover　16.4.19.1
出口侧轴套　discharge side sleeve　16.4.19.8
出口端部　outlet end　10.1.2.18
出口端温度计　discharge end thermometer　17.4.2.28
出口段　outlet portion　16.4.18.13
出口阀　outlet valve　23.1.2.34
出口管　outlet pipe　10.2.1.29
出口环　discharge ring　16.4.19.14
出口接头　nozzle head　16.4.27.32
出口接头衬套　nozzle head bushing　16.4.25.30
出口扩压器　outlet diffusor　25.3.1.28
出口外壳　outlet casing　25.3.1.27
出口斜槽　exit chute　3.3.1.7
　outlet sluice　12.5.2.7
出口斜料圈　exit damring　13.3.1.6
出口支架　outlet case　13.3.5.22
出料　outfeed　3.3.29.6
出料槽　gutter　5.5.1.11
出料格栅　discharge grate　2.2.87.3
出料管　discharge tube　11.5.5.7
出料口　outlet ports　3.3.17.8
出料输送机　outfeed conveyer　3.3.30.6
出气　gas out　3.4.4.2;17.2.14
出气孔　gas vent　14.1.11.7
出气口　discharge nozzle　18.1.2.6
出入口　access opening　5.4.1.21
出水　effluent　3.1.25.13;3.1.33.6;3.2.1.13
　Q　3.1.40.10
出水弯管　elbow for discharge　16.4.30.1
出液　drain　9.4.4.2
　liquid out　17.2.14.7
出液口　liquid outlets　9.3.2.20
出液溜槽　outlet launder　15.6.3.17
出油口　oil outlet　15.7.4.11
初步处理　preliminary treatment　3.1.37.2
初步物理单元操作　preliminary physical unit operations　3.2.1.6
初捣　premasher　2.3.12.42
初级分离　initial separation　8.6.3.9
初级绕阻　primary winding　25.4.3.5
初始流化床　incipiently fluidized bed　14.1.3.7
初始脱水　initial dewatering　9.3.9.1

初始装填线　original fill line　23.3.4.4
除尘　dust remover　2.2.108.16
　particulate removal　2.2.89.6
除尘器　dust extractor　2.3.12.40
除尘器平板盖　precipitator plate cover　10.3.1.4
除尘室　dust-removal chamber　2.2.88.18
除臭预热器　deodorizing preheater　3.3.24.18
除焦油　tar removal　2.2.89.6
除节器　knotter　2.3.11.13
除沫管　skimmer tube　5.5.1.52
除沫器　demister　2.2.74.8;5.3.3
　eliminator　20.2.3.18
　foam catcher　5.3.4.39
　foam separator　5.3.4.38
　mist eliminator　3.4.4.7
除沫器排液管　catch-all drain　7.2.11.3
除水器　entrainment eliminator　4.6.2.14
除甜罐　sweeten-off tank　6.3.1.6
除微粒　particulate removal　25.5.2.4
除雾器　mist eliminator　8.9.2.11
除油雾装置　oil spray arrester　18.1.4.13
除渣　de-sludging　15.6.3.30
　slag removal　2.2.31.10
储存　sequestration　2.2.3.16
储罐　storage tank　25.6.3.9
储料　storage　23.2.4
储料仓　storage building　23.2.4.6
储煤仓　coal storage　2.2.89.2
储煤斗　coal storage hopper　2.2.35.1
储气罐　air receiver　2.2.53.3
储液罐　reserve liquid tank　2.2.49.6
　hold tank　3.1.32.16
储油罐　oil-storage tank　2.1.7.12
储油柜　oil conservator　25.4.3.9
处理过的　treated　6.2.9.10
处理过的重质汽油　treated heavy gasoline　2.2.75.19
处理后出水　treated effluent　3.1.28.8
处理后的废液　treated effluent　3.1.29.5
处理后的空气　treated air　3.4.3.3
处理后的空气(废气)　treated air(effluent)　3.4.3.1
处理后的水　treated water　3.1.22.18
处理后的水出口(渗透)　treated water outlet(permeate)　3.1.29.7
处理后的中水　treated effluent　3.1.30.A.7;3.1.30.B.13
处理后废水　treated effluent[wastewater]　3.1.12.10

处理量　quantity　3.3.5.2
处理系统　disposal formation　3.1.24.8
处置　disposal　3.3.7;3.3.25.8
触变性　thixotropy　1
穿过式洗涤　thorough washing　9.1.1.11
穿流回转干燥器　through-flow rotary dryer　8.8.9
穿流筛板塔　perforated plate tower　5.1.2.13
穿流栅板　turbogrid tray　1
穿流塔盘　dual-flow tray　5.2.10.22
穿流循环　through-circulation　8.1.1.④
传导性电池　conductivity cell　6.2.1.8
传递函数　transfer function　1
传动齿轮　drive gear　11.2.1.6
传动带　driving belt　9.2.3.2
传动杆　actuator stem　22.1.9.6
传动机头　head motion　12.5.5.1
传动链　chain(drive)　11.2.6.4
传动链条　chain　13.5.1.17
传动链罩子　chain cover　9.2.5.2
传动皮带　drive belt　13.2.3.2
传动皮带轮　driven sheave　9.5.5.7　driving sheave　11.1.4.1
传动皮带轮键　key for driving sheave　11.1.4.3
传动小齿轮　(driving)pinion　11.1.4.6
传动轴　counter shaft　11.1.4.4
　　drive shaft　16.3.6.7
　　driving shaft　24.1.9.2
　　gear shaft　12.3.5.7
传动轴密封保持盖　countershaft seal retainer　11.1.4.2
传动轴套筒　countershaft housing　11.1.4.5
传动装置　drive　2.2.12.3
传动装置竖筒　column of driving device　4.6.2.15
传动装置轴座　drive socket　23.3.5.3
传动座　drive seat　19.1.1.6
传感层圆盘　sensing zone disc　9.5.13.16
pH 传感器　pH sensor　3.5.10.2
传感压力　sensing pressure　19.3.1.18
传感液泵　sensing liquid pump　9.5.13.18
传热　heat transfer　1
传热表面　heat transfer surface　14.3.1.19
传热膜系数　film heat transfer coefficient　1
传热设备　heat-transfer equipment　4
传热装置　heat-transfer equipment　4.4.3
传输线反应器　transfer line reactor　14.2.7.25
传送　travel　4.3.4.30
传统的滴流床生物过滤器示意图　schematic diagram of conventional trickle bed biofilter　3.4.1
传统的生物涤气器示意图　schematic diagram of conventional bioscubber　3.4.2
传统煤化工　traditional chemical processing of coal　2.2.1.22
传质单元数　number of(mass)transfer units（NTU）　1
传质区　mass transfer zone(MTZ)　1
船舶螺旋桨(推进式)　marine propeller　14.4.4.(c)
船坞泵　dock pump　16.1.2.3.51
船用泵　marine pump　16.1.2.3.50
喘振　surge　1
串联操作流程图　series operation flowchart　18.4.2
串塔合成甲醇工艺流程示意　methanol synthesis using a double-tower series　2.2.46
床 A　bed A　14.2.2.6
床 B　bed B　14.2.2.7
床层排料口　bed material drain　14.2.2.5
床层深度　bed depth　14.1.12.8
床层限制栅　hold down grate[grid]　5.4.1.6
床内空气冷却管　air-cooled tubes in bed　25.5.2.(b)
床内用过剩空气冷却　bed cooled by excess air　25.5.2.(c)
床内蒸汽冷却管　steam-cooled tubes in bed　25.5.2.(a)
床深　bed depth　3.1.2.7
创新的零排放电厂概念示意　simplified diagram of the advanced zero-emission power plant(AZEP)　2.2.114
吹玻璃工人　glassmaker　2.3.9.38
吹玻璃用铁管　blowing iron　2.3.9.39
吹风管嘴　tuyeres　2.2.14.11
吹灰装置　soot blower　4.2.4.17
吹空气卸饼　air blow for cake discharge　9.3.10.7
吹膜机　inflation film manufacturing machine　15.7.5
吹气箱　blow case　3.1.17.1
吹扫　purge　3.5.14.21
吹扫配件和组件　purge fittings ＆ devices　23.4.1.13
吹扫气　sweeping gas　2.2.42.3
吹制　blowing　2.3.9.28
吹制法　blowing processes　2.3.9.22-37
吹制过程　blow-and-blow process　2.3.9.22
吹制模　blow mould　2.3.9.26

垂直安装　installed vertically　15.5.2
垂直管（多孔管）聚并分离器　vertical tube；(perforated)coalescing separator　15.5.4.5
垂直管聚并　vertical tube coalescing　15.5.4
垂直管聚并分离器　vertical tube coalescing separator　15.5.1
垂直流动重负载板式除尘器　vertical-flow heavy-duty plateprecipitator　10.3.2
垂直流反应器　vertical-flow reactors　3.3.15
垂直剖分式离心泵　vertical split centrifugal pump　16.4.11
垂直筛板　vertical sieve tray(VST)　1
垂直重力张紧装置　vertical gravity takeup　23.2.1.8
垂直轴　vertical axis[shaft]　3.5.1.3
垂直轴流速计　vertical axis current meter　3.5.1.(a)
锤或冲击部件　hammers or impact members　11.2.9.1
锤或粉碎机　hammer mill　8.4.5.3
锤击器　knocker　8.8.1.5
锤磨　hammer mill　6.3.2.5
锤磨机　hammer mill　11.6.5
锤盘　hammer carriers　11.1.7.5
锤式粉碎机　hammer mill　8.4.5.(a)
锤式破碎机　hammer crusher[breaker]　11.1.5
锤碎　sledging crushing　11.1.2.3
锤砧块　anvil block　11.1.7.20
锤子　hammers　11.2.7.3
纯环己醇　pure cyclohexanol　2.3.8.18
纯甲醇　pure methanol　2.3.5.4
纯化装置　purifax unit　3.3.21.8
45°(唇形)抄板　45-deg.lip flight　8.8.1.18
90°(唇形)抄板　90-deg.lip flight　8.8.1.19
瓷弧鞍填料　ceramic berl saddle　5.3.2.(f)
瓷矩鞍填料　ceramic intalox saddle　5.3.2.(g)
磁测仪表　magnetic meter　3.5.18.(a)
磁场　field　16.5.4.7
磁分　magnetic separation　3.3.16.10
磁分离器　magnetic separator　3.3.16.30
磁粉控伤　mag.particle exam　21.1.2.15
磁感应分离器　induced-roll separator　12.6.11
磁感应元件　magnetic induction part　14.4.3.9
磁鼓　magnetic drum[roll]　12.6.9.6
磁鼓洗涤　drum wash　12.6.9.4
磁回路　magnetic-return circuit　16.5.4.4
磁力分离的富集物　magnetic concentrate　12.5.5.1
磁力分离器　magnetic separator　2.3.16.19;12.6.2
磁力轴承　magnetic bearing　9.5.21.10
磁铁　magnet　12.6.1.6
磁铁　magnetizing block　12.5.1.12
磁铁罩　magnet cover　14.4.3.4
磁铁座　magnet receiver　14.4.3.7
磁性　magnetics　12.6.4.8
磁性材料　magnetic materials　12.6.1.5
磁性富集物　magnetic concentrate　12.6.4.5
磁性鼓　magnetic drum　12.6.1.(a)
磁性鼓式分离器　magnetic drum separators　12.6.3
磁性体自动卸料　automatic magnetic discharge　12.6.10.1
磁性转轮　magnetic pulley　12.6.1
磁性转轮　magnetic pulley　12.6.1.1
磁选器　magnetic separators　11.6.3.18
磁选组　magnetic cobbers　11.6.3.10
次级绕阻　secondary winding　25.4.3.6
刺刀管换热器　bayonet-tube exchanger　4.2.5.9
从动齿轮　driven gear　16.3.1.6
从动滚　applicating rolls　8.7.2.3
从动螺杆　driven screw　12.3.2.11
从动轴　driven shaft　16.3.2.9
从空分装置来的氧气　O_2 from ASU　2.2.109.1
从煤锁斗来的煤　2.2.24.6　coal from lock hoppers
从填充剂中分离出蚯蚓（如需要）separation of earthworms from bulking agent（if required）　3.3.28.13
粗苯产出　production of crude benzol　2.3.14.41
粗苯罐　crude benzol tank　2.3.14.42
粗苯精制　crude benzol refining　2.2.1.7
粗糙度　roughness　1
粗分器浮选　rougher flotation　12.8.2.15
粗酚　crude phenol　3.1.5.8
粗酚罐　crude phenol tank　2.3.14.40
粗粉尘预况降器　dry precleaner　10.5.1.23
粗粉尘主流线　pattern of coarser dust main stream　10.1.2.24
粗粉溜槽　tailing chute　12.3.4.15
粗粉溜槽法兰　tailing chute flange　12.3.4.17
粗粉溜槽夹紧环　tailing elbow clamp ring　12.3.4.18
粗粉煤　crude dust　24.3.1.18
粗粉排出管　outlet duct for tailing　12.3.2.4

639

粗粉排出口　tailings discharge　12.3.1.12
粗粉锥筒　tailings cone　12.3.1.11
粗环己醇　raw cyclohexanol　2.3.8.16
粗甲醇　crude methanol　2.3.4.2;2.2.42.4;
　2.2.45.17;2.2.46.18;2.2.47.16;2.2.48.10;
　2.2.75.3
粗甲醇至精馏　crude methanol to rectification　2.2.44.17
粗甲醇至稳定塔　crude methanol to stabilization tower　2.2.44.18
粗甲酸甲酯　crude methyl formte　2.2.61.9
粗焦油罐　crude tar tank　2.3.14.32
粗颗粒　coarse particles　8.4.3
粗颗粒聚结　coarse coalescing　6.1.9.10
粗矿石储槽　coarse-ore bin　12.8.2.2
粗粒成品　coarse product　8.5.1.9
粗粒粉尘出口　coarse particles out　10.5.1.24
粗粒料　coarse material　12.4.2.14
粗料　coarse　12.2.3.8
粗滤器　strainer　3.1.19.13
粗螺旋　large spiral　4.3.3.(e)
粗螺旋窄板刮板　large spiral, hollow flight　4.3.3.(d)
粗煤气　crude gas　2.2.76.9
粗汽油　naphtha　2.1.9.19
　raw gasoline　2.1.1.34
粗汽油回收罐　naphtha recovery tank　2.1.9.24
粗汽油加氢处理装置　naphtha hydrotreater　2.1.9.36
粗汽油料　crude gasoline stock　2.2.91.8
粗碎　rubblizing　2.2.85.2
粗碎进料预处理　crushing, feed preparation　2.2.2.2
粗细粒子分层现象　stratification of particle sizes　13.3.3.(b)
粗乙醇来自加氢工序　crude ethanol from hydrogenation　2.2.59.2
粗乙酸　crude acetic acid　2.2.55.11
粗乙酸乙酯到加氢工序　crude ethyl acetate to hydrogenation process　2.2.59.3
粗纸　raw paper　2.3.11.29
粗制产品煤气　crude product gas　24.3.1.10
粗制合成气　raw synthesis gas　2.2.17.6
粗制煤气　crude gas　24.3.1.9
　raw gas　2.2.17.6
醋酸　acetic acid　6.1.3.8
醋酸盐　acetate　3.3.18.15
醋酸盐形成　acetate formation　3.3.18.22
催化　catalytic　2.2.28.3

催化重整　catalytic reforming　2.1.1.47
催化反应器的设计　Design of catalytic reactor　2.2.39
催化剂　catalyst　2.2.39.5;2.2.64.13
催化剂补充　catalyst makeup　14.2.7.4
催化剂层изgem料斗　catalyst level　14.3.3.26
催化剂沉积料斗　catalyst-settling hopper　14.3.2.28
催化剂冲洗段　catalyst purge section　14.3.3.32
催化剂出料　catalyst withdrawal　2.1.10.4
催化剂床　catalyst bed　3.4.5.4
催化剂床的冷却器　cooling unit for catalyst　2.2.76.12
催化剂分离塔　catalyst separator column　2.2.54.8
催化剂加入　catalyst addition　2.1.10.1
催化剂进口　catalyst inlet　14.3.3.28
催化剂进料　catalyst charging　14.2.7.6
催化剂冷却器　catalyst cooler　14.2.7.1
催化剂流化床　fluid catalyst bed　2.2.9.12
催化剂面　catalyst level　2.2.98.9
催化剂排放口　catalyst dropout　14.3.2.3
催化剂入口　catalyst input　14.3.2.1
催化剂循环　catalyst recirculation　2.3.5.14
　catalyst recycle　14.3.3.38
催化剂循环控制提升管　cat circulation control riser　14.3.4.8
催化剂循环率　catalyst circulation rate　14.3.3.(c)
催化加氢　catalytic hydrogenation　2.2.2.30
催化裂化　catalytic cracking　2.1.1.24
催化裂化原料　raw materials for catalytic cracking　2.1.1.38
催化裂化装置　catalytic cracking unit　14.2.6.(a)
催化流化床系统　catalytic fluidized-bedsystem　14.1.13
催化汽提塔　catalyst stripper　2.1.5.4
催化(H-coal)液化　catalytic(H-coal)liquefaction　2.2.90.5
催化蒸馏塔　catalytic distillation column　2.2.65.b
Evonik 催化蒸馏制甲基叔丁基醚(MTBE)工艺　Evonik catalytic distillation methyltbutyl ether(MTBE)process　2.2.65
催化裂化柴油　catalytic diesel oil　2.1.1.27
催化裂化汽油　catalytic cracked gasoline　2.1.1.26
脆性物料　friable material　23.1.7

萃残液　raffinate　6.1.12.13
萃取　extraction　6
萃取操作　extraction operation　6.1.1
萃取精馏塔　extractive distillation column　2.3.15.6
萃取器　extractor　2.3.8.37
Treybal 萃取器　Treybal extractor　6.1.15
萃取溶剂　extraction-solvent　6.1.3.15
萃取设备　extraction equipment　6.1
萃取塔　extraction tower[column]　5.1.1.7
萃取相　extractive phase　6.1.1.7
萃取液　extract　6.1.1.6
萃取蒸馏　extractive distillation　1
萃余相　raffinate phase　6.1.1.8
萃余液　extraction raffinate　6.1.1.9
　　raffinate　2.2.65.3
淬火车　quenching car　2.3.14.10
淬火塔　quenching tower　2.3.14.11
淬冷气体　quench gas　15.1.1.11
存储槽　holding tank　3.3.21.9
错流膜过滤工作原理　working principle of cross flow membrane filtration　3.1.29
错流湿式沉降分级器　crossflow wet-settling classifier　12.4.2
错流式喷淋吸收器　crossflow spray absorber　10.5.11.(b)

D

搭叠衬里　shiplap liner　11.2.2.3
搭接短管端　lap jt.stub end　21.1.1.4
打包机　contractor　3.3.5.20
大半径　wide pattern　17.6.2.2
大齿轮　girt gear　8.8.1.4
大齿轮盘　gear ring　12.3.5.29
大钩　hook　2.1.2.9
大环管　large ring　24.2.9.(d)
大颗粒　large particles　11.5.2.7
大孔筛板　large hole sieve tray　5.2.10.7
大粒度　oversize　11.6.4.8
大麦的接收　barley reception　2.3.12.41
大麦料斗　barley hopper　2.3.12.2
大麦提升机　barley elevator　2.3.12.33
大麦筒仓　barley silo　2.3.12.31
大气　atmosphere　19.4.1.2;3.3.25.14
　　atm.　2.3.17.10
大气冷凝器　barometric condenser　18.3.4 (b)
大气冷却器　atmospheric cooler　4.2.5.8
大气腿　barometric legs　18.3.5.14
大气中二氧化碳捕集系统的简化工艺流程图（填料塔方案）　simplified process diagram of an atmospheric CO_2 capture system (packed bed tower option)　3.4.6
大球　large balls　11.2.3.6
大容量聚并滤清器　high capacity coalescer filter　15.5.2
大头轴承盖　big end bearing　17.2.15.20
大物料　large material　3.3.16.7
大型泵　large pump sizes　18.2.4.4
大型容器　large scale vessel　5.3.3.14
代谢产物和多余的细胞产物　metabolic products and excess cell growth　3.1.47.10
带　belt　9.3.7.9
带粗介质反硝化塔的填料床反应器　PBR (packed bed reactor) system with coarse media denitrification columns　3.1.39
带电动机油压泵　hydraulic pump with motor　9.5.10.13
带定距管塔盘　trays with spacer tubes　5.2.4
带法兰大小头　flanged reducer　22.3.1.5
带法兰等径三通　flanged tee(straight size)　22.3.1.7
45°带法兰弯头　45° flanged elbow　22.3.1.4
90°带法兰弯头　90° flanged elbow　22.3.1.1
带箍管接头　coupling with band　22.3.2.17
带刮刀的转鼓真空过滤机　rotary-drum vacuum filter with scraper discharge　9.3.10
Eimco 带过滤机　Eimcobelt belt-discharge filter　9.3.7
带铰链端盖　hinged door　3.1.6.13
带颈对焊法兰　welding neck flange　22.4.2.(c)
带颈平焊法兰　slip-on weld hubbed flange　22.4.2.(b)
带孔转鼓　perforated basket　9.5.6.18
带螺纹三通　threaded tee　22.3.2.2
带螺纹四通　threaded cross　22.3.2.3
带螺纹异径三通接头　threaded reducing tee　22.3.2.5
带燃烧前二氧化碳捕集的整体煤气化联合循环简化流程　simplified IGCC (integrated gasification combined cycle) with pre-combustion CO_2 capture process.　2.2.117
带式　band　8.1.1.⑦
带式干燥机　band type drying machine　8.3.4
　　belt dryer[drier]　8.3
带式过滤机　belt filter　9.4.8
带式倾料器　belt tripper　23.2.4.3

带式输送机　belt conveyor　2.3.14.2
带式输送机卸料方式　belt conveyor discharge arrangements　23.2.3
带式输送系统　belt conveyor system　23.2
带式压榨机　belt filter press　9.6.2
带视镜人孔　manhole with sight glass　24.1.1.11
带同流换热器的催化氧化法分解挥发性有机化合物　VOC destruction by catalytic oxidation with recuperative heat exchanger　3.4.5
待处理的料液　liquid to be treated　2.2.56.11
待处理溶液　solution to be treated　6.2.4.1
待生催化剂　spent catalyst　2.2.72.2
待蒸馏的粗制品　crude distilland　5.5.1.1
待装料空桶　drum waiting to be filled　23.4.1.2
袋　bag　3.5.7.24
袋滤器　bag filters　10.2
袋式　bag type　10.2.1.(c)
袋式过滤器　cloth filters　10.2.1
bag filter　2.3.16.14；2.3.17.8
袋旋转器　bag turner　23.4.3.10
单程壳体　one pass shell　4.1.2.(f)
单纯重力式两产品系统　single-gravity two-product system　12.5.4.(a)
单缸　simplex　16.1.1.8
单缸卧式往复压缩机　simplex horizontal reciprocating compressor　17.2.1.1
单缸柱塞泵　simplex plunger pump　16.2.4
单轨梁　monorail beam　24.1.2.1
Fairmount 单辊破碎机　Fairmount single-roll crusher　11.1.2
单级　one stage　17.1.1.14
single stage　16.1.1.39
单级侧面进口蜗壳泵　single-stage end-suction volute pump　16.4.3
单级带网板锥形推料机　single-stage with conical pusher screen　9.5.6.(a)
单级离心泵　single stage centrifugal pump　16.4.1
单级离心式蒸气压缩机　single stage centrifugal vapor compressor　7.1.14.15
单级内旋风分离器　single-stage internal cyclone　14.1.5.(b)
单级喷射器　single-stage ejector　18.3.3
单级气流输送干燥器　single-stage pneumatic-conveyor dryer　8.4.1
单级推料离心机　traditional single-stage pusher　9.5.1.(m)

单级外旋风分离器　single-stage external cyclone　14.1.5.(c)
单级旋片式真空泵　single stage rotating blade vacuum pump　18.1.2
Swenson 单级再压缩式蒸发器　Swenson single-stage recompression evaporator　7.1.14
单晶　single crystal　1
单流塔盘　single pass tray　5.2.2
单氯苯　monochlorobenzene　2.3.8.9
单螺杆挤压机　single-screw extruder　13.2.3
单室　single-compartment　14.2.4
单室增稠器　unit thickener　15.6.1
单速横移刮刀自动离心机　single-speed automatic traversing knife　9.5.1.(h)
单位产量　per unit of output　8.1.1.35
单吸　single suction　16.1.1.34
单向阀　check valve　3.5.16.10
单向流动旋风分离器　uniflow cyclone　10.1.1.(a)
单相流　single-phase flow　2.1.9.48
单斜晶硫(液相)　S_8；liquid　2.1.9.48
单一刮刀可升降换向离心机　single-reversing knife rising knife　9.5.1.(f)
单一迷宫式密封　single labyrinth　16.4.8
单一污泥生物营养素脱除工艺　single-sludge biological nutrient removal process　3.2.4
单溢流塔板　cross-flow plate　5.2.6
单元操作　unit operation　1
单直径式　single diameter　8.8.5.(a)
单转子　one rotor　17.1.1.6
single rotor　16.1.1.19
单转子混合器　single rotor　12.1.3.(j)
单作用　single acting　16.1.1.7
单作用活塞　single-step-type cylinder　17.2.4
单作用气缸　single acting cylinder　17.2.2
淡化　desalination　25.1.1.16
淡水　fresh water　2.2.66.19；3.4.1.6
蛋白质工程　protein engineering　1
蛋形厌氧消化器　egg shaped anaerobic digester　3.2.6
氮储罐　nitrogen receiver　2.2.53.9
氮肥　nitrogen fertilizer　2.1.1.16
氮气　nitrogen　2.2.8.2；2.2.31.4；2.2.76.2；2.2.115.2
氮气　N_2　2.2.25.1；2.2.33.18；2.2.117,11；2.2.118,3；2.2.119.3
氮气洗涤　nitrogen wash　2.3.1.13

642

氮气压缩机　nitrogen compressor　17.1.2.4
氮洗　nitrogen wash　2.2.33.10
氮洗尾气　nitrogen wash tail gas　2.2.33.11
氮氧化物采样系统　nitrogen oxide sample train　3.5.15
当量比表面直径　Sauter mean diameter　1
挡板　baffle(plate)　10.2.1.30
  damper　11.1.9.12
  deflector　16.4.9.10
  retainer plate　10.4.2.6
  stop plate　11.1.7.28
挡板导架　screen guide　11.1.8.30
挡板螺钉　retainer plate screw　16.4.26.21
挡板塔　baffle tower　5.1.2.11
挡板凸缘　retainer lugs　10.4.2.4
挡片　baffle plate　18.2.3.11
挡圈　baffle ring　11.2.10.3
挡套　spacer sleeve　16.4.1.12
挡液环　flinger　16.4.12.23
  flinger(ring)　16.4.1.7
挡油环　oil retainer　17.3.2.13
  oil thrower　16.4.1.8,11
刀刃　knife edge　15.4.2.4
刀头锤子　knife hammer　11.1.5.9
(导电)芯线　conductor　25.4.5.1
导耳(导缘)　guide lug　19.1.5.13
导风锥　cone　10.3.4.15
  gas deflector cone　10.3.4.14
导轨　sliding way　18.1.1.7
导角逐步调节器　angle lead graduated adjustment　11.3.1.8
导流板　air deflectors　8.9.6.11
导流器　induced flow　10.2.3.18
导流筒　draft tube　6.1.12.7
  guide(flow)cylinder　18.2.2.5
导流筒-挡板结晶器　draft-tube-baffle crystallizer　7.2.6
导流叶片　vane　10.1.2.25
导轮　idler　9.5.11.3
导气浮选　induced-air flotation　3.1.13
导燃烧嘴　pilot burner　3.3.10.10
导热姆的泵回路系统　Dowtherm pumped-return system　25.6.3
导热姆的重力回路系统　Dowtherm gravity-return system　25.6.2
导热姆蒸发器　Dowtherm vaporizer　25.6.1
导丝轮　godet wheel　2.3.7.16
导温系数　thermal diffusivity　1
导线　conductor　25.4.5.1
  wires　10.3.5.15
导向板调节杆与螺钉　turnbeam adjusting level and screws　9.6.1.21
导向隔板　guide plate　16.4.18.12
导向环　guide ring　22.1.9.23
导向盘　guide　22.1.8.13
导向筛板　directional sieve tray　5.2.10.8
  slotted sieve tray　5.2.10.9
导向套　guide bushing　22.1.8.6
导向叶片　guide vane　16.4.18.8
  turning vanes　8.2.1.3
导向轴承　guide bearing　16.4.21.3
导叶　guide vane　16.4.30.2
导引轴承　guide bearing　15.6.1.14
捣棒　pestle　15.7.4.5
捣浆槽　mash tub　2.3.12.43
捣浆过程　mashing process　2.3.12.42~53
捣碎麦芽　mashing the malt　2.3.12.43
倒吊桶　inverted bucket　22.2.3.14
倒吊桶式蒸汽疏水阀　inverted bucket steam trap　22.2.5
倒卸式起重机　skip hoists　23.3.6
倒卸式起重机运行路径　skip-hoist paths　23.3.6.(d)
德国韦瑟灵公司甲醇制烯烃汽油厂的汽油模式(©Uhbe,埃克森美孚)　Gasoline mode of the methanol-to-olefins gasoline plant at wesseling,Germany(©Uhbe,Exxon Mobil)　2.2.76
德士古煤造气工艺　Texaco coal-gasification process　2.3.2
灯笼环　lantern ring　4.1.1.26
灯座　lampholder　25.4.1.60
等焓过程　isenthalpic process　1
等截面混合段　constant area mixing section　18.3.3.6
等径四通　cross straight size　22.3.1.13
等距　eq.sp.　10.4.2.3
等(理论)板高度　height equivalent of a theoretical plate(HETP)　1
等容过程　isochoric process　1
等熵　isentropic　20.18
等熵过程　isentropic process　1
等式约束　equality constraint　1
等温过程　isothermal process　1
等压过程　isobaric process　1
低含氮量产品　low N product　2.2.91.10
低含硫量产品　low S product　2.2.91.10
低合金钢　low-alloy steel　22.2.2.⑥
低聚反应　oligomerization　2.2.78.16
低空气速度　low air velocity　4.3.3.25
低密度球浮动床　floating bed of low-density sphere　10.5.4.16

643

低密度室 low-gravity compartment 12.5.2.17
低密质介质 low-gravity medium 12.5.2.12
低凝点油 low-freezing oil 2.1.1.22
低浓度废水 low-strenght wastewater 3.1.31.6
低热值 low-Btu 2.2.2.8
低热值燃料气 low-Btu fuel gas 2.2.104.1
低速搅拌器 low-speed agitator 14.4.4
低速厌氧工艺 low rate anaerobic process 3.2.7
低碳烯烃 low-carbon olefins 2.2.70.12
低投资成本 low capital cost 8.1.1.35
低温泵 cryogenic pump 16.1.2.1.10
低温变换 low temperature shift 2.3.2.12
低温操作螺旋管换热器 cryogenic-service spiral-tube exchanger 4.2.5.11
低温分离 cryogenic separation 2.2.104.11;2.2.78.15
low temperature separation 2.2.107.10
低温分离罐 cryogenic separation tank 2.2.74.10
低温罐 low-temperature vessels 25.6.4.6
低温甲醇洗 Rectisol wash 2.2.33.7
低温甲醇洗净化技术 （RECTISOL）中，还有氨合成以及氨与二氧化碳制成尿素的工艺（未层示）
Portion of a Rectisol plant desigmed to process syngas prior to ammonia synthesis and subsequent conversion of ammonia and CO_2 to urea 2.2.28
低温浆态床费托合成油工艺流程 Fischer-Tropsch low temperature bed oil synthesis process 2.2.74
低温区 low temperature zone 8.7.2.10
低温液氨洗 cryogenic liquid ammonia washing 2.2.38.10
低温贮罐 refrigerated storage tank 21.2.2
低压安全电流 low-voltage safety current 25.4.1.2
低压泵 low pressure pump 16.1.2.1.2
低压侧 low-pressure side 20.1.3.6
低压侧回油空腔 oil return cavity of low pressure side 19.3.2.14
低压分离器 low pressure separator 2.2.54.4
低压废热锅炉 low pressure waste heat boiler 2.2.34.9
低压锅炉给水加热器 feed water heater for low pressure boiler 2.2.34.10

低压回油线 low-pressure return line 2.1.
低压或第一级压缩机 low or first stage compressor 20.12.1
低压级压缩机 low-stage compressor 20.1.3.15
低压级蒸发器 low-stage evaporator 20.1.3.20
低压甲醇羰基化制乙酸流程简图 schematic of acetic acid synthesis from low pressure methanol carbon ylation 2.2.55
低压煤气总管 low-pressure gas main 2.3.14.44
低压气流 low pressure stream 17.2.2.12
低压气轮机 low-pressure turbine 25.3.2.9
低压气体 low pressure gas 19.3.2.18
低压汽缸 low-pressure cylinder 25.4.4.5
低压绕阻 low-voltage winding 25.4.3.6
低压吸收塔 low pressure absorbing tower 2.2.55.19
低压一级缸 low pressure cylinder 17.2.3.1
低压蒸气 low-pressure vapor 20.1.2.1
低压蒸汽 L.P.steam;light pressure 3.1.7.6
low pressure steam 2.2.43.4;2.2.111.15
低压蒸汽涡轮机和发电机 low pressure steam turbine & generator 2.2.111.16
低压蒸汽过热器 low pressure steam superheater 2.2.34.6,7
低真空 low vacuum 19.4.1.23
低蒸气压 low-vapor pressure 19.4.1.12
低酯副产物 low ester byproducts 2.2.58.2
低酯回收塔 low ester recovery column 2.2.58.12
低酯回收塔回流罐 low ester recovery tower reflux tank 2.2.58.7
滴流床 trickle bed 1
滴流床反应器 trickle reactor 14.3.1.(a)
滴流床加氢装置 trickle bed hydrotreater 2.2.79.7
滴流式过滤工艺示意图 schematic diagram of trickling filter process 3.1.47
涤气冷却器 scrubbing cooler 2.2.18.5
涤气器 scrubber 10.5.7;3.3.24.24
涤气器废水 spent scrubber water 10.5.13.11
涤气器用水 scrubber water 10.5.13.8
涤气水入口 scrubbing water inlet 10.5.1.4
涤气塔 scrubber 2.3.6.27;2.2.51.8
涤气液 scrubbing liquor 10.5.4.17
底板孔 floor openings 1
底泵壳 bottom case 16.4.28.27
底部跟踪辊 lower tracking roll 9.6.2.11

底部框架 bottom frame 12.2.1.15
底部框式卸料口 bottom-frame discharge 12.2.1.16
底部排料阀 bottom discharge valve 14.4.1.21
底部偏心块 eccentric bottom weight 12.2.1.17
底部驱动 bottom drive 9.5.2.12
底部驱动底部出料间歇式离心机 bottom drive bottom discharge batch basket 9.5.1.(c)
底部驱动间歇式带袋离心机 bottom drive batch basket with bag 9.5.1.(a)
底部驱动锥形混合干燥器 bottom drive conical mixer dryer 8.9.5
底部手工出料 bottom discharge manual 9.5.2.15
底产品 bottom product 5.5.1.25
底法兰 bottom flange 22.1.9.24
底封头 bottom head 9.2.5.29
底盖 bottom cover 18.2.1.11
底盖锁紧螺母 bottom cover securing nut 18.2.1.12
底货仓 lower hold 23.3.9.8
底基 substrate 15.2.3.1
底架 underframe 17.5.1.12
底料泵 sump pump 16.1.2.3.30
底流 underflow 1
底面涂布器 coater for the underside 2.3.11.34
底物 substrate 1
底卸式货车 hopper cars 23.3.8
底肘管 bottom bend 8.4.5.8
底座 base 11.3.1.7
　 bed 9.5.4.3
　 base bed 12.1.1.6
　 lower case 11.1.5.24
地板下的插座 underfloor sockets 25.4.1.22
地表面 ground surface 2.2.5.11
地表水层 surface water-bearing strata 3.1.24.3
地脚螺栓 anchor bolt 21.3.1.4
　 foundation bolt 11.6.3
地脚螺栓孔 fixing holes 15.5.5.2
地脚螺栓中心圆 anchor bolt circle 5.4.1.15
地脚螺栓座 base of stone bolt 5.3.4.35
地蜡 floor wax 2.1.1.84
地面干馏 surface retort 2.2.85.3
地面间接干馏 surface-indirect retorting 2.2.83
地面直接干馏 surface-direct retorting 2.2.82

地平 ground level 2.1.7.10
地平面 G.L.;ground level 4.6.1.3
地区加热 district heating 25.1.1.10
地上煤气化 surface-coal gasification 2.2.89
地下高压气化 high-pressure underground gasification 2.2.81
地下室排水泵 cellar drainage pump 16.1.2.3.9
地下水 water table 3.3.12.9
地下注入液体废物的深井 deep well used for the subsurface injection of liquid 3.1.24
地质构造 geological structure 25.2.4.1～12
第二沉降池 secondary settling tank 3.1.25.7
第二道湿毛布 second wet felt 2.3.11.18
第二道压榨 second press 2.3.11.20
第二反应器 second reactor 3.3.22.B
第二合成塔 second synthesis column 2.2.44.12
第二级 second stage 17.2.4.5
第二级高压蒸汽进口 high-pressure steam inlet,2nd stage 18.3.4.9
第二级混合喉管 second stage combining throat 18.3.4.12
第二级活塞 second stage piston 17.2.7.15
第二级活塞杆 second stage piston rod 17.2.9.16
第二级开口环 second stage snap ring 16.4.27.35
第二级排气阀 discharge valve of second stage 17.2.7.14
　 second stage discharge valve 17.2.9.18
第二级气缸 second stage cylinder 17.2.9.14
第二级吸气接头 second stage suction head 18.3.4.10
第二级吸入阀 second stage suction valve 17.2.9.15
第二级卸料阀 secondary discharge lock 23.1.13.4
第二级旋风分离器 secondary cyclone 23.1.13.2
第二级叶轮 second stage impeller 16.4.27.7
第二级蒸汽喷嘴 second stage steam nozzle 18.3.4.11
第二级蒸汽室 second stage steam chest 18.3.4.8
第二块板 plate No.2 15.2.3.7
第二缺氧区 secondary anoxic zone 3.1.35.5
第二热源 secondary heat 8.7.2
第二水分离器 second water separation column 2.2.34.15

645

第二再生层段　second regeneration zone　14.3.3.37

第三水分离器　third water separation column　2.2.34.16

第四水分离器　fourth water separation column　2.2.34.17

第五水分离器　fifth water separation column　2.2.34.18

第一次吹制　first blowing　2.3.9.24

第一沉降池　primary settling tank　3.1.25.4

第一道湿毛布　first wet felt　2.3.11.17

第一道压榨　first press　2.3.11.19

第一反应器　first reactor　3.3.22.2

第一合成塔　first synthesis column　2.2.44.4

第一级　first stage　17.2.4.2；2.4.2.3

第一级高压蒸汽进口　high-pressure steam inlet,1st stage　18.3.4.1

第一级混合喉管　first stage combining throat　18.3.4.7

第一级活塞　first stage piston　17.2.7.12

第一级活塞杆　first stage piston rod　17.2.9.6

第一级开口环　first stage snap ring　16.4.27.4

第一级排气阀　first stage discharge valve　17.2.9.9

第一级气缸　first stage cylinder　17.2.9.7

第一级吸气阀　first stage suction valve　17.2.7.3

第一级吸气接头　first stage suction head　18.3.4.5

第一级卸料阀　primary discharge lock　23.1.13.7

第一级旋风分离器　primary cyclone　23.1.13.8

第一级叶轮　first stage impeller　16.4.27.38

第一级叶轮挡圈　first stage impeller retaining collar[ring]　16.4.27.6

第一级蒸汽喷嘴　first stage steam nozzle　18.3.4.3

第一级蒸汽室　first stage steam chest　18.3.4.4

第一块板　plate No.1　15.2.3.6

第一水分离器　first water separation column　2.2.34.15

第一缺氧区　first anoxic zone　3.1.35.4

第一通道　first pass　8.3.3

第一再生层段　first regeneration zone　14.3.3.34

第1预热室　preheating compartment No1　14.2.5.8

第2预热室　preheating compartment No2　14.2.5.9

第3预热室　preheating compartment No3　14.2.5.10

典型的敞口池　typical open basin　3.1.46.10

典型的带辐射合成气冷却器的GE煤气化系统示意　schematic of typical GE gasification system with radiant syngas cooler　2.2.30

典型的带急冷室的GE煤气化系统示意　schematic of a typical GE coal gasification system with quench chamber　2.2.29

典型的GSD气化炉示意　schematic of a typical GSP gasifier　2.2.15

典型的GSP＋气化炉示意　schematic of a typical GSP plus gasifier　2.2.16

典型的壳牌气化工艺示意(MP,中压)　schematic of typical shell gasification process MP,medium pressure　2.2.31

典型的湿式涤气器示意图:(a)鼓泡塔;(b)填充床/塔;(c)喷雾塔;(d)文丘里涤气器　schematic diagram of typical wet scrubbers:(a)Bubble column;(b)Pack,bed/tower;(c)spray tower;(d)Venturi scrubber　3.4.4

典型的温克勒气化炉示意　schematic of a typicae Winkler gasifier　2.2.19

典型的整体煤气化联合循环工厂　A typical integrated gasification combined cycle plant　2.2.110

典型连续处理系统　typical continuous-treatment system　3.1.12

点火管　ignition tube　24.2.9.2

点火器　ignitor　3.5.21.5

点火装置　ignition source　3.3.11.8

点效率　point efficiency　1

电　electricity　2.2.39.4

电表　electricity meter　25.4.1.32

电表柜　electricity meter cupboard　25.4.9.31

电除尘器　cottrell precipitator　14.2.7.11　electrical precipitators　10.3

电磁阀　magnetic valves　5.5.4.9　solenoid valve　5.5.4.8

电磁搅拌式超高压反应釜　Electromagnetic stirring type superhigh-pressure autoclave　14.4.3

电磁排气阀　solenoid vent valve　3.5.22.4

电磁(式)　electromagnetic　16.1.1.51

电灯泡　electric light bulb　25.4.1.56

电动机 (electric)motor 4.5.1.1
电动机定子 motor stator 20.2.3.14
电动机驱动装置 motor drive 23.5.4.21
电动机支架 motor support column 16.4.27.21
电动机转子 motor rotor 20.2.3.13
电工刀 electrician's knife 25.4.1.63
电工器件 electric parts 25.4.1
电机 motor 10.2.1.20
电机侧联轴器锁紧螺母 motor half coupling lock nut 16.4.27.22
电机齿轮变速箱 gear head motor 5.5.2.2
电机/齿轮减速机组件 motor/gear reducer assembly 3.1.46.8
电机端 motor end 19.1.8.4
电机附加重块 auxiliary motor weight 11.3.2.1
电机机架 motor frame[stand] 16.4.28.20
电机减速传动装置 motor reducer drive 8.8.10.1
电机减速机 gear motor 9.2.1.3
 motor reducer variator 13.1.2.1
 motor with reducer 24.1.13.4
 reducer with motor 8.8.12.1
 variable speed reduction motor 13.5.1.2
电机壳体 motor casing 16.4.20.7
电机上端重块 upper motor weight 11.3.2.3
电机与减速机 motor and gear 3.1.14.20
电机支架 motor support 9.5.10.15
电机轴 motor shaft 16.4.28.23
电机座 motor base[frame] 9.5.4.16
电极 electrode 16.4.1.99
电加热器 electric heater 8.5.3.6
Petreco Cylectric 电聚结器 Petreco Cylectric coalescer 6.1.9
电缆 cable 25.4.5.8
电缆(接线)盒 cable box 25.4.7
电缆套 casing 25.4.5.4
电缆轴心 core 25.4.5.3
电烙铁 electric soldering iron 25.4.1.45
电力 electric power 25.1.1.8
电力 electricity 2.2.80.8;2.2.109.15;2.2.110.23;2.2.119.13
电力增强颗粒床过滤器 electrically augmented granular-bed filter 10.4.1
电力转换系统 power-conversion-system 25.1.1.18
电铃 electric bell 25.4.1.15
电铃按钮 bell push 25.4.1.2
电流 current 16.5.4.9
电路模拟板 matrix mimic board 25.4.2.8

电热辐射加热器 electrical radiant heater 5.5.1.31
电热元件 electric heating elements 24.1.7.12
电热蒸汽发生器 electric steam generator 3.1.6.3
电容法 capacitance 3.5.18.(e)
电渗析 electrodialysis 1
电石 calcium carbide 2.2.1.6
电视摄像机 television camera 25.2.5.6
电梯 vertical lift 2.3.8.36
电网 electricity grid 2.2.81.15
电压 voltage 25.4.1.41
电泳 electrophoresis 1
电源 power supply 3.4.7.1
电源接头 power connection 10.3.5.16
电源线 power line 25.4.1.23
电子放大器 electronic amplifier 12.7.5.11
电阻器 resistor 25.4.9.6
垫板 pad plate 21.3.3.1
垫片 gasket 10.4.2.2
 washer 17.2.9.19
吊车梁 beam 25.1.1.32
吊耳 lifting lug 4.1.1.36
吊杆 hanger rod 21.2.2.9
吊钩 lifting yoke 13.1.2.41
吊挂螺栓 hanger bolt 9.5.4.34
吊挂弹簧 hanger spring 9.5.4.35
吊架式带式输送机 gantry belt conveyor 23.3.9.3
吊柱 hoisting pillar 5.2.1.14
迭合 polyreaction 2.2.78.10
叠层织物填料 aminated cloth packing 19.2.3.5
叠合重整 polyforming 2.1.1.70
叠合汽油 polybenzine 2.1.1.71
叠装碟组 disk stack 9.5.15.9
碟式分离机转鼓 disk-centrifuge bowl 9.5.14
碟式离心机 disk centrifuge 1
碟形封头 torispherical head 21.1.1.63
碟组件 disc set 9.5.16.3
蝶阀式出口 butterfly outlet 23.3.8.(b)
蝶形 butterfly plate 22.1.4.9
丁二醇 BDO 2.1.1.10
丁炔二醇 butinediol 14.3.1.(a)
顶部出料 top discharge 9.5.2.14
顶部环板 top annular plate 9.5.4.22
顶部加料式双滚筒干燥器 top drum dryer with top-feed 8.7.1.(b)
顶部加热架 top heating shelf 8.2.2.5

647

顶部框架　top frame　12.2.1.13
顶部框式卸料口　top-frame discharge　12.2.1.12
顶部偏心块　eccentric top weight　12.2.1.4
顶部驱动　top drive　9.5.2.11
顶部驱动底部出料间歇式离心机　top drive bottom discharge batch basket　9.5.1.(b)
顶部装料阀和接管　top-fill valve & nozzle　23.4.1.20
顶层盘　top disc　9.5.12.6
顶封头　cover head　14.4.1.25
顶盖　cover　2.2.39.2;15.5.1.9
　cover(head)　9.2.7.11
　roof　10.2.1.8
顶盖垫片　packing for cover　9.2.1.20
顶盖加热管　cover heating　8.9.5.5
顶盖门　header door　24.1.2.14
顶盖升降控制杆　coven lift lever　24.1.7.10
顶盖支架　cover supporting bracket　24.1.7.9
顶盖座　block　9.2.1.21
顶滑车　crown blocks　2.1.2.4
(顶开)吸气阀卸荷器　Inlet valve unloader　17.2.12
顶面涂布器　coater for the top side　2.3.11.31
顶排液列管过滤机　top-outlet tubular filter　9.4.5
定常态　steady state　1
定距板　distance piece　11.1.5.17
　spacing plate　14.3.1.29
定距管　spacer tube　5.2.4.4
定距环　spacer ring　9.2.2.9
定距片　spacer　12.3.5.6
定距轴环　distance collar　16.4.18.7
定量　dosing　9.5.12.15
定量环　dosing ring　9.5.12.16
定量环室　dosing ring chamber　9.5.11.17
定量加料器　feed dosing　8.5.7.7
定量调节阀　batching valve　23.1.8.13
定时器　timer　10.2.3.8
定时取样装置　grab-sample setup　3.5.4.8
定态　steady state　1
定位环　locating ring　16.4.23.19
定位螺钉　set screw　19.1.1.12
定位销　dowel pin　22.1.8.4
　knock pin　22.2.1.2
定温控制阀　thermostat-controlled valve　25.6.4.10
定向　orientation　12.1.5.11
定向式　directional　23.3.2.13
定子　stator　15.7.6.13
定子线圈　stator coil　16.4.28.18
定子芯　stator core　16.4.28.21
动环　rotating seal ring　19.1.3.3
动力单元　power block　2.2.109.12
动力黏度　dynamic viscosity　1
动力气体　motive gas　18.3.3.1
动力升举机构　motorized lifting device　15.6.1.9
动力(式)　kinetic　16.1.1.29
　dynamic　17.1.1.3
动力涡轮机定子叶片　power turbine stator blades　25.3.1.25
动力涡轮机定子叶片托架　power turbine stator blade carrier　25.3.1.24
动力涡轮机径向止堆轴承　power turbine radial-axial bearing　25.3.1.29
动力涡轮机径向轴承　power turbine radial bearing　25.3.1.21
动力涡轮机转子　power turbine rotor　25.3.1.22
动力涡轮机转子叶片　power turbine rotor blades　25.3.1.26
动力涡轮可调转子叶片　power turbine adjustable rotor blades　25.3.1.23
动力装置　Power plant　27
　power unit　23.1.2.13
动力装置燃料　power-plant fuel　2.2.93.6
动量传递　momentum transfer　1
动态过滤机　dynamic filter　9.4.9
动态模拟　dynamic simulation　1
动压头　kinetic head　1
斗式提升机　bucket elevator　2.3.16.17;3.3.3.4
抖动器　kicker　8.3.4.22
独立的 MTG 反应器,烧去催化剂表面附着的焦质　offline MTG reactor burning coke off catalyst　2.2.77.2
毒品　toxic　8.1.1.28
堵缝垫圈的凹板压滤机滤板　calked-gasketed-recessed filter plate　9.1.3
堵头　plug　24.2.7.6
杜邦 PACT 生产工艺　DuPont's PACT process　3.1.20
杜邦公司乙二醇生产工艺　DuPont's ethylene glycol process　2.2.66
杜康旋风分离器　Duclone collector　10.1.2.(a)
镀金箔纸　metallized paper　25.4.8.2
端板　end plate　11.1.7.2
端部锤盘　end hammer carriers　11.1.7.27
端部剖视　sectional end elevation　10.3.1.19

端盖　bonnet　4.1.2.(b)
　　cover　9.4.4.12
　　end cap[cover]　13.3.5.28
端盖螺栓　end cover bolt　17.4.2.9
短臂　short arms　15.6.1.5
短管蒸发器　short-tube evaporator　7.1.3
短连杆　short connecting rod　16.2.3.38
断路器　circuit breakers　25.4.1.33
断续操作的剥离刮刀　peeling knives intermittently operated　4.3.1.13
断续器　contact maker　25.4.1.38，39
　　interrupter　25.4.9.5
煅烧　calcination　2.9.1.45
　　calcining　24.1.12.3
煅烧炉　calciner　3.4.6.14
煅烧炉燃料　calcining furnace fuel　14.2.1.7
煅烧室　calcining compartment　14.2.5.11
锻钢缸体　forged steel cylinder　16.2.4
锻钢制单作用高压气缸　forged-steel single-acting high-pressure cylinder　17.2.8
锻件　forging　21.1.2.7
堆焊硬质合金　surfacing hard alloy　22.1.8.18
堆积的介质(典型的)　dumped media(typical)　3.1.39.3
堆积角　angle of surcharge　23.2.1.15
堆聚　aggregate　3.3.1.17
堆料器　trimmer　23.3.9
堆密度　bulk density　1
堆置机　stacking machines　2.3.9.20
对比温度　reduced temperature　1
对苯二甲酸二甲酯 DMT　2.2.64.16
对苯二甲酸二甲酯塔 DMT column　2.2.64.10
对称膜　symmetric membrane　1
对称平衡式压缩机　balanced opposed reciprocating compressor　17.2.1.10
对二甲苯　p-xylene　2.2.64.12
对二甲苯氧化制苯二甲酸二甲酯　dimethyl terephthalate(DMT)production through p-xylene oxidation　2.2.64
对开挡圈　split shear ring　4.1.1.19
对开环　split shear ring　4.1.1.19
对开式端面密封　split end seal　19.1.5.10
对开轴瓦　bearing bushes in halves　11.1.7.22
对流　convection　8.1.1.5
对流带式　convection band　8.1.1.⑮
对流段炉管　heating tube in convection section　24.1.1.26
对流段外壳　shell of convection chamber　24.1.1.2
对流盘式　convection tray　8.1.1.⑭
对数速度　log velocity　14.1.1.2
对数压降　log pressure gradient　14.1.1.1
对于废气除臭和去除颗粒,有其它方案可选　alternatives available for exhaust gas deodorization and particulate removal　3.3.25.19
Pacific 多层床炉　Pacific multiple-hearth furnace　24.1.13
多层烘干机　multiple drying machine　2.3.7.31
多层流化床干燥器　multiple bed dryer　8.5.4.(b)
　　multistage fluized-bed dryer　8.5.1
多层生物固体焚烧炉耙臂详图(源于 BSP 环境技术公司)detail of rabble arm of multiple hearth biosolids furnace (courtesy of BSP Div. Envirotech Corp.)　3.3.26
多次循环的一次污水　primary effluent with recycles　3.1.35.2
多段　multiplex　16.1.1.11
多幅架帽　spider cap　11.1.4.15
多釜串联模型　tanks-in series model　1
多功能焚烧炉　multipurpose incinerator　3.3.5
多功能干馏釜　multifunction retort　2.2.84
多管式旋风分离器　multiclones　10.1.1.(b)
　　multiclone collector　10.1.2.(d)
多管塔　tube bundle column　5.1.2.6
　　multitubular column　5.1.2.6
多级　multistage　16.1.1.40
多级感应磁辊分离器　multiple induced-roll magnetic separator　12.6.4
Kerr-McGee 多级混合澄清器　Kerr-McGee multistage mixer-settler　6.1.12
多级离心泵　multistage centrifugal pump
　　multi-stage centrifugal pump　16.4.18
多级离心压缩制冷系统　multistage centrifugal system　20.1.6
多级推料离心机　traditional multi-stage pusher　9.5.1.(n)
多级烷化器　cascade alkylator　6.1.6.(a)
多级压缩机　multistage compressor　1
多级　multistage　9.5.6.(b)
多降液管筛板　multiple downcomer sieve tray　5.2.10.13
多聚体　polymer　3.1.12.8
多孔　multiple openings　21.1.1.37
多孔板　perforated plate　1；10.4.3.7
多孔壁　porous wall　15.4.1.4
多孔挡板　perforated baffle　15.6.5.6
多孔吊篮　perforated basket　9.5.5.21

649

多孔纺纱头 multi-holed spinneret 2.3.7.14
多孔管 perforated piping 3.3.20.10
多孔管分布器 perforated pipe distributor 5.3.4.11
多孔活塞 porous piston 7.2.14.3
多孔基质 porous matrix 15.3.1.7
多孔介质 porous medium 23.1.2.28
多孔金属滤布及支架 perforated metal cloth-support 9.4.2.9
多孔金属平板 flat metal perforated plates 14.1.8.(d)
多孔片 porous sheet 15.2.2.6
多孔套管 screened wellhead 3.1.24.11
多孔圆筒形排料筛 perforated cylindrical screen discharge 11.2.5.4
多孔支承板 porous support 6.3.3.9
多孔状进料分配管 porous-feed distributor tube 15.2.1.13
多孔状支撑片 porous backup disk 15.2.1.11
多列 multiplex 16.1.1.11
多流式泵 multiflow pump 16.1.2.2.60
多流体理论 poly-fluid theory 1
多路输送器 multiple conveyors 23.2.3.(e)
多室焚烧炉 Multiple-hearth incinerator 3.3.9
多室振打袋滤器 multi-compartment vibro bag filer 10.2.2
多速 multi speed 9.5.2.9
多膛炉 multiple-hearth furnaces 3.3.15.(c)
多膛燃烧炉 multiple-hearth furnaces 3.3.3
多通阀 splitter valve 8.4.1.17
多头喷嘴集管 spray manifold 10.5.4.8
多头卸料 multiple discharge 23.1.1.(b)
多相流 multiphase flow 1
多效蒸发 multiple-effect evaporation 1
多效蒸发器 multiple-effect evaporator 2.3.6.25
多效蒸馏法(海水淡化)厂示意 schematic of a multiple-effect distillation(MED) plant 2.4.3
多用电表 multiple meter 25.4.1.41
多用途电表 universal test meter 25.4.1.41
多元混合物 multicomponent mixture 1
多元离子交换过程 multicomponent ion-exchange process 6.2.7
多蒸汽喷嘴增压式喷射泵 booster ejector with multiple steam nozzles 18.3.2
多转子 multiple rotor 16.1.1.24
多组分混合物 multicomponent mixture 1
惰辊 idler roll 23.2.1.16

惰性粒子 inert particles 8.5.3.12
惰性粒子流化床干燥器 fluidized bed dryer with inert particles 8.5.3
惰性气体保护 protective inert-gas atmosphere 2.3.9.16
惰性气体补充口 inert-gas makeup 8.9.2.14
惰性气体供入装料室 inert-gas supply to burden chamber 13.5.2.4
惰性气体接口 inert-gas connection 4.3.4.2
惰性烃类液体 inert nydrocarbon liquid 2.2.39.d
惰性液体 inert liquid 2.2.9.4

E

鹅颈管 gooseneck 14.3.2.25
轭架 yoke 22.1.4.3
轭架螺栓 yoke bolting 22.1.4.7
Blake 颚式破碎机 Blake jaw crusher 11.1.1
颚式破碎机 jaw crusher 12.8.2.5
蒽醌二碘酸钠法脱硫器 Stretford desulfurizer 2.2.37.34
蒽醌二磺酸钠法脱硫装置 Stretford desulfurnation unit 2.2.23.14
耳式支座 lug support 21.3.3
耳式支座 support lugs 21.1.1.15
耳轴 trunnion 11.2.1.2
二次沉降池 secondary clarifier 3.1.2.13
二次澄清 secondary clarification 3.1.37.5
二次澄清器 secondary clarifier 3.1.8.8
二次出水 secondary effluent 3.1.44.9
二次处理 secondary treatment 3.1.27.7
二次底流污泥 underflow secondary sludge 3.2.1.17
二次电路(回路) secondary circuit[loop] 25.2.1.7
二次反应 secondary reaction 1
二次反应区[段] secondary reaction zone 3.2.6.17
二次废水 secondary effluent 3.1.8.9
二次分离 secondary separation 2.1.9.13
二次分离罐 secondary separation tank 2.1.9.16
二次焚烧室 secondary combustion chamber 3.3.12.3
二次活性污泥 secondary activated sludge 3.1.25.10
二次夹带 re-entrainment 1

二次浆化箱　repulping box　12.6.9.3
二次空气　secondary air　24.2.2.4
二次空气入口　secondary air inlet　14.1.9.7
二次空气入口管　secondary air inlet pipe　12.3.2.8
二次空气通口　secondary air port　3.3.12.5
二次扩压器　secondary augmentor　18.3.5.7
二次冷凝器　secondary condenser　3.1.6.10
二次钠泵　secondary sodium pump　25.2.1.8
二次浓缩污泥　secondary thickened sludge　3.1.26.(b).14
二次排料　secondary discharge　11.2.7.8
二次泡沫　secondary froth　2.1.9.15
二次喷射器　secondary ejectors　20.1.10.10
二次燃烧炉[器]　secondary burner　3.3.7.2
二次燃烧室　secondary combustion chamber　3.3.8
二次撕碎机　secondary shredder　3.3.16.22
二次污泥　secondary sludge　3.1.26.(b)13
二次涡流破碎器　secondary vortex breaker　10.4.3.18
二次污泥澄清器　secondary sludge clarifier　3.1.20.5
二次蒸汽　secondary steam　7.1.3.9
二次整形机　secondary shaper　13.2.4.25
二次转筒筛　secondary trommel　3.3.16.23
二道筛　secondary screen　2.3.10.58
二段　second stage　2.2.21.5
二段水煤浆　2nd stage coal water slurry　2.2.22.3
二段有氧化处理（硝化）　second-stage aerobic(nitrification)　3.1.40.6
二段转化　secondary reforming　2.3.3.8
二段转化炉　secondary reformer　2.3.2.9
二级缸　second stage cylinder　17.2.3.2
二级集尘器　secondary collector　8.4.1.5
二级排液口　secondary drain　10.1.4.12
二级喷射泵　second ejector　18.3.5.5
二级全混/增稠器 发酵罐　2-stage complete mix/thickener fermenter　3.2.10.(d)
二级相变　second-order phase transition　1
二级圆锥轧碎机　secondary cone crusher　12.8.2.8
二甲苯　dinethyl benzene　2.1.1.53
二甲醚　DME　2.2.51.9;2.2.70.14
二甲醚　dimethyl ether　2.2.1.18;2.2.49.10;2.2.54.16;2.2.50.9
二甲醚产品　DME product　2.2.52.7
二甲醚反应器　DME (dimethyl ether) reactor　2.3.4.1;2.2.51.6
二甲醚反应器进料/产物氢交换　DME reactor feed/product HX　2.2.51.4

二甲醚+水　DME+H_2O　2.2.77.4
二甲醚水溶液　DME agueous solution　2.2.70.10
二甲醚塔　DME (dimethylether) column　2.2.51.3;2.2.52.3
二甲醚塔　DME tower　2.2.49.8;2.2.50.7
二甲硫醚产品或循环（CH_3)$_2$S product or recycle　2.2.63.19
二硫化碳　carbon disulphide　2.3.7.8
二氯二氟甲烷　R-22; dichlorodifluoromethane　20.1.7.注2.
二氯乙烷　EDC(ethylene dichloride)　2.3.18.11
EDC二塔　EDC No.2 column　2.3.18.13
二氧化硫采样系统　sulfur dioxide sample train　3.5.14
二氧化碳　CO_2　2.2.32.7;2.2.33.17;2.2.67.9;2.2.68.9;2.2.70.8;2.2.71.22;2.2.80.12
二氧化碳　CO_2　2.2.109.13;2.2.114.14;2.2.116.14;2.2.117.7;2.2.118.15;2.2.119.8;2.3.19.10;3.1.47.9;3.3.29.3
二氧化碳　carbon dioxide　2.2.6.14
二氧化碳捕集　CO_2 capture　2.2.109.10
二氧化碳捕集方法简介：(a)燃烧前捕集；(b)燃烧后捕集；(c)富氧燃烧　Summary of CO_2 capture strategies. (a)pre-combustion capture, (b)post-combustion capture, and (c)oxyfuel combustion　2.2.119
二氧化碳捕集与压缩　CO_2 capture & compression　3.4.6.15
二氧化碳存储　CO_2 storage　2.2.117.13;2.118.17
二氧化碳，氮气　CO_2, N_2　2.2.68.5
二氧化碳，氮气，氢气，一氧化碳-CO_2, N_2, H_2, CO　2.2.38.7
二氧化碳耗尽的空气　CO_2-depleted air　3.4.6.1
二氧化碳解吸塔　CO_2 desorber　2.3.19.9
二氧化碳/汽轮机　CO_2/steam turbine　2.2.114.12
二氧化碳汽提塔　CO_2 stripper　2.2.118.16
二氧化碳去储存　CO_2 to storage　3.4.6.16
二氧化碳去压缩　CO_2 to compression　2.2.27.13
二氧化碳脱除　CO_2 removal　2.2.3.15
二氧化碳吸收塔　CO_2 absorber　2.3.19.8;2.2.32.4;2.2.118.14
二氧化碳压缩机　carbon dioxide compressor　17.1.2.8

F

发出的电力　generated power　3.1.23.14
发电　power generation　2.3.1.24

发电厂与电器件　power plant and electric parts　25.4
发电机　generator　25.2.1.15；2.2.110.14；2.2.113.8；2.2.114.11；2.2.116.9；2.2.117.19；2.2.118.12
　electric generator　2.2.111.7
发电站　power plant　2.2.108.3
发动机　engine　2.1.2.12
发酵　fermentation　1
发酵槽　fermentation vessel　2.3.13.8
发酵罐　fermenter　3.2.10.8
发酵罐混合液回流　fermenter mixed liquor return　3.2.10.10
发酵罐/增稠器　fermenter/thickener　3.2.10.12
发酵室　fermenting cellar　2.3.13.7
发酵温度计　fermentation thermometer　2.3.14.9
发射电极　emission electrode　10.3.5.19
发生器溢流管　generator overflow tube　20.1.8.4
发生氢气　generate hydrogen　2.2.21.12
发生蒸汽　steam generation　2.2.45.10
发运　to shipping　23.4.1.7
乏汽出口　steam exhaust　16.2.1.4
乏汽通　exhaust steam　14.3.3.3
阀　valve　13.3.5.25
Flapper 阀　Flapper valve　14.1.7.(a)
ICI 阀　ICI valve　14.1.7.(c)
J 阀　J valve　14.1.7.(b)
阀板　valve　12.3.4.7
阀瓣　valve clack　22.1.8.2
阀闭　valve closed　6.3.3.14
阀柄　valve handle　1.1.3.13
阀挡　valve guard　16.2.3.34
阀反向　valves reversed　6.2.4.8
阀盖　bonnet　22.1.2.6
　bonnet cover　22.1.10.1
　valve head　17.2.12.2
阀盖垫片　bonnet gasket　22.1.2.5
阀盖螺母　bonnet nut　22.1.2.17
　bonnet stud nut　22.1.8.7
阀盖螺栓　bonnet bolt　22.1.2.16
阀盖双头螺栓　bonnet stud bolt　22.1.8.8
阀杆　spindle　22.2.1.17
　stem　14.1.4.2
　valve rod〔spindle〕　12.3.4.8
阀杆衬套　stem sleeve　22.1.2.13
阀杆垫片　stem gasket　22.1.7.5
阀杆接头　stem connector　22.1.9.12
　valve-rod head　16.2.3.49

阀杆接头销　valve-rod head pin　16.2.3.48
阀杆接头销螺母　valve-rod head-pin nut　16.2.3.47
阀杆拉杆　valve-rod link　16.2.3.46
阀杆连杆接头　valve-rod link head　16.2.4.41
阀杆连接杆　valve-rod link　16.2.4.40
阀杆螺母　valve-rod nut　16.2.3.52
阀杆填料　stem packing　22.1.3.15
阀杆填料函压盖　valve-rod stuffing box gland　16.2.3.50
阀杆填料函压盖　valve-rod stuffing box gland　16.2.4.49
阀关闭　valve closed　22.1.10.11
阀环　valve ring　22.1.3.11
阀极　valve plate　17.2.12.8
阀开　valve open　6.3.3.13
阀门　valve　22.1
阀门手轮　valve handle　9.2.1.24
阀门紧螺钉　gate clamp serews　23.5.1.4
(阀门)压紧装置　pressing device　17.2.9.10
阀门组件　valve system　17.2.11
阀盘　valve disc　22.1.10.7
阀盘螺母　disc nut　22.1.10.4
阀片　valve plate　16.2.3.31
阀全开　valve fully open　22.1.10.13
阀升程限制器　lift stop of a valve　17.2.11.5
阀手轮　valve handle　22.1.3.3
阀弹簧　valve spring　16.2.3.35
阀体　valve body〔housing〕　22.1.10.2
阀体盖　body cap　22.1.7.4
阀体密封圈　body rings　22.1.2.2
阀调节杆　valve-adjusting rod　16.2.3.46
阀调节铰链　valve-adjusting pivot　9.3.3.41
阀芯　valve plug　22.1.4.16
阀罩　bonnet　22.1.8.11
阀针　needle　22.2.3.3
阀座　valve seat　16.2.3.37
阀座环　seat ring　22.1.9.22
阀座圈　valve(seat)ring　22.1.3.12
法兰　flange　22.4
法兰板　plate　16.4.4.1
法兰垫片　flange gasket　19.3.1.13
法兰盖　blind flange　22.4.2.(h)
法兰管件　flanged fittings　22.3.1
法兰接头　flanged only heads　22.5.1.(b)
法兰联接　flanged connection　22.3.1
法兰密封面型式　type of flange facing　22.4.3
法兰型式　flange type　22.4.2
翻拌器　turner　2.3.12.24

翻边开孔　flued openings　21.1.1.30
翻斗　skip　23.3.6.2
翻斗式抓斗　skip bucket　23.3.6.5
翻斗运输机　bucket conveyor　6.3.2.3
翻斗真空过滤机　Bird-Prayon tilting-pan filter　9.4.1
钒萃取　vanadium extraction　6.1.12.(c)
反冲洗供给　backwash supply　3.1.40.9
反萃取　reverse extraction　1
反萃取　stripping　1
反回信号盘　revertive signal panel　25.4.2.7
反馈控制　feedback control　1
反渗透　reverse osmosis　1
Permasep反渗透器　Permasep permeator　15.2.1
反时针方向转动　rotates counterclockwise　12.1.3.9
反洗　backwash　3.1.10.12
反洗泵　backwash pump　3.1.39.8
反洗空气　backwash air　3.1.38.8
反洗期　backwash period　6.2.9.(b)
反洗水　backwash water　3.1.41.8；3.1.38.9
反洗水去厂前区　backwash water to head of plant　3.1.39.4
反洗液　backwash　3.1.19.10
反洗液出口　backwash outlet　5.4.3.5
反洗液进口[入]　backwash inlet　5.4.3.4
反向空气净化(反吹)　reverse air cleaning　10.4.4.8
反向流空气分级器　reversed-current air classifier　11.6.9
反向皮托管　reverse-type Pitot tube　3.5.11.4
反硝化塔　denitrifica tion column　3.1.39.6
反应产物分离塔　reactor product splitter　2.2.51.10
反应程度　extent of reaction　1
反应澄清器　reactor-clarifier　15.6.3
反应动力学　reaction kinetics　1
反应段　reaction section　14.3.3.29
反应堆泵　reactor pump　16.1.2.3.38
反应堆建筑物　reactor building　25.2.2.29
反应堆压力容器　reactor pressure vessel　25.2.2.2
反应釜　reaction kettle　1
　　　 reactor　2.2.55.6
反应或吸收设备　reaction or absorption vessel　附录1.33
反应机理　reaction mechanism　1
反应级数　reaction order　1
反应进度　extent of reaction　1

反应井　reaction well　15.6.3.18
反应气体　reaction gas　3.5.22.1；2.2.67.17
反应器　reactor　2.2.52.4；2.2.54.1；2.2.56.1；2.2.61.2；2.2.63.8；2.2.66.5；2.2.69.4；2.2.72.9；2.2.74.2；2.2.76.11；2.3.16.21；2.3.17.4；2.3.18.4；2.3.19.2；3.4.7.6；14.3
Dorrco Fluosolids反应器　Dorrco Fluosolids reactor　14.2.4
MTO反应器　MTO reactor　2.2.68.4
反应器构型　reactor configurations　14.3.5
反应器壳体　reactor housing　2.2.39.3
反应器流出物　reactor effluent　2.3.4.12
反应器流出液冷却器　effluent cooler　2.3.4.19
反应器外壳　reactor casing　3.1.32.23
反应区　reaction zone　2.2.71.4
反应室　reaction chamber　14.3.5.8
反应型DTB结晶器　Swenson reaction type DTB crystallizer　7.2.8
反应液　reaction fluid　2.2.55.5
反应蒸馏　distillation with chemical reaction　1
反转　rev.；reverse　附录2
返回混合物　back-mix　8.1.1.11
返回空气　air return　11.6.9.15
返回口　return　8.8.11.6
返回输送带　conveyor return　8.3.3.4
H_2O返回吸收塔　H_2O return to absorbing tower　2.2.70.7
返混　back mixing　1
返氢气　return hydrogen　2.2.46.16；2.2.47.14
泛点　flooding point　1
方框图　block diagram　2.3.3
方头丝堵　square head plug　22.3.2.19
方形编织填料　square braided packing　19.2.3.1
方钻杆　kelly　2.1.2.14
芳烃分离与转化　separation and reforming of aromatics　2.1.1.49
芳香族　aromatics　6.1.3.13
防爆膜　rupture disk　3.1.6.5
防爆膜(板)　explosion panel　8.4.6.11
防尘密封滚动轴承　dusttight sealed roller bearings　8.8.11.2
防尘密封环　dust seal　12.3.5.10
防尘罩　dust cover　12.2.1.11
　　　 sanitary shield　23.3.8.6
防冲板　impingement baffle　4.1.1.29
防护环　protecting collar　16.4.23.4
防护罩　hood　10.5.18.3

防溅板　splash guard　11.7.1.1
防水胀保护器　protector for water swelling　16.4.28.26
防松环　lock ring　22.1.4.12
防涡流板　vortex breaker　5.4.1.14
防涡器　vortex breaker　5.3.4.41
防震安全帽　shock-resisting safety helmet　25.4.1.48
防止结块处理　pretreatment to prevent caking　2.2.101.6
防转短轴［棒］　anti-rotation stud［bar］　19.1.5.14
仿真　simulation　1
纺纱用丝束的整理　finishing of tow for spinning　2.3.8.57
放出　purge　7.1.14.6
放出气　released gas　2.2.43.6
放大　scale up　1
放电　charged　10.3.1.28
放电电极　discharge electrode　10.3.1.15
放电电子（负极）　discharge; negative electrons　10.3.1.2
放焦台　coke side bench　2.3.14.13
放净口　drain　9.4.4.11
放空　(air)vent　20.1.10.12
atmosphere　2.2.5.7
vent　2.1.7.9；2.2.55.18；3.1.42.3；3.3.20.15
vent air　3.1.32.14
放空 A　vent A　2.3.19.7
放空 B　vent B　2.3.19.13
放空管线　vent line　25.6.7.12
放空罐　blow tank　23.1.8.8
放空冷凝器　vent condenser　25.6.2.6
放空线　vent line　25.6.2.6
放气　gas release　3.3.18.1
放气阀　air cock　16.2.3.33
air vent valve　22.2.3.5
放气接口　vent connection　4.1.1.32
放气口　(air)vent　15.5.5.10
relief port　16.4.26.23
放气螺塞　plug for air vent　16.4.19.36
放热　heat release　24.2.8
放热反应　exothermic reaction　1
放泄阀　drain valve　15.5.5.4
飞灰　fly ash　10.3.5.5
飞灰　fly ash　3.3.8.13
飞灰涤气器　fly ash scrubber　10.5.10
飞灰控制器　fly-ash control　24.1.4
飞灰再注器　fly ash reinjector　14.2.2.9
飞溅布料式双滚筒干燥器　double drum dryer with splash-feed　8.7.1.(c)
飞轮　flywheel　11.1.1.5
非磁性材料　nonmagnetic materials　12.6.10.6
非磁性粗尾砂　non-magnetic coarse tailing　12.6.4.11
非磁性富集物　nonmagnetic concentrate　12.6.1.14
非磁性尾砂　non-magnetic tailing　12.6.4.6
非磁性细粒尾砂　nonmagnetic fine failings　12.6.4.10
非磁性渣料卸出　nonmagnetic tailings discharge　12.6.9.9
非催化流化床系统　non-catalytic fluidized-bed system　14.1.12
非定时操作　arbitrary operation　23.3.4.1
非定态　unsteady state　1
非定向定式　non-directional　23.3.2.14
非反应核模型　unreacted core model　1
非刚性　non-rigid　22.6.1.(c)
非火蒸汽锅炉　unfired steam boilers　21.1.2.17
非金属　nonmetal　22.6.2.2
非金属的　nonmetallic　19.3.3.(c)
非金属管系和非金属衬里管系　nonmetallic pipe and lined pipe systems　22.6.2.3
非均相反应　heterogeneous reaction　1
非均一流化床　heterogeneously fluidized bed　14.1.3.14
非能量资源回收　nonenergy resource recovery　3.3.1.15
非牛顿流体　non-Newtonian fluid　1
非平衡级模型　non-equilibrium stage model　1
非平衡式　unbalanced　23.3.2.6
非平衡系统　non-equilibrium system　1
非色散型红外线分析仪　non-dispersive infrared analyzer(NDIR)　3.5.11.27
非受压件　non pressure parts　21.1.1.38
非稳态　unsteady state　1
非洗涤板　non-wash plate　9.1.1.1
非正位移驱动装置　non-positive drive　23.3.2.3
非自吸式　non-self-priming　16.1.1.38
肥料　fertilizer　2.2.89.12
废催化剂　spent catalyst　2.2.98.5
waste catalyst　2.2.72.7
废固体颗粒　waste solids　2.2.8.7
废活性污泥　WAS; waste activated sludge　3.2.2.20；3.1.30.B.6；3.1.33.7；3.1.40.5
废空气管　exhaust-air tubes　8.7.2.8
废矿粉　fines discard　2.2.82.3

废料　refuse　2.2.101.4
waste　6.2.5.7
废料仓　waste bin　12.7.5.9
废料池　rejects chest　2.3.10.59
废料床　spent bed　25.5.3.11
废料堆　gravel pack　3.1.24.9
废料坑　waste sump　6.2.1.9
废流体　spent fluid　11.5.4.6
废黏土排放　spent clay to disposal　6.3.2.27
废气　exhaust　13.4.1.6；2.2.113.12
　exhaust air　3.1.16.8
　exhaust gas　24.1.12.10；3.1.45.7；3.3.24.9；3.3.25.11
　gaseous waste　2.2.4.11
　offgas　2.3.5.15；2.2.52.6；3.3.29.10
　tail gas　2.3.1.25
　waste gas　3.3.10.12；2.3.17.9
废气出口　exhaust air　8.9.1.4
　exhaust-gas outlet　8.8.9.7
　waste gas outlet　4.2.4.18
废气处理　waste gas treatment　3.4
废气冷却器　waste-gas cooler　14.2.7.10
废气排出　gaseous effluent　3.4.1.3；3.4.2.3
废气排放管　exhaust gas stack　25.2.2.17
废气燃烧炉　waste gas combustor　2.3.17.7
废气与灰分　exhaust and ash　3.3.27.15
废气预热器　economizer　2.2.81.13
废弃　sink　12.5.1.8
waste　6.2.5.4
废弃生物质送去固体处理　waste biomass to solids handling　3.1.36.10
废弃物　waste　2.2.66.12
废弃物加料　waste feed　3.3.2.1
废弃物进口　waste feed　3.3.12.10
废弃物进料　waste feed　3.3.3.6
废弃物调节器　waste conditioner　3.1.22.5
废热　waste heat　20.1.9.11
废热锅炉　W. H. B.；waste heat boiler　1；4.4
废热回收装置　waste heat recovery unit　2.2.46.8
废热炉气　hot kiln exhaust gases　11.6.4
废热气入口　waste gas inlet　4.2.4.13
废溶剂　waste solvent　3.1.6
废水　effluent　3.1.15.5。
　waste water　15.5.4.8；2.2.52.9；2.2.58.4；2.2.59.8；3.1.43.1
废水泵　waste water pump　2.3.10.63
废水槽　effluent flume　3.1.3.17
废水处理　waste water treatment　3.1
Reno-Sparks 废水处理厂流程图

Flowdiagram of Reno-Sparks wastewater treatment plant　3.1.37
废水处理厂研究中改进的化学辅助一次沉降流程图
Flow chart of CAPS (chemically assisted primary sedimentation) upgrading in the studied WWTP　3.1.27
废水管　effluent sewer　3.1.3.16
废水回收塔　wastewater recovery column　2.2.58.14；2.2.59.18
废水回收塔回流罐　reflux tank of waste water recovery tower　2.2.58.9；2.2.59.13
废水进　inlet wastewater　3.1.25.1；3.1.27.1
废水料槽　waste water feed tank　3.1.7.2
废水排出　waste water discharge　2.3.14.38
废水投入　waste water feed　15.2.4.1
废水溢流堰板　effluent weir and wall　3.1.3.15
Bardenpho 废水中除氮磷工艺（取自美国环境保护局）　Bardenpho process (source US EPA)　3.1.33
废酸　spent acid　6.1.5.13
废酸汽提塔　waste acid stripper　2.2.56.6
废炭/生物质　waste carbon/biomass　3.1.43.8
废污泥　sludge wastage　3.2.6.29
　waste sludge　3.1.32.22；3.1.34.12；3.1.44.8
废污泥氯化工艺系统简图（BIF 公司）Schematic of a waste sludge chlorination process system (BIF)　3.3.21
废污泥去污泥处理　waste sludge to sludge handling　3.2.10.6
废物　waste　3.2.6.23；2.3.15.13
废物处理　waste treatment　3
废物进口　waste entrance　3.3.11.12
废物垃圾燃料　fluff RDF　3.3.16.25
废物源　waste source　3.3.7.10
废盐矿　abandoned salt mine　25.2.4.1～12
废页岩　spent shale　2.2.82.18
废页岩固体颗粒　spent-shale solids　2.2.82.9
废页岩冷却段　spent-shale cooling zone　2.2.82.8
废页岩卸料装置　spent-shale discharge device　2.2.84.2
废液　barren liquor　6.2.6.2
　effluent　6.3.9.14
　liquid waste　3.1.22.7
　spent liquor　2.2.15.16
　waste(liquid)　3.3.10.8；3.1.47.5
　waste liquors　2.2.106.26

废液进料 liquid waste feed 3.3.11.5
废液 T 形出口管 T-pipe outlet for effluent 15.5.1.8
废油去锅炉 slop oil to boilers 2.3.6.2
废油页岩 spent shale 2.2.88.2
废再生液 waste regenerant 6.2.7.7
废蒸汽 exhaust steam 2.2.111.18
沸石催化剂 zeolite catalyst 1
沸石晶体 zeolite crystals 6.3.10.9
沸水系统 boiling water system 25.2.3
沸腾 boiling 25.5.1.9
沸腾泵 ebullition pump 2.1.10.11
沸腾表面 boiling surface 7.2.1.13
沸腾床催化反应器 ebullient-bed catalytic reactor 2.2.98.8
沸腾床加氢装置 ebullating bed hydrotreater 2.2.79.9
沸腾面 boiling surface 7.2.5.21
费-托合成 Fischer-Tropsch synthesis 2.2.2.18;2.2.78.9
SASOL-I 费-托合成过程 SASOL-I Fischer-Tropsch process 2.2.31
分辨率 resolution 1
分布极 distributor 3.1.42.6
分布器 bed plate 3.1.16.11
分布器栅板 distributor grid plate 2.1.10.9
分部装配 sub-assembly 6.2.8.14
分层罐 declamping tank 2.2.62.12
分层进料 layer-by-layer loading 12.1.4.(d)
分层碎石 graded stone 3.3.29.13
分层现象 stratification 13.3.3(b)
分程隔板 pass partition 4.1.1.31
分出的粗粉料槽 separated tailing trough 12.2.3.19
分隔流动 split flow 4.1.2.(h)
分级离心机叶片 classifying whizzer blades 11.2.11.5
分级器 dassifier 2.3.11.50
分级器出口 classifier outlet 11.5.4.7
分级器筒体 separator body chamber 12.3.2.5
分级器叶轮 classifier wheel 11.2.9.2
分级室 classification chamber 11.5.2.6
分级旋流器 classifying cyclone 11.6.3.16
分开 segregation 3.3.7.7
分开的组元 separated constituents 6.3.8.6
分块式塔板 segregative tray 5.2.10.28
分块式塔盘结构 segregative tray construction 5.2.7
分类 categorization 3.3.7.8
classification 16.1
分离 separation 2.2.85.8;2.2.36.11
分离仓 separation bin 2.2.4.9
分离槽 separating vessel 12.5.5.13
分离的水和 85% 的甲酸 separation of water and 85% formic acid 2.2.62.14
分离段 disengager 14.3.4.20
separation zone 7.1.2.6
分离固体 separating solids 12.8.1
分离罐 separation tank 2.2.49.7;2.2.50.5
分离机 separator 9.5.14.(a)
分离机转鼓部件 component parts of separator bowl 9.5.12
分离空间 disengaging space 14.1.12.6
freeboard 3.3.27.8
分离盘式 split pan 23.3.2.8
分离喷嘴 separation nozzle 15.4.2
分离器 separator 2.2.5.6;2.2.72.6
splitter 2.2.41.4;2.2.44.11;2.2.46.5;2.2.47.5;2.2.57.9;2.2.60.4
separation vessel 2.2.64.4
分离器流道 separator channel 3.1.3.8
分离器元件 separator cell 3.4.22.12
van Tongeren 分离器 van Tongeren cyclone 10.1.2.(c)
分离室 disengaging chamber 23.1.7.5
分离室 separator 7.1.3.7
分离设备 splitter 2.2.75.16
分离塔 separator column 2.2.54.7
separating column 2.2.76.17
分离液化产品 separate liquefaction products 2.2.90.14
分离因子 separation factor 1
分离与汽提器 separator and stripper 2.2.95.23
分离元件管 separation-element tube 15.4.3
分离装置 separating unit 2.2.61.6
分流板 breaker plate 15.7.2.4
dividing edge 12.7.4.12
分流阀 diverter valve 23.1.2.23
分流流动 divided flow 4.1.2.(j)
分流器 diverter 3.4.10.6
splitter 1
分馏 gradual distillation 2.3.8.4
分馏萃取 fractional extraction 6.1.10
分馏器 fractionator 2.2.60.2
分馏塔 fractionating column 5.1.1.2
fractionating tower 5.3.3.11
fractionator 2.1.5.12
C_2 分馏塔 C_2 fractionator 2.2.69.14
C_3 分馏塔 C_3 fractionator 2.2.69.16

分配管　distributor(pipe)　6.1.9.9
分配块　division bridge　9.3.2.26
分配器　distributor　18.4.2.11
分配室　channel section　21.1.1.52
分配送料器　distributing conveyor　23.2.5.9
分配头　rotary valve　9.3.2.21
分批　batch　8.1.1.2
分批进料　feeding in the batch　2.3.9.2
分批投料　batch charge　13.4.1.2
分批蒸馏　batch distillation　1
分散板　distributing plate　12.3.1.10
分散固体出口　divided-solids out　4.3.2.8
分散固体床层　divided-solids bed　8.9.6.1
分散固体进口　divided solids in　8.9.6.1
分散固体进料　divided solids feed　4.3.4.21
分散固体排料　divided-solids discharge　4.3.4.33
分散固体入口　divided-solids in　4.3.2.9
分散磨　dispersion mill　11.4.1
分散器　disperser breaks　1.1.13.3
分散设备　disperser　8.4.6.22
分析仪　analyzer　3.5.11.30
分选机　classifier　2.3.16.18
分子重新排列　molecular rearrangement　2.3.8.23
分子热力学　molecular thermodynamics　1
分子筛　molecular sieve　1
分子蒸馏　molecular distillation　5.5.1
Scott Smith 分子蒸馏釜　Scott Smith molecular still　5.5.3.14
分子蒸馏釜　molecular(distillation)still　5.5
酚　phenol　2.3.8.14
酚氢化　hydrogenation of phenol　2.3.8.16
酚醛塑料　phenolics　2.3.6
焚烧　incineration(process)　13.4.1(b)
焚烧段　combustion zone　3.3.9.4
焚烧炉　incinerator　2.3.6.28;3.4.5.5
焚烧炉-进料缓冲槽　incinerator-feed surge tank　2.3.6.26
焚烧炉-进料缓冲池　incinerator-feed surge pond　2.3.6.24
焚烧炉装置　incinerator facility　3.3.8
粉尘　dust　10.1.2.33
粉尘处理　dust disposal　10.4.4.2
粉尘分离器　dust separator　14.3.3.20
粉尘刮沟　share-off-dust channel　10.1.2.21
粉尘刮面板　dust shave-off　10.1.2.19
粉尘过滤器　dust filter　23.1.6.4
粉尘盒　dust bin　10.4.1.18
粉尘粒子　dust particle　10.3.1.24
粉尘料斗　dust hopper　10.1.2.31

粉尘流型　pattern of dust stream　10.1.2.20
粉尘排出(口)　dust discharge　10.2.1.28
粉尘取样管　dust sample line　10.3.5.10
粉尘收集器　dust collector　12.1.5.9
粉尘输送风机　ash conveyor fan　10.4.1.19
粉尘锁闭料斗　dust-tock hopper　10.3.5.13
粉尘筒　dust drum　8.8.6.5
粉尘与焦油　dust and tar　2.2.17.7
粉灰　fly ash　2.2.31.19
粉灰筒仓　fly ash silo　2.2.31.18
粉粒体的称量　weighing of bulk solids　23.5
粉粒体流化室　fluidized powder compartment　23.1.2.27
粉料　powder　12.1.5.13
粉料成品　fine product　8.5.1.4
粉煤　coal dust　25.3.1.17
粉煤燃烧　pulverized-coal firing　24.2.2
粉煤灰　fly ash　2.2.24.14;2.2.25.3;2.2.116.12
粉磨机　pulverizer　2.2.31.2
粉煤灰(循环的)fly ash(recycled)　2.2.25.15
粉末　dust　14.1.12.12
粉末分离器　dust separator　14.1.12.11
粉末活流　dust return　14.1.12.22
粉末活性炭　PAC(powdered activated carbon)　2.4.4.6
粉末活性炭处理活性污泥工艺　Powdered activated carbon activated sludge process（PACT）　3.1.43
粉末原料　powder stock　12.1.5.7
粉末转筒　dust rotor　8.5.6.22
粉碎　size reduction　11
粉碎后的煤　crushed coal　2.2.99.1
粉碎机　atomizer　11.1.8
　crusher　8.3.2.9
　disintegrator　2.3.10.5;3.3.22.5
　pulverizer　2.2.102.3;2.3.16.9
粉碎机底座　mill base　11.3.2.8
粉碎机盖　mill cover　11.3.2.6
粉碎机转子　pulverizer rotor　11.2.6.7
粉碎黏土　pulverized clay　6.3.2.10
粉碎黏土吹送机　pulverized clay blower　6.3.2.9
粉碎区　pulverizing zone　11.5.3.2
粉碎设备　breaking plant　13.2.4.22
粉碎室　grinding chamber　11.3.2.7
　milling chamber　11.5.5.2
粉碎纤维素板　shredding the cellulose sheets　2.3.7.6
粉碎装置　breaking device　12.1.3.3

粉体工程　powder technology　1
粉体技术　powder technology　1
粉状固体排料　granular solids discharge　4.3.4.27
粉状物料　fine material　3.3.16.8
风斗　plenum　4.5.1.14
风管　ducting　10.4.4.4
风机　fan　10.5.13.12;2.2.112.5;2.2.115.9;10.4.4.1
　　blower　8.3.1.4
风机电机　fan motor　8.2.1.1
风机空气　fan air　11.5.3.4
风机/排风系统　fan/venting system　3.3.20.4
风机外圈　fan ring　4.5.1.5
风机叶片　fan blade　12.3.5.19
风机叶片支架　fan blade arm　12.3.5.20
风口　tuyeres　3.3.27.12
风门　blast gate　10.1.2.13
风门调节器　shutter adjustment　12.3.1.8
风扇　fan　11.1.6.8
风扇轮　fan wheel　11.2.11.7
风扇叶轮　fan impeller　16.4.20.9
　　fan wheels　11.2.9.3
风扇罩　cowl　16.4.20.10
风送槽　blow tank　23.1.2.(f)
风筒　fan stack　4.6.2.1
风箱　wind box　24.2.4.13;3.3.27.2
封闭刮油环　seal wiper ring　19.1.7.2
封闭连接　gastight connections　13.5.2.2
封闭式传动度带通道　enclosed belt tunnel　10.5.1.16
封闭式电磁泵　hermetically sealed magnetic drive pump　16.1.2.4.7
封闭式透平制冷机　turbo refrigeration machine(closed) type　20.2.3
封闭圆桶　drum closing　23.4.1.6
封泥盖　lute cap　3.3.26.1
封填绝缘膏　filling compound　25.4.5.5
封头　head　21.4
　　shell cover　4.1.1.9
封头法兰　cover flange　14.4.1.17
　　shell cover flange　4.1.1.13
封头附件　head attachment　21.1.1.60
封头直边段　head skirt　21.4.1.10
封液环　lantern ring　16.4.2.9
　　seal cage　14.4.9.2
蜂鸣器　electric buzzer　25.4.1.15
敷设的　applied　21.1.1.11
弗克法玻璃拉制机　Fourcault glass-drawing machine　2.3.9.8

弗克法玻璃窑　glass furnace for the Fourcault process　2.3.9.1
伏(辊)工　coucher　2.3.11.49
伏特　volts;V　10.3.1.29
服务大楼　service building　25.2.2.16
氟化物采样装置　sampling apparatus for fluoride　3.5.16
浮槽　float bath　2.3.9.16
浮顶罐　floating-roof　21.2.1.3
浮动端壳体法兰　rear head end shell flange　4.1.1.11
浮动风门　floating damper　3.3.9.8
浮动管板裙　floating tube sheet skirt　4.1.1.22
浮动滑轮　float pulley　3.5.5.5
浮动角钢塔盘　floating angle tray　5.2.10.19
浮动筛板　floating sieve tray　5.2.10.10
浮动舌形塔盘　floating jet tray　5.2.10.16
浮阀塔　float valve tower　5.1.2.4
浮阀塔盘　floating valve tray　5.2.10.3
浮法玻璃制造法　float glass process　2.3.9.12
浮环　floating ring　19.3.2.1
浮环节流型密封　floating-type restrictive ring seal　19.3.1
浮环密封　floating ring seal　19.3.2
浮球　floating ball　22.2.3.9
浮球式冷凝水排除器　floating steam trap　22.2.4
浮桶　bucket　22.2.3.1
浮桶式蒸汽疏水阀　open bucket steam trap　22.2.2
浮头法兰　floating head flange　4.1.1.17
浮头盖　floating head cover　4.1.1.16
浮头勾圈　floating head backing device　4.1.1.18
浮头管板　floating tube sheet　4.1.1.15
浮头换热器　floating head heat exchanger　1
浮选　flotation　3.1.32.7
浮选澄清器　flotation clarifier　3.1.32.20
浮选单元　flotation units　3.1.13.1
浮选工厂　flotation plant　12.8.2
浮选机械　flotation machine　12.8.1.13
浮选加料　flotation feed　12.8.2.14
浮选物　float　3.1.15.6
浮选装置　flotation unit　3.1.15.4
浮油　free oil　3.1.3
浮渣层　scum layer　3.2.6.15;3.3.17.3
浮渣出口　scum port　3.2.6.6
浮渣挡板　scum baffle　15.6.5.13
浮渣导管　scum pipe　15.6.5.4

浮渣室和挡板　scum box and baffle　15.6.5.5
浮渣收集装置　scum collector　15.6.5.3
浮渣收集装置　scum collector　3.2.15.7
浮子阀　float valve　20.1.10.16
辐射段炉管　heating tube in radiant section　24.1.1.21
辐射段外壳　shell of radiation chamber　24.1.1.20
辐射防护屏　radiation shield　18.2.1.7
辐射强度　radiation intensity　1
辐射形叶片　spider vane　24.2.9.7
辐射锥　reradiating cone　24.1.2.5
辐射锥支架　cone support　24.1.3.11
俯视图　plan view　17.2.3
釜残液　residue　5.5.1.37
釜残液出口　residue outlet　5.5.2.15
釜底残留物　still bottoms　5.5.3.4
釜式浮头再沸器　kettle-type floating-head reboiler　4.1.1.(f)
釜式再沸器　kettle type reboiler　4.1.2.(k)
釜体　autoclave body　14.4.3.18
釜体　kettle　14.4.1.24
釜液　residue　1
釜支腿　autoclave leg　14.4.3.24
辅料　trains　13.3.5.5
辅助常开水管　auxiliary opening water　9.5.13.15
辅助萃取液　secondary-extract　6.1.2.4
辅助对流段　auxiliary convection section　24.1.2.4
辅助接点　auxiliary contact　25.4.9.7
辅助冷凝器　auxiliary condenser　18.3.5.4
辅助燃料　support fuel　3.1.22.4
辅助燃料气　auxiliary fuel gas　3.3.10.7
辅助燃气　support gas　3.1.22.3
辅助溶剂　secondary-solvent　6.1.2.5
辅助叶片　auxiliary blade　12.3.5.17
辅助叶片防磨板　auxiliary blade wear　12.3.5.18
辅助支承环板　subsupport plate ring　5.2.7.3
辅助支承角钢圈　subsupport angle ring　5.2.7.1
腐化物除臭　septage odor control　3.3.20
腐化物氯化系统示意图（美国 EPA）　Schematic of a septage chlorination system（US EPA）　3.3.22
腐蚀　corrosion　21.1.1.13
负荷　weight　3.5.1.1
负荷指示　load indicator　15.6.1.8
负压　neg.press.　3.2.6.12

负载　loadings　21.1.2.5
负载传感秤　load cell scale　23.4.1.4
附加处理　additional treatment　3.2.1.12
附加的系统　options　3.1.2.29
附加特征　special features　8.1.1（d）
复合肥混合　composting mix　3.3.29.4；3.3.30.3
复合回路循环混合系统　multiple-loop recirculation blending system　12.1.5
复合膜　composite membrane　1
复合样品　composite sample　3.1.2.27
复绕　rewinding　2.3.8.50
复式泵　plicate pumps　2.1.7.17
复式滤油器　duplex oil filter　2.1.7.16
复式内管　multiple intertube　24.2.2.5
复位按钮　resetting button　25.4.1.20
复位弹簧　return spring　16.2.8.5
复杂精馏塔　complex distillation column　5.1.1.3
副产品　by-product　2.2.94；2.2.54.15
CO_2副产品　CO_2 by-product　2.3.3.12
副产品气体（作为烘干机和电厂的燃料）　by-product gases（fuel for drier and power plant）　2.3.16.20
副产物　by-product　1
副产蒸汽　by product steam　2.2.73.13
副反应　side reaction　1
副十字头导板　hind guiding plate　17.2.15.19
副十字头导轨镶条　hind crosshead metal　17.2.15.21
副轴　counter shaft　11.1.3.9
副轴滚珠轴承箱　counter ball case　13.3.5.1
富的重溶剂　fat heavy solvent　6.1.9.2
富的轻溶剂　fat light solvent　6.1.9.19
富挥发性脂肪酸　VFA；volatile fatty acid　3.2.10.4
富挥发性脂肪酸发酵罐上清液　VFA rich fermenter supernatant　3.2.10.13
富挥发性脂肪酸一次排放物　VFA primary effluent　3.2.10.4
富集　enrichment　1
富集萃取液　pregnant strip solution　6.1.2.4
富集物　concentrate　12.6.4.15
富煤气　rich gas　2.2.81.4
富气　rich gas　2.3.1.22
富氢循环气　hydrogen-rich recycle gas　14.2.6.14
富吸收剂　rich absorbent　2.2.37.35
富相　rich phase　1
富氧空气　oxygen enriched air　2.2.53.8

富氧燃烧工艺中二氧化碳捕集示意 Schematic diagram of the oxyfuel combustion process for CO_2 capture 2.2.118
富液 full miscella 6.1.17.5
富液盘 flooded disk 10.5.9.2
腹板 wed plate 21.3.1.5

G

改质 upgrading 2.1.9.34
钙取代沸石 calcium-substituted zeolite 6.3.10.12
盖 cover 11.1.7.26
盖板 cover(ing)plate 14.4.3.15
盖帽 blank cap 14.4.3.2
干层树脂 lager of resin 6.2.2.(b)
干成品 dry product 8.4.5.6
干成品输送机 dry product conveyor 3.3.24.17
干磁鼓分离器 dry drum separator 12.6.10
干的油页岩分离器 dry oil-shale separator 2.2.88.11
干电池 dry cell battery 25.4.1.27
干封型 tank-Wiggins dry-seal type 21.2.1.4
干固体去除 dry solidsremoval 2.2.25.13;2.2.31.17
干级旋风分离器 dry-stage cyclone 8.4.1.36
干净的排水 clear effluent 3.1.42.10
干净油 clean and dry oil 15.5.5.9
干垃圾 dry rubbish 3.3.5.8
干垃圾格栅 dry rubbish grate 3.3.5.18
干料出口 dry discharge 8.9.1.8
干料分流管 dry divider 8.4.1.9
干馏 retort 2.2.82.4
retorting 2.2.85.17
干馏段 retorting zone 2.2.85.14
干馏器 retort 2.2.87
干毛布 dry felt 2.3.11.23
干煤粉 dry pulverized coal 2.2.105.1
干品出口 dry product discharge 8.5.6.10
干气 dry gas 2.1.1.15;2.2.69.18
干气体 dry gas 6.3.3.11
干球温度控制器 dry-bulb controller 20.2.2.4
干式分级 dry classification 12.3
干式冷却塔 dry cooling tower 25.2.2.19
干式气体流量计 dry-gas meter 3.5.7.11
干式真空泵 dry vacuum pump 9.3.1.8
干污泥 dried sludge 3.3.25.9
干物料 dry material 3.1.21.20
干物料卸出 dry product discharge 3.1.16.14
干燥 dry(ing) 2.2.101.16
干燥层 drying floor 2.3.12.16
干燥产品卸料输送器 dried material discharge conveyor 8.8.6.3
干燥产物 dry product 8.5.6.31
干燥氮气 dry nitrogen 2.2.53.10
干燥的聚酰胺(碎)片 dry polyamide chips 2.3.8.39
干燥段 drying zone 3.3.9.5
干燥段多孔板 perforated plate for drying 8.5.1.5
干燥粉末状固体加料斗 dry-powdered solids hopper 4.3.1.9
干燥粉碎机 flash drying mill 2.2.106.11
(干燥)过程 (dry)process 8.1.1.22
干燥机 dryer;drier 2.2.92.5
干燥剂 desiccant 6.2.8.11
干燥剂除湿器 desiccant dehumidifier 6.2.8.1
干燥剂冷却循环 desiccant cooling cycle 6.2.8
干燥剂盘 desiccant wheel 6.2.8.5
干燥器 dryer 13.1.1.6;8.1;2.2.71.2
drier 2.2.115.17
干燥器风管 drier duct 8.4.5.10
干燥器壳体 dryer shell 8.8.5.9
干燥器终端 end of dryer 8.8.5.11
干燥区 dry zone 9.3.10.13
干燥设备 drying equipment 8
干燥室 drying chamber 2.3.8.49
干燥塔 drying column 8.4.6.8
drying tower 2.2.69.10
干渣排放口 dry cake discharge 9.2.3.8
杆 bar 21.1.1.21
感液澄清盘 sensing liquid clarifying discs 9.5.13.17
感应线圈 coils 12.6.4.16
刚性 stiff 22.6.1.(a)
刚性方程 stiff equation 1
刚性防漏容器 rigid leak proof container 3.5.16.28
刚性刮板薄层蒸发器 thin-layer evaporator with rigid wiper blades 5.5.1.(b)
刚性刮片 rigid flights 15.6.1.24
刚性滚轴搬运器 rigid truck loader 23.4.2.14
刚性容器 rigid container 3.5.7.25
钢 steel 19.3.3.(a)
钢板 steel 14.2.5.5
钢板支承圈 plate support ring 5.2.7.9

钢表面覆盖不锈钢　stainless-clad steel　19.4.1.10
钢槽　steel tank　15.6.1.20
钢衬里　steel liner　11.2.1.3
钢衬水泥　cement-lined steel　22.6.2.③
钢带　steel belt[tape]　13.5.3.2
钢管　steel pipe　22.4.4.8
钢轨　steel track　23.3.7.8
钢拉线　steel draw-in wire　25.4.9.30
钢门　steel door　8.2.2.6
钢枪转子　pin rotor　11.2.10.8
钢栅板　steel grate　11.2.1.8
钢丝护套　steel wire armour　25.4.8.8
钢外壳　steel shell　3.3.26.3
钢圆桶　steel drums　23.4.1
钢制炉体　steel shell　24.1.2.11
钢质框架　steel casing　23.3.7.6
缸体　cylinder　11.4.1.11
缸体支承　cylinder support　17.2.9.20
杠杆键　lever key　16.2.4.34
高比转数泵　high-specific speed pump　16.1.2.2.21
高边雷蒙磨碎机　Raymond high-side mill　11.2.12
高纯氧气　HPO(high purity oxygen)　3.1.40.2
高沸点碳氢化合物　heavier hydrocarbons　14.3.3.(b)
高硅铸铁　high silicon cast iron　22.6.2.③
高硅铸铁泵　high-silicon cast iron pump　16.1.2.5.7
高合金钢　high-alloy;steels　22.6.2.⑮
高脚杯　goblet　2.3.9.41
高精度液体装料　high-precision liquid filling　23.4.1
高抗拉强度　high tensile strength　2.3.8.46
高卵石层　height shot gravel　3.1.19.11
高密度介质　high-gravity medium　12.5.2.11
高密度室　high-gravity compartment　12.5.2.16
高黏度三螺杆泵　high viscosity triple screw pump　16.1.2.1.11
高浓度废水　high-strength wastewater　3.1.31.1
高热值　HHV;high heat value;high-Btu　2.2.9.19
高热值管道气　high-Btu pipeline gas　2.2.89.2.1.7
高热值气体产品　high-Btu gas product　2.2.87.15
高热值燃料气　high Btu fuel gas　2.2.106.43

高速断面　high-velocity section　8.4.1.23
高速沟槽级　high velocity slot stage　10.5.1.8
高速空气　high-velocity air　8.7.2.9
高速空气阀　high speed air valves　12.7.4.9
高速离心泵　high speed centrifugal pump　16.1.2.2.16
高速率固体接触式反应澄清器　reactor-clarifier of the high-rate, solids-contact type　15.6.3
高速燃烧器　high-velocity burner　24.2.8
高位槽　head tank　3.1.32.15
elevated tank　2.2.57.3
高位槽来油　oil from elevated tank　19.3.2.16
高温变换　high temperature shift　2.3.3.9
高温分离罐　hot separator tank　2.2.74.11
高温罐　high-temperature vessel　25.6.4.3
高温核反应堆　high-temperature nuclear reactor　2.2.107.18
高温空气罩　high temperature air-cap　8.7.2.5
高温凝固物　hot break　2.3.13.2
高效率纤维雾净除器单元　H-E;brink mist eliminator element　10.4.2
高压泵　high lift pump　16.1.2.1.5
high pressure pump　16.1.2.1.4
高压侧　high-pressure side　20.1.3.5
高压侧回流孔　return opening of high pressure side　19.3.2.11
高压电　high-voltage　10.4.1.10
高压电极　high voltage electrode　10.3.4.10
高压电极重锤　electrode weight　10.3.4.9
高压电接入口　H.T.inlet　10.3.1.13
高压电缆　high voltage cable　10.3.1.6
高压电路支架　high-voltage support frames　10.3.4.4
高压电振打器传动轴　H.T.rapper shaft　10.3.1.16
高压放电极　high voltage discharge electrode　10.3.4.5
高压风机　high press.fan　11.6.4.15
高压釜　autoclave　2.3.8.12
高压供油线　high-pressure supply line　2.1.7.19
高压环形波纹管　high pressure toroidal bellows　22.5.1.(i)
高压级冷凝器　high-stage condenser　20.1.7.9
高压级压缩机　high-stage compressor　20.1.3.17

高压绝缘器 HV insulator 15.3.1.2
高压绝缘子室 high-voltage insulator compartment 10.3.4.18
高压煤气总管 high-pressure gas main 2.3.14.45
高压气流 high pressure stream 17.2.2.3
高压气体 high pressure gas 19.3.2.17
高压汽缸 high-pressure cylinder 25.4.4.3
高压汽轮机 high-pressure turbine 25.3.2.8
高压栅 high-voltage grid 10.4.1.7
高压吸收塔 high pressure absorbing tower 2.2.55.16
高压蒸气 high-pressure vapor 20.1.2.4
高压蒸汽 high-pressure steam 2.2.42.8;2.2.43.11;2.2.45.11;2.2.111.8
高压蒸汽涡轮机和发电机 high pressure steam turbine & generator 2.2.111.9
高压直流动力供应 high-voltage dc power supply 15.3.1.1
高扬程泵 high lift pump 16.1.2.1.5
高真空蒸馏装置 high vacuum distilling apparatus 18.5.1
膏糊状 paste 8.1.1
锆和锆合金 zirconium and zirconium alloys 22.6.2.㉕
格栅填料 grid packing 5.3.4.49
格栅支承 grid support 9.3.10.5
隔板 baffle 21.1.1.51
 closure plate 11.2.7.6
 diaphragm 17.4.2.18
 distance plate 8.5.1.16
 division plate 15.5.5.5
 partition wall 20.2.3.4
隔板真空接头 diaphragm vacuum connection 9.3.3.17
隔断阀 isolating valve 22.1.9.14
隔离环 seal cage 19.2.2.4
隔离室 distance piece 17.2.15.37
隔离竖筒 spacer column 16.4.27.9
隔离套筒 spacing tube 16.4.1.10
隔膜 diaphragm 16.2.7.4
隔膜泵 diaphragm(type)pump 16.2.7
隔膜阀 diaphragm valve 22.1.4
隔膜推杆 diaphragm push rod 16.2.8.7
隔热 thermal insulation 1
隔热层 thermal insulation 5.5.1.32
铬钒钢 chromium-vanadium steel 22.6.2.⑬
铬酸钠 sodium chromate 6.2.7.11
铬酸盐冲洗槽 chromate rinse tank 6.2.7.4
铬酸盐喷镀槽 chromate plating tank 6.2.7.3

4个碳(含C_4)以上的烃类产品 C_{4+} product 2.2.68.15;2.2.71.21
5个碳(含C_5)以上的烃类 C_{5+} 2.2.69.22;2.2.75.15
2个碳(含C_2)以下的烃类 C_{2-} 2.2.77.15
给料 feed 3.3.23.3
给料斗 feed hopper 11.5.5.3
给料段 feed section 15.7.2.2
给料管 conveyer pipe for material 12.3.2.1
给料链 feed chain 2.3.10.67
给水 feed water 2.2.110.8
给水泵 feed pump 16.1.2.3.3
给水分布孔 feed water distribution orifice 15.6.4.10
给水管路 feedwater line 25.2.1.12
给油口盖 oil hole cover 12.3.5.28
给油轴承 oil feed bearing 9.5.10.12
工具箱 tool case 25.4.1.49
工况 regime 14.1.1
工业废气 process gas 2.1.9.42
 waste gas 3.4.1.2;3.4.2.2
工业焚烧炉 commercial incinerator 24.1.4
 industrial incinerator 24.1.4
工业刮膜式分子蒸馏釜 industrial wiped-film stills 5.5.2
工业规模 large scale 8.1.1.25
工业锅炉 industrial boiler 3.3.1.13
工业过程 industrial processes 25.1.1.14
工业炉 industrial furnace 24.1
工业旋风分离器 commercial cyclone 10.1.1
工业用水 process water 2.2.107.19
工业用水平列管式过滤机 industrial horizontal tubular filter 9.4.4
工艺 process 15
Parsons工艺 Parsons Process 2.2.10
Petrosix工艺 Petrosix process 2.2.7
工艺过程 process 19.3.1.19
Udex工艺过程 Udex process 6.1.4
工艺过程机械 process machine 23.1.1.16
工艺过程料仓 process bin 23.1.6.11
工艺过程流体进出口 process fluid inlet and outlet 24.1.3.2
工艺过程气体 process gas 19.3.1.12
工艺过程蒸汽 process steam 24.3.1.11
工艺空气 process air 3.2.13.8;3.1.41.10
工艺冷却塔 process cooling towers 2.3.6.17
UOP/Hydro MTO工艺流程(生产聚合级烯烃) UOP/Hydro MTO process (production of polymer-grade olefins) 2.2.71

工艺气体　process gas　2.1.9.35
工艺容器　process vessels　21.2
工艺用水　process water　2.2.51.11
工艺蒸汽　process steam　2.2.108.5
工作表面　working face　12.6.10.4
工作池　working baths　2.3.9.5
工作流程图　operation process chart　18.4.1
工作喷嘴　operating nozzle　18.4.2.2
工作平台　working platform　2.1.2.5
工作期　service period　6.2.9.(a)
工作气体(空气)　operating gas(air)　18.4.2.1
工作室　operating chamber　18.4.1.5
　 working space　16.3.7.4
工作液　working fluid　18.2.3.4
工作液入口　working fluid inlet　25.5.1.11
工作液体　operating liquid[fluid]　19.1.3.2
工作原理图　functional diagram　18.2.3
工作蒸汽　operating steam　18.3.1.1
工作蒸汽入口　operating steam inlet　18.3.1.1
弓形板　segmental plate　5.2.2.3
弓形管段　pipe segments　22.5.1.(c)
Lurgi 公司甲醇合成反应系统流程　Lurgi Company methanol synthesis reaction process　2.2.42
Lurgi 公司甲醇合成和精馏工艺流程　Lurgi Company methanol synthesis and rectification process.　2.2.43
Davy 公司改进低压甲醇合成工艺流程　process flow for Davy company improved low pressure methanol synthesis　2.2.45
BASF 公司甲醇羰基化制乙酸工艺流程　BASF process of acetic acid synthesis from methanol carbonylation　2.2.54
公用工程　utility　3.3.1.13
功率燃气轮机　separate power turbines　25.3.3
供出空气　supply air　20.2.2.18
供料泵　supply pump　3.3.22.7
供料斗管线　feed hopper line　12.3.4.3
供料口　inlet　3.1.10.5
供料输送机　material feed conveyor　3.3.29.7
供料压缩机　feed compressor　2.2.48.2
Exxon 供氢溶剂过程　Exxon donor solvent process　2.2.97
供热水　heat supply　14.3.3
供入空气分布导管　air-supply distribution duct　4.3.4.16
供入热水　hot-water supply　8.9.6.7
供入液体　liquid supply　13.4.1.15

供入蒸汽　steam supply　14.3.3.2
供水　water supply　4.3.4.11
供水泵　water supply pump　16.1.2.3.2
供水管　supply water piping　10.5.1.27
　 water supply pipe　2.3.12.48
供水口　feedwater inlet　14.3.2.12
供氧　oxygen supply　3.1.46.5
供液　liquid supply　3.4.4.3
供液口　feed liquid inlet　9.5.3.27
供应　service　6.2.5.5
供应空气　supply air　6.2.8.8
供应氯　chlorine supply　3.3.22.13
供油器　service bin　2.1.8.3
供油与回油线　supply and return lines　2.1.7.1
拱顶罐　dome-roof　21.2.1.2
共沸物　azeotrope　1
共沸蒸馏　azeotropic distillation　1
共沸蒸馏塔　azeotropic distillation column　2.2.54.9
共混　blending　2.2.75.12
沟流　channel(ing)　14.1.3.18
构架　substructure cloumns　4.5.1.8
构件　components　18.2.4
箍环　shroud　11.2.10.3
鼓风机　(air)blower　2.3.16.8;12.1.5.1 fan　23.1.1.14
鼓风机　FD fan;forced-draft fan　25.5.3.21
鼓风机底座　blower base　8.3.4.9
鼓风式燃烧器　air-blown firebox　2.2.88.16
鼓泡　bubbling　14.1.1
鼓泡床　bubbling bed　14.1.2.3
鼓泡流化床　bubbling fluidized bed　2.2.18.10;14.1.3.15
鼓泡器　bubbler　3.5.18.(d)
鼓泡塔　bubble column　5.1.2.12
固氮(作用)　nitrogen fixation　1
固定鞍式支座　fixed supporting saddle　8.8.10.14
固定鞍座　fixed saddle　21.3.1.2
固定板　stationary plate　9.3.2.28
固定衬套　stationary sleeve　19.3.3.6
固定床　fixed bed　14.1.2.1
固定床反应器　fixed bed reactor　14.3.2.(c)
固定床合成　fixed bed synthesis　2.2.100.5
固定床加氢处理　fixed-bed hydrotreatment　2.2.99.10
固定床装置　fixed bed unit　14.3.4.(c)
固定磁铁　stationary magnet　12.6.1.9
固定磁性装置　stationary magnet assembly

663

12.6.9.8
固定刀片壳 stationary bladed shell 2.3.10.76
固定端板 fixed head 9.1.2.4
固定端端盖 stationary head bonnet 4.1.1.2
固定端法兰 stationary head flange 4.1.1.3
固定端盖 fixed end cover 4.2.1.5
固定端管板 stationary tube sheet 4.1.1.6
固定端管箱 stationary head channel 4.1.1.1
固定端壳体法兰 stationary head end shell flange 4.1.1.10
固定颚板 fixed-jaw plates 11.1.1.1
固定封头 stationary head 4.1.4.(n)
固定管板 fixed tube sheet 4.1.4.(m)
固定管板换热器 fixed-tube-sheet exchanger 4.1.1.(a)
固定锅体手把 lever for pan position 15.7.4.7
固定化技术 immobilization technology 1
固定化酶反应器 immobilized enzyme reactor 1
固定环 fixed ring 19.3.2.2
固定件 stationary element 8.8.5.13
固定角度 constant angle 9.5.2.16
固定壳 stationary shell 12.1.3.(h)
固定孔 fixing holes 15.5.5.2
固定料槽 fixed discharge chute 23.2.3.(e)
固定螺栓 clamping bolt 16.4.19.17
 setting bolt 14.4.3.12
固定膜生物电池 fixed-film bio-cell 3.1.44.3
固定模板 stationary die plate 15.7.3.7
固定倾料器 fixed trippers 23.2.3.(d)
固定栅板 retaining grid 10.5.4.15
固定式进口液分配器 stationary inlet distributor 6.3.8.3
固定式球阀 fixed ball valve 22.1.1.(f)
固定式梳状物组件 stationary comb 9.4.7.4
固定式套管换热器 fixed type double-pipe heat exchanger 4.2.2
固定式洗脱废液收集槽 stationary eluent waste collection 6.3.8.10
固定位置 fixed location 23.2.3.(f)
固定(旋流)叶片 fixed vane 10.5.4.11
固定永磁体组合体 stationary permanent-magnet assembly 12.6.4.21
固定于烟道上的织物 cloth fastened to breeching 8.8.5.18
固定轴套 stationary sleeve 19.1.7.8
固含量 total solids 3.3.4.3
固化成型 solidification 2.2.94.21
固化法造粒 pelletizing by solidification method 13.3.5.13.5
固态产品 solid product 2.2.102.24
固态电路质量控制器 solid-state weight controller 23.4.1.12
固体百分率 % solids 3.1.2.5
固体薄片 thin film of solid 9.3.8.5
固体产品 solid product 1.1.1.8
固体沉积 disposition of solid 9.1.1.10
固体持留空间 solid holding space 9.5.14.4
固体出口 solid outlet 4.3.2.15
固体处理系统 solid handling system 3.1.11
固体床层 bed of solids 8.9.6.5
固体混合与分离设备 mixing and separation equipment for solids 12
固体发酵工艺 solids fermentation process 3.2.10
固体废弃物 solids waste 3.1.8.3
固体废物 solid waste 3.3.16.1
固体废物处理 solid waste treatment 3.3
固体废物进料 solid waste feed 3.3.11.4
固体分离 solids separation 2.2.95.18
固体分离器 solid separation unit 2.2.24.10
固体锅炉燃料 solid boiler fuel 2.2.91.11
固体加料口 solid feed 6.1.17.4
固体加料器 solids feeder 15.1.1.8
固体接触式 solids-contact type 15.6.3
固体进口 solids in 4.3.2.13
固体进料 feed solid 15.1.1.1
 solid feed 14.1.12.1
固体进料斜槽 solid feed chute 13.3.1.1
固体(晶体)分布器 crystal distributor 9.5.6.23
固体聚集箱 solids collecting housing 9.5.10.17
固体颗粒 solid pastilles 13.5.3.3
固体颗粒缓冲仓 solids surge bin 2.2.4.2
固体颗粒加料孔 solids charge opening 4.3.1.1
>C_{30}(固体蜡) >C_{30}(solid wax) 2.2.73.11
固体粒子 solids 2.1.9.22
固体料面计 solid level 6.3.11.2
固体料面调节器 solid level controller 6.3.11.5
固体流量控制阀 solid flow-control valve 14.1.12.14
固体流量控制装置 solid flow control devices 14.1.6
固体流量调节阀 solids flow control valve

6.3.6.4
固体流通道 solids-flow channel 23.2.2.3
固体滤渣 solids cake 9.5.5.20
固体(滤渣)排出口 solids-discharge ports 9.5.5.11
固体煤粒立管 solid standpipe 2.2.23.4
固体排出 solid discharge 14.1.12.7
固体排出口 solids-discharge port 9.5.5.2
固体排料装置 solids discharge provision 4.3.2.4
固体平板输送表面 solid-plate conveying surface 4.3.4.28
固体取样 solids sampling 11.7
固体溶剂精制煤 solid solvent refined coal 2.2.91.2
固体速率增大 higher solid rate 14.1.1.6
固体物料混合机械 solid mixing machines 12.1.3
固体卸料 solids discharge 14.1.11
固体氧化物燃料电池 SOFC 2.2.113.3
固体氧化物燃料电池、燃气轮机和蒸汽轮机三结合联合循环示意图 schematic diagram of a triple combined cycle including SOFC(solid oxide fuel cell),gas and steam turbines 2.2.113
固体液体 solid liquid 2.2.98.13
固体用换热器 heat exchangers for solid 4.3
固体有机垃圾 solid organic rubbish 3.3.5.6
固位手柄 position retaining handle 23.3.8.9
固相化技术 immobilization technology 1
固相溶剂精制煤 solid SRC 2.2.91
固液分离 solid-liquid separation 1
固液分离器 liquid-solid separator 2.2.98.26
固液平衡 SLE;solid-liquid equilibrium 1
固液平衡 solid-liquid equilibrium(SLE) 1
固有频率输送器 natural frequency conveyor 23.1.7.2
故障诊断 failure diagnosis 1
刮板 blades 15.6.1.6
　flight 23.3.7.16
　scraper 13.3.5.26
　scraper blade 8.8.12.15
　wiper 8.9.1.6
刮板宽度 flight width 12.5.2.9
刮板冷却液 flight coolant 4.3.3.18
刮板轮毂 scraper boss 13.3.5.29
刮板输送机 flight conveyor 12.5.2.1
刮棒 scraper 15.7.4.3
刮刀 doctor blade[knife] 9.3.2.23
　scraper 13.5.1.11

scraper blade 9.5.3.10
scraping blade 8.7.2.1
刮刀杆 scraper bar 13.3.1.3
刮刀尖 scraper tip 9.3.3.29
刮刀片 scraper blade 9.3.3.30
刮刀调节器 scraper adjuster 9.3.3.31
刮刀涂布机 blade coating machine 2.3.11.29-35
刮刀下部卸料式 with bottom discharge by scraper 9.5.3.(a)
刮刀压板 scraper plate 9.3.3.32
刮刀轴承 scraper bearing 9.3.3.33
刮料板 stripper 23.3.7.14
刮料装置 scraper 9.5.4.29
刮面式换热器 scraped-surface exchanger 4.2.5.14
刮膜式分子蒸馏釜 wiped-film still 5.5.3
刮泥板链轮 flight scraper chain sprocket 3.1.3.9
刮泥板链条 flight scraper chain 3.1.3.11
刮油环 oil wiper rings 17.2.6.4
关键组分 key component 1
观察窗 observation window 24.1.13.8
观察孔 inspection port 12.2.1.9
　observation port 24.1.3.4
观察罩 inspection cover 14.1.11.6
　sight indicator 16.5.2.3
冠齿轮油池 crownwheel oil pan 9.6.1.4
冠齿轮罩 crownwheel shield 9.6.1.11
冠状软木瓶盖 crown cork closure 2.3.13.27
管 pipe 9.5.4.44
　tube 9.4.4.5
O管 O tube 11.5.2.12
P管 P tube 11.5.2.3
管板 header sheet 10.3.4.13
　pipe plate 9.3.3.13
管板支承板 tube sheet support 24.1.1.24
管程 side pass 1
　tube pass 1
　tube(side)pass 4.1.1.41
管程接管口 inner pipe nozzle 4.2.2.9
管道 tubing 18.4.1.11
管道泵 inline pump 16.4.20
　pipeline pump 16.1.2.3.52
管道气液分离器 in-line gas scrubber 5.3.9
管件 fittings 22
管接头 adapter 13.2.1.13
管接头加热器 adapter heater 13.2.1.9
管壳泵 tubular casing pump 16.1.2.2.28
管壳换热器 shell-and-tube heat exchanger

665

1;4.1
管路拐弯处　pipe bend　23.1.7
4管强制流袋包装机　four-tube force flow bag packer　23.4.2.1
4管强制流阀式袋包装机　four-tube force flow valve-bag packer　23.4.2
管坯　tubed　4.3.2.(c)
管式泵　piping pump　16.1.2.3.54
管式电除尘器　tubular precipitator　10.3.5
管式高速离心机　tubular-bowl centrifuge　9.5.11
管式炉　pipe still　6.3.2.18
tubular furnace　24.1.2
管式膜超滤　tubular membrane ultrafiltration　15.2.4
管式喷淋器　tubular sprayer　5.3.4.2
管式预冷器　pipe precooler　6.3.2.26
管式轴　tubular shaft　16.4.21.4
管束　tube bundle　1
管系　piping system　22.6.1
管系材料　pipe-systems materials　22.6.2
管线　pipeline　2.1.9.41
管线泵　in-line pump　16.1.2.3.53
管箱　channel section　21.1.1.52
header　4.5.1.10
管箱盖　channel cover　4.1.1.4
管箱接管　stationary head nozzle　4.1.1.5
管心距　tube pitch　4.1.5.5
管状圆筒　column pipe　16.4.25.17
管状圆筒定距块　column pipe spacer　16.4.25.19
管子　tube　21.1.1.50
管子排列形式　tube arrangement　4.1.5
管子随壳体回转　tubes rotate with shell　4.3.2.10
灌溉用泵　irrigation pump　16.1.2.3.7
灌液泵　priming pump　16.1.2.2.57
灌液漏斗　priming cup　16.4.19.21
罐区泵　tank farm pump　16.1.2.3.43
罐区潜水泵　tank farm submersible pump　16.1.2.3.44
罐装啤酒　beer can　2.3.13.25
光电倍增管　photomultiplier tube　12.7.5.12
光电二极管　photodiode　3.5.6.7
光电物料下落板凳设备　photocell dropout alarm　23.1.7.13
光滑式　plain　4.3.2.(a)
光室　optical chamber　12.7.4.7
Sortex 771M 光学分离器　Sortex 711M optical separator　12.7.4
光子分离器　optical separator　12.7

广义流态化　broad fluidization　14.1.3.24
generalized fluidization　1
规则的固体床层　regular-solids bed　4.3.3.1
规则填料　regular packing　5.3.4.45
规整填料　structured[arranged]packing　5.3.1.5
硅钢　carbon-silicon steel　22.6.2.⑩
硅胶　silica gel　3.5.11.11
硅胶干燥管　silica gel drying tube　3.5.14.11
硅胶管　silica-gel tube　3.5.7.6
硅酸钠　sodium silicate　6.3.10.2
硅铁泵　silicon iron pump　16.1.2.5.6
硅藻土过滤器　kieselguhr filter　2.3.13.4
硅镇静碳钢　silicon-killed carbon steel　22.6.3.⑭
R 轨道　R track　6.3.2.1
贵金属催化剂　precious metal catalysts　14.3.4.(c)
辊式输送机段　roller conveyor section　23.4.1.4
辊子　rollers　24.1.9.9
滚动团聚器　tumbling agglomerators　13.3.2
滚动轴承　trunnion roll bearings　8.8.4.(b)
滚轮　muller　12.1.3.10
muller turret　12.1.3.9
roll　9.3.5.7
滚轮切纸机　roll cutter　2.3.11.42
滚圈　riding ring　13.2.1.10
tire　12.1.1.11
doffer　8.3.4.23
滚筒　drum　8.1.1.⑧
revolving drum　8.7.1.1
滚筒干燥器　rotating drum dryer　8.7.1
滚筒回转筛　trommel　2.2.5.16
滚筒加料器　roll feeder　11.2.6.9
滚筒减速机　reducer for doffer　8.3.4.21
滚筒式造粒机　rolling drum granulator　13.3.1
滚柱式泵　roller-type pump　16.1.2.2.55
滚子　roller　4.2.4.8
锅　pan　15.7.4.6
锅炉　boiler　18.2.3.3
锅炉　combustor　25.1.2.11
锅炉厂专用燃烧炉　special burner on boiler plant　3.3.5.12
锅炉给水　boiler feed water　2.2.25.6;2.2.30.8;2.2.42.5;2.2.43.5;2.2.44.5;2.2.45.7;2.2.46.12;2.2.47.11;2.2.111.17;2.2.116.3;25.5.2.10
锅炉给水排污　boiler-feed water blowdown

2.3.6.30
锅炉循环泵 boiler circulating pump 16.1.2.3.5
锅体水平调节螺杆 adjusting screw for pan position 15.7.4.8
过程 process 20.1.9.5
SRC-Ⅰ过程 SRC-Ⅰ process 2.2.95
SRC-Ⅱ过程 SRC-Ⅱ process 2.2.96
过程加热 process heat 25.1.1.12
过程空气 process air 6.2.8.7
过程流体出口 process stream out 4.5.2.6
过程流体出口接管 process-fluid outlet nozzle 4.5.1.13
过程流体加入 process stream in 4.5.2.5
过程流体进口接管 process-fluid inlet nozzle 4.5.1.12
过程模拟 process simulation 1
过程气出口 process gas outlet 4.4.2.7
过程气体 process gas 19.1.7.12
过程热力学分析 thermodynamic analysis of process 1
过程系统工程 process system engineering 1
过程用溶剂 process solvent 2.2.95.16
过程优化 process optimization 1
过大颗粒 oversize particles 12.3.3.6
过渡半径 knuckle radius 21.1.1.62
过渡堵头 bridge blocks 9.3.10.15
过渡段扩散器 transition diffuser 16.4.23.8
过冷液体 subcooled liquid 20.2.4.24
过量焓 excess enthalpy 1
过量生物质 excess biomass 3.2.11.13
过量炭 excess char 3.3.15.16
过滤 filtration 1.1.37.14; 2.4.5.7
filtering 9.3.5.10
过滤板 filtration plate 13.2.3.13
过滤槽 lauter tun 2.3.12.50
过滤层 blanket 15.6.4.14
过滤层段 filter compartment 3.1.10.4
过滤带 filter belt 9.4.8.3
过滤袋隔离室 compartment isolated for bag cleaning 10.2.2.2
过滤罐 filter tank 9.3.3.38
过滤机 filter 12.8.1.15
过滤机盖支架 filter cover support 9.3.3.26
过滤机减速驱动装置 reduction drive-gear for filter 9.4.2.1
过滤机转鼓 filter drum 9.3.3.25
过滤接收器 filter receiver 23.1.1.11
过滤聚结器 filter coalescer 15.5.5
过滤离心机 filtration centrifuge 9.5.2.1
过滤麦芽汁 filtering off the wort 2.3.12.50

过滤膜捆 bundles 3.1.29.2
过滤母液 undiluted mother liquor 9.4.1.13
过滤器 filter 2.2.20.4; 2.2.117.8; 2.2.118.10; 12.8.2.19
strainer 16.4.25.1
sweetland filter 6.3.2.28
过滤器风机 filter fan 11.6.4.19
过滤器夹持器 filter holder 3.5.16.8
过滤器夹持器配置 filter holder location 3.5.16.7
过滤器筒体 filter cylinder 10.2.3.1
过滤器组 lauter battery 2.3.12.51
过滤筛 filter mesh 3.1.19.12
过滤设备 filtration equipment 9
过滤室贮斗 filter bouse 6.3.2.10
过滤收集器 filter receiver 23.1.6.8
过滤物料液面 level of material to be filtered 9.3.5.8
过滤叶片 filter leaf 9.2.2.7
过滤叶片接口 leaf nozzle 9.2.5.28
过滤叶片驱动电机 leaf driving motor 9.2.5.3
过滤叶片驱动装置 leaf driving unit 9.2.5.21
过滤叶片支架 leaf support 9.2.5.26
过滤叶片组件 filter leaf group 9.2.2.6
过热 superheating 25.5.1.10
过热低压蒸汽 superheated low pressure steam 2.2.43.2
过热器 superheater 2.2.81.10
过热区 superheat zone 20.2.4.10
过热蒸气 superheated steam[vapor] 25.5.3.6
过热蒸汽 superheated steam 2.2.31.14
过热蒸汽发生器 superheater-steam generator 25.5.1.(b)
过烧空气通道 overfire air port 3.3.12.9
过剩的污泥 excess sludge 3.1.28.9; 3.4.1.9
过剩空气 excess air 3.1.14.15
过剩蚯蚓(如果有的话) surplus earthworms(if any) 3.3.28.11
过剩蚯蚓可出售 surplus earthworms for sale 3.3.28.15
过失误差 gross error 1
过载断路器 overload release 9.5.5.12

H

哈斯贝克设计的氧燃烧动力装置 Oxy-com-

bustion power plant based on design by Haslbeck etal 2.2.115
海水 seawater 2.4.3.7
海水泵 sea water pump 16.1.2.3.13
(海水)多级闪速蒸馏工艺示意 Schematic of a multistage flash distillation process 2.4.2
海水进入 seawater in 2.4.2.7
含尘空气 dust laden air 10.1.3.13
含尘空气侧 dust air side 10.2.1.3
含尘空气入口 dust-laden air in 10.5.4.10
含尘气体 dust-laden gas 10.5.4.14
含尘气体进口 dust-laden gas inlet 10.2.2.1
含二氧化碳溶液 CO_2-loaded solution 2.2.27.5
含固形物 total solids 13.4.1.10
含氯的溶剂 solvents containing chlorine 3.1.21.2
含氢富气 hydrogen-rich gas 2.2.101.1
含湿量 moisture content 13.1.1
含石油气易燃残渣 wet flammable sludge 3.3.5.5
含水的环氧乙烷 ethylene oxide with water 2.3.19.17
含水甲醛 aqueous formaldehyde 14.3.1.1
C_2 含氧化合物 C_2 oxygen containing compounds 2.2.40.9
含油的水 oily water 2.3.6.12
含油废水 oily waste 3.1.13
含油废水进入 oily-water influent 3.1.14.19
含油水储存池 oily-water holding pond 2.3.6.10
含油污水 oily water sewer 3.1.1.2
含有二氧化碳的溶剂 CO_2 loaded solvent 2.2.32.2
含有硫化氢的溶剂 H_2S loaded solvent 2.2.32.9
焓 enthalpy 1
焓熵图 enthalpy-entropy diagram 1
焊缝接头局部检验 spot exam. of welded joint 21.1.2.11
焊缝系数 joint efficiency 21.1.2.3
焊接垫板 backing strip 21.1.1.46
焊接端结构 welding end construction 22.5.3.3
焊接环支承 welded ring support 5.3.4.22
焊接结构 weld const 21.1.2.7
焊接连接 welded connection 21.1.1.2
800 号合金专用板 alloy 800 production plate 4.4.2.12
好氧 AER;aerobic 3.2.2.2
好氧细菌 aerobic bacteria 1

合成 synthesis 2.3.3.15
合成氨 ammonia synthesis 2.3.2.15;2.2.33.13;
synthetic ammonia 2.2.1.13
合成氨厂 ammonia plant 2.3.2
合成氨工艺流程 ammonia synthesis process 2.2.38
合成氨装置 ammonia plant 2.3.3
合成的粗产物 synthetic crude 2.1.9.41
合成反应 synthesis 2.2.73.3
合成反应器 synthesis reactor 2.2.37.7;2.2.40.8;2.2.41.3;2.2.48.6
合成工序 synthesis process 2.2.45.8
合成气 synthesis gas 2.1.8.12
syngas 2.2.1.2;2.2.22.5;2.2.23.8;2.2.24.12;2.2.25.14;2.2.29.14;2.2.30.13;2.2.31.20;2.2.32.1;2.2.40.2;2.2.41.1;2.2.43.5;2.2.49.1;2.2.50.3;2.2.70.1;2.2.74.3
合成气变换单元工艺流程 Process of syngas conversion unit 2.2.34
合成气出口 syngas out 2.2.16.6
合成气加工 syngas processing 2.2.3.3
合成气进料 syngas feed 2.2.28.1
合成气净化(低温甲醇洗) syngas purification(Rectisol process) 2.2.67.7
合成气冷却器 syngas cooler 2.2.23.5;2.2.25.12;2.2.31.15
合成气溶液 ammonia stripper 2.3.6.13
合成气/循环气压缩机 synthesis gas/recycle gas compressor 2.2.43.2
合成气制乙醇工艺流程 process flow for ethanol synthesis from syngas 2.2.41
合成气压缩机 syngas compressor 2.2.44.2
合成气压缩机 synthesis gas compressor 2.3.3.14
合成轻油 synthetic light oil 2.2.74.12
合成燃料 synfuel 25.1.1.5
synthetic fuels 14.3.2
合成塔 synthesis converter 2.2.49.2
合成纤维 synthetic fibre 2.3.7
合成油产品路线 schematic drawing of oil product synthesis 2.2.73
合成油工艺流程 Gasoline synthesis process 2.2.78
合成原油 synthetic crude oil 2.2.30.11
syn-crude 2.2.2.34
合金风机 alloy fan 24.1.7.2
合金框 alloy basket 24.1.7.6
合金栅板 alloy grid plate 24.1.7.5
河水制备饮用水流程 Potable water production from a river 2.4.4

核动力站　nuclear power plant　25.2.2
核反应堆　nuclear reactor　25.2.2
核能　nuclear energy　25.1.1.2
核酸　nucleic acid　1
核糖　ribose　1
核糖核酸　ribonucleic acid(RNA)　1
褐煤　lignite　2.2.107.6
黑粒子　black particle　12.1.4.1
黑液过滤器　black liquor filter　2.3.10.28
黑液贮槽　black liquor storage tank　2.3.10.29
恒沸物　azeotrope　1
恒沸蒸馏　azeotropic distillation　1
恒温器　thermostat　8.3.4.15
恒温式疏水阀　thermostatic trap　22.2.6
恒压混合段　constant pressure mixing section　18.3.3.7
横杆　side rail　9.1.2.10
横刮板　cross flights　3.2.15.10
横梁式气体喷射支承板　beam-type gas-injection-support plate　5.3.4.24
横向流动　cross flow　4.1.2.(l)
横向送料器　cross conveyor　23.2.3.(c)
横向折流板　transverse baffles　4.1.1.28
横型分布器　trough-type distributor　5.3.4.13
横移刮刀　traversing knife　9.5.2.22
烘房内烘干　drying in heated room　2.3.7.24
烘干层　kilning floor　2.3.12.16～18
烘干机　drier　2.3.16.13
　　dryer　2.3.17.13
烘干窑　drying kiln　2.3.12.29
烘缸　drying cylinder　2.3.11.22
烘窑出气口　outlet from the kiln　2.3.12.19
红外线干燥器　infrared drier　2.3.11.32
宏观流体　macrofluid　1
虹吸管　siphon pipe[tube]　16.5.2
喉部　throat　18.3.1.7；3.4.4.13
喉部垫环　throat ring　16.4.20.22
喉部截面　throat section　3.5.4.6
后曝气　post-aeration　3.1.37.13
后处理　downstream processing　1
后端封头型式　Rear end head types　4.1.4
(后)副十字头导轨　hind counter guide　17.2.15.18
后盖　rear cover　16.4.18.14
后管　trail pipe　9.3.10.11
后级泵　secondary pump　18.1.2.10
后排气阀　secondary discharge valve　18.1.2.12
后加热段　post-heating　24.1.12.3
后加热区　back heat zone　13.3.1.17
后壳体　rear casing　16.5.5.6
后冷凝器　after condenser　18.3.4.26
后迷宫密封　rear labyrioth seal　17.4.2.26
后燃烧室　afterburner chamber　3.1.22.9
后弯叶片　backward-curved vane[blade]　16.4.5.3.1
后置冷凝器　after condenser　20.10.11
后置冷却器　after cooler　17.6.1.7
后轴承　rear bearing　9.3.3.3
呼唤按钮　call button　25.4.1.3
呼吸器　breather　25.4.3.10
弧度　rad.；radius　22.5.2.4
弧形搅拌器　agitator arc　9.3.3.34
弧形水盘　water paring disc　9.5.12.1
弧形油盘　paring disc　9.5.12.3
户内电话　house telephone　25.4.1.3
护板　screen　11.1.5.23
护板定位框　screen guide　11.1.5.11
护罩　cap　22.1.2.14
戽斗链轮再装料器　bucket-wheel reclaimer　23.2.5
滑板阀　slide valve　14.1.6.(a)
滑板式真空泵　sliding-vane vacuum pump　16.1.2.2.42
滑板型转鼓底　sliding bowl bottom　9.5.12.9
滑板闸门　slide gate　23.3.3.2
滑动鞍座　sliding saddle　21.3.1.3
滑动衬板　slip lining　9.5.4.12
滑动轮毂　slip boss　9.5.4.17
滑动托板　slip blade　9.5.4.13
滑动托板销　slip blade pin　9.5.4.14
滑动闸阀　slide gate　23.1.6.13
滑动轴承　sleeve bearing　17.4.2.10
滑阀　slide valve　16.2.3.54
滑阀杆　slide valve spindle　18.1.1.5
滑阀环　slide valve ring　18.1.1.4
滑阀式真空泵　slide valve vacuum pump　18.1.1
滑键　slide-key　16.2.5.9
滑块　slipper　17.2.15.5
滑块泵　shuttle block pump　16.1.2.2.66
滑片　sliding vane　16.3.5.1
滑片泵　sliding-vane pump　16.3.5
滑片式　sliding vane　17.1.1.8
滑片式压缩机　sliding vane compressor　17.3.1.(a)
滑移齿轮　flow gear　3.5.5.4
滑移凸轮　flow cam　3.5.5.1
化工工艺　chemical process　2.3
化工工艺流程图　flow diagram of chemical process　2
化工工艺用泵　process pump　16.1.2.3.36

669

化工机械　chemical machinery　1
化工机械工程　process equipment engineering　1
化工热力学　chemical engineering thermodynamics　1
化工塔类　chemical towers　1
化工系统工程　chemical[process] system engineering　1
化工用泵　chemical pump　16.1.2.3.35
化灰器　lime slaker　2.3.10.51
化石燃料　fossil fuel　25.1.1.4
化学瓷管　chemical-porcelain pipe　22.6.2.⑨
化学反应工程　chemical reaction engineering　1
化学工程　chemical engineering　1
化学工程学　chemical engineering science　1
化学工业　chemical industry　1
化学耗氧量　chemical oxygen demand (COD)　1
化学品　chemicals　2.2.89.11
化学平衡　chemical equilibrium　1
化学气相沉[淀]积　chemical vapor deposition(CVD)　1
化学识别　chemical identification　3.3.7.9
化学陶瓷管　chemical ware　22.6.2.④
化学调理　chemical conditioning　3.1.26.(a).6;3.1.26.(b).9
化学吸附　chemisorption　1
化学吸收　chemical absorption　1
化学洗涤器　chemical scrubber　3.3.25.13
化学需氧量　chemical oxygen demand (COD)　1
化学循环燃烧系统简图　simplified schematic diagram of chemical looping combustion system　2.2.80
化学药品　chemicals　3.2.11.19
化学药品添加　chemical-addition　3.1.43.4
化学原料　chemical feedstocks　2.2.93.10
化学制品　chemicals　3.3.1.3
环　ring　13.1.2.38
环槽密封面　ring type joint face　22.4.3.(e)
环段泵　ring-section pump　16.1.2.2.61
环焊缝接头　circumferential joints　21.1.1.47
环己酮　cyclohexanone　2.3.8.20
环己酮肟　cyclohexanoxime　2.3.8.22
环境空气　ambient air　10.4.1.5
环流　circulation　1
环流反应器　loop reactor　1
环隙开口　annuli open　14.1.8.(a)
环向开槽　circumferential groove　19.3.1.5
环行式　orbiting type　12.1.3.(f)
环形槽　annulus　9.3.2.27
环形床　annnular bed　6.3.8.8
环形风管　ring duct　8.4.5.25
ring-shaped path　8.4.5.(c)
Berks环形干燥器　Berks ring dryer　8.4.5.(c)
环形沟　ringed channel　5.3.1.14
环形管道　ring duct　8.4.6.10
环形加料提升机　loop-feed elevator　23.3.3.(c)
环形接头　toroidal　22.5.1.(f)
环形空间　annulus　13.4.1.20
环形气室　annular cell　17.3.3.8
环形收集溜槽　annular collection launder　15.6.3.25
环形下流　annular flow　23.3.4.5
环形堰板　ring dam　9.5.11.5
环形液体分离床　annular bed for liquid separation　6.3.8
环氧玻璃纤维增强支持管　fibreglass-reinforced epoxy support tube　15.2.4.3
环氧树脂管板　epoxy tube sheet　15.2.1.10
环氧乙烷　ethylene oxide　2.3.19.15
环氧乙烷解吸塔　EO desorber　2.3.19.11
环氧乙烷生产工艺　A process flow diagram to produce ethylene oxide　2.3.19
环氧乙烷洗涤塔　EO scrubber　2.3.19.6
环轴定位　retaining ring　16.4.19.29
环状的固体床层　annular solids bed　4.3.3.7
缓冲　buffer　2.2.56.7
缓冲衬板　breaker plate　13.3.1.10
缓冲挡板　impingement baffle　4.1.1.29
缓冲罐　knock-out drum　5.3.3.13
缓冲罐　surge tank　2.2.106.27;3.1.34.4
缓冲气体　buffer gas　19.1.7.15
缓冲弹簧　buffer spring　9.5.3.6
缓冲液　buffered　19.3.3.10
缓冲液进口　buffer inlet　19.3.1.17
缓冲装料斗　surge hopper　2.2.5.3
缓启闭手轮　temper handle　9.2.5.23
换热板组件　heat exchange plate pack　4.2.1.4
换热管　heat exchange tube　4.1.4.2
换热过程　heat exchange process　2.1.4.5
换热器　heat exchanger　2.2.108.7;2.2.46.3;2.2.47.2;2.2.110.7;3.3.25.15
换热器部件　heat-exchanger component　4.1.1
换热器单元　heater block　3.5.21.7
换向挡板　change over flap　8.4.5.15
换向阀　diverter valve　23.1.8.22
黄金分割法　golden section method　1

670

黄蜡布带　varnished-cambric tape　25.4.8.4
黄麻护层　jute serving　25.4.8.7
黄铁矿的硫　pyritic sulfur　2.2.2.33
黄铜板翅式换热器　brazed-plate-fin heat exchanger　4.2.5.3
簧片开关　reed switches　20.1.8.15
灰　ash　2.2.90.13
灰仓　ash bin　3.3.3.3
灰尘沉淀室　dust-settling chamber　2.3.10.6
灰分　ash　2.3.1.21；2.2.17.3
灰分排出　ash discharge　3.3.9.2
灰分去除　ash removal　2.2.19.9
灰锁斗　lock hopper　2.2.29.12；2.2.30.9
　　ash lock　2.2.13.6
灰分调理器　ash conditioner　3.3.3.2
灰分脱水　ash dewatering　2.3.1.31
灰分卸压　ash depressurization　2.2.24.5
灰坑　ash pit　24.1.4.2
灰筒　ash drums　3.3.8.14
灰渣　ash and residue　2.2.94.16
　　ash residue　2.2.97.14
　　char　2.2.22.7；2.2.23.7
挥发性馏分　volatile tops　5.5.3.7
回程　return runs　23.3.7.(b)
(回程)下托辊　return idlers　23.2.1.11
回风调节风门　R.A damper　20.2.2.7
回归分析　regression analysis　1
回卷台　rewind station　2.3.11.41
回流　backstreaming　18.2.4.4
　　reflux　14.3.4.29
　　return　25.6.7.7
回流的污泥　return sludge　3.1.32.17；3.1.34.10；3.1.44.5；3.1.45.3；3.1.46.6
回流导叶　return guide vane　20.2.1.7
回流分凝器　reflux dephlegmator　2.2.57.5
回流管　recycle tube　2.2.98.7
回流管口　reflux return nozzle　14.4.1.1
回流罐　reflux tank　2.2.76.16
回流活性污泥　RAS；return activated sludge　3.1.25.9；3.1.33.4；3.1.35.1；3.1.37.10；3.1.40.4；2.2.2.3
回流空气　return air　3.3.3.7
回流口　reflux inlet　5.2.1.2
回流冷凝器接管口　nozzle for reflux condenser　14.4.1.12
回流器　return passage　17.4.2.17
回流生物固体　return biosolids　3.1.31.8
回流室　return channel　20.2.3.11
回流弯道　return bend　17.4.2.16
回热器　regenerator　25.3.2.4
回收　recovering　2.3.14.37
回收处理　treatment for recovery　12.5.5.10
回收的粉末　recovered fines　2.2.19.8
回收的铬酸　recovered chromic acid　6.2.7.1
回收的溶剂　recovered solvent　6.3.7.10
回收的有机氯化物　recovered organic chloride　3.1.7.5
回收阀　reclaim valve　20.1.8.8
回收(率)　recovery　1
回收溶剂　recovered solvent　8.9.2.9
　　solvent recovery　8.9.2
EDC回收塔　EDC recovery column　2.3.18.15
回收盐　salt recovery　3.3.11.1
回收铀工艺　process for recovering uranium　5.4.3
回收装置　recovery plant　2.3.14.37
回水　return water　20.1.10.15
回水收集槽　collecting tank for backwater　2.3.10.26
回弯头　return bend　4.2.2.2
回吸收塔　back to the absorption tower　2.2.50.2
回油　returned oil　24.2.4.8
回油管　oil return pipe[tube]　18.2.3.9
回造气　back to gasification　2.2.50.1
回转　rotation　13.3.3.9
回转泵　rotating pump　16.3
回转的金属壳体　rotating metal shell　4.3.2.3
回转干燥器　rotary dryer[drier]　8.8；6.3.10.17
回转隔池　rotating cells　6.1.17.6
回转(公转)驱动装置　reversing rotational drive　8.9.5.15
回转过程　during rotation　13.3.3(b)
回转烘干机　rotary dryer　3.3.25.5
回转件　rotating element　8.8.5.12
回转壳体　rotating shell　8.8.5.16
回转破碎机　gyratory crusher　11.6.3.2
回转生物接触器　RBC；rotating biological contactor　3.2.11.4
回转式　rotary　17.1.1.5
回转式鼓风机　rotary blower　17.3
回转式滑片压气机　sliding-vane type of rotary compressor　17.3.3
回转式环形色谱仪　rotating annular chromatograph　6.3.8.11
回转式脱气器　rotary degassers　5.5.3.6
回转式液环压气机　liquid-piston type of rotary compressor　17.3.4
回转刷子　rotating brush　7.1.3.11
回转压缩机　rotary compressor　17.3.1

回转窑　rotary kiln　3.3.8.4
回转叶片加料器　rotary-vane feeder　12.1.5.16
回转圆筒混合器　rotary mixer　12.1.1
回转振动筛　gyrating screens　12.2.3
回转轴　rotary shaft　19.1.1.5
混合　mixing　12.1.1.18
混合 C_4　mixed C_4　2.2.67.20;2.2.69.21
混合 C_5　mixed C_5　2.2.67.21
混合槽　blending tank　3.1.8.2
　　mixing tank　2.3.10.35
Vitro 混合-澄清器　（用于铀萃取）Vitro uranium mixer-settler　6.1.13
混合床离子交换　mixed-bed ion exchange　6.2.9
混合段　mixing section　18.3.3.4
混合二甲酚　xylenols　2.2.91.9
混合釜壁　mixing kettle wall　4.3.1.7
混合喉管　combining throat　18.3.4.7
混合机械　mixing machine　15.7
混合甲酚　cresols　2.2.91.9
混合搅拌器　mixing agitator　6.3.2.12
混合料仓　blending bin　11.6.3.3
混合流式　ixed-flow　8.6.3.(b)
混合螺带螺旋　mixing-ribbon spirals　4.3.1.4
混合螺杆　mixing screw　12.1.1.26
混合盘管　mixing coil　2.2.57.8
混合器　mixer　16.5.1.4
Stratco 混合器　Stratco contactor　6.1.7
混合器　mixer　2.2.57.1;2.2.66.4;3.1.35.7;3.3.24.3;6.1.9.5
混合器出口　mixture outlet　24.2.7.8
混合器打碎物料　mixer breaks up material　11.4.3.2
混合器螺杆　mixer screw　8.9.5.9
混合器螺旋（自转）驱动装置　mixer screw drive unit　8.9.5.14
混合器支承面　mixer support level　8.4.1.10
混合区　zone of mixing　3.3.18.5
混合溶剂　component solvent　2.1.1.78
混合室　mixing chamber　15.6.4.15
混合树脂　mixed resin　6.2.9.1
混合/淘析槽　mixing/elutriation tank　3.2.10.2
混合物　mix　6.1.7.9
混合物料　mixed material　12.6.10.3
混合纤维素板　mixing cellulose sheets　2.3.7.2
混合研磨机　mix muller　6.3.10.15
　　mixing & grinding machine　15.7.4

混合液　mixed liquor　3.1.29.4;3.1.34.9;3.1.46.13;3.3.18.7
混合液流型　mixed liquor flow pattern　3.2.13.9
混合液去澄清器　mixed liquor to clarifier　3.1.45.8
混合液入口　mixed liquor inlet　3.1.29.1
混合液循环　MLR;mixed liquor recirculation　3.2.2.12
混合/造粒机　mixer/granulator　3.1.16.3
混合转鼓　mixing drum　9.6.2.4
混流泵　mixed flow pump　16.1.2.2.25
混流（式）　mixed flow　16.1.1.32
混流式叶轮　mixed-flow impeller　16.4.5.(c)
混凝土　concrete;conc.　3.1.4.3
混凝土板　concrete plate　12.2.3.22
混凝土基础　concrete foundation　21.2.2.1
混凝土屏蔽层　concrete shield　25.2.2.29
活动端板　movable head　9.1.2.7
活度系数　activity coefficient　1
活节　swivel　22.5.4.2
活门把手　valve handle　11.3.1.10
活塞　piston　16.2.1.14
活塞泵　piston pump　16.1.2.2.4
活塞底座　piston foot　16.5.6.4
活塞阀　piston valve　16.2.4.47
活塞杆　piston rod　16.2.1.15
活塞杆接头螺栓　piston-rod spool bolt　16.2.3.14
活塞杆填料　piston-rod packing　16.2.1.16
活塞杆填料函衬套　piston-rod stuffing box bushing　16.2.4.9
活塞环　piston ring　17.2.7.21
活塞流　plug flow　1
活塞密封环　piston packing　16.2.1.18
活塞受力　direction of forces on pistons　16.2.1
活塞推料离心机　pusher centrifuge　9.5.10
　　reciprocating-conveyor continuous centrifuge　9.5.7
活套法兰　loose type flange　21.1.1.5
活套靠背法兰　slip-on backing flange　4.1.1.20
活性沸石催化剂　active zeolite catalyst　14.3.4.(b)
活性分子筛颗粒　activated molecular sieve pellets　6.3.10.20
ABF(活性生物过滤设备)工艺流程图　ABF (activated bio filter) process flow diagram　3.1.44
活性衰减　decay of activity　1

活性炭　（active）carbon　3.1.19.7
活性炭补充　carbon make-up　3.1.18.3
活性炭出口　carbon out　6.3.5.6
活性炭床层表面　carbon bed surface　6.3.9.11
活性炭床层高度　carbon bed hight　6.3.9.1
活性炭管　charcoal tube　3.5.16.25
活性炭进入　carbon in　6.3.5.2
活性炭流量调节器　carbon flow controller　6.3.11.10
活性炭生物物理处理　biophysical treatment uses activated carbon　3.1.18
活性炭-污泥循环　carbon-sludge recycle　3.1.18.6
活性炭卸出口　carbon discharge　6.3.9.12
活性炭再生　carbon regeneration　3.1.18.1
活性炭再生用多层膛式炉　Multiple hearth furnace for carbon reactivation　6.3.5
活性炭柱　carbon columns　6.3.1.2
活性炭装填口　carbon charge　6.3.9.9
活性污泥　activated sludge　1
活性污泥池　activated sludge tank　3.4.2.6
活性污泥工艺　activated sludge process　3.2.2
活性污泥系统　activated-sludge system　3.1.8
活跃消解的污泥　actively digesting sludge　3.3.19.2.8.6
火车闷罐车　railroad boxcar　23.4.2.16
火管　firetube　25.6.5.1
火管式盐浴加热器　fire-tube salt-bath heater　25.6.5
火花（放电）间隙　spark gap　25.4.9.12
火墙　bridge wall　24.1.4.4
火焰　flame　8.1.1.29
火焰离子化检测器　flame-ionization detector　3.4.21
火焰通口　flame port　3.3.12.7
火焰预热器　fired preheater　2.2.29.3
货车　truck　23.4.2
货车联结点　car connection　23.1.2.18
货船油泵　cargo oil pump　16.1.2.3.26
霍尔电解质电导率检测器　Hall electrolytic conductivity detector　3.5.22

J

机动有轨车　railcar　23.1.7.3
机盖　cover　15.7.6.6
机架　frame　14.4.1.28
　stand leg　9.1.2.12
机架悬挂　spider suspension　11.1.4
机壳　enclosure　4.3.4.22
　housing　8.9.4.3
　main case　13.3.5.24
　shell　6.1.7.7
机（器）罩　machine hood　2.3.11.27
机身　machine body［frame］　11.1.8.13
机身顶盖　frame top cover　17.2.15.2
机身夹板　bag clamp　11.1.11.1
机身紧固螺栓　frame tie bolts　17.2.15.7
机体　main body　11.4.1.10
机筒　barrel　13.2.1.7
机箱　housing　8.8.12.14
　housing cap　8.8.12.9
机械　mechanical　8.1.1.32
机械浮动式　mechanical float　3.5.18.(c)
机械减振器　mechanical dash pot　23.3.2.4
机械浆料池　machine chest　2.3.10.86
机械接触式轴封　Mechanical contact shaft seal　19.1.7
机械密垫　mechanical seal［packer］　3.1.24.12
机械密封部件　mechanical seal components　19.1.1
机械驱动的　mech.operated　16.1.1.16
机械式　mechanical　23.3.2.11
机械式涤气器　mechanical scrubber　10.5.9
机械式空气分级器　mechanical air classifier　11.6.5.1
机械式喷嘴　mechanical spray thrower　5.3.4.6
机械添煤工业煅烧炉　mechanically fired industrial incinerator　24.1.5
机械压缩制冷　mechanical refrigeration　20.1.1.(a)
机械真空泵　mechanical vacuum pump　19.4.1.3
机械作用的　mechanically actuated　16.2.8
机座　base　12.2.7.13
　base support　13.3.3.3
　machine supporter　11.1.8.17
　machine-base　15.7.2.12
　under frame　12.2.5.6
机座曲轴箱　base crankcase　16.2.7.11
积分采样装置　integrated-sample setup　3.5.9
基板　base plate　21.3.3.3
基本合成天然气工艺　basic SNG process　2.2.34
基本Brayton或Joule循环　basic Brayton or Joule cycle　25.3.2.(a)

基本系统　basic system　3.1.2.30
基本型压缩机　basic compressor　17.4.3.5
基本原料　basic material　2.3.7.1
基本制冷系统　basic refrigeration system　20.1.1
基本制冷循环　basic refrigeration cycle　20.1.3
基础　base　21.3.1.1
基础环　basic ring　5.3.4.36
基础环板　base ring　21.3.1.2
基础螺栓　foundation bolt　9.2.5.6
基础螺栓孔　foundation bolt hole　17.2.15.27
基团活度系数　group activity coefficient　1
基因工程　genetic engineering　1
基于急冷反应器的甲醇合成工艺流程　Methanol synthesis flowsheet based on a quench reactor　2.2.48
基质　substrate　1
激磁绕组　exciting winding　16.5.4.8
激活电极　active electrode　12.7.2.1
激励铁催化剂　promoted iron catalyst　14.3.2.(c)
4级(除氮)工艺　4-stage(Nitrogen removal) process　3.1.33.1
5级(脱磷和脱氮)过程　5-stage(phosphorus and nitrogen removal)process　3.1.33.8
级间　interstage　18.3.4 (b)
级间挡板　inter stage baffle　3.2.11.6
级间隔板　stage baffle　3.1.45.9
级间冷凝器　intercondenser　18.3.4.21
级间冷却器　intercooler　17.2.7.16
级间迷宫密封　interstage labyrinth seal　17.4.2.20
级联式换热器　cascade heat exchanger　4.2.5.7
极化器电极　polarizer electrode　3.5.21.4
极面　pole face　16.5.4.2
极限开关　limit switch　9.5.6.19
极性起重机　polar crane　25.2.2.18
急冷罐[槽]　quench tank　3.1.17.7
急冷甲醇　chilled methanol　2.2.28.15
急冷气　quench gas　2.2.25.9;2.2.31.6;2.2.37.8;2.2.48.7
急冷室　quench chamber　2.2.29.11;3.3.8.7
quench cooler　2.2.13.4
急冷水　chilled water　20.1.10.26
quenching water　2.2.15.5
quench water　2.2.17.2;2.2.22.1;2.2.26.7
急冷水进口　quench-water inlet　14.1.11.1
急冷水进入　quench water in　2.2.16.3

急冷塔　quenching tower　2.2.69.7;2.2.71.2.2.72.11
急冷塔　quencher　2.3.18.7
EDC急冷塔　EDC quencher　2.3.18.16
急冷淘析器　quench elutriator　14.3.4.37
急冷液　quench liquid　2.2.13.3
集尘板(正极)　collecting(positive)plate　10.3.1.3
集尘袋套管　dust-bag sleeve　8.8.11.13
集尘电极管　collecting electrode-pipe　10.3.4.12
集尘斗　(dust)hopper　10.2.1.24
集尘空气　ash pickup air　10.3.5.3
集尘器　dust catcher　3.3.5.14
dust collector(D.C.)　12.6.4.19
dust trap　10.1.2.6
receiver　10.1.4.4
集尘器顶盖　cover of dust collector　12.1.1.4
集尘器　dust collector　10.4.4.7
集尘室　bag house　10.4.1.16
集电极　collector electrode　3.5.21.6
集粉器　dust collector　8.5.4.9
集管箱　manifold　8.8.7.(b)
集料吊斗　collection bucket　3.5.10.13
集料螺旋　collecting screw　8.4.5.14
集泥板　sludge collecting board　15.6.5.10
集泥斗　sludge-collecting hopper　3.1.3.3
集泥斗卸出口　sludge-collecting hopper　3.1.3.1
集泥机　sludge collector　15.6.5
集水槽　water collecting pond　4.2.3.5
集水孔　water collecting orifice　15.6.4.16
集水器　trap　17.6.2.9
集水箱　water-collecting tank　2.3.12.5
集中分离器　centralized separator　2.3.10.15
集中供热循环泵　central heating circulating pump　16.1.2.3.12
几何位置　geometries　10.5.11.(a)
己内酰胺　caprolactam　2.3.8.28
己内酰胺油　caprolactam oil　2.3.8.25
挤出机　extruder　15.7.5.2
挤缝　caulking　9.3.10.3
挤球　squeeze bulb　3.5.14.23
挤压　compaction　23.2.2.4
挤压泵　squeegee pump　16.1.2.2.64
挤压机　Extruder　15.7.2
挤压区　wedge zone　9.6.2.2
挤压造粒机　pellet extruder　6.3.10.16
挤压造粒设备　extrusion pelleting equipment　13.2
计量泵　constant rate pump　16.1.2.3.45

计量泵　metering pump　16.1.2.3.46
计量段　metering section　13.2.1.14
计量后进料　measured feed　23.3.3.8
计量及树脂清洗室　measuring and resin cleaning chamber　6.2.6.4
计量输送机[器]　metering conveyor　23.4.3.8
计算机辅助过程设计　computer aided process design(CAPD)　1
记录器　recorder　3.5.22.14
加部加强圈　external reinforcing ribs　8.2.2.10
加重皮带张紧轮　weigh idler(s)　23.5.4.15
加工　process　2.2.36.7
加工气流　process gas flow　15.4.1.2
加工气(逆流)流　process gas (countercurrent) flow　15.4.1.2
加工气体泵　process gas pump　15.4.1.6
加工烟道气　fluegas treating　2.3.1.30
加劳特油页岩干馏装置　Galoter oil-shale retorting system　2.2.88
加料　charge　14.3.1.20
　F;feed　12.4.1
　feed in　15.2.2.1
　feed inlet　8.5.7.5
加料泵　charge pump　16.1.2.3.27
　feed pump　18.5.1.9
加料斗　feed hopper　11.2.10.1
　hopper　15.7.2.7
加料端　filling end　2.3.9.2
加料段　feed section　13.3.1.16
　hopper　6.3.11.18
加料分布器　feed distributor　12.6.1.8
加料管线　fill line　25.6.7.13
加料罐　charging tank　6.3.2.14
　feed tank　8.5.7.6
加料喉管　feed throat　13.2.3.6
加料井　feed well　15.6.1.12
加料口　charge opening　24.1.12.1
　feed inlet　11.2.1.1
加料料仓　charge bin　2.2.20.5
加料流量计　feed flowmeter　5.5.3.4
加料螺旋　feed screw　11.2.10.2
加料器　feeder;fdr　11.2.11.1
加料器减速器　feeder speed reducer　11.2.8.4
加料裙罩　loading skirts　23.2.1.3
加料水　feed water　15.2.3.2
加料预热器　feed preheater　14.3.2.18
加料装置　feed system　3.3.7.20
加氯消毒　chlorination　2.4.5.8
加强垫板　stiffener plate　21.1.1.14

　stiffening pad　21.3.1.4
加强圈　reinforcement pad　21.1.1.3
　stiffener ring　21.2.2.10
　stiffening rings　21.1.1.42
加氢柴油　hydrogenated diesel fuel　2.1.1.75
加氢处理　hydrotreating　2.2.2.26
加氢反应　hydrogenating　2.2.66.13
加氢反应器　hydrogenation reactor　2.2.40.10;2.2.69.12;2.2.41.5
加氢精制　hydrofining　2.1.1.13
加氢裂化　hydrocracking　2.1.1.14
加氢裂化柴油　hydrocracking diesel oil　2.1.1.19
加氢裂化航煤　hydrocracking aviation fuel　2.1.1.18
加氢煤油　hydrogenated kerosene　2.1.1.75
加氢气化　hydro-gasification　2.2.2.15
加氢气化器　hydrogasifier　2.2.6.16
加氢汽油　hydrogenated gasoline　2.1.1.75
加权平均　weighted mean　1
加热　applying heat　20.1.2.10
加热管　heating tube　24.1.2.10
加热管束　heating tube bundle　7.1.8.3
加热管线　heating circuit　6.5.3.2
加热烘缸　heated drying cylinder　2.3.11.33
加热夹套　heating jacket　7.1.2.9
加热解吸器　heater resorber　14.2.3.8
加热介质　heating medium　15.1.1.4
加热介质进口　heating medium inlet　8.9.5.6
加热流体　heating fluid　4.3.3.23
加热炉　heating furnace　24.1.2
加热炉　furnace　2.2.61.1;3.3.24.20;3.3.25.3
加热面　heating wall　7.1.2.8
加热盘管　heating coil　25.6.4.4
加热器　heater　14.4.3.19;2.2.63.3;3.4.7.4
　heating element　7.2.5.16
　heating unit　2.2.76.5
加热器固定(双头)螺栓　heater retaining stud　18.2.4.17
加热器/冷却器　heater/cooler　13.2.3.7
加热区　heated area　3.5.11.7
加热筒　heating barrel　7.7.3.6
加热用蒸汽　steam for heating　6.3.7.6
加热元件　heating element　4.2.4.14
加热蒸馏塔　heated distilling column　5.5.1.9
加热蒸汽　heating steam　5.5.1.48
加热蒸汽　heating steam　7.2.15.11
加入粗柴油　gas oil charge　2.1.5.6
加入管　fill pipe　2.1.7.13
加速器输送机　accelerator conveyor　23.4.2.20
加炭　carbon addition　3.1.43.2

加压过滤膜　PFM　3.1.42.8
加压过滤器　pressure filter　3.1.2.16
加压空气　plenum air　3.3.8.15
加压流化床燃烧　pressurized-fluidized-bed combustion　25.5.2
加压流化床燃烧室　Pressurized fluidized-bed combustor　24.2.10
加压煤气　pressure gas　2.3.11.28
加压排液管　pressure drain tube　18.1.4.7
加压排液螺塞　pressure drain plug　18.1.4.12
加压燃烧室　pressurized combustor　25.5.2.3
加压水　H.W.; hydraulic water　12.4.1.2
加压吸附器　pressurized adsorber vessel　6.3.9
加压系统　pressurization system　3.1.15.3
加压下　pressurized　8.9.6.7
加压叶滤机　pressure leaf filter　9.2.1
加压叶片过滤机　pressure leaf filter　9.2.7
夹锤板　hammer plate　11.1.8.29
夹带剂　entrainer　15.1.1.3
夹带剂气体　entrainer gas　15.1.1.14
夹角　included angle　21.2.5
夹角之一半　half the included angle　21.2.5
夹紧环　clamp rings　12.2.1.7
夹紧螺栓　clamping bolt　13.1.2.34
夹紧装置　clamp　10.2.1.9
夹钳　tong　2.3.9.44
夹套　jacket　11.4.1.15
夹套附件　attachment of jacket　21.1.1.18
夹套冷却液　jacket coolant　4.3.3.11
夹套排液口　jacket drain　5.5.2.7
夹套容器　jacketed vessels　21.1.1.19
夹套式换热器　jacketed type exchanger　4.2.5.15
夹套蒸汽进口　steam inlets to jacket　8.8.10.15
甲板　deck　23.3.9.5
甲苯　methylbenzene　2.1.1.52
甲醇　methanol　2.2.1.14; 2.2.2.20; 2.2.28.2; 2.2.40.3; 2.2.43.16; 2.2.45.14; 2.2.50.10; 2.2.52.5; 2.2.54.12; 2.2.61.13; 2.2.62.2; 2.2.64.11; 2.2.65.2; 2.2.67.15; 2.2.68.1; 2.2.69.5; 2.2.71.12; 2.2.76.1; 3.1.37.8; 3.1.38.4; 3.1.39.2; 3.1.40.7
甲醇循环　methanol recycle　2.2.61.10
甲醇储罐　methanol storage tank　3.1.36.6; 3.1.38.1
甲醇萃取　methanol extraction　2.2.65.c
甲醇,二甲醚,碳氢化合物　methanol, dimethyl ether, hydrocarbons　2.2.3.19
甲醇法　rectisol process　2.3.1.12
甲醇法脱除酸性气体　rectisol acid-gas removal　2.2.104.3
ICI甲醇反应器　ICI methanol reactor　14.3.2.(a)
甲醇合成　methanol synthesis　2.2.2.17; 2.2.67.13
甲醇合成塔　methanol synthesis converter　2.2.46.1; 2.2.47.1
甲醇-甲酸甲酯分离　methanol-methyl formate separation　2.2.62.15
甲醇进料　methanol feed　3.1.36.4
甲醇进料泵　methanol feed pump　3.1.36.5; 3.1.38.2
甲醇汽化器　methanol vapouriser　2.2.51.5
甲醇塔　methanol column　2.2.64.3; 2.2.65.d
methanol tower　2.2.49.13; 2.2.50.8
甲醇羰基化　methanol carbonylation　2.2.40.4
甲醇脱氢制甲酸甲酯　Methyl formate production by dehydrogenation of methanol　2.2.61
甲醇循环　CH_3OH recycle　2.2.63,9
recirculated methanol　2.2.62.1
甲醇-氧化生成甲醛和水蒸气　formaldehyde + water vapor from methanol-oxidation　2.2.66.3
甲醇预加热器　methanol preheator　2.2.51.2
甲醇蒸气　methanol vapor　2.2.28.9
甲醇制低碳烯烃SDTO工艺原则流程　SDTO process principle for methanol to low-carbon olefins　2.2.70
甲醇制汽油的流化床工艺流程示意　Schematic flow diagram of the fluid-bed methanol-to-gasoline process　2.2.75
甲醇制烯烃　methanol to olefins　2.2.67.16
UOP/HYDRO甲醇制烯烃(MTO)工艺　UOP/HYDRO methanol to olefins(MTO) process　2.2.68
甲醇制烯烃(MTO)工艺流程示意图　Methanol to olefins(MTO) process　2.2.69
甲醇制烯烃（MTO）工艺装置简要流程　Schematic of methanol to olefins (MTO) process　2.2.72
甲基叔丁基醚　MTBE　2.2.65.5
甲硫醇　CH_3SH　2.2.63.6
甲硫醇产品　CH_3SH product　2.2.63.18
甲硫醇生产工艺　methyl mercaptane production process　2.2.63
甲酸甲酯　$HCOOCH_3$　2.2.61.11
甲酸甲酯　methyl formate　2.2.62.7
甲酸甲酯循环　recirculated methyl formate

2.2.62.11
甲烷　methane　2.2.68.19;3.3.18.16
甲烷　CH_4　2.2.71.23
甲烷合成天然气　methane SNG　2.2.104.13
甲烷化(反应)　methanation　2.2.103.7;2.2.36.10
甲烷化反应器　methanator　2.3.2.14
甲烷化作用　methanation　2.2.101.15
甲烷形成　methane formation　3.3.18.22
甲烷转化器　methanator　2.2.6.15;2.2.9.18;2.2.106.41
钾盐层　potash salt seam　25.2.4.9
钾盐床　potash salt bed　25.2.4.9
假想的相切圆　hypothetical tangent circle　11.5.1.4
架顶安全平台　crown safety platform　2.1.2.3
监挖井　monitoring well　3.3.12.1
检查　insp.;inspection　21.1.2.2
检查门　inspection door　10.5.4.12
检修孔　manhole　3.3.20.14
检修门　access door　10.2.2.3
剪钳　cutting plier　25.4.9.51
剪切销　shearing pin　11.1.7.25
剪切销套　shearing pin bush　11.1.7.24
减荷阀　unloaded valve　17.2.9.2
减黏裂化　visbreaking　2.1.1.33
减速齿轮　reduction gear　2.3.10.69
减速机　reducer　15.7.3.2
　reducing gear　9.5.4.11
减速机　speed reducer　14.4.1.16
减速机出油口　oil outlet for reducer　15.7.4.11
减速机底座　base for speed reducer　14.4.1.27
减速箱　reduction gear　15.7.4.13
减速装置　speed reducer　12.1.1.23
减小回流　reduce backstreaming　18.2.4.4
减压　pressure-let-down　2.2.90.9
减压残渣　vacuum residue　2.2.95.28
减压柴油　vacuum diesel oil　2.1.1.11
减压阀　pressure-relief valve　8.2.2.12
　relief valve　15.5.5.8
减压闪蒸　vacuum flash　2.2.92.18
减压塔　vacuum tower　2.1.4.2
减压渣油　vacuum residuum　2.1.1.12
减压蒸馏　vacuum still　1
　vacuum distillation　2.2.98.28
减振橡胶板　damping rubber plate　12.2.3.21
减振装置　vibration-stopper　24.1.1.22

简单升华　simple sublimation　15.1.1
简单洗涤　simple washing　9.1.1.12
简单蒸馏　simple distillation　1
简单制冷循环　simple refrigeration cycle　20.16.13
简化流程　simplified flow sheet　11.6.2
简捷法　shortcut method　1
碱　alkali　6.2.9.9;2.2.71.16
碱纤维素　alkali-cellulose　2.3.7.9
碱洗涤器　caustic scrubber　2.3.18.8
碱洗塔　caustic scrubber　2.2.69.9
　caustic washing tower　2.2.71.9
碱性纤维素碎片　alkali-cellulose crumbs　2.3.7.7
碱液　caustic solution　2.1.9.4
碱再生　caustic regeneration　6.2.9.(c)
间隔板条　division strip　9.3.3.28
间隔环　distance ring　13.1.2.29
间隔块　distance piece　11.1.8.21
间接传热设备　indirect heat-transfer equipment　4.3.4
间接过程加热器　indirect-process heater　25.5.1.(c)
间接加热回转　indirect rotary　8.1.1.⑨
间接加热连续操作圆盘干燥器　indirect-heated continuous plate dryer　8.9.4
间接加热式回转煅烧炉　gas-fired indirect-heat rotary calciner　8.8.8
间接燃烧　indirectly fired　3.3.15.(b)
间接液化　indirect liquefaction　2.2.20.11;2.2.1.15
间接蒸发冷却器　indirect evaporative cooler　6.2.8.10
间隙　clr.;clearance　8.8.5.4
间隙点　points of clearance　8.8.5.14
间隙调节器　clearance adjuster　11.4.1.20
　gap adjuster　15.7.6.16
间歇操作釜　batch kettle　5.3.3.2
间歇处理厂　batch treatment plant　6.3.2.13
间歇穿流循环　batch through-circulation　8.1.1.⑤
间歇法生产乙酸丁酯工艺流程　Process of intermittent production of butyl acetate　2.2.60
间歇混合沉降器　batch mixer-settler　6.3.3
间歇式带搅拌　agitated batch　8.1.1.②
间歇式顶喷嘴　intermittent overhead spray nozzle　10.3.4.2
间歇式滚轮混合器　batch muller　12.1.3.(g)
间歇式流化床造粒机　batch fluidezed bed granulator　13.4.1(a)

间歇式振动磨 batch vibrating ball mill 11.3.3
间歇投料 intermittent feed 3.2.5.a
间歇蒸馏 batch distillation 1
间歇蒸馏系统 batch-still system 3.1.6
间歇式自动离心机 automatic batch centrifugal 9.5.8
建筑物地面 building floor 4.3.4.17
溅料轮 splashing device 8.7.1.9
溅雾区 spray zone 10.5.1.20
键 key 13.1.2.12
将被氧化的工业废气(入口) waste gas to be oxidized(inlet) 3.4.5.2
浆 slurry 2.2.95.17
浆槽 vat 2.3.10.25
浆尖速度 tip speed 1
浆料 slurry 19.1.5
浆料反应器 slurry reactor 1
浆料混合池 mixing chest for stuff 2.3.11.1
浆料循环 slurry recycle 14.2.6.18
浆料液面 slurry level 9.3.9.9
浆面 slurry level 2.2.98.10
浆态床反应器 slurry bed reactor 2.2.50.4
浆态进料 slurry feed 14.2.4
浆液进料管 slurry feed pipe 15.5.3.1
浆液排出泵 slurry discharge pump 7.2.10.13
浆液循环 slurry recycle 14.3.4.28
浆状 slurry 8.1.1
浆状产品排出 product slurry discharge 7.2.8.17
浆状产品排出 product slurry discharge 7.2.9.6
浆状物 slurry 8.1.1.10
浆状物料面 pulp level 12.6.4.7
桨叶 impeller 6.1.5.7;6.1.7.4
 propeller 7.2.1.11;7.2.5.18
桨叶驱动器 propeller drive 7.2.1.16;7.2.5.22
降解产物 degradation products 3.2.11.8
降膜 falling film 15.4.1.9
降膜式分子蒸馏设备 operation of the falling film still 5.5.1.(a)
降膜式换热器 falling-film exchanger 4.2.5.12
降膜蒸发器 falling-film evaporator 7.1.5
降水管 water downcomer pipes 4.4.2.5
降温罐 rundown tank 6.3.2.32
降液板 downcomer apron 5.2.2.8
降液管 down spout 5.2.8.3
 downcomer 3.1.32.26;5.2.1.18

交变极性磁力分离器 alternating-polarity Ball-Norton separator 12.6.7
交叉编织填料 nterbraid packing 19.2.3.4
交叉进料皮带 cross feed belt 11.6.4.4
交叉型 inter-meshing type 16.4.8
交流电 AC electricity 2.2.113.2
交流电输入 A.C.input 10.3.1.9
交流发电机 alternator 2.2.81.14
胶板压榨 offset press 2.3.12.21
胶管泵 squeegee pump 16.1.2.64
胶囊化 encapsulation 1
胶体磨 colloid mill 11.4.2
焦化柴油 coking diesel oil 2.1.1.42
焦化厂 coking plant 2.3.10
焦化蜡油 coking waxy oil 2.1.1.43
焦化汽油 coking gasoling 2.1.1.41
焦化石脑油 coker naphtha 2.1.6.7
焦化塔馏出物 coker distillate 2.2.24.9
焦炉气 coke oven gas 2.2.1.5
焦煤卸料场 dumping of coking coal 2.3.14.1
焦气主管 gas-collecting main 2.3.10.17
焦炭 coke 2.1.1.44;2.2.1.3
焦炭产品 product coke 2.1.9.29
焦炭导槽 coke guide 2.3.14.9
焦炭罐 coke drum 2.1.6.2
焦炭去气化 coke to gasification 2.2.79.16
焦炭装载 coke loading 2.3.14.15
焦油的提取 coal tar extraction 2.3.14.10
焦油浆 tar/slurry 2.2.15.2
焦油去气化器 tar to gasifier 2.2.106.24
角封 angle seal 8.8.1.27
角钢圈 angle ring 5.2.7.3
角钢塔盘 angle bar tray 5.2.10.18
角加衬的耐磨轴台 bushed-angle pillow block 8.8.5.23
角接阀 angle valve 22.1.1.(c)
角接式升降止逆阀 angle lift check valve 22.1.1.(g)
角形避雷器 arcing horn 25.4.9.11
绞车 draw works 2.1.2.11
绞接膨胀节 Hinged expansion joint 22.5.3
绞链式筛网底 hinged screen bottom 6.1.17.3
铰接 hinge 22.6.1.1
铰接连接 hinged connection 22.5.4.1
铰接盘 hinged disc 22.1.10.5
铰接销 hinge pin 22.5.3.2
铰链 hinge 11.1.8.12
铰链盖 hinged lid 25.4.1.24
 snap lid 25.4.1.24

铰链杆　hinged lever　9.2.1.22
铰链刮料器　hinged plow　23.2.3.(f)
铰链盘式单向阀　hinged disc type check valve　22.1.10
铰链式地面插座　hinged floor socket　25.4.1.23
铰链销　hinge pin　22.1.10.6
铰链轴　shaft for hinged lever　9.2.1.23
搅拌　ploughs　25.1.11.10
　　stirring　1
搅拌薄膜蒸发装置　agitated thin-film evaporator　7.1.1
搅拌釜式　agitated kettle　4.3.1.(b)
搅拌釜式反应器　stirred tank reactor　14.3.1.(d)
搅拌过的污泥　blended sludge　3.3.25.18
搅拌机　mixer　3.3.28.4
　　blender　3.3.25.6
搅拌桨　propeller　14.4.1.18
搅拌结晶器　stirred type crystallizer　1
搅拌膜式蒸发器　agitated film evaporator　7.1.2
搅拌耙　agitator blade　24.1.13.11
搅拌耙臂　rabble arm　3.3.9.6
搅拌耙臂驱动装置　rabble-arm drive　3.3.9.12
搅拌耙齿　rabble teeth　6.3.5.5
搅拌器　agitator　14.4.3.22
　　blender　23.1.2.24
　　impeller　14.4.2
　　mixer　3.3.29.1
　　stirrer　11.3.4.5；3.4.2.5；3.4.7.5
搅拌器搂耙　agitator rakes　9.3.3.35
搅拌器驱动装置　mixer drive　3.1.45.6
搅拌器曲柄　agitator crank　9.3.3.18
搅拌容器　stirred vessel　13.2.4.20
搅拌式　agitated　8.1.1①
搅拌式反应器[釜]　agitated reactor　14.4
搅拌/絮凝　mixing/flocculation　2.4.5.4
搅拌叶片　mixing blade　12.1.1.14
　　propeller　6.1.5.20
搅拌装置　agitator　8.5.7.2
校核称量秤　check-weigher　23.4.3.7
校核称量输送器　check-weigh conveyor　23.4.2.7
校正单元　calibration unit　3.5.19.2
较细颗粒　finer particles　10.1.2.20

酵母　yeast　1
酵母纯培养设备　pure culture plant for yeast　2.3.13.6
阶式磨机　cascade mill　11.6.3.4
阶梯式气缸　step-type cylinder　17.2.4.7
阶梯形迷宫　stepped labyrinth　19.3.3.12
阶梯状平板　step plate　13.3.5.31
接触板　contacting element　6.1.18.1
接触过滤　contact filtration　6.3.2
接触接通器　contact maker　25.4.1.38,39
接触摩擦　rubbing contact　8.8.5.10
接触器　contactor　3.4.6.2
Higgins 接触器　Higgins contactor　6.2.4
接触弹簧　contact spring　25.4.8.28
接地　grounded　10.3.1.28
接地插头　earthed plug　25.4.8.9
接地金属电极　grounded metal electrode　15.3.1.5
接地收集电极　grounded receiving electrodes　10.3.1.27
接地双插座　earthed double socket　25.4.8.6
接地转子　earthed rotor　12.7.2.2
　　grounded rotor　12.7.2.2
接管　connecting pipe[tube]　21.2.4.4
接管接头　union nipple　22.2.3.24
接口　connection　9.3.2.29
接料盘　take-up(tray)　8.3.4.3
接收器　receiver　25.6.3.7
接收装置　receiver　23.1.8
接受器　receptacle　10.1.2.12
接受区　receiving area　3.3.16.2
接筒　extend neck　15.7.1.8
接头轴承　connector bearing　16.4.23.9
接头轴承　connector bearing　16.4.24.7
接头轴瓦　connector bearing　16.4.26.3
接吸入口　to suction　16.4.6.21
接线盒　terminal block　18.2.1.13
接线盒　terminal box　16.4.20.16
节流衬套　throttle bushing　19.1.6
节流挡板　cut damper　8.9.3.5
节流阀　throttle valve　25.6.2.5
　　throttling valve　20.1.2.15
节流过程　throttling process　1
节流降压室　relief chamber　16.4.6.14
节流降压装置　restricting orifice　16.4.6.22
节流圈　throttle ring　16.4.22.4

679

节能器 economizer 20.1.3.12
节涌 slugging 1
洁净合成气 clean syngas 2.2.36.7
洁净固体燃料 clean solid fuel 2.2.2.37
洁净焦炭 clean-coke 2.2.93
洁净焦炭实验厂 clean-coke demonstration plant 2.2.93.2
洁净气体燃料 clean gaseous fuels 2.2.2.13
洁净燃料制造 the production of clean fuels 2.2.2
洁净液体流出 clean liquid out 15.3.1.3
洁净液体燃料 clean liquid fuels 2.2.2.27
结构基础 structural foundation 8.8.11.9
结构示意图 structural representation 18.2.2
结构详图 construction details 21.1.1
结垢 fouling 1
  scale 1
结合的燃烧室 incorporated combustion chambers 25.3.1.17
结晶 crystallization 2.2.64.7
结晶槽 crystallization tank 6.3.10.4
结晶固体 crystalline solid 8.1.1.19
结晶浆液 crystal slurry 6.3.10.6
结晶器 crystallizer 5.3.3.7
结晶设备 crystallization equipment 7.2
结片机 flaker 13.5.1
截留 retention 1
截留物 retentate 3.1.29.6
截油挡板 oil retention baffle 3.1.3.14
截止阀 globe valve 22.1.3
  shutoff valve 2.1.7.5
  stop valve 17.6.1.12
截止式升降止逆阀 globe lift check valve 22.1.1.(h)
解聚装置 depolymerizer 2.2.102.13
解吸 stripping 1
解吸段 desorption section 6.3.7.4
解吸管 desorption tube 6.3.7.11
解吸塔 desorption column 5.1.1.6
  stripper 2.2.27.11
解吸因子 stripping factor 1
解吸用蒸汽 steam for desorption 6.3.7.7
  stripping steam 6.3.11.4
介电电泳 dielectrophoresis 15.3
介稳区 metastable region 1
介质 medium 12.5.2.18
  media 3.1.38.5
介质泵 medium pump 12.5.1.18
介质沉降槽 medium thickener 12.5.1.14
介质储槽 medium scmp 12.5.1.17
介质的液面 media level 12.5.2.3

介质返回 media return 10.4.1.20
介质/粉尘减少夹带室 media/dust deentrainment chamber 10.4.1.14
介质过滤床层 media filter bed 10.4.1.9
介质流动空气注入器 media flow air injector 10.4.1.6
介质流态化膨胀床 fluidized/expanded media bed 3.2.9.2
介质排出口 medium discharge 14.4.1.6
介质上部回流管 upper return nozzle for medium 14.4.1.13
介质上部入口 upper inlet for medium 14.4.1.14
介质提升气风机 media left air blower 10.4.1.4
介质洗涤 medium washing 9.3.7
介质下部回流管口 lower return nozzle for medium 14.4.1.11
介质下部入口管 lower inlet for medium 14.4.1.9
界面 interface 6.1.12.6
金刚石精矿 diamond concentrate 12.7.5.5
金刚石料仓 diamond bin 12.7.5.1
金属 metals 22.6.2.1;2.2.80.6
金属鲍尔环 metal Pall ring 5.3.2.(d)
金属箔螺旋卷制填料 metal foil spiral wrapped packing 19.2.2.10
金属捕集器 metal collector 3.3.2.2
金属底座 metallic foundation 25.3.1.1
金属碟 disk 9.5.15.8
金属定位块 metal retainer 19.3.1.10
金属阀 metal valve 16.2.4.20
金属法兰 metal flange 22.4.4.6
金属辐射锥 metal radiating cone 24.1.3.6
金属管路 metal systems 22.4.4.4
金属护罩 metal shroud 4.4.1.5
金属化纸 metallic paper 25.4.8.2
金属环矩鞍 metal Intalox saddle 5.3.2.(i)
金属壳体 metal 11.4.1.18
金属球形填料 metal tripak 5.3.2.(l)
金属筛板 metal screen 6.3.9.13
金属丝网 wire gauze 2.3.11.14
金属丝网通道 wire mesh walkway 10.2.1.6
金属丝支录圈 wire retainer 10.2.3.2
金属套筒 metal shroud 4.4.1.5
金属填料 metallic packing 19.2.3.9
金属填料高压填料函 high-pressure stuffing box using metallic packing 19.2.1
金属填料组件 complete metallic packing 17.2.15.33
金属桶 drums 3.3.5.9

680

金属线材绕组　wire winding　9.3.3.4
金属小室　metal chamber　19.3.1.3
金属芯填料　metal core packing　19.2.3.12
金属烟囱　metal stack　24.1.3.10
金属氧化物　MeO_x—metal oxide　2.2.80.3
金属氧化反应炉　MOR—metal oxidation reactor　2.2.80.2
金属与金属接触　metal-to-metal contact　18.2.4.1
筋板　web plate　5.2.2.5
　wed　21.3.3.2
紧定螺钉　set screw　19.1.5.9
紧固螺栓　clamping bolt　9.2.5.25
　closing bolt　9.2.5.10
紧固手轮　closing handle　9.2.5.9
紧密床层中央加料式塔结晶器　dense-bed center-fed column crystallizer　7.2.14
进厂污水　plant influent　3.1.37.1
进风口　air inlet port　8.5.4.8
　air intake　4.6.2.8
　air intake port　17.5.1.10
进空气　inlet air　3.3.24.16
进口　in　16.5.6.1
　inlet　10.1.1.1
　input　3.1.24.1
　intake　16.2.5.2
进口侧轴套　suction side sleeve　16.4.19.23
进口挡板　inlet baffle　3.2.15.11
进口挡料圈　inlet dam ring　13.3.1.2
进口导流叶片　inlet guide vane　17.4.2.13
进口段　inlet portion　16.4.18.1
进口管　inlet pipe　14.1.11.4
进口加压密封件　inlet pressure seal　6.3.8.4
进口壳体　suction casing　16.4.19.26
进口密封　inlet seal　8.8.1.24
进口密封件　inlet sealing　12.2.3.13
进口止逆阀　inlet check valve　16.2.7.1
进口锥体衬板　intake cone liner　12.3.4.5
进料　infeed　3.3.29.3;3.3.29.16
　feed inlet　7.2.8.2
　feed material　11.5.3.1
　feed-in　2.3.11.13
　fresh feed　2.1.6.4
　influent　6.2.5.3
　intake　9.4.4.20
　material feed　12.2.1.10
　material in　23.1.1.8
进料板　feed plate　11.1.3.1
进料泵　feed pump　13.4.1.10;25.5.3.2
　influent pump　3.1.39.1

进料仓　feed bins　11.6.4.1
进料槽　feed launder　12.4.2.9
　feed puddle　8.7.2.2
进料斗　feed funnel　9.5.6.9
进料阀　feed valve　9.5.6.21
进料分布器　feed distributor　9.5.10.2
进料管　feed pipe　12.3.1.1
　inlet pipe　15.6.3.10
进料管　spout feeder　11.2.4.1
进料滚压机　feed roller press　8.3.2.2
进料集管　feed manifold　11.5.1.8
进料计量泵　metering inlet pump　5.5.1.2
进料空气压缩机　feed air compressor　2.2.53.1
进料口　feed opening　11.1.7.1
　inlet　9.4.7.2
　material inlet　23.1.6.3
进料链　feed chain　2.3.10.69
进料螺旋　feed screw　11.2.9.6
进料气体　feed gas　14.2.3.1
进料输送机　infeed conveyer　3.3.30.7
进料器组件　feeder assembly　8.8.8.9
进料溶液　feed solution　7.1.14.1
进料入口　feed inlet　7.2.1.2
进料室　charging chamber　25.2.5.7
进料台　table feeder　11.1.6.1
进料调节管　adjustable pipe for feed flow　12.3.2.9
进料筒　charging funnel　13.1.2.20
进料头　inlet head　8.8.1.1
进料推料机　feed ram　25.1.5.1
进料斜槽　feed chute　8.8.1.2
进料压缩机　feed compressor　2.2.37.2
进料液槽　feed tank　18.5.1.7
进料液与残液热电偶　feed & residue thermocouples　18.5.1.5
进料与排放　feed and drain　9.3.7.7
进料预处理　feed preparation　2.3.5.3
进气　gas in　3.4.4.1;17.2.14.4
　inlet　17.3.3.1
　intake　10.4.3.14
　influent　3.4.3.4
进气端温度计　intake end thermometer　17.4.2.12
进气阀压罩　valve catcher case　17.2.7.2
进气分离器　inlet separator　6.3.3.3
进气管　inlet pipe　18.4.1.8
进气孔　inlet port　17.3.3.2
　inlet hole　18.4.1.4
进气口　air inlet　23.1.6.2

681

inlet 25.3.2.7
intake 17.2.4.1
suction inlet 18.2.2.1
suction port 17.2.15.35
进气喷射器 gas injectors 2.2.87.4
进气通道 inlet passage 17.2.7.11
进汽室 steam chest 16.2.3.55
进水 feedwater 25.5.3.1
　influent 3.2.1.1; 3.1.32.8; 3.1.33.2; 3.1.40.1; 3.1.42.1
进水管道 influent lines 3.1.32.25
进水加热器省煤器 feedwater heater economizer 25.5.3.3
进水口 feed water inlet 5.4.2.14
进水口工程 headworks 3.1.41.2
进水注入 influent injection 3.1.32.2
进液 liquid inlet 17.2.14.1
进油孔 oil intake port 19.3.2.13
近端 near end 4.3.4.27
浸解池 macerator 3.3.21.2
浸没布料式单滚筒干燥器 single drum dryer with dip-feed 8.7.1.(a)
浸没管束强制循环蒸发器 submerged-tube F.C.(forced-circulation)evaporator 7.1.8
浸没式燃烧炉 submerged combustor 3.3.10
浸泡 infusion 1
浸取 leaching 1
浸润 imbibition 1
浸湿风扇叶轮 wetted fan wheel 10.5.15.4
浸液 steep liquor 2.3.12.6
浸渍 infusion 1
浸渍层 steeping floor 2.3.12.28
经典流态化 classical fluidization 1
经验模型 empirical model 1
经乙酸乙酯加氢制备乙醇的工艺流程 process of ethanol production from hydrogenation via ethyl acetate 2.2.59
晶浆 crystal paste 7.2.12.8
　slurry 7.2.5.25
晶体生长区 crystal growth zone 7.2.10.5
晶种 seed crystal 1
晶种粒子 seed particles 13.3.4
精甲醇 refined methanol 2.2.49.11
精炼池 refining bath 2.3.9.4
精馏段 fractionation 14.2.7.18
精馏塔 distillation column 2.2.61.12; 2.2.62.8; 5.1.1.1
　fractionating column 5.1.1.2
　fractionating tower 5.3.3.11
　fractionator 14.3.4.30

rectification column 2.2.54.10
精磨机 refiner 2.3.10.84
精选矿 concentrate 12.8.1.10
精制 fine purification 2.1.1.83
　polishing 3.1.32.5
精制塔 refining tower 2.2.58.13
精制塔回流罐 reflux tank of refining tower 2.2.58.8
精制纤维素 refined cellulose 2.3.10.78
井管 tubing 2.1.2.20
井架底座 substructure 2.1.2.2
井口建筑物 minehead buildings 25.2.4.14
阱 trap 3.1.3.5.10
径流泵 radial pump 16.1.2.2.13
径流(式) radial flow 16.1.1.31
径向臂 radial arms 11.3.4.6
径向出口 outlet radial 15.6.3.6
径向反应器 radial flow reactor 1
径向活塞泵 radial piston pump 16.5.6
径向进口风门 radial inlet damper 4.6.24
径向溜槽 radial launders 15.6.3.11
径向球轴承 radial ball bearing 11.1.5.19
径向通气 radial vent 19.3.1.6
径向叶片 radial vane 16.4.5.3.2
径向轴承 radial bearing 13.3.5.7,8
净发电量 net power generation 2.2.107.13
净化 purif;purify 2.2.2.7
　purification 2.2.2.7; 2.2.78.5
EDC净化 EDC purification 2.3.18.D
VCM净化 VCM purification 2.3.18.F
净化的残渣 cleaning sludge 3.3.5.3
净化的流出液 clean effluent 15.5.4.2
净化的溶剂精制煤 clean SRC 2.2.94.20
净化合成气 purifier syngas 2.2.28.16;2.2.32.5
净化空气 cleaned air 10.1.3.12
净化空气侧 clean air side 10.2.1.33
净化空气出口 clean air out(let) 10.5.1.18
净化煤气 clean gas 2.2.19.6
　cleaned gas 10.5.4.19
净化气侧 clean gas side 10.2.1.26
净化气出口 clean gas outlet 10.3.4.17;10.5.1.2
净化气体 clean gas 6.3.7.15
净化气体出口 clean gas discharge 8.5.4.11
　clean gas out(let) 10.2.2.8
净化器 cleaner 12.8.2.1
Hi-eF净化器 Hi-eF purifier 10.4.3.(a)
净化器浮选 cleaner flotation 12.8.2.21
净化器排出物浮选 cleaner-scavenger flotation 12.8.2.17

净化水 clean water 3.1.14.6
treated water 15.5.4.11
净化水流出通道 clean-water outlet channel 3.1.4.4
净化水至换热器 purified water to heat exchanger 2.2.72.15
净化循环介质 cleaned/recycled media 3.2.9.4
净化指数 decontamination factor(DF) 1
净正吸压头 net positive suction head (NPSH) 1
静电场 electrostatic field 10.3.1.23
静电除尘器 electrostatic precipitator 10.3.1
静电分离器 electrostatic separator 12.7.1
静电极 static electrode 12.7.2.4
静电集尘器 electrostatic precipitator 2.3.14.20
静电加料器(＋) electrostatic feeder 12.7.2.10
静电筛 electrostatic sieve 12.7.3
静环 stationary seal ring 19.1.3.5
静密封圈 static seals 19.1.1.1
静态采样系统 stationary sample systems 3.5.10
静态混合器 static mixer 1
静压头 static head 1
静止的 stationary 19.3.3.4
静止环 stator ring 6.1.14.6
静止滤板 stationary filter plate 9.4.9.3
静止区 quiet zone 6.1.9.13
静轴 dead shaft 8.8.5.21
酒桶仓库 barrel store 2.3.13.16
就地干馏 in-situ retorting 2.2.94
局部剖面图 partial cross section 8.8.10.16
局部详图 detail drawing 12.3.5
局剖剖视图 cutaway view 10.2.1.(b)
矩形板 rectangular plate 5.2.2.2
矩形沉降槽 rectangular sedimentation tank 3.2.16
矩形澄清器 rectangular clarifier 15.6.2
聚丙烯 polypropylene 22.6.2 ④;2.2.67.25
聚丙烯产品 PP 2.2.67.26
聚并 coalescence 1
聚并分离器 coalescing separator 15.5.4
聚并过滤 coalescing processes 15.5
聚并滤清器 coalescer filter 15.5.2
聚并器元件 coalescer element 15.5.2.7
聚光透镜 collection leas 3.5.6.6
聚合物粉末 polymer powder 3.3.19.2
聚合物和污泥搅拌罐 polymer and sludge mixing tank 3.3.19.5
聚合物进料泵 polymer feed pump 3.3.19.4
聚合物溶解槽 polymer dissolving tank 3.3.19.7
聚集室 collection chamber 3.1.10.1
聚焦透镜 condenser len 3.5.6.2
聚结膜 coalescing membrane 6.1.8.3
聚结器 coalescer 6.1.5.14
聚式流化床 aggregative fluidized bed 14.1.3.13
聚水系统 underdrain 3.1.38.7
聚四氟乙烯采样管 Teflon sample line 3.5.16.22
聚四氟乙烯换热器 teflon heat exchanger 4.2.5.13
聚四氟乙烯密封环 polytetrafluoroethylene seal ring 19.4.1.20
聚酰胺 polyamide 2.3.8.35
聚酰胺(短)纤维 polyamide staple 2.3.8.61
聚酰胺丝线 polyamide thread 2.3.8.46
聚酰胺纤维 polyamide fibre 2.3.8
聚酰胺纤维包 bale of polyamide staple 2.3.8.62
聚酰胺锥形筒 polyamide cone 2.3.8.51
聚乙烯 PE;polyethylene 2.2.67.24;22.6.2.③
聚乙烯产品 PE 2.2.67.26
聚乙烯超高压压缩机 very high pressure compressor for polyethylene process 17.2.15
涓流床 trickle bed 1
卷取纸轴 delivery reel 2.3.11.28
卷筒纸 web 2.3.11.45
卷线轴 spool 2.3.9.55
卷压织物填料 rolled cloth packing 19.2.3.7
卷扬机 winding machine 23.3.6.1
wind-up unit 15.7.5.11
卷轴(螺旋)卸料 scroll discharge 9.5.2.6
绝对速度 absolute velocity 16.4.3.10
绝热顶盖 insulated cover 24.1.7.11
绝热过程 adiabatic process 1
绝热砌块 insulating block 24.1.7.8
绝热外壳 insulated housing 8.9.1.5
绝热罩 insulation cover 22.5.1.(h)
绝压 press. in.abs 20.7.4
绝缘柄 insulated handle 25.4.9.54
绝缘材料 dielectrics 12.7.2.6
绝缘(胶)带 insulating tape 25.4.9.34
绝缘膜浮动盖 floating insulated membrane cover 3.2.6.13

均方根误差 root-mean-square error 1
均四甲苯 durene 2.1.1.62
均压 equalization 3.1.37.6
均压箱 surge tank 3.5.14.15
均一流化床 homogeneously fluidized bed 14.1.3.10
均匀水膜 uniform waterfilm 4.2.3.3
均匀转化模型 uniform conversion model 1
菌膜 biofilm 3.2.11.18

K

卡箍 band 12.3.5.2
　　collar 10.2.3.3
卡诺循环 Carnot cycle 1
开并系统监控盘 monitoring controls for the switching systems 25.4.2.6
开放末端 open ends 15.2.1.9
开工 start 14.3.4.40
开工燃料烧嘴 startup burner 14.2.4.7
开工烧嘴 startup burner 2.2.24.3
开工烟囱 start-up stack 8.5.6.33
开工预热风箱的燃烧器 startup preheat burner for hot windbox 3.3.27.9
开工预热器 startup heater 14.2.7.22
开工/再生炉 startup/regen.furnace 2.3.4.8
开孔 opening 21.1.1.36
开口端 open end 4.3.2.7
开口搁架式液体分布器 open shelf liquid distributor 10.5.13.1
开口环 snap-ring 22.2.1.15
开口销 cotter 17.2.11.7
　　split pin 22.1.10.3
开路磨碎系统 open circuit grinding system 11.6.1
开始 start 9.3.8.3
开式 open 16.4.5.7
开式供水口 operating water supply 9.5.12.14
开式结构 open design 16.4.5.(d)
开式叶轮 open impeller 16.1.1.41
铠装泵 armoured pump 16.1.2.2.34
抗辐射屏障 anti-radiation screen 25.2.5.2
抗腐蚀材料 corrosion inhibitors 3.1.24.2
抗生物生长材料 biological-growth inhibitors 3.1.24.2
苛化搅拌器 causticizing agitators 2.3.10.49
苛化器 causticizer 3.4.6.6
苛性钠 caustic soda 2.3.7.3
苛性钠溶液 caustic soda solution 2.3.8.40

康菲气化炉 Conoco Phillips gasifier 2.2.21
康菲 E-Gas 气化炉 Conoco Phillips E-Gas gasifier 2.2.22
(颗粒)出口通道 chute 13.3.5.23
颗粒床 granule bed 13.3.1.5
颗粒阀 solids valve 14.1.10.3
颗粒轨迹 particle trajectory 23.3.1.4
颗粒活性炭吸附塔 GAC column 3.1.30.B.10
颗粒喷泉 particle fountain 13.4.1.22
颗粒去除 particulate removal 2.2.109.6
颗粒污泥 granular sludge 3.2.6.27
颗粒物采样仪器 Particulate-sample apparatus 3.5.11
颗粒载体 carrier particles 3.2.11.15
壳壁厚度 shell thickness 21.1.1.41
壳程 shell(side)pass 4.1.1.42
壳程接管口 shell nozzle 4.2.2.8
壳体 body 22.2.3.20
　　shell 10.3.4.11
壳体侧衬板 side liners for casing 11.1.7.8
壳体侧衬板螺栓 bolt for casing side liners 11.1.7.7
壳体接管 shell nozzle 4.1.1.12
壳体双头螺栓 casing stud 16.4.20.8
壳体终端 end of shell 8.8.5.2
可泵送的 pumpable 3.3.7.15
可泵送废料 pumpable waste 3.3.7.5
可泵送液体 pumpable liquid 3.1.22
可变角度转鼓 variable angle tumbler 9.5.2.17
可变水深 water depth variable 15.6.1.26
可变液位[面] variable level 14.3.4.11
可变蒸气空间贮罐 variable-vapor-space 21.2.1.5
可拆挡板 removable baffle 10.2.1.4
可拆的 removable 15.5.1.4
可拆法兰联接 demountable flange connexion 5.5.2.14
可拆盖板 removable cover 4.1.2.(d)
可拆管丝堵 removable tube plugs 4.5.1.9
可拆降液管 removable downcomer 5.2.2.10
可拆模板 movable die plate 15.7.3.11
可拆喷嘴组合件 removable jet assembly 18.2.4.5
可动隔板 movable partition 8.3.1.8
可锻铸铁 malleable cast-iron 22.6.2.②
可改变尺寸 variable sizes 11.5.2.3
可改变喉部开度 varying throat opening

可更换衬里 replaceable liners 11.2.8.6
可更换密封圈 renewable seal ring 22.1.2.3
可靠的深埋坑 secure burial pit 3.3.14
可靠性 reliability 1
可逆反应 reversible reaction 1
可逆输送器 reversible conveyor 23.4.2.13
可溶性废弃物 soluble waste 3.1.18.4
可溶性无机物质 SIM；soluble inorganic matter 3.2.1.4
可溶性有机物质 SOM；soluble organic matter 3.2.1.2
可视给油阀 sight feed valve 16.4.25.29
可调出口堰 adjustable outlet weir 3.1.4.6
可调挡板 adjustable damper 8.8.6.10
可调分流器 adjustable splitters 12.6.4.20
可调加料口 adjustable feed gate 24.1.5.2
可调减压阀 adjustable relief valve 17.2.14.6
可调角度风机 adjustable-pitch fan 8.2.1.2
可调进口堰 adjustable inlet weir 3.1.4.10
可调滤液排出口 adjustable filtrate ports 9.5.5.5
可调旁通阀 adjustable bypass valve 16.2.8
可调配风器 adjustable registers 24.2.4.12
可调式刮刀 adjustable unloader knife 9.5.5.13
可调式空气送风管 adjustable air-blast nozzles 8.2.1.5
可调式弹簧 adjustable spring 19.1.5.2
可调堰 adjustable dam 9.5.15.1
adjustable weir 15.6.1.21
可调堰板 adjustable weir 5.2.2.9
可调支承扁钢 adjustable support bar 11.1.7.18
可移动端盖 moveable end cover 4.2.1.3
可用能 exergy 1
可再生资源 renewable resources 1
克劳斯(二段脱硫)工艺 Clause process 2.2.117.15
克劳斯制硫装置 Claus sulfur plant 2.3.1.32
克劳斯转化器 Claus converter 2.1.9.46
克隆 clone 1
克努森数 Knudsen number 1
刻度间隔 span 3.5.11.22
刻石装置 stone-dressing device 2.3.10.70
坑道地段 mined portion 2.2.85.3
空分 air separation 2.2.78.4
空分装置 air separation plant 2.3.1.27
air separation unit 2.2.33.6；2.2.36.1；2.2.115.3；2.2.117.5；2.2.118.2
ASU 2.2.119.2
空回程 empty return run 23.3.7.7
空间加热 space heat 25.1.1.11
空间速率 space velocity(SV) 1
空间选择 space available 23.1.9.(d)
空冷换热器 air-cooled heat exchanger 4.5.1
空冷冷凝器 air-cooled condenser 3.5.11.29
空冷器 air cooler 2.2.44.10
空冷器循环系统 air-cooler recirculator 4.5.2
空气 air 2.2.33.5；2.2.45.6；2.2.64.14；2.2.67.1；2.2.68.2；2.2.69.3；2.2.71.13；2.2.72.1；2.2.80.1；2.2.110.12；2.2.111.3；2.2.113.5；2.2.114.6；2.2.115.1；2.2.116.1；2.2.117.6；2.2.118.1；2.2.119.1；2.1.7.5.11；2.3.17.3；3.1.28.4；3.1.32.9；3.1.47.6；3.3.13.12；3.3.25.1；3.3.29.14；3.3.29.5；3.3.30.9；3.4.6.3；3.4.6.12；10.3.5.4
atmospheric air 25.5.3.20
空气泵 air pump 16.1.2.2.50
空气闭锁器 airlock 8.4.1.8
空气冲击 air impingement 8.7.2
空气出口 air out(let) 10.2.1.16
空气出口管 air-exhaust duct 8.2.1.9
空气吹弧断路器 air-blast circuit breaker 25.4.9
空气导流装置 air deflectors 24.2.4.14
空气动力捕集器 aerodyne collector 10.4.3.(d)
空气动力管 aerodyne tube 10.4.3.(c)
空气(二级)喷射泵 air(secondary)ejectors 18.3.6.3
空气反吹 air blow 9.4.1.2
空气分布 air distribution 4.3.4.25
空气分布器 air distributor 24.2.10.1
空气分布器炉栅 air distribution grid 14.2.4
空气分布器栅板 air-distributor grating 8.3.2.8
空气分级器 air classifier 12.3.4
空气分类 air classification 3.3.16.13
空气分离 air separation 2.2.67.2
空气分配装置 air distribution 3.3.29.18
空气分散环 air dispersion ring 11.2.10.7
空气管 air pipe 16.5.1.3
空气管道 air lines 3.1.32.24
空气管路装置 air line installation 17.6.2
空气过滤器 air filter 8.4.6.3；2.2.53.5

685

空气环状入口　annular air inlets　11.2.9.4
空气混合室　air-mix chamber　8.4.1.31
空气加热器　air heater　25.3.2.16
空气接管　air connection　9.3.1.4
空气进口　air in(let)　10.2.1.2
　air intake　17.2.6.1
空气进口管　air-inlet duct　8.2.1.8
空气进口过滤器　air intake filter　17.6.1.1
空气进口壳体　inlet air casing　25.3.1.2
空气进入物料　air-into-material　23.1.8.①
空气孔　air vent　8.8.10.7
空气控制　air control　24.2.6
空气冷却器　air cooler　22.9.21
　atmospheric cooler　4.2.5.8
空气流　air flow　8.6.3.5;2.3.16.2
空气脉冲(发生)器　air pulser　6.1.16.6
空气脉冲填料塔　packed column with air pulser　6.1.16.(b)
空气排放　air out　6.2.9.8
空气泡　air bubble　22.2.3.17
空气喷枪　air lance　3.3.26.9
空气喷射冲击　air jets impinging　8.3.2
空气喷射器　air ejector　7.2.5.24
空气喷嘴　air jet　12.7.5.8
　air nozzle　23.1.8.20
空气歧管　air manifold　3.3.30.5
空气去除系统　air removal system　3.3.29.9
空气去循环　air to recycle　2.2.106.22
空气入口　air inlet　11.2.11.3
空气入口　air port　24.1.4.1
空气室　air compartment　23.1.2.29
　air housing　3.3.26.14
空气释放　air relief　6.2.1.4
空气输入系统　air input system　3.3.29.15
空气提升泵　air-lift pump　16.1.2.2.51
空气提升机　air lift　23.1.7.12
空气提升气流输送干燥器　air-lift pneumatic-conveyor dryer　8.4.3
空气提升涡流干燥作用区　airlift turbo drying action　8.4.1.25
空气提升压力损失　pressure drop across air lift line　14.3.3.(c)
空气调节器　air register　25.2.9.3
空气调节器　shutter　24.2.6.7
空气通道　air passage　17.2.5.11
空气消声器　air silencer　17.6.1.2
空气星形阀　airlock　10.2.3.13
空气循环　air circulation　12.3.1.5
空气循环风机　air circulating fan　8.3.3.1
空气压缩机　air compressor　2.3.3.5

空气,氧气,蒸汽　air,O_2,steam　2.2.24.1
空气-液滴混合物　air-droplet mixture　3.4.15
空气与物料　air and material　23.1.13.9
空气与物料排出口　air and material discharge　11.2.8.9
空气预热器　air preheater　4.2.4;2.3.16.7
空气增压压缩机　air booster compressor　2.3.4.7
空气罩密封圈　air hood collar seal　4.2.4.6
空气注入器　air injector　10.4.1.3
空腔　cavity　19.3.2.12
空时收率　space time yield(STY)　1
空速　space velocity(SV)　1
空调设备　air conditioning equipment　20.2.2.15
空桶储存仓　empty drum storage　23.4.1.1
空心活塞　hollow piston　17.2.10.5
空心螺旋　hollow screws　11.1.3.11
空心轴　hollow shaft　25.1.13.9
空心轴带轴承　hollow shaft with bearings　9.5.10.9
空心轴颈　hollow journal　11.2.4.11
空转锁定螺母　lost-motion locknut　16.2.4.38
空转调节螺母　lost-motion adjusting nut　16.2.4.39
孔　h1;hole　6.3.9.5
孔板　orifice　3.5.11.19
孔板测压计　orifice manometer　3.5.16.20
孔板塔　perforated plate tower　5.1.2.13
孔边缘　edges of orifices　10.5.1.10
孔流系数　discharge coefficient　1
空隙率　voidage　1
空隙速度　interstitial velocity　1
控制板　control board[panel]　25.4.2.1
　control panel　3.1.46.3
控制棒　control rod　25.2.2.4
控制棒驱动装置　control rod drive　25.2.2.3
控制阀　control valve　10.3.5.9
控制龙头　control tap　2.3.10.9
控制盘　control panel　15.7.5.4
控制室　control room　25.4.2
控制台　control console[desk]　25.4.2.9
　control desk　25.2.5.8
控制系统　control system　20.2.2.14
控制循环导流板　recycle control baffle　8.4.1.28
控制装置　control device　17.2.12.1
　controls　3.3.20.2
库存料输送器　inventory conveyor　23.4.2.

686

19

快速床　fast bed　14.1.2.5
快速流态化　fast fluidization　14.1.1
快速切断　quick disconnect　3.5.7.23
快速松脱接头　quick release joint　17.6.2.13
快中子增殖(反应)堆　fast-breeder reactor　25.2.1
快装管接头　quick connects female　3.5.16.31
宽刮刀　wiper blade　9.4.9.6
矿井　shaft　25.2.4.13
矿坑水　mine waters　5.4.3
矿石　ore　11.6.2.1
矿物废渣　mineral residue　2.2.92.24
框　fr.; frame　9.1.1.16
框架　frame　11.1.1.12
窥视和采光孔　viewing and lighting windows　13.5.2.7
窥视孔　viewing port　4.3.4.1
窥视视镜　peep sight glass　5.4.2.6
馈电端子(瓷套管)　feed-through insulator [terminal]　25.4.5.2
馈电屏　feeder panel　25.4.2.4
扩大断面　expanded sections　8.4.1.24
扩大管　expanding tube　8.4.3
扩大截面　diverging section　3.5.4.8
扩口壳体　flared shell　22.5.1.(c)
扩散泵　diffusion pump　18.2.1
扩散泵冷却盘管　diffusion pump cooling coils　5.5.2.12
扩散泵体　diffusion pump body　5.5.2.13
扩散分离过程　diffusional separation processes　15.4
扩散器　diffuser　11.2.11.8
扩散系数　diffusion coefficient　1
扩散装置　diffusion device　3.1.3.7
扩压板　diffuser plate　20.2.1.10
扩压管　diffuser　18.2.3.7
扩压器　augmentor　18.3.5.1
　　diffuser　16.4.27.5
扩压器出口　diffuser discharge　18.3.4.25
扩压器机体　diffuser body　18.3.1.5
扩压器进口　diffuser inlet　17.4.1.3;18.3.4.24
扩压器口环　diffuser ring　16.4.27.36
扩压器叶片　diffuser vanes　6.1.7.2
扩压型泵　diffuser-type pump　16.4.17
扩张角5°　5 deg divergence　23.2.2.6

L

垃圾　refuse　3.3.1.1
垃圾废料粉末　powder RDF　3.3.16.28
Silva Gas 垃圾气化系统工艺流程　Silva Gas refuse gasification system flowsheet　3.3.23
垃圾燃料　refuse-derived fuels(RDF)　3.3.16
拉杆　tension rod　11.1.1.14
　　tie rod　21.2.3.11
拉杆　tie-rod　5.2.4.3
拉杆螺栓　tie bolts　11.1.7.3
拉杆弹簧　tension rod spring　11.1.1.10
(拉杆弹簧)垫圈　washer　11.1.1.11
拉紧螺栓　staybolt　16.4.18.26
拉门　sliding door　3.3.12.47
拉伸　stretching　2.3.8.55
拉伸弹簧　extension spring　19.3.1.11
拉-推系统　pull-push system　23.1.1.(d)
拉瓦尔喷管　Laval nozzle　18.2.3.8
拉瓦尔喷嘴　Laval nozzle　1
拉乌尔定律　Raoult's law　1
拉西环　Raschig ring　5.3.2.(a)
拉西环或鞍形填料　rings or saddles　5.3.1.10
拉线开关　pull-switch　25.4.1.16
拉制机　drawing machines　2.3.9.7
喇叭形接头　flued heads　22.5.1.(e)
喇叭形进气口　bellmouth nozzle　23.1.7.14
喇叭形吸入口　suctional bell　16.4.27.2
喇叭形吸入罩　suction bell ring　16.4.27.40
蜡出口　wax outlet　14.3.2.14
蜡膏　slack wax　2.1.1.81
来自处理设施的空气　air from treatment facility　3.3.20.11
来自换热器　from heat exchanger　2.2.110.15
来自煤气化装置粗煤气　crude gas from coal gasification　2.2.34.21
来自污泥浓缩机　from sludge thickener　3.3.19.3
来源　source　3.3.5.1
莱茵岩盐　Leine rock salt　25.2.4.8
拦液圈　flinger ring　9.5.3.8
栏杆　hand rail　24.1.1.25
　　top rail　5.6.1.4
栏杆立柱　(hand rail) post　5.6.1.6
捞工　vatman　2.3.10.46
老化器　digester　3.1.2.14

687

雷蒙德闪蒸干燥器（源自雷蒙德分部、燃烧工程公司） Raymond flash dryer（courtesy of Raymond Division，Combustion Engineering Inc.） 3.3.24
雷蒙磨 Raymond mill 6.3.2.8
雷蒙闪急干燥器 Raymond flash dryer 8.4.5.(a)
雷诺数 Reynolds number(Re) 1
累积器 accumulator conveyor 23.4.3.5
累积输送器 accumulator conveyor 23.4.2.4
累计器 totalize 3.5.18.11
（累计）总量表 quantity meter 1
肋板 rid 21.3.1.7
泪孔 weep hole 5.2.1.20
类似较小容器 similar smaller containers 23.4.1
冷成品气 cold product gas 2.2.28.3
冷冻 refrigerating 2.3.3.16
冷冻干燥 freeze drying 1
lyophilization 1
冷冻剂 refrigerant 2.2.37.10
冷冻剂入口 refrigerant inlet 7.2.1.26
冷冻式干燥器 refrigerated dryer 2.2.53.4
冷冻压缩机 cooling compressor 2.3.12.22
冷激 quench 1
冷焦炭 cold coke 2.1.9.30
冷阱 cold trap 15.1.1.9
refrigerated trap 18.5.1.3
冷空气管 cold air tube 3.3.26.7
冷空气入口 cold air inlet 11.6.4.16
冷空气吸入口 cold air intake 4.6.1.1
冷流体 cold fluid 4.2.1.12
冷流体出口 cold fluid out 4.2.1.6
冷流体进入 cold fluid in 4.2.1.8
冷模试验 cold-flow model experiment 1
mockup experiment 1
冷凝 condensation 20.2.4.21
冷凝管 condenser tube 20.2.3.3
冷凝-冷却器壳体 con.-cooler shell 20.2.3.1
冷凝器 condenser;cond. 2.2.27.12;2.2.60.1;2.2.76.14;2.2.111.19;2.2.113.10;2.2.114.13;18.3.4
冷凝器回收热 condenser heat recovery 25.5.2.11
冷凝器水出口 condenser water outlet 5.5.2.19
冷凝器水进口 condenser inlet 5.5.2.16
condenser water inlet 18.3.6.2
冷凝器用水进口 condensing water in 20.1.8.3
冷凝扫掠蒸汽 condensed sweep vapor 15.4.1.9
冷凝水 condensate 3.1.6.2;2.4.3.2
冷凝水出 condensate out 2.4.2.2
冷凝水出口 condensate outlet 7.2.8.12
condensed water outlet 18.3.6.5
冷凝温度 condensing temp. 20.2.4.14
冷凝物 condensate 2.2.34.9
冷凝压力 cond.press. 20.1.7.2
condensing pressure 20.2.4.23
冷凝液 cnds.；condensate 2.2.9.14
冷凝液泵 condensate pump 20.1.10.8
冷凝液出孔 hole for condensate 5.5.1.45
冷凝液出口 condensate out 8.8.7.3
冷凝液排放 condensate drain 8.8.10.9
冷凝液柱高度范围 liquid heights in condensate 25.6.2
冷凝用水 condensing water 20.1.8.9
冷气体 cool gas 2.2.7.15
冷却 cool 2.2.115.14
cooling 2.2.53.2;2.2.53.6
冷却壁 cooling wall 2.2.16.4
冷却池 cooling tank 2.3.9.15
冷却的惰性烃液体 cooled inert hydrocarbon liquid 2.2.39.b
冷却涤气器 cooling scrubber 3.1.22.12
冷却段 cooling zone 3.3.9.3
冷却段多孔板 perforated plate for cooling 8.5.3.7
冷却负荷 cooling load 25.1.2.7
冷却工质 cooling medium 25.3.2.12
冷却管 cooling duct 20.2.1.19
冷却管线 cooling circuit 25.6.3.5
冷却滚轮 cooling roll 2.3.11.25
冷却滚筒 cooling cylinder 2.3.8.27
冷却剂泵 coolant pump 16.1.2.3.22
冷却剂出口 coolant outlet 7.2.1.10
冷却剂流通管 coolant flow passage 25.2.9
冷却夹套 cooling jacket 14.3.1.15;2.2.26.8
冷却结晶器 cooling crystoallizer 7.2.12
冷却介质 cooling medium 15.1.1.7
冷却空气 cooling air 13.1.1.5;3.3.26.13
冷却空气鼓风机 cooling air fan 3.3.9.1
冷却空气排出 cooling air discharge 3.3.9.9
冷却空气入口 cooling air in 25.1.13.1
冷却帘栅 cooling screen 2.2.26.5
冷却流体 cooling-fluid 13.5.2.9
冷却器 cooler 13.1.1.7;2.2.27.9;2.2.54.2;2.2.74.9

冷却器壁　condenser wall　15.4.1.8
冷却器管　cooler tube　20.2.3.7
冷却器组　cooler group　14.3.2.24
冷却区段　cooling zone　8.5.6.20
冷却圈　cooling ring　15.7.5.6
冷却设备　cooling device　18.2.3.5
冷却室　cooling compartment　14.2.5.12
冷却室盖　cooling compartment cover　16.4.20.23
冷却水　C.W;cooling water　2.2.9.7;2.2.15.4;2.2.34.23;2.2.37.9;2.2.45.13;2.2.46.4;2.2.47.9;2.2.48.8;2.2.110.19;3.4.7.9
冷却水泵　cooling water pump　16.1.2.3.12
冷却水补充　cold-water makeup　3.1.6.12
冷却水出口　cooling water out(let)　11.3.4.2
冷却水返回　cooling-water return　2.3.6.18
冷却水回路　CWR　2.2.34.24
冷却水夹套　water jacket for cooling　17.2.5.12
冷却水进出口　cooling water outlet and inlet　16.4.20.19
冷却水进口　cooling water in(let)　11.3.4.1
冷却水(流动)回路　cooling water flow circuit　25.2.1.6
冷却水入口　cooling water　7.2.5.6
冷却塔　cooling tower　25.2.2.19
冷却塔进料缓冲池　cooling-tower feed surge pond　2.3.6.9
冷却吸收器　cooling absorber　5.4.2.1
冷却系统　coolant system　25.2.3.5
cooling system　2.2.80.10
冷却箱　cooling box　6.3.2.30
冷却液膜　coolant film　4.3.3.10
冷却液收集装置　coolant-collecting plant　2.3.12.7
冷却油出口　cooling-oil outlet　14.3.2.21
冷却油进口　cooling-oil inlet　14.3.2.23
冷却装置　cooling unit　2.2.62.4
冷水　chilled water　20.1.10.17
cold water　8.9.2.8
冷水出口　chilled water out　20.1.8.13
冷水阀　chilled water valve　20.2.2.20
冷水供入　chilled water supply　20.2.2.21
冷水进口　cold-water inlet　4.3.2.16
冷水入口　chilled water in　20.1.8.14
冷水水箱　cold water tank　3.3.12.9
冷水循环泵　cold water circulating pump　20.1.10.27
冷水贮罐　cold-water reservoir　3.1.6.14
冷态粒子　cold shot　14.3.2.(b)

冷态粒子接管　cold shot nozzles　14.3.2.2
冷循环气　cool recycle gas　2.2.87.14
冷烟道气　cool flue gas　2.2.111.20
冷乙炔　cold acetylene　14.3.1.3
离合器　clutch　12.1.1.16
离析　segregates　23.1.3
离心泵　centrifugal pump　16.1.2.2.14
离心滗析器　centrifugal decanter　3.1.16.5
离心堆料器　centrifugal trimmer　23.3.9.7
离心分离萃取机　centrifugal extractor　6.1.18
离心分离器　centrifugal separators　10.1.4
离心风机　centrifugal fan　12.3.1.3
离心风机组件　components of a centrifugal fan　23.1.10
离心机　centrifuge　9.5;2.2.64.8
离心机传动装置　whizzer drive　11.2.12.5
离心机叶片　whizzer blades　11.2.8.7
离心空气分级　whizzed air classification　11.2.8
离心筛　centrifugal screen　2.3.10.22
离心式　centrifugal　17.1.1.12
离心式导流板　centrifugal deflector　7.2.14.9
离心式分离器　centrifugal separator　5.5.1.18
离心式分子蒸馏釜　centrifugal molecular still　5.5.1.(c)
离心式风机　centrifugal fan　7.5.1
离心式压缩机　centrifugal compressor　17.4.2
离心式压缩机　centrifugal compressor　20.1.3.7
离心式叶轮　centrifugal impeller　16.4.5.(a)
离心压缩机　centrifugal compressor　17.4
离心液精馏　centrifugate distillation　2.2.64.9
离心制冷系统　centrifugal refrigeration system　20.1.4
离子电极　ionic electrode　12.7.2.7
离子辐射　ionic bombardment　12.7.2
离子化单元　ioniying unit　10.3.1.21
离子交换槽　ion exchange tank　6.3.10.11
离子交换床　ion-exchange bed　3.5.22.9
离子交换器再生　ion-exchanger regeneration　6.2
离子交换设备　ion-exchange equipment　6.2
离子迁移膜　ITM（ion transport membranes）　2.2.114.3
理论板当量高度　height equivalent of a theoretical plate(HETP)　1

理论(塔)板 theoretical plate 1
理想气体 ideal gas 1
立管 stand pipe 2.2.24.7;14.2.6.10
立管冷凝器 vertical tube condenser 5.5.2.9
立井 shaft 25.2.4.13
立式泵 vertical pump 16.4.21
立式端吸离心泵 vertical-shaft end-suction pump 16.4.14
立式多级离心泵 vertical multistage centrifugal pump 16.4.27
立式干燥器 vertical type dryer 8.9
立式加压叶片过滤机 vertical type pressureleaf filter 9.2.6
立式扩压型排水泵 vertical wet-pit diffuser pump bowl 16.4.24
立式雷蒙磨 Raymond vertical mill 11.2.11
立式离心卸料加压叶滤机 centrifugal-discharge filter 9.2.3
立式连续干燥器 continuous vertical type dryer 8.9.3
立式螺杆混合器 vertical screw mixer 12.1.2
立式螺旋混合器 vertical screw 12.1.3.(f)
立式屏蔽泵 vertical can-type pump 16.1.2.2.33
立式气液分离器 vertical seperator 5.3.3.8
立式涡轮泵 vertical turbine pump 16.4.25
立式 U 形管水管锅炉 vertical U-tube watertube boiler 4.4.1
立式叶滤机 vertical pressure leaf filter 9.2.2
立式圆筒形压力罐 upright cylindrical pressure tank 9.2.5.31
立式轴流泵 Vertical axial flow pump 16.4.30
立式转子 vertical rotor 10.5.1.21
立轴式井泵 vertical spindle well pump 16.1.2.2.41
立柱 column 13.1.2.36
upright column 4.2.1.1
沥滤的蔡希斯坦残渣 residue of leached Zechstein 25.2.4.6
沥青 asphalt 2.1.1.86
bitumen 2.1.9.21
沥青进料 pitch feed 14.2.6.19
沥青纸 asphalted paper 25.4.8.6
砾石支撑层 support gravel 3.1.38.6
粒度 classified 12.3.3.7
particle size 1
粒度分布 particle size distribution 1
粒度分级 size classification 1

粒度分析仪 grain size analyzer 1
粒料 granules 13.3.4.5
粒料出口(固态)密封连接 burden outlet (solids state) gastight connection 13.5.2.8
粒料分级器 classifier 8.4.5.12
粒料分选器 classifier 8.4.1.14
粒状 granular;gran. 8.1.1
粒状活性炭 granulated active carbon 2.4.4.3
粒状介质过滤操作 granular-media filter operates on 3.1.10
粒状炉渣 granulated slag 2.2.26.11;24.3.1.22
粒子(的)收集 particle collection 15.3.1.7
粒子富集 particle concentrate 10.4.3.9
粒子计数器 particle counter monitors 3.5.6
连杆 connecting rod 16.2.5.6
pitman 11.1.1.8
连杆键 connecting rod key 16.2.3.43
连杆螺栓螺母 connecting rod bolts with nuts 17.2.15.26
连杆销 connecting rod pin 16.2.3.13
连杆组件 complete connecting rod 17.2.15.4
Y 连接 star connection[network]; Y-connection 25.4.3.2
连接板 connecting plate 5.2.2.11
连接卡圈 connect collar 11.1.5.2
连接孔 hole for connection 22.1.4.11
连接螺栓 coupling bolt 17.2.11.2
连通管 complete piping 16.4.27.26
连续 continuous 8.1.1.3
连续采样系统 continuous sample train for COCO 3.5.12
连续穿流-循环输送带式干燥器 continuous through-circulation conveyor dryer 8.3.3
连续穿流-循环式 continuous through-circulation 8.1.1.⑯
连续带式 continuous.band 8.1.1.⑲
连续法生产乙酸乙酯工艺流程 process of continuous production of ethyl acetate 2.2.57
连续回转过滤机 continuous rotary filter 9.3.1.5
连续搅拌釜式反应器 CSTR reactor 2.3.15.1
连续进料流 continuous feed stream 6.3.8.12
连续拉丝 continuous filament process 2.3.9.48

连续冷却盘管 continuous pipe cooling coils 6.3.2.30
Himsley 连续离子交换系统 Himsley continuous ion-exchange system 6.2.6
连续链式磨木机 continuous chain grinder 2.3.10.53
连续流动输送机 continuous-flow conveyors 23.3.3
连续流动组合取样器 continuous flow composite sampler 3.5.3.7
连续密闭循环涡轮转盘式干燥器 turbo-tray dryer in closed circuit for continuous drying 8.9.2
连续盘式 continuous tray 8.1.1.⑰
连续润滑油处理工厂 continuous L.O.treatment plant 6.3.2.15
连续式滚轮混合器 continuous muller 12.1.3.(h)
连续水平真空过滤机 continuous horizontal vacuum filter 9.4.2
连续投料带增稠 continuous feeds with thickening 3.2.5.(b)
连续型斗式提升机 continuous bucket elevator 14.3.3.21
连续研磨入口 series grinding inlet 11.3.1.
莲蓬头式喷淋器 shower nozzle thrower 5.3.4.3
联苯 diphenyl 5.5.3.5
联合生产 cogenerating 25.1.1
联合生产系统 cogeneration systems 25.1
联合循环电厂（来自合成气处理过程的蒸汽未显示） combined cycle plant (steam from syngas processing is not shown) 2.2.111
联接器 coupling 22.2.3.23
联轴器 coupling;cplg. 11.3.3.2
　shaft coupling 16.4.25.23
联轴器部件 complete coupling 16.4.27.20
联轴器键 coupling key 16.4.29.26
联轴器锁母 coupling lock nut 16.4.29.25
联轴器销衬套 coupling bushing 16.4.29.22
联轴器柱销 coupling pin 16.4.29.23
炼厂泵 refinery pump 16.1.2.3.37
炼厂废水 refinery wastewaters 3.1.3
炼厂加工 petroleum refinery 2.1.1
炼焦废液 coke waste 3.1.5
炼焦炉煤气的加工 coke-oven gas processing 2.3.14.16-45
炼焦炉煤气排出（口） discharge of gas from the coke ovens 2.3.14.16
炼焦炉组 battery of coke ovens 2.3.14.8
链 chain 13.1.2.42
链传动装置 chain-drive unit 8.8.12.2
链轮 chain wheel 13.5.1.16
链式输送机 chain conveyor 23.3.5
链输 feed-screw sprocket 9.3.3.1
链条防油防尘罩 oil and dusttight chain casing 8.8.11.4
链烷烃 paraffins 6.1.6.(a)
良好设计 well-designed 23.2.2.(b)
凉焦台 coke loaking bay 2.3.14.12
凉水池 cold water well 4.6.1.2
凉水塔 cooling tower 4.6
两侧进料 side-by-side loading 12.1.4.(c)
两端 both ends 4.3.4.34
两端进料 end to end loading 12.1.4.(a)
两端进气 suction in ends 17.5.4.2
两段湿式自磨机 two autogenous wet-grinding stages 11.6.3
两级电除尘器 two-stage electrical-precipitation 10.3.3
两级对置的单作用活塞 two-stage single-acting opposed piston 17.2.4
两级喷射泵 two-stage ejector 18.3.4.(a)
两级气流和笼式磨机气流输送干燥器 two-stage air stream and cage mill pneumatic-conveyor dryer 8.4.4
两级双作用空气压缩机 two-stage double acting air compressor 17.6.1.8
两级双作用压气机气缸 two-stage double-acting compressor cylinders 17.2.7
两级旋风分离接收器 two-stage cyclone receiver 23.1.13
两级再生器 two stage regenerator 14.3.4.18
两级逐级系统 two-stage cascade system 20.1.5
两级逐级循环（系统） two-stage cascade system 20.1.7
两级组合式旋风分离器 two-stage multiple cyclonic separator 10.1.3
两甲板间的空间 tween 23.3.9.4
两相流 two-phase flow 1
量纲分析 dimensional analysis 1
量子效应 quantum effect 1
料仓 bin 23.1.8.24
　feed bin 6.3.1.10
　feed hopper 12.7.5.6
　storage bin 23.3.4
料仓抖动器 bin activator 12.1.5.12

料仓弧形导流器　bin stream-deflector　23.
　　2.4.7
料仓松动器　bin activator　23.3.4.7
（料）袋　bag　9.5.2.18
料斗　bin　13.2.4.15
　　feed hopper　15.7.1.1
　　hopper　10.2.3.12；2.2.31.5；2.2.110.3
料斗衬板　feed hopper liner　12.3.4.4
料斗带式加料器　hopper-belt feeder　11.7.
　　1.12
料斗底阀　hopper valve　10.2.1.23
料斗管板　hopper tube sheet　10.1.2.34
料斗冷却夹套　hopper cooling jacket　13.2.
　　1.4
料斗支架　hopper base　11.1.5.4
料斗支座　hopper base　11.1.8.2
料化　bin　13.3.1.1
料将进口　slurry inlet　9.2.5.30
料浆加入　feed slurry　9.4.1.6
料浆样面　slurry level　9.3.10.6
料库　silo　23.1.2.3
料面高度　bin level　23.3.4.2
料面高度计　Hi-level　23.1.8.6
料面指示器　level indicator　8.9.3.3
料盘　discs　9.5.13.2
　　feed pan　8.7.1.5
　　plate　8.9.4.5
　　trays　8.2.1.6
料入口　feed inlet　9.1.1.7
料腿　dip leg　14.2.6.9
料液　feed　6.1.3.10
　　feed liquor　6.2.6.3
料液泵　feed pump　7.1.2.19
料液入口　liquor inlet　7.1.2.10
料液-塔底液换热器　feed-bottom exchanger
　　3.1.7.2
料液预热器　feed preheater　18.5.1.10
列管换热器　shell-and-tube heat exchangers
　　4.1
列管式反应器　shell-and-tube reactor　2.2.
　　70.3
裂化蒸气出口　cracked vapor outlet　14.3.
　　3.30
裂解气　cracked gas　14.2.7.12
裂解转筒　pyrolysis drum　2.2.5.19
EDC 裂解装置　EDC cracker．2.3.18.E
林德筛板　Linde sieve tray　5.2.10.12
林德循环　Linde cycle　1
临界点　critical point　20.2.4.20
临界流态化　minimum fluidization　14.1.3.8
临界指数　critical exponent　1

SOD 临氢重整装置　SOD hydroformer　14.
　　2.6.(f)
淋溶镁灰残留物　residue of leached Zech-
　　stein　25.2.4.6
淋水装置　water distribution　4.6.1.5
淋洗筛　rinse screen　12.5.5.19
淋洗水泵　rinsewater pump　12.5.5.11
灵活焦化　flexicoking　2.2.28.12
灵活焦化器　flexicoker　2.2.79.15
零点　zero　3.5.11.21
零泄漏　zero leakage　1
溜槽　launder　15.6.3.17
溜槽自动放液管　auto launder drain　15.6.
　　3.26
流变性质　rheological property　1
流程泵　process pump　16.1.2.3.36
流程图　diagrammatic flow sheet　2.1.1
　　flow charts　13.3.2
　　flow diagram　2
　　flow sheet　1
流程中的 MTG 反应器　online MTG reactor
　　2.2.77.7
流出　effluent　15.6.1.22
　　output　3.1.19.14
流出口　outfall　3.1.2.28
流出物　effluent　5.4.3.2
流出物放净口　effluent drain　9.4.4.10
流出物排出口　effluent discharge　9.4.4.8
流出液　effluent　3.1.1.8
流出液空冷器　effluent air cooler　2.3.4.16
流道　flow passage　16.4.15.18
流道口缝隙　slot for channel gate　3.1.3.4
流动不良　poor-flowing　23.2.2.1
流动方向　flow　16.5.4.6
流动模式　flow patterns　9.1.1
流动坡度　pitch with flow　17.6.2.7
流动网　flow screen　15.2.1.7
流化焙烧焚化炉　fluorides incinerator　3.3.4
流化床　fluid bed　8.1.1.⑥
　　fluidized bed　14.1.10.1；2.2.70.13
流化床反应器　fluid bed reactor　3.3.6.5；3.
　　1.36.7
　　fluidized bed-reactor　2.2.71.1
流化床反应器工艺　fluidized bed reactor
　　processes　14.3.4
流化床反应器中试装置　Pilot-plant fluid-
　　bed reactor system　2.3.5
流化床焚烧炉　fluid-bed incinerator　3.3.6
流化床干燥器　fluid bed dryer　3.1.16.15
　　fluidized bed dryer　8.5.5
　　fluidized bed dryer　2.2.107.7

692

流化床合成　fluid bed synthesis　2.2.100.9
Synthol 流化床合成燃料连续反应器系统　Synthol fluidized bed continuous reactor system　14.3.2.(d)
流化床焦化工艺　fluidized bed coking process　14.3.4.(d)
流化床焦化塔　fluidized-bed coker　2.1.9.31
流化床临氢重整装置　fluidized bed hydroformer　14.3.4.(c)
流化床煤干燥器　fluidized-bed coal dryer　8.5.2
流化床密封腿　fluidized-bed seal leg　14.1.10
流化床生物反应器 FBBR　FBBR（fluidized bed bioreactor）　3.2.12
流化床系统　fluidized-bed system　14；14.1.4
流化床旋风分离器　fluidized-bed cyclone　14.1.5
流化床与膨胀床工艺　fluidized bed and expanded bed process　3.2.9
流化床蒸汽发生器　fluidized-bed steamgenerator　14.2.2
流化床蒸汽发生系统　fluidized-bed steam-generation system　25.5.3
流化反应器　fluosolids reactor　3.3.4.14
流化过程　fluidization processes　14.2.6
流化焦化装置　fluidized coking unit　14.2.6.(h)
流化空气鼓风机　fluidizing air blower　3.3.4.17
流化空气入口　fluidizing-air inlet　3.3.27.1；25.5.1.2
流化气垫　fluidizing pad　23.3.8.(a)
流化室　fluidizing chamber　8.5.6.8
流化系统　fluidizing system　23.1.2.(e)
流化用鼓风机　fluidizing air blower　13.4.1.7
流化用气　fluidizing air　13.4.1.1
流化载气　fluidizing gas　2.2.30.2
流经叶轮与蜗壳　flow through impeller and volute　16.4.3.2
流空　empty　6.3.3.4
流量测量　flow measurement　3.5.3.4
流量计　flow meter　3.3.21.4；3.3.22.4
　rate meter　3.4.11.26
流量控制　flow control　3.5.19.5
流量控制和分流　flow control & splitting　3.1.44.4
流量调节部件　flow-adjustment feature　23.5.1.3
流量系数　discharge coefficient　1
流量显示器　flow indicator　6.2.1.12

流能磨　fluid-energy mill　11.5.4
Micronizer 流能磨（气体粉碎机）　Micronizer fluid-energy (jet) mill　11.5.1
流入　flow　3.5.10.5
inflow　3.3.30.1
influent　15.6.1.33
流入水平往返移动输送机　inflow traverse conveyor　3.3.29.8
流入物　influent　15.5.4.1
流入物进口　influent inlet　9.4.4.11
流入液　influent　5.4.3.1
流水管　flow tube　3.5.18.(b)
流速测量系统　Velocity-measurement system　3.5.2
流速计　current meter　3.5.1
流态化　fluidization　14.1.3
流态化焙烧　fluosolids roasting　13.4.1.11
流态化泵　fluidization pump　3.1.36.3
流态化初期阶段　incipient stage　4.3.3.26
流态化的吸收剂介质　fluidized absorbent media　3.1.42.4
流态化蝶形阀式出口　fluidizing butterfly outlet　23.3.8.(c)
流态化粉粒体　fluidized bulk powders　23.3.8.(b)
流态化接管　fluidizing connection　23.3.8.7
流态化空气　fluidized air　4.3.3.22
流态化空气管线　fluidizing air line　23.3.8.10
流态化气垫　fluidizing pad　23.3.8.3
流态化物料　fluidized product　8.5.6.19
流态化卸料口　fluidized outlets　23.3.8
流态空气风机　fluidizing air blower　3.3.2.8
流淌壁　flooded wall　10.5.13.2
流体出口　fluid outlet　11.5.1.9
liquid out　10.1.4.7
流体导管　fluid duct　16.5.4.5
流体动力学　fluid dynamics　1
流体分级系统　fluid fractionating system　18.2.4.12，13
流体静压试验　hydrostatic testing　21.1.2.15
流体密封槽　fluid-seal pot　14.1.7.(d)
流体盘管　flow coil　25.6.5.4
流体驱动的　fluid operated　16.1.1.15
流体入口　fluid inlet　11.5.1.1
流体速度　fluid velocity　14.1.2
流体压力总管　fluid pressure manifold　11.5.1.7
流体质点　fluid particles　16.4.15.13
流通式巴氏消毒器的基本特征　Essential

693

features of a flow-through pasteurizer 2.4.1
流线　flow line　16.4.3.3
flow　3.1.3.18
流向图　flow diagram　15.5.4.5
流型　flow pattern　3.2.15.8
硫　sulfur　2.2.92.13;2.2.109.11
硫铵产出　production of ammonium sulphate　2.3.14.35
硫保护反应器　sulfur guard reactors　2.2.106.39
硫,汞和微量元素脱除　sulphur, mercury and trace elements removal　2.2.110.11
硫化氢　hydrogen sulfide　2.2.94.12
H_2S　2.2.32.8;2.2.33.16;2.2.63.1;2.2.117.14;
硫化氢,氨,有机酸　H_2S, NH_3, organic acids　3.1.47.11
硫化氢,二氧化碳分离　H_2S, CO_2 separation　2.2.117.16
硫化氢,甲硫醇　H_2S, CH_3SH　2.2.63.7
硫化氢汽提塔　H_2S stripper　2.2.32.10
硫化氢洗涤塔　hydrogen sulphide scrubber　2.3.14.22
硫化氢循环　H_2S recycle　2.2.63.5
硫黄　sulfur　2.2.6.13;2.2.67.11
硫回收　sulfur recovery　2.2.36.8;2.2.67.10;2.2.95.22
硫回收装置　sulfur recovery unit　2.2.95.14
硫或硫酸　sulfur or sulfuric acid　2.2.36.9
硫加工设备　sulfur plant　2.2.102.22
硫酸　sulfuric acid　2.2.57.13
H_2SO_4　2.3.15.12
硫酸铵　ammonium sulphate　2.3.14.36
硫酸铵溶液　ammonium sulphate solution　2.3.8.26
硫酸产出　production of sulphuric acid　2.3.14.34
硫酸供应　sulphuric acid supply　2.3.14.33
硫酸钠颗粒产[成]品　sodium sulfate pellet product　2.3.4.16
硫酸烷烃化　sulfuric acid alkylation　6.1.6.(a)
硫酸盐纸浆厂　sulfate pulp mill　2.3.10.1-52
硫吸收塔　sulfur absorber　2.2.32.2
馏出液　distillate　5.5.1.47;2.2.79.11
馏出液出口　distillate out(let)　18.5.1.11
馏出液出料泵　distillate withdrawal pump　5.5.1.14
馏出液排出口　distillate drain　5.5.2.17

六角头丝堵　hex head plug　22.3.2.20
笼架　cage　15.6.3.22
笼式磨碎机　cage mill　8.4.1.1;3.3.24.1
漏斗　funnel　15.7.6.1
漏斗流料仓　funnel-flow bin　23.1.3
漏泄蒸汽通道　leakage steam path　25.4.4.8
炉　furnace　24.1.9.7
炉衬　furnace lining　8.8.8.5
炉床间隙调节器　hearth clearance adjustment　24.1.9.6
炉底灰　bottom ash　2.2.18.12;2.2.24.4
炉管支承架　heating tube support　24.1.1.19
炉管支承架底座　seat for heating tube support　24.1.1.18
炉灰　ash　24.3.1.8;2.2.117.10;2.2.118.9
boiler ash　2.3.1.26
炉灰分离器　ash separator　2.2.88.9
炉灰螺旋输送器　ash screw　24.3.1.13
炉灰热交换器　ash heat exchanger　2.2.88.3
炉灰拖运机　ash drag　24.1.5.9
炉内检查孔备用　spare nozzle for furnace inspection　24.1.1.10
炉排驱动缸　grate drive cylinder　24.1.5.5
炉气　kiln gas　11.6.4.22
炉腔内　in hearth　3.3.26.4
炉式工艺制炭黑流程　Furnace process(for carbon black production)　2.3.17
炉膛　hearth　6.3.5.1
炉膛底　pot with hearth　24.1.11.8
炉膛外　out hearth　3.3.26.2
炉体　furnace　8.8.8.4
炉条　grate　3.3.12.11
炉条下面空气通道　undergrate air port　3.3.12.10
炉渣　slag　2.2.15.8;2.2.25.7;2.2.29.7;2.2.30.10;2.2.31.11;2.2.109.3
炉渣猝灭槽　slag quench tank　24.3.1.21
炉渣急冷水　slag quench water　2.2.21.3
炉渣筛　slag screen　2.2.29.13
炉栅　grate　2.2.13.5
炉子　furnace;fce　24.1.5.3
炉子除尘室　kiln dust chamber　11.6.4.21
鲁奇干煤灰气化炉　Lurgi dry ash gasifier　2.2.13
鲁奇公司大型二甲醚制备过程　Lurgi's Mega DME process　2.2.51
鲁奇-鲁尔公司工艺　Lurgi-Ruhrgas process　2.2.4
鲁奇(煤气化)法　Lurgi process　2.2.8
鲁奇煤气化炉　Lurgi gasifier　2.2.11
铝　Al;aluminum　1.9.3.3.(a)

铝合金 aluminum alloys 22.6.2.⑳
铝酸钠 sodium aluminate 6.3.10.1
铝回收 aluminum recovery 3.3.16.11
绿麦芽 green malt 2.3.12.23
绿液预热器 green liquor preheater 2.3.10.43
氯 chlorine 2.3.8.7;2.3.18.2
氯丁橡胶垫圈 neoprene gasket 8.2.2.9
氯化钙 calcium chloride 6.3.10.10
氯化钠 sodium chloride 2.3.8.13
氯化氢 hydrogen chloride 2.3.18.6
氯化器 chlorinizer 3.3.21.7
  chlorinator 3.3.22.11
氯接触箱 chlorine contact chamber 3.1.34.13
氯气 chlorine 3.3.21.5;3.5.18.14
Cl$_2$ 3.1.34.15
氯气压缩机 chlorine gas compressor 17.1.2.3
氯乙烯单体 vinyl chloride monomer 2.3.18.20
氯乙烯单体生产工艺流程图 A process flow diagram to manufacture vinyl chloride monomer 2.3.18
氯乙烯单体塔 VCM column 2.3.18.18
滤板接缝 plate joint 9.1.3.2
滤饼 cake 9.1.3.6
  filter cake 9.1.3.9
  filtered cake 9.3.5.4
滤饼槽 cake recess 9.1.3.4
滤饼刮刀 scraper 9.3.9.11
滤饼清洗液 cake rinse 9.5.6.26
滤饼脱水 cake dewatering 9.4.1.1
滤饼洗涤 cake washing 9.3.9.7
滤饼卸出 cake dislodging and discharging 9.4.1.3
滤布 filter cloth 9.1.2.11
滤布干燥 cloth drying 9.4.1.5
滤布嵌条 calking strip 9.1.3.3
滤布上的薄层滤饼 thin cake on filter cloth 9.4.9.8
滤槽 filter tank 9.3.10.1
滤尘器 dust filter 11.6.4.17
滤(出)液 filtered liquid 9.1.3.14
滤袋 filter bag 10.2.1.34;2.3.16.10
  bags 10.4.4.6
滤后空气出口 filter air out 23.1.2.5
滤浆槽 slurry trough 9.3.2.16
滤浆进口 feed slurry 9.2.2.16
  slurry inlet 9.1.2.1
滤浆进料 feed slurry 9.5.5.8

滤浆入口 feed slurry inlet 9.2.5.4
滤框 frame 9.1.1.2
滤膜 filtration membrane 1
  membranes 3.1.30.A.5;3.1.31.7
滤膜托 filter holder 3.5.11.9
滤汽器[网] steam strainer 18.3.4.2
滤筛 strainer 3.3.10.54
滤室网 chamber screen 9.1.3.11
滤网 screen 12.3.5.8
  strainer 16.4.28.7
滤网架 screen holder 22.2.1.10
滤网通道 screen channel 3.1.34.2
滤芯 cartridge 15.5.2
滤叶 filter leaf 9.2.1.16
滤液 filtrate 3.1.9.11;3.3.19.10
滤液泵 filtrate pump 3.1.9.10
滤液槽 filtrate receiver 9.2.1.10
滤液出口 filtrate outlet 7.1.8.14
  liquid discharge 9.5.6.27
滤液集管 filtrate manifold 9.2.5.32
滤液排出壳体 filtrate housing 9.5.10.16
滤液排出口 filtrate discharge 9.5.5.10
滤液总管 manifold 9.2.5.19
滤油器丝堵 oil filler plug 18.1.4.3
滤渣 residua 2.2.33.11
滤渣出口 cake discharge 9.4.9.7
  sludge outlet 9.2.7.17
  solid outlet 9.5.4.9
滤渣干燥器 residue dryer 2.2.92.15
滤渣卸出 solids discharge 9.5.6.13
滤渣卸出(口) solids discharge 9.5.5.23
滤渣卸料 sludge discharge 3.1.9.8
卵石反应器 pebble reactor 3.3.3.(a)
卵石加料器 pebble feeder 14.3.3.11
卵石加入门 door for adding pebbles 14.3.3.10
卵石进料 pebble feed 11.6.3.13
卵石料斗 pebble hopper 14.3.3.6
卵石磨机 rock pebble mill 11.6.3.15
卵石提升机 pebble elevator 14.3.3.4
卵石填充床 marbles packed bed 10.5.8.4
卵石移出门 door for removing pebbles 14.3.3.9
乱堆塔填料 dumped tower packing 5.3.4.42
乱堆填料 dumped packing 1
伦纳德工艺:甲醇经甲酸甲酯制甲酸 Leonard process for indirect synthesis of formic acid via methyl formate 2.2.62
轮船装料系统 ship-loading system 23.3.9
轮盖 shroud 16.4.5.2

轮毂　hub　16.4.6.2
轮毂直径　hub diameter　16.4.6.11
轮盘　disk　15.7.6.3
罗茨泵　Roots pump　16.3.7
罗茨鼓风机　Roots blower　17.3.1.(c)
罗茨鼓风机　two-impeller type of rotary positive-displacement blower　17.3.5
罗茨式　Roots　17.1.1.10
螺钉　screw　13.1.2.6
螺杆　screw　11.1.5.16
螺杆泵　screw pump　16.1.2.2.6
螺杆离心(式)　screw centrifugal　16.1.1.52
螺杆式　screw　17.1.1.11
螺杆式挤压机　screw-type extruder　13.2.1.(a)
螺杆箱体　screw case　11.1.5.6
螺母　nut　13.1.2.4
螺塞　plug　13.3.5.13
螺栓　bolt　13.1.2.8
螺栓法兰　bolted flange　21.1.1.28
螺栓法兰圈　bolt ring　6.3.9.6
螺栓座　seat of bolt　21.3.1.3
螺丝起子　screwdriver　25.4.1.46
螺条混合器　ribbon mixer　12.1.3.(e)
螺纹法兰　threaded flange　22.4.2.(f)
螺纹管件　Threaded fittings　22.3.2
螺纹孔　tapping hole　5.4.2.18
45°螺纹弯头　45° threaded elbow　22.3.2.4
90°螺纹弯头　90° threaded elbow　22.3.2.1
螺线顶盖　helical top　10.1.2.3
螺旋　screw　9.4.2.6
螺旋板　helical plate　10.4.3.6
螺旋板换热器　spiral-plate exchanger　4.2.5.2
螺旋抄板　spiral flights　8.8.1.16
螺旋挡板　spiral baffle　15.2.3.3
螺旋灯头　screw base　25.4.1.59
螺旋干燥机　spiral dryer　8.9.7
螺旋给料机　screw feeder　2.2.18.2
螺旋管换热器　spiral-tube exchanger　4.2.5.10
螺旋加料器　screw feeder　11.2.5.2
螺旋桨　propeller　11.4.1.7
螺旋桨泵　propeller pump　16.1.2.2.23
螺旋桨键　propeller key　16.4.24.3
螺旋桨驱动装置　propeller drive　7.2.8.19
螺旋面刮板　helical flight　4.3.5.(e)
螺旋面式　helical　10.1.1.(c)
螺旋式　spiral　10.1.1.(e)
螺旋输送机[器]　screw conveyer　13.5.1.4
　helical conveyor　12.4.2.3

螺旋输送器轴承　bearing for screw conveyer　9.2.1.5
螺旋输送设备　spiral conveyor adaptations　4.3.3
螺旋送料机[器]　screw feeder　12.1.1.3
螺旋送料机齿轮　gear for screw feeder　12.1.1
螺旋脱水器　dewatering screw　3.1.17.5
螺旋形风帽　helicoid tuyere　10.4.3.17
螺旋形叶片　helical blades　13.2.4.18
螺旋压榨脱水系统示意图(Hollin 钢铁厂)　Schematic of the screw press dewatering system(courtesy of Hollin iron works)　3.19
螺旋叶轮泵　helical impeller pump　16.1.2.2.22
螺旋钻　augers　3.3.30.2
螺柱　studs　21.1.1.10
螺柱连接　studded connections　21.1.1.26
洛特塞萃取机　Rotocel extractor　6.1.17

M

麻花填料　twisted packing　19.2.3.3
码垛机　stacker　23.4.2.17
码头　pier　23.3.9.6
埋入地下　land disposal　3.3.1.16
麦芽成品收集漏斗　finished malt collecting hopper　2.3.12.20
麦芽浆　mash　2.3.13.10
麦芽浆蒸煮锅　mashing kettle　2.3.12.44
麦芽提升机　malt elevator　2.3.12.35
麦芽筒仓　malt silo　2.3.12.37
麦芽汁　wort　2.3.12.51
麦芽汁残渣的去除　wort break removal　2.3.13.1-5
麦芽汁冷却器　wort cooler　2.3.13.5
麦芽制造　malting　2.3.12
麦芽制造塔　malting tower　2.3.12.1
脉冲塔　pulsed column　5.1.2.7
满袋移去输送器　filled bag take-away conveyor　23.4.2.3
满货板移出输送器　filled-pallet take-away conveyor　23.4.3.2
满货板重力辊子输送器　filled-pallet gravity roll-conveyor　23.4.2.22
慢洗　slow rinse　6.2.1.5
慢洗供应　slow rinse supply　6.2.1.3
盲板法兰　blind flange　16.4.27.12
盲板法兰垫片　blind flange gasket　16.4.27.

11

盲盖　blind cover　17.2.15.3
毛布　felt　2.3.11.51
(毛)毡　felt　2.3.11.51
煤　coal　2.2.11.3;2.2.13.1;2.2.18.3.2.2.29.2;2.2.30.3;2.2.36,2.2.38.1;2.2.78.1;2.2.79.2;2.2.109.2;2.2.110.1;2.2.115.5
煤仓　service bunker　2.2.13.2;2.3.14.3;2.2.18.1
煤层　coal seams　2.2.81.17
煤称量给料机　coal gravimetric feeder　2.2.35.2
煤床　coal bed　2.2.11
煤分布器　coal distributor　2.2.17.8
煤分布器/搅拌器　coal distributor/stirrer　2.2.14.9
煤粉碎　coal pulverizer　2.2.92.5
煤干馏用焦化设备　coking plant for dry coal distillation　2.3.8.2
煤合成气　coal synthesis gas　14.3.2.(c)
煤化工　chemical processing of coal　2.2
煤化工技术路线　Techniques for chemical processing of coal　2.2.1
煤灰　ash　2.2.115.7
煤灰水　soot/water　2.2.15.7
煤加工制备　coal-handling and preparation　2.3.2.1
煤浆　coal slurry　2.2.21.1;2.2.102.5
　　slurry coal　2.2.90.3
煤浆泵　coal slurry pump　2.2.35.14
　　slurry pump　2.2.29.6
煤浆槽　slurry tank　2.2.29.5;2.2.30.6;2.2.79.4
　　coal slurry tank　2.2.35.13
煤浆槽搅拌器　mixer for coal slurry tank　2.2.35.12
煤浆产品　product slurry　2.2.102.15
煤浆出料槽　coal slurry discharge tank　2.2.35.10
煤浆出料槽搅拌器　mixer for coal slurry discharge tank　2.2.35.9
煤浆混合槽　slurry-mix tank　2.2.95.2
煤浆混合罐　slurry blend tank　2.2.105.4
　　slurry mixing tank　2.2.102.4
煤浆调和　slurry blend　2.2.92.9
煤浆/氧气　coal slurry/O_2　2.2.21.2
煤浆预热器　slurry preheater　2.2.94.5
煤浆预热溶解器　slurry preheater dissolver　2.2.92.10
煤浆制备　slurry preparation　2.2.2.29
煤浆制备单元示意　schematic of coal slurry preparation unit　2.2.35
煤浆制备罐　slurry makeup tank　2.2.94.3
煤焦化　coking　2.2.1.2
煤焦油　coaltar　2.2.1.4
煤进料系统　slurry feed system　2.2.105
煤经合成气制甲烷工艺流程　Process of methane production from coal via syngas　2.2.36
煤净化工厂　coal cleaning plant　12.5.5
煤,空气,蒸汽　coal, air, steam　2.2.23.1
煤气　gas　2.2.15.6;2.2.76.19
煤气抽出器　gas extractor　2.3.14.21
煤气出口　gas out let　2.2.14.5;2.2.26.9
煤气发生器　gas generator　2.2.108.11
煤气分布器(鼓泡器)　gas distributor (sparger)　3.4.4.4.
煤气干燥　gas drying　2.3.14.30
(煤)气化　gasification　2.3.1.7
煤化　coal gasification　2.2.1.11;2.2.36.3;2.2.40.1;2.2.67.4
煤气化炉　coal gasifier　24.3
煤气急冷　gas quench　2.2.14.6
煤气加工　gas processing　2.2.108.19
煤气净化　gas clean-up　2.2.109.9
煤气冷却　gas cooling　2.3.14.29
煤气冷却器　gas cooler　2.3.14.19
煤气喷燃器　gas jet burner　24.1.1.14
煤气去用户　gas to user　2.2.108.20
煤气燃烧器　gas burner　2.2.84.5
煤气压缩机　gas compressor　2.3.14.26
COED煤热解过程　COED coal pyrolysis　2.2.99
煤入口　coal in　24.2.10.4
煤锁斗　Rock hopper　2.2.13.7
煤塔　coal tower　2.3.14.5
煤塔输送机　coal tower conveyor　2.3.14.4
煤炭　coal　2.2.3.5;2.2.25.2;2.2.117.2;2.2.118.6
煤炭能源化工产业链　coal energy and chemical industrial chain　2.2.1.1
煤衍生液体　coal-derived liquid　2.2.2.28
煤,氧气,蒸汽进入　coal, oxygen, steam in　2.2.16.1
煤液化　coal liquefaction　2.2.1.20
煤油　kerosene　2.1.4.11
煤与灰的流化床　fluidized bed of coal and ash　24.2.10.5
煤预处理　coal preparation　2.2.98.2
煤预加工　coal preparation　2.2.101.5
煤渣　ash　2.2.11.12
　　residue　24.2.10.8

煤渣炉排操作气缸　dump grate operating cylinder　25.1.5.8
Exxon Donor 煤直接加氧制液体燃料的溶剂生产工艺　Exxon Donor Solvent process for direct hydrogenation of coal to produce liquid fuels　2.2.79
煤直接液化　direct-coal-liquefaction　2.2.90
煤制备　coal preparation　2.3.1.4
煤制聚烯烃示范项目工艺流程　process for a demonstration project of polyolefin produced from coal　2.2.67
煤制乙醇的主要技术路线　Techniques for ethanol synthesis from coal via syngas　2.2.40
酶法水解　enzymatic hydrolysis　1
酶反应动力学　enzymatic reaction kinetics　1
酶膜　enzyme membrane　1
霉菌　mould　1
美孚公司丙烯腈生产工艺流程图　Flow diagram of the Sohio process for the manufacturing of acrylonitrile　2.3.15
美国机械工程师协会规范　ASME code　21.1.2.20
美国石油学会　American Petroleum Institute；3.1.3
美国石油学会标准分离器　API separator 2.1.9.6
美国型　united States design　9.4.9.(b)
美国油页岩公司　TOSCO；the Oil Shale Corporation of America　2.2.5
美国油页岩公司Ⅲ工艺　TOSCO Ⅲ process　2.2.5
镁灰油页岩　Zechstein shale　25.2.4.12
门　door　11.1.7.12
门的（自动）启闭装置　door operating device　8.8.10.18
门框墙墩　gateway pier　3.1.3.6
锰钒钢　manganese-vanadium steel　22.6.2.⑨
锰钢　carbon-manganese steel　22.6.2.⑧
锰钼钒钢　manganese-molybdenum-vanadium steel　22.6.2.⑫
弥散烟囱　dispersion stack　3.1.22.14
迷宫材料　labyrinth materials　19.3.3.(a)
迷宫密封　labyrinth seal　16.4.20.6
迷宫密封螺母　labyrinth nut　11.1.8.22
迷宫式　labyrinth　16.4.8
迷宫式密封　labyrinth type seal　19.3.3
迷宫旋转型　rotating labyrinth type　19.3.3.(c)
密闭的纯氧活性污泥工艺流程图　Flow diagram for covered pure oxygen activated sludge process　3.1.45
密闭的中间轴　enclosed line shafting　16.4.25
密闭的中间轴轴承　enclosed line shaft bearing　16.4.25.21
密闭系统　seal system　2.2.7.19
密闭型刮板式输送提升机　drag-type enclosed conveyor-elevator　23.3.7
密封　seal　14.2.4.8
密封槽　seal trough　25.1.9.4
密封垫　seal packing　4.2.1.13
密封垫片　sealing gasket　9.1.3.1
密封动环　rotating seal ring　19.1.1.3
密封阀　sealing valve　23.1.8.14
密封盖　gastight cover　13.5.2.6
密封隔环　seal cage　16.4.27.18
密封管　seal pipe　14.1.11.5
密封盒　seal housing　19.3.1.14
sealing box　19.2.1.2
密封盒外表面　stuffing box area　19.1.5.15
密封环　seal ring　14.4.3.17
sealing ring　16.4.1.16
密封环隔圈　seal ring spacer　16.4.25.14
seal spacer　16.4.26.5
密封环组　seal ring set　17.2.9.36
密封件　seal　16.4.23.5
密封静环　stationary seal ring　19.1.1.2
密封煤斗　coal lock-hopper　2.2.14.7
密封气　seal gas　2.2.7.2
密封腔　seal(ing)cavity　19.1.3.1
密封圈　sealing ring　19.1.3.4
密封溶液泵　hermetic-solution pump　20.1.8.1
密封式磁力驱动泵　hermetically sealed magnetic drive pump　16.5.5
密封梳齿　seal labyrinth　19.3.2.9
密封套　gland sleeve　19.1.5.17
密封套O形圈　gland sleeve O-ring　19.1.5.16
密封腿　seal leg　10.4.1.2
密封效果　tighter seal　19.3.3.(b)
密封压盖　gland　19.1.1.9
seal cover　9.2.1.4
密封液管　seal piping[tubing]　16.4.9.16
密封液通入液封环　liquid seal to lantern ring　16.4.4.20
密封用油　seal oil　19.1.7.10
密封油排放管　seal oil drain line　19.1.7.18
密封渣斗　slag lock-hopper　2.2.14.15
密封制冷剂泵　hermetic-refrigerant pump

698

20.1.8.19
密封装置　seal　11.2.7.2
Fluo 密封装置　Fluo seal　14.1.9
密封装置　seal　4.2.4.3
密封组件　seal assembly　16.3.2.12
密相　dense phase　1
密相流化床　dense-phase fluidized bed　14.1.3.19
密相流料仓　mass-flow bin　23.1.2
密相流料斗　mass-flow hopper　23.5.1
描绘器　tracer　25.4.8.3
描绘器元件　tracer element　25.4.7.3
敏感性物料　sensitive product　8.1.1.30
铭牌　NP;name plate　22.2.1.3
铭牌位置　name plate set　22.2.1.18
命名规定　nomenclature　10.1.3.11
模拟　simulation　1
模拟重结晶法　simulated annealing　1
模盘　mould　2.3.11.48
模制耐腐蚀玻璃纤维容器　moulded corrosion-resistant fibreglass envelope　15.5.1.3
膜　membrane　15.2.2.5
膜分离过程　membrane processes　15.2
膜分离器　membrane separator　6.1.8
膜盒　diaphragm capsule　22.1.9.4
膜厚控制器　spreader　8.7.1.3
膜盘　diaphragm disk　22.1.9.2
膜片　diaphragm　22.1.9.1
膜生物反应器　membrane bioreactor　1
MBR(menbrane bioreactor)　3.1.30.A
膜生物反应器和等效的传统废水处理工艺比较　Comparison between MBR and equivalent traditional WWTP　3.1.30
膜式计量泵　diaphragm type metering pump　16.1.2.3.47
膜式泄压阀　diaphragm relief valve　2.1.7.20
膜头盖　diaphragm cap　22.1.9.3
膜渗透单元　membrane permeation module　15.2.2
膜组件　membrane module　1;3.1.28.7;3.1.29.8
摩擦材料　friction material　8.8.5.7
摩擦密封　friction seal　8.8.1.3
摩擦面　area of friction　8.8.5.17
摩擦曳力　frictional drag　23.2.2.5
磨棒装入口　roding hole　11.2.4.2
磨耗率　attrition rate　14.3.3.(c)
磨机　mill;mi.　11.6.4.6
磨料　attriters　14.3.4.32
磨煤机　coal pulverizer　2.2.110.2
　pulverizer　2.2.35.3

磨煤机出料槽泵　pulverizer discharge tank pump　2.2.35.11
磨木浆　groundwood　2.3.10.68
磨木浆厂　groundwood mill　2.3.10.53-65
磨木纸浆　groundwood pulp　2.3.10.77
磨砂玻璃插口　ground glass socket　3.5.14.30
磨砂玻璃锥　ground glass cone　3.5.14.31
磨石　grinding stone　2.3.10.71
磨碎　grind　2.2.90.2
磨碎干燥机　dryer grinder　3.1.21.8
磨碎机　mill　11.6.5.6
磨碎机传动装置　mill drive　11.2.12.1
磨碎机传动装置　mill drive　11.2.8.10
磨碎机料仓　mill feed bin　11.6.5.4
磨碎物料　pulverized material　11.2.6.8
磨细　size reduction　1
磨细的循环干物料　grinding recycled dry product　3.1.16.2
末端皮带轮　end pulley　23.2.3.(b)
末级叶轮　impeller(last stage)　16.4.19.15
模具　die　13.2.1.12
母线条　bus bar　10.3.5.18
母液出口　mother liquor outlet　7.2.1.6
母液回流　ML return　7.2.8.16
母液循环泵　ML recycle pump　7.2.8.18
母液循环管　ML circulating pipe　7.2.8.15
木板条　wood stave　9.3.3.27
木格栅　wood grid　5.3.2.(m)
木片装料机　chip packer　2.3.10.3
木质和衬木钢管　wood and wood-lined steel pipe　22.6.2.⑪
目　mesh　11.6.3.17
目标函数　objective function　1
钼钢　carbon-molybdenum steel　22.6.2.⑪
幕墙　curtain wall　3.3.12.14
幕墙通口　curtain wall port　3.3.12.2

N

纳米过滤　nanofiltration　1
纳氏泵　Nash pump　16.1.2.2.53
耐腐蚀泵　corrosion resistant pump　16.1.2.5.9
耐火材料　REFR;refractory　24.1.3.7;2.2.16.5
耐火材料衬里　refractory lining　2.2.14.10
耐火材料衬里　refractory lining　4.4.1.6
耐火层　refractory lining　2.2.84.4
耐火拱顶　refractory arch　3.3.27.3

耐火混凝土　refractory concrete　4.4.2.2
耐火绝热砖　refractory insulating brick　24.1.7.7
耐火砖　fire brick　24.1.7.4
　refractory brick　24.1.2.13
耐磨轴承　antifriction bearing　8.8.5.25
耐磨轴台　antifriction pillow block　8.8.5.22
耐磨板　wear plate　9.3.3.14
耐磨衬里　abrasion-resistant lining　11.3.1.14
耐磨环　wearing ring　16.4.22.21
耐磨性　erosion-resistant　10.5.9.3
耐磨性喉部衬环　erosion-resistant throat 10.5.9.3
耐热玻璃棉　Pyrex wool　3.5.14.4
耐酸泵　acid pump　16.1.2.3.28
萘　naphthalene　2.2.91.9
南温带板　south temperate zone plate　21.2.3.8
挠性构件　flexible member　16.1.1.22
挠性接头　flexible connection　12.2.3.7
　flexible joint　12.2.5.5,10
挠性进口法兰密封件　flexible inlet flange sealing　12.2.3.12
挠性联接　flexible connection　23.1.7.1
挠性联轴器　flexible coupling　15.7.2.9
挠性膜片　flexible diaphragm　22.1.4.4
挠性软管　flexible hose　17.6.2.12
挠性轴系　flexible shafting　16.4.21
挠性转子泵　Flexible rotor pump　16.3.4
挠性装车机　flexible car roader　23.4.2.18
内部分布器　internal distributors　14.3.2.(a)
内部管　internal pipes　9.3.2.22
内部换热器　internal heat exchanger　14.3.2.7
内部结构　internal structure　5.1.2
　internals　6.2.3
内部裙形挡板　internal skirt baffle　7.2.8.3
内部热交换器　internal heat transfer　14.1.4.(h)
内部双层壁　double interior wall　18.2.4.11
内部旋风分离器　internal cyclone　14.1.4.(c)
内部循环泵　internal recycle pump　3.1.35.8
内部循环流动　internal circulation　6.1.9
内侧　inboard　16.4.9.20
内插管式换热器　bayonet-tube exchanger 4.2.5.9
内衬板　lining plate　11.1.8.10
内浮头　internal floating head　4.1.4.9
内浮头换热器　internal-floating-head exchanger　4.1.1.(b)
内盖　inner cover　22.2.1.4
内管　inner pipe　4.2.2.4
内径　inside diameter　21.1.1.64
内壳　inner casing；injection tube　3.1.24.13
内壳层　inner shell　14.3.2.13
内壳体　inner shell　21.2.2.6
内扩散　internal diffusion　1
内离心分级器　internal whizzer classifier　11.2.12
内密闭室　inner closing chamber　9.5.13.9
内裙罩　inner skirt　23.1.13.3
内筒衬板　inside drum liner　12.3.4.12
内筒盖　inside drum cover　12.3.4.10
内循环器的驱动装置　internal circulator drive　7.2.9.8
内压管　int.pressure tube　21.1.1.50
内止点(死点)　inner dead point　17.2.2.7
内转鼓　inner drum　9.3.2.17
内转鼓空间　inner bowl space　9.5.16.6
内转角半径　inside-corner radius　21.4.1.9
内装挡板的混合容器　baffled mixing vessel　6.1.5
内装隔板的筒体　shell with diaphragm　8.8.1.20
内装式机械密封　internal mechanical seal　19.1.3
内装轴承　housed bearing　9.3.3.24
内锥体　inner cone　12.3.3.2
内阻层　intrinsic barrier　23.4.1.11
内钻孔　inner bore hole　3.1.24.7
能量衡算　energy balance　1
能量回收系统　energy-recovery systems　3.3.1
能量平衡　energy balance　1
能量守恒定律　law of conservation of energy　1
能伸缩唇形活塞　expanding lip piston　16.5.6.7
尼龙　nylon；nyl.　22.6.2.⑤
尼龙衬垫　nylon insert　19.1.5.12
泥饼　sludge cake　3.3.19.11
泥饼运输机　cake conveyor　3.3.19.12
泥浆包　mud drum　14.2.2.10
泥浆泵　slush pump　2.1.2.16
泥浆出口　slurry outlet　10.5.15.7
泥浆管　sludge pipe　15.6.1.35
泥浆坑　sludge sump　15.6.3.29
泥浆坑刮片　sludge sump scraper　15.6.3.27
泥浆料斗　sludge hopper　15.6.1.36
泥浆溢流　slimes overflow　12.4.2.4

泥状沉淀炉膛　sludge hearth　3.3.5.15
拟线性化　quasilinearization　1
逆流　countercurrent-flow　3.3.15.(a)　counterflow　6.2.2.(b)
Asahi 逆流离子交换过程　Asahi countercurrent ion-exchange process　6.2.5
逆流式　countercurrent　8.6.3.(a)
逆流隧道式干燥器　countercurrent tunnel dryer　8.3.1.(a)
逆流洗涤塔　counter-current flow scrubber　10.5.15
逆流直接加热式回转干燥器　countercurrent direct-heat rotary dryer　8.8.1
逆流转动型湿式鼓式磁力分离器　counter-rotation wet-drum separator　12.6.6
逆时针（方向）　counter-clockwise　17.5.2.9
逆时针（旋转）　counter clockwise; C. C. W.　17.3.3.7
逆向（反吹）脉冲纤维过滤器　reverse-pulse fabric filter　10.2.3
黏度　viscosity　1
黏胶短纤维（人造丝）　viscose rayon staple fibre　2.3.7.28～34
黏胶纺丝液　viscose spinning solution　2.3.7.10
黏塑性流体　viscoplastic fluid　1
黏弹性流体　viscoelastic fluid　1
黏土　clay　3.3.12.6
黏土衬砌　clay liner　3.3.12.7
黏土黏结剂　clay binder　6.3.10.13
黏土箱　clay bin　6.3.2.1
黏土贮存器　clay storage　6.3.2.4
黏土贮斗　clay storage bins　6.3.2.11
碾轮　runner　11.4.1.6
碾磨机　attrition mill　11.3.4　mill　2.2.29.4; 2.2.30.4
碾碎的煤　ground coal　2.2.19.3
酿酒捣碎车间　mashhouse　2.3.12.42～53
尿素合成　urea synthesis　2.2.33.14
尿素脱蜡　urea dewaxing　2.1.1.21
啮合开关　jog switch　8.8.11.8
镍基合金　nickel-base alloys　22.6.2.⑱
镍铜合金　nickel-copper alloys　22.6.2.⑲
凝并　coalescence　1
凝固槽　coagulating basin　2.3.7.15
凝固剂　coagulant　2.4.5.2
凝固物［粒］料　solidifying burden　13.5.2.3
凝结剂　coagulant　3.1.27.3
凝结剂加入管　coagulant feed pipe　15.6.4.21
凝结器　condenser　2.3.12.6
　condenser　25.2.1.16
凝结水　condensate water　2.2.45.16
凝结液贮存槽　storage tank for condensate　2.3.10.17
凝聚器　coagulator　15.6.4
　condenser　15.1.1.6
凝液　condensate　7.2.5.15
凝液出口　condensate outlet　7.1.3.5
　drain　7.1.2.12
凝液入口　condensate inlet　22.2.3.8
牛顿流体　Newtonian fluid　1
扭力扳手　torque wrench　23.4.1.8
纽约 Coxsackie 下水道污水处理厂工艺简图　Process schematic for Coxsackie sewage treatment plant，New York　3.1.34
农药废弃物　pesticide disposal　3.3.11
农业　agriculture　25.1.1.15
农业可利用的污泥　sludge for agricultural utilisation　3.1.26.(b).15
浓度（分布）剖面（图）　concentration profile　1
浓废液　concentrated liquid　3.3.10.13
浓集器　concentrator　10.1.3.6
浓集器机组　concentrator unit　10.1.3.5
浓集器集尘箱［斗］　concentrator dust hopper　10.1.3.8
浓浆至厂进液中　underflow to plant influent　3.1.26.(b).3; 3.1.26.(b).7
浓浆至厂流入液中　underflow to plant influent　3.1.26.(a).9
浓密机　thickener　1
浓铅　lead concentrate　12.8.1.1
浓溶液　concentrated solution　13.3.3.5
　strong solution　20.1.8.17
浓缩器　concentrator　2.3.10.48
　thickener　12.8.1.2
浓缩物（截留物）　concentrate（retentate）　3.1.29.9
浓缩液　concentrate　15.2.4.4
浓缩液出口　concentrate outlet　15.2.1.4
浓缩液排出泵　discharge pump for concentrated liquor　7.1.2.20
浓缩液滞留空间　concentrate space　9.5.16.5
浓相　dense phase　1
浓液　strong aqua　20.1.9.9
努塞特数　Nusselt number　1
暖空气　warm air　4.5.2.2

O

欧洲型　European design　9.4.9.(a)

P

耙　harrow　23.2.5.3
耙　rake　12.4.2.11
耙臂　rake arm　15.6.3.23
　rabble arm　3.3.26.5
耙臂齿　rabble arm teeth　3.3.26.10
耙臂固定销　arm bolding pin　3.3.26.8
耙子　rake　9.2.2.10
耙子驱动马达　rake drive motor　15.6.3.3
耙子使物料(向上)移动　motion caused by rake　12.4.2.15
帕里斯霍尔水槽　Parshall flumes　3.5.4
拍板　clappers　2.3.9.42
排出　discharge　16.2.6.2
　exhaust　3.5.19.11
排出端　discharge side　16.3.6.9
排出端泵盖　discharge cover　16.4.20.20
排出段　outlet portion　16.4.18.12
排出阀　discharge valve　16.2.3.36
排出阀螺塞　discharge-valve plug　16.2.4.26
排出(高压)气体　discharge gas　20.2.3.8
排出管　delivery pipe　20.2.3.6
　discharge pipe　16.2.1.13
排出空气(废气)　exhaust air　4.3.3.24
排出空气加热器　venting air heaters　4.3.4.22
排出孔　discharge port　17.3.3.6
排出口　discharge outlet　16.3.2.5
　discharge port　16.3.7.5
　outlet　12.1.1.17
排出口法兰　discharge flange　16.4.3.9
排出口接头　discharge head　16.4.25.24
排出口壳体　discharge casing　16.4.19.10
排出口直径　discharge dia.　16.4.3.18
排出气　purge gas　2.2.95.12
　vent gas　14.3.1.28;2.2.115.10
排出气体　off gas　25.5.1.7
排出气体与固体　purge gas and solid　10.1.1.6
排出腔　discharge chamber　16.3.2.4
　discharge manifold　16.2.1.11
排出通道　outlet passage　16.5.6.3
排出筒　discharge bowl　16.4.24.9
排出物　effluent　6.2.3.11
排出压力　discharge pressure　16.2.1.10
排出叶尖　discharge vane tip　16.4.3.12
排出叶片之边缘　discharge vane edge or tip　16.4.6.7
排出止逆阀　discharge check valve　16.2.7.2
排放　drain　3.1.19.6
　draining　6.3.3.11
排放槽　letdown tank　2.2.44.14
排放阀　drain valve　14.1.11.12
排放口　drain　15.3.1.10
　drain outlet　13.5.1.10
排放口管堵　drain plug　9.4.7.3
排放水　drain　6.2.9.6
　discharge water　2.2.49.12;2.2.50.11
排放丝堵　drain plug　22.2.3.7
排放物测定　emissions measurement　3.4
排放物去炉灰输送机　effluent to ash conveyor　2.3.6.29
排放液至澄清池　effluent to clarifier　3.2.13.3
排风管道　exhaust duct　8.3.4.17
排风机　exhaust blower　11.6.1.7
　exhaust fan　2.3.12.15
　vent fan　8.4.1.6
排风孔　exhaust fan　10.5.1.17
排干层　drained stratum　6.2.3.6
排灌站水泵　drainage station pump　16.1.2.3.8
排灰　ash disposal　2.2.89.16
排空　drain　25.6.7.6
　to ambient　6.2.8.6
排空气　air vent　9.4.4.21
排料　discharge　11.5.2.8
排料处　discharge　11.1.4.9
排料接口　metal for product discharge　11.1.8.18
排料口　discharge spout　11.4.3.8
排料口大小液压控制装置　hydraulic setting control　11.1.1.9
排料门　discharge door　15.7.1.6
排料门底座　door support　15.7.1.5
排料启闭器　door latch　15.7.1.4
排料启闭器气缸　latch cylinder　15.7.1.3
排料锥　discharge cone　11.1.7
排气　air discharge　17.2.6.7
排气　discharge　17.3.3.5
　discharge air　8.6.2.7
　exhaust;ex　10.4.3.15
　exhaust air　8.5.3.1
排气定时器　vent timer　3.5.22.3
排气阀　air valve　16.4.19.9
　air vent　22.2.3.10
　discharge valve　17.2.6.5
排气阀　vent valve　23.1.8.9

排气风机 vent fan 23.1.6.5
排气风门 exhaust damper 11.6.1.8
排气管 air vent 5.4.2.10
  discharge pipe 18.4.1.7,9
  vent 7.2.1.22
  vent nozzle 21.3.1.8
  vent pipe 24.1.6.2
排气管道 exhaust duct 8.4.6.23
排气管口 air vent nozzle 14.4.1.10
排气辊 breast roll 9.6.2.5
排气过滤器 vent filter 23.1.8.11
排气集管 exhaust header 8.4.6.12
排气接管 discharge nozzle 18.1.4.14
排气孔 discharge hole 18.4.1.6
  exhaust hole 11.1.8.7
  vent 10.3.5.7
  vent hole 22.2.3.15
排气口 air-exhaust 8.3.2.6
  delivery port 17.2.15.36
  discharge 17.2.4.6
  discharge nozzle 18.1.1.9
  discharge outlet 18.2.2.2
  exhaust 25.3.2.10
  vapor outlet 13.5.1.13
排气门 blast gate 10.4.4.3
排气室 evacuation chamber 2.2.84.9
排气通道 discharge passage 17.2.7.10
排气筒 exhaust barrel 11.1.8.32
排气温度计管 nozzle for thermometer of exhaust gas 24.1.1.1
排气旋塞 air cock 16.4.12.16
排气压力检查孔 inspection hole for exhaust gas pressure 24.1.1.28
排气烟囱 exhaust stack 3.3.11.7
排气烟道 exhaust-air stack 8.3.1.7
排气引风机 exhaust fan 23.1.7.7
排气装置 exhauster 10.1.3.6
排弃 disposal 12.5.5.10
排砂 sands discharge 12.4.2.7
排水 drainage 3.4.3.10
  effluent 3.1.43.6
排水泵 dewater pump 16.1.2.3.15
  wet-pit pump 16.1.2.3.14
排水表面 drainage surface 9.1.3.6
排水出口 drain outlet 12.5.2.5
排水口 water drain 4.3.4.12
排水流量测量 outfall flow measurement 3.5.18
排水去精制或过滤 effluent to polishing or filtration 3.1.39.5
排污 waste disposal 2.2.89.17

排污泵 faeces pump 16.1.2.3.16
排污管 drain 15.6.4.8
排污连接件 blowdown connection 4.4.2.4
排污螺塞 drain plug 22.2.1.11
排污旋塞 drain cock 15.5.2.2
排污至储罐 drains to storage tank 25.6.2.7
排泄口 outlet 3.3.17.7
排液 drain 20.1.10.13
排液沟槽 drainage channel 9.3.7.1
排液管 discharge spout 16.4.22.15;3.3.24.2
排液接口 drain connection 4.1.1.33
排液口 drain 10.3.4.8
  leakage fluid drain 16.4.20.24
  liquid outlet 9.5.4.2
排液筛网 effluent screen 3.2.13.11
排液弯管 discharge elbow 16.4.22.16
排油管 oil exit pipe 20.2.1.27
排油口 oil drain 18.2.4.15
  oil draw out 15.7.4.4
排渣阀 residual cake valve 9.2.2.13
排至大气 vent to atmosphere 8.4.1.5
盘管 coil(pipe);cl. 4.2.3.4
  tube coils 24.1.3.5
盘管式油罐加热器 tank coil heater 4.2.5.18
盘管托架 coil bracket 24.1.3.9
盘管支座 coil support 24.1.3.3
盘式 tray 8.1.1.③
盘式固化 pan granulation 13.3.5
盘式加料器 table feeder 14.1.6.(c)
盘式离心机 disk-type centrifuge 2.1.9.23
盘式淋洒器 disc spray thrower 5.3.4.8
盘式疏水阀 Disk type steam trap 22.2.2
盘式造粒机 dish granulator 13.4.3(a)
  pan granulator 13.3.5.2
盘式制动器 disc brake 9.5.10.10
盘梯 stairway 21.2.3.12
盘形阀 dick valve 22.2.1.5
旁路 by-pass 2.2.88.14;3.3.20.5;7.1.14.13
旁路阀 by-pass valve 2.1.7.21
旁路沟槽 by-pass channel 10.1.4.9
旁路调节阀 by-pass pressure-relief valve 25.6.7.5
旁通 by-passed 6.2.4.7
旁通阀 by-pass valve 3.5.11.15
旁通小孔 by-pass orifice 19.1.7.1
抛光杆 polish(ed) rod 2.1.2.27
抛起界面 strike level 7.2.11.8

抛射角　throw angle　23.3.1.7
抛扬分散器　dispersion sling　8.4.1.20
抛油环　flinger　16.4.4.8
泡点　bubble point　1
泡核沸腾　nucleate boiling　1
泡沫　froth　12.8.1.4
　head　2.3.13.31
泡沫流动接触器　froth flow contactor　3.1.42.5
泡沫塑料外套　foam encasement　3.5.14.28
泡罩　bubble-cap　5.2.8.11
泡罩(帽)塔盘　bubble cap tray　5.2.10.1
泡罩塔　bubble cap column;BPT　5.1.2.2
　bubble cap tray tower　5.2.8
培养基　culture medium　1
配合使用　used with simultaneous　23.5.1
配水系统　water distributing system　4.6.2.5
配有石墨活塞环和承磨导向环的活塞　piston equipped with carbon piston and wearing rings　17.2.10
配重له　counterweight　23.3.2.10
喷动床　spouted bed　1
喷动床造粒机　spouted bed granulators　13.4
喷动用气　spouting gas　13.2.1.19
喷镀漂洗水　plating rinse water　6.2.7
喷发　spout　13.2.1.21
喷放池　blow pit　2.3.10.13
喷放阀　blow valve　2.3.10.12
喷放罐　blow tank　2.3.10.11
喷管　shower header　9.2.5.17
喷淋泵　sprinkler pump　16.1.2.3.20
喷淋管　shower　9.2.5.24
　shower pipe　12.2.3.17
　spray pipe　2.3.10.72
喷淋密度　specific liquid rate　1
　spray density　1
喷淋器　sprayer　10.5.12
喷淋式盘管换热器　spray-type coil heat exchanger　4.2.3
喷淋水管　spray header　4.2.3.2
喷淋水洗设备　overhead spray washing plant　2.3.7.29
喷淋塔　spray tower　10.5.3
喷淋头　spray header　5.4.1.4
喷淋装置　spray apparatus[device]　5.3.4.1
喷燃器燃料罐　burner tanks　3.3.8.2
喷入　sprayedin　2.1.9.26
喷洒加料式双滚筒干燥器　double drum dryer with spray-feed　8.7.1.(d)
喷洒器　sprayer　10.5.12

喷洒室　spray chamber　4.3.4.10
喷洒塔　spray-column　10.5.11.(a)
喷散物　spray　8.6.3.7
喷射泵　ejector　18.3.5.3
　ejector pump　18.3
　jet ejector　18.4.2.6
　jet pump　16.1.2.2.45
喷射泵组件　injection pump assembly　16.2.7.8
喷射混合器　eductor mixer　2.2.95.3
喷射冷凝器　jet condenser　2.3.10.16
喷射器　eductor　3.3.22.3
　ejector　20.1.2.12
　injector　8.4.5.27
喷射器级　ejector stage　18.2.4.10
喷射器控制设备　ejector control unit　12.7.5.11
喷射式　jet　16.1.1.48
喷射式塔板　jet tray　5.2.10.15
喷射系统　ejection system　3.5.10.(c)
喷射用水　injected water　6.2.3.5
喷射增压式　ejector-boosted　16.1.1.48
喷射(蒸汽喷射)式制冷　ejector(steam-jet) refrigeration　20.1.1.(b)
喷射(蒸汽喷射)式制冷循环　ejector(steam-jet)refrigeration cycle　20.1.11
喷射嘴　jet nozzle　25.4.4.2
喷水管　spray pipe　1.1.7
喷水器　sprays　20.2.2.13
　water spray　12.7.4.1
喷水式冷却器　water-spray extended cooler　8.8.8
喷水用泵　spray pump　20.2.2.19
喷丝孔　spinneret holes　2.3.8.42
喷丝头　top of spinneret　2.3.8.41
喷头　shower nozzle　9.2.5.16
　spray head　8.9.2.6
　spray jet　10.5.1.26
　sprays　3.4.1.5
喷雾　spray　8.1.1.⑩;10.5.15.5
喷雾干燥器　spray dryer[drier]　8.6
喷雾干燥塔　spray drying tower　8.6.1
喷雾或喷射　spray or jet　3.4.4.14
喷注器　injector　11.5.4.3
喷嘴　jet　11.5.5.4;18.2.2.3
　nozzle;N　11.1.6.4
喷嘴　nozzles　8.6.3.4
　spray nozzle　12.1.3.1
喷嘴固定板　nozzle plate　18.3.1.4
喷嘴环轮　jet annular disk　11.5.5.5
喷嘴卸料　nozzle discharge　9.5.14.(b)

704

喷嘴卸料离心机　nozzle centrifuge　9.5.16
喷嘴组合件　jet assembly　18.2.4.5
盆　pan　13.3.1.4
盆式造粒机　pan granulator　13.3.4
膨胀　exp. ;expansion　20.2.4.7
膨胀波纹管　expansion bellows　8.4.5.11
膨胀槽　expansion tank　20.1.7.10
膨胀床　expanded bed　14.1.2.2
膨胀罐　expansion tank　2.2.43.13
膨胀机械　expansion device　20.1.3.19
膨胀接头　expansion joint　7.2.5.14
膨胀节　expansion bellows　8.4.5.11
　expansion joint　22.5;3.3.24.6
膨胀金属网填料　expanded metal packings　5.3.2.(o)
膨胀托辊　expansion rollers　8.8.10.11
膨胀液面　expanded level　2.1.10.7
批料凝固　batch solidification　13.5.2
皮带　belt　23.2.1.14
皮带的驱动电机　belt drive motor　12.7.4.4
皮带轮　belt pulley　7.1.2.27
皮带轮螺母　pulley nut　11.1.5.21
皮带输送机　belt conveyor　12.6.1.7
皮带运输机　belt conveyor　6.3.2.2
皮托测压管[计]　Pitot manometer　3.5.11.5
皮托管　Pitot tube　3.5.11.2
皮托管测压计　Pitot manometer　3.5.16.6
皮托管探测器　probe-pitot tube　3.5.7.18
皮托压力计　pitot manometer　3.5.7.5
啤酒杯　beer glass　2.3.13.30
啤酒厂　brewhouse　2.3.13
啤酒滤器　beer filter　2.3.13.15
啤酒酿造　brewing　2.3.12
啤酒酿造者　brewer　2.3.12.49
啤酒桶　beer barrel　2.3.13.17
啤酒箱堆　stack of beer crates　2.3.13.24
偏离的物流　deflected stream　12.7.4.11
偏摩尔量　partial molar quantity　1
偏心　eccentric　16.3.5.2
偏心变径接头　eccentric reducer　14.1.9.3
偏心滑块　eccentric slide block　16.2.5.20
偏心件　eccentric　11.1.3.7
偏心轮　eccentric　12.2.3.1
偏心式振动器　eccentric vibrator　12.2.3.2
偏心套筒　eccentric sleeve　16.2.5.21
偏心异径管　eccentric reducer　22.3.1.6
偏心重块　eccentric weight　11.3.3.4
偏心轴　eccentric shaft　11.1.1.7
偏置距离　offset　22.5.2.5
片状　sheet　8.1.1

片状产品输送带　product flaking belt　2.2.92.22
片状阀　flap valves　12.3.3.1
漂白　bleaching　2.3.7.21
撇出的油　skimmed-oil　3.1.14.4
撇沫板　skimmer　15.6.5.14
撇油槽　skimming hopper　3.1.14.8
撇油管　oil-skimming pipe　3.1.3.13
撇油器　oil skimmer　3.1.4.7
撇油器出口接头　oil skimmer outlet　15.5.1.10
贫的重溶剂　lean heavy solvent　6.1.9.17
贫的轻溶剂　lean light solvent　6.1.9.3
贫气　lean gas　14.2.3.3
贫气提升管　lean gas riser　5.4.2.9
贫溶剂　lean solvent　2.2.32.6
贫吸收剂　lean absorbent　2.2.37.36
贫相　lean phase　1
平板玻璃制造　sheet glass production　2.3.9.1～20
平板盖　flat head　21.1.1.48
平板式　flat plates　22.5.1.(a)
平板拖车　flat bed trailer　23.4.2.15
平底料仓　flat-floor building　23.2.4
平垫片　flat gasket　16.4.20.11
平封头　flat head　21.1.1.44
平盖　flat cover　11.1.7.10
平衡　equilibrating　6.3.3.8
平衡常数　equilibrium constant　1
平衡衬套　balance bush　16.4.28.6
平衡锤　balance weight　9.2.1.25
平衡锤操作杆　counter weighted operating levers　8.8.10.13
平衡釜　equilibrium still　1
平衡管　balance pipe　16.4.19.34
平衡环　balance ring　16.4.18.11
平衡活塞　balance piston　16.4.28.32
　balancing piston　17.4.2.25
平衡孔　balance hole　16.4.18.9
　balancing hole　16.4.15.8
平衡块　counterweight　23.3.6.4
平衡盘　balance disc　16.4.27.31
　balance disk　16.4.19.11
　balance runner　20.2.1.26
　balancing disk　16.4.6.18
平衡盘密封　balance seal　16.4.19.12
平衡盘上部　balancing disk head　16.4.6.20
平衡盘套筒　balance bush　16.4.19.13
平衡盘指针　pointer of balance disk　16.4.18.18
平衡式　balanced　23.3.2.7

平衡室　balance cylinder　20.13.25
　balancing chamber　16.4.6.23
平衡水池　equalization basin　3.1.1.4
平衡套　balance sleeve　16.4.18.10
平衡型　balanced　23.3.6.(c)
平衡型内装式机械密封　balanced internal mechanical seal　19.1.4
平衡重　counterweights　17.2.5.1
平衡重动作门　counter-weighted door　23.1.1.3
平键　flat key　16.2.5.8
平均停留时间　mean residence time　1
平料器　leveler　8.9.1.7
平面法兰　flat face flange　21.1.1.34
平面图　plan view　3.2.13.6
平推流　plug flow　1
平稳流化床　smoothly fluidized bed　14.1.3.11
平行进料　parallel feed　1
平直迷宫　straight labyrinth　19.3.3.(f)
平直型塔侧收集器　straight side wiper　5.3.4.17
平直叶开启蜗轮式　hub-mounted flatblade turbine　14.4.4.2
平直叶圆盘蜗轮式　disc-mounted flatblade turbine　14.4.4.1
评估　evaluating　3.1.19
屏蔽泵　canned motor pump　16.1.2.2.32
屏蔽好的容器　shielded container　25.2.5.10
瓶装啤酒　beer bottle　2.3.13.26
破碎　crushing　11.6
　disintegration　1
破碎板　breaking plate　11.1.7.16
破碎给料机　feed crusher　11.2.6.4
破碎机　disintegrator　8.4.5.32
破碎设备　crushing equipment　11.1
破碎室　crushing chamber　11.1.1.2
破碎头　crushing head　11.1.4.11
破碎物料　disintegrating the feed　8.4.5.(a)
破碎物料斜槽　crusher discharge chute　11.2.6.3
破损细粒用水　fines destruction water　7.2.9.11
　water addition for fines destruction　7.2.8.14
破旋器　swirl breaker　7.2.5.10
A-A 剖面　section A-A　14.3.3.23
剖面图　cross sectional view　8.8.10.19
　sectional view　12.1.5.11
A-A 剖视　view A-A　10.1.3.10
剖视图　sectional view　10.2.1.(a)

普遍(使用)方法　popular method　12.8.1
普朗特数　prandtl number　1
普通镍钢　plain nickel steel　22.6.2.⑦
普通平衡盘　simple balancing disk　16.4.9
普通型180°回弯头　medium pattern return bend　22.3.2.16
曝气　aeration　3.1.2.10；3.1.32.1；3.1.44.6
曝气池　aerated lagoon　3.1.6
　aeration basin　3.2.2.8
　aeration tank　3.1.34.8；3.1.43.3
曝气池盖　aeration tank cover　3.1.45.4
曝气池污泥　aeration basin　3.1.18.5
曝气罐　aeration tank　3.1.14.16
曝气井　aerated shaft　3.1.32
曝气(生化需氧量去除)　aeration(BOD removed)　3.1.37.4
曝气循环　aeration recycle　3.1.8.6

Q

漆布带　varnished-cambric tape　25.4.8.4
其他气体　other gas　2.2.3.9
其他燃油及化学品　other fuels and chemicals　2.2.73.12
歧管粒料分选器　manifold classifier　8.4.6.9
起燃室　ignition chamber　3.3.12.8
气　gas　2.2.28.4；2.2.77.10
气动顶部卸料式　with pneumatic top discharge　9.5.3.(b)
气动滑阀　air slide　11.6.4.12
气动排水阀　air actuated water dump valve　15.5.2.4
气动设备　pneumatic plant　2.3.12.12
气浮码垛器　air-flotation palletizer　23.4.3.3
气缸　compressor cylinder　17.2.5.10
　cylinder　16.2.6.9
气缸布置　cylinder arrangement　17.2.1
气缸衬套　cylinder liner　17.2.7.25
气缸盖　cylinder head　17.2.15.12
气缸盖侧　cylinder head end　17.2.2.1
气缸支承　cylinder support　17.2.15.34
气缸装配螺栓　cylinder assembling bolt　17.2.15.13
气缸座　cylinder bottom　17.2.15.32
气固分离器　gas-solids separator　14.2.3.5
气固分离设备　gas-solids separations equipment　10
气固接触方式　gas-solids contacting methods　8.6.3
气化　gasification　2.2.89.18；2.2.3.2；2.2.3.

气

10;2.2.78.3
气化的油 vaporized oils 3.3.15.9
气化段 gasification 24.3.1.4
气化反应器 gasification reactor 2.2.100.2
Koppers-Totzek 气化炉 Koppers-Totzek 24.3.3
气化和相关技术 Gasification and related technologies 2.2.3
气化剂 gasification agent 2.2.38.2
气化炉 gasifier 2.2.13.8;2.2.19.1;2.2.20.6;2.2.22.9;2.2.25.11;2.2.29.10;2.2.30.7;2.2.31.7;2.2.107.1;2.2.109.5;2.2.110.5;2.2.117.1;3.3.22.4
(气化炉)混合区 mixing zone 2.2.24.8;3.1.32.12
Lurgi 气化炉 Lurgi 25.3.1
Winkler 气化炉 Winkler 25.3.2
ECUST 气化炉示意 shematic of an ECUST gasifier 2.2.27
气化器主体 gasifier body 2.2.23.2
气化器 gasifier 2.2.106.23
vaporizer 3.1.46.7
气化器(如需要) vaporizer(if required) 3.3.21.6
气化物 vapors 2.2.82.11
气化作用 gasification 2.2.101.7
气(甲烷)相 gas(methane)phase 3.3.18.20
气冷反应器 gas-cooled reactor 2.2.42.6
气冷式高温反应堆 HTGR;high-temperature gas-cooled reactor 2.2.108.6
气冷式合成反应器 gas-cooled synthesis reactor 2.2.43.9
气力分级器 pneumatic classifier 12.3.2
气力提升器 air lift 2.1.2.4
气流 pneumatic 8.1.1.⑬
气流粉碎机 jet mill 11.5
Buttner-Rosin 气流干燥器 Buttner-Rosin pneumatic dryer 8.4.5.(b)
气流喷射方向 jet direction of airstream 11.5.5.8
气流式干燥器 pneumatic dryer 8.4
气流输送干燥管 pneumatic conveying drying pipe 8.5.1.14
气流输送干燥器 pneumatic conveying dryers 8.4.5
气轮机 turbine 25.3.2.2
气门 louvre damper 11.6.4.14
气密泵 airtight pump 3.5.7.9
气密阀 airtight valve 23.1.2.33
气密旋转出料器 rotary airlock 23.1.7.8
气密旋转加料器 rotary airlock 23.1.8.3

气升泵 gas-lift pump 16.5.1
气升(式) gas lift 16.1.1.49
气蚀 cavitation 1
气室 plenum chamber 14.1.12.3
气水分离器 gas-water separator 18.4.2.12
water knockout 6.3.3.8
气态进料 feed inlet gaseous 7.2.8.1
气提管 air-lift pipe 14.1.11.11
气体 gas 2.1.6.6;3.3.17.2;3.3.18.4
COED 气体 COED gas 2.2.30.6
气体产品 gas product 2.2.4.12
气体出口 air out 12.1.5.8
gas exit 24.1.5.4
gas out 10.1.2.32
vapo(u)r out 5.4.1.16
气体出口管 gas outlet duct 10.1.2.30
气体出口去烟道 gas outlet to stack 25.6.1.8
气体点火器 gas-fired lighter 24.2.1.4
气体分布板 gas distributor 14.1.12.4
gas distributor plate 8.5.6.18;3.4.4.10
气体分布器 gas distributor 14.1.8
气体分级 gas fractionation 14.2.3
气体分离 gas separating 2.1.1.63
gas separation 2.2.94.13
气体分离器 gas separation 2.2.94.2
气体分离装置 gas separation unit 2.2.119.9
气体干燥 gas drying 2.2.93.8
气体鼓泡器 gas spargers 3.2.6.9
气体接触区 gas contacting region 10.5.4.23
HCl 气体进口 HCl gas inlet 5.4.2.2
气体进口 gas in(let) 10.5.12.1
gas inlet port 17.2.5.15
气体进入 gas in 10.1.4.5
气体进入段 gas inlet section 10.5.10.6
气体精制 gas purification 2.2.94.13
气体净化 gas cleanup 2.2.100.4
gas purification 2.2.95.13
气体净化设备 gas cleaning unit 2.2.110.10
气体净化系统 gas cleaning system 2.2.107.11
气体净化装置 gas cleaning unit 2.2.104.4
气体冷却 gas cooling 2.3.1.9
气体冷却器 gas cooler 3.3.4.8
气体流 gas flow 10.4.2.1
气体流动方向 direction of gas flow 10.3.1.1
气体流化床 gas fluidized bed 14.1.3.16
气体流量 gas flow 3.1.2.15
气体流向 gas flow 10.3.1.26
气体密封 Gas seals 19.3

707

气体密封装置　air tight equipment　4.2.4.16
气体排出　gas out　10.1.4.6
气体排放　gas draining　3.4.7.8
气体喷射板　gas injection plate　5.4.1.9
气体切向入口　tangential gas inlet　10.5.4.5
气体取样口　vapo(u)r sample　5.4.1.19
HCl气体去吸收系统　HCl gases to absorption system　24.1.11.6
气体燃料　gaseous fuel　3.3.1.4
气体入口　gas inlet　10.3.1.11
气体释放空间　gas-disengaging plenum　2.2.87.5
气体提升管　gas lift line　6.3.7.13
气体烃　HC gas　2.2.95.11
　hydrocarbon gas　2.2.98.18
气体通道[路]　gas passage　17.2.13.3
气体透平　gas turbine　25.5.2.7
气体脱硫　gas desulfurization　2.3.2.4
气体吸入腔　suction gas chamber　19.1.7.11
气体洗涤　gas scrubbing　2.2.99.7
气体洗涤机　disintegrator　2.1.8.10
气体洗涤器　scrubber　3.3.2.15
气体洗涤塔　gas scrubber　2.2.29.15;5.4.2.16
气体循环压缩机　gas recycle compressor　2.3.4.20
气体/液体产品去分离器　gas/liquid product to separators　2.1.10.3
气体与催化剂混合物　gas and catalyst mixture　14.3.2.19
气体与污染油排放口　gas and contaminated oil drain　19.1.7.16
气体转化与调理　gas reforming and conditioning　2.2.103.1
气相甲醇　gaseous methanol　2.2.72.4
气相色谱　gas chromatograph　3.5.22.2
气相色谱系统　gas-chromatograph system　3.5.20
气压　gas pressure　6.3.3.1
气压试验　pneumatic test　21.1.2.15
气液分离罐　knock-out drum　5.3.3.13
气-液分离器　gas-liquid separator　2.3.4.13
　gas/liquid separators　2.2.94.9
气液混合管　air and liquid mixed tube　16.5.1.5
气-液接触器　gas/liquid contactor　3.5.22.7
气液平衡　gas-liquid equilibrium;GLE　1
气源　air source　3.1.41.11
气载产品　airborne product　12.3.3.3
气载原料　airborne material　12.3.3.7
气闸　air lock　11.6.4.5;10.4.4.5

汽镇入口孔　air ballast inlet hole　18.1.4.9
汽包　drum　25.5.3.4
　steam drum　14.2.2.11;2.2.44.7;2.2.46.7;2.2.47.7;2.2.74.4
汽包排污　steam drum blow-off　2.2.44.6
汽缸端的放液阀　drain valve for steam end　16.2.3.8
汽缸放液阀　steam cylinder drain valve　16.2.4.2
汽缸支腿　steam-cylinder foot　16.2.3.6
汽缸支座　steam-cylinder foot　16.2.4.5
汽轮发电机　turbogenerator　25.2.2.14
汽轮发电机组　turbogenerator set　25.2.2.13
汽轮机　steam turbine　25.2.1.14;2.2.113.9;2.2.114.10;2.2.118.11
　turbine　2.2.80.7
汽轮机监控盘　turbine monitoring panel　25.4.4.1
汽蚀余量　NPSH;net positive suction head　1
汽水分离器　trap　1
汽提　(steam)stripping　1;5.4.4
汽提段　stripper　14.3.4.19
汽提工艺水　stripped water　3.1.7.7
汽提工艺水冷却器　stripped-water cooler　3.1.7.8
汽提过程　stripping processes　3.1.7
汽提后煤浆　stripped slurry　2.2.95.27
汽提气冷却后作为浓缩水回炼　cooled stripping gas as concentrated water back to refining　2.2.72.13
汽提器　stripper　2.2.95.23
汽提塔　stripper　2.3.15.4;2.2.72.14;3.1.7.9;2.2.63.12
　stripping column　2.3.19.15;5.1.1.8
　stripping tower　5.1.1.4
汽提蒸汽　stripping steam　14.2.6.12
汽液平衡　vapor-liquid equilibrium(VLE)　1
汽液平衡比　vapor-liquid equilibrium ratio　1
汽油　ga.;gaso.;gasoline;gasolene　2.1.1.74
　gasoline　2.2.75.13
汽油和柴油　gasoline and diesel　2.2.1.21
器尾　breeching　8.9.2.4
卡口灯泡　light bulb with bayonet fitting　25.4.9.69
卡口灯座　bayonet fitting　25.4.1.65
卡子　clamp　5.3.4.30
　grip　5.3.4.29
牵紧螺母　sleeve nut　16.4.18.16
牵拉浮头式封头　pull through floating head　4.1.4.(r)
铅包皮　lead sheath　25.4.5.6

708

铅玻璃窗　lead glass window　25.2.5.3
铅封　lead seal　22.1.8.17
铅封管　lead-in pipe　25.4.5.7
铅护套　lead sheath　25.4.5.6
铅塞　lead plug　3.1.3.1
前池　forebay　3.1.3.5
前端固定封头型式　Front end stationary head types　4.1.2
前管　lead pipe　9.3.10.10
前级泵　fore-pump　5.5.2.20
前级（泵）接头　backing connection　18.2.1.16
前级管线冷凝器　backing line condenser　18.2.4.8
前级机械泵　mechanical backing pump　18.5.1.1
前级排气阀　backing stage discharge valve　18.1.2.11
前级真空泵　backing pump　18.1.2.9
前加热区　front heat zone　13.2.1.18
前进方向　travel　12.5.2.4
前壳体　front casing　16.5.5.3
前轮和驱动装置　head pulley and drive　23.2.1.9
前迷宫密封　front labyrinth seal　17.4.2.11
前弯式叶片　forward curved blade　17.3.3.10
前弯叶片　forward curved vane　16.4.5.3.3 forward-curved-blade　17.5.1.
（前）主十字头导轨　front counter guide　17.2.15.28
钳压密封　clipper seal　16.4.26.18
潜热　latent heat;lat. ht.　20.2.4.5
潜热流　Q_L(latent heat flow)　2.4.1.10
潜水泵　submersible motor pump　16.4.28
潜水（电机）泵　submersible motor pump　16.1.2.2.38
浅床　shallow bed　1
浅碟形封头　shallow dished head　21.4.1.2
嵌装式接地插座　flush-mounted earthed socket　25.4.1.5
嵌装式摇杆开关　flush-mounted rocker switch　25.4.1.4
强化喷嘴　energizing nozzles　11.5.4.8
强制润滑供油接头　force feed lubricator connection　17.2.6.6
强制通风室　plenum chamber　8.2.1.4
强制洗涤液　strong wash liquor　9.4.1.12
强制循环油泵　forced oil circulation pump　15.7.2.11
强制循环有挡板表面冷却结晶器　forced-circulation baffle surface-cooled crystallizer　7.2.1
强制循环（蒸发）结晶器　forced-circulation (evaporative) crystallizer　7.2.5
墙固定板　wall casting　24.2.6.3
羟胺　hydroxylamine　2.3.8.21
羟基化作用　hydroxylation　24.2.8
敲打器连杆　beater connecting rod　10.2.1.12
敲击机构　tapping mechanism　10.2.1.18
桥　bridge　23.4.2.11
桥式支撑机构　bridge-supported mechanism　15.6.1
壳牌煤气化炉　the shell coal gasifier　2.2.25
壳牌装置　shell unit　14.2.6.(e)
切变的介质和回流生物质　sheared media & return biomass　3.1.36.12
切断阀　shut-off valve　17.6.2.3
切换杆　changer lever　9.2.1.11
切换装置　switch station　23.1.1.9
切换装置　switch station　23.1.5.12
切碎机　shredder　3.3.2.4
切向二次空气　tangential secondary air　24.2.3.4
切削槽式管件　cut-groove type fitting　22.3.1.23
切纸机　cutter　2.3.11.44
亲和势　affinity　1
亲水分离膜　hydrophilic separatory membranes　6.1.8.1
青铜　bronze　19.3.3.(a)
青铜　bronze　22.6.2.㉒
轻部分(成分)　light fraction　15.2.4.3
轻成分聚集洗矿槽　light fraction collection launder　15.5.3.8
轻催化裂化粗柴油　light catalytic gas oil　2.1.5.9
轻的矿石粒子　light mineral ripple　15.5.3.9
轻负荷夹套式结构　light-duty jacketed construction　4.3.4.(c)
轻击分离器　flick separator　10.4.3.(b)
轻减压粗柴油　LVGO；light vacuum gas oil　2.1.4.17
轻焦化粗柴油　LCGO；light coking gas oil　2.1.6.8
轻馏出物　light distillate　2.1.4.25
轻馏分　light distillate　2.2.98.26 light ends　2.2.51.7
轻馏分油　light distillate　2.2.25.18
轻石脑油　light naphtha　2.1.1.16
轻相流　light-phase effluent　9.5.14.2

轻型钢锯　junior hacksaw　25.4.9.52
轻液　LL;light liquid　6.1.15.1
轻液出口　light liquid outlet　6.1.18.5
轻液分布管　light liquid dispersion pipe　6.1.18.12
轻液体　light liquid　6.1.14.3
轻液循环　LLR;light liquid recycle　6.1.15.3
轻油　light liquid oil　2.2.6.7
轻质矿石收集池　light mineral collection tank　15.5.3.5
轻质气　light gas　2.2.75.8
轻质汽油　light gasoline　2.2.75.18;2.2.77.9
轻质物　lt.ends　2.2.90.8
轻质页岩油　light shale oil　2.2.7.17
轻质有机氯化物　lighter organic chlorides　3.1.7
轻组分　light components　15.5.4.10
　　lights　2.3.18.21
轻组分和苯　light and benzene　6.1.3.11
氢　hydrogen　14.3.1.14
氢-煤过程　H-coal process　2.2.98
氢冷却器　hydrogen cooler　25.4.4.7
氢气　hydrogen　2.2.8.1;2.2.41.6
　H_2　2.2.117.12;2.2.109.14;2.2.68.16;2.2.33.9;2.2.43.17;2.2.61.7;2.2.67.18
氢气,电力　hydrogen,electric power　2.2.3.17
氢气和二氧化碳　hydrogen and carbon dioxide　2.2.39.c
氢气生产装置　hydrogen plant　2.2.6.21
氢气压缩机　hydrogen gas compressor　17.1.2.5
氢气,一氧化碳,二氧化碳,硫化氢,氮气等　H_2,CO,CO_2,H_2S,N_2,etc　2.2.38.4
氢循环　hydrogen recycle　2.2.6.9
氢回收　hydrogen recovery　2.2.98.16
氢氧化钙　$Ca(OH)_2$　3.4.6.7
氢氧化钠　caustic soda;sodinm hydroxide　3.1.12.6
　NaOH　2.2.41.12;3.4.6.4
氢氧化物　caustic　3.4.6.3
倾析　decantation　1
倾析器　decanter　2.2.10.10
倾斜壁卷轴卸料离心机　inclined wall scroll discharge　9.5.1.(1)
倾斜滚光壁自动卸料离心机　inclinedtumblingwall self-discharge　9.5.1.(k)
倾斜角　slope　23.2.2.8
倾斜转鼓卸料　inclined bowl discharge　9.5.2.13
清华大学液相一步法二甲醚工艺流程　tsing hua university one-step liquid phase dimethylether(DME) synthesis, 2.2.49
清洁气体出口　clean gas outlet　10.5.15.3
清理门　cleanout door　11.2.6.6
清理爪　cleaning finger　23.3.7.2
清母液　clarified M.L.　7.2.5.27
清扫　CO;cleanout　3.1.6.13
清扫口　cleanout door　3.3.12.1
清水　treated water　15.6.4.17
清水泵　clean water pump　16.1.2.3.1
清水池　clear well　2.4.5.9;3.1.41.6
清水导管　treated water conduit　15.6.4.18
清洗板　scavenger plate　9.4.4.16
清洗机　cleaning machine　2.3.12.36
清洗滤布　cloth cleaning　9.4.1.4
清洗气　purge　8.9.2.12
清洗物　scavenger　9.4.4.17
清液　clarified liquid　9.4.7.1
　clear filtrate　9.4.2.11
清液　clear liquid　5.2.6.3
清液管　washing pipe　15.6.4.3
氰化氢　HCN　2.3.15.15
氰化氢塔　HCN column　2.3.15.5
蚯蚓床　worm beds　3.3.28.8
蚯蚓采集器　earthworm harvester　3.3.28.9
蚯蚓工艺园　diagram of an earthworm process　3.3.28
球阀　ball valve;BV　22.1.7
球加热器　ball heater　3.3.16.29
球面碟形盖　spherically dished cover　21.4.1.6
球面封头　spherical cover　21.1.1.28
球磨　ball mill　3.3.16.26
球磨机　ball mill　11.6.2.17
球磨机衬里　ball mill liners　11.2.2
球墨铸铁　ductile cast iron　21.1.2.7
球体　ball　22.1.1.3
球心阀　globe valve;Gl.V.　22.1.3
球心型阀盘　globe type disc　22.1.1.2
球形度　sphericity　1
球形分离器　spherical separator　5.3.3.12
球形填料　ball packing　5.2.9.4
球形止逆阀　ball check　16.5.6.2
　ball check valve　3.5.16.30
球形贮罐　spherical storage tank　21.2.3
球轴承箱　ball bearing case　11.1.5.13
球状物　balls　2.2.5.8
曲柄　crank　16.2.4.35
曲柄键　crank key　16.2.4.37
曲柄销　crankpin　16.2.3.41
曲柄销螺母　crankpin nut　16.2.3.40

曲径式雾沫分离器　zig-zag entrainment separator　10.5.8.2
曲线形产品出口　curved product outlet　11.1.1.4
曲线砧座　curved anvil　11.1.2.2
曲折因子　tortuosity　1
曲轴　crank shaft　17.2.5.3
曲轴侧　crank end　17.2.2.8
驱动齿轮　drive gear　11.2.3.2
　driving gear　11.1.4.7
驱动磁铁　driving magnet　16.5.5.8
驱动端视图　drive end view　8.8.10.20
驱动辊　drive roll　9.6.2.9
驱动滚轮　driving rolls　12.1.1.19
驱动控制器　drive control　15.6.1.8
驱动轮　driving sheave　6.1.18.10
　driving wheel　11.1.7.23
驱动马达　drive motor　15.6.1.2
驱动器　drive　8.8.1.29
驱动器装配体　drive assembly　8.8.1.13
驱动轴　drive shaft　12.1.1.7
驱动装置　drive　13.3.3.2
　drive unit　15.6.1.32
　driver　24.1.8.4
　driver unit　7.1.11.9
驱动装置侧　driver half　16.4.9.24
驱动装置组件　drive assembly　13.3.1.8
取热器　heat remover　2.2.69.1
取样　sample　3.5.11.20
取样阀　sample valve　3.1.19.9
取样管线　sampling line　3.5.3.2
取样口　sample out　3.1.19.8
取样瓶　sample bottle　3.5.3.6
取样歧管　sample manifold　3.5.19.1
取汁龙头　broaching tap　2.3.13.14
去包衣(涂膜)　coating　13.1.1.10
去磁煤　demagnetizing coat　12.5.5.7
去磁线圈　demagnetizing coil　12.5.1.16
去除空气　air removal　3.3.29.19
去处理流程　to treatment process　3.3.20.6
去低温甲醇洗　to Rectisol process　2.2.34.22
去放空　to vent　2.2.110.21
去离子水　deionized water　6.2.7.10
去气化炉　to gasifier　2.2.35.18
去清水池　to clear well　3.1.38.11
去热交换器　to heat exchanger　2.2.110.20
去生物反应器　to bioreactor　3.1.30.A.6
去湿器　moisture separator　17.6.1.9
去脱碳塔　to decarbonization tower　2.2.49.4
去往火炬　to torch　2.2.56.9
去尾渣池　to tailings pond　2.1.9.7

去污指数　decontamination factor(DF)　1
去下道工序　to process　3.1.9.12
去硝化污水　demitrified effluent　3.1.38.10
去蒸汽轮机　to steam turbine　2.2.110.6
全部的　complete　2.2.104.9
全尺寸生物滤池设计　full-scale biofilter design　3.4.3
全回流　total reflux　1
全混　perfect mixing　1
全混发酵罐　complete mix fermenter　3.2.10.(b)
全凝器　complete condenser　2.2.57.6
全平面垫片　full face gasket　21.1.1.1
缺氧　ANX；anoxic　3.2.2.11
缺氧池　anoxic cells　3.1.34.3
缺氧/好氧消化　anoxic/aerobic digestion　3.2.5
缺氧区　anoxic zone　3.1.33.10
裙式挡板　skirt baffle　7.2.1.4
裙式支座　skirt support　21.3.1
　support skirt　21.1.1.24
裙罩　skirt　23.2.2.7
裙座　skirt support　5.2.1.7
　support skirt　21.1.1.24

R

燃尽段　burned-out zone　2.2.5.12
燃料　fuel　14.2.5.2；2.2.26.1；2.2.80.5；2.2.111.1；2.2.119.6；3.3.25.2
燃料棒　fuel rods　25.2.1.4
燃料仓库　fuel storage　25.2.2.8
燃料储存器　fuel storage　3.3.23.2
燃料管道通发电站　pipeline fuel to power plant　2.2.102.12
燃料喷枪　fuel gun　3.3.27.10
燃料气　fuel gas　2.2.97.13；2.2.43.8；2.2.45.9；3.3.23.9
燃料气体　fuel gas　14.3.3.22
燃料燃烧器　fuel burner　3.3.3.10
燃料氧化反应炉　FOR-fuel oxidation reactor　2.2.80.4
燃料油　fuel oil　2.2.91.5；2.2.79.12
燃料装卸吊车　fuel-handling hoists　25.2.2.7
燃气　fuel gas　2.2.119.12
燃气出口孔　gas-outlet holes　24.2.9.4
燃气发生器透平定子叶片　gas-generator turbine stator blades　25.3.1.13
燃气发生器透平定子叶片托架　gas-generator turbine stator blade carrier　25.3.1.14

燃气发生器透平转子　gas-generator turbine rotor　25.3.1.7
燃气发生器透平转子叶片　gas-generator turbine rotor blades　25.3.1.15
燃气阀　gas valve　24.2.6.2
燃气供入　gas supply　24.2.9.1
燃气管线　gas line　24.2.6.1
燃气环形管　gas ring　24.2.9.5
燃气进口　gas inlet　24.2.7.2
燃气轮机　gas-turbine　25.3.1；2.2.110.13；2.2.113.7；2.2.114.8；2.2.117.18
燃气轮机循环　gas-turbine cycles　25.3.2
燃气器　gas burner　8.4.1.11
燃烧产物　combustion product　24.2.10.7
燃烧段　combustion zone　2.2.82.7
燃烧焚化炉　burner incinerator　3.3.10.5
燃烧炉　burner　2.1.9.28；2.2.20.3；3.3.25.16　furnace　2.2.117.17；2.2.118.5　combustion furnace　8.8.1.22　combustion chamber　3.1.21.22　combustor　2.2.113.6；2.2.119.10；3.3.23.11
燃烧炉进料槽　burner feed tank　3.3.7.19
燃烧炉膛　furnace　3.3.3.17
燃烧气(体)　combustion gas　14.3.3.(a)
燃烧器　burner　14.3.4.36；25.2　3.3.27.13；25.1.16.3
燃烧器本体　burner body　24.2.6.9
燃烧器喉管　burner throat　24.2.6.5
燃烧器胶结水泥　burner cement　24.2.6.12
燃烧室　combustion chamber　14.3.3.17　combustor　25.3.2.19　fuel combustor　25.3.2.11
燃烧室主壳体　combustion-chamber main casing　25.3.1.12
燃烧送风机　burner ventilator　2.3.12.17
燃烧效果　combustion result　24.2.8
燃烧用空气　combustion air　25.3.2.14
燃油　fuel oil　2.3.16.5
燃油或燃气器　oil burner　8.4.1.11
绕长轴旋转的双锥混合器　double-cone revolving around long axis　12.1.3.(d)
绕线中心滑动盒　wiring center dolly box 9.3.3.20
绕线链轮　wiring sprocket　9.3.3.2
绕线式灯丝　coiled-coil filament　25.4.1.58
热　heat　2.2.114.5
热泵　heat pump　16.1.2.3.41
热成品气　warm product gas　2.2.28.4
热处理　H. T.；h. tr.；heat treatment　21.1.2.1
热传导　conduction　8.1.1.4

热电偶　TC；thermocouple　3.3.27.5；14.3.2.4
热电偶套　sleeves for thermocouples　2.2.39.1
热电偶套管　thermowell　5.4.2.11
热动力式疏水阀　thermodynamic trap　22.2.1
热反射器　heat reflectors　18.2.4.18
热废页岩　hot spent shale　2.2.5.22
热分解　PTGL processes　3.3.1.11　pyrolysis　3.3.15.3
热分解过程　thermal pyrolysis　3.3.1
热风发生器　hot air producer　8.9.3.7
热风炉　furnace　8.6.1.6
热风通入口　hot air compartment　3.3.26.6
热风室　hot air blast hole　8.5.1.6
热辐射损失　radiation loss　18.2.4.6
热负载[荷]　heat load　20.1.10.28
热工作液　hot working fluid　25.5.1.12
热固体颗粒　hot solids　2.2.4.6
热固性树脂　thermoset resin　15.7.3.(b)
热管　heat pipe　4.8
热过程(SRC & EDS)　thermal (SRC & EDS)　2.2.90.6
热合成气　hot syngas　2.2.112.2
热回收　heat recovery　2.4.3.5
热回收级　heat recovery stages　2.4.2.4
热回收蒸汽发生器　HRSG(heat recovery steam generator)　2.2.111.14；2.2.112.8；2.2.114.9
热回收装置　heat-recovery unit　2.2.4.10　thermal recovery unit　2.3.3.19
热机　heat engines　25.1.1
热交换　heat exchange　2.2.117.20
热交换夹套　heat transfer-jacketed body　12.1.3.(c)
热交换器　heat exchanger　20.1.8.2；2.2.76.13；2.3.19.3；2.4.1.4
热交换器用来预热工业废气流　heat exchange to preheat the waste gas stream　3.4.5.6
热焦炭　hot coke　14.3.4.34
热解　pyrolysis　2.2.2.22
热解气　pyrolysis gas　2.2.99.4
热解容器　pyrolysis vessel　2.2.7.5
热空气　hot air　2.2.106.12
热空气出口　hot-air out(let)　4.3.5.3
热空气导管　hot air duct　8.6.1.7
热空气进入　hot air inlet　3.1.16.13
热空气喷射器　hot air jets　8.3.2.4
热空气室　hot air box　8.4.6.5　hot-air chamber　8.8.9.9
热扩散系数　thermal diffusivity　1

热力学第二定律　second law of thermodynamics　1
热力学第一定律　first law of thermodynamics　1
(热力学)环境　surroundings　1
热力学平衡　thermodynamic equilibrium　1
热力学特性函数　thermodynamic characteristic function　1
热力学一致性检验　thermodynamic consistency test　1
热沥青　hot bitumen　2.1.9.26
热沥青储罐　hot-bitumen storage　2.1.9.25
热量　heat　4.3.5.5
thermal　2.2.107.18
热量传递　heat transfer　1
热裂化　thermal cracking　2.1.1.28
热裂化柴油　thermal diesel oil　2.1.1.31
热裂化汽油　thermal gasoline　2.1.1.30
热流体　hot fluid　4.2.1.11
热流体出口　hot fluid out　4.2.1.9
热流体进入　hot fluid in　4.2.1.7
热敏电阻探针　thermistor probes　7.2.14.8
热气出口　heated air out　24.1.13.13
热气导管　hot gas duct　3.3.24.22
热气入口　hot gas inlet　14.3.3.16
热气体　hot gas　2.2.7.14;3.3.25.4
热气体出口　hot gas outlet　4.2.4.12
热气体管道　hot gas duct　8.4.1.13
热气体生成和熔解过程　PTGL processes　3.3.1.11
热气循环甲烷转化器　hot gas recycle methanator　2.2.106.42
热球　hot balls　2.2.5.15
热燃气套管　hot gas casing　25.3.1.11
热容　heat content　20.2.4.1
热熔断器　thermal cut-out　18.2.4.9
热输送　heat transport　25.6
热水泵　hot water pump　16.1.2.3.11
热水槽　hot water tank　2.3.10.18
热水池　hotwell　18.3.6.6
热水箱　hot water tank　2.3.12.10
热塑防潮电线　thermoplastic moisture-proof cable　25.4.1.42
热塑连接器的插座板　strip of thermoplastic connectors　25.4.1.29
热塑性树脂　themroplastic resin　15.7.3.(a)
thermoplastics　22.6.2.①
热耗损　heat rejection　2.4.3.6
热耗损级　heat rejection stage　2.4.2.6
热损失　heat loss　2.4.1.9
热炭　hot char　3.3.15.19

热通量　heat flux　1
热稳定性　thermal stability　1
热烟道气　hot flue gas　2.2.5.1;2.2.111.6
热油循环系统　hot-oil circulating system　25.6.7
热源　heat source　8.5.6.3
热载体　heating medium　7.1.2.11
heat carrier　2.2.70.2
热载体泵　heat transfer media pump　16.1.2.3.42
热载体出口　medium outlet　24.1.1.17
热载体出口总管　header of thermal medium outlet　25.1.1.16
热载体分离器　heat-carrier separator　2.2.88.13
热载体机械循环　mechanical recirculation of heat carrier　3.3.15.(d)
热载体加热炉　thermal medium heater　24.1
热载体入口　medium inlet　24.1.1.4
热载体入口总管　header of thermal medium inlet　24.1.1.3
热载体蒸发器　thermal medium evaporator　7.1.12
热载体制备室　heat-carrier preparation chamber　2.2.84.9
人工吹制　hand-blown　2.3.9.41
人工吹制的高脚杯　hand-blown goblet　2.3.9.41
人工的　manual　9.5.2.19
人工启动开关　manual start switch　23.5.2.19
人工智能　artificial intelligence　1
人孔　man door　10.2.1.5
man hole;M. H.　10.1.2.26
manway　5.3.1.18
人孔法兰　manhole flange　8.9.5.1
人孔盖　manhole cover;m. c.　11.2.1.5
人孔盖板　manhole cover plate　21.1.1.29
人孔梯子　ladder to manhole　24.1.1.12
人行道　walkway　15.6.3.2
人员专用道　manway manway　25.2.2.12
人造丝　viscose rayon　2.3.7.34
人字齿轮箱　herring bone gearbox　15.7.2.10
任选的　optional　11.2.7.6
任意型式法兰　optional type flange　21.1.1.27
日光灯管　fluorescent tube　25.4.1.61
日光灯管座　bracket for fluorescent tubes　25.4.1.62
日期　data　21.1.2.16
容积　volume;Vol.　21.2.5

容积泵　positive displacement pump　16.1.2.2.1
容积传氧系数　volumetric oxygen transfer coefficient　1
容积式泵　positive displacement pump　5.5.3.11
容积式压缩机　positive displacement compressor　17.2.11
容积式　positive displacement　17.1.1.2
容量　capacity　12.1.5.17
容量控制阀　capacity-control valve　20.1.8.11
容量瓶　volumetric flask　2.3.11.3
容器　vessel　9.2.1.15
容器法兰　vessel flange　5.2.4.12
容器高　vessel hight　6.3.9.2
容器规范范围　code termination of vessel　21.1.1.4
容器加热管　vessel heating　8.9.5.8
容器内活塞流复合肥生物反应器:(a)圆筒形;(b)矩形;(c)隧道式　Plug-flow in-vessel composting bioreactors:(a)cylindrical;(b)rectangular;(c)tunnel　3.3.29
容器内的搅拌(混合)复合肥生物反应器:(a)圆形;(b)矩形　Agitated(mixed)in-vessel composting bioreactors:(a)circular;(b)rectangular　3.3.30
容器上段　vessel upper section　8.9.5.10
容器外壳　outer tank shell　11.3.1.13
容器下段　vessel lower section　8.9.5.11
容器直径　vessel diameter　6.3.9.3
溶剂　solvent　2.2.90.16
溶剂泵前储罐　solvent pumping tank　3.1.5.5
溶剂储存　solvent reservoir　3.5.22.11
溶剂萃取　solvent extraction　1
溶剂回收塔　solvent recovery column　3.1.5.12
溶剂回收蒸馏　solvent-recovery distillation　2.2.92.19
溶剂回收蒸馏釜　solvent recovery still　2.2.92.16
溶剂加氢　solvent hydrogenation　2.2.97.2
溶剂加氢　solvent hydrogenation　2.2.90.17
溶剂精制煤　solvent refined coal;SRC　2.2.2.36
溶剂精制煤产品　SRC product　2.2.93.5
溶剂精制煤产品仓　solvent-refined-coal product storage　2.2.92.21
溶剂精制煤工艺　SRC process　2.2.94
溶剂精制煤实验厂　SRC demonstration plant　2.2.93.4

溶剂汽提塔　solvent stripping column　3.1.5.6
溶剂容器　solvent container　2.2.64.6
溶剂溶解　dissolution in solvent　2.2.2.31
溶剂脱沥青　solvent deasphalting　2.1.1.36
溶剂脱油　SDO;solvent deoiling　2.1.1.82
溶剂循环　recycle solvent　3.1.5.10
溶解　dissolve;dis.　2.2.90.4
溶解槽　dissolving tank　2.3.10.37
溶解度　solubility　1
溶解器　dissolver　2.2.94.6
溶解氧分析仪　D.O.analyzer　3.1.46.9
溶解氧传感器　D.O.probe　3.1.46.11
溶解固体　dissolved solids　3.3.18.13
溶解氧量　dissolved oxygen　3.1.2.8
溶气浮选　dissolved-air flotation　3.1.1.5
溶气浮选系统　dissolved-air flotation system　3.1.14
溶脱器元件　stripper elements　15.5.2.8
溶液　solution;sol.　20.1.2.17
溶液回流　solution return　7.2.10.10
溶液聚合　solution polymerization　1
溶液流动期　solution flow period　6.2.4.7
溶液雾滴　solution sprays　13.3.3.6
溶质　solute　1
熔断器座　fuse holder　25.4.1.35
熔化玻璃　molten glass　2.3.9.50
熔化玻璃坩埚　glasshouse pot　2.3.9.46
熔化池　melting bath　2.3.9.14
熔化的锡　molten tin　2.3.9.17
熔化罐　melting pot　2.3.8.30
熔化聚酰胺　melting the polyamide　2.3.8.41
熔解的玻璃原料　god　2.3.9.40
熔融的玻璃料滴　gob of molten glass　2.3.9.23
熔融段　melting section　7.2.19.7
熔融硅石　fused silica　22.6.2.⑩
熔融化学品　molten chemical　13.5.3.1
熔融石英　fused quartz　22.6.2.⑩
熔隔石英　fused quartz　22.6.2.⑪
熔融塑料　molten plastics　13.2.1.(a)
熔融物料　melt　13.1.1
熔融物料　melt　13.5.1.14
melted burden　13.5.2.1
熔融物料加入　feed melt　13.3.4.1
熔融物料热电偶　melt thermocouple　13.2.1.11
熔融硝酸铵　ammonium nitrate melt　13.1.1.1
熔盐　molten salt　3.3.11.11
熔盐除沫器　molten-salt demister　3.3.11.8
熔盐焚烧炉结构　molten-salt incinerator

design 3.3.11
熔盐室 molten-salt chamber 3.3.11.3
熔盐位面控制器 molten-salt level control 3.3.11.2
熔渣急冷室 slag quench-chamber 2.2.14.14
熔渣气化炉 slagging gasifier 2.2.14
融化 thawing 1
柔度 flexibility 22.6.1
柔性 flexibility 1
柔性材料 flexible materials 8.8.5.19
柔性联接结构 flexible connection 23.3.1.3
柔性石墨填料 flexible graphite packing 19.2.3.13
柔性织物摩擦密封 flexible cloth rubbing seal 8.8.5.15
蠕动(泵) peristaltic 16.1.1.47
蠕动泵 peristaltic pump 16.1.2.2.65
乳化塔 packed column filled with liquid 5.1.2.9
乳化液进口 emulsion inlet 6.1.8.4
入口 inlet 10.1.2.8
入口导流叶片 suction vane 20.2.1.3
入口点 point of entrance 16.4.3.4
入口管 inlet pipe 10.2.1.31
入口管板 inlet tube sheet 10.2.1.35
入口斜槽 inlet sluice 12.5.2.8
入口叶片齿轮 suction vane gear 20.2.1.2
入炉热空气 warm air to furnace 25.5.3.13
软材料 soft materials 19.3.3.(c)
软膏状物 soft paste 8.1.1.12
软管 flexible tube 23.1.2.19
软蜡 soft wax 2.2.74.14
软填料 soft-packed 19.2.2
软填料填函箱 soft-packed stuffing box 16.4.20.18
锐孔 orifice 10.2.3.6
锐孔板 orifice 3.5.16.19
锐孔型分布器 orifice-type distributor 5.3.4.12
润滑 lubrication;lub. 22.5.3.2
润滑环密封垫 lantern gland 16.2.4.16
润滑式干式摩擦 lubricated or dry friction 8.8.5.1
润滑塑料填料 lubricated plastic packing 19.2.3.14
润滑油 LO;lubricating oil 2.1.1.77
润滑油泵 lubricating oil pump 16.1.2.3.25
润滑油料 raw lubricating oil 2.1.1.37
润滑脂杯 grease cup 9.5.4.26
润滑脂通入压盖 grease to gland 16.4.4.19

润滑脂(油)杯 grease(oil)cup 16.4.9.8
润湿周边 wetted perimeter 1
弱碱液 weak liquor 3.3.4.6

S

洒水式冷却器 water-spray cooler 8.8.8.3
塞焊 plug welds 21.1.1.20
三次空气 tertiary air 24.2.2.3
三次扩压器 tertiary augmentor 18.3.5.1
三缸 triplex 16.1.1.10
三缸立式压缩机 triplex vertical compressor 17.2.1.4
三辊榨糖机 three-roll sugar mill 9.6.1
三级缸 third stage cylinder 17.2.3.3
三级冷凝器 three-stage condenser 2.3.5.9
三级流化床造粒机 three-stage fluidized bed granulator 13.4.1.(c)
三甲苯 trimethylbenzene 2.1.1.61
三角连接 delta connection 25.4.3.3
三脚插头 three-pin plug 25.4.1.67
三脚架 tripod 2.3.11.6
三列 triplex 16.1.1.10
三螺杆泵 Three-screw pump 16.3.3
三氯化铁 $FeCl_3$ 3.1.34.16
三通阀 3-way valve 23.1.1.4
diverter valve 12.1.5.5
三通滑槽 three-way chute 2.3.12.34
三通旋塞 3-way stop cock 3.5.14.32
三通插头 three-phase plug 25.4.1.14
三相点 triple point 1
三相发电机 three-phase generator 25.4.4.6
(三相)高压电缆 high voltage cable(for three-phase current) 25.4.8
三相流态化 three-phase fluidization 1
三芯电线 three-core cable 25.4.1.44
三眼插座 three-pin socket 25.4.1.66
三足式转鼓离心机 three-column basket centrifuges 9.5.3
伞齿轮 bevel gear 13.3.5.14
伞形喷嘴 bevel jet 18.2.3.1
散式流化床 particularly fluidized bed 14.1.3.9
散式流态化 particulate fluidization 1
散装仓 bulk storage 23.5.1.1
散装储存货架 bulk STG 2.3.17.14
散装填料 dumped packings 5.3.2
散装运输车辆 bulk transporting vehicle 2.3.16.15
散热 thermal insulation 18.2.4.11

散热片式冷却器　plate-fin cooler　4.2.5.4
扫掠蒸气　sweep vapor　15.4.1.5
扫掠蒸气锅炉　sweep vapor boiler　15.4.1.1
刹车片　brake lining　8.8.5.7
砂　sand　2.1.9.12

砂封　sand seal　3.3.26.11
砂浆产品　S;sand product　12.4.1
砂粒过滤器　sand filter　3.1.1.7;3.1.30.B.9
砂滤器　sand table　2.3.11.13
　　sand-bed filter　1
砂箱　grit chamber　3.1.25.3;3.1.27.4
砂子　sand　3.3.23.7
砂子和木炭　sand & char　3.3.23.8
砂子进口　sand feed　3.3.27.6
砂子流化床　fluidized sand bed　3.3.27.7
筛　screen;Scr.　3.3.16.24
D.S.M.筛　D.S.M.screen　11.6.3.8
筛板　perforated plate　6.1.16.2
　　perforated tray　5.2.10.6
　　screen deck　12.2.1.5
　　screen plate　12.2.3.16
　　sieve　5.2.6
　　sieve tray　5.2.10.5
筛板固定环　screen retaining ring　9.5.10.5
筛板塔　perforated-plate column　6.1.16.2(a)
　　sieve tower　3.3.8.10
　　sieve tray column　5.1.2.3
筛分　screen analysis　1
　　screening　3.3.16.9
　　screen　13.3.5.4
筛分机械　screening machine　12.2
筛架　screen frame　12.2.5.7
筛浆机　screen　2.3.10.57
筛孔板　sieve plate　10.5.1.12
筛孔直径　sieve diameter　1
筛上物料　oversize　12.7.2.14
筛　screen　2.4.5.1;3.1.28.2;3.1.30.A.3;3.1.30.B.3;10.2.1.7
筛网分离器　mesh separator　7.1.14.8;7.2.1.21
筛网击打器　screen beater　10.2.1.13
筛网洗涤液　mesh wash　7.1.14.7
筛析　screen analysis　1
筛下物料　undersize　12.7.2.13
筛选机组　screening unit　13.2.4.17
筛摇架　screen cradle　12.2.3.14
筛余粗料　rejects　11.6.4.13
筛子　screen　2.2.30.11
筛组　screen pack　13.2.3.1

栅板　grate　2.2.12.8
栅筐　grate basket　11.1.7.19
栅筛　grizzly　12.8.2.4
栅条　grid bars　11.1.7.15
　　grids　10.2.1.15
栅条架　grid frames　11.1.7.14
闪急干燥器　flash dryer　8.5.7
Strong-Scott闪急干燥器　Strong-Scott flash dryer Strong-Scott　8.4.2
闪蒸　flash　2.2.92.12;2.2.115.11
　　flash evaporation　1
闪蒸槽　flash tank　2.1.10.21;2.2.44.12;2.2.46.9;2.2.47.8
闪蒸槽蒸发器　flash tank-evaporator　20.1.10.25
闪蒸分离器　flash separator　2.2.98.20
闪蒸罐　flash drum　2.1.4.8
　　flash tank　2.2.56.2
闪蒸气　flash gas　20.1.3.9
　　flashed vapour　2.2.55.9
　　flash steam　2.2.47.15;2.2.44.16;2.2.46.17
闪蒸气体压缩机　flash gas compressor　2.2.28.10
闪蒸室　flash chamber　7.1.8.6
扇尾式　fantail　24.2.2.1
扇形　segmented　19.3.1.10
扇型往复压缩机　semiradial reciprocating compressor　17.2.1.8
商用分离元件管　commercial separation-element tube　15.4.3
熵产生　entropy generation　1
熵增原理　principle of entropy increase　1
上半部　upper half　16.9.13
上部跟踪辊　upper tracking roll　9.6.2.7
上部空气分布罩　upper air hood　4.2.4.7
上部滤带冲洗　upper belt wash　9.6.2.6
上部滤带张紧器　upper belt tensioning　9.6.2.8
上部喷嘴　upper jet　18.2.1.4
上部气室　upper plenum　10.2.3.14
上部提升筒　upper lift drum　6.3.11.19
上部压紧胶辊　nipping roll(upper)　15.7.5.8
上部圆筒　top bowl　16.4.25.11
上部圆筒管　top column pipe　16.4.25.22
上部圆筒连接器支座　top bowl connector bearing　16.4.25.15
上部圆筒支架　top bowl bearing　16.4.25.13
上部重块　upper weight　11.3.1.12
上部轴　top shaft　16.4.25.25
上部轴套管　top shaft tube　16.4.25.32

上层构架　superstructure　15.6.1.1
上层卵石床　upper pebble bed　14.3.3.18
上层筛　upper screen　12.2.3.6
上垫片　upper gasket　16.4.27.16
上端盖　upper end cover　16.4.28.9
上段轴　upper shaft　14.4.3.8
上浮产品　float product　12.5.5.26
上浮物　float　12.5.2.13
上浮物料排出　float-material drain　12.5.5.19
上盖　upper case　11.1.5.10
upper head　16.2.7.3
上盖板　top cap　9.6.1.13
上盖开关　top cap key　9.6.1.14
上辊法兰　top roll flange　9.6.1.15
上（滑动）轴承　upper metal　9.5.4.31
上极板　top crown　21.2.3.4
上壳体　upper casing　16.4.12.20
上流式流化床处理系统流程图　Flow diagram of upflow fluidized-bed system　3.1.36
上流式流化床去硝化　upflow fluidized-bed denitrification　3.1.37.9
上喷嘴密封帽　sealed top jet cap　18.2.4.3
上倾式出风　top angular down　17.5.2.8
上清液　supernatant　3.2.2.16；3.3.17.4；3.3.18.8
上清液出料　supernatant withdrawal　3.2.6.8
上清液回厂　supernatant return to plant　3.3.21.10
上清液去除　supernatant removal　3.3.18.10
上栅板　top grating　5.2.9.3
上升管　riser pipes　4.4.2.9
upstack　11.5.2.5
上水平出风　top horizontal　17.5.2.7
上套筒　upper sleeve　16.2.5.19
上套筒环　upper sleeve ring　16.4.27.37
上筒　top bowl　16.4.26.6
上筒轴瓦　top bowl bearing　16.4.26.7
上涂料　sizing　2.3.9.53
上托架　bracket　16.4.28.24
上（晚）壳灰岩层　upper muschelkalk　25.2.4.2
上悬式转鼓离心机　top-suspended basket centrifugal　9.5.6
上仰角出风　top angular up　17.5.2.6
上一层塔板　plate above　5.2.6.1
上圆筒　upper cylinder　10.1.2.9
上圆锥　upper cone　10.1.2.16
上轴承　bearing(upper)　12.3.5.1

top bearing　4.2.4.2
上轴承　upper bearing　9.2.2.4
上轴承盖　top bearing cover　12.3.5.30
upper ball cover　13.3.5.10
上轴承螺母　top bearing shaft nut　12.3.5.32
上轴承锁紧螺母　top bearing nut lock　12.3.5.31
上轴承调节螺母　top bearing adjusting nut　12.3.5.3
上轴承箱　top bearing housing　12.3.5.4
上轴套　upper shaft sleeve　16.4.27.24
烧碱　caustic soda　2.3.7.3
烧碱塔　caustic tower　2.3.18.19
烧结　sintering　1
烧结的多孔不锈钢板　porous sintered stainless plate　4.3.3.27
烧瓶　flask　3.5.14.26
烧瓶阀　flask valve　3.5.14.19
烧瓶屏罩　flask shield　3.5.14.27
烧嘴　burner　24.1.11.2；2.2.26.3；3.3.24.19
勺式温度计　ladle-type thermometer　2.3.12.53
构斗　dipper　3.5.10.14
构斗系统　dipper systems　3.5.10.(d)
少量的 CO，CO_2，N_2，H_2　little amount of CO，CO_2，N_2，H_2　2.2.38.9
舌形板　jet tray　1
舌形塔盘　jet tray　5.2.10.15
蛇管式换热器　spiral-tube exchanger　4.2.5.10
舍弃的蚯蚓可用于土地利用的土壤改良剂　castings for land untilization as soil amendment　3.3.28.12
设备（巴氏消毒器）　device(pasteurizer)　2.4.1.8
设备法兰　vessel flange　5.2.4.12
设备远距控制　FRC；facility remote control　6.3.3.12
设备支座　structural support　10.2.1.21
设计温度　design temperature　21.1.2.9
设计压力　design pressure　21.1.2.9
O 射流　O jet　11.5.2.11
P 射流　P jet　11.5.2.2
射流反应器　jet reactor　1
射流技术　fluidics　17.6.3.20
射流孔　jet orifice　11.5.1.11
射流器进料口　feed injector　11.5.1.5
射流轴线　jet axes　11.5.1.3
X 射线源　X-ray source　12.7.5.3
射线照相检验　radiographic exam　21.1.2.11
伸缩式装料长杆阀　retractable filling lance

717

velve 23.4.1.19
深翅片式 deep-pinned type 4.3.2.(d)
深沟球轴承 deep groove ball bearing 16.4.20.15
深井泵 borehole pump 16.1.2.2.40
deep well pump 16.1.2.2.39
cryogenic pump 16.1.2.1.10
深埋坑 burial pit 3.3.14
深水层 deep water bearing strata 3.1.24.16
渗滤 percolation 1
渗透系数 osmotic coefficient 1
渗透液 permeate 15.2.1.12
渗透液流出 permeate out 15.2.2.2
渗析 dialysis 1
渗余物 retentate 1
渗余液流出 retentate out 15.2.2.3
升程限制器 stop plate 17.2.12.9
升华 sublimation 15.1
升华器 sublimer 15.1.1.2
升华顶储罐 lifter-roof 21.2.1.4
升降刮刀 rising knife 9.5.2.20
升降刮刀自动离心机 automatic basket, rising knife 9.5.1.(d)
升降机 vertical lift 2.3.8.36
升降螺杆 lifting screw 22.1.3.4
升膜蒸发器 climbing-film evaporator; upward-flow evaporator 7.1.9
升气管 vapor riser 5.2.8.6
升压喷射器 booster ejector 20.1.10.22
生产规模 scale of production 8.1.1 (c)
生产过程料仓 process bins 23.1.8.23
生产(能)力 plant capacity 3.1.6
生产排序 scheduling of production 1
生长因子 growth factor 1
生成热 heat of formation 1
生成水 product water 3.1.18.8
生化操作 biochemical operation 3.2.1.8
生化分离 biochemical separation 1
生化工程 biochemical engineering 1
生胶混炼机 Banbury mixer 15.7.1
生物催化反应 biocatalytic reaction 1
生物电池 biocell 1
生物电池提升工段 bio-cell lift station 3.1.44.2
生物反应器 bioreactor 3.2.10.4;3.1.28.5;3.1.30.A.4;3.1.30.B.4;3.1.31.4
生物反应器外设膜组件的膜生物反应器系统 MBR (membrane bioreactor) process system with membrane module situated out side the bioreactor 3.1.28
生物曝气过滤系统流程图 Flow diagram for a biological aerated filter(BAF) system 3.1.41
生物工程 bioengineering 1
生物固体流化床焚烧炉截面图 Cross section of fluidized bed biosolids incinerator 3.3.27
生物过滤(BF$_s$) biofiltration(BF$_s$) 3.1.25.8
生物化学废水处理系统 biochemical wastewater treatment systems 3.2
生物化学转化 biochemical conversion 3.3.1.5
生物技术 biotechnology 1
生物降解 biodegradation 3.1.32.3
生物粒子 bioparticle 3.2.11.17
生物流动床—颗粒活性炭系统 Fluidized bed biological(FBB)—GAC(granular activated carbon)system 3.1.42
生物膜 biofilm 3.2.11.18
biological film 3.1.47.1
生物曝气过滤/生物炭系统 BAF/biocarbone system 3.1.41.5
生物气 biogas 3.2.6.11
生物制剂 biological agent 1
生物质 biomass 3.2.1.9;2.2.3.7;2.2.31.1;3.1.42.9;3.3.23.1
生物质燃料 biomass fuels 25.1.1.3
生物质收集系统 biomass harvesting system 3.1.42.7
生物转化 biotransformation 1
生长控制泵 growth control pump 3.1.36.9
生长控制装置 growth control device 3.1.36.8
声响式 sonic 3.5.18.(f)
绳索 coil 9.3.7.4
绳索过滤机 coil filter 9.3.6
绳索绕面转鼓 strings returning to drum 9.3.5.6
绳索卸料过滤机 string-discharge filter 9.3.5
省煤[热]器 econ.;economizer 2.2.81.13
剩余的生物固体 excess biosolids 3.1.31.9
剩余的污泥（减少了） excess sluage (reduced) 3.1.30.A.8
剩余焓 residual enthalpy 1
剩余冷凝液去深井 excess condensate to deep well 2.3.6.31
剩余沥青 residual bitumen 2.1.9.5
剩余酸性气体 remaining acid gas 2.2.89.7
剩余炭 residual char 2.2.107.1
剩余样品回流管线 excess sample return line 3.5.3.8

剩余蒸汽　surplus steam　2.1.8.7
失活　deactivation　1
失效活性炭储仓　spent carbon storage　3.1.17.4
失效活性炭增稠器　spent carbon thickener　3.1.18.9
施密特数　Schmidt number　1
施真空　vacuum applied　9.3.10.8
湿壁降膜吸收塔　Falling film type absorber　5.4.2
湿壁塔　wetted wall column　5.1.2.14
湿磁鼓分离器　wet drum separator　12.6.9
湿度控制器　humidity controller　20.2.2.5
湿法涤气器　wet scrubbing　2.2.31.21
湿含量测定系统　Moisture-sample train　3.5.7
湿灰　wet ash　2.3.1.34
湿式旋风分离器　wet-stage cyclone　8.4.1.35
湿加料器　wet feeder　8.4.1.34
湿进料　wet feed　8.4.1.33
湿进料混合器　wet feed mixer　8.4.1.2
湿井　wet well　3.1.8.5
湿空气氧化作用　wet-air oxidation　3.1.23
湿煤　wet-feed　8.8.9.1
湿煤仓　wet-coal bin　8.5.6.26
湿气体　wet gas　2.1.5.11
湿球温度　wet bulb temperature;WB　1
湿区　wet zone　20.2.4.8
湿式涤气塔　wet scrubber　2.2.118.13
湿式分级器　wet classifiers　12.4
湿式分级设备　wet classification machine　12.4.1
湿式机器　wet machine　12.7.4.1
湿式静电除尘器　wet cottrell　10.3.4
湿水蒸气进入　wet steam in　10.4.3.1
湿污泥输送机　wet sludge conveyor　3.3.24.4
湿物料　wet feed　8.1.1.8
　　wet material　8.3.1.1
湿物料加[进]入　wet feed　8.4.5.1
湿物料进[入]口　wet product inlet　8.9.6.9
湿氧化实验室设备　wet oxidation lab equipment　3.4.7
湿蒸汽　wet steam[vapor]　10.1.4.13
十二烷基苯　dodecylbenzene　14.3.1.27
十二烷基苯磺酸盐　dodecylbenzene sulfonate　14.3.1.30
十字格环　cross partition ring　5.3.2.(b)
十字流过滤　cross-flow filtration　9.4.9
十字头　crosshead　16.2.5.22
十字头侧金属填料　metallic packing of crosshead side　19.2.1.9
十字头衬瓦　crosshead liner　17.2.9.27
十字头滑道　crosshead guide　16.2.6.6
十字头滑块板　crosshead slipper with lower frame　17.2.15.17
十字头螺栓螺母　crosshead bolts with nuts　17.2.15.6
十字头上体　upper crosshead frame　17.2.15.8
十字头销　crosshead pin　17.2.15.24
十字头组件　complete crosshead frame　17.2.15.5
十字支架　cross stand　16.2.3.42
石膏　gypsum　2.2.115.19
石灰石　limestone　25.5.3.9
石灰石加料管　limestone feed pipe　14.2.2.2
石灰石煤浆　coal/limestone slurry　2.2.116.4
石灰石,水　limestone,water　2.2.115.21
Fluosolids石灰窑　Fluosolids lime kiln　14.2.5
石蜡　paraffin wax　2.1.1.79
$C_{20}\sim C_{30}$(石蜡级重油)　$C_{20}\sim C_{30}$(paraffin level heavy oil)　2.2.73.10
石棉绳　asbestos cord　5.2.4.11
石棉水泥　asbestos cement　22.6.2.①
石棉填料　asbestos packing　19.2.3.16
石棉橡胶盘根(填)　asbestos rubber packing　19.2.3.15
石墨泵　graphite pump　16.1.2.5.8
石墨活塞环　carbon piston ring　17.2.10.2
石墨块体换热器　graphite-block exchanger　4.2.5.6
石墨密封块　carbon seal　19.3.1.7
HSR石脑油　HSR naphtha　2.1.4.14
LSR石脑油　LSR naphtha　2.1.4.16
石脑油　naphtha　2.1.1.6;2.2.79.10;2.2.5.11;2.2.95.8;2.2.103.5
$C_5\sim C_{12}$(石脑油及汽油)　$C_5\sim C_{12}$(naphtha and gasoline)　2.2.73.8
石脑油进口　naphtha feed　14.3.4.25
石脑油进料　naphtha feed　14.2.6.15
石脑油馏分　naphtha fraction　2.2.93.3
石英　quartz　3.5.14.4
石英喷嘴　quartz jet　3.5.21.3
石油化工产品　chemicals from petroleum　1
石油化工基本原料　petrochemical building block　2.2.91
石油加工　petrochemical processing　2.1
石油焦　petrol coke　2.1.1.46
　　petcoke　2.2.3.6
石油馏分转化　conversion of petroleum

fractions 14.3.4
石油气 petrogas 2.1.1.29
石油钻探 crude oil drilling 2.1.3
石子介质 stone media 3.1.47.4
实心轴 solid shaft 9.4.9.9
实验流程 the flow sheet of the experimental 8.5.3
食品工业 food industry 17.6.3.26
食物 food 3.2.11.11
示意图 schematic diagram 12.6.14
sketch 13.3.1
事先警告 early warning 3.1.2.2
试剂管 chemical pipes 15.6.3.1
试剂加入管 chemical feed piping 15.6.3.12
试压泵 pressure test pump 16.1.2.3.49
视镜 sight glass 15.5.2.5;3.3.27.14
视孔 peep hole 14.4.1.4
sight hole 24.1.1.6
视密度 apparent density 1
适度间隙 moderate clearance 19.3.1.4
适合在厂内循环/再利用 suitable for recycle/reuse within the plant 3.1.30.A.9
适应性 flexibility 1
suitability 8.1.1 (d)
室 chamber 9.4.4.1
室间密封装置 compartment seal 9.3.7.6
室内 room 20.2.2.3
室内冷凝液排出口 chamber condensate drain 5.5.2.18
室排水口 chamber drain 8.2.2.8
室外空气 O.A;outdoor air 20.2.2.24
室外空气湿球恒温极限值 outdoor air W.B. limit thermostat 20.2.2.11
释放器 disengager 2.3.5.7
收尘板 collector plates 10.3.1.28
收集单元 collecting unit 10.3.1.22
收集电极 collecting electrode 10.3.1.17
收集器 accumulator 2.2.5.20
collector;ocllr 23.1.1.10
收集器的移动 collector travel 15.6.1.28
收集室 collector housing 10.2.3.10
收集型再分布器 wiper redistributors 5.3.4.15
收集轴 collector shaft 15.2.3.9
收率 yield 1
收缩板 constriction plate 14.1.12.4
收缩截面 converging section 3.5.4.5
手柄 grip 11.1.8.4
handle 12.3.4.9
手动泵 hand pump;HP 16.1.2.4.5
手动操作间歇式 batch manual 9.5.2.4

手动干料分流器 manual dry divider 3.3.24.5
手动压紧丝杠 hand screw 9.1.2.8
手动闸门调节螺旋 manual gate-adjustment screw 23.5.4.18
手工顶部卸料式 with manual top discharge 9.5.3.(c)
手工造纸 papermaking by hand 2.3.11.46~51
手孔盖 handhole cover 16.4.9.30
手轮 hand wheel 22.1.2.12
handle 15.7.4.17
手轮螺母 wheel nut 22.1.3.1
手摇泵 hand pump 16.1.2.4.5
手摇高压泵 high pressure manual pump 16.1.2.4.9
守时装置 timing unit 9.5.13.7
受料槽 vat 12.2.3.20
受料筐板条 bars for receiving basket 11.1.7.29
受料筐端部板条 end bars for receiving basket 11.1.7.30
受内压半球形封头 int.pressure hemispherical head 21.1.1.39
受内压碟形封头 int.pressure torispherical head 21.1.1.63
受内压椭圆形封头 int pressure ellipsoidal head 21.1.1.6
受内压锥形封头 int.pressure conical head 21.1.1.57
受外压碟形封头 ext.pressure torispherical head 21.1.1.63
受外压折边锥形封头 ext.pressure toriconical head 21.1.1.25
受压半球形封头 pressure hemispherical head 21.1.1.39
受压容器 pressure vessels 21.1.2.10
受压外壳 pressure shell 2.2.14.4
受液盘 receiving pan 5.2.1.17
seal pot 5.2.2.6
受液器 receiver 20.1.9.3
枢轴刮片 pivoted flight 15.6.1.37
枢轴装配体 trunnion roll assembly 8.8.1.10
梳齿座 comb 19.3.2.10
疏水泵 drainage pump 16.1.2.3.10
疏水阀 trap 7.1.2.15
疏水阀排水 trap to drain 3.1.6.1
疏水器 trap 1
输出螺旋 discharge screw 8.4.6.15
输出设备 output device 3.5.20.7
输电系统 discharge system 10.3.1.5

输电线路　transmission line transmission line　25.2.1.10
输送　delivery　2.3.9.37
transportation　3.3.7.12
输送表面　conveying surface　8.9.6.2
输送槽　trough　23.3.1.5
输送床　transport　14.1.4.(g)
输送床反应器　entrained-bed reactor　3.3.15.(d)
输送床气化炉　the transport gasifier　2.2.24
输送带　conveyor (belt)　12.7.4.6
输送反应器　transport reactor　3.3.15.(d)
(输送)分离高度　TDH; transport disengaging height　1
输送风机　conveying fan　8.6.2.10
Exxon输送管催化裂化装置　Exxon transfer-line catalytic-cracking unit　14.2.9
输送管线　conveying line　23.1.2.35
输送和传热表面　conveying and heat-transfer surface　4.3.4.5
输送机　accumulator conveyor　23.4.3.5
conveyor　3.3.2.3
输送机用电机　motor for conveyer　8.3.4.19
输送机用减速机　reducer for conveyer　8.3.4.18
输送皮带　belt　8.5.6.24
输送器　conveyor　8.7.1.8
输送器驱动装置　conveyor drive　9.5.5.1
输送设备　conveying; equipment　23
输送式提升机　conveyor-elevator　23.3.3.(b)
输送系统　conveying system　8.9.4.4
输液管　transfer pipe　3.1.10.6
熟化　maturing　2.3.7.7
熟化层　curing floor　2.3.13.18
熟化窑　curing kiln　2.3.12.30
鼠洞　rathdes　23.1.3
树胶球　gum ball　12.2.5.8
树脂　resin　13.2.1.2
树脂泵　resin pump　6.2.4.4
树脂床　resin bed　6.2.2.(b)
树脂阀　resin valves　6.2.4.5
树脂混合　resin mixing　6.2.9.(e)
树脂粒子　resin particles　6.3.3.3
树脂料斗　resin hopper　6.2.5.6
树脂移动期　resin movement period　6.2.4.9
树脂转移路线　resin transfer lines　6.2.6.7
竖的引管排水口　drip leg drain　17.6.2.10
竖管　standpipe　2.1.2.13
竖式炉　shaft furnace　3.3.15.(a)
竖式石灰窑　shaft furnace for lime production　24.1.12

数据记录系统　data recording system　3.5.19.10
数据进口压轮　data-entry thumb wheels　23.4.1.16
数据收集　data acquisition　3.5.18.13
数字控制器(温度,压力和搅拌速度)　digital controller (temperature, pressure and stirring speed)　3.4.7.2
数字式质量选定机和显示器　digital weight selector & display　23.4.1.17
甩板泵　swash plate pump　16.1.2.2.54
甩油环　oil slinger　19.3.2.7
甩油环甩油环　thrower　16.4.20.26
双瓣阀　double flap valve　3.3.24.23
双层筛分离器　two-deck screen separator　12.2.2
双程壳体带纵向隔板　two pass shell with longitudinal baffle　4.1.2.(g)
双床脱离子系统　two-bed deionizing system　6.2.1
双床TSA系统　two-bed TSA system　6.3.4
双端面密封　double mechanical seal　19.1.8
双阀　double valves　16.1.1.36
双分隔流动　double split flow　4.1.2.(i)
双缸　duplex　16.1.1.9
双缸立式往复压缩机　duplex vertical reciprocating compressor　17.2.1.3
双缸往复式压缩机　duplex reciprocating compressor　17.2.9
双鼓干燥器　double-drum dryer　8.7.2
双辊破碎给料机　double roll feed crusher　11.2.6.2
双滚筒磨式　double-drum mill　4.3.1.(c)
双滚筒真空干燥器　vacuum double drum dryer　8.7.1.(e)
双活塞泵　duplex piston pump　6.3.2.16
双级合成甲醇工艺流程示意　Flow chart for two-stage synthesis of methanol　2.2.47
双级内旋风分离器　two-stage internal cyclone　14.1.5.(b)
双级外旋风分离器　two-stage external cyclone　14.1.5.(d)
双级旋片式真空泵　double stage rotating blade vacuum pump　18.1.3
双浆混合器[机]　double paddle mixer　8.4.5.18
DP(双浆)型结晶器　Escher-Wyss or Tsukushima DP (double propeller) crystallizer　7.2.10
双壳球罐　spherical double-wall tank　21.2.2.4

双列　duplex　16.1.1.9
双列对称平衡压缩机　double row balanced opposed compressor　17.2.1.9
双列刮板输送机　flight chain conveyer; double lined　15.6.5.9
双流塔盘　two pass tray　5.2.3
双螺杆泵　two-screw pump　16.3.2
双螺杆加料器　twin-screw feeder　12.1.5.15
双螺杆压缩机　double screw compressor　17.3.3.1.(d)
双螺旋环　double spiral ring　5.3.2.(c)
双螺旋搅拌器　double spiral agitator　8.8.10.8
双膜理论　two-film theory　1
双缺氧区系统　dual anoxic zone system　3.1.35
双台式　double deck　23.3.2.9
双筒混合器(V型)　twin shell(vee)mixer　12.1.3.(b)
双头螺栓　SB;stud bolt;stud　19.2.2.6
双头螺柱　stud end　13.1.2.2
双蜗壳　double volute　16.4.9.31 double-volute casing　16.4.14
双吸　double suction　16.1.1.35
双吸离心泵　double suction centrifugal pump　16.4.12
双吸排水泵　double-suction wet-pit pump　16.4.23
双吸叶轮　double-suction impeller　16.4.8
双吸叶轮部件　parts of a double-suction impeller　16.4.6
双向分流器　two-way diverter　23.1.6.6
双叶型旋转正位移式(容积式)鼓风机　two-impeller type of rotary positive-displacement blower　17.3.5
双闸进料器　double-gate airlock　23.1.7.9
双支管弯头　double branch elbow　22.3.1.12
双直径式　two diameter　8.8.5.(b)
双轴布置　two-shaft arrangement　25.3.3
双轴锤式破碎机　Double shaft hammer crusher　11.1.7
双转子　two rotor　17.1.1.7
双转子混合器　twin rotor　12.1.3.(i)
双锥混合器　double cone mixer　12.1.3.(a)
双锥空气分级器　double-cone air classifier　12.3.3
双锥形混合器　double cone mixer　12.1.4.(f)
双锥型回转式真空干燥器　rotating(double-cone)vacuum dryer　8.8.11
双作用　double acting　16.1.1.6
双作用泵缸　double-acting liquid end　16.2.2

双作用循环搅拌叶片　double-acting circulation propeller　7.2.10.4
双作用压缩机气缸与活塞　double-acting piston and compressor cylinder　17.2.6
双作用蒸汽往复泵　duplex acting steam-driven reciprocating pump　16.2.3
水　water　2.1.9.8;2.2.18.6;2.2.27.2;2.2.28.7;2.2.29.1;2.2.30.14;2.2.32.11;2.2.33.2;2.2.57.16;2.2.64.15;2.2.68.7;2.2.74.6;2.2.75.7;2.2.76.20;2.2.77.12;2.2.80.13;2.2.115.15;2.3.17.12;2.3.18.10;3.3.19.1;
水　H_2O　2.2.63.10;2.2.65.4;2.2.70.5;2.2.71.11;2.2.114.15;3.4.6.9;
(水)泵房　pump room　2.3.12.11
水表面　water surface　3.5.4.4
水槽　water tank　4.3.2.5
水层　agueous layer　2.2.60.3
水厂泵　waterworks pump　16.1.2.3.6
水出口　water discharge　18.3.4.20 water out(let)　10.5.4.2
水处理过程概述　An over view on water treatment process　2.4.5
水锤　water hammer　1
水滴　water droplet　10.5.1.10
水分布器　water distributor　5.4.2.15
水分离器　moisture trap　9.3.1.7
水分离塔　water splitter　2.2.28.5
水分分离器　moisture separator　17.6.1.9
水分控制　moisture control　3.3.16.31
水封　barometric seal　9.3.1.10 water seal　6.2.3.10
水封管　sealing water pipe　16.4.19.19
水封管接头　sealing water pipe connection　16.4.19.20
水管　water tubes　20.1.10.6
水管线　water line　25.6.6.3
水环泵　water-ring pump　18.4.1
水环真空泵　water-ring vacuum pump　18.4
水灰　water ash　3.3.8.11
水或营养物进料　water or nutrient feed　3.4.3.9
水加入　water in　7.2.8.6
水加入口　feed water inlet　25.5.1.6
水夹套　water jacket;jw;jacket of water;Wat.jack　2.2.12.12
水检测探头　water detection probes　15.5.2.2
水解反应器　hydrolysis reactor　2.2.62.10
水进口　water in　20.1.7.14
水冷反应器　water-cooled reactor　2.2.42.7
水冷凝器　water condenser　2.2.80.11
水冷器　water cooler　2.2.44.3

水冷却　water-cooled　13.5.3.2
水冷式单级双作用压缩机　single-stage, double-acting water-cooled compressor　17.2.5
水冷式合成反应器　water-cooled synthesis reactor　2.2.43.10
水冷式填料压盖　water cooled gland　16.4.11.12
水冷塔　water cooling tower　2.2.47.4;2.2.46.4
水冷稳定板　water cooling stabilizer panel　15.7.5.7
水力半径　hydraulic radius　1
水力头　hydraulic head　6.1.7.3
水力旋流分离器　hydrocyclone　2.2.18.8
水力旋流器　hydroclone　2.2.98.25
水流　stream flow　3.5.5
水流方向　water direction　15.6.5.16
水炉渣混合浆　slag/water slurry　2.2.22.8
水炉渣浆　slag/water slurry　2.2.21.7
水煤浆　coal/water slurry　2.2.29.8
水煤浆,氧气　coal/water slurry, oxygen　2.2.17.1
水煤气变换炉　WGS　2.2.117.9
水煤气废热锅炉　water gas waste heat boiler Ⅰ　2.2.34.1;2.2.34.8
水膜　water film　4.3.2.1
水泥　concrete　3.1.24.5
水泥槽　concrete tank　15.6.1.17
水泥壳泵　concrete casing pump　16.1.2.5.4
水泥窑　cement kiln　3.3.1.14
水盘　water pond　10.3.4.6
水喷淋或空气喷雾　water shower or air spray　3.3.19.8
水喷射泵　water jet pump;WJP　16.1.2.4.3
水喷头　water spray　14.2.4.5
Sparkler水平板式过滤机　Sparkler horizontal plate filter　9.4.6
水平层叠的元件　horizontal element stacks　15.5.2
水平底面　level floor　3.5.4.2
水平管式蒸发器　horizontal-tube evaporator　7.1.7
水平管束　horizontal tube bundle　7.1.3.14
水平环板　level ring　9.5.12.4
水平回转式　horizontally gyrated　12.2.3
水平坑道　level　25.2.5.1
水平面　water level　15.6.4.19
水平倾斜输送式提升机　horizontal-inclined conveyor-elevator　23.3.3.(e)
水平输送机　horizontal conveyor　23.3.3.(a)
水平轴　horizontal axis　3.5.1.5
水平轴流速计　horizontal axis current meter　3.5.1.(b)
水平贮槽式　horizon tal-tank type　4.3.1.(a)
水平转鼓混合器　horizontal drum mixer　12.1.3.(c)
水气变换　water gas shift　2.2.3.14
水汽壁　waterwalls　25.5.3.7
水去循环　water to recycle　2.2.37.5
水入口　water inlet　10.5.4.1
水,石脑油等　water and naphtha, etc　2.2.36.13
水竖管　water leg　17.6.2.9
水塔　water column　2.2.52.8
水体　water body　25.1.1.13
水位　water level　24.1.5.7
水位记录仪　water-level recorders　3.5.5
水位探测器　water level detector　15.5.5.6
水洗　scrubbing　1
washing　2.3.7.19
水洗塔　water scrubber　2.2.69.8;2.2.72.10
水下泵　underwater pump　16.1.2.2.36
水下电机泵　underwater motor pump　16.1.2.2.37
水相　aqueous　6.1.12.1
water-phase　6.1.8.7
水相萃残液　aqueous raffinate　6.1.2.2
水相萃取液　aqueous strip solution　6.1.2.5
水相料液　aqueous feed　6.1.12.8
水相流出物　aqueous effluent　6.1.12.13
水循环　water recycle　2.2.27.4
水,乙酸　H_2O,AcOH　2.2.41.10
水银泵　mercury pump　16.1.2.3.32
水银扩散泵　mercury diffusion pump　16.1.2.2.49
水蒸气　steam;STM　2.1.5.5;2.4.3.1
steam　2.3.6.22
water vapor　7.1.14.9
水蒸气出口　water vapor outlet　6.3.2.22
水蒸气进入　steam in　2.4.2.1
水蒸气蒸馏　steam distillation　1
水总管　water header　7.1.2.23
顺时针(旋转)　clockwise(CW)　17.3.3.9
顺时针转动　rotates clockwise　17.3.3.8
顺序阀　sequencing valve　23.1.2.11
说明打印机　label printer　23.4.1.9
丝包的洗涤　washing of yarn packages　2.3.8.48
丝饼柔软处理　treating of cake to give fila-

丝 ments softness 2.3.7.22
丝堵 plug 13.1.2.11
丝束的烘干 drying of tow 2.3.8.58
丝束的起皱 crimping of tow 2.3.8.59
丝网滤叶截面图 wire filter leaf 9.1.4
丝网支承 wire mesh support 5.3.4.23
斯塔斯弗矿层（钾盐层，钾盐床） Stassfurt seam 25.2.4.9
斯塔斯弗盐 Stassfurt salt 25.2.4.10
斯托克斯直径 Stokes diameter 1
撕剪机 shear shredder 3.3.16.3
撕碎机 shredder 3.3.16.6
死区 non-flowing region 23.2.2.1
四插座转接器 four-socket（four-way）adapter 25.4.9.8
四级缸 fourth stage cylinder 17.2.3.4
四级压缩机 four-stage compressor 17.2.3
四连插座 four-socket［way］adapter 25.4.1.8
伺服电动机 servomotor 9.5.6.1
松动器 bin activator 23.3.4
松节油分离器 turpentine separator 2.3.10.14
松紧装置 take up unit 15.7.5.9
  take-up 15.6.1.25
松套法兰 lap joint flange 22.4.2.(g)
送料板 feed plate 23.3.7.15
送料槽 feeding trough 11.1.8.19
送料螺旋 feed-screw sprocket 9.3.3.1
送料盘 feed tray 12.2.1.6
速度（分布）剖面（图） velocity profile 1
塑料鲍尔环 plastic Pall ring 5.3.2.(e)
塑料泵 plastic pump 16.1.2.5.1
（塑料）电缆夹（头） cable clip 25.4.1.40
塑料管 plastic pipe 22.6.2.⑭
塑料矩鞍 plastic intalox saddle 5.3.2.(h)
塑料球形填料 plastic tripak 5.3.2.(k)
塑料填料 plastic packing 4.6.1.4
塑模旋转装置 die turning gear 15.7.5.3
酸 acid；A 2.3.6.3
酸蛋 acid egg 16.1.2.2.52
  blow case 16.1.1.17
酸管 acid tube 5.4.2.17
酸化 acidification 3.3.18.19
酸碱值 pH 3.1.2.3
酸进 acid in 6.1.7.6
酸稀释 acid dilution 6.2.1.3
酸相 acid phase 3.3.18.17
酸性气 acid gas 2.2.67.8
酸性气体吸收器 acid-gas absorber 2.2.106.37

酸再生 acid regeneration 6.2.9.(d)
随动件 follower 22.3.1.16
随机过程 random process 1
随机控制 stochastic control 1
碎浆机 pulper 2.3.10.80
碎粒返回 chip return 11.6.3.14
碎煤燃烧 pulverized coal combustion 2.2.115.6
碎片容器 chip container 2.3.8.40
碎石 gravel 3.3.20.9
碎渣机 slag crusher 2.2.31.9
隧道（式）干燥器 tunnel dryer 8.3
榫槽密封面 tongue-groove seal contact face 22.4.3.(d)
榫面法兰 tongued flange 22.4.3
缩核模型 shrinking core model 1
缩径管接头 reducing coupling 22.3.2.7
索曼塔盘 Thormann tray 5.2.10.2
索特平均直径 Sauter mean diameter 1
锁紧板 locking plate 9.6.1.7
锁紧棒 locking bars 11.1.7.11
锁紧垫圈 lock washer 16.4.20.4
锁紧螺母 clamping nut 12.3.5.14
  lock(ing)nut 11.1.5.18

T

塔板 tray 1
  tray sheet 5.2.4.22
塔板圈 tray ring 5.2.4.19
（塔）板效率 plate efficiency 1
塔底残留物 bottoms 2.1.4.28
塔底产品 bottom product 5.3.1.1
塔底重油浆 heavy bottom slurry 2.2.28.10
塔顶产品 top product 6.3.11.16
塔顶馏出物 overhead 2.1.4.23
塔节 sectional tower shell 5.3.4.33
塔节高度 height of sectional tower shell 5.2.4.15
塔内件 tower internals 5.3.4
塔内径 tower inside diameter 5.4.1.2
  tower internal diameter 5.2.4.13
塔盘 tray 5.2.1.4
  tray 5.2.8.2
  tray 6.3.7.14
塔盘板 tray sheet 5.2.2.12
塔盘紧固件 compact set of tray 5.3.4.27
塔盘类型 type of trays 5.2.10
塔盘圈 tray ring 5.2.4.5
塔盘支承 tray stiffer 5.2.8.5

塔盘支持圈　tray support ring　5.2.8.4
塔器　columns　21.3.1.10
塔器类型　type of towers　5.1
塔体　tower body　4.6.1.6
　　tower shell　5.2.1.9
塔外壳　shell　5.2.8.1
塔中部液体入口　intermediate feed　5.2.8.9
台面截面　deck section　4.3.4.8
太阳能　solar energy　25.1.1.1
太阳能电力　solar power　2.4.1.6
钛和钛合金　titanium and titanium alloys　22.6.2.㉔
泰勒标准筛　tyler standard sieve　1
弹簧　spring　14.4.3.10
弹簧安全泄压阀　spring safety-relief valve　22.1.8
弹簧导座　spring retainer　19.1.7.4
弹簧垫圈　spring washer　22.1.10.9
弹簧调整螺母　spring adjustment nut　19.1.5.4
弹簧调整螺栓　spring adjustment bolt　19.1.5.3
弹簧调整装置　spring adjustor　22.1.9.9
弹簧组件　spring assembly　12.2.1.3
弹簧座　spring retainer　16.2.8.6
　　spring seat　17.2.12.4
弹力环　elastic ring　17.2.10.3
弹性挡环　snap ring　15.2.1.3
炭　char　2.2.106.3
炭分子筛容器　carbon sieve container　2.2.53.7
炭黑工艺流程图　carbon black process diagram　2.3.16
炭黑流　carbon black flow　2.3.16.1
炭黑原料　carbon-black　2.2.91.9
炭加热器　char heater　3.3.15.18
炭料罐　carbon feed tank　6.3.1.4
炭料浆罐　char slurry tank　2.2.106.18
炭滤饼　char cake　2.2.106.6
炭气化　char gasification　3.3.15.4
炭/生物质循环　carbon/biomass recycle　3.1.43.7
探头　probe　3.5.11.3
碳钢　carbon steel；CS　21.1.2.7
碳化段　carbonization　24.3.1.3
碳化硅炉拱　carborundum arch　25.1.11.5
碳化塔　carbonizer　2.2.108.13
碳化钨柱塞　tungsten carbide plunger　17.2.15.9
碳氢化合物　hydrocarbons　6.1.7
碳氢化合物气体　hydrocarbon gases　2.2.94.11
碳石墨动环　rotating carbon ring　19.1.7.7
碳石墨刮壁器　carbon wipers　5.5.2.10
碳酸钙　$CaCO_3$　3.4.6.8
碳酸钠　Na_2CO_3　3.4.6.5
碳4烃　C_4-hydrocarbons　2.2.65.1
糖化作用　saccharification　1
糖真空结晶器　vacuum pan for crystallization of sugar　7.2.11
糖汁进料　juice feed　6.3.1.1
Thermix 陶瓷管分离器　Thermix ceramic tube　10.1.4.(b)
陶瓷化工泵　ceramic process pump　16.1.2.5
陶瓷筒　porcelain　14.4.3.20
淘析　sluice separation　1
淘析区　elutriation zone　7.2.10.14
淘析腿　elutriation leg　7.2.1.17
淘析液加入　elutriation liquid feed　7.2.10.12
套叠式斜槽　telescoping chute　23.3.9
套管　casing　2.1.2.18
　　thimble　23.1.13.10
套管换热器　double-pipe exchanger　4.2.5.17
套管冷却结晶器　votator apparatus　1
套管式纵向翅片换热器　double-pipe longitudinal finned exchanger　4.2.5.19
套环　lantern ring　4.1.1.26
套扣开孔（螺纹孔）　threaded openings　21.1.1.59
套圈　collar　13.3.3.4
套筒　bushing containing　2.3.9.50
　　mantle　11.1.3.4
　　skirt　16.5.6.6
套筒端　bushing tips　2.3.9.51
特定气-液过程反应器　reactors for specific liquid-gas processes　14.3.1
特勒花环填料　Teller Bosette packing　5.3.2.(j)
特殊高压密封端盖　special high pressure closure　4.1.2.(e)
特殊输送式干燥器　special conveyor dryer　8.3.2
特殊橡胶圈密封　seal by special rubber ring　16.3.6.3
特殊形状　special form　8.1.1.34
特种　special　16.1.1.46
特种泵　special pump　16.5
腾涌　slugging　14.1.3.21
梯蹬　rung　5.6.1.9
梯度　gradient　1

梯子　ladder　8.3.4.14;27.4.1.1
提抽管道　extraction line　3.1.32.18
提纯段　rectifying section　6.3.11.14
提纯区　purification zone　7.2.14.7
提纯装置　purification unit　2.2.6.11
提高细颗粒速率　higher fine powder rate　14.1.2.6
提馏　stripping　1
提馏段　stripping section　1
提馏塔　stripping tower　5.1.1.4
提浓塔　concentration tower　2.2.58.11
提浓塔回流罐　reflux tank of concentration tower　2.2.58.6
提取　extraction　2.3.14.39
　withdrawal　3.1.32.6
提升抄板　lifting flights　8.8.1.12
提升磁体　lifting magnet　12.6.1.(a)
提升高度　delivery lift　16.5.1.8
提升管　lift line　14.2.3.7
　lift pipe　2.2.4.4
　riser　14.3.2.20;2.2.24.9
提升管反应器　riser reactor　14.3.4.16
提升机　elevator　11.6.4.9
提升机基线　base line of elevator　14.3.3.8
提升机械　lifting mechanism　12.4.2.10
提升气体　lift gas　6.3.11.7
提升气体返回口　lift gas return　6.3.11.1
提升式(卸")阀板　removable valve plate　9.5.5.22
提升台　lift platform　2.3.11.40
提升整平机　elevating flattener　23.4.3.9
提叶器　elevator　11.6.5.8
提质方案　options for upgrading　2.2.104
体积流量　volumetric flow rate　1
体积流速[率]　volumetric flow rate　1
替代天然气,费-托法合成烃类化合物　substitute natural gas, Fischer-Tropsch hydrocarbons　2.2.3.20
天车　crown blocks　2.1.2.4
天然气　natural gas　2.2.3.8;2.2.15.1;2.2.45.1;2.2.78.7;2.2.102.6;2.3.17.5;
　NG(natural gas)　2.2.113.13;2.2.114.2;3.4.6.13
天然气泵　natural-gas pump　10.3.5.1
天然气去管网　natural gas to pipeline network　2.2.36.12
添加剂　additive　2.2.35.15
添加剂泵　additive pump　2.2.35.8
添加剂槽　additive tank　2.2.35.7
添加剂地下槽　underground tank for additives　2.2.35.6

甜水　sweet water　6.3.1.7
填充床涤气器　packed-bed scrubber　10.5.8
填料床反应器系统流程图　Flow diagram of packed bed reactor system　3.1.38
填充管　fill pipe　14.1.9.1
填充剂(如需要)　bulking agent(if required)　3.3.28.3
填充介质　packing media　3.4.4.6
填角焊缝　fillet welds　21.1.1.61
填料　packing　16.2.6.8
填料衬环　throat ring　16.4.20.22
填料封液环　lantern ring　16.4.11.10
填料隔圈　packing ring　16.4.12.13
填料函　gland packing　16.4.19.24
　packing(box)　14.4.3.14
　SB;stuffing box　16.2.5.4
填料函衬套　stuffing box bushing　16.4.27.15
填料函法兰　packing box flange　4.1.1.23
填料函盖　stuffing box cover　16.4.9.32
填料函盖圈　stuffing box cover ring　16.4.9.37
填料盒　packing box　16.4.30.7
　packing case　19.2.1.3
　stuffing box　2.1.2.26
填料环　packing ring　16.4.19.7
填料类型　type of packing　19.2.3
填料密封　packing seal　7.1.3.13
填料润滑装置部件　packing lubricator assembly　22.1.9.13
填料塔　packed column　5.3.1;3.4.1.1
　packed tower　5.3.3.6
填料卸出孔　packing draw-off hold　5.3.4.26
填料卸出口　unloading connection　5.4.1.18
填料压盖　gland　14.4.3.13
　packing　17.2.13.6
　packing flange　22.1.2.10
　packing follower　16.4.25.30
　packing follower ring　4.1.1.25
　packing gland　14.4.3.16
　stuffing box gland　16.4.20.25
填料压盖螺栓　gland bolt　16.4.27.19
填料压盖双头螺柱　packing flange stud　22.1.2.9
填料压盖随动件　gland follower　16.4.4.6
填料压环　packing follower　22.1.9.16
　packing press ring　5.3.4.25
填料压紧装置　packing take-up device　16.2.6.7
填料支承　packing support　5.3.4.21
填料支承板　packing support plate　5.3.4.20
填埋打包机　contractor for dumping　3.3.5.21
挑选　option　3.3.16.4
条形浮阀塔盘　rectangular valve tray　5.2.

10.4
调和　blend　2.1.1.85
调和汽油　blended gasoline　2.1.1.60
调节 H_2/CO 比值　adjust H_2/CO ratio　2.2.36.5
调节板　damper　8.2.1.9
调节池　conditioning tank　3.1.28.3;3.1.30.A.2;3.1.30.B.2
调节阀　adjustable gate　23.5.1.5
　control valve　22.1.9
调节风机　control blower　14.3.4.3
调节环　adjusting ring　22.1.8.5
调节剂　conditioning agent　1.1.11.4
调节空气入口　control air inlet　14.3.4.21
调节轮　adjustment wheel　11.4.3.6
调节螺栓　adjusting bolt　5.2.4.16
调节器本体　governor body　24.2.7.10
调节器隔膜　governor diaphragm　24.2.7.3
调节手轮　adjust handle　15.7.6.17
调节位置　in adjustment　11.4.3.5
调节位置指示器　damper position indicator　10.5.4.3
调节蜗杆　adjusting worm　16.2.5.15
调节蜗轮　adjusting wormgear　16.2.5.16
调节 pH 值　pH adjust　3.1.31.2
调节座　adjusting seat　16.2.5.18
调理过的污泥　conditioned sludge　3.3.22.10
调速器　velocity adjustor　6.1.9.8
调优操作　evolutionary operation(EVOP) 1
调整刀片间隙　adjusting the clearance between the knives　2.3.14.74
调整轮　adjusting wheel　11.1.6.6
调准盘　aligning disc　12.7.4.3
调准皮带　aligning belts　12.7.4.5
贴标签　labeling　3.3.7.1
铁　Fe;ferrum;iron　11.6.3.19
铁轭　yoke　25.4.3.14
铁金属　ferrous metal　3.3.16.12
铁矿石富集器　iron mine concentrator　11.6.3
　mine iron-ore concentrator　11.6.2
铁路货车　rail car　23.1.2.7
铁素体不锈钢　ferritic stainless steel　22.6.2.⑰
铁芯　core　16.5.4.3
铁栅筛　bar screens　3.1.25.2;3.1.27.2
烃基化　hydrocarbylation　2.1.1.56
烃类　hydro-carbon　2.2.2.21
停留时间　hold-up　8.4.3
　residence time　2.2.66.7
停树脂泵　resin pump stopped　6.2.4.7

挺杆　tappet　16.2.4.36
通大气　to atmosphere　11.6.4.20
通道　channel;path　9.1.2.3
　passage　18.1.2.8
　tunnel　24.2.6.11
通道板　man way plate　5.2.2.1
通断控制　on-off control　1
通风管道　ventilation shaft　2.3.12.14
　plenum　3.3.30.10
通风机　fan　8.5.6.28
通过量　throughput　1
通级冷却器　intercooler　17.2.4.3
通量　flux　1
通流截面　section through　17.4.1
通气格子板　air vane　12.3.4.14
通气管　vent pipe　5.4.1.23
通气孔　vent hole　21.3.1.7
通气　vent　23.1.5.8
通前级管　backing tube　18.2.1.15
通信线　communication lines　25.4.1.23
同步齿轮　synchromesh gear　17.3.2.7
同步传动齿轮　synchronizing gear　16.3.6.6
同步器　synchro　25.4.6
同心筛网　concentric screens　10.4.2.9
同心收集器　concentric collector　11.5.1.12
同心圆环　concentric ring　14.1.8.(a)
铜导体　copper conductor　25.4.1.43
铜的冷却盘管　copper cooling-coils　18.2.4.7
铜铬铁矿　Cu-chromite　2.2.66.13
铜合金　copper alloys　22.6.2.⑳
铜芯线　copper conductor　25.4.9.43
统计模型　statistical model　1
统装　bulk　3.3.7.2
桶顶　dome of the tun　2.3.12.45
桶口灰尘罩　bung dust caps　23.4.1.10
筒仓　silo　23.1.8.10;2.2.31.3
筒仓　silo　3.1.21.10
筒体　barrel　15.7.2.1
　cylindrical shell　21.2.4.1
　shell　11.2.4.8
　tank　16.4.27.8
筒体半径　radius of cylinder　21.2.5
筒体长度　length of cylinder　21.2.5
筒体衬板　shell liner　11.2.4.9
筒体冷却器　barrel cooler　15.7.2.1
筒体直径　diameter of cylinder　21.2.5
筒尾　breeching　8.8.5.5
筒尾密封　breeching seals　8.8.1.7
筒形泵　barrel pump　16.1.2.2.31

筒形泵壳　barrel casing　16.4.27
筒形过滤机　cartridge filter　9.4.7
筒锥形螺旋卸料离心机　cylindrical-conical helical-conveyor centrifuge　9.5.5
投入产出　input-output　1
投资收益率　rate of return on investment (ROI)　1
透明圆盘　transparent disk　11.7.3.7
透平机燃料　turbine fuel　2.2.91.8
透平膨胀机　turboexpander　3.1.23.15
透平驱动泵　turbine driven pump　16.1.2.4.2
透平制冷压缩机　turbo refrigerator　20.2.1
透气管　air breather　13.1.1.27
透析　dialysis　1
凸轮　lobe　16.3.6.1
凸轮泵　cam pump　16.1.2.2.63
　lobe pump　16.1.2.2.10
凸轮-活塞泵　cam-and-piston pump　16.1.2.62
凸轮(式)　lobe　16.1.1.26
凸面法兰　male flange　22.4.3.
凸叶(式)　lobe　16.1.1.26
凸缘双头螺柱　flange stud bolt　22.1.2.11
突(平)面　raised face;RF　22.4.3.(b)
突然扩大　sudden enlargement　1
突然缩小　sudden contraction　1
$p\text{-}h$ 图　$p\text{-}h$ diagram　20.2.4
图解法　graphical method　1
图像分析仪　quantimet　1
涂布的卷筒纸　reel of coated paper　2.3.11.35
涂覆聚四氟乙烯的金属筛网　PTFE coated metal mesh　15.5.5.13
涂铝聚酯袋　aluminized mylar bag　3.5.17.32
土地表层　earth cover　3.3.12.2
土壤过滤器　soil filter　3.3.20.7
土壤调理剂　soil conditioner　3.3.1.6
土质　soil　3.3.12.8
湍动气化床　turbulent fluidized bed　1
湍流　turbulent flow　14.1.1
湍流床　turbulent bed　14.1.2.4
湍球塔　turbulent ball column　5.2.9
团块再循环轨迹　pad recirculation line　3.2.13.13
团块再循环汽提泵　airlift pump for pad recirculation　3.2.13.10
团粒　aggregate　23.3.4.3
推环　thrust ring　19.1.1.8
推进溜板　pusher ram　2.3.14.7
推进区　drive zone　2.2.85.24
推力轴承　thrust bearing　16.2.5.13
推力轴承衬瓦　thrust metal　16.4.28.15

推力轴承滑道　thrust runner　16.4.28.16
推料板　push out plate　23.3.7.3
推料活塞　pusher piston　9.5.10.8
推料轴　pusher shaft　9.5.10.7
推挽式起重螺旋　push-pull jacking screws　9.6.1.16
退火窑　annealing lehr　2.3.9.18
退纸装置　unwind station　2.3.11.39
托范式离心罐　Topham centrifugal pot [box]　2.3.7.17
托辊　spaced idlers　23.2.1.4
托架　bracket;brkt　9.5.4.15
托架板　yoke plate　19.1.5.1
托普索公司二甲醚合成工艺(两步合成法)　Topsфe DME synthesis process (two pot synthesis)　2.2.52
拖曳槽型稠密介质分离器　drang-tank-type dense-media separatory vessel　12.5.2
脱胺塔　amine scrubber　2.1.9.44
脱苯　debenzoling　2.3.14.27
脱丙烷塔　depropanizer　2.2.68.13;2.2.71.6
脱尘板　skimmer　10.1.4.8
脱除溶剂　solvent removing　6.1.1.5
脱除酸性气体　acid gas removal　2.2.101.12
脱除微量硫　trace sulfur removal　2.2.101.13
脱除微量有机物　trace organics removal　2.2.101.14
脱氮　denitrification　3.1.25.5
脱氮的污水　denitrified effluent　3.1.36.11
脱丁烷塔　debutanizer　2.2.75.9
脱二氧化碳　CO_2 removal　2.2.68.8;2.3.2.13
脱酚后液体排出　dephenolized effluent　3.1.5.7
脱过热器　desuperheater　20.1.7.13
脱甲烷塔　demethanizer　2.2.68.18;2.2.71.5
脱焦油　tar remover　2.2.108.14
脱空气　air take-off　20.1.10.9
脱硫　desulfurization　3.3.3.6
　desulfurization　2.1.9.49;2.2.38.6
　desulphurization　2.3.14.28
　desulphurizing　2.3.7.20
　sulfur removal　2.2.103.4
脱硫脱碳　desulfurization and decarbonization　2.2.36.6
脱硫塔　desulfurization tower　2.2.45.3
脱气　remove gases　2.2.90.7
　de-gassing　3.1.32.4
脱气槽　deaerating tank　13.2.4.19
脱气器　degassers　5.5.3
脱气室　degassing chamber　5.5.1.7

脱气塔　degassing tower　2.2.54.6
脱轻塔　off with light constituents tower　2.2.55.10
脱轻组分塔　light component removal column　2.2.56.3
脱氢　dehydrogenation　2.1.1.57
脱溶剂　solvent removal　2.2.2.32
脱湿器　mist eliminator　10.5.4.18
脱水　dewater　12.5.1.7
　dewatering　3.1.2.21;3.3.28.2
　dehydration　2.2.68.1
　water removal　2.2.68.6
脱水罐　dewatering tank　6.3.1.9
脱水后生物固体流去处置　dewatered biosolids flow to disposal　3.1.26.(a).8;3.1.26.(b).12
脱水器　dehydrator　2.3.18.3
脱水塔　dehydration tower　2.2.55.12;2.2.56.4
脱C_1塔　de-C_1 tower　2.2.69.13
脱C_2塔　de-C_2 tower　2.2.69.11
脱C_3塔　de-C_3 column　2.2.69.15
脱C_4塔　de-C_4 column　2.2.69.17
脱CO_2塔　CO_2 removal tower　2.2.70.9
脱碳　decarbonizing　2.3.3.11
　decarbonization　2.2.38.8
脱碳塔　decarbonization tower　2.2.50.6
脱烷塔　paraffins removal column　2.2.56.5
脱盐设备　desalter　2.1.4.7
脱盐水加热器　desalinated water heater　2.2.34.11
脱乙烷塔　deethanizer　2.2.68.12;2.2.71.8
脱液极边缘　skimmer edge　10.1.4.10
脱油沥青　detarring asphalt　2.1.1.39
椭圆形封头　ellipsoidal head　21.1.1.6
　elliptical head　5.4.1.1

W

挖斗　digging buckets　23.2.5
瓦楞状隔板　corrugated spacer　15.2.2.7
瓦斯油　G.O.;gas oil　14.2.6.17
瓦斯油加氢处理装置　gas oil hydrotreater　2.1.9.40
外包绝热纤维的水管口密套　ferrule wrapped with insulating fibre　4.4.2.13
外保温层　external insulation　4.4.2.6
外部　exterior　8.2.2.13
外部密封的浮头管板　externally sealed floating tubesheet　4.1.4.(t)
外部热交换　external heat transfer　14.1.4.(j)
外部收集粉尘　dust collected externally　14.1.5.(e)
外部调节的　externally adjustable　10.5.13.3
外部旋风分离器　external cyclone　14.1.4.(b)
外部贮存　outside storage　6.3.2.33
外部贮存罐　outside storage tanks　6.3.2.34
外侧　outboard　16.4.9.9
外侧环形表面　outer annular surface　15.5.3.4
外重整炉　external reformer　2.2.113.4
外浮头端盖　floating head cover-external　4.1.1.21
外盖　outer cover　22.2.1.1
外观检验　visual exam　11.1.2.11
外管　outer pipe　4.2.2.6
外界环境冷空气　cold ambient air　4.5.2.1
外界空气　outside air　20.2.2.9
外径　O.D.;outside diameter　21.1.1.64
外壳　casing　10.2.1.1
　shell　10.1.2.29
　surface casing　3.1.24.18
外壳板　outside casing plate　9.5.4.20
外壳衬板　outside casing liner　12.3.4.11
外壳体　housing　9.5.3.7
外扩散　external diffusion　1
外流凝结器　outflow condenser　2.3.12.4
外密闭室　outer closing chamber　9.5.13.8
外燃式热机　external-combustion heat engine　25.1.1.7
外填料函浮头换热器　outside-packed floating-head exchanger　4.1.1.(d)
外填料函式浮头　outside packed floating head　4.1.4.(p)
外筒　bowl　16.4.26.11
外筒轴瓦　bowl bearing　16.4.26.12
外压管　ext.pressure tube　21.1.1.50
外运输送器　outgoing conveyor　23.2.5.5
外罩　cover　13.5.1.7
外止点　end of stroke　17.2.2.4
外置轴承　outboard bearing　8.8.10.3
外轴　external shaft　13.1.2.21
外转鼓　outer drum　9.3.2.19
外装式机械密封　external mechanical seal　19.1.2
外锥体　outer cone　12.3.3.5
外钻孔　outer bore hole　3.1.24.4
弯管　turn tube　13.1.2.37

弯管壳体泵　elbow casing pump　16.1.2.2.29
弯曲壁　curved wall　15.4.2.6
弯头　tailing cone elbow　12.3.4.16
180°弯头　return bend　17.6.2.2
弯叶开启涡轮式　hub-mounted curved-blade turbine　14.4.4.3
完成　finish　9.3.8.4
烷基化　alkylation　2.1.1.72;2.2.75.10
烷基化产品　alkylate product　6.1.5.16
烷基化汽油　gasoline alkylate　2.1.1.73
烷烃　paraffin　6.1.5.18
烷烃原料　paraffin feed　6.1.5.9
万向接泵　universal-joint pump　16.1.2.2.67
万向接头　universal joint　16.4.21.5
万用表　universal test meter　25.4.1.41
网　screen　9.4.4.13
网波纹填料　corrugated wire gauze packing 1
wire-web packing　5.3.4.47
网孔塔盘　perform tray　5.2.10.20
网屏　screen　14.3.3.12
网形连接　mesh connection　25.4.3.3
往复泵　reciprocating pump　16.2
往复加料盘　reciprocating feed tray　6.3.6.5
往复炉箅　reciprocating grates　24.1.5
往复式　reciprocating　17.1.1.4
往复式刮刀　reciprocating scraper　13.3.3.7
往复式活塞杆　reciprocating piston rod　9.5.6.2
往复式平台刮刀　reciprocating floor scraper 13.3.4.3
往复式压缩机　reciprocating compressor　17.2
往复推料　reciprocating push discharge　9.5.2.8
(往复)运动　motion　16.2.1.7
往复轴　reciprocating shaft　19.2.1.1
往上流再生设备　upflow regenerated unit 6.2.3
危险废弃物　hazardous wastes　3.3.7
危险品　hazard　8.1.1.26
　hazardous material　3.3.12.4
威尔逊密封　Wilson seal　19.4.1.13
微波干燥　microwave drying　1
微粉出口通道　dust chute　13.3.5.18
微粉流道　dust chutes　10.1.3.9
微胶囊　microcapsule　1
微(孔过)滤　microfiltration　1;3.1.30.B.11
微粒控制设备　particulate control device　2.2.24.13
微滤　microfiltration　1
微生物　microorganism　1
Dustex 微型(斜管)收集器组　Dustex minia-

ture collector assembly　10.1.2.(e)
Dustex 微型旋风分离器　Dustex cyclone tube　10.1.2.(f)
维修　repair　21.1.2.13
尾部叶片　tail vane　3.5.1.2
尾管　tail piece　10.1.2.11
　tail pipe　18.3.6.4
尾轮　tail pulley[whee]　23.2.1.1
尾气　offgases　14.3.1.11
　tail gas　14.3.2.26;2.2.54.13;2.2.55.15;2.2.74.7;2.2.78.13
　tail gases　2.3.15.14
尾气去重整或作燃料气　tail gas to reformer or fuel gas　2.2.73.6
尾气吸收塔　off gas absorber　2.2.52.5
尾气洗气塔　tail gas scrubber　2.2.54.5
尾砂　tailings　12.5.5.3
尾砂出口　tailing outlet　12.8.1.7
尾油　tail oil　2.1.1.20
尾渣　tailings　11.6.5.2
未变形位置　undeflected position　22.5.2.1
未处理煤气　raw gas　2.2.107.5
未反应的气体,以及甲醇和水蒸气　unreacted gases,and methanol and water vapors 2.2.39.a
未反应之合成气　unreacted syngas　2.2.70.4
未过滤液体　unfiltered liquid　9.1.3.7
未加工褐煤　raw lignite　2.2.107.2
未加工煤气　raw gas　2.2.104.5
未加工气体　raw gas　2.2.102.19
未加工页岩　raw shale　2.2.88.6
未加工页岩加料　raw shale feed　2.2.82.10
未经处理的废水(原废水)　raw wastewater 3.1.30.A.1
未精制纤维素　unrefined cellulose　2.3.10.77
喂送输送机[器]　feed conveyor　23.4.3.4
温度　temp.;temperature　20.1.7.4;3.1.2.9
温度 T_1　T_1(temperature 1)　2.4.1.2
温度 T_p　T_p(temperature p)　2.4.1.5
温度(分布)剖面(图)　temperature profile 2.2.85.18
温度计接管口　nozzle for thermometer　14.4.1.15
温度记录控制器　TRC;temperature recording controller　3.1.6.17
温度记录仪　TR;temperature recorder　3.1.6.8
温度控制电子元件　temperature-control electronics　3.5.20.8
温度显示器　temperature indicator　8.5.3.7
温克勒气化炉　The Winkler gasifier　2.2.18

温控器　temperature controller　3.5.22.5
温热小球　warm balls　2.2.5.21
温水出口　warm-water outlet　4.3.2.14
文丘里涤气器　Venturi scrubber　10.5.13
文丘里涤气蒸发器　Venturi scrubber evaporator　3.3.4.10
文丘里(管)　Venturi(tube)　10.5.13.7
文丘里管喉部　Venturi throat　3.3.8.12
文丘里喉管　Venturi throat　10.5.13.3
文丘里进料　Venturi feed　23.1.1.(c)
文丘里洗涤器系统　Venturi-scrubber system　10.5.14
紊流　turbulent flow　1
稳定的残渣与生物质　stable residue and biomass　3.2.1.19
稳定后的污泥　stabilized sludge　3.1.26.(a).5
稳定后的一次污泥　stabilized primary sludge　3.1.26.(b).7
稳定杆　sway bar　21.2.2.11
稳定水池　holding equalization　3.1.2.1
稳定器　stabiliser　2.2.63.13
稳定性分析　stability analysis　1
稳定状态　stable state　1
稳态　stable state　1
　　　　steady state　1
涡流　eddy current[flow]　10.1.2.20
　　　vortex　10.1.2.4
涡流返混混合器　turbulizer-backmixer　8.4.1.19
涡流分离器　whirlpool separator　2.3.13.2
涡流扩散　eddy diffusion　1
涡流屏　vortex shield　10.1.2.5
涡轮　turbine　15.6.3.13
涡轮泵　turbine pump　16.1.2.2.17
涡轮发电机　turbine generator　2.2.119.11
涡轮发电机组　turbogenerator unit　25.4.4
涡轮风机　turbo-fan　8.9.1.2
涡轮干燥器　turbo-dryer　8.9.2.1
涡轮混合搅拌器　turbo-mixer agitators　6.1.9.18
涡轮混合器　turbine　12.1.3.(k)
涡轮机　turbine　2.2.81.11;2.2.111.5;2.2.116.8
涡轮搅拌器　turbine agitator　1
涡轮驱动马达　turbine drive motor　15.6.3.4
涡轮式　turbine　24.1.2.9.(c)
涡轮式搅拌器　turbine impeller　14.4.2.(a)
涡轮转盘式干燥器　turbo-tray dryer　8.9.1.7
涡旋式离心机　scroll-type centrifuge　2.1.9.20

涡旋式输料器　scroll conveyor　9.5.5.4
蜗杆　worm　11.1.8.15
蜗杆传动蜗轮　worm-drive gear　9.3.3.44
蜗杆轴　worm shaft　9.3.3.47
蜗壳　volute　1
蜗壳泵　volute casing pump　16.1.2.2.27
蜗壳喉部　volute throat　4.3.4.13
蜗壳卸料口　volute race discharge　9.5.10.17
蜗壳形转子　spiral-shaped rotor　15.7.1.2
蜗轮　worm gear[wheel]　16.2.5.11
蜗轮操作　worm-gear-operated　8.8.11.14
蜗轮减速器　worm speed reducer　11.1.5.14
蜗轮箱　worm case　11.1.8.16
蜗形机壳　volute casing　20.2.1.8
卧式泵　horizontal pump　16.1.2.2.30
卧式单级单吸离心泵　horizontal single-stage single suction centrifugal pump　16.4.2
卧式单级双吸蜗壳泵　horizontal single-stage double-suction volute pump　16.4.13
卧式对置式往复压缩机　horizontal opposed reciprocating compressor　17.2.1.11
卧式多室流化床干燥器(冷却器)　horizontal multi comparfment fluidized bed dryer [cooler]　8.5.2
卧式刮壁(膜)型蒸发器　horizontal wiped-film evaporator　7.1.11
卧式混流泵　horizontal mixed flow pump　16.4.29
卧式加压叶片过滤机　horizontal pressure leaf filter　9.2.4
卧式气液分离器　horizontal seperator　5.3.3.5
卧式容器　horizontal vessel　21.3.1.1
卧式圆筒形容器　horizontal cylindrical vessel　21.2.4
卧式蒸汽分离器　horizontal steam separator　10.1.4.(d)
卧式贮罐　horizontal tank　21.2.5
污垢　fouling[scale]　1
污泥　sludge　3.1.14.1;2.4.5.5;3.3.21.11;3.1.26.(a).3
污泥泵　sludge pump　16.1.2.3.31
污泥泵吸入管　sludge pump suction pipe　3.1.3.2
污泥槽池　sludge pit　3.1.4.1
污泥沉淀池　sludge settling tank　3.4.1.5
污泥沉积室　sludge collector　15.6.4.13
污泥出口　sludge port　9.5.12.19
污泥储存池　sludge storage　3.3.21.1
污泥储存罐　sludge storage tank　3.3.19.6
污泥的两级厌氧消化示意图　Schematic dia-

gram of two-stage anaerobic digestion of sludge 3.3.18
污泥管 sludge pipe 15.6.5.2
污泥回流 return sludge 3.1.20.4
sludge return 3.3.18.3
污泥加入 feed sludge 3.2.2.14
污泥降解 digesting sludge 3.3.17.5
污泥接受器 sludge hopper 3.2.15.4
污泥进料 sludge feed 3.3.6.6
feed sludge 3.3.25.7
污泥进料泵 sludge feed pump 3.3.19.9;3.3.21.3
污泥进入 sludge inlet 3.3.9.10
污泥老化罐 sludge conditioning tank 3.1.11.5
污泥排出管 sludge drawoff 3.3.18.11
污泥入口 sludge inlet 3.3.27.4;3.3.17.1;3.3.18.2
污泥收集器 sludge collectors 3.2.15.3
污泥增稠罐 sludge thickening tank 3.1.11.1
污泥制备 sludge preparation 3.1.16.6
污泥贮斗 sludge hopper 15.6.5.17
污染物分析器 pollutant analyzer 3.5.19.9
污染物监控系统 pollutant monitoring system 3.5.19
污水 polluted water 2.2.74.13
sewage 19.1.5
waste water 3.1.20.2
污水泵 sewage pump 16.1.2.3.18
污水池 waste 3.1.12.12
污水处理厂 sewage treatment plant 3.1.21.4
污水处理厂所用装置 wastewater treatment plant showing instrumentation 3.1.2
污水处理装置 wastewater-treatment facility 3.1.1
污水污泥处理方法挑选 waste-sludge disposal choices 3.1.20.8
污水污泥处理系统 waste-sludge disposal systems 3.1.20.7
污水污泥脱水和活性炭再生 waste-sludge dewatering and carbon regeneration 3.1.20.9
无挡板夹套 unbaffled jacket 4.3.3.6
无二氧化碳烟气 CO_2-free flue gas 2.2.116.11
无二氧化碳烟气排放到大气 CO_2-free flue gas to atmosphere 2.2.27.7
无粉尘带出 no dust carryover 4.3.3.25
无规范 no code 21.1.1.49
无级变速驱动装置 variable speed drive 5.5.3.15
无菌操作 sterile operation 1

无量纲数群 dimensionless group 1
无平衡块 unterweighted 23.3.6.(a)
无水氨 anhydrous ammonia 2.3.1.15
无水乙醇产品 absolute alcohol product 2.2.59.6
无损检验 non-destructive testing 21.1.2.15
无效能 anergy 1
无泄漏联结 leak-free connections 3.5.2.4
无油润滑气缸 non-lubricated cylinder 17.2.10
炕 anergy 1
五级逆流级联 five-stage countercurrent cascade 6.1.10
五级蒸汽喷射泵 five-stage steam ejector 18.3.5
物理单元操作 physical unit operation 3.2.1.10
物料仓 material source 23.1.2.14
物料槽 feed tank 8.5.3.10
物料层 lager of product 3.1.16.10
material bed 2.2.5.4
物料抽提机构 material extraction mechanism 3.3.29.12
物料处理工序 materials-handling sequence 3.3.7
物料床 material bed 23.2.5.8
物料床层 bed of material 4.3.3.26
物料管线 material line 23.1.2.8
物料衡算 material balance 1
物料加入 material input 11.5.2.1
物料接收器 material receiver 23.1.2.20
物料进口 materal in 8.4.1.30
product feed 8.5.6.17
物料进口法兰 product inlet flange 8.9.5.4
物料进入空气中 Material-into-air 23.1.8.③
物料平衡 material balance 1
物料去除系统 material removal system 3.3.29.11
物料收集器 material destination 23.1.2.6
物料竖筒 solid column of material 23.3.7.5
物料卸出 material removal 3.3.30.4
物流 stream 1
物品 product 20.1.3.21
物位传感器 level sensors 23.1.2.21
雾滴 sprays 13.3.1.4
雾化 atomized 10.5.1.10
雾化空气 atomizing air 3.3.10.9
雾化喷嘴 atomizing nozzle 13.4.1.23
spraying nozzle 8.6.1.2

雾化器 atomizer 8.6.2.4
atomizing device 13.1.1.2
雾沫 spray 10.5.8.5
雾沫分离板 entrainment separator plate 10.5.4.21
雾沫分离级 entrainment separation stage 10.5.1.3
雾沫分离器 entrainment separator 10.5.1.19
雾沫夹带 entrainment 1
雾沫净除器单元 brink mist eliminator element 10.4.2
雾水盆 spray pan 20.2.2.15

X

西门子气化炉 Siemens gasifier 2.2.26
吸尘装置 dust extractor 2.3.10.1
吸除麦芽装置 corn removal suction 2.3.12.38
吸附 adsorbing[adsorption] 6.3.3.4
吸附段 adsorption section 6.3.7.3
吸附罐 adsorption tank 6.2.5.2
吸附剂流 adsorbent flow 6.3.7.1
吸附剂载气 adsorbent carrier gas 6.3.7.8
吸附器 adsorber 3.1.17.2
吸附设备 adsorption equipment 6
Purasiv HR 吸附塔 Purasiv HR adsorber vessel 6.3.7
吸附柱 adsorption column 6.2.6.1
吸气 inlet[suction] 18.3.1.3
吸气阀 inlet[suction]valve 17.2.12.11
吸气口 suction(nozzle) 18.1.1.8
吸气室 suction chamber 18.3.1.4
吸取物料 pick up the solid 23.1.1.(d)
吸取装置 pickup device 23.1.1.1
吸热 absorbing heat 20.1.2.9
吸热反应 endothermic reaction 1
吸热管 heat-absorption tube 24.2.10.2
吸入 suction 16.2.6.1
吸入侧轴承压盖 suction side bearing cover 16.4.19.28
吸入端 suction side 16.3.6.10
吸入端(泵)盖 suction cover 16.4.1.2
吸入阀螺塞 suction-valve plug 16.2.4.27
吸入管 suction duct 20.2.3.17
  suction line 2.1.7.25
  suction pipe 16.2.1.12
吸入接头 suction head 16.4.26.15
吸入接头盖帽 suction head cap 16.4.26.14
吸入接头管塞 suction head plug 16.4.26.17

吸入接头套圈 suction head ring 16.4.9.38
吸入接头支承面[轴承] suction head bearing 16.4.26.16
吸入口 suction 16.4.21.8
  suction eye 16.4.6.4
  suction inlet 16.3.2.1
  suction port 16.3.7.1
吸入口法兰 suction flange 16.4.3.7
吸入口盖 suction cover 16.4.9.39
吸入口机壳 suction casing 20.2.1.6
吸入口接头 suction head 16.4.25.3
吸入口壳体 suction cover 16.4.29.1
吸入口直径 eye diameter 16.4.6.10
  suction diameter 16.4.3.17
吸入气体 suction gas 20.1.3.14
吸入腔 suction chamber[manifold] 16.3.2.2
吸入筒 suction bowl 16.4.24.1
吸入压力 suction pressure 16.2.1.9
吸入叶尖 suction vane tip 16.4.3.11
吸入叶片之前缘 suction vane edge or tip 16.4.6.3
吸入支管 suction branch 18.3.5.11
吸收 absorption 1;20.1.2.21
吸收工段 absorbing section 2.2.56.8
吸收剂 sorbent 25.5.2.2
吸收剂再生器 absorbent regenerator 2.2.106.33
吸收甲醇 absorbed methanol 2.2.55.4
吸收硫 sulfur absorption 2.2.103.5
吸收器 absorber 20.1.2.14
吸收热 heat of absorption 1
CO$_2$吸收塔 CO$_2$ absorber 5.4.1
  absorption 2.2.102.23;2.2.27.6;2.2.61.8;2.2.63.11;2.2.15.3
  absorbing tower 2.2.70.6
  absorption column 5.1.1.5
  absorption tower 2.2.49.3
吸收循环 absorption cycle 20.1.8
吸收制冷 absorption refrigeration 1
吸水滚轮 suction roll 2.3.11.16
吸水喇叭口 suction bell 16.4.23.14
吸水喇叭口轴承 suction bell bearing 16.4.23.15
吸液 imbibition 1
吸引板 attracting plate 12.7.2.9
吸引栅 attracting grid 12.7.2.10
吸油管 oil attracting tube 15.5.1.4
析出段 elutriating leg 7.1.14.4
析叶开启涡轮式 pitched bladed turbine 14.4.2.(b)

烯烃　olefin　6.1.5.19
　olefin hydrocarbon　2.2.1.17
烯烃分离　olefin separation　2.2.67.19
$C_2 \sim C_4$（烯烃及液化气）　$C_2 \sim C_4$（olefin hydrocarbon and liquefied petroleum gas　2.2.73.7）
烯烃原料　olefin feed　6.1.5.10
稀的　diluted　12.6.4.1
稀介质泵　dilute-medium pump　12.5.5.28
稀介质储槽　dilute-medium sump　12.5.5.27
稀苛性碱　dilute caustic　6.2.1.7
稀溶液　weak solution　20.1.8.16
稀释的硫酸　diluted H_2SO_4　2.2.66.2
稀释罐　dilution tank　2.1.9.17
稀释过的泡沫　diluted froth　2.1.9.18
稀释液　diluted liquid　3.3.10.4
稀释液贮槽　storage tank for the weak liquor　2.3.10.45
稀酸　dilute acid　6.2.1.5
稀洗液浓缩器　concentrator for the weak wash liquor　3.3.10.44
稀相　dilute phase　1
稀相流化床　dilute[lean]-phase fluidized bed　14.1.3.20
稀液　weak aqua　20.1.9.8
稀渣排放　slurried cake discharge　9.2.3.6
熄火蒸汽吹入管　steam inlet for snuffing　25.1.1.13
Tol-洗氨塔　Tol-ammonia washing tower　2.2.34.20
洗涤　rinse　12.5.1.6
　scrubbing　1
　wash　9.4.8.2
　washing　9.1.1
洗涤板　wash plate　9.1.1.3
洗涤层　washing floor　2.3.12.3
洗涤后的气体　scrubbed gas　3.1.22.13
洗涤后排出气　scrubbed exhaust　10.5.13.9
洗涤(溶)剂　wash solvent　2.2.92.17
洗涤篮筐　wash basket　9.5.10.2a
洗涤水　wash[washing]water　12.4.1.1
洗涤水泵　wash water pump　9.3.1.1
洗涤水出口　washwater outlet　6.2.6.5
洗涤水流出　wash-water outflow　15.5.3.5
洗涤水喷头　wash spray　9.3.1.3
洗涤水入口　washing water inlet　9.2.5.15
洗涤水水管支架　support for drip-piping wash　9.4.2.3
洗涤水水幕（水帘）　wash-water curtain　15.5.3.3
洗涤水总管　wash header　9.3.7.7

洗涤器　scrubber　3.3.25.17
洗涤塔　column washer　1
　washing tower　5.1.1.9
洗涤塔／冷却塔　washer/cooler　2.1.8.8
洗涤系统　flushing system　10.3.6.30
洗涤液　scrubbing agents　2.3.14.37
　scrubbing liquid　2.2.11.5
　wash liquid　3.1.9.2
　wash solvent　6.1.3.9
　washings　1
洗涤液接管　washing nozzle　9.2.1.28
洗涤液入口　scrubbing liquid inlet　10.5.15.2
洗涤油罐　scrubbing oil tank　2.3.14.43
洗出的煤　washed coal　2.2.93.1
洗瓶车间　bottle-washing plant　2.3.13.18
洗瓶机　bottle-washing machine　2.3.13.19
洗气塔　scrubber　2.2.49.5
洗气叶片　scrubbing vanes　10.5.15.6
洗水　wash water　9.1.1.14
洗水路径　path of wash water　9.1.1.9
洗水入口　wash inlet　9.1.1.6
　washwater inlet　6.2.6.6
洗提剂[液]　eluate　6.2.6.11
洗提器　elutriator　10.3.5.6
洗提柱　elution column　6.2.6.9
洗脱液出口　eluate exit　6.3.8.9
　eluent inlet　6.3.8.2
洗脱液流　eluent streams　6.3.8.5
洗液　wash (liquor)　9.4.8.5
洗液出口　wash discharge[removal]　9.5.6.14
洗液分布器　wash distributor　9.3.9.3
洗液管　wash pipe　9.5.6.10
洗液喷嘴　spray nozzle　9.5.6.12
洗液入口　wash inlet　9.5.5.19
洗液通真空系统的接口　wash vacuum connection　9.3.2.29
洗油罐　wash oil tank　2.2.106.28
系杆　tie rod　8.8.5.33
F-13系统　F-13 system　20.1.7.17
F-22系统　F-22 system　20.1.7.18
系统风机　system fan　8.4.1.15
系统控制装置　system controls　23.1.2.25
细胞培养　cell culture　1
细粉　fines　13.1.1.9
细粉回流　fines return　2.2.18.9
细粉排出管　outlet duct for fine material　12.3.2.2
细粉物料　fine materials　23.1.12
细粉锥筒　fines cone　12.3.1.6
细金属粉　fine metal powders　12.7.3

细(颗)粒 fines 12.5.1.1
细颗粒 fine particle 14.1.3.1
细颗粒分离 fine particle separation 11.5.2.9
细颗粒聚结 fine particle coalescing 6.1.9.11
细颗粒流态化相图 fluidization phase diagram for a fine powder 14.1.1
细颗粒排弃 fines discard 2.2.82.16
细矿进料器 fine ore feeder 12.8.1.6
细矿石储槽 fine ore bin 12.8.2.9
细粒 undersize 13.3.3.10
细粒流 fines stream 7.2.9.12
细粒溶解槽 fines dissolving tank 7.2.9.10
细粒随溢流而出 overflow and fines 12.4.2.13
细料 fines 12.2.3.9
细流加料闸门 dribble gate 23.5.2.6
细盐 fine salt 7.2.12.7
下部夹套 lower jacket 14.4.1.22
下部空气集气罩 lower air hood 4.2.4.11
下部滤带冲洗 lower belt wash 9.6.2.12
下部滤带张紧器 lower belt tensioning 9.6.2.13
下部喷嘴 lower jet 18.2.1.5
下部提升筒 lower lift drum 6.3.11.8
下部压紧胶辊 nipping roll(lower) 15.7.5.10
下部重块 lower weight 11.3.1.9
下层卵石床 lower pebble bed 14.3.3.14
下层筛 lower screen 12.2.3.10
下沉的产品 sink product 12.5.5.25
下沉物 sink 12.5.2.14
下沉物料淋洗筛 sink-material rinse screen 12.5.5.18
下沉物料提升机 elevating conveyor for sink material 12.5.5.12
下垫片 lower gasket 16.4.27.33
下端盖 bottom end cover 16.4.28.13
下端开口的下降管 open-end downcomer 24.1.6.5
下段轴 low shaft 14.4.3.21
下分配板 lower distribution plate 12.3.5.15
下盖 lower head 16.2.7.6
下(旱)考依波(泥灰岩及砂岩) lower Keuper 25.2.4.1
下(旱)壳灰岩层 lower Muschelkalk 25.2.4.4
下板 bottom crown 21.2.3.9
下降管 downcomer 3.3.10.2
downstack 11.5.2.10
下壳体 lower casing 16.4.12.4
下料 material discharge 10.2.3.17
下流式填充床系统(反硝化) downflow packed-bed-system(denitrification) 3.1.40.8
下流式Ⅱ型催化裂化装置 downflow model Ⅱ catalytic-cracking unit 14.2.8
下漏 underflow 1
下(面)摇臂轴 lower rock shaft 16.2.3.44
下倾角出风 bottom angular down 17.5.2.2
下栅板 lower grating 5.2.9.6
下水管 downcomer 14.2.2.3
下水平出风 bottom horizontal 17.5.2.3
下套筒 lower sleeve 16.2.5.10
下套筒轴承 lower sleeve bearing 16.4.27.3
下托架 bottom bracket 16.4.28.14
下仰角出风 bottom angular up 17.5.2.4
下一层塔板 plate below 5.2.6.10
下游处理 downstream processing 1
下圆锥 tower cone 10.1.2.14
下轴承 lower bearing 11.1.2.9
下轴承盖 lower ball cover 13.3.5.12
下轴承箱 lower bearing housing 12.3.5.11
下轴承箱盖 lower bearing housing cap 12.3.5.12
下轴套 lower shaft sleeve 16.4.27.13
Hookers Point 先进废水处理厂流程图 Flow diagram of Hookers Point advanced wastewater treatment plant 3.1.40
纤维接触元件 fibrous contacting element 10.5.11.3
纤维素黄酸酯 cellulose xanthate 2.3.7.9
纤维填充床涤气器 fibrous-bed scrubber 10.5.11
纤维填充物 fiber packing 10.4.2.10
纤维状 fibrous 8.1.1
纤维状固体 fibrous solid 8.1.1.20
闲置中 standby 6.3.2.20
显热流 Q_s(sensible heat flow) 2.4.1.7
现代催化裂化装置 modern FCC unit 14.3.4.(b)
现代煤化工 modern chemical processing of coal 2.2.1.23
现代物料输送方法 modern material handling 23.3.2
现代整体煤气化联合循环(IGCC)工厂的简化框图 simplified block diagram of modern integrated gasification combined cycle (IGCC)plant 2.2.109
现有供应线 existing supply line 23.4.1.18
现有装置 existing plant 2.3.2.11

735

限界流体　grenzanhydrite　25.2.4.11
限流环　stripper　16.4.15.12
限位波形膨胀节　constrained-bellows expansion joint　22.5.4
限位杆　limit rod　22.5.3.1
线路通断测试器　continuity tester　25.4.1.55
线芯　core　25.4.5.3
线性化电路　linearizer　3.5.18.4
相对挥发度　relative volatility　1
相配的上下磨石　matched upper and lower stones　11.4.3.3
相切圆　tangent circle　11.5.1.4
相通　through to　16.3.2.4
厢式干燥器　shelf dryer　1
　tray and compartment dryer　8.2
　tray dryer　8.2.1
箱式燃烧器　burner box　2.2.111.2
箱体　casing　10.2.1.1
详图　detail; detail drawing　10.5.1.11
向后传送　travel away　4.3.4.19
向前传送　travel forward　4.3.4.18
向上通风装置　upward ventilator　25.2.5.9
向心泵　centripetal pump　9.5.13.3
项目评审技术　PERT; project evaluation and review technique　1
相平衡　phase equilibrium　1
相图　phase diagram　1
橡胶衬里泵　rubber-lined pump　16.1.2.5.3
橡胶垫　rubber support　12.2.3.18
橡胶管　rubber pipe　22.6.2.⑮
橡胶密封圈　rubber seal ring　19.4.1.6
橡胶圈　rubber ring　17.2.12.10
橡胶塑料加工成型机械　forming machine for rubber and plastics　15.7
橡胶叶轮　neoprene vane　16.3.4.1
削片机　chipper　2.3.10.1
消除真空用阀　vacuum break valve　8.2.2.3
消毒　disinfecton　3.1.25.12; 3.1.30.B.12; 3.1.37.15
消防泵　fire-fighting pump　16.1.2.3.19
消弧角　arcing horn　25.4.9.11
消化器　slaker　3.4.6.10
消化器进液　digester feed　3.2.6.4
消化污泥　digested sladge　3.3.17.6; 3.3.18.9
消沫器　foam breaker　5.3.4.40
消声器　silencer　3.1.9.14
消旋叶片　antispin vane　10.5.4.7
消振器　vibration-stopper　14.4.1.20
硝化　nitrification　3.1.37.7
硝化污水　nitrified effluent　3.1.36.1; 3.1.38.5

硝化循环（400％Q）　nitrified recycle（400％Q）　3.1.33.3
硝酸　nitric acid　14.3.1.9
硝酸铵颗粒　ammonium nitrate pill　13.1.1.10
硝酸铵造粒塔　prilling tower for ammonium nitrate　13.1.1
销毁　destroying　3.3.5
销轴垫板　pin plate　11.1.8.28
销子　pin　13.1.2; 13.1.1.2.32
小车　car (truck)　24.1.8.5
小齿轮　pinion　12.3.5.26
小齿轮轴　pinion shaft　12.3.5.24
小齿轮轴内侧轴承　inside pinion bearing　12.3.5.25
小齿轮轴外侧轴承　pinion shaft outside bearing　12.3.5.23
小环管　small ring　24.2.9.(b)
小螺旋粗轴式　small spiral large shaft　4.3.3.(b)
小批量　small scale　8.1.1.23
小气泡　minute bubble　3.1.13.3
小球（small）　ball　2.2.5.8
小球加热器　ball heater　2.2.5.14
小球提升机　ball elevator　2.2.5.9
小设备　smaller unit　15.5.2
小弹簧　minispring　19.1.1.7
小头轴承盖　small end bearing　17.2.15.25
小型断路器　miniature circuit breaker　25.4.1.33
小型断路器（熔断器）　miniature circuit breaker　25.4.1.19
小型隔焰炉　small muffle furnace　24.1.10
小型工业焚烧炉　small industrial [commercial] incinerator　24.1.4
小型鼓泡器　midget bubbler　3.5.14.8
小型焊接管件　small welded fitting　21.1.1.58
小于　11.6.2.6
楔形挡板衬里　wedge bar liner　11.2.2.1
楔形紧固件　wedge compact set　5.3.4.28
楔形闸板　wedge gate　22.1.1.1
斜壁自动卸料离心机　inclined wall self-discharge　9.5.1.(i)
斜槽　chute　23.4.2.12
斜槽（固体排出）　chute (solids discharge)　9.5.6.22
斜槽口　sluice opening　12.5.2.2
斜底槽　sloping bottom tank　12.4.2.12
斜孔塔板　inclined hole tray　5.2.10.17
斜面　incline　23.3.5
斜面坡度1:4　slope 1/4　3.4.4.1
斜盘式泵　swash plate pump　16.1.2.2.54

斜盘式止逆阀　tilting-disk check valve　22.1.1.(k)
斜坡水槽　flume　3.5.3.1
携带湿分的废气　moisture-carrying exhaust air　8.9.6.12
泄漏检测孔　leak detection port　16.2.7.14
泄漏途径　leakage path　19.3.3.3
泄压阀　bleed off valve　3.1.19.5
pressure-relief valve　2.1.7.17
泄压装置　relief device　21.1.2.12
泄液　drain　10.4.3.8
卸饼　discharge　9.3.7
卸出槽　outlet pocket　15.6.3.7
卸料　discharging　9.3.5.9
material discharge　23.3.5.4
卸料槽　discharge chute　23.2.1.7
卸料池　dump chest　2.3.10.81
卸料端　discharge end　23.3.7.(a)
卸料阀　discharge gate　25.1.12.5
discharge lock　23.1.13.4
dumping valve　14.1.9.9
卸料方式　discharge arrangement　23.2.3
卸料辊子　discharge roll　9.3.7.2
卸料机械装置　discharge mechanism　2.2.7.6
卸料口　discharge opening　23.3.8.1
discharge outlet　8.8.1.28
discharge point　3.3.3.5
discharge port　24.1.13.3
卸料门　discharge door　8.8.10.12
卸料装置控制气缸　unloader control cylinder　9.5.6.20
卸料装置用的液压泵　unloader hydraulic pump　9.5.6.16
卸料锥　discharge cone　15.6.1.18
卸渣　discharge sludge　9.4.2.12
卸渣管法兰垫片　packing for discharge cover　9.2.1.26
卸渣管法兰盖　discharge cover　9.2.1.27
蟹爪式阀罩　crab claw　17.2.12.7
向心泵　centripetal pump　9.5.16.7
锌的浮选　zinc flotation　12.8.1.14
新鲜的碳氧化物　fresh carbon oxide　2.2.66.9
新鲜的原料气　fresh feed gas　2.2.37.1;2.2.48.1
新鲜活性炭补充　virgin-carbon makeup　3.1.20.1
新鲜甲醇　fresh methanol　2.2.55.17
新鲜空气　fresh-air　8.3.1.2
新鲜空气鼓风机　fresh air fan　3.1.21.1
新鲜空气加入　fresh air in　8.4.6.1
新鲜空气入口　fresh air inlet　8.3.4.10
新鲜料　fresh feed　14.3.2.17
新鲜淋洗水　fresh rinse water　12.5.5.17
新鲜气　fresh syngas　2.2.44.1
新鲜溶剂进口　fresh solvent in　6.1.17.1
新鲜溶液　fresh liquor　2.2.9.13
新鲜水　fresh water　2.2.35.16
新鲜酸　fresh acid　6.1.5.11
新鲜油进料　fresh-oil feed　14.2.7.14
新鲜蒸气进口　fresh vapor inlet　14.3.3.27
信号处理　signal processing　3.5.19.7
信号处理系统　signal processor　3.5.20.6
(信号)传感体积　sensing volume　3.5.6
信号灯　lamp[singal light]　3.5.6.12
信息流图　information flow diagram　1
星形阀　star valve　14.1.6.(b)
星形给料器　rotating segment feeder　11.6.1.5
行程标尺　travel indicator scale　22.1.9.11
行程长度　stroke length　17.2.2.11
行程指针　travel indicator　22.1.9.10
形成区　form zone　9.3.10.16
T形锤子　T-shaped hammer　11.2.5.3
V形阀　V-belt　12.3.4.2
V形带传动的产品泵　V-belt driven product pump　7.2.9.7
V形带轮　V-belt pulley　12.3.5.21
U形管管束　U-tube bundle　4.1.4.(s)
U形管换热器　U-tube heat exchanger　4.1.1.(c)
U形管压差计　U-gauge pressure　8.5.3.11
O形环[圈]　O-ring　22.4.4.5
O形环密封接头　O-ring sealed joint　18.2.4.1
U形流塔盘　reverse flow tray　5.2.10.26
U形螺栓　U-bolt　4.2.2.1
O形密封环　O-ring seal　16.2.7.5
O形密封圈　O-ring seal　18.2.1.3
O-ring　13.1.2.24,30,39
T形喷射器　injector tee　23.1.8.1
V形皮带　V-belt　12.1.1.22
S形塔盘　S-shape tray　5.2.10.14
U形弯管　U-bend　14.3.4.26
形状系数　shape factor　1
Lorain型衬里　Lorain liner　11.2.2.2
Mikro型锤磨机　Mikro-Pulverizer hammer mill　11.2.5
Exxon Ⅳ型催化裂化(FCC)装置　Exxon Model Ⅳ fluid catalytic cracking(FCC)unit　14.3.4.(a)

SOD Ⅱ 型催化裂化装置　SOD model Ⅱ catalytic cracking unit　14.2.6.(a)
SOD Ⅲ 型催化裂化装置　SOD model Ⅲ catalytic cracking unit　14.2.6.(b)
SOD Ⅳ 型催化裂化装置　SOD model Ⅳ catalytic cracking unit　14.2.6.(c)
Ⅰ型催化裂化装置　Model Ⅰ catalytic-cracking unit　14.2.7
V 型带传动　V-belt drive　4.5.1.2
V 型带轮　V-pulley　15.7.4.16
M 型对称平衡式压缩机　M-type balanced opposed reciprocating compressor　17.2.1.10
L 型阀　L-valve　14.1.6.(f)
Sirocco D 型分离器　Sirocco type D collector　10.1.2.(b)
Mikro 型粉磨机　Mikro-pulverizer　11.2.6
Mikro-ACM 型粉磨机的截面图　Section of Mikro-ACM pulverizer　11.2.10
Mikro 型粉碎机　Mikro-atomizer　11.2.9
L 型固定轮　L type fixed ring　19.3.2.2
T 型轨　tee rail　15.6.1.27
P 型环形干燥器　P-type ring dryer　8.4.6
V 型混合器　V-type mixer　12.1.4.(e)
Mannheim 型机械盐酸炉　Mannheim-type mechanical hydrochloric acid furnace　24.1.11
M 型胶体磨　Model M colloid mill　11.4.3
A 型搅拌器驱动装置　type A agitator drive　9.3.3.21
JO 型聚四氟乙烯密封环　JO type polytetrafluoroethylene seal ring　19.4.1.20
型块　matrix　8.1.1.16
Gayco 型离心分离器　Gayco centrifugal separator　12.3.1
V 型皮带　V-belt　15.7.4.10
V 型皮带轮　V-pulley　11.1.5.15
S 型皮托管　type-s Pitot tube　3.5.2.2
Rietz 型破碎机　Rietz disintegrator　11.2.7
Trost 型气流粉碎机　Trost jet mill　11.5.2
45°Y 型三通（等径）　45°Y-branches (straight size)　22.3.2.6
T 型栅板　grids of T bars　14.1.8.(c)
Marcy 型栅板式连续球磨机　Marcy grate-type continuous ball mill　11.2.1
Majac 型射流粉碎机　Majac jet pulverizer　11.5.3
Z 型输送式提升机　Z-type conveyor-elevator　23.3.2.(b)
H 型四级压缩机　four-corner four-stage compressor　17.2.3

H 型往复压缩机　H-type reciprocating compressor　17.2.1.12
L 型往复压缩机　L-type reciprocating compressor　17.2.1.5
V 型往复压缩机　V-type reciprocating compressor　17.2.1.6
W 型压缩机　W-type compressor　17.2.1.7
Index V 型永久磁铁组件　Index V magnet assembly　12.6.10.5
L 型油杯　oil cup(L type)　15.7.4.12
RA 型在线分离器　type RA line separator　10.4.3.(e)
A 型转鼓驱动装置　type A drum drive　9.3.3.39
性能系数　coefficient of performance(COP)　1
修整工具　trimming tool　2.3.9.43
(袖珍)手电筒　pocket torch　25.4.1.26
溴化锂吸收循环　lithium bromide absorption cycle　20.1.8
虚线轮廓线　dotted outline　15.5.5.3
需氧　AER；aerobic　2.2.2.2
需氧的　aerobic　3.1.47.3
需氧的生物反应器　aerobic biological reactor　3.1.20.3
需氧区　aerobic zone　3.1.33.11；3.1.35.6
需要附加的工序　additional step required　3.1.30.B.8
序贯模块法　sequential modular approach　1
序列关联　serial correlation　1
絮凝　flocculation　1；2.4.4.1
絮凝剂　coagulant　3.1.12.5
flocculant　3.1.27.5
絮凝剂制剂　flocculant preparation　3.1.16.4
絮凝污泥　flocculant sludge　3.2.6.26
絮凝作用　flocculation　3.1.12.9
蓄热式换热器　regenerative heat exchanger　2.2.113.11
悬臂　arm　9.2.5.22
悬臂支承销　arm base pin　13.3.5.20
悬浮　suspension　1
悬浮室　suspension chamber　7.2.4.23
悬浮固体　suspended solid　3.1.13；3.3.18.12
悬浮聚合　suspension polymerization　1
悬浮生长反应器　suspended growth reactor　3.2.13.4
悬浮式燃烧器　suspended combustor　2.2.16.2
悬浮液　susp.；suspension　1；
悬挂式支座　support bracket　4.1.1.37
悬挂轴衬　suspension bushing　11.1.4.13

悬架支承部件　bracket support　16.4.2.6
悬筐　basket　9.5.6.4
悬挂重物　suspension weight　10.3.5.22
旋柄控制拉紧　knob operates take-up　23.3.7.12
旋风涤气器　cyclone spray scrubber　10.5.4
旋风分离管　cyclone tube　10.1.2.27
旋风分离机组　cyclonic separator unit　10.1.3.2
旋风分离器　cyclone (separator)　2.2.19.7；2.2.76.2；3.3.24.7；3.3.23.6；3.3.25.10；10.1 dust rotor　8.5.6.22
oycione　2.2.37.16
van Tongeren 旋风分离器　van Tongeren cyclone　10.1.4.(c)
旋风分离器固体回流密封装置　cyclone solids-return seal　14.1.7
旋风集尘器　cyclone dust collector　24.2.10.6
旋风炉　cyclone furnace　24.2.3
旋风汽提器　cyclone stripper vessel　14.2.7.24
旋风器　cyclone　24.2.2.8
旋风器出渣孔　cyclone slag-tap hole　24.2.3.5
旋风除尘器　cyclone　10.4.3.11
旋流塔盘　rotating stream tray　5.2.10.21
旋片　blade　18.1.2.2
旋启式止逆阀　swing check valve　22.1.1.(l)
旋塞　cock　11.4.1.5
plug cock　22.1.1.(d)
旋塞阀　plug valve　22.1.1.6
旋涡(泵)　regenerative　16.1.1.45
旋涡泵　peripheral pump　16.1.2.2.19
vortex pump　16.1.2.2.18
旋涡式喷淋器　spinner thrower　5.3.4.7
旋液分离器　cyclone　12.8.2.11
hydraulic cyclone　3.3.6.7
hydroxy clone　13.1.8.11
旋转板　rotating plate　9.3.2.24
旋转杯　rotating cup　3.5.1.4
旋转臂　rotating arm　24.1.11.10
旋转表面(钢)　rotating surface (steel)　19.3.3.5
旋转部件　rotating element　9.3.5.12
Liprotherm 旋转床膜蒸发器　the Liprotherm rotating thin film evaporator　5.5.1.(d)
旋转磁鼓　rotating drum　12.6.9.5
旋转刀片的锥部　rotating bladed cone　2.3.

10.75
旋转的迷宫　rotating labyrinth　19.3.3.(d)
旋转阀　rotary valve　23.1.1.7
旋转阀进料器　airlock feeder　23.1.6.1
旋转方向　(direction of) rotation　16.3.5.7
旋转刮板臂　rotating scraper arm　3.1.14.2
旋转刮板真空泵　rotary moving blade vacuum pump　18.1.4
旋转刮刀　rotary knife　9.5.2.21
旋转刮刀单速自动离心机　single-speed automatic rotary knife　9.5.1.(g)
旋转刮刀自动离心机　automatic basket rotary knife　9.5.1.(e)
旋转管式撇油器　rotary pipe skimmer　15.5.1.6
旋转管束干燥机　rotating tube bundle dryer　8.8.12
rotary tube dryer　2.2.107.3
旋转混合器轴　impeller shaft　12.8.1.9
旋转加料器　rotary feeder　11.2.8.2
旋转接头　rotary joint　8.8.12.3
旋转壳体　revolving [rotating] shell　12.6.1.10
旋转壳体(泵)　rotating casing (pump)　16.5.3
旋转离心机　revolving whizzer　11.2.12.6
旋转炉栅　rotating grate　2.2.18.11
旋转滤叶　rotating filter leave　9.4.9.2
旋转泥浆管　rotary hose　2.1.2.13
旋转撇油器叶片　rotating skimmer blade　3.1.14.9
旋转气塞阀　rotary air lock　23.1.2.4
旋转气体密封　rotary gas seal　8.8.4.(a)
旋转气闸　rotary air lock　10.4.1.17
旋转筛　rotary screen　2.3.10.2
旋转筛浆机　rotary sorter　2.3.10.23
旋转闪蒸干燥机　spin flash dryer　1
旋转式采样管　rotary type sampling tube　15.6.4.1
旋转式定位开关　drill-cast rotary switch　25.4.1.18
旋转式空气闭锁器　rotary airlock　8.4.1.27
旋转式入口气流调节器　swinging inlet damper　10.5.4.4
旋转式叶片泵　rotary vane type pump　16.1.2.2.11
旋转膛式炉　rotary-hearth furnace　25.1.9
旋转凸轮泵　rotary lobe pump　16.3.6
旋转凸(形)栅　rotating grate　2.2.11.10
旋转尾端运送器　slewing boom　23.2.5.6
旋转卸料阀　airlock discharger　23.1.6.7
旋转星形加料器　rotating segment feeder

2.2.87.6
旋转叶轮　rotating disk　9.4.9.10
旋转叶片　swirl vane　10.1.1.4
旋转元件　rotating element　9.4.9
旋转圆筒体　revolving cylinder　12.6.10.2
旋转真空过滤机　rotary vacuum filter　3.1.11.6
旋转蒸发器机体　revolving vaporizer body　5.5.1.40
旋转蒸汽接头　rotary steam joint　8.8.7.1
旋转炉床　rotary hearth　24.1.9.8
旋转轴　rotating shaft　19.4.1.5
旋转轴密封　sealing of rotating shaft　19
选矿池进料分离器　bath feed separator　15.5.3
选余物　tailing　11.6.3.11
选择器　selector　3.2.2.7
选择器活性污泥工艺　selector activated sludge(SAS)process　3.2.3
选择性催化还原　SCR(selective catalytic reduction)　2.2.111.13
选择性控制　selective control　1
选择性脱除硫化氢和二氧化碳的 SELEXOL 工艺流程　SELEXOL process for the selective removal of H_2S and CO_2　2.2.32
选择性系数　selectivity coefficient　1
循环　recirculation[recycle]　3.2.11.12
recycle　2.2.52.2;2.2.75.4;2.2.114.1;3.1.42.2
循环 CO_2　recycle CO_2　2.2.106.14
循环 EDC　recycle EDC　2.3.18.14
循环泵　circulating pump　16.1.2.3.39
circulation pump　7.2.12.5
recirculation pump　25.5.3.12;3.3.22.1
循环泵　recycle pump　2.3.4.15
循环槽　circulating vessel　12.5.5.20
recycle tank　3.3.4.11
循环床　circulating bed　14.1.4.(f)
循环催化剂　regenerated catalyst　2.2.98.11
循环催化剂提升管　cat circulation riser　14.2.6.11
循环萃取溶剂　recycle extraction-solvent　6.1.3.6
循环的蚯蚓,吃剩的污泥颗粒和填充剂(如果使用的话)　recycled earthworms, uneaten sludge particles, and(if used)bulking agent.　3.3.28.10
循环的填充剂　bulking agent for recycle　3.3.28.14
循环的热载体　recirculating heat carrier　3.3.15.17

循环二氧化碳　recycle CO_2　2.2.112.4
循环返回的细粉末　recycle fines　13.4.3.8
循环风机　circulating[recirculation]fan　12.3.1.2
循环固体　recycled solids　2.2.29.3;2.2.30.5
循环管　circulating pipe　7.2.5.11
circulation pipe　7.2.10.6
circulation tube　6.1.7.8
recirculation duct　8.4.5.13
循环管道再沸器　circulation pipe to reboiler　5.3.1.2
循环管线　circulating line　6.3.2.23
循环合成气　recycle syngas　2.2.24.2
循环回流管　circulating loop　2.1.7
循环活性污泥　RAS　3.1.30.B.5
循环机　circulator　2.2.46.6;2.2.47.6
循环急冷水　circulating quench-water　2.2.14.2
循环(快速)床　circulating(fast)bed　14.1.4.(e)
循环冷却　circulating[recycle]cooling　4.3.3.23
循环料　recycle　14.3.2.17
循环流　cycle-flow　23.3.4.8
循环流动　recycle-flow　3.1.14
循环流化床煅烧炉　circulating-fluid-bed calciner　14.2.1
循环煤浆　recycle slurry　2.2.95.26
循环母液　circulating mother liquor　6.1.17.8;2.2.55.3
循环气　recycle gas　14.3.4.24;2.2.37.4;2.2.45.2;2.2.48.4;2.2.73.1;2.2.77.5
循环气入口　recycle gas inlet　2.2.84.8
循环气压缩机　recycle gas compressor　2.3.18.9
recycle syngas compressor　2.2.44.9
循环氢气　recycle H_2　2.2.79.5
recycle hydrogen　2.2.94.7
循环溶剂　recycle solvent　2.2.95.15;2.2.79.5
循环式挡板集尘器　recirculating baffle collector　10.4.4
循环水　circulated water　2.2.66.15
recirculating water　3.4.1.8
recycle water　3.1.14.23
循环酸　recycle acid　6.1.5.12
循环筒　recirculation drum　15.6.3.19
循环污泥　recycled sludge　3.1.12.7
sludge recycle　3.2.10.5
循环物料　recirculated material　8.4.5.20
recycle　13.3.4.2

循环洗涤液 weak wash liquor 9.4.1.11
循环系统 circulation system 11.3.4.3
循环旋风分离器 recycle cyclone 2.2.23.3
循环压缩机 recycle compressor 2.2.25.10;2.2.37.3;2.2.41.2;2.2.42.2;2.2.48.3;2.2.74.1;2.2.79.6
循环盐水 recycle brine 2.4.2.8
循环盐水系统 circulating salt system 25.6.6
循环液流 recycle flow 3.1.36.2
循环油 recycling oil 2.1.10.10
循环重馏分 recycle heavy distillate 2.2.98.30

Y

压板 clamp plate 5.2.2.16
压差传感器 differential-pressure cell 3.4.18.3
压-吹法 press-and-blow process 2.3.9.30
压盖 bonnet 22.2.3.21
  gland 13.1.2.40
压盖螺母 gland nut 16.4.4.12
压盖螺栓 gland bolt 19.2.1.8
压光滚轮 calender roll 2.3.11.38
压光机 calender 2.3.11.36
压焓图 pressure-enthalpy diagram 1
压紧底部现场封口机 pinch-bottom field closer 23.4.3.11
压紧螺母 gland nut 15.7.6.10
压紧螺栓 compression bolt 4.2.1.10
压紧螺丝 compression screw 22.1.8.16
压紧器螺钉 catcher screw 17.2.7.9
压紧纤维填料 fibrous compression packing 19.2.3.8
压紧装置 pressing device 17.2.11.9
压力 p; pressure 20.2.4.2
压力变送器 pressure transmitter 3.5.18.6
压力表 PI; pressure indicator 3.1.6.19
  pressure ga(u)ge 14.4.3.11
压力表接管头 pressure gage connection 6.1.8.2
压力侧金属填料 metallic packing of pressure side 19.2.1.10
(压力)打包机 baling press 2.3.7.33
压力罐 overhead tank 2.1.7.6
压力罐用 for overhead tanks only 2.1.7.6
压力降 pressure gradient 14.1.2
压力接口 pressure tap 3.3.27.11
压力控制泵 pressure control pump 3.3.22.9
压力密封盖 pressure-tight cover 4.3.4.3

压力平衡弹簧 pressure equalizing spring 11.1.2.4
压力容器 pressure vessel 21.1
压力容器规范 pressure vessel code 21.1.2.21
压力式防水注油器 pressure type waterproof of lubricating device 15.6.4.12
压力式过滤机 pressure filter 2.2.92.16
压力试验 pressure testing 21.1.2.15
压力室 force chamber 16.2.3.32
压力水出口 pressure, water outlet 2.2.26.4
压力水进口 pressure, water inlet 2.2.26.6
压力调节阀 pressure regulating valve 6.2.1.2
  pressure-adjusting valve 15.7.2.5
压力系统 pressure 23.1.2.(a)
压力真空表接管口 nozzle for compound pressure and vacuum gauge 14.4.1.8
压力-真空系统 pressure-vacuum 23.1.2.(c)
压力-真空卸料与输送系统 pressure-vacuum unloading and transfer 23.1.2.(d)
压滤机[器] press filter 3.1.2.18
压片机 sheeting machine 13.2.4.21
压气机 compressor 25.3.2.1
压 press ring 5.2.4.10
压入配合 press fit 8.8.5.26
压水反应堆 pressurized-water reactor 25.2.2
压水系统 pressurized-water system 25.2.2.20
压送的气体 compressed gas 17.1.2
压送系统 pressure system 23.1.1.(b)
压碎机 crusher 13.3.5.8
压缩 compression 2.2.101.17
压缩段 compression section 13.2.1.15
压缩工序 compression process 2.2.45.12
压缩机 compression section 15.7.2.2
  compressor 2.2.20.1;2.2.31.13;2.2.63.4;2.2.68.10;2.2.76.15;2.2.77.16;2.2.111.4;2.2.115.16;2.2.116.2;3.1.32.10;17.1.1.17
压缩机出口 exhaust port 20.2.1.9
压缩机定子 compressor stator 25.3.1.9
压缩机进口导叶 compressor inlet guide vane 25.3.1.3
压缩机径向止推轴承 compressor radial-axial bearing 25.3.1.4
压缩机径向轴承 compressor radial bearing 25.3.1.18
压缩机静叶片 compressor stator blade 25.3.1.6

压缩机空气进口　air inlet to compressor　17.2.9.1
压缩机气缸　compressor cylinder　17.2.11.10
压缩机/燃气轮机/发电机　compressor/gas turbine/generator　2.2.112.3
压缩机吸入口　suction port　20.2.1.1
压缩机转子叶片　compressor rotor blade　25.3.1.8
压缩空气　compressed air　10.4.1.15
压缩空气出口　pressurizing air delivery　17.2.9.21
压缩空气罐　compressed-air tank　25.4.9.1
压缩空气排出下沉产品　air sink removal　12.5.4.(b)
压缩空气入口　compressed air inlet　11.5.5.6
压缩空气洗涤装置　compressed air washing unit　2.3.12.3
压缩空气站　compressed air installation　17.6
压缩空气、蒸汽或气体　compressed air, steam, or gas　11.5.3.3
压缩空气注入井　compressed air injection well　2.2.85.20
压缩空气总管　compressed air manifold　10.2.3.9
压缩因子　compressibility factor　1
压缩制冷　compression refrigeration　1
压榨滚轮　press roll　2.3.11.37
压榨机　size press　2.3.11.24
压榨区　press zone　9.6.2.3
压榨设备　expression equipment　9
压榨速率　expression rate　1
压制　pressing　2.3.9.33
压制模　press mould　2.3.9.34
压制品　press　8.1.1.13
亚稳区　metastable region　1
氩弧焊接　argon arc welding　19.4.1.9
烟囱　stack　2.2.110.22;2.2.111.21;2.2.116.6;3.3.24.10;3.4.5.7
烟囱壁　stack wall　3.5.11.6
烟囱连接段　connected stack　24.1.1.30
烟囱气体　stack gas　25.5.3.17
UOP 烟囱式装置　UOP stacked unit　14.2.6.(d)
烟道　breeching[stack]　8.8.1.8
烟道连接　stack connection　8.8.8.6
烟道气　flue[stack]gas　14.2.6.6
flue gas　2.2.118.7
烟道气放空　flue gas to atmosphere　2.3.1.28
烟道气(废气)出口　flue gas outlet　14.3.3.19
烟道气排出　flue gas out　3.3.9.7
烟道气去静电除尘器　flue gas to electrostatic precipitator　2.2.88.15
烟道气去烟囱　flue gas to stack　2.3.6.32
烟灰　soot　2.3.1.18
烟气　fumes　3.1.22.8
flue gas　2.2.27.1;2.2.71.15;2.2.116.5
烟气去储存待用　flue gas to sequestration　2.2.115.13
烟气去旋风分离器　stack gas to cyclone　14.2.5.4
烟气脱硫　flue gas desulfurization　2.2.115.20
烟气循环　flue gas recycle　2.2.115.8
烟气与尘粒　flue gas and dust particle　25.5.3.14
烟气至热量回收系统　flue gas to heat recovery system　2.2.72.8
延长线　extension lead　25.4.1.10
延长线插头　extension plug　25.4.1.11
延长线插座　extension socket　25.4.1.12
延迟焦化　delayed coking　2.1.1.40
延迟焦化装置　delayed coker　2.2.93.8
delayed-coking unit　2.1.6
延时曝气　extend aeration　1
研究与开发　R&D;research and development　1
研磨　grinding　11.6
研磨锤　grinding hammer　11.2.11.2
研磨、干燥、进料　grinding, drying, & feeding　2.2.25.4
研磨环　grinding ring　11.2.12.4
研磨介质　grinding media　11.3.1.4
grinding medium　11.3.4.4
milling medium　11.3.8
(研磨)介质挡板　media retainer　11.3.1.11
研磨设备　grinding equipment　11.2
研磨室　grinding chamber　11.3.1.2
研磨水泵　grinder pump with water　2.2.35.5
研磨水槽　water tank with grinder　2.2.35.4
研磨物　grind　8.1.1.17
研磨元件　(grinding) element　11.2.11.2
盐　salt　1.1.11.3
盐饼　salt cake　24.1.11.7
盐饼贮槽　salt cake storage tank　2.3.10.36
盐出口　salt outlet　7.2.12.13
盐急冷室　salt quenching chamber　3.3.11.13
盐矿　salt mine　25.2.5
盐水　brine　2.4.3.3
盐水排放　brine discharge　2.4.2.9
盐酸出口　hydrochloric acid outlet　5.4.2.5
盐酸塔　HCl column　2.3.18.17
盐悬浮液　salt suspension　7.2.12.14

盐液面　salt level　25.6.5.2
颜色指示器　colour indicator　25.4.9.37
衍生的　derived　25.1.1.5
掩埋　landfill　3.3.16.14
厌氧　ANA；anaerobic　3.2.2.10
厌氧的　anaerobic　3.1.47.2
厌氧区　anaerobic zone　3.1.33.9
厌氧污泥上流过滤层生物反应池　upflow anaerobic sludge blanket bioreactor　3.2.2.8
厌氧污泥消化池　anaerobic sludge digestion tank　3.1.26.(a).4；3.1.26.(b).5
厌氧细菌　anaerobic bacteria　1
厌氧消化系统示意图　schematic diagram of anaerobic digestion system　3.3.17
堰　weir　15.5.1.7
堰板　cut-off wall　15.6.4.7
weir　4.1.1.38
堰高　weir height　1
堰-上升管分布器　weir-riser distributor　5.3.4.14
堰式溢流　weir type overflow　11.3.2.4
扬程　head　1
height　1
lift　1
扬水高度　delivery lift　16.5.2.5
扬析　elutriation　1
扬析环　elutriation ring　12.3.2.7
阳离子交换器　cation exchanger　6.2.7.5
阳离子交换树脂床层　cation-exchange resin bed　6.2.2.(a)
阳离子型　cation　6.2.9.3
阳螺杆　male screw　17.3.2.3
氧　oxygen　2.2.32.2
氧化　oxidation　8.1.1.33
氧化氮吸收　nitrogen oxide absorption　14.3.1.(e)
氧化反应器　oxidation reactor　2.2.64.1
氧化钙　CaO　3.4.6.11
氧化焦炭　anode coke　2.2.91.11
氧化沥青　oxidized asphalt　2.1.1.76
氧化区　oxidation zone　3.1.32.11；3.4.53
氧化铁过滤器　iron oxide filter　3.3.20.13
氧化铁/木屑混合物　Fe_2O_3/woor chip mixture　3.3.20.12
氧化（脱氢作用）　oxidation (dehydrogenation)　2.3.8.19
氧化/硝化池　oxidation/nitrification tank　3.1.25.6
氧化锌反应器　ZnO reactor　2.2.8.5
氧化用空气　combustion air　2.1.5.1
氧氯化　oxychlorination　2.3.18.3

氧气　oxygen　2.2.30.1；2.2.33.4；2.2.67.3；2.3.18.3
氧气　O_2　2.2.25.5；2.2.29.9；2.2.112.1；2.2.114.4；2.2.115.4；2.2.117.4；2.2.119.5；2.2.118.4；2.3.19.4；3.1.47.7.8
氧气发生器　oxygen generator　3.1.46.1
氧气供气　oxygen feed gas　3.1.45.1
氧气耗尽的空气　O_2 depleted air　2.2.80.9；2.2.114.7
氧气扩散　O_2 diffusion　3.1.46.12
氧气/空气　oxygen/air　2.2.19.2
氧气/空气/蒸汽　oxygen/air/steam　2.2.110.4
氧气瓶　oxygen bottle　3.4.7.3
氧气压缩机　oxygen compressor　17.1.2.6
氧气/蒸汽　O_2/steam　2.2.26.2
oxygen/steam　2.2.31.8
样品存储器　sample container　3.5.10.7
样品进入　sample in　3.1.19.4
样品提取　samples extraction　3.4.7.7
窑　kiln　2.2.87.1
窑炉　kiln　3.3.7.22
窑炉燃烧器　kiln burner　3.3.7.21
摇臂　rocker arm　10.2.1.10
摇臂轴　rocker shaft　10.2.1.11
摇瓶培养　shake-flask culture　1
遥控板　remote-control panel　23.4.1.14
遥控板内打印机　printer recessed in panel　23.4.1.15
遥控风门机构　damper with remote control device　24.1.1.29
遥控手动风门　remote manual dampers　3.3.24.14
冶金焦炭　metallargical coke　2.2.93.12
叶轮　propeller　16.4.24.2
vane wheel　18.4.2.10
叶轮衬套　impeller bushing　16.4.25.6
叶轮磁体　impeller magnet　16.5.5.5
叶轮端　impeller[propeller] end　19.2.2.14
叶轮键　impeller key　16.4.9.28
叶轮紧固螺母　impeller nut　16.4.29.2
叶轮空距环　impeller spacer　17.4.2.21
叶轮口环　impeller ring　16.4.9.36
叶轮宽度　impeller width　16.4.3.20
叶轮类型　types of impeller　16.4.5
叶轮轮盖　shroud　16.4.6.1
叶轮轮毂　impeller hub　16.4.5.4
叶轮螺母　impeller nut　16.4.1.17
叶轮迷宫密封　impeller labyrinth　20.2.1.5
叶轮叶片　impeller vane　16.4.5.3

743

叶轮直径　impeller diameter[diam]　16.4.
  6.9
叶轮轴　impeller shaft　6.1.9.7
叶轮轴键　impeller key　16.4.12.17
叶轮轴联轴器　impeller shaft coupling
  16.4.26.1
叶片　blade[vane]　17.5.1.8
叶片安装底架　blade retainer　12.3.5.16
叶片紧固器　leaf fastener　9.2.7.7
叶片控制电机　vane control motor
  20.2.3.16
叶片(式)　vane　16.1.1.20
页岩(层)　shale　3.1.24.6
页岩干馏段　shale retorting zone　2.2.82.6
页岩淤浆　shale slurry　2.2.6.21
页岩油　shale oil　2.2.4.14
页岩预热段　shale preheating zone　2.2.82.5
液氨　liquid ammonia　2.2.37.11;2.2.38.15;
  14.3.1.25
液泵　liquor pump　2.3.10.33
液滴　drop　13.5.3.1
液滴防护屏　drop shield　18.2.4.16
液滴分离器　(water)drop separator　2.1.8.11
液封槽　(liquid)seal pot　5.2.1.21
液封环　lantern ring　16.4.12.15
  seal cage　16.4.19.22
液封筒　liquid seal pot　10.4.2.12
液缸底盖　liquid cylinder foot　16.2.4.18
液缸活塞杆　liquid piston rod　16.2.3.15
液缸活塞纤维填料环　liquid piston fibrous
  packing ring　16.2.3.24
液固萃取　liquid-solid extraction　1.
液化　liquefaction　2.2.97.7;3.3.18.18
液化产品　liquefaction product　2.2.21.15
液化气　liquefied gas　2.1.1.25
液化气泵　liquefied gas pump　16.1.2.33
液化石油气　LPG;liquefied petroleum gas
  2.1.1.1;2.2.77.8
液环泵　liquid ring pump　16.1.2.2.44
液环鼓风机　liquid ring compressor
  17.3.1.(b)
液环(式)　liquid-ring　16.1.1.28
液蜡　liquid paraffin　2.1.1.23
液力千斤顶　hydraulic jack　9.2.7.14
液流　liquid flow　5.2.6.6
  liquid stream　5.2.1.15
液流方向　flow　16.4.6.17
液流途径　developed path　16.4.3.8
液面　liquid level　12.4.2.5
  liquid tevel　10.5.4.26
  water level　15.6.1.31

液面测定仪　level detectors　2.1.10.2
液面计　liquid level gauge　21.2.4.5
液面计接口　liquid level connection　4.1.1.39
液面指示管　level indication pipe　13.1.2.18
液-气相界面　liquid-vapor interface　2.2.9.6
液态　liquid state　13.5.2.1
液态产品　liquid product　2.2.6.18
液态过冷　liquid subcooling　20.2.4.17
液态化床　fluidized-bed　4.3.3.(e)
液态金属　liquid metal　19.4.1.8
液态金属转动轴密封　rotating shaft seal for
  liquid metal　19.4.1
液态空速　LHSV;liquid hourly space velocity
  1
液态炉渣　liquid slag　24.3.1.20
液态碳氢化合物[烃]　liquid hydrocarbon
  2.3.5.18
液态锡基合金　liquid tin-base alloy　19.4.1.
  16
液体　liquids　2.2.79.14;3.4.4.8
液体出口　liquid outlet　5.2.9.7
  liquid out　10.5.12.4
液体的容积　liquid volume　21.2.5
液体顶热器　liquor preheater　2.3.10.8
液体分布　liquid-distribution　10.5.11.(a)
液体分布器　liquid distributor　14.3.1.4
液体分配器　liquid dispenser　2.2.39.6
液体和蒸汽　liquid and vapor　20.2.4.9
液体环　liquid-ring　18.4.1.3
液体或冷凝液接口　liquid or condensate
  connection　4.3.1.8
液体或蒸汽夹套　liquid or steam jacket　4.
  3.1.2
液体甲醇　liquid methanol　2.2.28.8
液体接口　liquid connection　4.3.1.6
液体进　liquid in　3.4.4.9
液体进口　liquid in[inlet]　10.5.12.2
液体进口与锥形机座组合体　combined
  liquid inlet and cone support　10.5.10.1
液体进料(口)　liquid feed　3.3.1.813
液体进入　influent　6.3.9.7
液体静电洁净器　electrostatic liquid cleaner
  15.3.1
液体流动比例抽样自动装置　automatic
  device proportions samples to flow　3.5.3
液体流化床　liquid fluidized bed　14.1.3.
  12
液体路线　liquid path　25.6.2.2
液体排出　liquid draw off　9.5.5.14
液体排放口　liquid drain　3.4.4.5
液体排泄　liquid drainage　10.4.2.13

744

液体喷射焚烧过程　liquid injection incineration　3.1.22
液体燃料　liquid fuel　3.3.1.9
液体入口　liquid feed　5.2.8.12　liquid inlet　10.5.11.2;3.4.4.12
液体深度　depth of liquid　21.2.5
液体渗透探伤　liquid penetration exam　21.1.2.15
液体收集器　liquid collector　5.3.1.15
液体通道　liquid path　15.3.1.8
液体吸收蒸气　absorb vapor in liquid　20.1.2.22
液体压力　pressure of liquid　20.1.2.23
液体堰　liquid dam　13.5.2.5
液体再分布器　liquid redistributor　5.3.4.10
液烃　hydrocarbon liquids　2.2.77.11
液位　liquid level　2.2.39.7
液位控制澄清　LC-Fining　2.2.22.3
液位控制器　LC;level controller　3.1.6.20
液下泵　submerged pump　16.1.2.35
液下搅拌叶片（任选）　submerged propeller (optional)　3.1.45.10
液相进料　liquid feed　2.3.5.2
液相区　liquid zone　20.2.4.16
NKK液相一步法二甲醚工艺流程　NKK one-step liquid phase dimethylether synthesis　2.2.50
液压泵　hydraulic pump　15.7.3.10
液氧储罐（备用）　LOX storage(stand-by)　3.1.46.2
液压传动的高压小流量隔膜压气机　high-pressure, low-capacity compressor having a hydraulically actuated diaphragm　17.2.14
液压夯锤　hydraulic ram　3.3.29.17
液压和气动式　hydraulic and pneumatic　23.3.2.5
液压滑块（式）　hydraulic ram　16.1.1.50
液压马达　hydraulic motor　15.7.3.3
液压设备　hydraulic plant　2.3.12.13
液压流体　hydraulic fluid　16.2.8.10
液液平衡　LLE;liquid-liquid equilibrium　1
液柱静压头　hydrostatic head　1
液柱压力计　manometer　1
一般沉降　typically sedimentation　3.2.1.10
一般流程　generalized flow diagram　2.2.90
一般注释　general note　21.1.2
一次沉降[澄清]池　primary clarifier　3.1.2.6
一次沉降池　primary settler　3.1.34.5　primary settling tank　3.1.27.6
一次澄清　primary clarification　3.1.37.7

一次底流污泥　underflow primary sludge　3.2.1.15
一次电路[回路]　primary circuit[loop]　25.2.2.22
一次反应区　primary reaction zone　3.2.6.16
一次分离　primary separation　2.1.9.1
一次分离罐　primary separation tank　2.1.9.9
一次和二次污泥　primary and secondary sludge　3.1.26.(a).1
一次活化槽　activated primary tank　3.2.10.(a)
一次空气　primary air　24.2.3.2
一次空气与煤　primary air and coal　24.2.2
一次扩压器　primary augmentor　18.3.5.10
一次冷却剂泵　primary coolant pump　25.2.2.5
一次钠泵　primary sodium pump　25.2.1.5
一次浓缩污泥　primary thickened sludge　3.1.26.(b).4
一次排放液　primary effluent　3.2.10.11
一次泡沫　primary froth　2.1.9.14
一次污泥　primary sludge　3.2.10.9;3.1.25.11;3.1.26.(b).1;3.1.27.8;3.1.34.6
一次污泥和二次污泥传统的和厌氧消化的联合处理　Combined treatment (anaerobic digestion) and traditional disposal of the primary and secondary sludge　3.1.26.(a)
一次污泥循环　primary sludge recycle　3.2.10.15
一次污泥厌氧消化和二次污泥在农业中的直接再利用　Anaerobic digestion of the primary sludge and direct reuse of the secondary sludge in agriculture　3.1.26.(b)
一次污水　primary effluent　3.1.34.7
一次污水进　primary effluent feed　3.1.44.1
一次性包装　disposable pack　2.3.13.28
一次旋流叶片　primary whirling vane　25.2.4.6
一次整形机　primary shaper　13.2.4.23
一段　first stage　2.2.21.4
一段水煤浆,氧气　1st stage coal/water slurry,O_2　2.2.22.2
一段有氧处理(去除五天生化需氧量)　first-stage acrobic(BOD_5 removal)　3.1.40.3
一段转化　primary reforming　2.3.3.7
一段转化炉　primary reformer　2.3.2.8
一级发酵罐/增稠器　single-stage fermenter/thickener　3.2.10.(c)
一级集尘箱[斗]　primary dust hopper　10.1.3.7

一级排液挡板　primary discharge baffle　10.1.4.11
一级喷射泵　first ejector　18.3.5.3
一氯二氟甲烷　R-12・chlorod　20.2.4
一氯三氟甲烷　R-13（chlorotrifluoromethane）　20.1.7.注 1.
一维模型　one-dimensional model　1
一氧化碳　CO　2.2.56.12；2.2.62.3　carbon monoxide　2.2.54.11
一氧化碳变换反应炉　CO-shift reactor　2.2.109.8
一氧化碳+氢气　$CO+H_2$　2.2.67.12
一氧化碳循环　CO cycle　2.2.66.8
一乙醇胺法脱除烟气中的 CO_2　Mono-ethanol-amina（MEA）based process CO_2 removal from flue gas　2.2.27
一乙醇胺溶液　MEA solution　2.2.27.8
仪表接口　instrument connection　4.1.1.34
移出热量　removing heat　20.1.2.6
移动床　moving bed　14.1.3.25
移动床催化裂化装置　moving bed catalytic cracker　14.3.3.(c)
移动床反应器　reactor with moving bed　14.3.3
移动床吸附器　moving bed sorber　6.3.11
移动床系统　pittsburgh moving-bed system　6.3.1
移动搅拌分段式反应器　moving-stirred-bed staged reactor　3.3.15.(c)
移动式吊架　traveling gantry　23.3.9.2
移动式卸料器　traveling tripper　23.2.3.(c)
移动式重锤　floating weight　15.7.1.9
移动填充床反应器　moving-packed-bed reactor　3.3.15.(a)
移走砂粒　to grit removal　3.1.8.4
遗传工程　genetic engineering　1
乙苯　ethylbenzene　2.1.1.54
乙醇　ethanol　2.2.40.11；2.2.57.14；2.2.59.1
95%乙醇　95%EtOH　2.2.41.3
乙烯　ethylene　2.2.67.22；2.2.68.20
乙醇酸　glycolic acid+H_2O　2.2.66.1
乙醇塔　ethanol tower　2.2.59.16
乙醇塔回流罐　reflux tank of ethanol tower　2.2.59.11
乙醇、乙酸　ethanol, acetic acid　2.2.58.1
乙二醇　ethylene glycol　2.2.66.14
乙二醇干燥器　glycol dryer　2.2.9.20
乙腈　acetonitrile　2.3.15.16
乙腈净化塔　acetonitrile purification column　2.3.15.7
乙炔　acetylene　2.2.1.9

乙炔饱和塔　acetylene saturator　2.2.68.17
乙炔进料　acetylene feed　14.3.1.2
乙酸　acetic acid　2.2.1.19；2.2.40.5；2.2.54.14；2.2.57.12；2.2.59.2
乙酸成品　CH_3COOH final product　2.2.56.10
乙酸乙酯产品　ethyl acetate products　2.2.58.3
乙酸乙酯、甲醇　AcOEt, MeOH　2.2.41.11
乙酸乙酯生产的工艺流程　ethyl acetate production process　2.2.58
乙酸乙酯塔　ethyl acetate tower　2.2.59.15
乙酸乙酯塔回流罐　reflux tank of ethyl acetate tower　2.2.59.10
乙酸乙酯与乙醇　ethyl acetate and ethanol　2.2.59.5
乙酸酯　acetate　2.2.40.6
乙酸酯化　acetic acid esterification　2.2.40.7
乙烷　C_2^0　2.2.69.24
ethane　2.2.71.18
乙烯　$C_2^=$　2.2.69.19
ethylene　2.3.18.1；2.2.71.17；2.3.19.5
乙烯精馏塔　ethylene fractionator　2.2.71.10
乙烯压缩机　ethylene compressor　17.1.2.9
已干产品排出口　discharge for dried product　8.9.6.14
已破碎的页岩区　fractured shale zone　2.2.85.19
已燃　combustion　2.2.85.23
已燃空气回流　combustion-air return　3.3.9.11
已洗净的瓶子　cleaned bottle　2.3.13.21
已卸出滤饼　discharged filter cake　9.3.9.6
已预热油页岩　preheated shale　2.2.5.5
已增浓的料浆　thickened slurry　9.4.9.5
异丁烷　isobutane　2.1.1.67
异丁烯　isobutene　2.1.1.68
异径管　flanged reducer　22.3.1.5
异径螺纹弯头　reducing street elbow　22.3.9
异径三通　street tee　22.3.2.11
45°异径弯头　45° street elbow　22.3.2.10
90°异径弯头　90° reducing elbow　22.3.1.3
抑制剂　inhibitors　2.3.19.1
易燃废液　flammable liquid waste　3.3.5.7
易熔元件　fuse cartridge　25.4.1.36
易碎物料　friable material　23.1.1.(c)
逸出　runoff　5.5.1.43
逸度系数　fugacity coefficient　1

意大利 Avellino 废水处理厂研究中的水处理工艺 Water processing flow diagram considered in the study of wastewater treatment plant in Avellino, Italy 3.1.25
意大利 Avellino 废水处理厂研究中的固体处理工艺流程图 Solid processing flow diagram considered in the study of wastewater treatment plant in Avellino, Italy 3.1.26
溢流 over flow 25.6.7.11
溢流板 overflow plate 8.5.1.15
溢流产品 O; overflow product 12.4.1
溢流管 overflow tube 20.1.8.4
overflow well 14.3.4.5
溢流口 overflow 9.2.5.14
溢流口液面控制 overflow level control 7.2.9.3
溢流溜槽 overflow launder 15.6.1.10
溢流溜槽出口 overflow launder outlet 15.6.1.4
溢流水 water overflow 2.2.26.10
溢流卸出 overflow discharge 12.6.9.1
溢流堰 overflow plate 15.6.5.8
overflow weir 12.4.2.8
weirs 3.5.5
溢流堰板 weir plate 8.5.6.21
溢流堰调节环 adjustable weir ring 10.3.4.7
溢流至澄清器 overflow to clarifier 3.1.11.2
溢流肘管 flooded elbow 10.5.13.4
因次分析 dimensional analysis 1
阴离子交换器 anion exchanger 6.2.7.8
阴离子型 anion 6.2.9.2
阴螺杆 female screw 17.3.2.4
引出管孔 port for outgoing pipe 21.3.1.9
引出线 lead wire 16.4.28.11
引风机 blower 23.1.1.5
exhauster 10.2.3.15
fan 13.4.1.5
ID fan; ID-induced draft 25.5.3.18
induced draft fan 3.1.21.16; 3.3.24.12
引上 drawn upward 2.3.9.10
引射器本体 inspirator body 24.2.7.7
饮用水去消毒和配送 potable water to disinfection and distribution 2.4.4.7
40 英寸的管道 40″duct 3.4.3.7
英国 Davy 公司甲醇合成工艺流程 Davy methanol synthesis process 2.2.44
英(国)热单位 Btu; british thermal unit 20.2.4
英国天然气协会 British Gas Corp 2.2.19.1
荧光灯管 fluorescent tube 25.4.1.61
营养素 nutrient 3.2.11.10

应变仪负荷传感器 strain-gauge load cell 23.5.3.9
应急旁路排风管 emergency bypass stack 3.3.3.9
硬的膏状物 hard paste 8.1.1.16
硬(颗粒) hard 8.1.1
硬蜡 hard wax 2.2.74.15
㶲 availability; exergy 1
永磁体干细粉磁力分离器 dry-fines permanent magnet separator 12.6.8
用过的催化剂 spent catalyst 14.1.12.19
用过的催化剂料斗 spent-catalyst hopper 14.2.7.13
用过的活性炭 spent carbon 6.3.1.8
用(油)页岩 spent shale 2.2.4.1
用河水冷却 cooling with water from river 24.2.3.5
用离心机和带式压滤机脱水 dewatering with centrifuge and belt-filter press 3.1.26.(a).7; 3.1.26.(b).10
用热设备 user 24.6.7.4
用于合成氨/尿素厂的两级低温甲醇洗工艺流程 2-Stage Rectisol wash for ammonia/urea plant 2.2.33
用于高温温克勒气化炉的加压进料系统 pressurized feed system for the high temperature Winkler gasifier 2.2.20
油 2.2.15.3; 2.3.17.2; 3.3.6.4
COED 油 COED oil 2.2.30.9
油杯 lubricator 22.1.2.15
oil cup 15.7.4.1
油杯盖 (oil)cap 16.4.29.13
油泵 lube pump 6.3.2.15
oil pump 2.3.16.4
油标 oil gauge 13.3.5.15
油槽 oil pan[reservoir] 9.5.10.15
油层 oil layer 3.1.4.9
油产品 oil product 2.2.6.17
油出口 oil out 6.1.9.14
油分离器 oil separator 18.2.3.6
油封 oil seal 11.1.5.3
油封环 oil seal 16.4.29.4
oil sealed ring 13.1.2.25
油封圈 oil seal 16.4.12.5
oil seal ring 16.4.22.8, 9
油封式旋转机械真空泵 rotary oil sealed mechanical pump 18.1
油封式正位移式打液泵 positive displacement oil-sealed liquor pump 5.5.3.2
油封填料函 oil-sealed stuffing box 5.5.2.4
油供入 oil feed 14.3.4.13

747

油管   oil pipe   12.3.5.9
   tubing   2.1.2.24
油罐吸入加热器   tank suction heater   2.1.
   7.14
油和煤气   oil and gas   2.2.85.16
油回收与过滤   oil recovery and filtration
   2.2.99.5
油或气体燃烧器   oil or gas burner   24.1.3.1
油急冷回收   recovery quench for oil   3.3.15.15
油孔   oil-hole   19.2.2.10
油扩散泵   oil diffusion pump   18.2.2
油漆废料   paint residue   3.3.5
油气   oil vap   2.2.84.14
油气分离器   oil(gas)seperator   5.3.3.3
油气化法   KOPPERS-TOTZEK   2.1.8
油气去冷凝工段   oil vapor to condensation
   section   2.2.88.17
油气收集   oil-vapor collection   2.2.4.9
油腔   oil reservoir(sump)   16.4.29.10
油燃烧器   oil burner   24.1.12.7
油润滑   oil lubrication   16.4.25
油沙   oil sand   2.1.9.2
油沙加工   processing of oil sand   2.1.9
油水分离器   oil/water separator   3.1.3
油位计   oil gauge   25.4.3.11
油位视镜   oil level sight glass   18.1.4.11
油箱   oil conservator   25.4.3.9
油压泵   hydraulic pump   9.5.10.13
油压表   oil pressure gauge   15.7.3.15
油页岩   oil shale   2.2.84.13
油-页岩半焦化室   oil-shale semicoking chamber   2.2.84.11
油页岩分布器   shale distributor   2.2.7.4
油页岩加工   oil-shale process   2.2.84
油页岩加料   shale feed   2.2.4.5
油页岩矿   oil-shale mine   2.2.82.1
油页岩矿柱   shale pillar   2.2.85.4
油预热器   oil preheater   2.3.16.6;2.3.17.1
油增压泵   oil booster pump   18.2.3
油脂   fat   14.3.1.2
油脂杯   grease cup   16.4.12.11
油脂的连续加氢反应   continuous hydrogenation of fat   14.3.1.(c)
油脂间歇加氢   batch hydrogenation of fat   14.3.1.(d)
油珠   oil globule   3.1.4.8
铀的液-液萃取   liquid-liquid extraction of uranium   6.1.2
游动床涤气器   fluidized-bed scrubber   10.5.4.5
游动滑车   travelling block   2.1.2.8
游离油   free oil   15.5.4.4

有臭味废气   odorous exhaust air   3.1.21.3
有盖玻璃皿   petri dish   2.3.11.7
有盖的熔化玻璃坩埚   covered glasshouse pot   2.3.9.46
有机废液   organic waste   3.1.23
有机酸   organic acids   3.3.18.14
有机污染物   organic pollutant   3.1.47.7
有机物与烃类的采样系统   sampling system for organic and hydrocarbon   3.5.17
有机相   organic phase   6.1.2.1
有机相流出物   organic effluent   6.1.12.10
有帽螺钉   cap screw   10.4.2.7
有气味的废气   odorous exhaust   10.5.13.6
有限元法   finite element method   1
有效能   availability;exergy   1
有效区   active area   5.2.6.7
有效因子   effectiveness factor   1
有氧燃烧与带有碳捕集和封存的整体煤气化联合循环的组合工艺   A combination of oxy-combustion and IGCC with CCS（carbon capture and sequestration）   2.2.112
有氧消化过的污泥   aerobically digested sludge   3.3.28.1;3.3.28.7
有载换接器   on-load tap changer   25.4.3.13
釉面陶土排污管   vitrified-clay sewer pipe   22.6.2.⑤
淤浆泵   slurry pump   14.2.4.1
淤浆出口   slurry out   10.5.11.6
淤浆进料枪   slurry feed gun   14.2.4.4
淤泥   sludge   1
淤泥泵   slurry pump   16.1.2.3.55
淤泥滤饼去处理   sludge cake to disposal   2.3.6.5
淤渣   sludges   3.3.1.1
余热锅炉   HRSG   2.2.110.16
余隙阀   clearance valve   17.2.13.2
余隙腔[容积]   clearance pocket   17.2.13.1   17.2.2.14
余隙调节   clearance control   17.2.13
隅角导向板   corner deflector   11.2.16.12
雨水积池   storm-water retention   3.1.1.1
预称量秤   preweigh scale   23.4.3.12
预成型的膏状物   preformed paste   8.1.1.15
预成型物   preform   8.1.1
预处理   preprocessing   3.3.1.2
预处理（曝气等）   pre-treatment（aeration etc.）   3.3.28.6
预处理设施   pretreatment facilities   3.3.20.8
预敷层过滤介质   precoat filter medium   9.3.8.6
预敷层转鼓真空过滤机   vacuum precoat

filter 9.3.8
预覆盖 precoated 9.1.4
预覆盖层 precoat 9.1.3.10
预混燃烧器 premix burner 24.2.7
预加机械载荷 mechanical preloading 19.3.4.2
预加氯消毒 prechlorination 2.4.5.3
预净化旋风分离器 pre-cleaner cyclone 2.2.23.6
预拉伸 preliminary stretching 2.3.8.45
预冷(却)器 precooler 13.3.5.3
预热 preheat 2.2.90.4;2.2.79.8
预热的气体 preheated gas 3.4.5.1
预热段 preheating 24.1.12.2
预热管 preheating tube 6.3.7.12
预热后的蒸气 preheated vapor 3.1.21.21
预热器 preheater 14.2.1.5;2.2.54.3;2.3.17.6
预热燃烧器 preheat burner 3.3.11.10
预热室 preheating compartment 14.2.5.8
预热用蒸汽 steam for preheating 3.1.23.6
预湿筛 pre-wet screen 12.5.1.2
预调节筛 preconditioning screen 12.5.5.6
预涂过滤系统 precoat filtration system 3.1.9
预涂助滤剂 precoat material 3.1.9.3
预涂助滤剂泵 precoat pump 3.1.9.1
预涂助滤剂混合槽 precoat mix tank 3.1.9.4
预涂助滤液 precoat liquid 3.1.9.5
预雾沫分离器 primary entrainment separator 10.5.4.22
预选料器 presorter 2.3.10.21
预置开关 initializing switch 23.5.3.1
预重整 pre-reforming 2.2.45.4
原废水 raw wastewater 3.1.28.1;3.1.30.B.1;3.1.45.2
原废水流入或一次水流出 influent raw wastewater or primary effluent 3.1.46.4
原废液 raw waste 3.1.12.4
原腐化物 raw septage 3.3.20.1;3.3.22.6
原料 feedstock 1
　feedstocks 2.2.3.1
　raw material 1
HC原料 HC feed 6.1.7.5
原料CO input material CO 2.2.55.1
原料甲醇 input material methanol 2.2.55.2
　methanol feedstock 2.2.51.1
原料进给装置 batch feeder 2.3.9.13
原料进口 feed inlet 12.2.6.11
原料煤 feed coal 2.2.14.8

原料煤 feed stock coal 2.2.33.1;2.2.35.17
原料煤 rau shuff coal 2.2.67.5
原料气 feed[raw]gas 2.2.37.6;2.2.48.5;14.1.12.16
原料气压缩机 feed gas compressor 2.3.3.4
原料入口 raw material inlet 14.4.1.7
原料水 raw water 6.2.1.1
原料液 feed solution 6.1.1.1
原料油 feed stock oil 2.3.16.3
原料油加入 raw oil charge 14.3.4.1
原煤 raw coal 2.2.92.8
原煤贮仓 raw coal storage hopper 2.2.37.17
原配料仓 existing hopper 23.1.9.3
原水 raw influent 3.2.10.1
原污水 raw wastewater 3.1.23.2
原型试验 prototype experiment 1
原油 crude(oil) 2.1.1.3
原油常压塔 crude atmospheric tower 2.1.4
原油初馏炼厂 crude-oil-topping petroleum refinery 3.1.1
原油加入 crude charge 2.1.4.3
原油与再沸器 crude and reboiler 2.1.4.6
原油蒸馏装置 crude unit 2.1.2
原子能动力站 atomic power plant 25.2.2
圆底长颈烧瓶 round-bottom boiling flask 3.5.14.29
圆环 ring 16.4.4.2
圆环多活塞 circumferential piston 16.1.1.27
圆环挤压机 ring extruders 13.2.1.(b)
圆盘板组件 disk 9.4.7.5
圆盘离心机系统 disk-centrifuge system 11.7.5
圆桶操作人员 drumming operator 23.4.1.5
圆筒 bowl 16.4.25.10
圆筒管 column pipe 16.4.23.13
圆筒或简单压力容器 drum or simple pressure vessel
圆筒连接器 column coupling 16.4.25.20
圆筒形玻璃冷凝器 cylindrical glass condenser 5.5.1.10
圆筒形管式加热炉 cylindrical-furnace tubular heater 24.1.3
圆筒形机壳 cylindrical shell 11.3.3.6
圆筒形油杯 solenoid oiler 16.4.25.28
圆筒支架 bowl bearing 16.4.25.7
圆筒状金属壳 cylindrical metal shell 15.3.1.4
圆头管 blunt pipe 22.2.9.(a)
圆形 circular 24.2.2.7
圆形沉降槽 circular sedimentation tank 3.2.15
圆形进口挡板 circular inlet baffle 3.2.15.2

749

圆形平膜 flat circular diaphragm 16.2.9
圆形清理球 ball cleaner(s) 12.2.3.11
圆形燃烧器 circular burner 24.2.1
圆周（式） peripheral 16.1.1.44
圆周速度 peripheral speed 1
圆锥 cone 12.3.4.13
圆锥破碎机 cone crusher 11.6.2.7
圆锥体 cone 19.1.5.8
圆锥体内 O 形密封圈 cone O-ring 19.1.5.7
圆锥筒 bowl 11.1.3.3
圆锥筒衬里 bowl liner 11.1.3.2
圆锥形壳体 cone body 12.1.1.28
圆嘴钳 round-nose plier 25.4.1.50
远端 far end 4.3.2.7
月牙形块 crescent 16.3.4.3
越界流体 grenzanhydrite 25.2.4.11
匀化 homogenization 1
匀化器 homogenizer 15.7.6
允差（公差） tolerance 21.1.2.19
运动黏度 kinematic viscosity 1
运输车 hauling 3.3.3.1
运送 dispatch 2.3.8.6
  handling 23.4
运行良好 perform well 23.3.5
运载 carrying 23.3.7.(b)
运渣带 conveyor 9.6.2.10

Z

杂质泵 contaminated medium pump 12.5.1.15
杂质排出口 tramp discharge 8.4.1.21
杂质脱除 impurity removal 2.2.3.12
再堆放 reclaiming 23.2.5.4
再堆放运送器 reclaim conveyor 23.2.5.2
再沸器 reboiler 1;2.2.27.10;2.3.19.12
再沸器返回口 reboiler return 5.2.1.8
再分布器 redistributor 5.3.1.12
再浮选 reflotation 11.2.6.2
再混合 note-remix 23.3.4.6
再活化和活性炭处理系统 reactivation and carbon-handling system 3.1.17
再活化空气加热器盘管 reactivation air heater coil 6.2.8.4
再加热 reheating 2.3.9.27
  reheat 2.2.115.18
再加热盘管 reheat coil 20.2.2.22
再加热器 reheater 20.2.2.16
再净化器浮选 recleaner flotation 12.8.2.16

再磨碎球磨机 regrind ball mill 11.6.2.20
再热燃烧室 reheating combustion chamber 25.3.2.6
再生 regenerating[regeneration] 6.3.3.5
  regenerative 16.1.1.45
再生泵 regenerative pump 16.1.2.2.20
再生泵操作 regeneratine turbine pump 16.4.16
再生催化剂 regenerated catalyst 14.1.12.18
  regenerating catalyst 2.2.72.3
再生催化剂料斗 regenerated catalyst hopper 14.2.7.3
再生罐 regeneration tank 6.2.5.8
再生过程 regeneration process 6.2.3.11
再生过的活性炭 regenerated carbon 6.3.1.5
再生后的炭 revivified char 6.3.12.11
再生回热循环的燃气轮机 regenerative-cycle gas turbine 25.3.3
再生活性炭储仓 reactivation carbon storage 3.1.17.3
再生或处置 regeneration or disposal 3.1.43.9
再生剂 regenerant 6.2.4.3
再生剂进口 regenerant inlet 6.2.3.1
再生加热炉 reactivation furnace 3.1.17.6
再生空气 regeneration air 6.2.8.1
再生炉 regenerating furnace 6.3.1.11
再生气 regenerating gas 14.1.12.15
再生气冷却器 regen.gas cooler 2.3.4.10
再生气体 regeneration gas 6.3.3.2
  regenerated gas 2.2.76.1
再生气体加热器 regen gas heater 6.3.3.10
再生气体冷却器 regen gas cooler 6.3.3.7
再生气体压缩机 regen gas compressor 6.3.3.6
再生器 regenerator 14.1.12.20;2.2.68.3;2.2.69.2;2.2.71.4;2.2.72.5;2.2.75.2;2.2.76.1
再生设备 regeneration equipment 6.3.4
再生石灰 reconverted lime 2.3.10.52
再生塔 desorption column 5.1.1.6
  regeneration tower 2.2.59.17
再生塔回流罐 reflux tank of regeneration tower 2.2.59.12
再生循环 regeneration cycle 6.2.3.3
再生/循环换热器 regen./recycle exchanger 2.3.4.9
再生循环压缩机 regen.recycle compressor 2.3.4.11
再生液 regenerant solution 6.2.2.(b)

750

再生乙二醇　regenerated glycol　2.2.59.7
再循环　recycle　11.6.3.9
再循环泵　recirculating pump　2.2.36.5
再循环澄清器　recycle clarifier　9.5.14.(b)
再循环的淋洗水泵　recycle-rinsewater pump　12.5.5.11
再循环的中间富集物　recycle-medium concentrate　12.5.5.8
再循环风机　recirculating fan　8.9.2.5
再循环管　recirculation pipe　7.2.5.5
再循环管道　recirculating duct　4.3.4.25
再循环淋洗水　recycled rinse water　12.5.5.24
再循环热空气　recirculated hot air　4.5.2.3
再循环水流径　recirculation water flow path　25.2.3.1
再装料操作　reclaiming　23.2.4
在高压釜内聚合　polymerization in the autoclave　2.3.8.33
在线　on-line　1
载点　loading point　1
载荷传感器　load cells　23.5.2.13,14
载气　carrier gas　14.1.10.4
载气供给与控制　carrier-gas supply and control　3.5.20.1
载气进口　carrier gas inlet　8.9.6.16
载热体　heat carrier　1
heating medium　1
载体　supporter　1
载体加入　carrier in　15.2.2.8
载体流出　carrier out　15.2.2.9
载运的物料　material in carrying run　23.3.7.9
脏空气　dirty air　3.4.4.11
脏空气入口　contaminated air inlet　10.5.1.14
脏空气入口　dirty air inlet　10.5.1.25
脏气体侧　dirty gas side　10.2.1.27
脏气体进口　dirty gas inlet　10.4.1.11;10.5.15.1
脏气体入口　dirty gas in　10.5.11.5
脏水泵　dirty water pump　16.1.2.3.17
脏水出口　dirty water outlet　10.5.1.6
dirty water outlet　10.5.4.9
脏液体入口　dirty liquid in　15.3.1.9
脏油　dirty oil　15.5.5.14
脏油进口　dirty oil inlet　15.5.5.1
造粒　pelletizing[solidifies]　1
造粒板　marume plate　13.3.5.27
造粒机　granulator[pelletizers]　4.3.2.23
11.6.3.18
pelletizer　2.3.16.12;2.3.17.11
造粒喷头组件　prill spray assembly　13.1.2
造粒设备　granulation equipment　13
pelletizing plant　13.2.4
造粒塔　prill tower　2.2.92.20
造粒塔　prilling tower　13.1
造粒筒　prill bucket　13.1.2.15
造粒筒支承套　support prill bucket　13.1.2.19
造粒用水　granulation water　2.3.16.11
造气　gasification　2.2.38.3
造气装置　gasification unit　2.3.2.2
造纸　papermaking　2.3.11
造纸机料箱　breast box of a paper machine　2.3.11.11
centrifugal cleaners　2.3.11.11
造纸机生产线　paper machine production line　2.3.12.13～28
噪声水平　noise level　1
增稠　thickening　3.2.1.18
增稠池　thickener　3.1.2.12
增稠固体相　concentrated solid phase　9.5.14.6
增稠器　thickener　1
增稠区　thickening zone　7.2.10.1
增稠与脱水　thickening & dewatering　3.2.1.20
增浓器　densifier　12.5.1.13
增强环　reinforcing ring　22.5.1.(h)
增强热固性树脂　reinforced thermosetting resin　22.6.2.⑥
增湿器　humidifier　6.2.8.2;3.4.3.8
增湿雾化器　humidification spray　10.5.1.5
增压　pressurization　3.1.14
增压泵　booster pump　16.1.2.1.7
增压泵机体　booster body　18.3.1.6
增压流化床燃烧　pressurized fluid bed combustion　2.2.116.10
增压喷射泵　booster ejector　18.3.6.1
憎水分离膜　hydrophobic separatory membrane　6.1.8.5
甑式多室焚烧炉　retort multiple-chamber incinerator　3.3.12
渣排口　slag tap　2.2.14.3
渣斗　slag hopper　2.2.110.9
渣油　residual oil　2.1.1.32
渣油加氢裂化反应器　Resid hydrocracker reactor　2.1.10
渣油进料　residuum feed　14.3.4.27
渣水处理　slag water treatment　2.2.31.12
闸板　gate(disc)　22.1.2.4
榨汁挡板　juice shield　9.6.1.2
榨汁盘　juice pan　9.6.1.1

751

窄孔　close opening　11.6.2.2
毡垫圈　felt washer　9.3.3.43
辗辊　roll　4.3.4.23
胀塑性流体　dilatant fluid　1
沼气至贮气柜　biogas to gasometer　3.1.26.(a).10;3.1.26.(b).6
照明设备　illumination　3.1.16.9
照相机　camera　12.7.4.8
罩　cap　14.4.3.1
罩子　cap　10.5.11.2
遮环　shroud ring　6.1.12.11
折边　flanging　21.4.1.3
折边标准碟形封头　flanging standard dishes head　21.4.1.3
standard dishes head　21.4.1.3
折边平封头　flanging-only head　21.4.1.1
折边锥形封头　toriconical head　21.1.1.25
折叠编织填料　laid packing　19.2.3.2
折叠织物填料　hooded cloth packing　19.2.3.6
折流板　baffle　12.3.1.4
折流环　deflector ring　10.1.1.5
针(形)阀　needle valve　3.5.11.25
针状高压电极　needle-point HV electrode　15.3.1.6
珍珠岩　pearlite　21.2.2.7
真空表　vacuum gauge　3.5.16.12
真空表接口　vacuum gage connection　18.3.1.8
真空成熟槽　vacuum ripening tank　2.3.7.11
真空带式　vacuum band　8.1.1.⑪
真空分布器　vacuum distributor　9.4.1.14
真空管线　vacuum line　23.1.2.9
真空罐　vacuum hopper[tank]　23.1.2.10
真空回转干燥器　vacuum rotary dryer　8.8.10
真空计　vacuum gauge　3.5.7.7
真空接口[头]　vacuum connection　8.8.11.1
真空精馏　vacuum distillation　2.2.66.10
真空精馏塔　vacuum distillation column　2.2.94.17
真空连续过滤　continuous vacuum filtration　9.3.1
真空连续回转过滤机　continuous rotary vacuum filter　9.3.2
真空盘架式干燥器　vacuum-shelf dryer　8.2.2
真空盘式　vacuum tray　8.1.1.⑫
真空平衡管线　vacuum balancing line　18.5.1.4
真空闪蒸罐　vacuum flash tank　2.2.102.10
真空设备　vacuum equipment　18

真空受液罐　vacuum receiver　9.3.1.11
真空顺序系统　vacuum sequencing system　23.1.2.12
真空系统　vacuum(system)　18.5
真空下聚合　polymerization under vacuum　2.3.8.54
真空箱　vacuum box　2.3.11.15
真空-压力和物料接收器　vacuum-pressure & material receiver　23.1.2.20
真空源　vacuum sourcer　15.1.1.10
真空蒸馏　vacuum distillation　1;2.2.79.13
真空装置　vacuum equipment[unit]　7.2.8.11
真实就地干馏　true in-situ retorting　2.2.87
真实气体　real gas　1
砧面垫板　anvil plate　11.2.9.8
振打棒　rapper bar　10.3.1.14
振打电机　vibro motor　10.2.2.5
振打机构　rapping mechanism　10.3.5.15
振打机构(管)　rapping mechanism(tube)　10.3.5.11
振打机架　vibro frame　10.2.2.4
振荡输送机　reversible shuttle conveyor　23.2.3.(b)
振动的动力装置　vibratory power unit　4.3.4.13
振动隔离体　vibration isolator　4.3.4.15
振动加料器　vibrating feeder　12.7.4.2
振动流化床　vibrated fluidized bed　1
振动盘式干燥器　vibrating tray dryer　8.9.6
振动器　shaker[vibrator]　8.9.6.8
振动球磨机　vibratory ball mill　11.3.2
振动筛　vibrating screen　1
振动输送机　vibrating conveyor　23.3
振动输送机　vibratory-conveyor adaptation　4.3.4
振动输送式　vibrating-conveyor type　13.5.2
振动送料机　shaking feeder　8.9.3.6
振动斜壁自动出料离心机　inclined vibrating wall self-discharge　9.5.1.(j)
振动卸料　vibratory discharge　9.5.2.7
振动研磨机　vibro-energy mill　11.3.1
蒸发　evaporation[evaporate]　2.3.6.19
Oslo蒸发结晶器　Oslo evaporative crystallizer　7.2.6
蒸发结晶器　evoporalor cryslaltaer　7.2.13
蒸发器　evaporator　20.1.2.8;2.2.55.8;2.2.61.3;2.2.76.8;3.3.22.12
vaporizer　7.2.1.20
蒸发器壁　evaporator wall　5.5.2.3

752

蒸发器鼓内最高与最低液面　maximum and minimum liquid levels in vaporizer drum　25.6.2
蒸发器或进料缓冲槽　evaporator or feed surge tank　2.3.6.21
蒸发器-进料缓冲池　evaporator-feed surge pond　2.3.6.23
蒸发器壳体　vapor body　7.1.14.5
蒸发器型式　evaporator type　7.1.13
蒸发室　evaporation［flash］chamber　7.2.10.3
蒸发温度　evap.temp.　20.1.7.3
evaporating temp.　20.2.4.6
蒸发压力　evap.press.　20.1.7.3
蒸解锅　digester　2.3.10.7
蒸馏　distillation　2.2.97.8
蒸馏工序　distillation process　2.2.45.15
蒸馏塔　distillation column［tower］　20.1.9.1；2.2.70.11；2.3.19.16
蒸馏液　distilland　5.5.1.11
蒸馏液出料泵　distilland withdrawal pump　5.5.1.13
蒸馏液进料口　distilland feed　5.5.2.5
蒸馏与吸收设备　Distillation and absorption equipments　5
蒸馏柱　still　2.2.98.21
蒸气　vapo（u）r　5.2.1.16；2.4.3.8
蒸气产品　vapor product　2.1.9.32
蒸气出口　vapo（u）r outlet　5.2.1.1
vapor out　10.4.3.5
蒸气出口通冷凝器　vapor outlet to condenser　5.3.1.19
蒸气分离器　vapor separator　6.3.2.21
蒸气风机　vapor fan　3.3.24.13
蒸气降温器　vapor desuperheater　7.1.14.12
蒸气进口　vapo(u)r feed［inlet］　5.3.1.7
蒸气进料　vapor feed　2.3.5.1
蒸气进入　vapor in　10.4.3.4
蒸气冷凝段　vapor condensation zone　2.2.85.15
蒸气流　vapor flow　2.4.3.9
vapor steam　18.2.3.12
蒸气路线　vapor path　25.6.2.1
蒸气室主体　vapor body　7.2.8.4
蒸气压缩　vapor compression　20.1.2.19
蒸气域　vapo(u)r space　5.5.1.37
蒸气罩　vapor hood　8.7.1.10
蒸汽　steam［vapour］　2.2.3.13；2.2.18.7；2.2.25.8；2.2.28.6；2.2.30.12；2.2.37.5；2.2.44.8；2.2.46.13；2.2.57.15；2.2.61.9；2.2.74.5；2.2.78.2；2.2.109.7；2.2.110.17；2.2.116.7；2.2.118.8；2.3.18.5；3.3.23.5；14.2.6.5
蒸汽包　steam collector　14.3.2.16
蒸汽重整　steam reforming　2.2.97.5；2.2.3.11
蒸汽重整炉　steam reformer　2.2.107.17
蒸汽出口　steam out［outlet］　10.1.4.14
蒸气出口法兰　vapor flange　8.9.5.2
蒸汽发生器　steam generator　2.2.22.4；2.2.108.8
蒸汽发生系统　steam-generation system　25.5
蒸汽阀　steam valve　20.2.2.24
蒸汽分配盘　steam manifold　8.8.6.2
蒸汽干燥器　steam dryer　18.3.5.12
蒸汽缸　steam cylinder　16.2.1.1
蒸汽缸端盖　steam cylinder head　16.2.4.1
蒸汽缸活塞　steam piston　16.2.3.4
蒸汽管　steampipe　2.3.10.64
蒸汽管路　steam line　25.2.1.11
蒸汽管式回转干燥器　steam-tube rotary dryer　8.8.6
蒸汽管线　steamline　25.6.6.1
蒸汽锅炉　steam boiler　2.3.1.16
蒸汽恒温控制阀　thermostatically controlled steam valve　15.5.5.11
蒸汽活塞　steam piston　16.2.4.4
蒸汽活塞杆　steam piston rod　16.2.3.7
蒸汽活塞杆接头　steam piston spool　16.2.3.12
蒸汽活塞杆螺母　steam piston-rod nut　16.2.4.3
蒸汽活塞杆锁紧螺母　steam piston-rod jam nut　16.2.4.11
蒸汽活塞杆填料函衬套　steam piston-rod stuffing box bushing　16.2.3.9
蒸汽活塞杆填料函压盖　steam piston-rod stuffing box gland　16.2.3.11
蒸汽活塞杆填料函压盖衬套　steam piston-rod stuffing box gland lining　16.2.3.10
蒸汽活塞环　steam piston ring　16.2.3.5
蒸汽活塞螺母　steam piston nut　16.2.3
蒸汽集管箱部面　section through steam manifold　8.8.6.7
蒸汽加热鼓　steam-heated drum　8.7.2.4
蒸汽加热管　steam-heated tube　8.8.6.9
蒸汽加热器　steam heater　14.3.2.10
蒸汽加热装置　steam-heating unit　8.3.2.5
蒸汽加入　steam inlet　7.2.11.5
蒸汽接口　vapor connection　4.3.1.3
蒸汽接头　steam joint　8.8.7.(b)
蒸汽解吸段　steam section　6.3.11.12
蒸汽进口　steam in［inlet］　6.3.6.2

蒸汽进口旋转接头　rotary joint for steam inlet　8.8.10.2
蒸汽进入　steam in[inlet]　16.2.1.3
蒸汽＋空气或氧气　steam＋air or O_2　2.2.18.4
蒸汽轮机　steam turbine　2.2.110.18；2.2.117.21
蒸汽盘管　steam coil　10.3.4.3
蒸汽泡　steam bubble　22.2.3.16
蒸汽喷射　steam jet　20.1.2.20
蒸汽喷射泵　steam ejector　18.3.5
steam jet　16.1.2.4.4
steam-jet ejector　18.3.1
蒸汽喷射冷却单元　steam jet water cooling unit　20.1.10.19
蒸汽喷射系统　steam-jet system　18.3.6
蒸汽喷射制冷循环　steam-jet refrigeration cycle　20.1.1
蒸汽喷嘴　steam nozzle　18.3.1.2
蒸汽去汽提段　steam to stripping section　14.2.7.19
蒸汽入口　steam[vapor]inlet　7.1.3.2
蒸汽式空气加热器　steam air heater　8.4.1.32
蒸汽室　steam chest　16.2.4.48
steam dome　18.3.1.3
蒸汽室顶盖　steam-chest cover　16.2.3.53
steam-chest head　16.2.4.50
蒸汽/水　steam/water　4.4.2.8
蒸汽/水出口　steam/water out　4.4.1.4
蒸汽缩颈管　steam neck　8.8.6.1
蒸汽调节器　steam regulator　7.1.2.18
蒸汽透平[涡轮机]　steam turbine　25.5.2.8
蒸汽涡轮机/发电机/冷凝器　steam turbines/generator/condenser　2.2.112.7
蒸汽压力　steam pressure　16.2.1.5
蒸气压缩循环　vapor-compression cycle　20.2.1
蒸汽/氧气　steam/oxygen　2.2.13.9；2.2.19.4
蒸汽总管　steam manifold　8.8.7.4
蒸煮麦芽汁用的锅炉　hop boiler for boiling the wort　2.3.12.52
蒸煮液贮槽　storage tank for the cooking liquor　2.3.10.46
整袋顶部吊出卸料式　with top discharge by bag withdrawal　9.5.3.(d)
整合　integration　2.2.93
整块式塔板　intact tray　5.2.10.27
整流高压器机组　transformer rectifier set　10.3.1.10

整体法兰　integral type flange　22.4.2.(d)
整体复合板　integrally clad plate　21.1.1.12
整体盖板　integral cover　4.1.2.(b)
整体接头　integral fitting　15.5.1.2
整体器壁　solid wall　9.5.14.(a)
整体式法兰　integral type flange　21.1.1.53
整体式防喷溅挡板　integral splash baffle　18.2.4.2
整体式风机　integral fan　25.2.9.8
整装填料　structured packing　1
正常液面　normal level　14.3.4.23
正丁烷　normal butane　2.1.1.66
正丁烯　n-butene　2.1.1.69
正方形排列　square pitch　4.1.5.1
Kellogg正流式装置　Kellogg orthoflow cat 14.2.6.(g)
正三角形排列　equilateral triangle pitch　4.1.5.3
正视　elevation　8.8.10.16
正视图　face view　13.3.3.(a)
正位移驱动装置　positive drive　23.3.2.1
正压输送　positive-pressure conveying　23.1.5
正压输送系统　positive pressure system　23.1.8
正装料圆桶　drum filling　23.4.1.3
支座　stay　21.1.1.21
支撑板　spider plate　6.1.17.12
支撑表面　stayed surface　21.1.1.21
支撑顶　tank-supported roof　21.2.1.1
支撑螺栓　stay bolts　21.1.1.22
支撑体　support　3.1.29.3
支承板　support(ing)plate　5.5.1.12
支承板固定螺钉　abutment plate mounting screw　19.1.5.18
支承扁钢　support bar　11.1.7.13
支承带　support belt　9.4.8.7
支承格栅　support grid　5.3.1.16
支承管　support column　16.4.22.7
支承滚轮　supporting rolls　12.1.1.20
支承结构　supporting construction　25.1.13.2
支承绝缘体　supporting insulators　10.3.6.24
支承绝缘子　support insulator　10.3.4.1
支承块　support piece　4.2.2.8
支承面板　abutment plate　19.1.5.19
支承平面　support level　8.4.1.7
支承轮　support wheel　12.2.4.21
支承输电系统的绝缘体　discharge system support insulator　10.3.1.5
支持板　support plate　15.2.3.5

支持圈 supporting ring 5.2.2.4
支耳 side lug 9.1.1.5
支杆 arm 9.5.4.4
支杆销 arm pin 9.5.4.5
支管 branch main 17.6.2.6
支管为异径的斜三通 reducing lateral on branch 22.3.1.10
支梁固定板 minor beam support clamp 5.2.7.2
支架 mounting hinge 24.2.4.11
　support 9.5.11.14
　supportor 13.5.1.3
　yoke 21.1.1.31
支架绝缘子 support insulator 25.4.9.4
支架螺栓 yoke bolts 22.1.3.6
支链(分子) branched (molecules) 14.3.4.(c)
支腿 leg 21.3.3.4
　leg 9.2.5.7
支线 branch circuit 2.1.7.2
支轴销 fulcrum pin 16.2.4.30
支柱 support 4.2.2.7
　support column 21.2.3.7
　support leg 5.2.4.18
　supporter 9.5.4.37
支柱帽 supporter cap 9.5.3.33
支座 support 21.3
　supporting block 21.2.4.6
支座板 base plate 13.3.5.19
支座法兰 support flange 8.9.5.7
支座焊接装配体 bracket-welded assembly 8.8.5.29
知识库 knowledge base 1
织物 cloth 8.8.5.19
脂族化合物 aliphatics 6.1.3.12
执行机构弹簧 actuator spring 22.1.9.5
直动泵 direct acting steam pump 16.1.2.4.6
直接过程加热器 direct-process heater 25.5.1.(d)
直接加热回转 direct rotary 8.1.1.⑱
直接加热式并流回转干燥器 direct-heat cocurrent rotary dryer 8.8.3
直接加热式回转干燥器 direct-heat rotary-dryer 8.8.2
直接接触冷冻结晶器 direct-contact-refrigeration crystallizer 7.2.2
直接接触式冷却器 direct contact cooler 2.2.27.3
直接氯化 direct chlorination 2.3.18.1
直接明火 direct firing 21.1.2.10
直接排放到大气中 direct discharge to atmosphere 3.3.25.12
直接燃烧 directly fired 3.3.15.(a)
直接燃烧空气加热器 direct-fired air heater 10.3.5.2
直接使用蒸汽 direct steam use 25.1.2.5
直接脱硫 direct desulfurization 2.2.2.38
直接氧化 direct oxidation 14.3.3.(a)
直接液化 direct liquefaction 2.2.89.12
直接蒸发冷却器 direct evaporative cooler 6.2.8.9
直径 diameter;di. ;dia. ;diameter 13.3.3.(a)
直径100英尺钢罐 100′DIA steel tank 3.4.3.2
直立挡板 vertical-slot baffle 3.1.3.7
直立升降式止逆阀 vertical lift check valve 22.1.1.(m)
直立式(径向)抄板 radial flights 8.8.1.17
直链分子 straight chain molecules 14.3.4.(c)
直列式多室焚烧炉 inline multiple-chamber incinerators 3.3.13
(直流变交流)逆变器 inverter 2.2.113.1
直流电操作的电磁泵 d.c. electromagnetic pump 16.5.4
直流电输出 D.C.output 10.3.1.7
直馏柴油 straight-run diesel oil 2.1.1.9
直馏煤油 virgin kerosene 2.1.1.8
直馏汽油 distilled gasoline 2.1.1.7
直馏重柴油 straight-run heavy diesel fuel 2.1.1.10
直燃式回转烘干机的工艺流程 flowsheet of direct-fired rotary dryer 3.3.25
直梯 ladder 5.2.1.11
直凸轮 straight-lobe 17.3.3.11
直线段衬里 straight lining 11.1.8.6
直线型往复压缩机 straight-line reciprocating compressor 17.2.1.2
直叶瓣 straight-lobe 17.3.3.11
直叶片 straight-blade 17.5.1.8.1
pH值传感元件 pH sensor 3.1.12.3
止动环 snap ring 16.4.28.1
止动螺钉 set screw 16.4.6.30
止回阀 COV; check-out valve; NRV; non-return valve; backpressure valve 3.1.14.18
止逆阀 check valve 16.2.7.7
　non-return valve 22.1.10
止逆阀顶丝 check valve spindle 16.4.28.1
止逆阀罩 check valve case 16.4.28.37
止推垫圈 thrust washer 16.4.1.14,18
止推辊 thrust roll 8.8.5.31

755

止推辊轴　thrust roll shaft　8.8.5.29
止推辊轴瓦　thrust roll bushing　8.8.5.30
止推滚动轴承　thrust roll bearing　8.8.5.32
止推滚珠轴承　thrust ball bearing　13.3.5.5
止推环　thrust collar　16.4.24.11
止推轴承　thrust bearing　13.2.3.4
止推轴承定位环　thrust collar　20.2.1.17
止推轴承调节垫　thrust pad　20.2.1.16
纸幅　web　2.3.11.30
纸浆泵　pulp pump　16.1.2.3.29
(纸)浆槽　vat　2.3.11.47
纸浆废液　paper mill waste　13.2.1.(b)
纸浆干燥机　pulp-drying machine　2.3.10.61
纸浆水泵　pulp water pump　2.3.10.55
纸料池　stuff chest　2.3.10.85
指示孔　tell tale holes　21.1.1.17
指示盘　indicator dial　15.6.4.2
指针式真空表　dial-type vacuum ga(u)ge　8.2.2.4
酯干燥塔　ester drying tower　2.2.57.10
酯化　esterification　2.2.66.11
酯化反应器　esterification reactor　2.2.64.2
酯化反应塔　esterification reactor tower　2.2.57.4
酯化釜　esterification reactor　2.2.60.5
酯化塔　esterification tower　2.2.58.10;2.2.59.14
酯化塔回流罐　reflux tank of esterification tower　2.2.58.5;2.2.59.9
酯塔　ester column　2.2.64.5
酯蒸出塔　ester distilled tower　2.2.57.7
至饱和塔　to desaturation tower　2.2.43.15
至大气　to atmosphere　3.3.16.18
至火炬排放　to torch　2.2.55.20
制备　preparation;prep.;prepare　2.2.2.2
制备分子筛吸附剂　making molecular sieve adsorbents　6.3.10
制锭　pastillization　13.5.3
制锭操作　operation of pastillization　13.5.3
制动轮　brake pulley　9.5.4.6
制动器　dray　9.5.11.11
制动手柄　brake handle　9.5.4.30
制动闸　brake hand　9.5.4.8
制动闸衬　brake lining　9.5.4.7
制动轴　brake axle　9.5.4.32
制浆　pulping　2.3.10
制冷　refrigeration;refr.;refrigerate　20
制冷机　refrigeration machine　20.2.2.23
制冷剂　refrigerant　20.1.8.18;2.2.28.12
制冷剂回流管　refrigerant return pipe　20.2.3.8
制冷剂节流孔板　refrigerant reducing orifice　20.2.3.9
制冷系统　refrigeration system　20.1
制冷循环　refrigeration cycle　1
制麦芽　preparation of malt　2.3.12.1～41
制瓶机　bottle-making machine　2.3.9.21
制氢　hydrogen making　2.1.1.2
hydrogen production　2.2.1.8
Monsanto/BP　制乙酸工艺流程　Monsanto/BP process for acetic acit production　2.2.56
制造厂　fact.;factory;manufacturing plant　3.3.7.11
质量　qua;quality;weight　3.1.2.26
X-质量　X-quality　20.2.4.27
质量扩散筛　mass diffusion screen　15.4.1.3
质量扩散塔　mass-diffusion column　15.4.1
质量流量　mass flow rate　1
质量流率　mass flow rate　1
致死物质　lethal substance　21.1.2.4
滞流　laminar flow　1
streamline flow　1
中部横断面　exploded cross section at center　15.2.3.9
中部框架　middle frame　12.2.1.14
中部框式卸料口　middle-frame discharge　12.2.1.8
中等尺寸物料　intermediate-size material　3.3.16.27
中等规模　medium scale　8.1.1.24
中段　intermediate portion　16.4.18.2
中放射性废料　storage of medium-active waste　25.2.5
中等热值可燃气体　medium-BTU gas　3.3.23.10
中国化工学会　Chemical Industry and Engineering Society of China　1
中合金钢　intermediate-alloy steel　22.6.2.⑤
中和　neutralize　2.3.8.24
中和池　neutralization pond　3.1.2.4
中和器　neutralizer　2.3.15.2
中级品　middling　12.6.4.14
中继透镜　relay lens　3.5.6.4
中间泵壳　middle case　16.4.28.31
中间产物　intermediate-product　1
middlings　2.1.9.10
中间催化裂化粗柴油　intermediate catalytic gas oil　2.1.5.8
中间垫片　intermediate gasket　16.4.27.10
中间富集物　medium concentrate　12.5.5.8

中间固定板　stationary intermediate plate　9.3.2.25
中间焦化粗柴油　MCGO；intermediate coking gas oil　2.1.6.9
中间壳体　intermediate casing　25.3.1.16
middle casing　16.4.19.31
中间冷凝器　inter condenser　20.1.10.14
中间冷却器　intercooler　17.6.1.4
side cooler　1
中间馏分油　middle distillate　2.2.94.19
中间排气式隧道式干燥器　center exhaust tunnel dryer　8.3.1.(c)
中间平台　intermediate platform　2.1.2.5
中间试验装置　pilot plant　1
中间网状螺旋　holo-flite intermeshed screws　4.3.3.23
中间温度的液体　intermediate temperature liquid　20.1.3.13
中间物　middling　12.5.2.15
中间轴　line shaft　16.4.23.12
Porcupine 中间轴式　Porcupine medium shaft　4.3.3.(c)
中壳灰岩层　middle Muschelkalk　25.2.4.3
中空纤维　hollow fiber　15.2.1.6,8
中空纤维型　hollow-fiber module　15.2.1
中空纤维型 Permasep 反渗透器　hollow-fiber module Permasep permeator　15.2.1
中空纤维组件　hollow-fiber module　1
中馏分　middle distillate　2.2.95.24
中热值　medium-Btu　2.2.2.10
中热值燃料气　med-Btu fuel gas　2.2.20.15
中试装置　pilot system　3.1.19
pilot-scale　10.3.5
中水排放　effluent discharge　3.1.31.10
中体导杆　intermediate rod　17.2.15.15
中体导向装置　intermediate guide　17.2.15.14
中体十字头框　indermediate crosshead frame　17.2.15.16
中温　medium temperature　8.7.2.6
中温换热器/蒸汽过滤器　intermediate temperature heat exchanger/steam filter　2.2.34.3
中温一氧化碳变换　intermediate-temperature CO-shift　2.3.2.3
中心分隔壁　center dividing wall　16.4.6.8
中心管　center pipe　22.2.3.6
中心管盖　center tube cover　11.3.2.5
中心加热管　center heating tube　8.8.10.9
中心降液管　center downcomer　5.2.2.13
ACF 中心流动货车　ACF center-flow　23.3.8.(b)
中心套管　central casing　25.3.1.19
中心(星形)轮　center spider　9.3.3.8
中心型组合阀　central valve　17.2.15.11
中心轴齿轮传动装置　center shaft gear drive　3.3.26.12
中性导体　neutral conductor　25.4.1.13
中(性)点　neutral point　25.4.20.4
中性亚硫酸盐半化学废液　neutral sulfite semi-chemical waste liquor　3.3.4
中压　moderate pressure　2.2.100.2
中压泵　medium pressure pump　16.1.2.1.3
中压锅炉给水加热器　feed water heater for intermediate pressure boiler　2.2.34.2;2.2.34.13
中压汽缸　medium-pressure cylinder　25.4.4.4
中压蒸气　MP steam　2.2.31.16
intermediate pressure steam　2.2.46.11;2.2.47.10;2.2.111.10
中压蒸汽涡轮机和发电机　intermediate pressure steam turbine & generator　2.2.111.11
中央挡盘　core buster disk　10.5.4.6
中央降液管　central downcomer　7.1.3.4
中央进气　suction in center　17.4.3.3
中央空调机组与控制系统　central-station air-conditioning unit　20.2.2
中央空调机组与控制系统　central-station air-conditioning unit and control system　20.2.2
中央控制室　central control room　2.3.12.25
中央塔　center column　15.6.3.20
中央循环管　central downcomer　7.1.3.4
中央圆盘　central disc　15.5.3.7
中央圆筒　center column　11.3.1.15
中圆筒　middle cylinder　10.1.2.10
中圆锥　middle cone　10.1.2.15
中轴套　intermediate shaft sleeve　16.4.27.17
终端沉降池　final settler　3.1.34.11
终端澄清器　final clarifier　3.1.30.B.7;3.1.33.5
终端环　end ring　16.4.28.19
终端控制池　final control basin　3.1.31.8
终端速度　terminal velocity　1
钟帽式除沫器　bell cap　7.1.8.8
钟罩型再分布器　bell cap redistributor　5.3.4.16
重柴油　heavy gas oil　2.1.4.13
重成分　heavy fraction　15.4.2.5
重锤导向筒　weight cylinder　15.7.1.10
重催化裂化粗柴油　heavy catalytic gas oil

757

2.1.5.7
重芳烃　heavy aromatic　2.1.1.55
重芳烃分离　heavy aromatics separation　2.1.1.59
重负荷夹套式　heavy-duty jacketed　4.3.4.(a)
重减压粗柴油　HVGO；heavy vacuum gas oil　2.1.4.18
重介质泵　heavy-medium pump　12.5.5.16
重介质储槽　heavy-medium storage sump　12.5.5.22
重介质返回　heavy-medium return　12.5.5.15
重介质分离　heavy-media　12.5.1
重介质循环　heavy-medium circulation　12.5.5.14
重介质溢流　heavy-medium overflow　12.5.5.23
重矿石粒子　heavy mineral ripple　15.5.3.2
重力过滤器　gravity filter　3.1.2.17
重力盘　gravity disc　9.5.12.2
重力渗滤区　gravity drainage zone　9.6.2.1
重力输送器　gravity conveyor　23.4.3.1
重力下料管　gravity spouts　23.1.6.10
重力卸料　gravity discharge　23.1.7.4
重力增稠罐　gravity thickener tank　3.1.26.(a).2；3.1.26.(b).2
重量传感元件　weight-sensing elements　23.5.4.24
重馏出物　heavy distillate　2.1.4.26
重馏分　heavy distillate　2.2.95.25
heavies　2.2.69.25
重石脑油　heavy naphtha　2.1.1.17
重相流　heavy-phase effluent　9.5.14.3
重型环状固体床层　heavy annular solids bed　4.3.3.12
重型锰钢锤　heavy manganese steel hammer　11.1.7.4
重型止推轴承　heavy-load thrust bearing　15.7.2.8
重液　HL；heavy liquid　6.1.15.2
重液泵　heavy liquor pump　2.3.10.34
重液分布管　heavy liquid dispersion pipe　6.1.18.7
重液入口　heavy liquid inlet　6.1.18.4
重液体　heavy liquid　6.1.14.2
重液循环　HLR；heavy liquid recycle　6.1.15.4
重油　heavy oil　2.2.6.20
mazut　2.1.1.35
重质矿石收集池　heavy mineral collection tank　15.5.3.6

重质汽油　heavy gasoline　2.2.77.13
重质汽油处理反应器　HGT（heavy gasoline treating reactor）　2.2.77.14
重质汽油处理器　HGT unit　2.2.75.17
重质页岩油　heavy shale oil　2.2.7.13
重质液体　heavy liquids　2.2.90.11
重组分　heavies　2.3.18.21
heavy components　15.5.4.9
周边圆环固定板　peripheral ring clamps　5.2.7.10
轴　spindle　9.5.15.7
轴衬　bush　9.2.1.17
轴衬套　gear shaft liner　12.3.5.13
轴承A　shaft sleeve A　16.4.18.15
轴承部件A　bearing assembly A　16.4.18.21
轴承部件B　bearing assembly B　16.4.18.19
轴承衬套　bearing bushing　16.4.24.4
轴承底盖　cover　13.1.2.17
轴承垫片　bearing sheet　16.4.19.4
轴承定距环　bearing spacer　16.4.9.33
轴承定位环　bearing collar　13.3.5.9
轴承盖　bearing cap　16.4.25.5
轴承盖止动螺钉　bearing cap set screw　16.4.26.19
轴承架　bearing bracket　16.4.19.5
bearing frame　8.8.12.12
bearing pedestal　16.4.22.13
bearing stand　16.4.4.18
轴承间隔块　bearing distance piece　11.1.8.23
轴承紧固螺母　bearing nut　16.4.29.16
轴承螺母　bearing nut　11.1.5.20
轴承迷宫密封　bearing labyrinth　20.2.1.13
轴承上盖　cover　13.1.2.23
轴承甩油环　bearing oiling rings　11.1.7.21
轴承锁紧螺母　bearing lock nut　16.4.9.4
bearing nut　13.3.5.6
轴承外侧　bearing outboard　16.4.9.7
轴承箱　ball case　11.1.8.27
bearing box　16.4.22.11
bearing housing　16.4.9.27
轴承箱端盖　bearing cover　16.4.9.19
bearing end cover　16.4.9.6
轴承箱盖　bearing cover　16.4.12.8
轴承压盖　ball cover　13.3.5.2
bearing cover　16.4.22.10
轴承座　bearing box　17.5.1.2
bearing holder　16.4.12.1
bearing pedestal　16.3.6.5
bearing seat　16.4.29.18
bearing support　25.3.1.5
轴承座套　bearing sleeve　16.4.23.7

轴导套 shaft guide 14.4.3.6
轴的密封套管 shaft enclosed tube 16.4.23.10
轴端螺母 shaft end nut 17.2.9.31
轴封 shaft seal(ing) 7.1.2.4
轴封环 oil seal ring 17.3.2.14
轴盖 shaft cap 16.4.28.3
轴管固定件 shaft tube stabilizer 16.4.25.18
轴键 impeller key 16.4.19.18
轴颈座调节螺钉 trunnion block adjusting screw 9.6.1.22
轴连接件 shaft coupling 16.4.23.11
轴连接链 shaft coupling link 9.3.3.40
轴联轴器 shaft coupling 9.3.3.22
轴流泵 axial flow pump 16.1.2.2.24
轴流风机 axial flow fan 4.5.1.4
轴流式 axial flow 17.1.1.13
轴流式低压头循环泵 axial flow low head circulating pump 12.5.5.21
轴流式叶轮 axial flow impeller 16.4.5.(d)
轴流送风机 axial flow fan 20.2.3.12
轴螺母 shaft nut 16.4.6.29
轴迷宫圈 shaft collar 16.4.9.34
轴内的冷却液池 coolant pool in shaft 4.3.3.9
轴套 shaft protecting sleeve 16.4.20.12
  shaft sheave 19.3.2.5
  sleeve 16.4.28.12
  spindle sleeve 16.4.22.5
轴套 B shaft sleeve B 16.4.18.23
轴套定位螺母 shaft nut 16.4.19.6
轴套管 shaft tube 16.4.25.34
轴套管压力支承块 shaft tube tension bearing 16.4.25.27
轴套结构 sleeve construction 16.4.10
轴套螺母 shaft sleeve nut; sleeve nut 16.4.16.22
轴瓦 bushing 8.8.5.28
  sleeve bearing 16.4.28.17
轴向间隙 axial clearance 16.4.6.24
轴向伸长 axial extension 22.5.2.3
轴向压缩 axial compression 22.5.2.2
肘板 toggle plates 11.1.1.13
绉状金属箔卷制填料 metal foil crinkled and twisted packing 19.2.3.11
骤扩 sudden enlargement 1
骤冷 quench 1
骤冷塔 quench tower 2.2.117.3
骤缩 sudden contraction 1
珠磨机 bead mills 11.3
烛型过滤器 candle filter 2.2.22.6; 2.2.23.9

逐级计算法 stage-by-stage method 1
逐级冷凝器 cascade condenser 20.3.16
逐级系统 cascade system 20.1.5
逐级循环(系统) cascade system 20.1.7
主操作阀 main operating valve 25.1.4.8.2
主澄清器 primary clearifier 3.1.41.3
主动齿轮 drive gear 11.1.2.1
  driving gear 16.3.1.4
主动链轮 drive sprocket 23.3.7.1
主动螺杆 driving screw 16.3.2.10
主动 V 型皮带轮 main pulley 11.1.5.22
主动轴 driving shaft 16.3.2.8
主阀 main valve 3.5.11.16
主分馏塔 main fractionator 2.1.6.3
主供蒸汽 main steam supply 18.3.5.9
主级萃取溶剂 primary extraction-solvent 6.1.2.6
主甲烷化反应器 main methanation reactor 2.2.14.8
主进料皮带 main feed belt 11.6.4.3
主开关 master switch 25.4.1.2
主空气风机 main air blower 14.2.7.21
  main air blower 14.3.4.2
主空气入口 main air inlet 14.3.4.22
  primary air inlet 14.1.9.8
主冷凝器 main condenser 18.3.5.2
  primary condenser 20.1.10.5
主梁 major beam 5.2.2.14
主梁固定板 major beam clamp 5.2.7.5
主流向排气装置 main flow to exhauster 10.1.3
主流向旋风分离机组 main flow to cyclonic separator unit 10.1.3.1
主炉膛出渣孔 primary furnace slag-tap hole 24.2.3.6
主煤气冷却器 primary gas cooler 2.2.24.11
主喷射器 primary ejector 20.1.10.3
主汽轮机 main turbine 25.3.2.18
主入口截止阀 main-inlet shutoff valve 2.1.7.2
主十字头导板 front guiding plate 17.2.15.30
主十字头(导轨)镶条 front crosshead metal 17.2.15.29
主塔 primary column 2.2.28.13
主体加料阀门 bulk gate 23.5.2.5
主物流 main stream 12.7.4.10
主循环泵 primary circulating pump 16.1.2.3.40

主压气机　main compressor　25.3.2.17
主要部件(图)　main parts　18.3.4.(a)
主要沉降池　primary settler　3.1.31.3
主要磁极　primary magnet pole　12.6.11.3
主要的有机物料　mainly organic fraction　3.3.16.19
主要零部件　major components　23.2.1
主要排料口　primary discharge　11.2.7.7
主要燃烧室　primary combustion chamber　24.1.4.3
主要设备配置图　basic equipment arrangements　23.1.1
主要设备与流程　basic equipment arrangement　8.5.4.(a)
主蒸汽管道　prime steam line　25.2.2.11
主轴　head shaft　15.6.1.30
　shaft　20.2.1.18
　spindle　16.4.22.6
主轴承　main bearing　11.2.4.4
主轴键　main shaft key　9.5.4.41
主轴冷却风机　shaft cooling air fan　3.3.26.15
主轴螺母　shaft nut　20.2.1.4
主轴迷宫密封　shaft labyrinth　20.2.1.11
主轴轴承　main shaft bearing　9.2.1.2
助燃风机　combustion air fan　3.3.24.21
助燃空气　combustion air　14.3.3.24
助燃空气预热器　combustion air preheater　3.3.24.15
助燃气　combustion air　3.1.22.2
贮仓　silo　23.1.6.14
贮槽　storage　14.3.4.39
　storage sump　16.5.2.6
　storage tank　2.3.6.15
贮存　storage　2.2.106.6
贮存室　storage chamber　25.2.4.15
　storage chamber　25.2.5.4
贮斗　hopper　3.2.6.7
　storage hopper　2.2.92.7
贮罐　storage　21.2
　storage tank　3.3.4.1
贮浆池　stuff chest　2.3.11.1
(贮酒的)大地窖　lager cellar　2.3.13.12
贮料层段　storage compartment　3.1.10.7
贮料斗　inventory hopper　10.4.1.1
贮气罐　air receiver　17.6.1.13
　gas holder　2.3.14.25
　receiver　17.6.2.1
贮气柜　gasometer　3.1.26.(b).8
贮液器　hopper　2.3.5.13
贮运器　storage　3.1.22.6
注入　filling　6.3.3.6

注入接头　filling connection　25.6.7.8
注入孔　injection port　19.1.7.15
注入液体　fluid charge　18.2.1.9
注射泵　injection pump　2.2.105.6
注射器　injector　2.2.106.13
注水口旋塞　feed water plug　16.4.28.25
注塑成型机　Injection moulding machine　15.7.3
注塑缸　injection cylinder　15.7.3.4
注油管　oil filler pipe　19.2.1.4
　oil filler point　19.2.1.5
注油口　oil filling port　15.7.4.15
　oil inlet　15.7.4.2
注油器　lubricating device　6.1.18.11
驻波原理　standing wave principle　15.5.3
柱　column　3.5.20.3
柱塞　plunger　16.2.4.29
柱塞泵　plunger pump　16.1.2.2.3；16.2.6
柱塞后十字头　rear cross head of plunger　16.2.4.23
柱塞螺母　plunger nut　16.2.4.13
柱塞前十字头　front cross head of plunger　16.2.4.12
柱塞(式)　plunger　16.1.1.5
柱塞式计量泵　plunger type metering pump　16.2.5
柱塞填料压盖衬套　plunger gland lining　16.2.4.15
柱塞填料压盖法兰　plunger gland flange　16.2.4.14
柱式支架　column support　21.3.4
铸造蒸汽集管箱　cast steam manifold　8.8.7.(a)
抓斗　bucket　23.3.6.3
砖洞道　brick tunnel　4.3.2.6
转化　conversion　1
转化反应器　conversion reactors　2.2.75.5；2.3.4.5
C_4 转化反应器　C_4 conversion reactor　2.2.69.6
转化釜　conversion reactor　2.2.55.7
转化率　conversion　1
转角　angular rotation　22.5.2.4
转角正方形排列　rotated square pitch　4.1.5.2
转角正三角形排列　rotated equilateral triangle pitch　4.1.5.4
continued below　3.1.37.12
转向　direction of rotation　16.4.15.14
　rotation　9.3.10.17
转向器　diverter　23.4.2.10

转动表面　rotating surface　19.3.3.7
转动除沫器　revolving scum skimmer　15.6.1.23
转动方向　rotation　8.8.6.8
转动耙　rotating rake　15.5.3.11
转动筛分机　rotax-screen　12.2.5
转鼓　basket　9.5.10.4
  bowl　9.5.16.2
  drum　12.1.1.13
  revolving drum　8.7.1.1
  rotating bowl　9.5.11.8
  rotating drum　12.6.4.3
  rotor　6.1.18.13
  tumbler　12.1.3.2
转鼓齿轮　drum gear　12.1.1.10
转鼓导管　drum piping　9.3.3.6
转鼓底　basket bottom　9.5.4.48
转鼓端板　closed drum head　9.3.3.11
转鼓阀　bowl valve　9.5.13.11
转鼓辐条　drum arm　9.3.3.7
转鼓毂　basket hub　9.5.17.37
转鼓过滤机　drum filter　9.3.9
转鼓混合器　drum mixer　12.1.4.(b)
转鼓连接管　drum nipple　9.3.3.10
转鼓清洗螺塞　plug for rotor cleaning　6.1.18.6
转鼓外壳　drum shell　9.3.10.12
转鼓外罩　bowl hood　9.5.12.5
  rotor cover　6.1.18.14
转鼓洗液　drum wash　12.6.4.4
转鼓型V型和双锥型混合器　drum-V-type and double-cone mixers　12.1.4
转鼓旋转方向　drum rotation　12.6.4.18
转鼓轴　drum shaft　9.3.3.9
转鼓主体　bowl body　9.5.12.10
转轮　pulley　12.6.1.(b)
转盘　disc　13.3.1.2
  rotary table　2.1.2.15
  rotor　11.4.1.21
  rotor disk　6.1.14.5
Peterson转盘沉降器　Peterson roto-disc clarifier　9.4.3
转盘萃取器　rotating extractor　1
转盘萃取塔（RDC）　rotating-disk（RDC）extractor　6.1.14
转盘阀　rotary valve　9.3.2.21
转盘雾化器　disk atomizer　8.6.3.8
转盘造粒机　rotating dish granulator　13.3
转筒　rotating drum　15.6.3.21
转筒混合机　tumbler mixer　2.1.9.3

转筒筛　trommel　3.3.16.5
  trommel screen　14.3.3.7
转筒式稠密介质分离器　revolving-drum-type dense-media separatory vessel　12.5.3
转筒真空过滤机　Oliver continuous vacuum drum filter　9.3.4
  rotary vacuum drum filter　9.3
转速表传感器　tachogenerator　23.5.4.19
转栅驱动器　grate drive　2.2.12.9
转叶泵　swinging vane pump　16.1.2.2.9
转轴　rotating shaft　16.4.23.1
  shaft　12.3.2.12
转轴固定螺钉　rotary shaft set screw　15.7.6.8
转轴密封装置　shaft seal　7.1.11.8
转子　rotor　11.1.7.6
转子导流锥体　rotor guide cone　12.3.2.11
转子端板　rotor shoe　23.1.12.5
转子端板拉紧螺栓　rotor shoe tension bolt　23.1.12.9
转子流量计　rotameter　8.5.3.2
转子驱动装置　rotor drive　8.9.6.13
转子外壳　rotor casting　23.1.11.4
  rotor housing　16.5.3.7
转子下轴承　lower rotor bearing　5.5.1.24
转子芯　rotor core　16.4.28.22
转子叶片　rotor blades　12.3.2.10
转子（叶片间）狭缝空气通道　air way slits in rotor　12.3.2.6
转子轴承　rotor bearing　5.5.1.16
转子组件　rotor assembly　17.4.2.22
装袋　bagging　13.1.1.10
装袋机　bag-packaging machine　23.5.3
装袋器　sacker　2.3.12.39
装货设备　loading facilities　23.4.2
装夹模具用气缸　die clamping cylinder　15.7.3.12
装料　filling　23.2.5.8
  material loading　23.3.5.2
装料端　loading end　23.3.7.(c)
装料孔　charge opening　11.3.1.1
装料口　charge opening　8.8.11.11
  loading points　23.3.3.1
装料门　charging door　3.3.12.9
装料排出口　burden discharge　4.3.1.5
装料区　charging zone　2.2.84.12
装料室　burden chamber　3.3.4.4
  charging chamber　25.2.5.7
装料筒　charging domes　8.8.10.6
装料向前传送　burden travel forward　4.3.

装料移动　burden travel　4.3.2.12
装料嘴　loading spout　23.1.2.30
装煤车　larry car　2.3.14.6
装配体　assembly　8.8.5.29
装配图　assembly　8.8.4.(c)
装瓶　bottling　2.3.13.22
装桶　drummed　3.3.7.3
装箱　packaging　3.3.7.4
装置并联　units in parallel　14.3.3.(a)
状态变量　state variable　1
状态点　state point　6.2.8.12
状态方程　equation of state(EOS)　1
撞击板　impingement plate　10.5.1.11
撞击板式涤气器　impingement-plate scrubber　10.5.1
撞击挡板级　impingement baffle plate stages　10.5.1.7
撞击器　impinger　3.5.14.9
撞击室　impact chamber　11.5.2.4
雅波　standing wave　3.5.4.3
锥底球形晶浆罐结晶器　conispherical magma crystallizer　7.2.7
锥顶罐　cone-roof　21.2.1.1
锥固定器　spud holder　24.2.6.8；24.2.7.5
锥塞　spud　24.2.6.4
锥形稠密介质分离器结构　dense-media cone vessel arrangements　12.5.4
锥形的中间零件　conical intermediate piece　25.3.1.20
锥形短头　cone shorthead　11.6.2.11
锥形阀　cone valve　14.1.6.(e)
锥形封头　conical head　21.1.1.57
锥形刮板　cone scraper　15.6.1.19
锥形罐　conical tank　12.1.3.7
锥形机座　cone support　10.5.10.1
锥形截面　cone sections　8.4.3
锥形进口套管　conical inlet casing　25.3.1.5
锥形扩散器　diffuser cone　3.1.14.10
锥形料堆　conical pile　3.2.3.(a)
锥形磨浆机　conical refiner　2.3.10.27
锥形瓶　conical flask　8.5.3.5
锥形球磨机　conical(ball) mill　11.2.3 conical mill　11.6.9.2
锥形绕线机　cone-winding machine　2.3.7.26
锥形栅板　conical grate　11.2.3.7
锥形(烧)瓶　erlenmeyer flask　2.3.11.2
锥形丝筒　cone form　2.3.7.25
锥形碎纸机　cone breaker　2.3.10.82
锥形同心环　concentric rings in the form of a cone　14.1.8.(b)

锥形头　conical head　11.1.3.5
锥形推料脱水离心机　conical pusher with de-watering cone　9.5.1.(o)
锥型塔侧收集器　cone side wiper　5.3.4.18
锥型支持板　pyramid support　5.3.4.19
准备压制的位　post ready for pressing　2.3.11.50
准静态过程　quasi-static process　1
准球形封头　torispherical head　21.4.1.4
拙打机构　shaking mechanism　10.2.1.25
拙劣设计　poorly designed　23.2.2.(a)
子系统　subsystem　1
自撑顶　tank-self-supported roof　21.2.1.2
自导式喷雾涤气器　self-induced spray scrubber　10.5.6
自定心滚柱轴承　self-aligning roller bearing　11.1.1.6
自动　automatic　23.5.2.20
自动袋放置器　automatic bag placer　23.4.3.13
自动电压调节器　automatic voltage regulator　25.4.4
自动对中法兰　self-centering flange　18.3.1.7
自动阀　automatic valve　9.3.3.15
自动阀法兰　automatic valve flange　9.3.3.45
自动阀环　automatic valve ring　9.3.5.3
自动放泄阀　automatic drain valve　17.6.1.10
自动记录　record　3.5.18.10
自动加料　self-feeding　23.3.7.10
自动截切机　automatic cutter　2.3.9.19
自动控制　automatic control　15.5.5.4
自动码垛　automatic palletizing　23.4.2
自动码垛机　automatic palletizer　23.4.2.21
自动排污管　automatic sludge discharge pipe　15.6.4.4
自动调节风门　automatic dampers　3.3.24.11
自动调节锥体　automatically adjustable cone　10.5.10.4
自动温度控制器　ATC；automatic temperature controller　15.7.3.14
自动卸料转鼓　self-cleaning bowl　9.5.13
自动循环式　cyclic automatic　9.5.2.5
(自动)压紧装置　closing device　9.1.2.9
自动圆盘定位器　automatic disk positioner　10.5.9.5
自动转换堆放器　tripper stacker　23.2.5.9
自动装袋吊具　automatic bag hanging　23.4.3
自发过程　spontaneous process　1
自流出料　gravity discharge　23.1.7.4

自流排放旋塞　gravity drain plug　18.1.4.8
自流式料斗　gravity-flow hopper　23.1.1.2
自清洗式真空密封卸料阀　self-wiping vacuumtight discharge valve　8.8.11.14
自然抽风筒　natural-draft stack　8.8.6.11
自然通风(凉水)塔　atmospheric tower　4.6.1
自然循环火管锅炉　natural circulation firetube boilers　4.4.2
自上往下送风　down blast　17.5.2.1
自身梁式塔板　self-beam tray　5.2.10.29
自位轴承　self-aligning bearings　15.6.1.29
自吸泵　self-priming pump　16.1.2.2.58
自吸式　self-priming　16.1.1.37
自下往上送风　up blast　17.5.2.5
自由沉降　free sedimentation　1
自由浮动介质　free-floating media　3.2.13.12
自由空间　freeboard　14.1.12.9;2.2.18.5
自由流动泵　free-flow pump　16.1.2.2.59
自由流动产品　free-flowing products　23.5.1
自由流动的颗粒　free-flowing granular　8.1.1.19
自由式球阀　free ball valve　22.1.1.(e)
自由移动　free movement　22.6.1.(d)
综合性示意图　generalized schematic　15.1.1
总产量　total capacity　2.2.88.1
总传热系数　overall heat transfer coefficient　1
总传质单元高度　height of overall transfer unit　1
总废物处理系统　total waste-treatment system　3.1.21
总管　main　17.6.2.5
manifold　9.1.3.13
总计表　totalizing meter　6.2.1.11
总进料　feed-all　8.4.1.18
总开关　master switch　25.4.2.2
总流程　general systems description　2.82.20
总压头　total head　16.4.3.15
纵缝接头　longitudinal joints　21.1.1.16
纵向漩涡　longitudinal vortex　16.4.15.20
纵向折流板　longitudinal baffle　4.1.1.30
走道　manway manway　25.2.2.12
walkway　15.6.5.12
阻挡液　barrier　19.3.3.10
阻漏环　casing ring　16.4.9.12
liner ring　16.4.29.6
阻塞　jamming　23.2.2.4
组合　configuration　25.1.1
组合干燥-磨碎系统　combined drying-grinding system　11.5.2.4

组合流动模型　composite flow model　1
组合取样管线　composite sample line　3.5.3.5
组合取样器　composite sampler　3.5.18.12
组合式迷宫　combination labyrinth　19.3.3.(e)
组合式润滑油控制装置　plug in type lubrocontrol unit　17.6.2.8
钻杆　drill pipe　2.1.2.6
drilling pipe　2.1.2.19
钻井　well　2.1.2.17
钻井机械　drilling rig　2.1.2.1
钻头　drilling bit　2.1.2.21
钻油井用的钢丝绳　drilling cable　2.1.2.7
最大混合度　maximum mixedness　1
最大开度　maximum opening　11.1.3.12
最大室外空气调节风门　maximum O.A. damper　20.2.2.8
最大许用操作压力　max.allowable working pressure　21.1.2.8
最大许用应力值　max.allow.stress value　21.1.2.14
最高水平面　maximum water level　15.6.3.14
最高液位　max liquid level　5.4.1.13;2.1.10.5
最后成形　final shaping　2.3.9.36
最后拉伸　final stretching　2.3.8.47
最后脱水　final dewatering　9.3.9.8
最小回流比　minimum reflux ratio　1
最小开度　minimum opening　11.1.3.13
最小流化速度　minimum fluidizing velocity　1
最小室外空气手动调节风门　minimum O.A.damper;manual　20.2.2.10
最小压头损失　minimum head loss　3.5.4
最终产品　end-product　13.5.3.3
final product　12.1.5.6
finished product　11.5.3.5
最终产品仓　finished product bin　11.6.5.7
最终处置　ultimate disposal　3.2.1.14
最终处置场地　ultimate disposal site　3.1.11.5
最终富集物　final concentrate　12.8.2.20
最终甲烷化反应器　final methanation reactor　2.2.8.9
最终尾砂　final tailing　12.8.2.22
左侧　near side　10.3.6.9
作功　work　20.1.2.2
座体　body　21.3.1.6